GENERAL RELATIVITY

GENERAL RELATIVITY

Robert M. Wald

The University of Chicago Press
Chicago and London

The painting reproduced on the cover of the paperback version of this book is René Magritte's *Les Belles Réalités*, © by ADAGP, Paris, 1984. The transparency was furnished by Galerie Isy Brachot, Bruxelles-Paris.

The University of Chicago Press, Chicago 60637
The University of Chicago Press, Ltd., London

Printed in the United States of America

14 13 12 11 15 16 17

Library of Congress Cataloging-in-Publication Data

Wald, Robert M.
General relativity.

Bibliography: p. 473
Includes index.
1. General relativity (Physics) I. Title.
QC173.6.W35 1984 530.1'1 83-17969
ISBN 0-226-87033-2 (pbk.)

♾ The paper used in this publication meets the minimum requirements of the American National Standard for Information Sciences—Permanence of Paper for Printed Library Materials, ANSI Z39.48-1992.

CONTENTS

PREFACE

This book is intended to provide a thorough introduction to the theory of general relativity. It is intended to serve as both a text for graduate students and a reference book for researchers. These two goals are somewhat contradictory, and to the extent that they are, part I of the book emphasizes the first goal. It treats the topics usually covered in introductory relativity courses: basic differential geometry, Einstein's equation, gravitational radiation, the standard cosmological models, and the Schwarzschild solution. More emphasis is placed on the second goal in part II of the book, which treats a wide variety of advanced topics. However, even here I have attempted to explain all the basic ideas at an introductory level.

If I were teaching a one term introductory course on general relativity, I would cover most of the material of part I together with much of appendices B and C. For a full year course, I then would choose several chapters from part II as the basis for the material covered in the second term. For example, chapters 8 and 9 and parts of chapter 12 could comprise a one term course on global methods. Chapter 7, supplemented by current literature, could serve as the basis for a course on methods for obtaining solutions. Chapter 10 and appendix E, supplemented by further reading, could be used for a course on the dynamics of general relativity. Chapter 12 (supplemented by background material from chapters 8, 9, and 11) and chapter 14 could comprise a course on the classical and quantum properties of black holes. It should be noted that the chapters in part II of the book are largely independent of each other and, for the most part, can be read out of sequence with the following major exceptions: prior reading of chapter 8 is essential for chapter 9, and chapter 8 together with parts of chapters 9 and 11 are essential for the first two sections of chapter 12.

One of the most difficult issues which arises during the writing of a book on general relativity is where in the book to present the rather substantial amount of mathematical material that is needed. Much of this material (e.g., tensor calculus and curvature) is required even for the formulation of general relativity. Some material (e.g., Lie derivatives and Killing fields) could be avoided initially but soon becomes necessary to make the discussion clearer and to simplify computations. Finally, some of the mathematical material (e.g., many of the theorems on topological spaces) is not really needed until part II of the book. If all this material were presented at the beginning of the book, it would comprise a truly formidable obstacle to learning general relativity. On the other hand, if the mathematical results were introduced only "as needed" in the later chapters, the mathematical discussion would become greatly fragmented and these fragments would interrupt the discussion of physical issues. The best solution I could find to this problem was to put all the mathematical material essential for the formulation of general relativity into chapters 2 and 3, and then to put the remaining mathematical topics into appendices A, B, and C. In this way, the reader can get to chapter 4 without unnecessary detours, but the discussion of all the mathematical topics remains intact and can be referenced as needed in the text. Thus, it should be emphasized that *appendices A, B, and C are an essential part of this book*. The results derived in appendices B and C are used in many places throughout the book, and the definitions and results on topological spaces which are compiled in appendix A are referred to frequently in chapters 8 and 9.

One other somewhat unusual organizational feature of the book is that the Lagrangian and Hamiltonian formulations of general relativity also have been put in an appendix. Often, the Hamiltonian formulation is presented in conjunction with the initial value formulation of general relativity, but since the statements and proofs of the initial value results do not rely upon the Hamiltonian formulation, I found it logically clearer to discuss them independently. This left the Lagrangian and Hamiltonian formulations as topics which are unlinked to the material in the other chapters, too short to comprise a whole chapter on their own, and too important to omit. Thus, they ended up being treated in appendix E.

Problems are given at the end of each chapter in text. There is significant variation in the amount of thought and computation required to solve the problems, but there are very few trivial, "mechanical" exercises and none which are, in my opinion, inordinately difficult (i.e., I think I can solve them). Part of my purpose in giving some of the problems (particularly in the second half of the book) was to introduce important side points for which I did not want to make a detour in the text. Hence, even the reader who is determined not to do any exercises may still wish to read the problems.

I have benefited from numerous interactions with many colleagues while planning and writing this book. The influence of Robert Geroch should be apparent to readers familiar with his viewpoints on general relativity. Some of the arguments used in chapter 3 are adopted directly from the notes from a course we taught jointly in 1975. I particularly wish to thank colleagues who took the time and trouble to read parts (and, in a few cases, all) of the book and send me their suggestions for improvements. These include Abhay Ashtekar, Arvind Borde, S. Chandrasekhar, David Garfinkle, John Friedman, Robert Geroch, James Hartle, James Isenberg, Bernard Kay, Karel Kuchař, Liang Can-bin, Roger Penrose, Michael Turner, and William Unruh. Additional thanks are due David Garfinkle for checking most of the equations.

I wish to thank Susan Lancaster and Roxy Boersma for typing drafts of the manuscript and Fred Flowers for typing the final product on a word processor. Support by NSF grant PHY 80 26043 to the University of Chicago during the writing of this book is gratefully acknowledged. Finally, I wish to thank my wife, Veronica, for the considerable amount of patience displayed during the three years it took me to write this book.

NOTATION AND CONVENTIONS

In this book we shall follow the sign conventions of Misner, Thorne, and Wheeler (1973). In particular, we use metric signature $- + + +$, we define the Riemann tensor by equation (3.2.3), and we define the Ricci tensor by equation (3.2.25). However, we shall make one important exception to these conventions. We choose to use the metric signature $- + + +$ because it is generally much more convenient than the alternative choice $+ - - -$ in that it induces a positive definite (rather than negative definite) metric on spacelike hypersurfaces. Unfortunately, for the reasons explained in chapter 13, it is much more convenient to use the metric signature $+ - - -$ for the treatment of spinors. Furthermore, the standard references on spinors all use this signature. *Hence, in chapter 13—and only in chapter 13—we will change our metric signature convention to* $+ - - -$. The confusion which might result from this should be minimized by the fact that the equations of chapter 13 are written in spinor notation, so the reader need only remember to change the sign of the metric when transcribing equations from spinor notation to tensor notation for use elsewhere in the book. With regard to these changes of sign, it is useful to note that the derivative operator, ∇_a, associated with the metric is unaffected by a sign change of the metric. Hence, the Riemann tensor with index structure $R_{abc}{}^d$ also is unaffected, since it is defined purely in terms of ∇_a. (However, it should be noted that some authors define the Riemann tensor with sign opposite that of equation [3.2.3]; see Misner, Thorne, and Wheeler (1973) for a table of sign conventions.) Similarly, it is conventional to take the stress-energy tensor T_{ab} and Maxwell field tensor F_{ab} to be unaffected by a change of metric signature. However, each raising or lowering of an index on $R_{abc}{}^d$, T_{ab}, F_{ab}, and any other tensor results in a change of sign.

Throughout most of this book, we shall use "geometrized units," where the gravitational constant, G, and the speed of light, c, are set equal to one. However, for the convenience of the reader we have restored the G's and c's in section 5.4 and in many of the formulas elsewhere in the book where observational predictions are made. A conversion table from "geometrized" to "nongeometrized" units is given in appendix F.

Our notation differs from standard conventions in one important respect. Most relativity texts use an index notation for components of tensors. Usually, greek indices are used to denote space or time components of a tensor, while latin indices are used to denote purely spatial components, although in some references (e.g., Landau and Lifshitz 1962) these conventions are reversed. This index notation provides an extremely efficient scheme for denoting tensor operations such as contraction, covariant differentiation, and the taking of outer products. However, this standard notational convention suffers from the serious drawback that it is impossible to distinguish a relation between tensors from a relation which holds only for tensor components with respect to a specially chosen basis. We shall overcome this difficulty by employing an *abstract index notation* discussed by Penrose (1968) and Penrose and Rindler (1984) and used extensively by Geroch. In our notation, latin indices on a tensor do *not* represent components but are part of the notation for the tensor itself, much like the arrow used to denote a vector in ordinary three-dimensional space. Thus, in this book any equation involving tensors which employs latin indices represents a relation between tensors; the taking of basis components need not even be contemplated. The complete rules for interpreting the notation are given in section 2.4. On the other hand, greek indices on a tensor represent components, as in the usual convention. Any equation employing greek indices is a relation between tensor components and, usually, holds only with respect to a specially chosen basis. Unfortunately, in our notation we cannot denote purely spatial tensor components without introducing yet another alphabet. However, only rarely in this book do equations arise which hold only for spatial components, and in such cases we simply shall state explicitly for which components a given equation applies.

For the benefit of the reader who is not well versed in mathematical notation, we list below the definitions of some of the standard mathematical symbols used frequently in the text:

$\cup$	$A \cup B$ denotes the union of sets A and B
$\cap$	$A \cap B$ denotes the intersection of sets A and B
$\subset$	$A \subset B$ denotes that A is a subset of B
$-$	$B - A$ denotes the complement in B of the set A
$\in$	$p \in A$ denotes that p is an element of A
$\{\mid\}$	$\{p \in A \mid Q\}$ denotes the set consisting of those elements p of the set A which satisfy condition Q
$\times$	Cartesian product; $A \times B$ is the set $\{(a, b) \mid a \in A \text{ and } b \in B\}$
$\emptyset$	the empty set
$\mathbb{R}$	the set of real numbers
$\mathbb{R}^n$	the set of n-tuples of real numbers
$\mathbb{C}$	the set of complex numbers
$\mathbb{C}^n$	the set of n-tuples of complex numbers
$: \rightarrow$	$f: A \rightarrow B$ denotes that f is a map from the set A to the set B
$\circ$	$f \circ g$ denotes the composition of maps $g: A \rightarrow B$ and $f: B \rightarrow C$, i.e., for $p \in A$ we have $(f \circ g)(p) = f[g(p)]$.
$[\ ]$	$f[A]$ denotes the image of the set A under the map f, i.e., the set $\{f(x) \mid x \in A\}$.
C^n	the set of n-times continuously differentiable functions
C^∞	the set of infinitely continuously differentiable (i.e., smooth) functions

PART I

FUNDAMENTALS

In addition, a number of symbols defined in the book appear frequently and are not always redefined each time they are used. Hence, for the convenience of the reader we list these symbols below, together with the section of the book where they are defined:

$\bar{S}$	the closure of the set S (appendix A)
$\text{int}(S)$	interior of the set S (appendix A)
$\dot{S}$	boundary of the set S (appendix A)
$\pounds_v$	Lie derivative with respect to the vector field v^a (appendix C)
$\mathcal{F}$	the set of smooth functions from a manifold M into $\mathbb{R}$ (section 2.2)
V_p	tangent space at point p of a manifold (section 2.2)
V_p^*	dual space to V_p (section 2.3)
$I^+(S)$	chronological future of the set S (section 8.1)
$J^+(S)$	causal future of the set S (section 8.1)
$D^+(S)$	future domain of dependence of the closed, achronal set S (section 8.3)
$H^+(S)$	future Cauchy horizon of the closed, achronal set S (section 8.3)
$\mathscr{I}^+$	future null infinity (section 11.1)
i^0	spatial infinity (section 11.1)

The symbols $I^-(S)$, $J^-(S)$, $D^-(S)$, $H^-(S)$, and $\mathscr{I}^-$ are defined as above with "past" replacing "future." $D(S)$ denotes $D^+(S) \cup D^-(S)$, and $H(S)$ denotes $H^+(S) \cup H^-(S)$. Finally, round and square brackets around tensor indices denote, respectively, symmetrization and antisymmetrization, as defined by equations (2.4.3) and (2.4.4).

ONE

INTRODUCTION

1.1 Introduction

General relativity is the theory of space, time, and gravitation formulated by Einstein in 1915. It is often regarded as a very abstruse and difficult theory, partly because the new viewpoint it introduced on the nature of space and time takes some effort to get used to since it goes against some deeply ingrained, intuitive notions, and partly because the mathematics required for a precise formulation of the ideas and equations of general relativity (namely, differential geometry) is not familiar to most physicists. Although it has been universally acknowledged as being a beautiful theory, the potential relevance of general relativity to the rest of physics has not been universally acknowledged and, indeed, probably for this reason, the subject has lain nearly dormant during much of its history.

Strong interest in general relativity began to be revived starting in the late 1950s, particularly by the Princeton group led by John Wheeler and the London group led by Herman Bondi. Although it is difficult to determine the reasons for trends in physics, two developments—relating general relativity to other areas of physics and astronomy—have contributed greatly to the sustained interest in general relativity which has continued since then. The first is the astronomical discovery of highly energetic, compact objects—in particular, quasars and compact X-ray sources. It is likely that gravitational collapse and/or strong gravitational fields play an important role here, and if so, general relativity would be needed to understand the structure of these objects. The modern theory of gravitational collapse, singularities, and black holes was developed beginning in the mid-1960s largely in response to this impetus.

A second factor promoting renewed interest in general relativity is the realization that although gravitation may be too weak to play an important role in laboratory experiments in particle physics, nevertheless it is of great importance to our further understanding of the laws of nature that a quantum theory of gravitation be developed. In order to make progress toward this goal, a deeper understanding of some aspects of the classical theory of gravitation—general relativity—may be needed. Interest in this program has been greatly strengthened by the prediction of quantum particle creation in the gravitational field of a black hole, as well as by advances in the study of gauge theories in particle physics.

But even aside from the potential impact of general relativity on astronomy and on other branches of physics, the theory in its own right makes many remarkable statements concerning the structure of space and time and the nature of the gravitational field. After one has learned the theory, one cannot help feeling that one has gained some deep insights into how nature works.

The purpose of this book is to present the theory of general relativity. We will take a more modern, geometrical viewpoint than Einstein had, and we will, of course, discuss the recent advances and developments, but the essential content of the theory is the one Einstein gave over half a century ago. We begin in this chapter by discussing the structure of space and time and the basic ideas of relativity theory from an intuitive, physical point of view. More complete introductory discussions are given by Geroch (1978*a*) and Wald (1977*a*). The remainder of this book will be devoted to making these ideas mathematically precise and exploring their consequences.

1.2 Space and Time in Prerelativity Physics and in Special Relativity

Perhaps the greatest obstacle to understanding the theories of special and general relativity arises from the difficulty in realizing that a number of previously held basic assumptions about the nature of space and time are simply wrong. We begin, therefore, by spelling out some key assumptions about space and time. In both the past and modern viewpoints, space and time have at least the following structure in common. We can consider space and time ($\equiv$ spacetime) to be a continuum composed of *events,* where each event can be thought of as a point of space at an instant of time. Furthermore, all events (or, at least, all events in a sufficiently small neighborhood of a given event) can be uniquely characterized by four numbers: in ordinary language, three numbers for the spatial position and one for the time. As will be discussed in chapter 2, a mathematically precise statement of these ideas is that spacetime is a four-dimensional manifold.

However, prior to relativity theory it was believed that spacetime had the following additional structure: Given an event p in spacetime, there is a natural, observer-independent notion of events occurring "at the same time" as p. More precisely, given two events p and q, one of the following three mutually exclusive possibilities must hold: (1) It is possible, in principle, for an observer or material body to go from event q to event p, in which case one says q is to the past of p. (2) It is possible to go from p to q, in which case one says q is to the future of p. (3) It is impossible, in principle, for an observer or material body to be present at both events p and q. In prerelativity physics, events in the third category are assumed to form a three-dimensional set and define the notion of simultaneity with p, as is illustrated in Figure 1.1.

The belief that the causal structure of spacetime has the character shown in Figure 1.1 turns out to be wrong. In special relativity theory the above classification of the causal relationships between events still holds. The crucial difference is that events in category (3) form more than a three-dimensional set; the causal relation between p and other events has the structure sketched in Figure 1.2. The events in category (3) can be further subdivided as follows: (i) Events that lie on the boundary of the set of points to the future of p. These events cannot be reached by a material particle

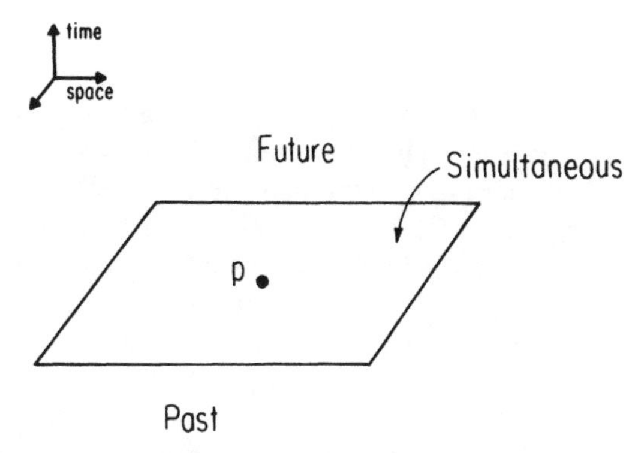

Fig. 1.1. A diagram showing the causal structure of spacetime in prerelativity physics. Given an event p, all other events in spacetime either are to the future of p, to the past of p, or simultaneous with p. The simultaneous events form a three-dimensional surface in spacetime.

starting at p but can be reached by a light signal emitted from p. They form the "future light cone" of p (a three-dimensional set). (ii) Events on the past light cone of p, defined similarly. (iii) Events in category (3) which are on neither the past nor the future light cone. These events are said to be spacelike related to p and comprise a four-dimensional set.

A key fact closely related to the above is that in special relativity there is no notion of absolute simultaneity; there are no absolute three-dimensional surfaces in spacetime as in Figure 1.1. As we shall see below, an observer still can define a notion of which events occur "at the same time" as a given event—thus defining a three-dimensional surface in spacetime—but the notion he gets depends upon his state of motion. (On the other hand, the light cones of Fig. 1.2 *are* absolute surfaces.) The notion that there is absolute simultaneity is a deeply ingrained one. The fact that there is no such notion is one of the most difficult ideas to adjust to in the theory of special relativity.

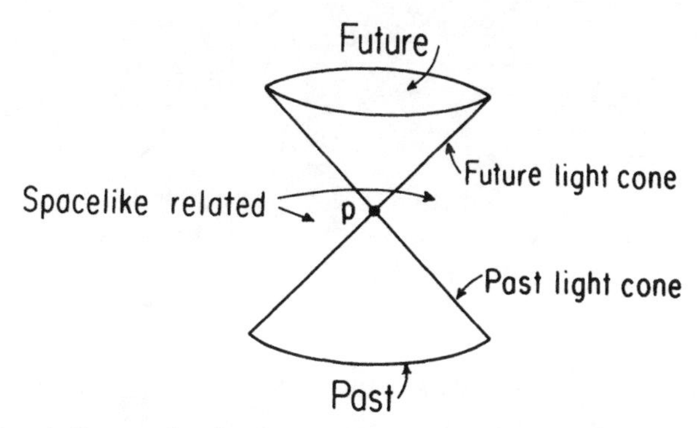

Fig. 1.2. A diagram showing the causal structure of spacetime in special relativity. The "light cone" of p rather than a "surface of simultaneity" with p now plays a fundamental role in determining the causal relationship of p to other events.

In special relativity (as in prerelativity physics) one has the notion of inertial, "nonaccelerating" motion, namely the motion a material body would undergo if subjected to no external forces. An inertial observer can label the events of spacetime in the following manner. He can build himself a rigid frame and label the grid points of the frame with the Cartesian coordinates x, y, z of the (assumed Euclidean) geometry of the frame. He can then have a clock placed at each grid point and can synchronize each clock with his by a symmetrical procedure, e.g., by making sure that a given clock and his give the same reading when they receive a signal sent out in a symmetrical manner by an observer stationed halfway between the two. (Because the causal structure of spacetime is that of Figure 1.2, not Figure 1.1, synchronization is not a trivial issue.) The observer may carry the grid, complete with synchronized clocks, in a nonrotating manner. Each event in spacetime can now be labeled with the coordinates x, y, z of the grid point at which the event occurred and the reading t of the (synchronized) clock at that event. The labels t, x, y, z assigned to events in this manner are referred to as global inertial coordinates.

If two such inertial observers go through this procedure, one may compare the coordinate labels they assign to events. In prerelativity physics (where the same labeling procedure works, the only difference being that clock synchronization *is* trivial), if observer O labels an event p with coordinates t, x, y, z and O' moves with velocity v in the x-direction, passing observer O at the event labeled by $t = x = y = z = 0$, the coordinate labels that O' assigns to event p are

$$t' = t, \tag{1.2.1}$$

$$x' = x - vt, \tag{1.2.2}$$

$$y' = y, \tag{1.2.3}$$

$$z' = z. \tag{1.2.4}$$

In special relativity, however, the labeling by O' will be related to that of O by a Lorentz transformation,

$$t' = (t - vx/c^2)/(1 - v^2/c^2)^{1/2}, \tag{1.2.5}$$

$$x' = (x - vt)/(1 - v^2/c^2)^{1/2}, \tag{1.2.6}$$

$$y' = y, \tag{1.2.7}$$

$$z' = z, \tag{1.2.8}$$

where c is the speed of light. Equation (1.2.5) shows that the notion of simultaneity determined by O (namely, t = constant) differs from that determined by O' (t' = constant), as illustrated in Figure 1.3.

1.3 The Spacetime Metric

In the previous section, we gave a prescription for how an inertial observer O can label the events in spacetime with global inertial coordinates t, x, y, z. However, a fundamental tenet of special relativity is that there are no preferred inertial observers. As seen above, a different inertial observer using the same procedure assigns differ-

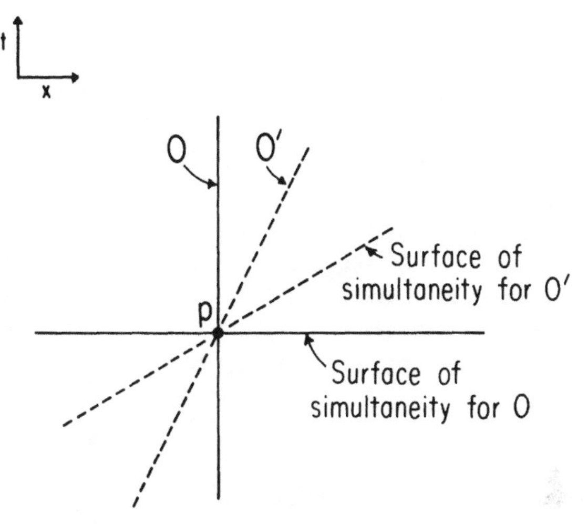

Fig. 1.3. A spacetime diagram illustrating the fact that in special relativity the inertial observers O and O' disagree over the definition of simultaneity with event p.

ent labels t', x', y', z' to the events in spacetime. Thus, the coordinate labels themselves do not have intrinsic significance since they depend as much on which observer does the labeling as they do on the properties of spacetime itself. It is of great interest to determine what quantities have absolute, observer-independent significance, i.e., truly measure intrinsic structure of spacetime. This is equivalent to determining what functions of global inertial coordinates are independent of the choice of inertial frame.

In prerelativity physics the answer is the following. The time interval Δt between two events has absolute significance; all observers will agree on the value of Δt. Furthermore, the spatial interval $|\Delta \vec{x}|$ between two simultaneous events is observer independent. However, these quantities (or functions of them) are the only ones with absolute significance. For example, observers moving with nonzero relative velocity will disagree over the spatial interval between nonsimultaneous events.

In special relativity neither the time interval nor the space interval between "relatively simultaneous" events (i.e., events determined to be simultaneous by a particular observer) has absolute significance. The quantity which is observer independent is the spacetime interval, I, defined by

$$I = -(\Delta t)^2 + \frac{1}{c^2}[(\Delta x)^2 + (\Delta y)^2 + (\Delta z)^2]. \tag{1.3.1}$$

Indeed, the Poincaré transformations (the set of all possible transformations between global inertial coordinates) consist precisely of the linear transformations which leave I unchanged. The spacetime interval I and functions of I are the only observer-independent quantities characterizing the spacetime relationships between events.

What is truly remarkable about the expression for I is that it is quadratic in the coordinate differences, just like the distance function in Euclidean (i.e., flat, positive definite) geometry. Indeed, the only difference is the minus sign in front of $(\Delta t)^2$, allowing I to become zero or negative. We shall refer to I as the metric of spacetime in analogy to an ordinary Euclidean metric. (More precisely, the metric of spacetime in special relativity will be defined later to be a tensor field associated with the formula for the spacetime interval between two "infinitesimally nearby" events; see eq. [4.2.2] below.) As we shall see in chapter 3, this difference in metric signature makes very little difference in the mathematical analysis of metrics. In particular definitions of geodesics ("straightest possible lines") and curvature carry through in the same way for metrics with the signature of I as for ordinary positive definite metrics. It is interesting to note that, as discussed more fully in chapter 4, the paths in spacetime of inertial observers in special relativity are geodesics of the spacetime metric, and the curvature associated with I is zero, i.e., the spacetime geometry in special relativity is flat.

1.4 General Relativity

Prior to special relativity, the prerelativity notions of space and time pervaded—among many other things—the formulation of the laws of physics. When these notions were overthrown, the task remained of modifying and reformulating physical laws to be consistent with the spacetime structure given by the theory of special relativity. Maxwell's theory of electromagnetism was already consistent with special relativity. Indeed, its incompatibility with prerelativity notions of spacetime structure unless preferred inertial frames were introduced led directly to the discovery of special relativity. Newton's theory of gravitation is not consistent with special relativity since it invokes notions of instantaneous influence of one body on another, but it might be thought that one could simply modify it to fit within the framework of special relativity.

However, two key ideas motivated Einstein *not* to follow this path but rather to seek an entirely new theory of spacetime and gravitation—a theory that revolutionized our notions of space and time every bit as much as special relativity already had done.

The first idea is that *all* bodies are influenced by gravity and, indeed, all bodies fall precisely the same way in a gravitational field. This fact, known as the *equivalence principle*, is expressed in the Newtonian theory of gravitation by the statement that the gravitational force on a body is proportional to its inertial mass. Because motion is independent of the nature of the bodies, the paths of freely falling bodies define a preferred set of curves in spacetime just as in special relativity the paths in spacetime of inertial bodies define a preferred set of curves, independent of the nature of the bodies. This suggests the possibility of ascribing properties of the gravitational field to the structure of spacetime itself. As already mentioned in the previous section, the paths of inertial bodies in special relativity are geodesics of the spacetime metric. Perhaps, then, the paths of freely falling bodies are always geodesics, but the spacetime metric is not always that given by special relativity. What we think of as a gravitational field would then not be a new field at all, but rather would correspond to a deviation of the spacetime geometry from the flat geometry of special relativity. We shall discuss these ideas further in chapter 4.

The second much less precise set of ideas which motivated the formulation of general relativity goes under the name of *Mach's principle*. In special relativity as in prerelativity notions of spacetime, the structure of spacetime is given once and for all and is unaffected by the material bodies that may be present. In particular, "inertial motion" and "nonrotating" are not influenced by matter in the universe. Mach as well as a number of earlier philosophers and scholars (in particular, Riemann) found this idea unsatisfactory. Rather, Mach felt that all matter in the universe should contribute to the local definition of "nonaccelerating" and "nonrotating"; that in a universe devoid of matter there should be no meaning to these concepts. Einstein accepted this idea and was strongly motivated to seek a theory where, unlike special relativity, the structure of spacetime is influenced by the presence of matter.

The new theory of space, time, and gravitation—general relativity—proposed by Einstein states the following: The intrinsic, observer-independent, properties of spacetime are described by a spacetime metric, as in special relativity. However, the spacetime metric need not have the (flat) form it has in special relativity. Indeed, curvature, i.e., the deviation of the spacetime metric from flatness, accounts for the physical effects usually ascribed to a gravitational field. Furthermore, the curvature of spacetime is related to the stress-energy-momentum tensor of the matter in spacetime via an equation postulated by Einstein. In this way, the structure of spacetime (as embodied in the spacetime metric) is related to the matter content of spacetime, in accordance with some (but not all!) of Mach's ideas. Thus far, the predictions of general relativity have been found to be in excellent agreement with experiments and observations (see section 6.3 below and Will 1981).

Most of the remainder of this book is devoted to exploring the consequences of this theory. Our first task, however, is to give a precise, mathematical expression to the ideas discussed in this chapter. To begin with, we must give a precise formulation of the notion that spacetime is a four-dimensional continuum. This will be accomplished with the definition of a manifold given in section 2.1. We must then introduce the basic mathematical framework needed to discuss curved geometry: vectors and tensors (2.2), the metric (2.3), derivative operators (3.1), curvature (3.2), and geodesics (3.3). Almost all of the discussion we shall give applies equally well to the differential geometry of ordinary surfaces (positive definite metric) as to the geometry of spacetime (metric of Lorentz signature). After development of these mathematical tools and techniques, we will then be in position to begin our study of general relativity in chapter 4.

Problem

1. *Car and garage paradox:* The lack of a notion of absolute simultaneity in special relativity leads to many supposed paradoxes. One of the most famous of these involves a car and a garage of equal proper length. The driver speeds toward the garage, and a doorman at the garage is instructed to slam the door shut as soon as the back end of the car enters the garage. According to the doorman, "the car Lorentz contracted and easily fitted into the garage when I slammed the door." According to

the driver, "the garage Lorentz contracted and was too small for the car when I entered the garage." Draw a spacetime diagram showing the above events and explain what really happens. Is the doorman's statement correct? Is the driver's statement correct? For definiteness, assume that the car crashes through the back wall of the garage without stopping or slowing down.

TWO

MANIFOLDS AND TENSOR FIELDS

In this chapter we lay the foundations for a precise, mathematical formulation of general relativity by obtaining some basic properties of manifolds and tensor fields. As defined in section 2.1, an n-dimensional manifold is a set that has the local differential structure of $\mathbb{R}^n$ but not necessarily its global properties. In section 2.2 we define tangent vectors as directional derivative operators acting on functions defined on a manifold. We obtain there some important properties of coordinate bases of the tangent space and tangents to curves. Tensors are introduced in section 2.3, and the notion of a metric is defined. Finally, in section 2.4 we introduce the abstract index notation for tensors, which we shall use throughout the remainder of this book. We will use a fair number of standard mathematical symbols in this chapter, and the reader unfamiliar with these symbols should consult the section "Notation and Conventions" at the beginning of this book.

2.1 Manifolds

As mentioned in the previous chapter, our experience tells us that spacetime is a "four-dimensional continuum" in the sense that it requires four numbers to characterize an event. In prerelativity physics as well as in special relativity it is assumed that this is globally true, i.e., that all events in spacetime can be put into one-to-one continuous correspondence with the points of $\mathbb{R}^4$. However, in general relativity we will be solving for the spacetime geometry and we do not wish to prejudice in advance any aspects of the global nature of spacetime structure. Our situation is very similar to that of hypothetical investigators of the structure of the surface of Earth prior to the explorations of Columbus and Magellan. Such investigators might notice that in their vicinity they can characterize positions on the surface of the Earth by two numbers. However, they would be making a serious error is they were to extrapolate from this fact to the conclusion that the entire collection of points on the surface of the Earth can be put into one-to-one correspondence with points of $\mathbb{R}^2$ in a continuous manner. Thus, what is needed as a mathematical basis for beginning the investigation of spacetime structure (as well as the surface of the Earth) is a precise notion of a manifold, that is, a set in which the vicinity of every point "looks like" $\mathbb{R}^n$ but which may have quite different global properties.

In the case of the Earth, our investigators might be aware that its surface "lives" in the higher dimensional Euclidean space $\mathbb{R}^3$ of all space points (at least, according to prerelativity notions of space and time). Thus the study of two-dimensional surfaces embedded in $\mathbb{R}^3$ would provide an adequate mathematical framework to analyze the structure of the Earth's surface, and one could avoid making an abstract definition of manifolds. However, in general relativity, spacetime itself does not (as far as we know) naturally live in a higher dimensional Euclidean space, so an abstract definition is much more natural. Indeed, such a definition turns out to be extremely useful even for the study of ordinary surfaces in $\mathbb{R}^3$.

Before defining the notion of a manifold we remind the reader that an *open ball* in $\mathbb{R}^n$ of radius r centered around point $y = (y^1, \ldots, y^n)$ consists of the points x such that $|x - y| < r$, where

$$|x - y| = \left[\sum_{\mu=1}^{n} (x^\mu - y^\mu)^2\right]^{1/2} .$$

An *open set* in $\mathbb{R}^n$ is any set which can be expressed as a union of open balls. This notion of open set makes $\mathbb{R}^n$ a topological space in the sense discussed in appendix A.

Basically, a *manifold* is a set made up of pieces that "look like" open subsets of $\mathbb{R}^n$ such that these pieces can be "sewn together" smoothly. More precisely, an *n-dimensional, C^∞, real manifold M* is a set together with a collection of subsets $\{O_\alpha\}$ satisfying the following properties:

(1) Each $p \in M$ lies in at least one O_α, i.e., the $\{O_\alpha\}$ cover M.

(2) For each α, there is a one-to-one, onto, map $\psi_\alpha : O_\alpha \to U_\alpha$, where U_α is an open subset of $\mathbb{R}^n$.

(3) If any two sets O_α and O_β overlap, $O_\alpha \cap O_\beta \neq \emptyset$ (where $\emptyset$ denotes the empty set), we can consider the map $\psi_\beta \circ \psi_\alpha{}^{-1}$ (where $\circ$ denotes composition) which takes points in $\psi_\alpha[O_\alpha \cap O_\beta] \subset U_\alpha \subset \mathbb{R}^n$ to points in $\psi_\beta[O_\alpha \cap O_\beta] \subset U_\beta \subset \mathbb{R}^n$ (see Fig. 2.1). We require these subsets of $\mathbb{R}^n$ to be open and this map to be C^∞, i.e., infinitely continuously differentiable. (Since we are dealing here with maps of $\mathbb{R}^n$ into $\mathbb{R}^n$, the advanced calculus notion of C^∞ functions applies.)

Each map ψ_α is generally called a *chart* by mathematicians and a *coordinate system* by physicists. We shall use these terms interchangeably. In order to prevent one from defining new manifolds by merely deleting or adding in a coordinate system, it is convenient also to require in the definition of M that the cover $\{O_\alpha\}$ and chart family $\{\psi_\alpha\}$ is maximal, i.e., all coordinate systems compatible with (2) and (3) are included. The definition of C^k or analytic manifolds is the same as above with the appropriate change in requirement (3). To define a complex manifold, one merely replaces $\mathbb{R}^n$ by $\mathbb{C}^n$ above.

We can define a topology on the manifold M by demanding that all the maps ψ_α in our maximal collection be homeomorphisms (see appendix A for definitions). Indeed, it is perhaps more natural to proceed by defining a manifold to be a topological space satisfying the above properties, with each ψ_α a homeomorphism. (We have not done so simply in order to avoid introducing the machinery of topological spaces in the main text.) Viewed as topological spaces, we shall consider in this book only manifolds which are *Hausdorff* and *paracompact;* these terms are explained in appendix A.

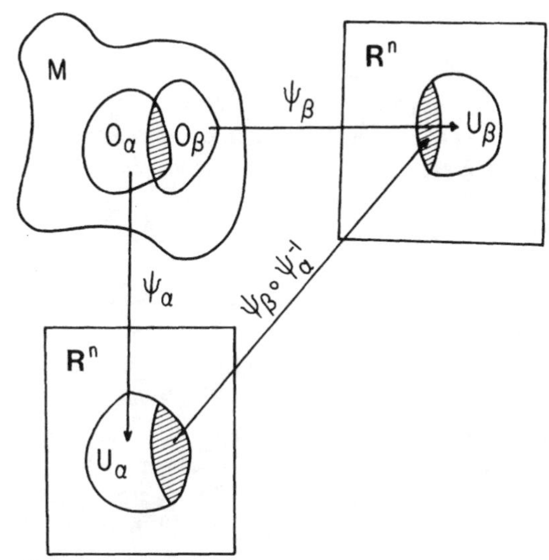

Fig. 2.1. An illustration of the map $\psi_\beta \circ \psi_\alpha^{-1}$ arising when two coordinate systems overlap.

Euclidean space, $\mathbb{R}^n$, provides a trivial example of a manifold, which can be covered by a single chart ($O = \mathbb{R}^n$, ψ = identity map). A more interesting example of a manifold is the 2-sphere S^2,

$$S^2 = \{(x^1, x^2, x^3) \in \mathbb{R}^3 \,|\, (x^1)^2 + (x^2)^2 + (x^3)^2 = 1\} \quad .$$

The entire 2-sphere S^2 cannot be mapped into $\mathbb{R}^2$ in a continuous, 1–1 manner, but "pieces" of S^2 can, and these can be "smoothly sewn together." For example, if we define the six hemispherical open sets $O_i^\pm$ for $i = 1, 2, 3$ by

$$O_i^\pm = \{(x^1, x^2, x^3) \in S^2 \,|\, \pm x^i > 0\} \quad ,$$

then $\{O_i^\pm\}$ covers S^2. Furthermore, each $O_i^\pm$ can be mapped homeomorphically into the open disk $D = \{(x, y) \in \mathbb{R}^2 \,|\, x^2 + y^2 < 1\}$ in the plane via the "projection maps" $f_1^+ : O_1^+ \to D, f_1^- : O_1^- \to D$, etc., defined by $f_1^+(x^1, x^2, x^3) = (x^2, x^3)$, etc. The overlap functions $f_i^\pm \circ (f_j^\pm)^{-1}$ can be checked to be C^∞ in their domain of definition (problem 1). Thus, S^2 is a two-dimensional manifold. In a similar manner, the n-dimensional sphere S^n is seen to be a manifold.

Given two manifolds M and M' of dimension n and n', respectively, we can make the product space $M \times M'$ consisting of all pairs (p, p') with $p \in M$ and $p' \in M'$ into an $(n+n')$-dimensional manifold as follows. If $\psi_\alpha : O_\alpha \to U_\alpha$ and $\psi'_\beta : O'_\beta \to U'_\beta$ are charts, we define a chart $\psi_{\alpha\beta} : O_{\alpha\beta} \to U_{\alpha\beta} \subset \mathbb{R}^{n+n'}$ on $M \times M'$ by taking $O_{\alpha\beta} = O_\alpha \times O'_\beta$, $U_{\alpha\beta} = U_\alpha \times U'_\beta$, and setting $\psi_{\alpha\beta}(p, p') = [\psi_\alpha(p), \psi'_\beta(p')]$. It is easily checked that the chart family $\{\psi_{\alpha\beta}\}$ satisfies the properties required to define a manifold structure on $M \times M'$. Most manifolds we will consider in this book can be expressed as products of Euclidean space $\mathbb{R}^n$ with spheres S^m.

With the structure on manifolds given by the coordinate systems, we may now define the notion of differentiability and smoothness of maps between manifolds. Let

M and M' be manifolds and let $\{\psi_\alpha\}$ and $\{\psi'_\beta\}$ denote the chart maps. A map $f: M \to M'$ is said to be C^∞ if for each α and β, the map $\psi'_\beta \circ f \circ \psi_\alpha^{-1}$ taking $U_\alpha \subset \mathbb{R}^n$ into $U'_\beta \subset \mathbb{R}^{n'}$ is C^∞ in the sense used in advanced calculus. If $f: M \to M'$ is C^∞, one-to-one, onto, and has C^∞ inverse, f is called a *diffeomorphism* and M and M' are said to be *diffeomorphic*. Diffeomorphic manifolds have identical manifold structure.

2.2 Vectors

The concept of a vector space is undoubtedly familiar to most readers. In prerelativity physics it is assumed that space has the natural structure of a three-dimensional vector space once one has designated a point to serve as the origin; the natural rules for adding and scalar multiplying spatial displacements satisfy the vector space axioms.[1] In special relativity, spacetime similarly has the natural structure of a four-dimensional vector space. However, when one considers curved geometries (as we do in general relativity), this vector space structure is lost. For example, there is no natural notion of how to "add" two points on a sphere and end up with a third point on the sphere. Nevertheless, vector space structure can be recovered in the limit of "infinitesimal displacements" about a point. It is this notion of "infinitesimal displacements" or tangent vectors which lies at the foundation of calculus on manifolds. Therefore, we will devote considerable attention below to giving a precise mathematical definition of this concept.

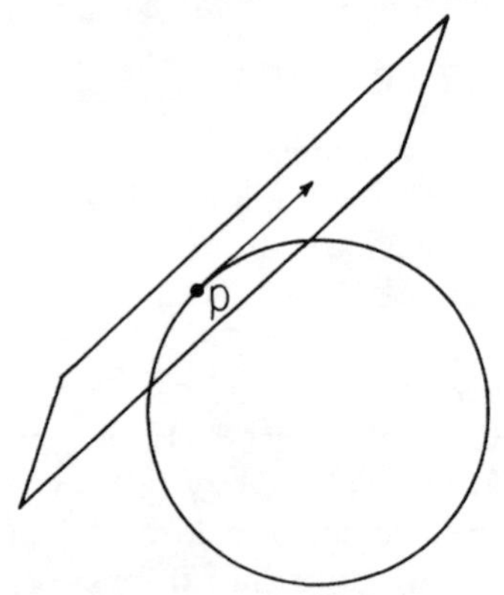

Fig. 2.2. The tangent plane at point p of a sphere in $\mathbb{R}^3$.

For manifolds like the sphere, which arise naturally as surfaces embedded in $\mathbb{R}^n$, the intuitive notion of a tangent vector at point p is a vector lying in the tangent plane illustrated in Figure 2.2. For manifolds embedded in $\mathbb{R}^n$, this idea can be made mathematically precise. However, in many situations—most importantly in general relativity—one is given a manifold without an embedding of it in $\mathbb{R}^n$. Thus, it is important (and, in the long run, much more useful) to define a tangent vector in a way that refers only to the intrinsic structure of the manifold, not to its possible embeddings in $\mathbb{R}^n$.

1. See, e.g., Royden (1963) for the list of vector space axioms.

Such a definition is provided by the notion of a tangent vector as a directional derivative. In $\mathbb{R}^n$ there is a one-to-one correspondence between vectors and directional derivatives. A vector $v = (v^1, \ldots, v^n)$ defines the directional derivative operator $\sum_\mu v^\mu(\partial/\partial x^\mu)$ and vice versa. Directional derivatives are characterized by their linearity and "Leibnitz rule" behavior when acting on functions. Thus on a manifold M let $\mathcal{F}$ denote the collection of C^∞ functions from M into $\mathbb{R}$. We define a tangent vector v at point $p \in M$ to be a map $v:\mathcal{F} \to \mathbb{R}$ which (1) is linear and (2) obeys the Leibnitz rule:

(1) $v(af + bg) = av(f) + bv(g)$, for all $f, g \in \mathcal{F}$; $a, b \in \mathbb{R}$;
(2) $v(fg) = f(p)v(g) + g(p)v(f)$.

Note that (1) and (2) imply that if $h \in \mathcal{F}$ is a constant function, i.e., $h(q) = c$ for all $q \in M$, then $v(h) = 0$, since from (2) we have $v(h^2) = 2cv(h)$ whereas from (1) we have $v(h^2) = v(ch) = cv(h)$.

Though it may not be obvious at first glance, this definition does indeed make precise and give intrinsic meaning to the concept of an "infinitesimal displacement." In the first place, it is easy to see that the collection, V_p, of tangent vectors at p has the structure of a vector space under the addition law $(v_1 + v_2)(f) = v_1(f) + v_2(f)$ and scalar multiplication law $(av)(f) = av(f)$. A second vital property of V_p is given by the following theorem:

THEOREM 2.2.1. *Let M be an n-dimensional manifold. Let $p \in M$ and let V_p denote the tangent space at p. Then* $\dim V_p = n$.

Proof. We shall show that $\dim V_p = n$ by constructing a basis of V_p, i.e., by finding n linearly independent tangent vectors which span V_p. Let $\psi: O \to U \subset \mathbb{R}^n$ be a chart with $p \in O$ (see Fig. 2.3). If $f \in \mathcal{F}$, then by definition $f \circ \psi^{-1}: U \to \mathbb{R}$ is C^∞. For $\mu = 1, \ldots, n$ define $X_\mu: \mathcal{F} \to \mathbb{R}$ by

$$X_\mu(f) = \frac{\partial}{\partial x^\mu}(f \circ \psi^{-1})\bigg|_{\psi(p)}, \tag{2.2.1}$$

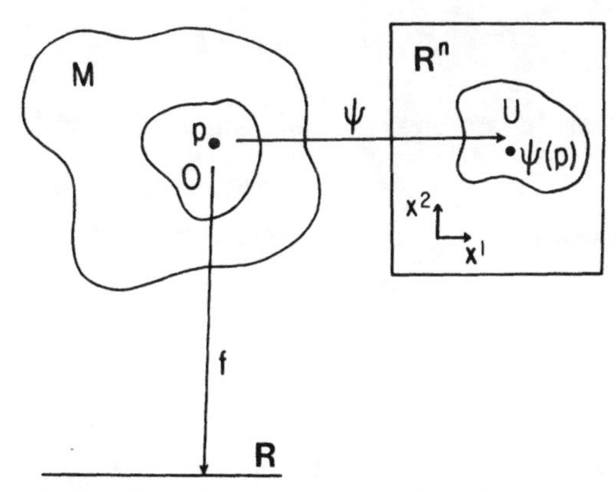

Fig. 2.3. A diagram illustrating the definition of the directional derivatives, X_μ, used in theorem 2.2.1.

where $(x^1, \ldots, x^n)$ are the Cartesian coordinates of $\mathbb{R}^n$. Then $X_1, \ldots, X_n$ are tangent vectors, and it is easily seen that they are linearly independent. To show that they span V_p we make use of the following result from advanced calculus (see problem 2): If $F: \mathbb{R}^n \to \mathbb{R}$ is C^∞, then for each $a = (a^1, \ldots, a^n) \in \mathbb{R}^n$ there exist C^∞ functions H_μ such that for all $x \in \mathbb{R}^n$ we have

$$F(x) = F(a) + \sum_{\mu=1}^{n} (x^\mu - a^\mu) H_\mu(x) \quad . \tag{2.2.2}$$

Furthermore, we have

$$H_\mu(a) = \left. \frac{\partial F}{\partial x^\mu} \right|_{x=a} \quad . \tag{2.2.3}$$

We apply this result here, letting $F = f \circ \psi^{-1}$ and $a = \psi(p)$. Then, for all $q \in O$ we have

$$f(q) = f(p) + \sum_{\mu=1}^{n} [x^\mu \circ \psi(q) - x^\mu \circ \psi(p)] H_\mu(\psi(q)) \quad . \tag{2.2.4}$$

Let $v \in V_p$. We wish to show that v is a linear combination of $X_1, \ldots, X_n$. To do so, we apply v to f, using equation (2.2.4), the linearity and Leibnitz properties of v, and the fact that v applied to a constant [such as $f(p)$] vanishes. We obtain

$$\begin{aligned} v(f) = v[f(p)] &+ \sum_{\mu=1}^{n} \left\{ [x^\mu \circ \psi(q) - x^\mu \circ \psi(p)] \Big|_{q=p} v(H_\mu \circ \psi) \right. \\ &+ \left. (H_\mu \circ \psi) \Big|_p v[x^\mu \circ \psi - x^\mu \circ \psi(p)] \right\} \\ &= \sum_{\mu=1}^{n} [H_\mu \circ \psi(p)] v(x^\mu \circ \psi) \quad . \end{aligned} \tag{2.2.5}$$

But by equation (2.2.3), $H_\mu \circ \psi(p)$ is just $X_\mu(f)$. Thus, for all $f \in \mathcal{F}$, we have

$$v(f) = \sum_{\mu=1}^{n} v^\mu X_\mu(f) \quad , \tag{2.2.6}$$

where the coefficients v^μ are the values of v applied to the function $x^\mu \circ \psi$,

$$v^\mu = v(x^\mu \circ \psi) \quad . \tag{2.2.7}$$

Thus, we have expressed an arbitrary tangent vector v as a sum of the X_μ,

$$v = \sum_{\mu=1}^{n} v^\mu X_\mu \quad , \tag{2.2.8}$$

which completes the proof. □

The basis $\{X_\mu\}$ of V_p introduced in the proof of theorem 2.2.1 is called a *coordinate basis* and is of considerable importance in its own right. Frequently, one denotes X_μ as simply $\partial/\partial x^\mu$. Had we chosen a different chart, ψ', we would have obtained a different coordinate basis $\{X'_\nu\}$. We can, of course, express X_μ in terms of the new basis $\{X'_\nu\}$. Using the chain rule of advanced calculus, we have

$$X_\mu = \sum_{\nu=1}^{n} \left. \frac{\partial x'^\nu}{\partial x^\mu} \right|_{\psi(p)} X'_\nu \quad , \tag{2.2.9}$$

where x'^ν denotes the νth component of the map $\psi' \circ \psi^{-1}$. Hence, from equations (2.2.8) and (2.2.9) we find that the components v'^ν of a vector v in the new coordinate basis are related to the components v^μ in the old basis by

$$v'^\nu = \sum_{\mu=1}^{n} v^\mu \frac{\partial x'^\nu}{\partial x^\mu} \quad . \tag{2.2.10}$$

Equation (2.2.10) is known as the *vector transformation law*.

A *smooth curve, C,* on a manifold M is simply a C^∞ map of $\mathbb{R}$ (or an interval of $\mathbb{R}$) into M, $C:\mathbb{R} \to M$. At each point $p \in M$ lying on the curve C we can associate with C a tangent vector $T \in V_p$ as follows. For $f \in \mathcal{F}$ we set $T(f)$ equal to the derivative of the function $f \circ C:\mathbb{R} \to \mathbb{R}$ evaluated at p, i.e., $T(f) = d(f \circ C)/dt$. Note that the above coordinate basis vector X_μ associated with a chart ψ is the tangent to the curves on M obtained by keeping all coordinate values except x^μ constant. Notice also that when we choose a coordinate system ψ, the curve C on M will get mapped into a curve $x^\mu(t)$ in $\mathbb{R}^n$. Then, for any $f \in \mathcal{F}$, we have

$$T(f) = \frac{d}{dt}(f \circ C) = \sum_\mu \frac{\partial}{\partial x^\mu}(f \circ \psi^{-1})\frac{dx^\mu}{dt} = \sum_\mu \frac{dx^\mu}{dt} X_\mu(f) \quad . \tag{2.2.11}$$

Thus, in any coordinate basis, the components T^μ of the tangent vector to the curve are given by

$$T^\mu = \frac{dx^\mu}{dt} \quad . \tag{2.2.12}$$

In the discussion above, we fixed a point $p \in M$ and considered the tangent space, V_p, at p. At another point $q \in M$ we could, of course, define V_q. It is important to emphasize that, given only the structure of a manifold, there is no natural way of identifying V_q with V_p; that is, there is no way of determining whether a tangent vector at q is "the same" as a tangent vector at p. In chapter 3, we shall see that when additional structure is given (namely, a connection or derivative operator on the manifold), one has a notion of "parallel transport" of vectors from p to q along a curve joining these points. However, if the curvature is nonzero, the identification of V_p with V_q obtained in this manner will depend on the choice of curve.

A *tangent field, v,* on a manifold M is an assignment of a tangent vector, $v|_p \in V_p$, at each point $p \in M$. Despite the fact that the tangent spaces V_p and V_q at different points are different vector spaces, there is a natural notion of what it means for v to vary smoothly from point to point. If f is a smooth (C^∞) function, then at each $p \in M$, $v|_p(f)$ is a number, i.e., $v(f)$ is a function on M. The tangent field v is said to be *smooth* if for each smooth function f, the function $v(f)$ is also smooth. Since the coordinate basis fields X_μ are easily verified to be smooth, it follows that a vector field v is smooth if and only if its coordinate basis components, v^μ, are smooth functions.

In the heuristic discussion above, we described tangent vectors as "infinitesimal displacements." We shall now show that precise meaning can be given to this idea. Let M be a manifold. A *one-parameter group of diffeomorphisms* ϕ_t is a C^∞ map from $\mathbb{R} \times M \to M$ such that for fixed $t \in \mathbb{R}$, $\phi_t : M \to M$ is a diffeomorphism and for all $t, s \in \mathbb{R}$, we have $\phi_t \circ \phi_s = \phi_{t+s}$. (In particular, this last relation implies that $\phi_{t=0}$ is the identity map.) We can associate to ϕ_t a vector field v as follows: for fixed $p \in M$, $\phi_t(p): \mathbb{R} \to M$ is a curve, called an *orbit* of ϕ_t, which passes through p at $t = 0$. Define $v|_p$ to be the tangent to this curve at $t = 0$. Thus, associated to a one-parameter group of (finite) transformations of M is a vector field, v, which can be thought of as the infinitesimal generator of these transformations.

Conversely, given a smooth vector field, v, on M we can ask if it is possible to find *integral curves* of v, that is, a family of curves in M having the property that one and only one curve passes through each point $p \in M$ and the tangent to this curve at p is $v|_p$. The answer is yes: If we pick a coordinate system in a neighborhood of p as in the proof of theorem 2.2.1, we see that the problem of finding such curves reduces to solving the system,

$$\frac{dx^\mu}{dt} = v^\mu(x^1, \ldots, x^n) \tag{2.2.13}$$

of ordinary differential equations in $\mathbb{R}^n$, where v^μ is the μth component of v in the coordinate basis $\{\partial/\partial x^\mu\}$. Such a system of equations has a unique solution given a starting point at $t = 0$, and thus every smooth vector field v has a unique family of integral curves (see, e.g., Coddington and Levinson 1955). Given the integral curves, for each $p \in M$ we define $\phi_t(p)$ to be the point lying at parameter t along the integral curve of v starting at p. Except for potential problems arising from the possibility that the integral curves of v may extend to only a finite value of the curve parameter, ϕ_t will be a one-parameter group of diffeomorphisms.

Finally, we note that given two smooth vector fields v and w it is possible to define a new vector field, $[v, w]$, called the *commutator* of v and w, by

$$[v, w](f) = v[w(f)] - w[v(f)] \tag{2.2.14}$$

(see problem 3). We note that the commutator of any two vector fields X_μ and X_ν occurring in a coordinate basis vanishes. (This fact follows directly from the definition of the coordinate basis given in the proof of theorem 2.2.1 together with the equality of mixed partial derivatives in $\mathbb{R}^n$.) Conversely, given a collection $X_1, \ldots, X_n$ of nonvanishing vector fields which commute with each other and are linearly independent at each point, one can always find a chart for which they are the coordinate basis vector fields (see problem 5).

2.3 Tensors; the Metric Tensor

Given the notion of displacement vectors, the notion of tensors arises when one considers other quantities of interest. It turns out that many quantities have a linear (or multilinear) dependence on displacements. Consider, for example, a measurement of the magnetic field (say, in the context of prerelativity physics). For each probe orientation, a number is recorded: the magnetic field strength in that given direction. Since there is an infinite number of possible orientations of the probe, in principle an infinite number of readings would be needed to determine the magnetic field. However, this is not necessary because the magnetic field strength has a linear dependence on the probe orientation. All that is required is readings in three linearly independent probe directions; the reading in any other probe direction is equal to a linear combination of these readings.

This fact gives rise to the notion of the magnetic field as a vector or, more precisely, a dual vector. One could define a dual vector as a collection of three numbers (i.e., the probe readings) associated with a basis of spatial displacement vectors (the three independent probe directions) which transform in an appropriate manner when the basis is changed. However, we will give below a simpler and more direct definition of a dual vector as a linear map from spatial displacement vectors into numbers. We have defined the magnetic field as a dual vector here, but, as we shall see at the end of this section, because space has a metric defined on it, we can naturally associate to any dual vector an ordinary (spatial displacement) vector.

In a similar manner, other quantities that occur in physics have a similar linear dependence on spatial displacement vectors but may be functions of more than one such vector. For example, for an ordinary material body in equilibrium, consider the plane with normal vector $\vec{n}$ passing through a point p in the body. At p, one could measure the force per unit area, F, in the $\vec{l}$-direction exerted on the matter on one side of the plane by the matter on the other side. One finds that F depends linearly on the choice of $\vec{n}$ and $\vec{l}$. Thus, although there is an infinite number of possible choices of $\vec{n}$ and $\vec{l}$, the value of F for any $\vec{n}$ and $\vec{l}$ can be calculated by knowing 3×3 numbers, namely the values F takes when $\vec{n}$ and $\vec{l}$ point in basis directions. This motivates the definition of a tensor which will be given below: A tensor is a multilinear (i.e., linear in each variable) map from vectors (or dual vectors) into numbers. The tensor which maps the pair of vectors $(\vec{n}, \vec{l})$ into the value of F is known as the *stress tensor* of the material body at p.

We give now the precise mathematical definition of tensors and discuss their properties. Let V be any finite-dimensional vector space over the real numbers. (The case of prime interest for us is the tangent space, $V = V_p$.) Consider the collection, V^*, of linear maps $f: V \to \mathbb{R}$. If one defines addition and scalar multiplication of such linear maps in the obvious way, one gets a natural vector space structure on V^*. We call V^* the *dual vector space* to V, and elements of V^* are called *dual vectors*. If $v_1, \ldots, v_n$ is a basis of V, we can define elements $v^{1*}, \ldots, v^{n*} \in V^*$ by

$$v^{\mu*}(v_\nu) = \delta^\mu{}_\nu \quad , \tag{2.3.1}$$

where $\delta^\mu{}_\nu = 1$ if $\mu = \nu$ and 0 otherwise. (This defines the action of $v^{\mu*}$ on the basis elements; its action on an arbitrary vector $v \in V$ is determined by this and linearity.) It follows directly (problem 6) that $\{v^{\mu*}\}$ is a basis of V^*, called the *dual basis* to the basis $\{v_\mu\}$ of V. In particular, this shows that dim V^* = dim V. The correspondence $v_\mu \longleftrightarrow v^{\mu*}$ gives rise to an isomorphism between V and V^*, but this isomorphism depends on the choice of basis $\{v_\mu\}$, so there is no natural way of identifying V^* with V (unless more structure is given on V, such as a preferred basis or, as described below, a metric).

We now can apply the above construction starting with the vector space V^*, thereby obtaining the *double dual* vector space to V, denoted V^{**}. A vector v^{**} in V^{**} is a linear map from V^* into $\mathbb{R}$. However, V^{**} is naturally isomorphic to the original vector space V. To each vector v in V we can associate the map in V^{**} whose value on the vector $\omega^* \in V^*$ is just $\omega^*(v)$. In this way, we obtain a one-to-one linear map of V into V^{**} which must be onto since dim V = dim V^{**}. Thus, taking the double dual gives nothing new; we can naturally identify V^{**} with the original vector space V. This identification will be assumed in the discussion below.

We now are ready to define the notion of a tensor. Let V be a finite dimensional vector space and let V^* denote its dual vector space. A *tensor, T, of type (k, l)* over V is a multilinear map

$$T: \underbrace{V^* \times \ldots \times V^*}_{k} \times \underbrace{V \times \ldots \times V}_{l} \to \mathbb{R} \quad .$$

In other words, given k dual vectors and l ordinary vectors, T produces a number, and it does so in such a manner that if we fix all but one of the vectors or dual vectors, it is a linear map on the remaining variable.

Thus, according to the above definition, a tensor of type $(0, 1)$ is precisely a dual vector. Similarly, a tensor of type $(1, 0)$ is an element of V^{**}. However, since we identify V^{**} with V, a tensor of type $(1, 0)$ is nothing more than an ordinary vector. Because of the identification of V^{**} with V, we may view tensors of higher type in many different (though, or course, equivalent) ways. For example, a tensor T of type $(1, 1)$ is a multilinear map from $V^* \times V \to \mathbb{R}$. Hence, if we fix $v \in V$, $T(\cdot, v)$ is an element of V^{**}, which we identify with an element of V. Thus, given a vector in V, T produces another vector in V in a linear fashion. In other words, we can view a tensor of type $(1, 1)$, as a linear map from V into V, and vice versa. Similarly, we can view T as a linear map from V^* into V^*.

With the obvious rules for adding and scalar multiplying maps, the collection $\mathcal{T}(k, l)$ of all tensors of type (k, l) has the structure of a vector space. Because of the multilinearity property, a tensor is uniquely specified by giving its values on vectors in a basis $\{v_\mu\}$ of V and its dual basis $\{v^{\nu*}\}$ of V^*. Since there are n^{k+l} independent ways of filling the slots of a tensor of type (k, l) with such basis vectors (where n = dim V = dim V^*), the dimension of the vector space $\mathcal{T}(k, l)$ is n^{k+l}.

We now introduce two simple but important operations on tensors, which will be used commonly in what follows. The first is called *contraction* with respect to the ith (dual vector) and jth (vector) slots and is a map $C: \mathcal{T}(k, l) \to \mathcal{T}(k-1, l-1)$ defined as follows. If T is a tensor of type (k, l), then

$$CT = \sum_{\sigma=1}^{n} T(\ldots, v^{\sigma*}, \ldots; \ldots, v_\sigma, \ldots) \quad , \tag{2.3.2}$$

where $\{v_\sigma\}$ is a basis of V, $\{v^{\sigma*}\}$ is its dual basis, and these vectors are inserted into the ith and jth slots of T. [Note that the contraction of a tensor of type $(1, 1)$, viewed as a linear map from V into V, is just the trace of the map.] The tensor CT thus obtained is independent of the choice of basis $\{v_\mu\}$, so the operation of contraction is indeed well defined (see problem 6).

The second operation on tensors is the outer product. Given a tensor T of type (k, l) and another tensor T' of type (k', l'), we can construct a new tensor of type $(k+k', l+l')$ called the *outer product* of T and T' and denoted $T \otimes T'$, by the following simple rule. Given $(k+k')$ dual vectors $v^{1*}, \ldots, v^{k+k'*}$ and $(l+l')$ vectors $w_1, \ldots, w_{l+l'}$, we define $T \otimes T'$ acting on these vectors to be the product of $T(v^{1*}, \ldots, v^{k*}; w_1, \ldots, w_l)$ and $T'(v^{k+1*}, \ldots, v^{k+k'*}; w_{l+1}, \ldots, w_{l+l'})$.

Thus, one way of constructing tensors is to take outer products of vectors and dual vectors. A tensor which can be expressed as such an outer product is called *simple*.[2] If $\{v_\mu\}$ is a basis of V and $\{v^{\nu*}\}$ is its dual basis, it is easy to show that the n^{k+l} simple tensors $\{v_{\mu_1} \otimes \cdots \otimes v_{\mu_k} \otimes v^{\nu_1*} \otimes \cdots \otimes v^{\nu_l*}\}$ yield a basis of $\mathcal{T}(k, l)$. Thus, every tensor T of type (k, l) can be expressed as a sum of simple tensors in this collection

$$T = \sum_{\mu_1, \ldots, \nu_l = 1}^{n} T^{\mu_1 \cdots \mu_k}{}_{\nu_1 \cdots \nu_l} v_{\mu_1} \otimes \cdots \otimes v^{\nu_l*} \quad . \tag{2.3.3}$$

The basis expansion coefficients, $T^{\mu_1 \cdots \mu_k}{}_{\nu_1 \cdots \nu_l}$, are called the *components* of the tensor T with respect to the basis $\{v_\mu\}$. Note that we follow the standard convention in the notation for components of using superscripts for labels μ_i associated with vectors and subscripts for labels ν_j associated with dual vectors.

In terms of components, we have the following formulas for contraction and outer product. Suppose the tensor T has components $T^{\mu_1 \cdots \mu_k}{}_{\nu_1 \cdots \nu_l}$ as in equation (2.3.3). Then, the contraction, CT, of T has components given by

$$(CT)^{\mu_1 \cdots \mu_{k-1}}{}_{\nu_1 \cdots \nu_{l-1}} = \sum_{\sigma=1}^{n} T^{\mu_1 \cdots \sigma \cdots \mu_{k-1}}{}_{\nu_1 \cdots \sigma \cdots \nu_{l-1}} \quad . \tag{2.3.4}$$

If T' has components $T'^{\mu'_1 \cdots \mu'_{k'}}{}_{\nu'_1 \cdots \nu'_{l'}}$, then the outer product $S = T \otimes T'$ has components given by

$$S^{\mu_1 \cdots \mu_{k+k'}}{}_{\nu_1 \cdots \nu_{l+l'}} = T^{\mu_1 \cdots \mu_k}{}_{\nu_1 \cdots \nu_l} T'^{\mu_{k+1} \cdots \mu_{k+k'}}{}_{\nu_{l+1} \cdots \nu_{l+l'}} \quad . \tag{2.3.5}$$

The above discussion applies to an arbitrary finite-dimensional vector space V. Let us now consider the case of prime interest for us, where V is the tangent space, V_p, at point p of a manifold M. In this case, V_p^* is commonly called the *cotangent space* at p and vectors in V_p^* are called *cotangent vectors*. We also commonly refer to vectors in V_p as *contravariant* vectors and vectors in V_p^* as *covariant* vectors. As discussed in section 2.2, given a coordinate system, we can construct a coordinate basis $\partial/\partial x^1, \ldots, \partial/\partial x^n$ of V_p. The associated dual basis of V_p^* is usually denoted as $dx^1, \ldots, dx^n$. [Thus, we stress that dx^μ is merely the symbol for the linear map

2. In many references, a tensor of type $(0, l)$ which can be expressed as the totally antisymmetric part of a simple tensor (see eq. [2.4.4] below) also is referred to as simple.

defined by $dx^\mu(\partial/\partial x^\nu) = \delta^\mu{}_\nu$.] If we change coordinate systems, we already showed that the components $v'^{\mu'}$ of a vector v in the new basis are related to its components v^μ in the old basis by the vector transformation law,

$$v'^{\mu'} = \sum_{\mu=1}^{n} v^\mu \frac{\partial x'^{\mu'}}{\partial x^\mu} \quad . \tag{2.3.6}$$

If ω_μ denotes the components of a dual vector ω with respect to the dual basis $\{dx^\mu\}$, then from equations (2.3.1) and (2.3.6) it follows that under a coordinate transformation its components become

$$\omega'_{\mu'} = \sum_{\mu=1}^{n} \omega_\mu \frac{\partial x^\mu}{\partial x'^{\mu'}} \quad . \tag{2.3.7}$$

In general, the components of a tensor T of type (k, l) transform as

$$T'^{\mu'_1 \cdots \mu'_k}{}_{\nu'_1 \cdots \nu'_l} = \sum_{\mu_1, \ldots, \nu_l = 1}^{n} T^{\mu_1 \cdots \mu_k}{}_{\nu_1 \cdots \nu_l} \frac{\partial x'^{\mu'_1}}{\partial x^{\mu_1}} \cdots \frac{\partial x^{\nu_l}}{\partial x'^{\nu'_l}} \quad . \tag{2.3.8}$$

Equation (2.3.8) is known as the *tensor transformation law*.

In other treatments, equation (2.3.8) often is used as the defining property of a tensor. The definition we have given here has the advantage that it generally is much easier to define a quantity as a tensor by displaying it as a multilinear map on vectors and dual vectors than it is to display it as a collection of numbers associated with a coordinate system which changes according to equation (2.3.8) when we change coordinate systems. In fact, as we shall illustrate throughout this book, it is rarely worthwhile to introduce a basis and take components of a tensor at all, let alone to worry about how these components change under a change of basis.

An assignment of a tensor over V_p for each point p in the manifold M is called a *tensor field*. The notions of smoothness of a function and of a (contravariant) vector field v were already defined in section 2.2. A covariant vector field ω is said to be smooth (C^∞) if for each smooth vector field v, the function $\omega(v)$ is smooth. A tensor T of type (k, l) is said to be smooth if for all smooth covariant vectors fields $\omega^1, \ldots,$ ω^k and smooth contravariant vector fields $v_1, \ldots, v_l$, $T(\omega^1, \ldots, \omega^k; v_1, \ldots, v_l)$ is a smooth function. The notion of a tensor field being C^k is defined similarly.

We now introduce the notion of a metric. Intuitively, a metric is supposed to tell us the "infinitesimal squared distance" associated with an "infinitesimal displacement." As discussed above in section 2.2, the intuitive notion of an "infinitesimal displacement" is precisely captured by the concept of a tangent vector. Thus, since "infinitesimal squared distance" should be quadratic in the displacement, a metric, g, should be a linear map from $V_p \times V_p$ into numbers, i.e. a tensor of type $(0, 2)$. In addition the metric is required to be symmetric and nondegenerate. By *symmetric*, we mean that for all $v_1, v_2 \in V_p$ we have $g(v_1, v_2) = g(v_2, v_1)$. By *nondegenerate*, we mean that the only case in which we have $g(v, v_1) = 0$ for all $v \in V_p$ is the case $v_1 = 0$. Thus, a *metric*, g, on a manifold M is a symmetric, nondegenerate tensor field of type $(0, 2)$. In other words, a metric is a (not necessarily positive definite) inner product on the tangent space at each point.

In a coordinate basis, we may expand a metric g in terms of its components $g_{\mu\nu}$ as

$$g = \sum_{\mu, \nu} g_{\mu\nu} dx^{\mu} \otimes dx^{\nu} \quad . \tag{2.3.9}$$

Sometimes the notation ds^2 is used in place of g to represent the metric tensor, in which case we write equation (2.3.9) as

$$ds^2 = \sum_{\mu, \nu} g_{\mu\nu} dx^{\mu} dx^{\nu} \quad , \tag{2.3.10}$$

where, following standard practice, we have omitted writing the outer product sign between dx^{μ} and dx^{ν}. The notation of equation (2.3.10) conveys the intuitive flavor of a metric as representing "infinitesimal squared distance."

Given a metric g, we always can find an *orthonormal basis* $v_1, \ldots, v_n$ of the tangent space at each point p, i.e., a basis such that $g(v_{\mu}, v_{\nu}) = 0$ if $\mu \neq \nu$ and $g(v_{\mu}, v_{\mu}) = \pm 1$ (see problem 7). There are, of course, many other orthonormal bases at p, but the number of basis vectors with $g(v_{\mu}, v_{\mu}) = +1$ and the number with $g(v_{\mu}, v_{\mu}) = -1$ are independent of choice of orthonormal basis (problem 7). The number of $+$ and $-$ signs occurring is called the *signature* of the metric. In ordinary differential geometry, one usually deals with *positive definite* metrics, i.e., metrics with signature $+ + \ldots +$. On the other hand, the metric of spacetime has a signature $- + + +$. Positive definite metrics are called *Riemannian;* metrics with signatures like those on spacetime (one minus and the remainder plus) are called *Lorentzian*.

As defined above, at each point $p \in M$ a metric g is a tensor of type $(0, 2)$ over V_p, i.e., a multilinear map from $V_p \times V_p \to \mathbb{R}$. However, we can also view g as a linear map from V_p into V_p^* via $v \to g(\cdot, v)$. Because of the nondegeneracy of g, this map is one-to-one and onto. In particular the inverse map exists. Thus, we can use g to establish a one-to-one correspondence between vectors and dual vectors. Indeed, given a metric g we could use this correspondence to entirely circumvent the necessity of introducing dual vectors. Normally this is done and accounts for why the concept of a dual vector is not more familiar to most physicists. However, in general relativity we shall be solving for the metric of spacetime; since the metric is not known from the start, it is essential that we keep the distinction between vectors and dual vectors completely clear.

2.4 The Abstract Index Notation

In the previous section we introduced the notion of tensors and defined a number of operations that can be performed on them. However, if one performs even the simplest manipulations, serious notational problems arise for the following reasons: (1) Tensors of high type are functions of many vectors and dual vectors. In operations such as contraction one has to keep track of which slots are involved. Introduction of a new symbol to denote, say, a particular contraction of a given tensor becomes extremely cumbersome and can make simple operations appear very complicated.

(2) As mentioned above, a given tensor can be viewed in a variety of equivalent ways. It is important that a simple, consistent notational scheme be developed so that the same expressions are written down regardless of the viewpoint taken.

A notation which solves the above problems and is used in most relativity texts as well as most older differential geometry texts is the following. As noted in section 2.3, if we introduce a basis, we can characterize a tensor by its components $T^{\mu_1 \cdots \mu_k}{}_{\nu_1 \cdots \nu_l}$. The notation consists of writing all equations in terms of these components. This solves problems (1) and (2), since one has unique, simple expressions for operations such as contraction and outer products in terms of components.

However, this component notation has a serious drawback. If we do not specify how the basis we use is to be chosen, the equations we write down will be true tensor equations, having basis-independent meaning. However, in some cases it will be convenient to use a particular type of basis, e.g., a coordinate basis adapted to the symmetries of a particular spacetime. If we do this, then the equations we write down for the tensor components may be valid only in this basis. It is important to make a clear distinction between equations that hold between tensors and equations for their components that hold only in a special basis. However, this distinction is blurred by the component notation.

We shall use a notation, the *abstract index notation,* which in practice is merely a slight modification of the component notation. It has all the advantages of the component notation but avoids the above drawback. The idea is *not* to introduce a basis but to use a notation for tensors that mirrors the expressions for their basis components (had we introduced a basis). The rules are as follows. A tensor of type (k, l) will be denoted by a letter followed by k contravariant and l covariant, lowercase latin indices, $T^{a_1 \cdots a_k}{}_{b_1 \cdots b_l}$. Thus, for example, $T^{abc}{}_{de}$ denotes[3] a tensor of type $(3, 2)$. The latin indices here should be viewed as reminders of the number and type of variables the tensor acts on, *not* as basis components. Any lowercase latin letters can be placed in any slot, but in any equation the same letter must be used to represent the same slot on both sides of the equation. Mirroring the component expression, equation (2.3.4) (but omitting the summation sign), we denote the contraction of a tensor by using the same letter as for the tensor but repeating the index on the contracted slots. Thus, $T^{abc}{}_{be}$ denotes the tensor of type (2, 1), obtained by contracting $T^{abc}{}_{de}$ with respect to the second contravariant and first covariant slots. The outer product of two tensors is denoted by simply writing them adjacent to each other (and omitting the $\otimes$ sign). Thus, $T^{abc}{}_{de} S^f{}_g$ denotes the tensor of type $(4, 3)$ obtained by taking the outer product of $T^{abc}{}_{de}$ and $S^a{}_b$.

Using the index notation, one only can write down true tensor equations, since no basis has been introduced. If we were to introduce a basis, one could of course take components and write equations for them. To distinguish between equations for components and the (very similar looking) tensor equations in the index notation, we

3. More precisely, we may view $T^{abc}{}_{de}$ as consisting of the tensor T and the elements a, b, c, d, e of a labeling set which mark the "slots" of this tensor. See Penrose and Rindler (1984) for further discussion.

adhere to the following conventions. Component labels in the component notation always will be denoted with greek letters as has been done above. Thus, for example, $T^{\mu\nu\lambda}{}_{\sigma\rho}$ denotes a basis component of the tensor $T^{abc}{}_{de}$. Given any tensor equation expressed in the index notation, the corresponding equation (with greek letters replacing latin ones in the superscripts and subscripts) holds for the basis components in any basis. Conversely, for any equation relating basis components which is a true tensor equation (i.e., is valid independently of how the basis is chosen), the corresponding tensor equation in the index notation is valid.

Thus, the distinction between the index notation and the component notation is much more one of spirit (i.e., how one thinks of the quantities appearing) than of substance (i.e., the physical form the equations take). The main advantages of the index notation are that one is not forced to introduce a basis unnecessarily and one is assured that all equations written in the index notation are equations holding between tensors, since only true tensor equations can be expressed in this notation. In the cases where one wishes to write a nontensorial equation for the basis components in a particular basis, the component notation may, of course, still be used. In this manner a clear distinction can be seen in the notation between true tensor equations and equations for components holding in a particular basis.

Additional notational rules apply to the metric tensor, both in the index and component notations. Since a metric g is a tensor of type $(0, 2)$, it is denoted g_{ab}. If we apply the metric to a vector, v^a, we get the dual vector $g_{ab}v^b$. It is convenient to denote this vector as simply v_a, thus making notationally explicit the isomorphism between V_p and V_p^* defined by g_{ab}. The inverse of g_{ab} (which exists, as remarked at the end of section 2.3, on account of the nondegeneracy of g_{ab}) is a tensor of type $(2, 0)$ and could be denoted as $(g^{-1})^{ab}$. It is convenient, however, to drop the inverse sign and denote it simply as g^{ab}. No confusion arises from this since the upper position of the indices distinguishes the inverse metric from the metric. Thus, by definition, $g^{ab}g_{bc} = \delta^a{}_c$, where $\delta^a{}_c$ (viewed as a map from V_p into V_p) is the identity map. If we apply the inverse metric to a dual vector ω_a, we denote the resultant vector $g^{ab}\omega_b$ as simply ω^a. In general, raised or lowered indices on any tensor denote application of the metric or inverse metric to that slot. Thus, for example, if $T^{abc}{}_{de}$ is a tensor of type $(3, 2)$, then $T^a{}_b{}^{cde}$ denotes the tensor $g_{bf}g^{dh}g^{ej}T^{afc}{}_{hj}$. This notation is self-consistent since the tensor resulting from the successive raising and lowering of a given index is identical to the original tensor. Furthermore, the notation also is self-consistent when applied to the metric itself, since $g^{ab} = g^{ac}g^{bd}g_{cd}$, i.e., g^{ab} *is* the tensor g_{ab} with its indices raised.

The index notation may also be used to express the symmetry properties of tensors. A tensor T_{ab} of type $(0, 2)$ takes a pair of vectors (v^a, w^a) into a number $T_{ab}v^aw^b$. We may wish to consider the new tensor obtained by interchanging the order in which the tensor T_{ab} acts on this pair of vectors, i.e., the tensor which takes (v^a, w^a) into $T_{ab}v^bw^a$. In the index notation this new tensor is denoted T_{ba}. Thus, for example, the equation $T_{ab} = T_{ba}$ says that the tensor T_{ab} is symmetric. Similar notational rules apply to any pair of covariant or contravariant indices of tensors of higher type.

It is convenient to introduce a notation for the totally symmetric and totally antisymmetric parts of tensors. If T_{ab} is a tensor of type $(0, 2)$, we define

$$T_{(ab)} = \frac{1}{2}(T_{ab} + T_{ba}) \quad , \tag{2.4.1}$$

$$T_{[ab]} = \frac{1}{2}(T_{ab} - T_{ba}) \quad . \tag{2.4.2}$$

More generally, for a tensor $T_{a_1 \cdots a_l}$ of type $(0, l)$ we define

$$T_{(a_1 \cdots a_l)} = \frac{1}{l!} \sum_{\pi} T_{a_{\pi(1)} \cdots a_{\pi(l)}} \quad , \tag{2.4.3}$$

$$T_{[a_1 \cdots a_l]} = \frac{1}{l!} \sum_{\pi} \delta_{\pi} T_{a_{\pi(1)} \cdots a_{\pi(l)}} \quad , \tag{2.4.4}$$

where the sum is taken over all permutations, π, of $1, \ldots, l$ and δ_π is $+1$ for even permutations and -1 for odd permutations. Similar definitions apply for any group of bracketed covariant or contravariant indices; e.g., we have

$$T^{(ab)c}{}_{[de]} = \frac{1}{4}[T^{abc}{}_{de} + T^{bac}{}_{de} - T^{abc}{}_{ed} - T^{bac}{}_{ed}] \quad . \tag{2.4.5}$$

A totally antisymmetric tensor field $T_{a_1 \cdots a_l}$ of type $(0, l)$,

$$T_{a_1 \cdots a_l} = T_{[a_1 \cdots a_l]} \quad , \tag{2.4.6}$$

is called a *differential l-form*. Some properties of differential forms are obtained in appendix B. If one is dealing strictly with differential forms, it is sometimes convenient to drop the index notation and denote an l-form $T_{a_1 \cdots a_l}$ as simply T. However, except for some isolated instances of dealing with differential forms and a few cases where the index notation can be confusing, such as with commutators and Lie derivatives (see appendix C), we will use the index notation throughout the book.

Problems

1. *a*) Show that the overlap functions $f_i^{\pm} \circ (f_j^{\pm})^{-1}$ are C^∞, thus completing the demonstration given in section 2.1 that S^2 is a manifold.

b) Show by explicit construction that two coordinate systems (as opposed to the six used in the text) suffice to cover S^2. (It is impossible to cover S^2 with a single chart, as follows from the fact that S^2 is compact, but every open subset of $\mathbb{R}^2$ is noncompact; see appendix A.)

2. Prove that any smooth function $F : \mathbb{R}^n \to \mathbb{R}$ can be written in the form equation (2.2.2). (Hint: For $n = 1$, use the identity

$$F(x) - F(a) = (x - a) \int_0^1 F'[t(x - a) + a]\, dt \quad ;$$

then prove it for general n by induction.)

3. *a*) Verify that the commutator, defined by equation (2.2.14), satisfies the linearity and Leibnitz properties, and hence defines a vector field.

b) Let X, Y, Z be smooth vector fields on a manifold M. Verify that their commutator satisfies the Jacobi identity:

$$[[X,Y],Z] + [[Y,Z],X] + [[Z,X],Y] = 0 \quad .$$

c) Let $Y_1, \ldots, Y_n$ be smooth vector fields on an n-dimensional manifold M such that at each $p \in M$ they form a basis of the tangent space V_p. Then, at each point, we may expand each commutator $[Y_\alpha, Y_\beta]$ in this basis, thereby defining the functions $C^\gamma{}_{\alpha\beta} = -C^\gamma{}_{\beta\alpha}$ by

$$[Y_\alpha, Y_\beta] = \sum_\gamma C^\gamma{}_{\alpha\beta} Y_\gamma \quad .$$

Use the Jacobi identity to derive an equation satisfied by $C^\gamma{}_{\alpha\beta}$. (This equation is a useful algebraic relation if the $C^\gamma{}_{\alpha\beta}$ are constants, as will be the case if $Y_1, \ldots, Y_n$ are left [or right] invariant vector fields on a Lie group [see section 7.2].)

4. *a*) Show that in any coordinate basis, the components of the commutator of two vector fields v and w are given by

$$[v, w]^\mu = \sum_\nu \left(v^\nu \frac{\partial w^\mu}{\partial x^\nu} - w^\nu \frac{\partial v^\mu}{\partial x^\nu} \right) \quad .$$

b) Let $Y_1, \ldots, Y_n$ be as in problem 3(*c*). Let $Y^{1*}, \ldots, Y^{n*}$ be the dual basis. Show that the components $(Y^{\gamma*})_\mu$ of $Y^{\gamma*}$ in any coordinate basis satisfy

$$\frac{\partial (Y^{\gamma*})_\mu}{\partial x^\nu} - \frac{\partial (Y^{\gamma*})_\nu}{\partial x^\mu} = \sum_{\alpha,\beta} C^\gamma{}_{\alpha\beta} (Y^{\alpha*})_\mu (Y^{\beta*})_\nu \quad .$$

(Hint: Contract both sides with $(Y_\sigma)^\mu (Y_\rho)^\nu$.)

5. Let $Y_1, \ldots, Y_n$ be smooth vector fields on an n-dimensional manifold M which form a basis of V_p at each $p \in M$. Suppose $[Y_\alpha, Y_\beta] = 0$ for all α, β. Prove that in a neighborhood of each $p \in M$ there exist coordinates $y_1, \ldots, y_n$ such that $Y_1, \ldots, Y_n$ are the coordinate vector fields, $Y_\mu = \partial/\partial y^\mu$. (Hint: In an open ball of $\mathbb{R}^n$, the equations $\partial f/\partial x^\mu = F_\mu$ with $\mu = 1, \ldots, n$ for the unknown function f have a solution if and only if $\partial F_\mu/\partial x^\nu = \partial F_\nu/\partial x^\mu$. [See the end of section B.1 of appendix B for a statement of generalizations of this result.] Use this fact together with the results of problem 4(*b*) to obtain the new coordinates.)

6. *a*) Verify that the dual vectors $\{v^{\mu*}\}$ defined by equation (2.3.1) constitute a basis of V^*.

b) Let $v_1, \ldots, v_n$ be a basis of the vector space V and let $v^{1*}, \ldots, v^{n*}$ be its dual basis. Let $w \in V$ and let $\omega \in V^*$. Show that

$$w = \sum_\alpha v^{\alpha*}(w) v_\alpha \quad ,$$

$$\omega = \sum_\alpha \omega(v_\alpha) v^{\alpha*} \quad .$$

c) Prove that the operation of contraction, equation (2.3.2), is independent of the choice of basis.

7. Let V be an n-dimensional vector space and let g be a metric on V.
a) Show that one always can find an orthonormal basis $v_1, \ldots, v_n$ of V, i.e., a basis such that $g(v_\alpha, v_\beta) = \pm\delta_{\alpha\beta}$. (Hint: Use induction.)
b) Show that the signature of g is independent of the choice of orthonormal basis.

8. *a*) The metric of flat, three-dimensional Euclidean space is

$$ds^2 = dx^2 + dy^2 + dz^2 \quad .$$

Show that the metric components $g_{\mu\nu}$ in spherical polar coordinates r, θ, ϕ defined by

$$r = (x^2 + y^2 + z^2)^{1/2} \quad ,$$

$$\cos\theta = z/r \quad ,$$

$$\tan\phi = y/x$$

is given by

$$ds^2 = dr^2 + r^2\, d\theta^2 + r^2 \sin^2\theta\, d\phi^2 \quad .$$

b) The spacetime metric of special relativity is

$$ds^2 = -dt^2 + dx^2 + dy^2 + dz^2 \quad .$$

Find the components, $g_{\mu\nu}$ and $g^{\mu\nu}$, of the metric and inverse metric in "rotating coordinates," defined by

$$t' = t \quad ,$$

$$x' = (x^2 + y^2)^{1/2} \cos(\phi - \omega t) \quad ,$$

$$y' = (x^2 + y^2)^{1/2} \sin(\phi - \omega t) \quad ,$$

$$z' = z \quad ,$$

where $\tan\phi = y/x$.

THREE

CURVATURE

Our intuitive notion of curvature arises mainly from two-dimensional surfaces which are embedded in ordinary three-dimensional Euclidean space. We normally think of a surface as curved because of the way it bends in $\mathbb{R}^3$. In chapter 9, we will capture this notion by defining the *extrinsic* curvature of a surface embedded in a higher dimensional space. However, our interest here is to investigate the curvature of spacetime. Our spacetime manifold M with spacetime metric g_{ab} is not naturally embedded (at least so far as we know) in a higher dimensional space. Thus we wish to develop an *intrinsic* notion of curvature that can be applied to any manifold without reference to a higher dimensional space in which it might be embedded.

Such a notion of curvature can be defined in terms of parallel transport. On a surface such as a plane (Fig. 3.1) or sphere (Fig. 3.2), we have an intuitive notion (which will be made mathematically precise below) of what it means to keep a vector "pointing in the same direction" (but always in the tangent space of the manifold) as one moves it along a path. On the plane, if one parallel-transports a vector around any closed path, the final vector always coincides with its initial value. However, this is not so on the sphere. The vector shown in Figure 3.2 comes back rotated with respect to its initial value when carried along the path shown. This basic idea allows us to characterize the plane as flat, the sphere as curved, and more generally allows us to characterize the curvature of any manifold intrinsically once we are told how to "parallel transport" vectors along curves.

An alternative characterization of curvature also can be given as follows. A geodesic is a curve whose tangent is parallel-transported along itself, that is, it is a "straightest possible" curve. A space will be curved if and only if some initially parallel geodesics fail to remain parallel, i.e., Euclid's fifth postulate fails.

Given only the manifold structure of space, we do not have a natural notion of parallel transport. The reason is that the tangent space V_p and V_q of two distinct points p and q are *different* vector spaces and hence there is no way of saying that a vector

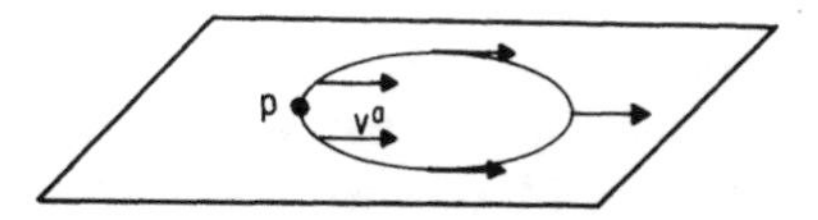

Fig. 3.1. The parallel transport of a vector, v^a, around a closed curve in the plane. The vector v^a always "comes back" pointing in the same direction as it did initially.

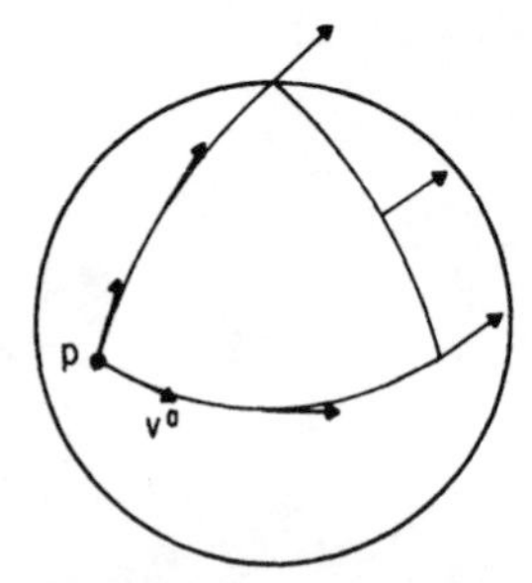

Fig. 3.2. The parallel transport of a vector, v^a, around a closed curve on the sphere. In the case shown here of a closed curved composed of three mutually orthogonal segments of great circles, the vector v^a comes back rotated by 90°.

at p is the same as a vector at q. Thus, the definition of parallel transport requires more than just the manifold structure. It is not difficult to convince oneself that a notion of how to parallel-transport vectors should be equivalent to the knowledge of how to take derivatives of vector fields. If we know how to parallel-transport vectors along a curve, we can define the derivative of a vector field in the direction of the curve; similarly, given a notion of derivative, we can define a vector to be parallel transported if its derivative along the given curve is zero. It turns out to be most convenient to work directly with the notion of a derivative operator, and we shall do so in this chapter. The failure of a vector to return to its original value when parallel transported around an infinitesimal closed curve translates into the lack of commutativity of derivatives. Thus, the notion of curvature can be defined in terms of the failure of successive differentiations on tensor fields to commute. This is the route we shall follow in section 3.2.

From where does this extra structure needed to define parallel transport or a derivative operator arise? We will show in section 3.1 that given a metric (of any signature), there is a unique definition of parallel transport which preserves inner products of all pairs of vectors. Thus, the existence of a metric gives rise to a unique notion of parallel transport and, thus, to an intrinsic notion of the curvature of the manifold. This is the notion of the curvature of a spacetime (M, g_{ab}) in which we are interested. However, it is more convenient to proceed by giving a general definition of the notions of derivative operator, parallel transport, and curvature before specializing to the case where they arise from a metric, and we shall proceed in this manner.

Derivative operators and parallel transport are defined in section 3.1, and curvature is defined in section 3.2. In much of our discussion in these sections, we shall follow closely the treatment given in unpublished notes of Geroch. Geodesics are introduced in section 3.3, and the geodesic deviation equation—which characterizes curvature in terms of the failure of initially parallel geodesics to remain parallel—is derived. Finally, section 3.4 discusses methods for computing curvature.

3.1 Derivative Operators and Parallel Transport

A *derivative operator*, ∇, (sometimes called a *covariant derivative*) on a manifold M is a map which takes each smooth (or merely differentiable) tensor field of type (k, l) to a smooth tensor field of type $(k, l + 1)$ and satisfies the five properties listed below. In the index notation, if $T^{a_1 \cdots a_k}{}_{b_1 \cdots b_l} \in \mathcal{T}(k, l)$, we denote the tensor field resulting from the action of ∇ on T by $\nabla_c T^{a_1 \cdots a_k}{}_{b_1 \cdots b_l}$. It is often notationally convenient to attach an index directly to the derivative operator and write it as ∇_a, although this is to some extent an abuse of the index notation since ∇_a is not a dual vector. Expressed in the index notation, the five conditions required of a derivative operator are

1. Linearity: For all A, $B \in \mathcal{T}(k, l)$ and α, $\beta \in R$,

$$\nabla_c(\alpha A^{a_1 \cdots a_k}{}_{b_1 \cdots b_l} + \beta B^{a_1 \cdots a_k}{}_{b_1 \cdots b_l}) = \alpha \nabla_c A^{a_1 \cdots a_k}{}_{b_1 \cdots b_l} + \beta \nabla_c B^{a_1 \cdots a_k}{}_{b_1 \cdots b_l}$$

2. Leibnitz rule: For all $A \in \mathcal{T}(k, l)$, $B \in \mathcal{T}(k', l')$,

$$\begin{aligned} \nabla_e[A^{a_1 \cdots a_k}{}_{b_1 \cdots b_l} B^{c_1 \cdots c_{k'}}{}_{d_1 \cdots d_{l'}}] &= [\nabla_e A^{a_1 \cdots a_k}{}_{b_1 \cdots b_l}] B^{c_1 \cdots c_{k'}}{}_{d_1 \cdots d_{l'}} \\ &\quad + A^{a_1 \cdots a_k}{}_{b_1 \cdots b_l} [\nabla_e B^{c_1 \cdots c_{k'}}{}_{d_1 \cdots d_{l'}}] \quad . \end{aligned}$$

3. Commutativity with contraction: For all $A \in \mathcal{T}(k, l)$,

$$\nabla_d(A^{a_1 \cdots c \cdots a_k}{}_{b_1 \cdots c \cdots b_l}) = \nabla_d A^{a_1 \cdots c \cdots a_k}{}_{b_1 \cdots c \cdots b_l}$$

4. Consistency with the notion of tangent vectors as directional derivatives on scalar fields: For all $f \in \mathcal{F}$ and all $t^a \in V_p$

$$t(f) = t^a \nabla_a f \quad .$$

5. Torsion free:[1] For all $f \in \mathcal{F}$,

$$\nabla_a \nabla_b f = \nabla_b \nabla_a f \quad .$$

The fifth condition is sometimes dropped, and indeed there are theories of gravitation where it is not imposed. However, in general relativity the derivative operator is assumed to satisfy condition 5, and, unless otherwise stated, all derivative operators considered in this book will be assumed to be torsion free.

It is worth noting that the conditions 4 and 5 together with the Leibnitz rule allow us to derive a simple expression for the commutator of two vector fields v^a, w^b in terms of any derivative operator ∇_a. Applied to any smooth function f, we have

$$\begin{aligned} [v, w](f) &= v\{w(f)\} - w\{v(f)\} \\ &= v^a \nabla_a (w^b \nabla_b f) - w^a \nabla_a (v^b \nabla_b f) \\ &= \{v^a \nabla_a w^b - w^a \nabla_a v^b\} \nabla_b f \quad . \end{aligned} \tag{3.1.1}$$

Thus we have

$$[v, w]^b = v^a \nabla_a w^b - w^a \nabla_a v^b \quad . \tag{3.1.2}$$

1. If condition 5 is not imposed, it can be shown that there must exist a tensor $T^c{}_{ab}$ antisymmetric in a and b such that $\nabla_a \nabla_b f - \nabla_b \nabla_a f = -T^c{}_{ab} \nabla_c f$ (see problem 1). $T^c{}_{ab}$ is called the *torsion tensor*, and thus our condition 5 states that the torsion tensor vanishes.

Our first important task is to show that derivative operators exist. Let ψ be a coordinate system and let $\{\partial/\partial x^\mu\}$ and $\{dx^\mu\}$ be the associated coordinate bases. Then in the region covered by these coordinates we may define a derivative operator, ∂_a, called an *ordinary derivative*, as follows. For any smooth tensor field $T^{a_1 \cdots a_k}{}_{b_1 \cdots b_l}$ we take its components $T^{\mu_1 \cdots \mu_k}{}_{\nu_1 \cdots \nu_l}$ in this coordinate basis and define $\partial_c T^{a_1 \cdots a_k}{}_{b_1 \cdots b_l}$ to be the tensor whose components in this coordinate basis are the partial derivatives $\partial(T^{\mu_1 \cdots \mu_k}{}_{\nu_1 \cdots \nu_l})/\partial x^\sigma$. All five conditions follow immediately from the standard properties of partial derivatives. Indeed, by the equality of mixed partial derivatives, the fifth condition holds for all tensor fields, not just scalar fields. Thus, given a coordinate system ψ, we can construct an associated derivative operator ∂_a. Of course, a different choice of coordinate system ψ' will yield a different derivative operator ∂'_a, that is, the components of the tensor $\partial_c T^{a_1 \cdots a_k}{}_{b_1 \cdots b_l}$ in the new (primed) coordinates will *not* be equal to the partial derivatives of the primed components of $T^{a_1 \cdots a_k}{}_{b_1 \cdots b_l}$ with respect to the primed coordinates. Thus, a given ordinary derivative operator is coordinate dependent, i.e., it is not naturally associated with the structure of the manifold.

How unique are derivative operators? By condition (4), any two derivative operators ∇_a and $\tilde{\nabla}_a$ must agree in their action on scalar fields. To investigate their possible disagreements on tensors of the next highest rank, let ω_b be a dual vector field and consider the difference $\tilde{\nabla}_a(f\omega_b) - \nabla_a(f\omega_b)$ for an arbitrary scalar field f. By the Leibnitz rule we have

$$\begin{aligned}\tilde{\nabla}_a(f\omega_b) - \nabla_a(f\omega_b) &= (\tilde{\nabla}_a f)\omega_b + f\tilde{\nabla}_a\omega_b - (\nabla_a f)\omega_b - f\nabla_a\omega_b \\ &= f(\tilde{\nabla}_a\omega_b - \nabla_a\omega_b) \quad , \end{aligned} \tag{3.1.3}$$

where we have used property (4) again. Now at a point p, $\tilde{\nabla}_a\omega_b$ and $\nabla_a\omega_b$ each depend on how ω_b changes as one moves away from p. However, equation (3.1.3) shows that their difference $\tilde{\nabla}_a\omega_b - \nabla_a\omega_b$ depends only on the value of ω_b at point p. To see this, we suppose that ω'_b equals ω_b at p and show that we get the same answer if we replace ω_b by ω'_b. By problem 2 of chapter 2, it follows that since $\omega'_b - \omega_b$ vanishes at p we can find smooth functions, $f_{(\alpha)}$, which vanish at p and smooth dual vector fields, $\mu_b^{(\alpha)}$, such that

$$\omega'_b - \omega_b = \sum_{\alpha=1}^{n} f_{(\alpha)}\mu_b^{(\alpha)} \quad . \tag{3.1.4}$$

Hence, using equation (3.1.3), at point p we have

$$\begin{aligned}\tilde{\nabla}_a(\omega'_b - \omega_b) - \nabla_a(\omega'_b - \omega_b) &= \sum_\alpha \{\tilde{\nabla}_a(f_{(\alpha)}\mu_b^{(\alpha)}) - \nabla_a(f_{(\alpha)}\mu_b^{(\alpha)})\} \\ &= \sum_\alpha f_{(\alpha)}\{\tilde{\nabla}_a\mu_b^{(\alpha)} - \nabla_a\mu_b^{(\alpha)}\} = 0 \end{aligned} \tag{3.1.5}$$

since each $f_{(\alpha)} = 0$ at p. Thus, we have

$$\tilde{\nabla}_a\omega'_b - \nabla_a\omega'_b = \tilde{\nabla}_a\omega_b - \nabla_a\omega_b \quad , \tag{3.1.6}$$

which proves our assertion.

Thus, we have shown that $\tilde{\nabla}_a - \nabla_a$ defines a map of dual vectors at p (as opposed to dual vector fields defined in a neighborhood of p) to tensors of type $(0, 2)$ at p. By property (1), this map is linear. Consequently $(\tilde{\nabla}_a - \nabla_a)$ defines a tensor of type $(1, 2)$ at p, which we will denote as $C^c{}_{ab}$. Thus, we have shown that given any two derivative operators $\tilde{\nabla}_a$ and ∇_a there exists a tensor field $C^c{}_{ab}$ such that

$$\nabla_a \omega_b = \tilde{\nabla}_a \omega_b - C^c{}_{ab} \omega_c \quad . \tag{3.1.7}$$

This displays the possible disagreements of the actions of ∇_a and $\tilde{\nabla}_a$ on dual vector fields.

A symmetry property of $C^c{}_{ab}$ follows immediately from condition (5). If we let $\omega_b = \nabla_b f = \tilde{\nabla}_b f$, we find

$$\nabla_a \nabla_b f = \tilde{\nabla}_a \tilde{\nabla}_b f - C^c{}_{ab} \nabla_c f \quad . \tag{3.1.8}$$

Since both $\nabla_a \nabla_b f$ and $\tilde{\nabla}_a \tilde{\nabla}_b f$ are symmetric in a and b, it follows that $C^c{}_{ab}$ must also have this property

$$C^c{}_{ab} = C^c{}_{ba} \quad . \tag{3.1.9}$$

Equation (3.1.9), of course, need not hold if the torsion-free requirement is dropped.

The difference in the action of $\tilde{\nabla}_a$ and ∇_a on vector fields and all higher rank tensor fields is determined by equation (3.1.7), the Leibnitz rule, and property (4). For every vector field t^a and one-form field ω_a, property (4) tells us that

$$(\tilde{\nabla}_a - \nabla_a)(\omega_b t^b) = 0 \quad . \tag{3.1.10}$$

On the other hand, by the Leibnitz rule, we have

$$(\tilde{\nabla}_a - \nabla_a)(\omega_b t^b) = (C^c{}_{ab} \omega_c) t^b + \omega_b (\tilde{\nabla}_a - \nabla_a) t^b \quad . \tag{3.1.11}$$

Thus, index substituting on contracted indices, we find

$$\omega_b [(\tilde{\nabla}_a - \nabla_a) t^b + C^b{}_{ac} t^c] = 0 \tag{3.1.12}$$

for all ω_b. This implies

$$\nabla_a t^b = \tilde{\nabla}_a t^b + C^b{}_{ac} t^c \quad . \tag{3.1.13}$$

Continuing in a similar manner, we can derive the general formula for the action of ∇_a on an arbitrary tensor field in terms of $\tilde{\nabla}_a$ and $C^c{}_{ab}$. For $T \in \mathcal{T}(k, l)$ we find

$$\nabla_a T^{b_1 \cdots b_k}{}_{c_1 \cdots c_l} = \tilde{\nabla}_a T^{b_1 \cdots b_k}{}_{c_1 \cdots c_l} + \sum_i C^{b_i}{}_{ad} T^{b_1 \cdots d \cdots b_k}{}_{c_1 \cdots c_l}$$
$$- \sum_j C^d{}_{ac_j} T^{b_1 \cdots b_l}{}_{c_1 \cdots d \cdots c_l} \quad . \tag{3.1.14}$$

Thus, the difference between the two derivative operators ∇_a and $\tilde{\nabla}_a$ is completely characterized by the tensor field $C^c{}_{ab}$. Conversely, it is not difficult to check that if $\tilde{\nabla}_a$ is a derivative operator and $C^c{}_{ab}$ is an arbitrary smooth tensor field which is symmetric in its lower indices, then ∇_a defined by equation (3.1.14) will also be a derivative operator. This shows that there is a great deal of freedom involved in the choice of a derivative operator, as on an n-dimensional manifold $C^c{}_{ab}$ has $n^2(n+1)/2$ independent components to be specified at each point.

The most important application of equation (3.1.14) arises from the case where $\tilde{\nabla}_a$ is an ordinary derivative operator ∂_a. In this case, the tensor field $C^c{}_{ab}$ is denoted $\Gamma^c{}_{ab}$ and called a *Christoffel symbol*. Thus, for example, we write

$$\nabla_a t^b = \partial_a t^b + \Gamma^b{}_{ac} t^c \quad . \tag{3.1.15}$$

Since we know how to compute the ordinary derivative associated with a given coordinate system, equation (3.1.15) (and, more generally, eq. [3.1.14] with ∂_a and $\Gamma^b{}_{ac}$ replacing $\tilde{\nabla}_a$ and $C^b{}_{ac}$) tells us how to compute the derivative ∇_a if we know $\Gamma^b{}_{ac}$. Note that, as defined here, a Christoffel symbol is a tensor field associated with the derivative operator ∇_a and the coordinate system used to define ∂_a. However, if we change coordinates, we also change our ordinary derivative operator from ∂_a to ∂'_a and thus we change our tensor $\Gamma^c{}_{ab}$ to a new tensor $\Gamma'^c{}_{ab}$. Hence the coordinate components of $\Gamma^c{}_{ab}$ in the unprimed coordinates will not be related to the components of $\Gamma'^c{}_{ab}$ in the primed coordinates by the tensor transformation law, equation (2.3.8), since we change tensors as well as coordinates.

Given a derivative operator ∇_a we can define the notion of the parallel transport of a vector along a curve C with a tangent t^a. A vector v^a given at each point on the curve is said to be *parallelly transported* as one moves along the curve if the equation

$$t^a \nabla_a v^b = 0 \tag{3.1.16}$$

is satisfied along the curve. More generally, one can define the parallel transport of a tensor of arbitrary rank by

$$t^a \nabla_a T^{b_1 \cdots b_k}{}_{c_1 \cdots c_l} = 0 \quad . \tag{3.1.17}$$

Choosing a coordinate system and using equation (3.1.15), we can express equation (3.1.16) as

$$t^a \partial_a v^b + t^a \Gamma^b{}_{ac} v^c = 0 \quad , \tag{3.1.18}$$

or, in terms of components in the coordinate basis and the parameter t along the curve,

$$\frac{dv^\nu}{dt} + \sum_{\mu, \lambda} t^\mu \Gamma^\nu{}_{\mu\lambda} v^\lambda = 0 \quad . \tag{3.1.19}$$

This shows that the parallel transport of v^a depends only on the values of v^a on the curve, so we may consider the parallel transport properties of vectors defined only along the curve as opposed to vector fields. Furthermore, from the properties of ordinary differential equations it follows that equation (3.1.19) always has a unique solution for any given initial value of v^a. Thus, a vector at a point p on the curve uniquely defines a "parallel transported vector" everywhere else on the curve. We may use this notion of parallel transport to identify (i.e., map into each other) the tangent spaces V_p and V_q of points p and q if we are given a derivative operator and a curve connecting p and q. The mathematical structure arising from such a curve-dependent identification of the tangent spaces of different points is called a *connection*. Conversely, one could start with the general notion of a connection and use it to define the notion of derivative operator.

As we have seen above, given only the manifold structure, many distinct derivative operators are possible and no one of them is naturally preferred over the others. However, we show now that if one is given a metric g_{ab} on the manifold, a natural choice of derivative operator is uniquely picked out. This is because the metric gives rise to a natural condition which we may impose on the notion of parallel transport. Given two vectors v^a and w^a, we demand that their inner product $g_{ab}v^a w^b$ remain unchanged if we parallel-transport them along any curve. Thus we require

$$t^a \nabla_a (g_{bc} v^b w^c) = 0 \tag{3.1.20}$$

for v^b and w^c satisfying equation (3.1.16). Using the Leibnitz rule, we obtain

$$t^a v^b w^c \nabla_a g_{bc} = 0 \quad . \tag{3.1.21}$$

Equation (3.1.21) will hold for all curves and parallelly transported vectors if and only if

$$\nabla_a g_{bc} = 0 \quad , \tag{3.1.22}$$

which is the additional condition we wish to impose on ∇_a. That this equation uniquely determines ∇_a is shown by the following theorem.

THEOREM 3.1.1. *Let g_{ab} be a metric. Then there exists a unique derivative operator ∇_a satisfying $\nabla_a g_{bc} = 0$.*

Proof. Let $\tilde{\nabla}_a$ be any derivative operator, e.g., an ordinary derivative operator associated with a coordinate system. We attempt to solve for $C^c{}_{ab}$ so that the derivative operator determined by $\tilde{\nabla}_a$ and $C^c{}_{ab}$ will satisfy the required property. We will prove the theorem by showing that a unique solution for $C^c{}_{ab}$ exists.

By equation (3.1.14), $C^c{}_{ab}$ must satisfy

$$0 = \nabla_a g_{bc} = \tilde{\nabla}_a g_{bc} - C^d{}_{ab} g_{dc} - C^d{}_{ac} g_{bd} \tag{3.1.23}$$

that is,

$$C_{cab} + C_{bac} = \tilde{\nabla}_a g_{bc} \tag{3.1.24}$$

By index substitution, we also have

$$C_{cba} + C_{abc} = \tilde{\nabla}_b g_{ac} \quad , \tag{3.1.25}$$

$$C_{bca} + C_{acb} = \tilde{\nabla}_c g_{ab} \quad . \tag{3.1.26}$$

We add equations (3.1.24) and (3.1.25) and then subtract equation (3.1.26). Using the symmetry property of $C^c{}_{ab}$, equation (3.1.9), we find

$$2C_{cab} = \tilde{\nabla}_a g_{bc} + \tilde{\nabla}_b g_{ac} - \tilde{\nabla}_c g_{ab} \quad ; \tag{3.1.27}$$

that is,

$$C^c{}_{ab} = \frac{1}{2} g^{cd} \{ \tilde{\nabla}_a g_{bd} + \tilde{\nabla}_b g_{ad} - \tilde{\nabla}_d g_{ab} \} \quad . \tag{3.1.28}$$

This choice of $C^c{}_{ab}$ solves equation (3.1.22) and is manifestly unique, which completes the proof. □

Thus, a metric g_{ab} naturally determines a derivative operator ∇_a. For the remainder of this book, when a metric is present we will always choose the derivative operator to be this natural one. Furthermore, equations (3.1.14) and (3.1.28) tell us how to compute ∇_a in terms of any other derivative operator $\tilde{\nabla}_a$. In particular, in terms of an ordinary derivative operator the Christoffel symbol is

$$\Gamma^c{}_{ab} = \frac{1}{2} g^{cd}\{\partial_a g_{bd} + \partial_b g_{ad} - \partial_d g_{ab}\} \quad , \tag{3.1.29}$$

and thus the coordinate basis components of the Christoffel symbol are

$$\Gamma^\rho{}_{\mu\nu} = \frac{1}{2} \sum_\sigma g^{\rho\sigma}\left\{\frac{\partial g_{\nu\sigma}}{\partial x^\mu} + \frac{\partial g_{\mu\sigma}}{\partial x^\nu} - \frac{\partial g_{\mu\nu}}{\partial x^\sigma}\right\} \quad . \tag{3.1.30}$$

Thus, we can compute $\Gamma^c{}_{ab}$ (and thence ∇_a) by taking partial derivatives of the coordinate basis components of the metric.

3.2 Curvature

In the previous section we showed that given a derivative operator, there exists a notion of how to parallel transport a vector from p to q along a curve C. However, the vector in V_q which we get by this parallel transport procedure starting from a vector in V_p will, in general, depend on the choice of curve connecting them. As already indicated in the discussion at the beginning of this chapter, we can use the path dependence of parallel transport to define an intrinsic notion of curvature. In this section we carry out this program by first defining the Riemann curvature tensor in terms of the failure of successive operations of differentiation to commute when applied to a dual vector field. Then we show that this tensor is directly related to the path-dependent nature of parallel transport; specifically, the failure of a vector to return to its original value when parallel transported around a small closed loop is governed by the Riemann tensor. In the next section we will show that the Riemann tensor also fully describes the other characterization of curvature mentioned above: the failure of initially parallel geodesics to remain parallel.

Let ∇_a be a derivative operator. Let ω_a be a dual vector field and let f be a smooth function. We calculate the action of two derivative operators applied to $f\omega_a$,

$$\begin{aligned}\nabla_a\nabla_b(f\omega_c) &= \nabla_a(\omega_c\nabla_b f + f\nabla_b\omega_c)\\ &= (\nabla_a\nabla_b f)\omega_c + \nabla_b f\nabla_a\omega_c + \nabla_a f\nabla_b\omega_c + f\nabla_a\nabla_b\omega_c \quad .\end{aligned} \tag{3.2.1}$$

If we subtract from this the tensor $\nabla_b\nabla_a(f\omega_c)$, the first three terms of the right-hand side of equation (3.2.1) will cancel the corresponding terms of the expression for $\nabla_b\nabla_a(f\omega_c)$ and we obtain the simple result,

$$(\nabla_a\nabla_b - \nabla_b\nabla_a)(f\omega_c) = f(\nabla_a\nabla_b - \nabla_b\nabla_a)\omega_c \quad . \tag{3.2.2}$$

By exactly the same reasoning as given above in the discussion of derivative operators (see eq. [3.1.3]), it follows that the tensor $(\nabla_a\nabla_b - \nabla_b\nabla_a)\omega_c$ at point p depends only on the value of ω_c at p. Consequently, $(\nabla_a \nabla_b - \nabla_b \nabla_a)$ defines a linear map from dual vectors at p to type $(0, 3)$ tensors at p; i.e., its action is that of a tensor of type $(1, 3)$. Thus, we have shown that there exists a tensor field $R_{abc}{}^d$ such that for all dual vector fields ω_c, we have

$$\nabla_a \nabla_b \omega_c - \nabla_b \nabla_a \omega_c = R_{abc}{}^d \omega_d \quad . \tag{3.2.3}$$

$R_{abc}{}^d$ is called the *Riemann curvature tensor*.

We first show that $R_{abc}{}^d$ is directly related to the failure of a vector to return to its initial value when parallel transported around a small closed curve. We can conveniently construct a small closed loop at $p \in M$ by choosing a two-dimensional surface S through p and choosing coordinates t and s in the surface [with the coordinates of p chosen, for simplicity, to be $(0, 0)$]. Consider the loop formed by moving Δt along the $s = 0$ curve, followed by moving Δs along the $t = \Delta t$ curve, and then moving back by Δt and Δs as illustrated in Figure 3.3. Let v^a be a vector at p (not necessarily tangent to S) and let us parallel transport v^a around this closed

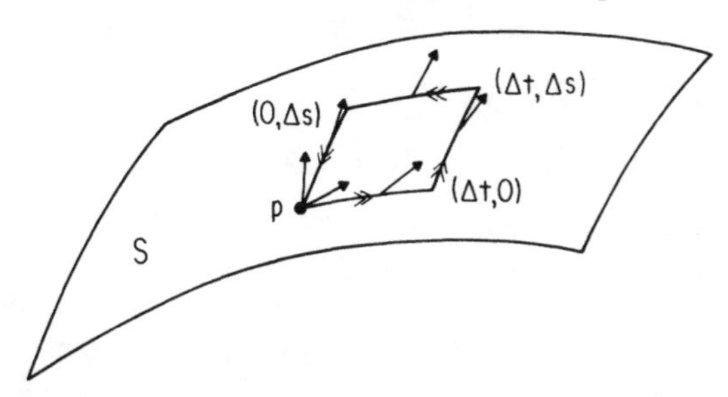

Fig. 3.3. The parallel transport of a vector v^a around a small closed loop. As derived in the text, to second order in Δt and Δs, the change in v^a is governed by the Riemann tensor at p.

loop. It is easiest to compute the change in v^a when we return to p by letting ω_a be an arbitrary dual vector field and finding the change in the scalar $v^a \omega_a$ as we traverse the loop. For small Δt the change, δ_1, in $v^a \omega_a$ on the first leg of the path is

$$\delta_1 = \Delta t \frac{\partial}{\partial t}(v^a \omega_a)\bigg|_{(\Delta t/2, 0)} \quad , \tag{3.2.4}$$

where, by evaluating the derivative at the midpoint, we have ensured that this expression is accurate to second order in the displacement Δt. We may rewrite δ_1 as

$$\begin{aligned} \delta_1 &= \Delta t \; T^b \nabla_b (v^a \omega_a) \big|_{(\Delta t/2, 0)} \quad , \\ &= \Delta t \; v^a T^b \nabla_b \omega_a \big|_{(\Delta t/2, 0)} \quad , \end{aligned} \tag{3.2.5}$$

where T^b is the tangent to the curves of constant s and $T^b \nabla_b v^a = 0$ by equation (3.1.16). Similar expressions hold for the changes δ_2, δ_3, and δ_4 on the other parts of the path. The two "Δt variations," δ_1 and δ_3, combine to yield

$$\delta_1 + \delta_3 = \Delta t \{ v^a T^b \nabla_b \omega_a \big|_{(\Delta t/2, 0)} - v^a T^b \nabla_b \omega_a \big|_{(\Delta t/2, \Delta s)} \} \quad , \tag{3.2.6}$$

and δ_2 and δ_4 combine similarly. Since the term in brackets vanishes as $\Delta s \to 0$, this shows that to first order in Δt and Δs, the total change in $v^a \omega_a$ (and thus the total change in v^a) vanishes; i.e., parallel transport *is* path-independent to *first order* in Δt and Δs. To calculate the second order change in $v^a\omega_a$, we need to evaluate the term in brackets in equation (3.2.6) to first order. We do this by the following procedure: We consider the curve $t = \Delta t/2$ and imagine parallel transporting v^a and $T^b\nabla_b\omega_a$ along this curve from $(\Delta t/2, 0)$ to $(\Delta t/2, \Delta s)$. Now to first order in Δs, v^a at $(\Delta t/2, \Delta s)$ equals the parallel transport of v^a at $(\Delta t/2, 0)$ along this curve since, as remarked above, parallel transport is path-independent to first order. On the other hand, to first order, the term $T^b\nabla_b\omega_a$ at $(\Delta t/2, \Delta s)$ will differ from the parallel transport of that quantity from $(\Delta t/2, 0)$ by the amount $\Delta s S^c\nabla_c(T^b\nabla_b\omega_a)$, where S^c is the tangent to the curves of constant t. Hence, the term in brackets is just this quantity contracted with v^a. Thus, to second order in $\Delta t, \Delta s$, we find

$$\delta_1 + \delta_3 = -\Delta t\, \Delta s\, v^a S^c\nabla_c(T^b\nabla_b\omega_a) \quad , \tag{3.2.7}$$

where, to this accuracy, we may evaluate all tensors at p. Adding the similar contribution for δ_2 and δ_4, we find the total change in $v^a\omega_a$ is

$$\begin{aligned}\delta(v^a\omega_a) &= \Delta t\, \Delta s\, v^a\{T^c\nabla_c(S^b\nabla_b\omega_a) - S^c\nabla_c(T^b\nabla_b\omega_a)\} \\ &= \Delta t\, \Delta s\, v^a T^c S^b(\nabla_c\nabla_b - \nabla_b\nabla_c)\omega_a \\ &= \Delta t\, \Delta s\, v^a T^c S^b R_{cba}{}^d\omega_d \quad ;\end{aligned} \tag{3.2.8}$$

where in the second line we used the fact that the coordinate vector fields T^a and S^b commute (see the end of section 2.2 and eq. [3.1.2]) and we used the definition of the Riemann tensor equation (3.2.3) in the last step. But equation (3.2.8) can hold for all ω_a if and only if the total change in v^a (accurate to second order in Δt and Δs) is

$$\delta v^a = \Delta t\, \Delta s\, v^d T^c S^b R_{cbd}{}^a \quad . \tag{3.2.9}$$

This is the desired result. It shows that the Riemann tensor indeed directly measures the path dependence of parallel transport.

Using equation (3.2.3) we may—by a procedure completely analogous to the derivation of equation (3.1.14)—express the action of the commutator of derivative operators on arbitrary tensor fields in terms of the Riemann tensor. To find the expression for a vector field t^a, we let ω_a be a dual vector field. Then using property 5 of derivative operators together with the Leibnitz rule and equation (3.2.3), we find

$$\begin{aligned}0 &= (\nabla_a\nabla_b - \nabla_b\nabla_a)(t^c\omega_c) \\ &= \nabla_a(\omega_c\nabla_b t^c + t^c\nabla_b\omega_c) - \nabla_b(\omega_c\nabla_a t^c + t^c\nabla_a\omega_c) \\ &= \omega_c(\nabla_a\nabla_b - \nabla_b\nabla_a)t^c + t^c(\nabla_a\nabla_b - \nabla_b\nabla_a)\omega_c \\ &= \omega_c(\nabla_a\nabla_b - \nabla_b\nabla_a)t^c + t^c\omega_d R_{abc}{}^d \quad .\end{aligned} \tag{3.2.10}$$

Thus, we obtain

$$(\nabla_a\nabla_b - \nabla_b\nabla_a)t^c = -R_{abd}{}^c t^d \quad . \tag{3.2.11}$$

By induction, for an arbitary tensor field $T^{c_1 \cdots c_k}{}_{d_1 \cdots d_l}$ we find

$$(\nabla_a \nabla_b - \nabla_b \nabla_a) T^{c_1 \cdots c_k}{}_{d_1 \cdots d_l} = -\sum_{i=1}^{k} R_{abe}{}^{c_i} T^{c_1 \cdots e \cdots c_k}{}_{d_1 \cdots d_l} + \sum_{j=1}^{l} R_{abd_j}{}^{e} T^{c_1 \cdots c_k}{}_{d_1 \cdots e \cdots d_l} \quad . \tag{3.2.12}$$

Next we establish four key properties of the Riemann tensor:

1. $R_{abc}{}^{d} = -R_{bac}{}^{d}$. (3.2.13)
2. $R_{[abc]}{}^{d} = 0$. (3.2.14)
3. For the derivative operator ∇_a naturally associated with the metric, $\nabla_a g_{bc} = 0$, we have

$$R_{abcd} = -R_{abdc} \quad . \tag{3.2.15}$$

4. The Bianchi identity holds:

$$\nabla_{[a} R_{bc]d}{}^{e} = 0 \quad . \tag{3.2.16}$$

Property (1) follows trivially from the definition of $R_{abc}{}^{d}$, equation (3.2.3). To prove property (2), we note that for an arbitrary dual vector field, ω_a, and any derivative operator ∇_a, we have

$$\nabla_{[a} \nabla_b \omega_{c]} = 0 \quad . \tag{3.2.17}$$

This equation can be proven from equation (3.1.14), substituting an ordinary derivative ∂_a for $\tilde{\nabla}_a$, and using the commutativity of ordinary derivatives and the symmetry of $C^c{}_{ab} = \Gamma^c{}_{ab}$, equation (3.1.9). (In differential forms notation, it is the statement that $d^2 \boldsymbol{\omega} = 0$ [see appendix B].) Thus, we have

$$0 = 2\nabla_{[a} \nabla_b \omega_{c]} = \nabla_{[a} \nabla_b \omega_{c]} - \nabla_{[b} \nabla_a \omega_{c]} = R_{[abc]}{}^{d} \omega_d \tag{3.2.18}$$

for all ω_d, which proves property (2).

Property (3) follows from equation (3.2.12) applied to the metric g_{ab}. We find

$$0 = (\nabla_a \nabla_b - \nabla_b \nabla_a) g_{cd} = R_{abc}{}^{e} g_{ed} + R_{abd}{}^{e} g_{ce} = R_{abcd} + R_{abdc} \quad , \tag{3.2.19}$$

which yields property (3). It follows from properties 1, 2, and 3 that the Riemann tensor also satisfies the following useful symmetry property (see problem 3):

$$R_{abcd} = R_{cdab} \quad . \tag{3.2.20}$$

Finally, to prove the Bianchi identity, property (4), we apply the commutator of derivative operators to the derivative of a dual vector field. We obtain, using equation (3.2.12),

$$(\nabla_a \nabla_b - \nabla_b \nabla_a) \nabla_c \omega_d = R_{abc}{}^{e} \nabla_e \omega_d + R_{abd}{}^{f} \nabla_c \omega_f \quad . \tag{3.2.21}$$

On the other hand, we have

$$\nabla_a (\nabla_b \nabla_c \omega_d - \nabla_c \nabla_b \omega_d) = \nabla_a (R_{bcd}{}^{e} \omega_e) = \omega_e \nabla_a R_{bcd}{}^{e} + R_{bcd}{}^{e} \nabla_a \omega_e \quad . \tag{3.2.22}$$

If we antisymmetrize over a, b, and c in equations (3.2.21) and (3.2.22), the left-hand sides become equal. Equality of the right-hand sides yields

$$R_{[abc]}{}^{e}\nabla_{e}\omega_{d} + R_{[ab|d|}{}^{f}\nabla_{c]}\omega_{f} = \omega_{e}\nabla_{[a}R_{bc]d}{}^{e} + R_{[bc|d|}{}^{e}\nabla_{a]}\omega_{e} \quad , \tag{3.2.23}$$

where the vertical bars indicate that we do not antisymmetrize over d. The first term on the left-hand side vanishes by equation (3.2.14) while the second terms on both sides cancel each other. Thus, we obtain, for all ω_e,

$$\omega_{e}\nabla_{[a}R_{bc]d}{}^{e} = 0 \quad , \tag{3.2.24}$$

which yields property (4).

It is useful to decompose the Riemann tensor into a "trace part" and a "trace free part." By the antisymmetry properties (1) and (3), the trace of the Riemann tensor over its first two or last two indices vanishes. However, its trace over the second and fourth (or equivalently, the first and third) indices defines the *Ricci tensor*, R_{ac},

$$R_{ac} = R_{abc}{}^{b} \quad . \tag{3.2.25}$$

By equation (3.2.20), R_{ab} satisfies the symmetry property

$$R_{ac} = R_{ca} \quad . \tag{3.2.26}$$

The *scalar curvature*, R, is defined as the trace of the Ricci tensor:

$$R = R_{a}{}^{a} \quad . \tag{3.2.27}$$

The "trace free part" is called the *Weyl tensor*, C_{abcd}, and is defined for manifolds of dimension $n \geq 3$ by the equation

$$R_{abcd} = C_{abcd} + \frac{2}{n-2}(g_{a[c}R_{d]b} - g_{b[c}R_{d]a}) - \frac{2}{(n-1)(n-2)}R\, g_{a[c}g_{d]b} \quad . \tag{3.2.28}$$

The Weyl tensor satisfies the symmetry properties (1), (2), and (3) of the Riemann tensor as well as being trace free on all its indices. It also behaves in a very simple manner under conformal transformations of the metric (see appendix D) and for this reason is sometimes called the *conformal tensor*.

Contraction of the Bianchi identity (3.2.16) leads to an important equation satisfied by R_{ab}. We find

$$\nabla_{a}R_{bcd}{}^{a} + \nabla_{b}R_{cd} - \nabla_{c}R_{bd} = 0 \quad . \tag{3.2.29}$$

Raising the index d with the metric and contracting over b and d, we obtain

$$\nabla_{a}R_{c}{}^{a} + \nabla_{b}R_{c}{}^{b} - \nabla_{c}R = 0 \tag{3.2.30}$$

or

$$\nabla^{a}G_{ab} = 0 \quad , \tag{3.2.31}$$

where

$$G_{ab} = R_{ab} - \frac{1}{2}R\, g_{ab} \quad . \tag{3.2.32}$$

The tensor G_{ab} is called the *Einstein tensor*. It appears in Einstein's equation, and the twice contracted Bianchi identity, equation (3.2.31), plays an important role in ensuring consistency of Einstein's equation (see chapters 4 and 10).

3.3 Geodesics

Intuitively, geodesics are lines that "curve as little as possible"; they are the "straightest possible lines" one can draw in a curved geometry. Given a derivative operator, ∇_a, we define a *geodesic* to be a curve whose tangent vector is parallel propagated along itself, i.e. a curve whose tangent, T^a, satisfies the equation

$$T^a \nabla_a T^b = 0 \quad . \tag{3.3.1}$$

Actually, in order to satisfy the intuitive requirement that the curve be "as straight as possible," one might require only that the tangent vector to the curve point in the same direction as itself when parallel propagated, and not demand that it maintain the same length. This would yield the weaker condition,

$$T^a \nabla_a T^b = \alpha T^b \quad , \tag{3.3.2}$$

where α is an arbitrary function on the curve. However, it is easy to show that given a curve which satisfies equation (3.3.2) we can always reparameterize it so that it satisfies equation (3.3.1) (see problem 5). Thus, there is no true loss of generality in considering only curves which satisfy equation (3.3.1). A parameterization which yields equation (3.3.1) is called an *affine parameterization*. Thus, our definition of a geodesic requires it to be affinely parameterized. (Some other references apply the term geodesic to any curve satisfying eq. [3.3.2].)

We can get some insight into the nature of the geodesic equation by writing out the components of this equation in a coordinate basis. In a coordinate system ψ, the geodesic is mapped into a curve $x^\mu(t)$ in $\mathbb{R}^n$. By equation (3.1.19), the components, T^μ, of T^a in this coordinate basis satisfy

$$\frac{dT^\mu}{dt} + \sum_{\sigma,\nu} \Gamma^\mu{}_{\sigma\nu} T^\sigma T^\nu = 0 \quad . \tag{3.3.3}$$

However, by equation (2.2.12), the components T^μ are simply

$$T^\mu = \frac{dx^\mu}{dt} \quad . \tag{3.3.4}$$

Thus, in a coordinate basis, the geodesic equation becomes

$$\frac{d^2 x^\mu}{dt^2} + \sum_{\sigma,\nu} \Gamma^\mu{}_{\sigma\nu} \frac{dx^\sigma}{dt}\frac{dx^\nu}{dt} = 0 \quad . \tag{3.3.5}$$

Equation (3.3.5) is a coupled system of n second order ordinary differential equations for the n functions $x^\mu(t)$. From the theory of ordinary differential equations, it is known that a unique solution of equation (3.3.5) always exists for any given initial value of x^μ and dx^μ/dt. This means that *given* $p \in M$ *and any tangent vector,* $T^a \in V_p$*, there always exists a unique geodesic through p with tangent* T^a. Notice that the solutions of the equations of motion in ordinary mechanics share this property: Given an initial position and velocity, a unique solution exists.

Indeed, the existence and uniqueness of geodesics allow us to use them to construct coordinate systems that are very convenient for some computational purposes. Let $p \in M$. We define a map, called the *exponential map,* from the tangent space V_p to M by mapping each $T^a \in V_p$ into the point in M lying at unit affine parameter from p along the geodesic through p with tangent T^a. For large T^a one might encounter a singularity before the affine parameter $t = 1$ is reached. Also geodesics may cross, thereby making the exponential map fail to be one-to-one. However, one can show that there always exists a (sufficiently small) neighborhood of the origin of V_p on which the exponential map is defined and is one-to-one (see, e.g., Bishop and Crittenden 1964). Since V_p is an n-dimensional vector space, we may identify it with $\mathbb{R}^n$, and hence use the exponential map to give us a coordinate system, called *Riemannian normal coordinates at p*. These coordinates have the property that all geodesics through p get mapped into straight lines through the origin of $\mathbb{R}^n$. From equation (3.3.5) it follows that in this coordinate system the Christoffel symbol components $\Gamma^\mu{}_{\sigma\nu}$ vanish at p. This fact makes Riemannian normal coordinates particularly useful if one is performing calculations at a given point.

In the case where the derivative operator ∇_a arises from a metric g_{ab} a second type of coordinate system, called *Gaussian normal coordinates,* or *synchronous coordinates,* often is useful for calculations in situations where one is given a *hypersurface* S, i.e., an $(n - 1)$-dimensional embedded submanifold of the n-dimensional manifold M (see appendix B). At each point $p \in S$, the tangent space $\tilde{V}_p$ of the manifold S can be naturally viewed as an $(n - 1)$-dimensional subspace of the tangent space V_p of M. Thus, there will be a vector $n^a \in V_p$, unique up to scaling, which is orthogonal (with respect to the metric g_{ab}) to all vectors in $\tilde{V}_p$. This vector, n^a, is said to be *normal* to S. In the case of a Riemannian metric, n^a cannot lie in $\tilde{V}_p$; in the case of a metric of indefinite signature, n^a could be a null vector, $g_{ab}n^a n^b = 0$, in which case it does lie in $\tilde{V}_p$ and S is said to be a *null hypersurface* at point p. If S is nowhere null, we may normalize n^a by the condition $g_{ab}n^a n^b = \pm 1$. Gaussian normal coordinates may be defined for any non-null hypersurface as follows (see Fig. 3.4). For each $p \in S$ we construct the unique geodesic through p with tangent n^a. We choose arbitrary coordinates $(x^1, \ldots, x^{n-1})$ on (a portion of) S and label each point in a neighborhood of (that portion of) S by the parameter t along the geodesic on which it lies and the coordinates $x^1, \ldots, x^{n-1}$ of the point $p \in S$ from which the geodesic emanated. The geodesics emanating from S may eventually cross or run into singularities, but in a (sufficiently small) neighborhood of each $p \in S$, the map $q \to (x^1, \ldots, x^{n-1}, t)$ defines the chart we wished to construct.

Gaussian normal coordinates satisfy the important property that the geodesics remain orthogonal to all the hypersurfaces S_t defined by the equation $t =$ constant. This is true by construction for the hypersurface $S_0 = S$. To show that this remains true for all S_t in the domain of validity of the construction, it suffices to show that the geodesic tangent field n^a remains orthogonal to all of the coordinate basis fields

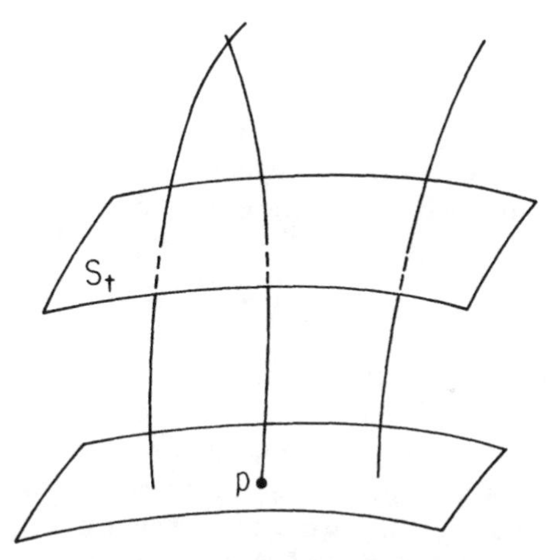

Fig. 3.4. The construction of Gaussian normal coordinates starting from the hypersurface S. The geodesics orthogonal to S eventually may cross, but until they do, Gaussian normal coordinates are well defined and the surfaces, S_t, of constant t remain orthogonal to the geodesics.

$X_1^a, \ldots, X_{n-1}^a$ which generate the tangent space to S_t. Denoting by X^a any one of these fields, we have

$$\begin{aligned} n^b\nabla_b(n_a X^a) &= n_a n^b \nabla_b X^a \\ &= n_a X^b \nabla_b n^a \\ &= \frac{1}{2} X^b \nabla_b (n^a n_a) \\ &= 0 \quad , \end{aligned} \tag{3.3.6}$$

where the first equality comes from the geodesic equation, the second from the fact that n^a and X^b—being elements of a coordinate basis on M—commute, the third follows directly from the Leibnitz rule, and the last follows from the fact that the normalization condition $n^a n_a = \pm 1$ on S is preserved by parallel transport so that $n^a n_a$ is constant on M. Since $n_a X^a = 0$ on S, equation (3.3.6) shows that this condition is preserved off of S, which yields the desired result.

A further property of geodesics of a derivative operator arising from a metric is that they extremize the length of curves connecting given points as measured by the metric. For a smooth (or merely differentiable) curve C on a manifold M with Riemannian metric g_{ab}, the *length*, l, of C is defined by

$$l = \int (g_{ab} T^a T^b)^{1/2} \, dt \quad , \tag{3.3.7}$$

where T^a is the tangent to C and t is the curve parameter. For a metric of Lorentz signature $- + + \cdots +$, a curve is said to be *timelike* if the norm of its tangent is everywhere negative, $g_{ab}T^aT^b < 0$; it is said to be *null* if $g_{ab}T^aT^b = 0$; it is said to be *spacelike* if $g_{ab}T^aT^b > 0$. For spacelike curves, the length may again be defined by equation (3.3.7); for null curves the length is zero; for timelike curves, we change the sign in the square root and use the term *proper time*, τ, rather than length,

$$\tau = \int (-g_{ab}T^aT^b)^{1/2}\, dt \quad . \tag{3.3.8}$$

The length of curves which change from timelike to spacelike is not defined. Note that since a geodesic tangent is parallel transported and thus its norm is constant, geodesics in a Lorentz manifold cannot change from timelike to spacelike or null. Note also that the length (or proper time) of a curve does not depend on the way in which the curve is parameterized. If we define a new parameterization $s = s(t)$, the new tangent will be $S^a = (dt/ds)T^a$ and the new length will be

$$l' = \int [g_{ab}S^aS^b]^{1/2}\, ds = \int [g_{ab}T^aT^b]^{1/2} \frac{dt}{ds}\, ds = l \quad . \tag{3.3.9}$$

We wish now to derive the condition on a curve which makes it extremize the length between its endpoints, i.e., wish to find those curves whose length does not change to first order under an arbitrary smooth deformation which keeps the endpoints fixed. We will perform the calculation of the change in length of a curve under a small deformation by choosing a chart and working in $\mathbb{R}^n$. (In chapter 9 we shall repeat this calculation without introducing coordinates and will calculate also the second variation of arc length.) For definiteness, we consider a spacelike curve. Writing equation (3.3.7) in the coordinate basis, we have

$$l = \int_a^b \left[\sum_{\mu,\nu} g_{\mu\nu} \frac{dx^\mu}{dt}\frac{dx^\nu}{dt} \right]^{1/2} dt \quad , \tag{3.3.10}$$

where $C(a) = p$ and $C(b) = q$ are the endpoints of the curve. The extremization problem for l is mathematically identical to the extremization problem for the action in Lagrangian mechanics. The variation in l is

$$\delta l = \int_a^b \left[\sum_{\mu,\nu} g_{\mu\nu} \frac{dx^\mu}{dt}\frac{dx^\nu}{dt} \right]^{-1/2} \sum_{\alpha,\beta} \left\{ g_{\alpha\beta} \frac{dx^\alpha}{dt}\frac{d(\delta x^\beta)}{dt} + \frac{1}{2} \sum_\sigma \frac{\partial g_{\alpha\beta}}{\partial x^\sigma} \delta x^\sigma \frac{dx^\alpha}{dt}\frac{dx^\beta}{dt} \right\} dt \quad . \tag{3.3.11}$$

Without loss of generality (since length is parameterization independent), we may assume that the original curve was parametrized so that

$$g_{ab}T^aT^b = 1 = \sum_{\mu,\nu} g_{\mu\nu} \frac{dx^\mu}{dt}\frac{dx^\nu}{dt} \quad .$$

With this choice of parameterization, the extremization condition is

$$0 = \int_a^b \sum_{\alpha,\beta} \left\{ g_{\alpha\beta} \frac{dx^\alpha}{dt} \frac{d(\delta x^\beta)}{dt} + \frac{1}{2} \sum_\sigma \frac{\partial g_{\alpha\beta}}{\partial x^\sigma} \frac{dx^\alpha}{dt} \frac{dx^\beta}{dt} \delta x^\sigma \right\} dt$$

$$= \int_a^b \sum_{\alpha,\beta} \left\{ -\frac{d}{dt}\left(g_{\alpha\beta} \frac{dx^\alpha}{dt} \right) + \frac{1}{2} \sum_\lambda \frac{\partial g_{\alpha\lambda}}{\partial x^\beta} \frac{dx^\alpha}{dt} \frac{dx^\lambda}{dt} \right\} \delta x^\beta dt \quad . \qquad (3.3.12)$$

(No boundary terms occur in the integration by parts since δx^β vanishes at the endpoints.) Equation (3.3.12) will hold for arbitrary δx^β if and only if

$$-\sum_\alpha g_{\alpha\beta} \frac{d^2 x^\alpha}{dt^2} - \sum_{\alpha,\lambda} \frac{\partial g_{\alpha\beta}}{\partial x^\lambda} \frac{dx^\lambda}{dt} \frac{dx^\alpha}{dt} + \frac{1}{2} \sum_{\alpha,\lambda} \frac{\partial g_{\alpha\lambda}}{\partial x^\beta} \frac{dx^\alpha}{dt} \frac{dx^\lambda}{dt} = 0 \quad . \qquad (3.3.13)$$

Using our formula for $\Gamma^\sigma{}_{\alpha\lambda}$, equation (3.1.30), we see that equation (3.3.13) is just the geodesic equation (3.3.5). (Had we not chosen the parameterization $g_{ab}T^aT^b = 1$ for C, we would have obtained the geodesic equation in the form [3.3.2].) Thus, a curve extremizes the length between its endpoints if and only if it is a geodesic.

An identical derivation shows that the curves which extremize proper time between two points are precisely the timelike geodesics. These derivations also show that the geodesic equation (with affine parameterization) can be obtained from variation of the Lagrangian,

$$L = \sum_{\mu,\nu} g_{\mu\nu} \frac{dx^\mu}{dt} \frac{dx^\nu}{dt} \quad . \qquad (3.3.14)$$

In many cases, the most efficient way of computing the Christoffel symbol $\Gamma^\mu{}_{\sigma\nu}$ in a given coordinate basis is to start with the Lagrangian, equation (3.3.14), write down the corresponding Euler-Lagrange equations, and read off $\Gamma^\mu{}_{\sigma\nu}$ by comparison with equation (3.3.5).

On a manifold with a Riemannian metric, one can always find curves of arbitrarily long length connecting any two points. However, the length will be bounded from below, and the curve of shortest length connecting two points (assuming the lower bound in length is attained) is necessarily an extremum of length and thus a geodesic. Thus, the shortest path between two points is always a "straightest possible path." However, a given geodesic connecting two points need not be the shortest path between them. For a manifold with a Lorentz metric, given two points that can be connected by a timelike curve, one can always find timelike curves of arbitrarily *small* proper time connecting the points (see Fig. 9.5 of chapter 9). In some spacetimes the proper time of timelike curves connecting the two given points need not be bounded from above; but if a curve of greatest proper time exists, it must be a timelike geodesic. Again, however, a given geodesic need not maximize the proper time between two points. In chapter 9 we shall introduce the notion of conjugate points along a geodesic and will show that their absence is the necessary and sufficient condition for a geodesic to be a local maximum of proper time (or, in the Riemannian case, a local minimum of length) between two points.

Our final task is to derive the geodesic deviation equation, the equation which relates the tendency of geodesics to accelerate toward or away from each other to the curvature of the manifold. This gives another characterization of curvature, and it also plays an important role in motivating Einstein's equation (see section 4.3) and in arguments relating to the singularity theorems (see chapter 9).

Let $\gamma_s(t)$ denote a smooth one-parameter family of geodesics (see Fig. 3.5), that is for each $s \in \mathbb{R}$, the curve γ_s is a geodesic (parameterized by affine parameter t);

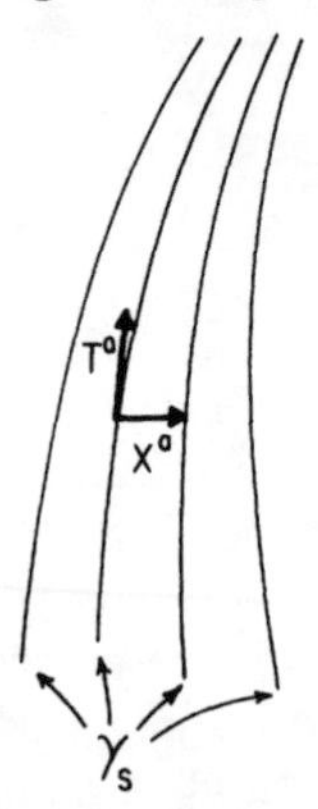

Fig. 3.5. A one-parameter family of geodesics γ_s, with tangent T^a and deviation vector X^a.

and the map $(t, s) \rightarrow \gamma_s(t)$ is smooth, is one-to-one, and has smooth inverse. Let Σ denote the two-dimensional submanifold spanned by the curves $\gamma_s(t)$. We may choose s and t as coordinates of Σ. The vector field $T^a = (\partial/\partial t)^a$ is tangent to the family of geodesics and, thus, satisfies

$$T^a \nabla_a T^b = 0 \quad . \tag{3.3.15}$$

The vector field $X^a = (\partial/\partial s)^a$ represents the displacement to an infinitesimally nearby geodesic, and is called the *deviation vector*. There is "gauge freedom" in X^a in the sense that X^a changes by addition of a multiple of T^a under a change of the affine parameterization of the geodesics $\gamma_s(t)$, $t \rightarrow t' = b(s)t + c(s)$ (see problem 5). It is worth noting that in the case where the geodesics arise from the derivative operator associated with a metric g_{ab}, X^a always can be chosen orthogonal to T^a. Namely, by re-scaling t by an s-dependent factor, we may ensure that $g_{ab}T^aT^b$ (which is automatically constant along each geodesic) does not vary with s. Since X^a and T^a are coordinate vector fields, they commute:

$$T^b \nabla_b X^a = X^b \nabla_b T^a \quad ; \tag{3.3.16}$$

so by the same calculation as equation (3.3.6), we see that X^aT_a is constant along each geodesic. By further reparameterizing each $\gamma_s(t)$ by adding a constant (depending on s) to t, we may ensure that the curve $C(s)$ comprising the points with $t = 0$ is orthogonal to all the geodesics. Thus, with this affine parameterization of $\gamma_s(t)$ we have $X_aT^a = 0$ at $t = 0$ and hence $X^aT_a = 0$ everywhere.

The quantity $v^a = T^b \nabla_b X^a$ gives the rate of change along a geodesic of the displacement to an infinitesimally nearby geodesic. Thus, we may interpret v^a as the relative velocity of an infinitesimally nearby geodesic. Similarly, we may interpret

$$a^a = T^c \nabla_c v^a = T^c \nabla_c (T^b \nabla_b X^a) \tag{3.3.17}$$

as the relative acceleration of an infinitesimally nearby geodesic in the family. We now shall derive an equation which relates a^a to the Riemann tensor. We have

$$\begin{aligned} a^a &= T^c \nabla_c (T^b \nabla_b X^a) \\ &= T^c \nabla_c (X^b \nabla_b T^a) \\ &= (T^c \nabla_c X^b)(\nabla_b T^a) + X^b T^c \nabla_c \nabla_b T^a \\ &= (X^c \nabla_c T^b)(\nabla_b T^a) + X^b T^c \nabla_b \nabla_c T^a - R_{cbd}{}^a X^b T^c T^d \\ &= X^c \nabla_c (T^b \nabla_b T^a) - R_{cbd}{}^a X^b T^c T^d \\ &= -R_{cbd}{}^a X^b T^c T^d \quad . \end{aligned} \tag{3.3.18}$$

Equation (3.3.18) is known as the *geodesic deviation equation*. It yields the final characterization of curvature which we sought: We have $a^a = 0$ for all families of geodesics if and only if $R_{abc}{}^d = 0$. Thus, some geodesics will accelerate toward or away from each other (and, in particular, some initially parallel geodesics—i.e., goedesics with $v^a = T^b \nabla_b X^a = 0$ initially— will fail to remain parallel) if and only if $R_{abc}{}^d \neq 0$.

3.4 Methods for Computing Curvature

In section 3.2 we defined the Riemann curvature tensor by proving existence of a tensor field which satisfies equation (3.2.3) for all dual vector fields ω_a. However, this existence argument does not tell us directly how to calculate $R_{abc}{}^d$ given a metric g_{ab}. Since the ability to calculate curvature is crucial for solving Einstein's equation of general relativity, we devote this section to methods for calculating $R_{abc}{}^d$.

3.4a Coordinate Component Method

To calculate the curvature by the coordinate component method, we begin by choosing a coordinate system. We express the derivative operator ∇_a in terms of the ordinary derivative ∂_a of this coordinate system and the Christoffel symbol $\Gamma^c{}_{ab}$, as discussed in section 3.1. For a dual vector field ω_a, we have

$$\nabla_b \omega_c = \partial_b \omega_c - \Gamma^d{}_{bc} \omega_d \quad , \tag{3.4.1}$$

and thus,

$$\begin{aligned} \nabla_a \nabla_b \omega_c = &\ \partial_a (\partial_b \omega_c - \Gamma^d{}_{bc} \omega_d) \\ &- \Gamma^e{}_{ab} (\partial_e \omega_c - \Gamma^d{}_{ec} \omega_d) \\ &- \Gamma^e{}_{ac} (\partial_b \omega_e - \Gamma^d{}_{be} \omega_d) \quad . \end{aligned} \tag{3.4.2}$$

Hence, equation (3.2.3) may be expressed as

$$R_{abc}{}^d \omega_d = [-2\, \partial_{[a} \Gamma^d{}_{b]c} + 2\Gamma^e{}_{c[a} \Gamma^d{}_{b]e}] \omega_d \quad , \tag{3.4.3}$$

where we have used the commutativity of ordinary derivatives and the symmetry, equation (3.1.9), of ${\Gamma^c}_{ab}$. Since equation (3.4.3) holds for all ω_d, we may "cancel" ω_d from both sides to obtain the desired formula for ${R_{abc}}^d$. Taking the components of the tensors appearing in this equation in the coordinate basis associated with our chart, we obtain the formula

$$ {R_{\mu\nu\rho}}^\sigma = \frac{\partial}{\partial x^\nu} {\Gamma^\sigma}_{\mu\rho} - \frac{\partial}{\partial x^\mu} {\Gamma^\sigma}_{\nu\rho} + \sum_\alpha ({\Gamma^\alpha}_{\mu\rho} {\Gamma^\sigma}_{\alpha\nu} - {\Gamma^\alpha}_{\nu\rho} {\Gamma^\sigma}_{\alpha\mu}) \quad , \tag{3.4.4} $$

where we have used the definition of the ordinary derivative as the partial derivative of components with respect to the coordinates.

Thus, to calculate ${R_{abc}}^d$ starting from g_{ab}, we first obtain the components, $g_{\mu\nu}$, of the metric in our coordinate basis. We then calculate ${\Gamma^\sigma}_{\mu\nu}$ by equation (3.1.30) (or by reading them off from the components of the geodesic equation [3.3.5] calculated from the Lagrangian, equation [3.3.14]). Finally we calculate the components ${R_{\mu\nu\rho}}^\sigma$ via equation (3.4.4). The Ricci tensor components then may be obtained by contracting equation (3.4.4):

$$ \begin{aligned} R_{\mu\rho} &= \sum_\nu {R_{\mu\nu\rho}}^\nu \\ &= \sum_\nu \frac{\partial}{\partial x^\nu} {\Gamma^\nu}_{\mu\rho} - \frac{\partial}{\partial x^\mu} \left(\sum_\nu {\Gamma^\nu}_{\nu\rho} \right) + \sum_{\alpha,\nu} \left({\Gamma^\alpha}_{\mu\rho} {\Gamma^\nu}_{\alpha\nu} - {\Gamma^\alpha}_{\nu\rho} {\Gamma^\nu}_{\alpha\mu} \right) \quad . \end{aligned} \tag{3.4.5} $$

We take this opportunity to point out several useful facts for calculations in coordinate bases. We may write the coordinate basis components, $g_{\mu\nu}$, of the metric as a matrix. The components, $g^{\mu\nu}$, of the inverse metric g^{ab} will then be the inverse of the matrix $(g_{\mu\nu})$. We define g to be the determinant of $(g_{\mu\nu})$,

$$ g = \det(g_{\mu\nu}) \quad . \tag{3.4.6} $$

Then, as discussed in appendix B, the natural volume element on the manifold M induced by g_{ab} is $\sqrt{|g|}\, dx^1 \ldots dx^n$.

A simple formula may be derived for the contracted Christoffel symbol ${\Gamma^a}_{ab}$. From equation (3.1.30) we have

$$ {\Gamma^a}_{a\mu} = \sum_\nu {\Gamma^\nu}_{\nu\mu} = \frac{1}{2} \sum_{\nu,\alpha} g^{\nu\alpha} \frac{\partial g_{\nu\alpha}}{\partial x^\mu} \quad . \tag{3.4.7} $$

But, using the formula for the inverse of a matrix, it is not difficult to show that

$$ \sum_{\nu,\alpha} g^{\nu\alpha} \frac{\partial g_{\nu\alpha}}{\partial x^\mu} = \frac{1}{g} \frac{\partial g}{\partial x^\mu} \quad . \tag{3.4.8} $$

Thus, we have

$$ {\Gamma^a}_{a\mu} = \frac{1}{2} \frac{1}{g} \frac{\partial g}{\partial x^\mu} = \frac{\partial}{\partial x^\mu} \ln \sqrt{|g|} \quad . \tag{3.4.9} $$

This is a useful formula since $\Gamma^a{}_{ab}$ appears in the expression for the Ricci tensor R_{ab} as well as in the formula for the divergence of an arbitrary vector field, T^a, in terms of its coordinate basis components. Indeed, using equation (3.4.9), we have

$$\begin{aligned}\nabla_a T^a &= \partial_a T^a + \Gamma^a{}_{ab} T^b \\ &= \sum_\mu \frac{1}{\sqrt{|g|}} \frac{\partial}{\partial x^\mu} (\sqrt{|g|}\, T^\mu) \quad .\end{aligned} \tag{3.4.10}$$

Coordinate basis methods have the advantage of providing a straightforward mechanical procedure for calculating curvature. One is normally presented a metric in terms of its components in a coordinate basis, so all one must do to obtain the curvature is to perform the partial differentiations and summations given in equations (3.1.30) and (3.4.4). On the other hand, even in the simplest cases, these straightforward computations are extremely laborious, and the "nongeometrical" nature of the calculations often makes it difficult to detect algebraic errors.

3.4b Orthonormal Basis (Tetrad) Methods

While coordinate basis methods have the advantage of providing a straightforward procedure for calculation of ∇_a and $R_{abc}{}^d$, for many purposes it is advantageous to use an orthonormal basis in tensor calculations. A coordinate basis $\{\partial/\partial x^\mu\}$, of course, it is not orthonormal except for the trivial case of flat spacetime in Cartesian coordinates. Thus, we may wish to introduce a "nonholonomic," i.e., noncoordinate, orthonormal basis of smooth vector fields $(e_\mu)^a$, satisfying

$$(e_\mu)^a (e_\nu)_a = \eta_{\mu\nu} \quad , \tag{3.4.11}$$

where $\eta_{\mu\nu} = \text{diag}(-1, \ldots, -1, 1, \ldots, 1)$. (Here the greek indices μ,ν have the range $1, \ldots, n$ and label the vector of the basis; the latin a is the usual index of the index notation designating that e_μ is a vector.) In four dimensions—and sometimes more generally—$\{(e_\mu)^a\}$ is called a *tetrad*. Equation (3.4.11) implies the useful relation

$$\sum_{\mu,\nu} \eta^{\mu\nu} (e_\mu)^a (e_\nu)_b = \delta^a{}_b \quad , \tag{3.4.12}$$

where $\delta^a{}_b$ is the identity map on vectors and $\eta^{\mu\nu}$ is the inverse of $\eta_{\mu\nu}$ (so that $\eta^{\mu\nu} = \eta_{\mu\nu}$). Equation (3.4.12) can be proven by verifying that it gives equality when contracted with an arbitrary basis vector $(e_\sigma)^b$.

The calculation of curvature by tetrad methods has a sufficiently different appearance from the coordinate basis methods that it is worth pointing out explicitly three key ingredients that must be used in determining the curvature of a metric: (1) The derivative operator is compatible with the metric, $\nabla_a g_{bc} = 0$. (2) The derivative operator is torsion free (property 5 of section 3.1). (3) The Riemann tensor is related to the derivative operator by equation (3.2.3). In the coordinate basis methods, ingredient (2) is expressed by equation (3.1.9), ingredient (1) (given (2)) is expressed by equation (3.1.29), and ingredient (3) is expressed by equation (3.4.4). As shown below ingredients (1) and (2) are expressed in quite different ways when using an orthonormal basis.

We begin by defining the *connection 1-forms*, $\omega_{a\mu\nu}$, by

$$\omega_{a\mu\nu} = (e_\mu)^b \nabla_a (e_\nu)_b \quad . \tag{3.4.13}$$

The components $\omega_{\lambda\mu\nu}$ of $\omega_{a\mu\nu}$ are called the *Ricci rotation coefficients*,

$$\omega_{\lambda\mu\nu} = (e_\lambda)^a (e_\mu)^b \nabla_a (e_\nu)_b \quad . \tag{3.4.14}$$

The orthonormality of $\{(e_\mu)^a\}$ implies

$$\begin{aligned} \omega_{a\mu\nu} &= (e_\mu)^b \nabla_a (e_\nu)_b \\ &= -(e_\nu)^b \nabla_a (e_\mu)_b \\ &= -\omega_{a\nu\mu} \quad , \end{aligned} \tag{3.4.15}$$

where the compatibility condition $\nabla_a g_{bc} = 0$ was used. Conversely, equation (3.4.15) together with equation (3.4.11) implies $\nabla_a g_{bc} = 0$. Thus, in the orthonormal basis approach, ingredient (1) is expressed by the simple condition

$$\omega_{a\mu\nu} = -\omega_{a\nu\mu} \quad . \tag{3.4.16}$$

(Note, in contrast, that the symmetry of $\Gamma^\alpha{}_{\mu\nu}$ stems from ingredient [2].) The antisymmetry of the Ricci rotation coefficients as compared with the symmetry of the Christoffel symbol indicates a significant potential advantage of this approach: There are $n^2(n+1)/2$ ($= 40$ when $n = 4$) components $\Gamma^\lambda{}_{\mu\nu}$ but only $n^2(n-1)/2$ ($= 24$ when $n = 4$) components $\omega_{\lambda\mu\nu}$.

The Riemann tensor can be expressed in terms of the Ricci rotation coefficients in the following manner. The components of R_{abcd} in our orthonormal basis are

$$\begin{aligned} R_{\rho\sigma\mu\nu} &= R_{abcd}(e_\rho)^a (e_\sigma)^b (e_\mu)^c (e_\nu)^d \\ &= (e_\rho)^a (e_\sigma)^b (e_\mu)^c (\nabla_a \nabla_b - \nabla_b \nabla_a)(e_\nu)_c \quad . \end{aligned} \tag{3.4.17}$$

However, we have

$$(e_\mu)^c \nabla_a \nabla_b (e_\nu)_c = \nabla_a \{(e_\mu)^c \nabla_b (e_\nu)_c\} - [\nabla_a (e_\mu)^c][\nabla_b (e_\nu)_c] \quad . \tag{3.4.18}$$

Furthermore, we have

$$\begin{aligned} [\nabla_a (e_\mu)^c][\nabla_b (e_\nu)_c] &= [\nabla_a (e_\mu)^f] \delta^c{}_f [\nabla_b (e_\nu)_c] \\ &= \sum_{\alpha,\beta} \eta^{\alpha\beta} [\nabla_a (e_\mu)^f](e_\alpha)^c (e_\beta)_f [\nabla_b (e_\nu)_c] \quad . \end{aligned} \tag{3.4.19}$$

Thus, from the definition of the connection one forms, equation (3.4.13), we obtain

$$\begin{aligned} R_{\rho\sigma\mu\nu} = (e_\rho)^a (e_\sigma)^b \{ &\nabla_a \omega_{b\mu\nu} - \nabla_b \omega_{a\mu\nu} \\ &- \sum_{\alpha,\beta} \eta^{\alpha\beta} [\omega_{a\beta\mu} \omega_{b\alpha\nu} - \omega_{b\beta\mu} \omega_{a\alpha\nu}] \} \quad . \end{aligned} \tag{3.4.20}$$

We may rewrite equation (3.4.20) in terms of the Ricci rotation coefficients as

$$R_{\rho\sigma\mu\nu} = (e_\rho)^a \nabla_a \omega_{\sigma\mu\nu} - (e_\sigma)^a \nabla_a \omega_{\rho\mu\nu}$$

$$-\sum_{\alpha,\beta} \eta^{\alpha\beta} \{\omega_{\rho\beta\mu}\, \omega_{\sigma\alpha\nu} - \omega_{\sigma\beta\mu}\, \omega_{\rho\alpha\nu} + \omega_{\rho\beta\sigma}\, \omega_{\alpha\mu\nu} - \omega_{\sigma\beta\rho}\, \omega_{\alpha\mu\nu}\} \quad , \quad (3.4.21)$$

where the last two terms compensate for taking the components of $\omega_{a\mu\nu}$ inside the derivative in the first two terms. Since $\omega_{\sigma\mu\nu}$ is a scalar, the derivative ∇_a in equation (3.4.21) may, of course, be replaced by an ordinary derivative ∂_a. Thus, equation (3.4.21) tells us how to compute the curvature tensor in terms of $\omega_{\sigma\mu\nu}$. The components of the Ricci tensor may then be computed via the formula

$$R_{\rho\mu} = \sum_{\sigma,\nu} \eta^{\sigma\nu} R_{\rho\sigma\mu\nu} \quad . \quad (3.4.22)$$

Equation (3.4.20) or equation (3.4.21) fully expresses ingredient (3) in the calculation of curvature. Thus, only ingredient (2)—the torsion-free condition on ∇_a—remains to be expressed in this approach. There are two different procedures which may be employed to do this. The first begins by noting that the torsion-free condition allowed us to express the commutator of two vector fields in terms of the derivative operator ∇_a via equation (3.1.2). Conversely, if equation (3.1.2) holds for all vector fields in a basis, it implies the torsion-free property of ∇_a. Thus, we may express ingredient (2) by the *commutation relations* of the basis vector fields,

$$(e_\sigma)_a [e_\mu, e_\nu]^a = (e_\sigma)_a \{(e_\mu)^b \nabla_b (e_\nu)^a - (e_\nu)^b \nabla_b (e_\mu)^a\} = \omega_{\mu\sigma\nu} - \omega_{\nu\sigma\mu} \quad . \quad (3.4.23)$$

This yields the $n^2(n-1)/2$ equations needed to solve for $\omega_{\sigma\mu\nu}$ (see problem 8).

An alternative procedure is to note that from the definition of the connection one-forms, it follows that

$$\nabla_{[a}(e_\sigma)_{b]} = \sum_{\mu,\nu} \eta^{\mu\nu} (e_\mu)_{[a} \omega_{b]\sigma\nu} \quad , \quad (3.4.24)$$

as can be verified by contracting equation (3.4.24) with arbitrary basis vectors $(e_\rho)^a (e_\lambda)^b$. However, the torsion-free condition implies that the antisymmetrized derivative of a one-form is independent of derivative operator, and thus may be replaced by an ordinary derivative ∂_a,

$$\partial_{[a}(e_\sigma)_{b]} = \sum_{\mu,\nu} \eta^{\mu\nu} (e_\mu)_{[a} \omega_{b]\sigma\nu} \quad . \quad (3.4.25)$$

Conversely, the validity of equation (3.4.25) for all basis vectors implies that ∇_a is torsion free. Thus, equation (3.4.25) is an alternate expression of ingredient (2).

Equations (3.4.20) amd (3.4.25) can be expressed more elegantly in the notation of differential forms (see appendix B). We drop the dual vector index a on $\omega_{a\mu\nu}$ and $(e_\mu)_a$ and use boldface letters to designate forms. For convenience, we also raise and lower greek indices with $\eta^{\mu\nu}$ and $\eta_{\mu\nu}$, e.g., we have

$$\boldsymbol{\omega}_\nu{}^\mu = \sum_\sigma \eta^{\mu\sigma}\, \boldsymbol{\omega}_{\nu\sigma} \quad . \quad (3.4.26)$$

Then equation (3.4.25) becomes, in differential forms notation,

$$de_\sigma = \sum_\mu e_\mu \wedge \omega_\sigma{}^\mu \quad . \tag{3.4.27}$$

Furthermore, equation (3.4.20) can be written as

$$R_\mu{}^\nu = d\omega_\mu{}^\nu + \sum_\alpha \omega_\mu{}^\alpha \wedge \omega_\alpha{}^\nu \quad , \tag{3.4.28}$$

where $R_{\mu\nu}$ is the two-form corresponding to $R_{ab\mu\nu}$. Equations (3.4.27) and (3.4.28) are sometimes referred to as the *equations of structure*.

Thus, we have basically two procedures for calculating curvature with orthonormal bases. The first is the straightforward approach by which one solves for the Ricci rotation coefficients from the commutation relations, equation (3.4.23), and substitutes in equation (3.4.21). The second approach—first discussed in the context of calculations in general relativity by Misner (1963)—uses equation (3.4.27) (≡ eq. [3.4.25]) to determine the connection one-forms, from which the curvature may be calculated by equation (3.4.28) (≡ eq. [3.4.20]). The major advantage of the second approach is that it is relatively easy to calculate the left side of equation (3.4.27). In simple cases it is possible to guess the solution for $\omega_{a\mu}{}^\nu$ without further calculations. In this way, a great deal of labor (such as that required to solve eq. [3.4.23] or, worse yet, the labor required to calculate $\Gamma^\alpha{}_{\mu\nu}$) can be avoided. We will illustrate this in chapter 6. Note that in both approaches, one will still introduce a coordinate system when it comes time to do the calculations. The aim of tetrad methods is not to avoid the introduction of a coordinate system but rather to use a more convenient basis for the expansion of tensors than that provided by the coordinate basis.

A variant of the "straightforward approach"—motivated by spinor methods (see chapter 13) and specially adapted to the case of a Lorentz metric in four dimensions—was introduced by Newman and Penrose (1962). Instead of choosing an orthonormal basis, one chooses a "null tetrad" consisting of two null vectors l^a and n^a with $l^a n_a = -1$; one completes the basis by picking two orthonormal spacelike vectors x^a and y^a orthogonal to l^a and n^a and defines the complex vector $m^a = (x^a + iy^a)/\sqrt{2}$. The analogs of the 24 real Ricci rotation coefficients of an orthonormal tetrad are 12 complex quantities called *spin coefficients*. They are related to the null tetrad by the analog of equation (3.4.13) and the curvature is given in terms of them by the analog of equation (3.4.21). The precise equations are given by Newman and Penrose (1962), Pirani (1965), and others. The Newman-Penrose approach has proven very useful for many applications in general relativity.

Which method—coordinate basis, orthonormal basis, or null tetrad—should one use to do calculations? There is no universal answer. In many simple cases the second orthonormal basis approach—using equations (3.4.27) and (3.4.28)—saves considerable labor over the other approaches, provided one can solve equation (3.4.27) for the connection one-forms by inspection. In situations where a null vector plays a geometrically distinguished role—specifically in algebraically special spacetimes (see chapter 7)—the Newman-Penrose approach is likely to be the most useful. Furthermore, in some cases where there is a great deal of symmetry, it may be possible to adapt the coordinate system to the symmetries and make the coordinate basis method the preferred approach. Finally in general situations—without symmetries or geometrically preferred vectors to serve as natural basis vectors—all methods are likely to be about equally laborious.

Finally, we note that in this section we have presented the coordinate and tetrad methods in the context of calculating the Riemann tensor, given the metric. However, the equations obtained, of course, are equally applicable in the much more frequently encountered situation where (part of) the curvature tensor is given (e.g., via Einstein's equation, as discussed in the next chapter) and we wish to find the metric. Furthermore, even in cases where we do not wish to calculate the curvature or solve for the metric, the above equations often directly provide useful relations between the metric and its curvature, particularly in cases where the coordinate system or tetrad is well adapted to the properties of the spacetimes under consideration.

Problems

1. Let property (5) (the "torsion free" condition) be dropped from the definition of derivative operator ∇_a in section 3.1

a) Show that there exists a tensor $T^c{}_{ab}$ (called the *torsion tensor*) such that for all smooth functions, f, we have $\nabla_a \nabla_b f - \nabla_b \nabla_a f = -T^c{}_{ab} \nabla_c f$. (Hint: Repeat the derivation of eq. [3.1.8], letting $\tilde{\nabla}_a$ be a torsion-free derivative operator.)

b) Show that for any smooth vector fields X^a, Y^a we have

$$T^c{}_{ab} X^a Y^b = X^a \nabla_a Y^c - Y^a \nabla_a X^c - [X, Y]^c \quad .$$

c) Given a metric, g_{ab}, show that there exists a unique derivative operator ∇_a with torsion $T^c{}_{ab}$ such that $\nabla_c g_{ab} = 0$. Derive the analog of equation (3.1.29), expressing this derivative operator in terms of an ordinary derivative ∂_a and $T^c{}_{ab}$.

2. Let M be a manifold with metric g_{ab} and associated derivative operator ∇_a. A solution of the equation $\nabla_a \nabla^a \alpha = 0$ is called a *harmonic function*. In the case where M is a two-dimensional manifold, let α be harmonic and let ϵ_{ab} be an antisymmetric tensor field satisfying $\epsilon_{ab} \epsilon^{ab} = 2(-1)^s$, where s is the number of minuses occurring in the signature of the metric. Consider the equation $\nabla_a \beta = \epsilon_{ab} \nabla^b \alpha$.

a) Show that the integrability conditions (see problem 5 of chapter 2 or appendix B) for this equation are satisfied, and thus, locally, there exists a solution, β. Show that β also is harmonic, $\nabla_a \nabla^a \beta = 0$. ($\beta$ is called the harmonic function *conjugate* to α.)

b) By choosing α and β as coordinates, show that the metric takes the form

$$ds^2 = \pm \Omega^2(\alpha, \beta)[d\alpha^2 + (-1)^s d\beta^2] \quad .$$

3. *a*) Show that $R_{abcd} = R_{cdab}$.

b) In n dimensions, the Riemann tensor has n^4 components. However, on account of the symmetries (3.2.13), (3.2.14), and (3.2.15), not all of these components are independent. Show that the number of independent components is $n^2(n^2-1)/12$.

4. *a*) Show that in two dimensions, the Riemann tensor takes the form $R_{abcd} = Rg_{a[c}g_{d]b}$. (Hint: Use the result of problem 3(*b*) to show that $g_{a[c}g_{d]b}$ spans the vector space of tensors having the symmetries of the Riemann tensor.)

b) By similar arguments, show that in three dimensions the Weyl tensor vanishes identically; i.e., for $n = 3$, equation (3.2.28) holds with $C_{abcd} = 0$.

5. *a*) Show that any curve whose tangent satisfies equation (3.3.2) can be reparameterized so that equation (3.3.1) is satisfied.

b) Let t be an affine parameter of a geodesic γ. Show that all other affine parameters of γ take the form $at + b$, where a and b are constants.

6. The metric of Euclidean $\mathbb{R}^3$ in spherical coordinates is $ds^2 = dr^2 + r^2(d\theta^2 + \sin^2\theta\, d\phi^2)$ (see problem 8 of chapter 2).

a) Calculate the Christoffel components $\Gamma^\sigma{}_{\mu\nu}$ in this coordinate system.

b) Write down the components of the geodesic equation in this coordinate system and verify that the solutions correspond to straight lines in Cartesian coordinates.

7. As shown in problem 2, an arbitrary Lorentz metric on a two-dimensional manifold locally always can be put in the form $ds^2 = \Omega^2(x,t)[-dt^2 + dx^2]$. Calculate the Riemann curvature tensor of this metric (*a*) by the coordinate basis methods of section 3.4*a* and (*b*) by the tetrad methods of section 3.4*b*.

8. Using the antisymmetry of $\omega_{a\mu\nu}$ in μ and ν, equation (3.4.15), show that

$$\omega_{\lambda\mu\nu} = 3\omega_{[\lambda\mu\nu]} - 2\omega_{[\mu\nu]\lambda} \quad .$$

Use this formula together with equation (3.4.23) to solve for $\omega_{\lambda\mu\nu}$ in terms of commutators (or antisymmetrized derivatives) of the orthonormal basis vectors.

FOUR

EINSTEIN'S EQUATION

In this chapter we shall give a mathematically precise formulation of the ideas sketched in the introduction. We begin by giving an exposition of the elementary topic of the geometry of space and the spatial tensorial character of physical laws in prerelativity physics. The discussion of special relativity which follows will completely parallel this discussion, with "spacetime" replacing "space." Our next (and most important) task will be to formulate general relativity and provide a motivation for Einstein's equation, which relates the geometry of spacetime to the distribution of matter in the universe. Finally, we will examine general relativity in the limit where gravity is weak. We will show that in appropriate circumstances general relativity reproduces the predictions of the Newtonian theory of gravity. We also will show that in general relativity, gravity has dynamical degrees of freedom which can be excited by the motion of matter, i.e., that general relativity predicts the existence of gravitational radiation.

4.1 The Geometry of Space in Prerelativity Physics; General and Special Covariance

In prerelativity physics it is assumed that space has the manifold structure of $\mathbb{R}^3$. It is further assumed that the association of points of space with elements (x^1, x^2, x^3) of $\mathbb{R}^3$ can be achieved in a natural manner by construction of a "rigid rectilinear grid" of metersticks. The coordinates of space obtained in this manner are referred to as *Cartesian coordinates*. Many different systems of rigid grids (i.e., many Cartesian coordinate systems) are possible—specifically they can be put in one-to-one correspondence with elements of the six-parameter group of rotations and translations of $\mathbb{R}^3$. Thus, the Cartesian coordinates (x^1, x^2, x^3) of a point in space do not, in themselves, have any intrinsic meaning. However, the distance, D, between two points, x and $\bar{x}$, defined in terms of Cartesian coordinates by

$$D^2 = (x^1 - \bar{x}^1)^2 + (x^2 - \bar{x}^2)^2 + (x^3 - \bar{x}^3)^2 \qquad (4.1.1)$$

is independent of the choice of Cartesian coordinate system and thus can be viewed as describing an intrinsic property of space.

This formula for the distance between two points gives rise to a metric of space h_{ab} (in the sense of section 2.3) in the following manner. According to equation (4.1.1), the distance between two "nearby" points is

$$(\delta D)^2 = (\delta x^1)^2 + (\delta x^2)^2 + (\delta x^3)^2 \tag{4.1.2}$$

This suggests that the metric of space is given by (in the notation of eq. [2.3.10])

$$ds^2 = (dx^1)^2 + (dx^2)^2 + (dx^3)^2 \quad , \tag{4.1.3}$$

or, in the index notation, in the Cartesian coordinate basis, we have

$$h_{ab} = \sum_{\mu,\nu} h_{\mu\nu}(dx^\mu)_a(dx^\nu)_b \tag{4.1.4}$$

with $h_{\mu\nu} = \text{diag}\,(1, 1, 1)$. This definition of h_{ab} is independent of choice of Cartesian coordinate system; i.e., the same tensor field h_{ab} would result if we had chosen a different Cartesian system.

Let us examine the geometry determined by equation (4.1.4). Since the components of the metric in the Cartesian coordinate basis are constants, the ordinary derivative operator of this coordinate system satisfies

$$\partial_a h_{bc} = 0 \quad . \tag{4.1.5}$$

Consequently, this ordinary derivative *is* the derivative operator associated with h_{ab} and thus $\Gamma^a{}_{bc}$ vanishes for this coordinate system (as also can be seen directly from eq. [3.1.30]). Since ordinary derivatives commute on all tensors, by equation (3.2.3) the curvature vanishes, i.e., the metric h_{ab} is flat. By equation (3.3.5), the geodesics of space are precisely those curves which are "straight lines" when expressed in Cartesian coordinates, i.e., the curves whose Cartesian coordinates are linearly related to the affine parameter. Consequently, there is a unique geodesic connecting any given pair of points, and the length, equation (3.3.7), of this geodesic is given by equation (4.1.1). Thus, our metric, equation (4.1.4), reproduces the distance formula which motivated its definition.

Thus, our assumptions about space in prerelativity physics have led to the statement that space is the manifold $\mathbb{R}^3$ which possesses a flat Riemannian metric. Conversely, given that space is $\mathbb{R}^3$ with a flat Riemannian metric, we can derive all our initial assumptions. We can use the geodesics of the flat metric to construct a Cartesian coordinate system, using the fact that initially parallel geodesics remain parallel because the curvature vanishes. The distance formula, equation (4.1.1), will hold, and hence one may lay down metersticks over the Cartesian coordinate lines to construct a "rigid rectilinear grid." Thus, everything we have said thus far in this section is encapsulated in the statement that *space is the manifold* $\mathbb{R}^3$ *with a flat Riemannian metric defined on it.*

Let us consider, now, quantities of physical interest in space. All experiments in physics measure numbers, so all quantities of physical interest must eventually be reducible to numbers. However, many quantities of interest—such as the magnetic field or the stress tensor mentioned at the beginning of section 2.3—require the additional specification of a basis of vectors in order to produce numbers. A very general class of quantities of interest are maps of vectors and dual vectors into numbers. Since any such analytic map can be Taylor-expanded as a sum of multilinear maps, we see that tensor fields—i.e., multilinear maps of vectors and dual vectors into numbers—encompass an extremely wide class of quantities. Indeed, the generality of the mathematical notion of tensor fields appears to be great enough so that essentially all[1] quantities which one considers in physics can be viewed as tensor fields. The laws of physics governing these quantities can be expressed as tensor equations, i.e., equalities between tensor fields defined on space.

An important principle—which goes under the name of *general covariance*—applies to the form of the laws of physics in the prerelativity description of space as well as in special relativity and general relativity. The principle of general covariance in this context states that the metric of space is the only quantity pertaining to space that can appear in the laws of physics. Specifically, there are no preferred vector fields or preferred bases of vector fields pertaining only to the structure of space which appear in any law of physics. This idea played an important role in motivating the formulation of special relativity and general relativity, but the principle is rather vague because the phrase "pertaining to space" does not have a precise meaning.

Historically, there has been a great deal of discussion concerning the tensor nature of the laws of physics and the principle of general covariance, so it is worthwhile to remark here upon other formulations of these ideas. In many treatments, it is assumed that a coordinate system has been chosen and the equations of physics have been written out in component form using the coordinate basis. (Note that, in our viewpoint, this already assumes that the laws of physics are tensor equations.) Now, if our formulation of general covariance were violated, say by the existence of a preferred vector field v^a, it would be possible to adapt a coordinate system to this vector field so that, say, $(\partial/\partial x^1)^a$ equals v^a. If we wrote out the components of an equation of physics in such an adapted coordinate system *without explicitly incorporating* v^a *into the equation* but rather substituting everywhere the components $v^\mu = (1, 0, \ldots, 0)$, we would find, of course, that the form of the equation is not preserved when we make a coordinate transformation which violates our condition $(\partial/\partial x^1)^a = v^a$. Thus in these treatments it would be concluded that in this example the equations are not preserved under general coordinate transformations. Furthermore, it would be concluded that the equations are not tensor equations because the components fail to transform according to equation (2.3.8) under coordinate transformations. However, in our viewpoint, the "nontensorial" nature of the equations can be attributed to a failure to *explicitly* incorporate the extra geometrical structure into the equation. When this is done, the equation will have a tensorial character, but our formulation of the principle of general covariance will be violated by the appearance of additional quantities pertaining to space.

A good example of an implication of the principle of general covariance which illustrates this difference in viewpoint is the following statement: A Christoffel symbol $\Gamma^c{}_{ab}$ cannot appear (by itself, in undifferentiated form) in any law of physics. From our viewpoint, $\Gamma^c{}_{ab}$ is a tensor, but it is equivalent to specifying an ordinary

1. Spinor fields arise in physics and comprise a more general class of objects than tensor fields; see chapter 13.

derivative, ∂_a. However, this ordinary derivative is an additional geometric quantity pertaining to space which is not derivable from the metric (unless it coincides with the derivative operator satisfying eq. [3.1.22], in which case $\Gamma^c{}_{ab} = 0$), so the appearance of $\Gamma^c{}_{ab}$ violates our formulation of general covariance. From the other viewpoint, $\Gamma^c{}_{ab}$ cannot appear because, as already remarked in section 3.1, its coordinate components do not transform according to equation (2.3.8) under general coordinate transformations.

The laws of prerelativity physics also obey the principle of *special covariance*. The meaning of this principle can be explained as follows. The metric of space, equation (4.1.4), has a nontrivial group of isometries (see appendix C), namely the six parameter group of rotations and translations of $\mathbb{R}^3$ together with the discrete parity symmetry. Consider a family, O, of observers stationed in space who make measurements on a physical field. Consider another family, O', of observers obtained by "acting" on O with an isometry. The principle of special covariance says that any physically possible set of measurements obtained by O also is a physically possible set of measurements for O'. It implies the existence of an action of the isometry group on the states of the physical fields being measured. The importance of special covariance lies mainly in the fact that in appropriate contexts, one can use this group action to derive, or at least motivate, the possible laws governing the physical fields. This will be illustrated in chapter 13, where—in the context of special relativity rather than prerelativity physics—special covariance will be used to motivate the dynamical laws satisfied by fields of mass m and spin s.

The principle of special covariance is closely related to the principle of general covariance. Suppose an object of physical interest is described by a tensor field $T^{a\cdots}{}_{b\cdots}$. If general covariance holds, then the equations governing $T^{a\cdots}{}_{b\cdots}$ should involve only $T^{a\cdots}{}_{b\cdots}$, the metric h_{ab}, and the quantities determined by h_{ab}, such as its derivative operator. Furthermore, all quantities measurable by O should be expressible as scalars resulting from contracting $T^{a\cdots}{}_{b\cdots}$ and its derivatives with the basis vector fields $(e_\alpha)^a$ associated with O. Let ϕ be an isometry. Then if the above assumptions hold, the action of ϕ on $T^{a\cdots}{}_{b\cdots}$, defined in appendix C, will produce a tensor field $\phi^* T^{a\cdots}{}_{b\cdots}$ which will satisfy the equations for the physical field and will yield the same measurable quantities for O' as $T^{a\cdots}{}_{b\cdots}$ yields for O. Thus, for physical quantities describable by tensor fields and satisfying tensor equations, special covariance of the physical laws under isometries follows, in essence, from general covariance. (However, note that the formulation of special covariance given above does not require that the physical quantities be describable as tensor fields, and, indeed, in chapter 13 we will use special covariance to motivate the introduction of spinor fields.) On the other hand, the principle of general covariance may be viewed as being, essentially, a formulation of the idea of special covariance which is applicable in the absence of isometries.

Special covariance of the laws of physics for tensor fields implies that if we pick a coordinate system and write out the coordinate component equations *without explicitly incorporating the metric into the equation* but rather substituting everywhere its coordinate components $h_{\mu\nu}$ [e.g., diag (1, 1, 1) in Cartesian coordinates], then the form of the equations will be preserved under the coordinate transformations corresponding to the isometries, for the simple reason that the numerical values of the components $h_{\mu\nu}$ remain unchanged under these transformations. Thus, special covariance can be viewed as expressing the invariance of the component form of equations under a "special" group of coordinate transformations (namely, those corresponding to isometries), while general covariance can be viewed as expressing a (quite different type of) invariance of the equations under general coordinate transformations. Historically these formulations of special and general covariance played a large role in discussions of relativity theory, and, indeed, the names "special" and "general" relativity derive from these formulations of special and general covariance.

4.2 Special Relativity

We have already described in section 1.2 the major revolution in our notions of space and time brought about by the special theory of relativity. Here we will reformulate these ideas in a manner closely analogous to the discussion of the previous section.

In special relativity, it is assumed that spacetime has the manifold structure of $\mathbb{R}^4$. As already mentioned in section 1.2, it is assumed that there exist preferred families of motion in spacetime, referred to as "inertial" or "nonaccelerating" motions. Inertial observers, it is further assumed, can set up a rigid grid of metersticks, synchronize clocks placed at the gridpoints, and label each event in spacetime by the grid coordinates x^1, x^2, x^3 and the clock reading t at an event. This map of spacetime into $\mathbb{R}^4$ is called a *global inertial coordinate system*. Many different global inertial coordinate systems are possible—specifically, the different systems can be put into one-to-one correspondence with elements of the 10-parameter Poincaré group. Thus, the labels at $t = x^0, x^1, x^2, x^3$ of a given event do not have intrinsic meaning. However, as already mentioned in section 1.3, the spacetime interval I between two events x and $\bar{x}$ defined (in units where $c = 1$) by

$$I = -(x^0 - \bar{x}^0)^2 + (x^1 - \bar{x}^1)^2 + (x^2 - \bar{x}^2)^2 + (x^3 - \bar{x}^3)^2 \tag{4.2.1}$$

has the same value for all global inertial coordinate systems and thus can be viewed as representing an intrinsic property of spacetime.

In parallel with our discussion of the previous section, equation (4.2.1) suggests that we define the *metric of spacetime* η_{ab} by

$$\eta_{ab} = \sum_{\mu,\nu=0}^{3} \eta_{\mu\nu}(dx^\mu)_a(dx^\nu)_b \tag{4.2.2}$$

with $\eta_{\mu\nu} = \text{diag}(-1, 1, 1, 1)$, where $\{x^\mu\}$ is any global inertial coordinate system. Again, the tensor field η_{ab} thereby obtained is independent of the choice of global inertial coordinates. Again, the ordinary derivative operator, ∂_a, of the global inertial coordinates satisfies

$$\partial_a \eta_{bc} = 0 \tag{4.2.3}$$

and thus is the derivative operator associated with the spacetime metric. Again, since ordinary derivatives commute, the curvature of η_{ab} vanishes. The geodesics of η_{ab} are those curves which are straight lines when expressed in global inertial coordinates. In particular, the timelike geodesics of η_{ab} are precisely the world lines of inertial observers in spacetime.

Thus, the theory of special relativity asserts that *spacetime is the manifold* $\mathbb{R}^4$ *with a flat metric of Lorentz signature defined on it*. Conversely, the entire content of special relativity as we have presented it thus far is contained in this statement, since, given $\mathbb{R}^4$ with a flat Lorentz metric, we can use the geodesics of this metric to construct global inertial coordinates, etc.

Quantities of physical interest in special relativity again are represented by tensor fields, but now quantities which in prerelativity physics were viewed as spatial tensors are recognized in special relativity to comprise spacetime tensors. In the context of special relativity, the principle of general covariance states that the spacetime metric, η_{ab}, is the only quantity pertaining to spacetime structure which can appear in any physical law. This principle is believed to apply to the laws of physics in special relativity with one important modification. Experiments demonstrating parity violation and (indirectly) demonstrating the failure of time reversal symmetry have shown that two further aspects of spacetime structure can appear in physical laws: the time orientation and space orientation of spacetime. By the "time orientation" we mean a continuous choice throughout spacetime of which half of the light cone represents the future direction, and which half represents the past direction (see Fig. 1.2, and see the beginning of chapter 8 for further discussion of this notion in the context of curved spacetime). By the "space orientation" we mean a continuous choice of "right handed" versus "left handed" orthonormal triads of spacelike vectors at each point, or equivalently, the specification of a continuous, nonvanishing totally antisymmetric tensor field $e_{abc} = e_{[abc]}$ on spacetime satisfying $t^a e_{abc} = 0$ for some timelike vector field[2] t^a. Thus, these two additional aspects of spacetime structure apparently do enter the laws of physics. Similarly, the laws of physics in special relativity are believed to satisfy the principle of special covariance with respect to the *proper* Poincaré transformation, i.e., with respect to the translations, rotations, and boosts of spacetime, but not with respect to time reflections, parity transformations, or their composition. Although these latter "improper" transformations are isometries, they do not preserve all the relevant structure of spacetime since they reverse the time and/or space orientation of spacetime.

Let us describe more explicitly the formulation of some of the laws of physics in special relativity. We have already mentioned that curves are classified as timelike, null, or spacelike according to whether the norm $\eta_{ab}T^aT^b$ of their tangent, is, respectively, negative, zero, or positive. Special relativity asserts that the paths in spacetime of material particles are always timelike curves. (This fact is, of course, simply a reformulation of the familiar statement that "nothing can travel faster than

2. Note that if t^a is chosen to be continuous, thus defining a time orientation, then $\epsilon_{abcd} = -t_{[a}e_{bcd]}$ yields a spacetime orientation in the sense of appendix B. Any two of (i) a time orientation, (ii) a space orientation, and (iii) a spacetime orientation of Minkowski spacetime determines the third.

the speed of light.") We may parameterize timelike curves by the *proper time* τ defined by

$$\tau = \int (-\eta_{ab} T^a T^b)^{1/2} \, dt \quad , \tag{4.2.4}$$

where t is an arbitrary parameterization of the curve (with increasing t corresponding to "forward in time"), and T^a is the tangent to the curve in this parameterization. According to special relativity, τ is precisely the time which would elapse on a clock carried along the given curve. Different timelike curves connecting the same pair of events may have different elapsed times (the "twin paradox"), just as different paths between two points of space can have different lengths. As discussed in section 3.3, the maximum elapsed time between two events is given by geodesic (i.e., inertial) motion.

The tangent vector u^a to a timelike curve parameterized by τ is called the *4-velocity* of the curve. It follows directly from the definition of τ that the 4-velocity has unit length,

$$u^a u_a = -1 \quad . \tag{4.2.5}$$

As already mentioned above, a particle subject to no external forces will travel on a geodesic; i.e., its 4-velocity will satisfy the equation of motion,

$$u^a \partial_a u^b = 0 \quad , \tag{4.2.6}$$

where ∂_a is the derivative operator associated with η_{ab}, i.e., the ordinary derivative operator of a global inertial coordinate system. If forces are present, the right-hand side of equation (4.2.6) will be nonzero (see, e.g., eq. [4.2.26] below).

All material particles have an attribute known as "rest mass" m, which appears as a parameter in the equations of motion when forces are present. The energy-momentum 4-vector, p^a, of a particle of mass m is defined by

$$p^a = m u^a \quad . \tag{4.2.7}$$

The *energy* of a particle as measured by an observer—present at the site of the particle—whose 4-velocity is v^a is defined by

$$E = -p_a v^a \quad . \tag{4.2.8}$$

Thus, in special relativity, energy is recognized to be the "time component" of the 4-vector p^a. For a particle at rest with respect to the observer (i.e., $v^a = u^a$), equation (4.2.8) reduces to the familiar formula $E = mc^2$ (in our units with $c = 1$). Since the spacetime metric, η_{ab}, is flat and thus parallel transport is path independent, we may define the energy of a particle as measured by an observer who is *not* present at the site of the particle to be the energy measured by the observer who *is* at the site of the particle and has 4-velocity parallel to that of the distant observer.

Continuous matter distributions in special relativity are described by a symmetric tensor T_{ab} called the *stress-energy-momentum* tensor. For an observer with 4-velocity v^a, the component $T_{ab} v^a v^b$ is interpreted as the energy density, i.e., the mass-energy per unit volume, as measured by this observer. For normal matter, this quantity will be nonnegative,

$$T_{ab}v^a v^b \geq 0 \quad . \tag{4.2.9}$$

If x^a is orthogonal to v^a, the component $-T_{ab}v^a x^b$ is interpreted as the momentum density of the matter in the x^a-direction. If y^a also is orthogonal to v^a, then $T_{ab}x^a y^b$ represents the x^a–y^a component of the stress tensor of the material defined near the beginning of section 2.3 above. Thus, the stress tensor of prerelativity physics is combined with the energy and momentum densities to form the stress-energy-momentum tensor of special relativity. Following standard practice, we often shall abbreviate the term "stress-energy-momentum tensor" as "stress-energy tensor" or even "stress tensor."

A *perfect fluid* is defined to be a continuous distribution of matter with stress-energy tensor T_{ab} of the form

$$T_{ab} = \rho u_a u_b + P(\eta_{ab} + u_a u_b) \quad , \tag{4.2.10}$$

where u^a is a unit timelike vector field representing the 4-velocity of the fluid. According to the above interpretation of T_{ab}, the functions ρ and P are, respectively the mass-energy density and pressure of the fluid as measured in its rest frame. The fluid is called "perfect" because of the absence of heat conduction terms and stress terms corresponding to viscosity.

The equation of motion of a perfect fluid subject to no external forces is simply

$$\partial^a T_{ab} = 0 \quad . \tag{4.2.11}$$

Writing out equation (4.2.11) in terms of ρ, P, and u^a, and projecting the resulting equation parallel and perpendicular to u^b, we find:

$$u^a \partial_a \rho + (\rho + P)\partial^a u_a = 0 \quad , \tag{4.2.12}$$

$$(P + \rho)u^a \partial_a u_b + (\eta_{ab} + u_a u_b)\partial^a P = 0 \quad . \tag{4.2.13}$$

In the nonrelativistic limit, $P << \rho$, $u^\mu = (1, \vec{v})$, and $v\, dP/dt << |\vec{\nabla} P|$, these equations become

$$\frac{\partial \rho}{\partial t} + \vec{\nabla} \cdot (\rho \vec{v}) = 0, \tag{4.2.14}$$

$$\rho \left\{ \frac{\partial \vec{v}}{\partial t} + (\vec{v} \cdot \vec{\nabla})\vec{v} \right\} = -\vec{\nabla} P \quad , \tag{4.2.15}$$

so we see that equation (4.2.11) reduces to the conservation of mass, equation (4.2.14), and Euler's equation (4.2.15).

Equation (4.2.11) has an important physical interpretation. Consider a family of inertial observers with parallel 4-velocities v^a, so that $\partial_b v^a = 0$. According to the above interpretation of T_{ab}, the quantity

$$J_a = -T_{ab}v^b \tag{4.2.16}$$

represents the mass-energy current density 4-vector of the fluid as measured by these observers. Equation (4.2.11) implies

$$\partial^a J_a = 0 \quad . \tag{4.2.17}$$

Using Gauss's law (see appendix B) equation (4.2.17) implies that over the three-dimensional boundary, S, of any four-dimensional spacetime volume V, we have

$$\int_S J_a n^a dS = 0 \quad , \tag{4.2.18}$$

where n^a is the unit normal whose direction is defined in appendix B. Applying this to the volume shown in Figure 4.1, we see that the energy change in the fluid in this volume (i.e., the contribution to the integral, eq. [4.2.18], from the top and bottom parts of S) equals the time integrated energy flux into the volume (i.e., the contribution from the "side" part of S). Thus, equation (4.2.18) implies conservation of energy. Conversely, conservation of energy as measured by all inertial observers requires equation (4.2.11). Thus, more generally, conservation of energy implies that equation (4.2.11) must hold for all continuous matter distributions, not just for perfect fluids.

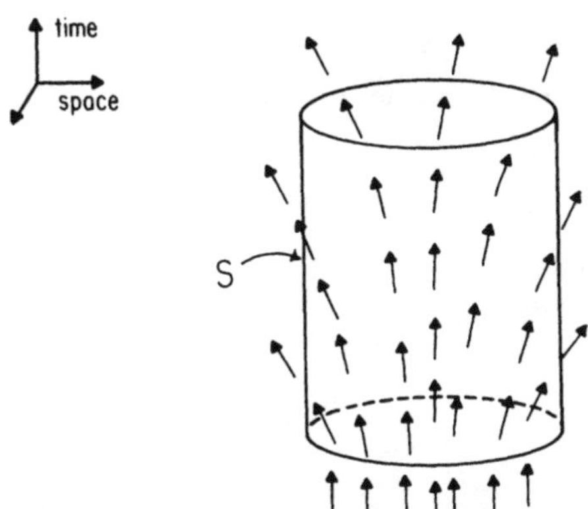

Fig. 4.1. A spacetime diagram showing the mass-energy current density J^a in a volume V of spacetime.

To illustrate the description of fields in special relativity, we will give two examples: the scalar field and the electromagnetic field. Although no classical scalar field exists in nature, it is instructive for many purposes to consider a field ϕ satisfying the Klein-Gordon equation,

$$\partial^a \partial_a \phi - m^2 \phi = 0 \quad . \tag{4.2.19}$$

The stress-energy tensor of this scalar field is[3]

$$T_{ab} = \partial_a \phi \partial_b \phi - \frac{1}{2} \eta_{ab} (\partial^c \phi \partial_c \phi + m^2 \phi^2) \quad . \tag{4.2.20}$$

3. General prescriptions for determining the stress-energy tensor of a field are discussed below, in appendix E.

Again, T_{ab} satisfies the energy condition, equation (4.2.9), and is conserved, equation (4.2.11), by virtue of the field equation (4.2.19).

In prerelativity physics, the electric field $\vec{E}$ and magnetic field $\vec{B}$ each are spatial vectors. In special relativity these fields are combined into a single spacetime tensor field F_{ab} which is antisymmetric in its indices, $F_{ab} = -F_{ba}$. Thus F_{ab} has six independent components. For an observer moving with 4-velocity v^a, the quantity

$$E_a = F_{ab}v^b \tag{4.2.21}$$

is interpreted as the electric field measured by that observer, while

$$B_a = -\frac{1}{2}\epsilon_{ab}{}^{cd}F_{cd}v^b \tag{4.2.22}$$

is interpreted as the magnetic field, where ϵ_{abcd} is the totally antisymmetric tensor of positive orientation with norm $\epsilon_{abcd}\epsilon^{abcd} = -24$ (see appendix B) so that in a right-handed orthonormal basis we have $\epsilon_{0123} = 1$.

In terms of F_{ab}, Maxwell's equations take the simple and elegant form,

$$\partial^a F_{ab} = -4\pi j_b \quad , \tag{4.2.23}$$

$$\partial_{[a}F_{bc]} = 0 \quad , \tag{4.2.24}$$

where j^a is the current density 4-vector of electric charge. Note that the antisymmetry of F_{ab} implies that

$$0 = \partial^b\partial^a F_{ab} = -4\pi\partial^b j_b \quad . \tag{4.2.25}$$

Thus, Maxwell's equations imply $\partial^b j_b = 0$, which, by the same argument as given above for J_a, states that electric charge is conserved. The equation of motion of a particle of charge q moving in the electromagnetic field F_{ab} is

$$u^a\partial_a u^b = \frac{q}{m}F^b{}_c u^c \quad , \tag{4.2.26}$$

which reformulates the usual Lorentz force law in terms of F_{ab}.

The stress-energy tensor of the electromagnetic field is

$$T_{ab} = \frac{1}{4\pi}\left\{F_{ac}F_b{}^c - \frac{1}{4}\eta_{ab}F_{de}F^{de}\right\} \quad . \tag{4.2.27}$$

Again, T_{ab} satisfies the energy condition, equation (4.2.9), and if $j^a = 0$, we have $\partial^a T_{ab} = 0$ by virtue of Maxwell's equations. If $j^a \neq 0$, then the stress-energy T_{ab} of the electromagnetic field alone is not conserved, but the total stress-energy of the field *and* the charged matter is still conserved.

By the converse of the Poincaré lemma (see appendix B), equation (4.2.24) implies that there exists a vector field A^a (called the *vector potential*) such that

$$F_{ab} = \partial_a A_b - \partial_b A_a \quad . \tag{4.2.28}$$

In terms of A^a, Maxwell's equations become

$$\partial^a(\partial_a A_b - \partial_b A_a) = -4\pi j_b \quad . \tag{4.2.29}$$

We have the gauge freedom of adding the gradient $\partial_a \chi$ of a function χ to A_a since, by equation (4.2.28), this leaves F_{ab} unchanged. By solving the equation

$$\partial^a \partial_a \chi = -\partial^b A_b \tag{4.2.30}$$

for χ, we may make a gauge transformation to impose the Lorentz gauge condition,

$$\partial^a A_a = 0 \quad , \tag{4.2.31}$$

in which case, using the commutativity of derivatives in flat spacetime, equation (4.2.29) becomes

$$\partial^a \partial_a A_b = -4\pi j_b \quad . \tag{4.2.32}$$

We may seek solutions of Maxwell's equations of the form of a wave oscillating with constant amplitude,

$$A_a = C_a \exp(iS) \quad , \tag{4.2.33}$$

where C_a is a constant vector field (i.e., constant norm and everywhere parallel to itself) and the function S is called the *phase* of the wave. To yield a solution with $j^a = 0$, the phase must satisfy (from eq. [4.2.32])

$$\partial^a \partial_a S = 0 \quad , \tag{4.2.34}$$

$$\partial_a S \partial^a S = 0 \quad , \tag{4.2.35}$$

and (from eq. [4.2.31]) ,

$$C_a \partial^a S = 0 \quad . \tag{4.2.36}$$

Now, for any function f on any manifold with a metric, the vector $\nabla^a f$ is normal (i.e., orthogonal) to the surfaces of constant f, since for any vector t^a tangent to the surface we have $t^a \nabla_a f = 0$. Equation (4.2.35) states that the normal $k^a = \partial^a S$ to the surfaces of constant S is a null vector, $k_a k^a = 0$. We call such a surface a *null hypersurface*. Note that null hypersurfaces satisfy the property that their normal vector is tangent to the hypersurface. Differentiation of equation (4.2.35) yields

$$\begin{aligned} 0 &= \partial_b(\partial_a S \partial^a S) \\ &= 2(\partial^a S)(\partial_b \partial_a S) \\ &= 2(\partial^a S)(\partial_a \partial_b S) \\ &= 2k^a \partial_a k_b \quad ; \end{aligned} \tag{4.2.37}$$

i.e., the integral curves of k^a are null geodesics. Indeed, since the derivation of equation (4.2.37) from equation (4.2.35) is valid in curved spacetime as well (with ∇_a replacing ∂_a everywhere), it follows that all null hypersurfaces in a Lorentz spacetime are generated by null geodesics.[4] The *frequency* of the wave (i.e., minus

4. The derivation of equation (4.2.37) is appropriate to the case in which one has a function S for which all the hypersurfaces of constant S are null. For a single null hypersurface, the vanishing of $\epsilon^{abcd} k_c \nabla_d (k_e k^e)$ shows that k^a is tangent to a (non-affinely parameterized) null geodesic.

the rate of change of the phase of the wave) as measured by an observer with 4-velocity v^a is given by

$$\omega = -v^a \partial_a S = -v^a k_a \quad . \tag{4.2.38}$$

The most important solutions of the form, equation (4.2.33), are the plane waves,

$$S = \sum_{\mu=0}^{3} k_\mu x^\mu \quad , \tag{4.2.39}$$

where $\{x^\mu\}$ are global inertial coordinates and k_μ are constants (and thus k^a is a constant vector field). From the theory of Fourier transforms, it follows that all well behaved solutions of Maxwell's equations which go to zero sufficiently rapidly at large spacial distances can be expressed as superpositions of plane waves.

Our analysis, above, suggests that electromagnetic disturbances, i.e., light signals, propagate on null geodesics, equation (4.2.37). This supposition is indeed correct: The retarded Green's function for Maxwell's equations has support confined to the past light cone (see, e.g., Jackson 1962), so the electromagnetic radiation from a source which reaches event p depends only on what the source does on the past light cone of p. Thus, the terminology of "light cone," which we have been using all along, is justified by Maxwell's equations; light does indeed propagate along the light cone.

4.3 General Relativity

Maxwell's theory is a remarkably successful theory of electricity, magnetism, and light which is beautifully incorporated into the framework of special relativity. Therefore, one might expect that the next logical step would have been to develop a new theory of the other classical force, gravitation, which would generalize Newton's theory and make it compatible with special relativity in the same way that Maxwell's theory generalized Coulomb's electrostatics. However, Einstein chose an entirely different path and instead developed general relativity, a new theory of spacetime structure and gravitation. As already mentioned in the introductory chapter, the equivalence principle and Mach's principle provided the primary motivation for formulating a new theory.

To see the relevance of the equivalence principle—that all bodies fall the same way in a gravitational field—to developing a new viewpoint on gravitation, consider how one measures the electromagnetic field in special relativity. The first step is to set up "background observers" who are not subject to electromagnetic forces (i.e., they are electrically neutral, have no magnetic dipole moment, etc.) or any other forces. These observers are called inertial and satisfy the geodesic equation of motion (4.2.6). The next step is to release a charged test body. The world line of this body will satisfy equation (4.2.26), and by observing the deviation from inertial motion (for sufficiently many test bodies) we can determine F_{ab}.

If we apply this procedure to gravitation, we are immediately faced with a serious problem: By the equivalence principle, we have no way of "insulating" an observer or body from the gravitational force, so we have no simple, direct physical procedure for constructing inertial observers in the sense used for electromagnetism. Any observer will move in exactly the same way as a test body, so we have no natural "background motion" to compare with the test body. Thus we have no simple, direct way of measuring the gravitational force field. It is, of course, possible that complicated procedures for accomplishing these constructions and measurements may exist. If special relativity were correct, one could construct inertial observers by spacetime measurements, i.e., the (flat) spacetime metric could be measured using clocks and metersticks, and its geodesics determined. Inertial observers would need to be equipped with rocket engines, but aside from that the gravitational force field could be determined in the same way as for electromagnetism. The equivalence principle would then be viewed as a peculiar quirk of the gravitational force law, just as it is viewed in standard treatments of Newtonian theory.

The basic framework of the theory of general relativity arises from considering the opposite possibility: that we cannot in principle—even by complicated procedures—construct inertial observers in the sense of special relativity and measure the gravitational force. This is accomplished by the following bold hypothesis: *The spacetime metric is not flat, as was assumed in special relativity. The world lines of freely falling bodies in a gravitational field are simply the geodesics of the (curved) spacetime metric*. In this way, the "background observers" (geodesics of the spacetime metric) automatically coincide with what was previously viewed as motion in a gravitational force field. As a result we have no meaningful way of describing gravity as a force field; rather, we are forced to view gravity as an aspect of spacetime structure. Although absolute gravitational force has no meaning, the relative gravitational force (i.e., tidal force) between two nearby points still has meaning and can be measured by observing the relative acceleration of two freely falling bodies. This relative acceleration is directly related to the curvature of spacetime by the geodesic deviation equation (3.3.18).

How does this viewpoint of general relativity that there is no such thing as gravitational force square with the well known "fact" that there is a gravitational force field at the surface of the Earth of 980 cm s^{-2}? Recall that in the standard Newtonian viewpoint this gravitational force on an object placed on the Earth's surface is balanced by the force the surface exerts, leaving the body in equilibrium, i.e., "at rest." In the viewpoint of general relativity, the only force acting on the body is the force of the surface of the Earth. On account of this force, the body accelerates (i.e., deviates from geodesic motion) at the rate of 980 cm s^{-2}. Nevertheless, it remains in a stationary state, because in the curved spacetime geometry in the vicinity of the Earth, the orbits of time translation symmetry differ from the geodesics of the metric. We could use the time translation symmetry of this example to define a preferred set of background observers. We then could define the gravitational force field of the Earth to be minus the acceleration a body must undergo in order to remain stationary. Thus, in this case a well defined meaning can be assigned to gravity as a force field. However, in the absence of time translation symmetry—e.g., in a case where there are several massive bodies in relative motion—there exists no natural set of curves whose comparison with geodesics could be used to define gravitational force. (This remark also applies to an observer in a nonstationary laboratory near the Earth who would not be aware of the time translation symmetry.) Thus, although we may meaningfully speak of the gravitational force field of the Earth, this is a very special notion, applicable only to situations with time translation symmetry. In general situations, there are no preferred background observers, and only the tidal force—the relative acceleration of nearby geodesics—is well defined.

Thus, in general relativity, we do not assert that spacetime is the manifold $\mathbb{R}^4$ with a flat metric η_{ab} defined on it. This is a possibility corresponding to no tidal forces, i.e., no gravitational field, but it is not the only possibility. The framework of general relativity permits the Lorentz metric, g_{ab}, of spacetime to be curved. Indeed, it asserts that spacetime must be curved in all situations where, physically, a gravitational field is present. Since we are allowing curved geometries, it is also much more natural to allow spacetime to have a manifold structure M other than $\mathbb{R}^4$. Hence, in general relativity we place no *a priori* restriction on the spacetime manifold. The final crucial feature of general relativity is Einstein's equation which relates the spacetime geometry to the matter distribution. We will discuss this equation below, but first we discuss the nature of the laws of physics under the new framework of spacetime structure given by general relativity: *Spacetime is a manifold M on which is defined a Lorentz metric* g_{ab}.

The laws of physics in general relativity are governed by two basic principles: (1) the principle of general covariance—already discussed in the previous two sections—which states that the metric, g_{ab}, and quantities derivable from it are the only spacetime quantities that can appear in the equations of physics; (2) the requirement that equations must reduce to the equations satisfied in special relativity in the case where g_{ab} is flat. As previously mentioned, the first principle is imprecise because the term "spacetime quantity" is not well defined. As will be illustrated below, these two principles alone do not uniquely determine the laws of physics in general relativity. However, together with simplicity and aesthetics, they serve as guides which, in many cases, lead directly to natural candidates for physical laws.

Since the basic framework of general relativity modifies that of special relativity only in that it allows the manifold to differ from $\mathbb{R}^4$ and the metric to be nonflat, we may continue to represent physical quantities by the same type of tensor fields as in special relativity. Thus, in general relativity, particle motion continues to be represented by a timelike (in the metric g_{ab}) curve; perfect fluids are still described in terms of a 4-velocity u^a, a density ρ, and a pressure P; the electromagnetic field is represented by an antisymmetric tensor F_{ab}. Only the equations satisfied by these fields need to be amended. The above two principles suggest the following simple rule: In the equations holding in special relativity, replace everywhere the metric η_{ab} of special relativity by g_{ab} and correspondingly replace the derivative operator ∂_a associated with η_{ab} by the derivative operator ∇_a associated with g_{ab}. This rule for, in effect, coupling particles and fields to gravity is closely analogous to the "minimal coupling" rule $p_a \rightarrow p_a - eA_a$ for coupling to electromagnetism. However, as we shall illustrate below, this rule is not entirely free of ambiguity.

Thus, in general relativity, we again define the 4-velocity, u^a, of a particle to be the unit tangent (as measured by g_{ab}) to its world line. A free particle satisfies the geodesic equation of motion,

$$u^a \nabla_a u^b = 0 \quad , \tag{4.3.1}$$

where ∇_a is the derivative operator associated with g_{ab}. If the acceleration[5] $a^b = u^a \nabla_a u^b$ of the particle is nonvanishing, we say that a force $f^b = ma^b$ acts on the particle, where m is its (rest) mass. For example, if the particle has (rest) mass m and charge q, and is placed in an electromagnetic field F_{ab}, it satisfies the Lorentz force equation,

$$u^a \nabla_a u^b = \frac{q}{m} F^b{}_c u^c \quad , \tag{4.3.2}$$

where indices are raised and lowered by g_{ab}, i.e., $F^b{}_c = g^{bd} F_{dc}$. (Again, we emphasize that there is no natural flat metric defined on spacetime and, by the principle of general covariance, only g_{ab} can enter the equations.) The 4-momentum of the particle is defined by

$$p^a = mu^a \quad . \tag{4.3.3}$$

The energy of the particle as determined by an observer who is present at the event on the particle's world line at which the energy is measured is again

$$E = -p_a v^a \tag{4.3.4}$$

where v^a is the 4-velocity of the observer. However, there is one important difference here: Because spacetime is curved, there is no well defined notion of vectors at different points being parallel; parallel transport is curve dependent. Thus, there is no natural "global family" of inertial observers, and a given observer cannot, in general, define the energy of a distant particle.

In general relativity, continuous matter distributions and fields again are described by a stress-energy tensor T_{ab}. The stress tensor of a perfect fluid is given by

$$T_{ab} = \rho u_a u_b + P(g_{ab} + u_a u_b) \quad , \tag{4.3.5}$$

and it satisfies the equations of motion

$$\nabla^a T_{ab} = 0 \quad , \tag{4.3.6}$$

which yield

$$u^a \nabla_a \rho + (\rho + P) \nabla^a u_a = 0 \quad , \tag{4.3.7}$$

$$(P + \rho) u^a \nabla_a u_b + (g_{ab} + u_a u_b) \nabla^a P = 0 \quad . \tag{4.3.8}$$

However, the interpretation of equation (4.3.6) is altered now. A family of observers is represented by a unit timelike vector field v^a. If one could find such a vector field which is covariantly constant, i.e., $\nabla_a v_b = 0$—or for which merely $\nabla_{(a} v_{b)} = 0$—then we would have $\nabla^a (T_{ab} v^b) = 0$. Applying the curved spacetime version of Gauss's law (see appendix B), we again would obtain strict conservation of energy in the form (4.2.18) for the energy-momentum four-vector $J_a = -T_{ab} v^b$ measured by the observers represented by v^b. However, in curved spacetime in general one no longer can find a v^a satisfying $v^a v_a = -1$ and $\nabla_{(a} v_{b)} = 0$. (Indeed, the equation $\nabla_{(a} v_{b)} = 0$ is Killing's equation and holds if and only if v^a generates a one-parameter

5. The (absolute) acceleration a^b should be clearly distinguished from the relative acceleration of geodesics discussed in section 3.3.

group of isometries ⌊see appendix C⌋.) Thus the argument fails that equation (4.3.6) implies strict energy conservation. Physically, this makes sense because the gravitational tidal forces can do work on the fluid and may increase or decrease its locally measured energy.[6] However, if one considers a spacetime region of dimension small compared with radii of curvature, then, physically, the tidal forces can do little work and the energy of the fluid should be approximately conserved. But over this small spacetime region it *is* possible to find vector fields with $\nabla_b v^a \approx 0$, and thus equation (4.3.6) does yield approximate conservation of energy as measured by these observers. Thus, equation (4.3.6) may be interpreted as a local conservation of material energy over small regions of spacetime. On account of this interpretation, we expect equation (4.3.6) to hold for all matter and fields, not just for perfect fluids.

The most natural generalization of the equation satisfied by a Klein-Gordon scalar field to curved spacetime is given by our "minimal substitution" rule $\eta_{ab} \rightarrow g_{ab}$, $\partial_a \rightarrow \nabla_a$,

$$\nabla^a \nabla_a \phi - m^2 \phi = 0 \quad . \tag{4.3.9}$$

The stress tensor of the field is

$$T_{ab} = \nabla_a \phi \nabla_b \phi - \frac{1}{2} g_{ab} (\nabla_c \phi \nabla^c \phi + m^2 \phi^2) \tag{4.3.10}$$

and satisfies $\nabla^a T_{ab} = 0$. We should point out, however, that there are many other possible generalizations of equation (4.2.19) which are consistent with the two basic principles stated above. For example, the equation

$$\nabla^a \nabla_a \phi - m^2 \phi - \alpha R \phi = 0 \quad , \tag{4.3.11}$$

where α is a constant, is such a generalization. Indeed, equation (4.3.11) with $\alpha = 1/6$ arises naturally on account of its conformal invariance properties (see appendix D).

Maxwell's equations in curved spacetime become

$$\nabla^a F_{ab} = -4\pi j_b \quad , \tag{4.3.12}$$

$$\nabla_{[a} F_{bc]} = 0 \quad . \tag{4.3.13}$$

The electromagnetic stress tensor again is given by equation (4.2.27) with g_{ab} replacing η_{ab},

$$T_{ab} = \frac{1}{4\pi} \left\{ F_{ac} F_b{}^c - \frac{1}{4} g_{ab} F_{de} F^{de} \right\} \quad . \tag{4.3.14}$$

Again, equation (4.3.13) allows us to introduce a vector potential A_a (at least

6. One might hope to recover an energy conservation law by including the stress-energy of the gravitational field as can be done in Newtonian theory. However, in general relativity there exists no meaningful local expression for gravitational stress-energy and thus there is no meaningful local conservation law which leads to a statement of energy conservation. Nevertheless, as will be discussed in chapter 11, a conserved total energy of an isolated system can be defined, even though there is no local expression for energy density.

locally). However, Maxwell's equations for A_a in the Lorentz gauge, contains an explicit curvature term resulting from the commutation of derivatives in the derivation of equation (4.2.32); we find

$$\nabla^a \nabla_a A_b - R^d{}_b A_d = -4\pi j_b \quad . \tag{4.3.15}$$

This illustrates an important deficiency of our minimal substitution rule. Had we minimally substituted in Maxwell's equations in the form (4.2.32), we would have been led to equation (4.3.15) without the Ricci tensor term. In this instance, we can decide in favor of equation (4.3.15) over the alternative equation without the $R^d{}_b A_d$ term because equation (4.3.15) implies current conservation, $\nabla_a j^a = 0$ (problem 1), while the alternative equation conflicts with it. However, this example shows that "minimal substitution" by itself is not a unique prescription.

In situations where the spacetime scale of variation of the electromagnetic field is much smaller than that of the curvature, one would expect to have solutions of Maxwell's equations of the form of a wave oscillating with nearly constant amplitude, i.e., solutions of the form

$$A_a = C_a e^{iS} \quad , \tag{4.3.16}$$

where derivatives of C_a are "small." Substituting equation (4.3.16) into equation (4.3.15) with $j_b = 0$ and neglecting the "small" term $\nabla_b \nabla^b C_a$ as well as the Ricci tensor term yields the condition

$$\nabla_a S \nabla^a S = 0 \quad ; \tag{4.3.17}$$

i.e., we find again that the surfaces of constant phase are null, and thus (by the same argument as given above for flat spacetime) $k_a = \nabla_a S$ is tangent to null geodesics. This suggests that, in this approximation (known as the *geometrical optics approximation*), light travels on null geodesics, a suggestion which can be confirmed by studies of the Green's function.

We now have described how general relativity treats gravitation in terms of curved spacetime geometry and we have illustrated the nature of the laws of physics in this new framework of spacetime structure. The remaining ingredient of general relativity is the equation satisfied by the spacetime metric. It is here that Mach's principle comes into play. Rather than prescribe the spacetime geometry in advance, general relativity asserts that the spacetime geometry is influenced by the matter distribution in the universe, in accordance with some of Mach's ideas (see section 1.4). In this way the spacetime metric now becomes not only a background arena on which the laws of physics are staged but also a dynamical variable which responds to the matter content of spacetime, as must be the case if the spacetime geometry is to describe gravity.

What equation describes the relation between spacetime geometry and the matter distribution? An important clue is provided by the comparison of the description of tidal force in Newtonian gravity and general relativity. In the Newtonian theory, the gravitational field may be represented by a potential, ϕ, and the tidal acceleration of two nearby particles is given by $-(\vec{x} \cdot \vec{\nabla})\vec{\nabla}\phi$, where $\vec{x}$ is the separation vector of the particles. On the other hand, in general relativity, from equation (3.3.18) the tidal acceleration of two nearby particles is given by $-R_{cbd}{}^a v^c x^b v^d$, where v^a is the 4-velocity of the particles and x^a is the deviation vector. This suggests that we make the correspondence,

$$R_{cbd}{}^a v^c v^d \longleftrightarrow \partial_b \partial^a \phi \quad . \tag{4.3.18}$$

However, Poisson's equation tells us that

$$\nabla^2 \phi = 4\pi\rho \quad , \tag{4.3.19}$$

where ρ is the mass (i.e., energy) density of matter and we remind the reader that we use units where $G = c = 1$ here and throughout the text. Furthermore, as we have discussed above, in special and general relativity the energy properties of matter are described by a stress-energy tensor T_{ab}, and we have the correspondence

$$T_{ab} v^a v^b \longleftrightarrow \rho \quad , \tag{4.3.20}$$

where v^a is the 4-velocity of the observer.

The correspondences (4.3.18) and (4.3.20) together with equation (4.3.19) suggest that we have $R_{cad}{}^a v^c v^d = 4\pi T_{cd} v^c v^d$, which suggests the field equation $R_{cd} = 4\pi T_{cd}$. Indeed, this equation was originally postulated by Einstein. However, it has a serious defect. As discussed above, the stress tensor satisfies $\nabla^c T_{cd} = 0$. On the other hand, the contracted Bianchi identity, equation (3.2.31), tells us that $\nabla^c (R_{cd} - \frac{1}{2} g_{cd} R) = 0$. Hence equality of R_{cd} and $4\pi T_{cd}$ would imply $\nabla_d R = 0$, i.e., that R, and hence $T = T^a{}_a$, is constant throughout the universe. This is a highly unphysical restriction on the matter distribution, and it forces us to reject this equation, as Einstein quickly realized (Einstein 1915*b*).

However, this difficulty also suggests its resolution. If instead we consider the equation

$$G_{ab} \equiv R_{ab} - \frac{1}{2} R g_{ab} = 8\pi T_{ab} \quad , \tag{4.3.21}$$

then there is no longer a conflict between the Bianchi identity and local conservation of energy; indeed, the Bianchi identity implies local energy conservation if equation (4.3.21) holds. Furthermore, the correspondences which motivated the previous equation are not destroyed. Taking the trace of equation (4.3.21), we find

$$R = -8\pi T \quad , \tag{4.3.22}$$

and thus,

$$R_{ab} = 8\pi (T_{ab} - \frac{1}{2} g_{ab} T) \quad . \tag{4.3.23}$$

In situations where Newtonian theory should be applicable, the energy of matter as measured by an observer who is roughly "at rest" with respect to the masses will be much greater than the material stresses (in units where $c = 1$), so we have $T \approx -\rho = -T_{ab} v^a v^b$. Thus, in this case, equation (4.3.23) still leads to $R_{ab} v^a v^b \approx 4\pi T_{ab} v^a v^b$.

Equation (4.3.21) is the desired field equation of general relativity. It was written down by Einstein in 1915 and is known as *Einstein's equation*. The entire content of general relativity may be summarized as follows: *Spacetime is a manifold M on which there is defined a Lorentz metric g_{ab}. The curvature of g_{ab} is related to the matter distribution in spacetime by Einstein's equation (4.3.21).*

Most of the rest of this book is devoted to studying the solutions of Einstein's equation and their physical properties. However, before concluding this section, we make three brief remarks concerning the nature of this equation. The first remark concerns its mathematical character. If we choose a coordinate system and express the coordinate basis components $R_{\mu\nu}$ in terms of $g_{\mu\nu}$, we see from section 3.4*a* that $R_{\mu\nu}$ depends on derivatives of $g_{\mu\nu}$ up to second order, and is highly nonlinear in $g_{\mu\nu}$ (although it is linear in the second derivatives of $g_{\mu\nu}$). Thus, Einstein's equation is equivalent to a coupled system of nonlinear second order partial differential equations for the metric components $g_{\mu\nu}$. For a metric of Lorentz signature, these equations have a hyperbolic (i.e., wave equation) character (see chapter 10). That we have the correct number of equations and unknowns to permit a good initial value formulation will be shown in chapter 10. Some methods for solving Einstein's equation will be discussed in chapter 7.

The second remark concerns how one should view Einstein's equation. In one sense, Einstein's equation (4.3.21) is analogous to Maxwell's equation (4.2.32) with the stress tensor T_{ab} serving as the source of the gravitational field in much the same way as the current j_a serves as a source of the electromagnetic field. However, there is an important difference. It makes sense to solve Maxwell's equation by specifying j_a first, and then finding A_a. One could try to solve Einstein's equation by specifying T_{ab} first and then finding g_{ab}. However, this does not make much sense because until g_{ab} is known, we do not know how to physically interpret T_{ab}; indeed, the formulas for T_{ab} for fluids and the fields considered above explicitly contain the metric. Thus, in general relativity, one must solve simultaneously for the spacetime metric and the matter distribution. This feature contributes to the difficulty of solving Einstein's equation when sources are present.

The final remark concerns the equations of motion of matter. As we have presented the theory, the equations of motion of particles, continuous matter, and fields are postulated first, and then Einstein's equation relating the matter distribution to the curvature of spacetime is given. However, Einstein's equation implies the relation $\nabla^a T_{ab} = 0$, and this relation contains a great deal of information on the behavior of matter. Indeed, for a perfect fluid, the relation $\nabla^a T_{ab} = 0$ is the entire content of the equations of motion. Thus for a fluid we may economize our assumptions by merely postulating the form of T_{ab}; the equations of motion of the fluid are already contained in Einstein's equation. Notice that for a perfect fluid with $P = 0$, i.e., a fluid composed of grains of "dust" which exert no forces upon each other, the fluid equation of motion (4.3.8) implied by $\nabla^a T_{ab} = 0$ tells us that the individual dust particles move on geodesics. More generally, it can be shown (Fock 1939; Geroch and Jang 1975) that the relation $\nabla^a T_{ab} = 0$ implies that any sufficiently "small" body whose self-gravity is sufficiently "weak" must travel on a geodesic. Thus, Einstein's equation alone actually implies the *geodesic hypothesis* that the world lines of test bodies are geodesics of the spacetime metric. This demonstrates an important self-consistency of Einstein's equation with the basic framework of general relativity. Note however, that bodies which are "large" enough to feel the tidal forces of the gravitational field will deviate from geodesic motion. The equations of motion of such bodies also can be found from the condition $\nabla^a T_{ab} = 0$ (Papapetrou 1951; Dixon 1974).

4.4 Linearized Gravity: The Newtonian Limit and Gravitational Radiation

The aim of this section is to treat the approximation in which gravity is "weak." In the context of general relativity this means that the spacetime metric is nearly flat. In practice, this is an excellent approximation in nature except for phenomena dealing with gravitational collapse and black holes and phenomena dealing with the large scale structure of the universe.

We will systematically develop the theory of small gravitational perturbations of an arbitrary solution in chapter 7. For the present, we simply shall assume that the deviation, γ_{ab}, of the actual spacetime metric

$$g_{ab} = \eta_{ab} + \gamma_{ab} \tag{4.4.1}$$

from a flat metric η_{ab} is "small." (Since there is no natural positive definite metric on spacetime, there is no natural norm by which "smallness" of tensors can be measured. An adequate definition of "smallness" in this context is that the components $\gamma_{\mu\nu}$ of γ_{ab} be much smaller than 1 in some global inertial coordinate system of η_{ab}.) We mean by "linearized gravity" that approximation to general relativity which is obtained by substituting equation (4.4.1) for g_{ab} in Einstein's equation and retaining only the terms linear in γ_{ab}.

We denote by ∂_a the derivative operator associated with the flat metric η_{ab}. In order not to have γ_{ab} hidden in a raised or lowered index, it is convenient to raise and lower tensor indices with η_{ab} and η^{ab} rather than g_{ab} and g^{ab}. We will adopt this notational convention for the remainder of this section with one exception: The tensor g^{ab} itself will still denote the inverse metric, not $\eta^{ac}\eta^{bd}g_{cd}$. It should be noted that in the linear approximation we have

$$g^{ab} = \eta^{ab} - \gamma^{ab} \tag{4.4.2}$$

since the composition of the right-hand sides of equations (4.4.1) and (4.4.2) differs from the identity operator only by terms quadratic in γ_{ab}.

The linearized Einstein equation can be obtained in a straightforward manner as follows. In a global inertial coordinate system, to linear order in γ_{ab} the Christoffel symbol is

$$\Gamma^c{}_{ab} = \frac{1}{2}\eta^{cd}(\partial_a \gamma_{bd} + \partial_b \gamma_{ad} - \partial_d \gamma_{ab}) \quad . \tag{4.4.3}$$

To linear order in γ_{ab}, the Ricci tensor (3.4.5) is

$$\begin{aligned} R^{(1)}_{ab} &= \partial_c \Gamma^c{}_{ab} - \partial_a \Gamma^c{}_{cb} \\ &= \partial^c \partial_{(b}\gamma_{a)c} - \frac{1}{2}\partial^c\partial_c \gamma_{ab} - \frac{1}{2}\partial_a\partial_b \gamma \quad , \end{aligned} \tag{4.4.4}$$

where $\gamma = \gamma^c{}_c$. Hence, the Einstein tensor to linear order is

$$G^{(1)}_{ab} = R^{(1)}_{ab} - \frac{1}{2}\eta_{ab}R^{(1)}$$

$$= \partial^c\partial_{(b}\gamma_{a)c} - \frac{1}{2}\partial^c\partial_c\gamma_{ab} - \frac{1}{2}\partial_a\partial_b\gamma - \frac{1}{2}\eta_{ab}(\partial^c\partial^d\gamma_{cd} - \partial^c\partial_c\gamma) \quad . \tag{4.4.5}$$

This expression can be simplified by defining

$$\overline{\gamma}_{ab} = \gamma_{ab} - \frac{1}{2}\eta_{ab}\gamma \quad . \tag{4.4.6}$$

In terms of $\overline{\gamma}_{ab}$, the linearized Einstein equation is found to be

$$G^{(1)}_{ab} = -\frac{1}{2}\partial^c\partial_c\overline{\gamma}_{ab} + \partial^c\partial_{(b}\overline{\gamma}_{a)c} - \frac{1}{2}\eta_{ab}\partial^c\partial^d\overline{\gamma}_{cd} = 8\pi T_{ab} \quad . \tag{4.4.7}$$

As discussed in detail in appendix C, there is a gauge freedom in general relativity corresponding to the group of diffeomorphisms: If $\phi: M \to M$ is a diffeomorphism of spacetime, the metrics g_{ab} and $\phi^* g_{ab}$ represent the same spacetime geometry, where ϕ^* is the map on tensor fields induced by ϕ (see appendix C). In the linear approximation, this implies that two perturbations γ_{ab} and γ'_{ab} represent the same physical perturbation if (and only if) they differ by the action of an "infinitesimal diffeomorphism" on the flat metric η_{ab}. As discussed in section 2.2, an "infinitesimal diffeomorphism" is generated by a vector field, ξ^a, and, as discussed in appendix C, the change in a tensor field induced by such an infinitesimal diffeomorphism defines the Lie derivative. This means that γ_{ab} and $\gamma_{ab} + \pounds_\xi \eta_{ab}$ describe the same physical perturbation. From appendix C, we see that we can express $\pounds_\xi \eta_{ab}$ in terms of the flat derivative operator ∂_a as

$$\pounds_\xi \eta_{ab} = \partial_a \xi_b + \partial_b \xi_a \quad . \tag{4.4.8}$$

This means that linearized gravity has a gauge freedom given by

$$\gamma_{ab} \to \gamma_{ab} + \partial_a\xi_b + \partial_b\xi_a \tag{4.4.9}$$

which is closely analogous to the electromagnetic gauge freedom $A_a \to A_a + \partial_a \chi$. This gauge freedom of γ_{ab} also can be derived without employing the machinery of appendix C from the tensor transformation law (2.3.8). According to equation (2.3.8), the components of γ_{ab} and $\gamma_{ab} + \partial_a\xi_b + \partial_b\xi_a$ differ, to first order, merely by a coordinate transformation and, hence, represent the same physical perturbation.

We may use this gauge freedom to simplify the linearized Einstein equation. By solving the equation

$$\partial^b\partial_b\xi_a = -\partial^b\overline{\gamma}_{ab} \tag{4.4.10}$$

for ξ^a, we can make a gauge transformation, equation (4.4.9), to obtain

$$\partial^b\overline{\gamma}_{ab} = 0 \quad , \tag{4.4.11}$$

which is the analog of the Lorentz gauge condition. In this gauge, the linearized Einstein equation simplifies to become

$$\partial^c \partial_c \bar{\gamma}_{ab} = -16\pi T_{ab} \tag{4.4.12}$$

and is closely analogous to Maxwell's equation (4.2.32).

In vacuum ($T_{ab} = 0$) equations (4.4.11) and (4.4.12) are precisely the equations written down by Fierz and Pauli (1939) to describe a massless spin-2 field propagating in flat spacetime (see chapter 13). Thus, in the linear approximation, general relativity reduces to the theory of a massless spin-2 field. The full theory of general relativity thus may be viewed as that of a massless spin-2 field which undergoes a nonlinear self-interaction. It should be noted, however, that the notion of the mass and spin of a field require the presence of a flat background metric η_{ab} which one has in the linear approximation but not in the full theory, so the statement that, in general relativity, gravity is treated as a massless spin-2 field is not one that can be given precise meaning outside the context of the linear approximation.

4.4a The Newtonian Limit

The theory of general relativity may have great aesthetic appeal, but this does not mean that its predictions are in accord with nature. We know that the Newtonian theory of gravitation gives excellent predictions under a wide range of conditions. Thus, the first crucial test of general relativity is that its predictions reduce to those of Newtonian gravity under the circumstances when Newtonian theory is known to be valid—specifically, when gravity is weak, the relative motion of the sources is much slower than the speed of light c, and the material stresses are much smaller than the mass-energy density (in units where $c = 1$).

When gravity is weak, the linear approximation to general relativity should be valid. The assumptions about the sources then can be reformulated more precisely as follows: There exists a global inertial coordinate system of η_{ab} such that

$$T_{ab} \approx \rho t_a t_b \quad , \tag{4.4.13}$$

where $t^a = (\partial/\partial x^0)^a$ is the "time direction" of this coordinate system. (Eq. [4.4.13] asserts that T_{ab} has only a "time-time" component; the neglect of the "time-space" components is essentially the statement that velocities [and thus, momentum densities] are small while the neglect of the "space-space" components is the statement that the stresses are small.) Since the sources are "slowly varying," we expect the spacetime geometry to change slowly as well, and thus we seek solutions of equation (4.4.12) where the time derivatives of $\bar{\gamma}_{ab}$ are negligible.

With these assumptions, the components of equation (4.4.12) in our global inertial coordinate system become

$$\nabla^2 \bar{\gamma}_{\mu\nu} = 0 \tag{4.4.14}$$

for all μ, ν except $\mu = \nu = 0$, while

$$\nabla^2 \bar{\gamma}_{00} = -16\pi\rho \quad , \tag{4.4.15}$$

where ∇^2 denotes the usual Laplace operator of space. The unique solution of equation (4.4.14) which is well behaved at infinity is $\bar{\gamma}_{\mu\nu} = 0$. (The solutions $\bar{\gamma}_{\mu\nu}$ = constant are also permissible, but they can be eliminated by a further gauge transformation.) Thus, in the Newtonian limit our solution for the perturbed metric γ_{ab} is

$$\gamma_{ab} = \overline{\gamma}_{ab} - \frac{1}{2}\eta_{ab}\overline{\gamma} = -(4t_a t_b + 2\eta_{ab})\phi \quad , \tag{4.4.16}$$

where $\phi \equiv -\frac{1}{4}\overline{\gamma}_{00}$ satisfies Poisson's equation,

$$\nabla^2 \phi = 4\pi\rho \quad . \tag{4.4.17}$$

The motion of test bodies in this curved spacetime geometry is governed by the geodesic equation,

$$\frac{d^2 x^\mu}{d\tau^2} + \sum_{\rho,\sigma} \Gamma^\mu{}_{\rho\sigma}\left(\frac{dx^\rho}{d\tau}\right)\left(\frac{dx^\sigma}{d\tau}\right) = 0 \quad , \tag{4.4.18}$$

where $x^\mu(\tau)$ is the world line of the particle in global inertial coordinates. For motion much slower than the speed of light, we may approximate $dx^\alpha/d\tau$ as $(1,0,0,0)$ in the second term, and the proper time τ may be approximated by the coordinate time t. Thus, we find

$$\frac{d^2 x^\mu}{dt^2} = -\Gamma^\mu{}_{00} \quad . \tag{4.4.19}$$

From our solution, equation (4.4.16), we have, for $\mu = 1, 2, 3$:

$$\Gamma^\mu{}_{00} = -\frac{1}{2}\frac{\partial \gamma_{00}}{\partial x^\mu} = \frac{\partial \phi}{\partial x^\mu} \quad , \tag{4.4.20}$$

where, again, time derivatives of ϕ have been neglected. Thus, the motion of test bodies is governed by the equation,

$$\vec{a} = -\vec{\nabla}\phi \quad , \tag{4.4.21}$$

where $\vec{a} = d^2\vec{x}/dt^2$ is the acceleration of the body relative to global inertial coordinates of η_{ab}.

Equations (4.4.17) and (4.4.21) are, of course, the basic equations of Newtonian gravity, and thus general relativity does indeed reduce to Newtonian gravity in the appropriate limit. Note, however, that although the predictions of general relativity agree with those of Newtonian gravity, the underlying viewpoint is radically different. In the Newtonian viewpoint, the Sun creates a gravitational field that exerts a force upon the Earth, which, in turn, causes it to orbit the Sun rather than move in a straight line. In the general relativistic viewpoint, the mass-energy of the Sun produces a curvature of the spacetime geometry. The Earth is in free motion (no forces act upon it) and it travels on a geodesic of the spacetime metric; but because spacetime is curved, it orbits the Sun. From the Newtonian viewpoint, the Earth undergoes acceleration; from the general relativistic viewpoint, it is the inertial observers of the flat metric, η_{ab}, who must accelerate.

It is instructive to examine the predictions of linearized gravity when the lowest order effects of the motion of the sources are taken into account. If we continue to neglect stresses, the stress energy tensor is approximated to linear order in velocity by

$$T_{ab} = 2t_{(a}J_{b)} - \rho t_a t_b \quad , \tag{4.4.22}$$

where $J_b = -T_{ab}t^a$ is the mass-energy current density 4-vector. The linearized Einstein equation again predicts that the space-space components of $\bar{\gamma}_{ab}$ satisfy the source free wave equation, but the space-time and time-time components now satisfy

$$\partial^a \partial_a \bar{\gamma}_{0\mu} = 16\pi J_\mu \quad . \tag{4.4.23}$$

Thus, $A_a \equiv -\frac{1}{4}\bar{\gamma}_{ab}t^b$ satisfies precisely Maxwell's equations in the Lorentz gauge with source J_a. If again we assume that the time derivatives of $\bar{\gamma}_{ab}$ are negligible, then the space-space components of $\bar{\gamma}_{ab}$ vanish, and we find that to linear order in the velocity of the test body, the geodesic equation now yields (problem 3)

$$\vec{a} = -\vec{E} - 4\vec{v} \times \vec{B} \quad , \tag{4.4.24}$$

where $\vec{E}$ and $\vec{B}$ are defined in terms of A_a by the same formulas as in electromagnetism. This is identical to the Lorentz force equation of electromagnetism (with $q = m$) except for an overall minus sign and a factor of 4 in the "magnetic force" term. Thus, linearized gravity predicts that the motion of masses produces magnetic gravitational effects very similar to those of electromagnetism.

One final, somewhat troublesome point deserves further comment. Above, we showed that general relativity reduces to Newtonian gravity in an appropriate limit, but, strictly speaking, we went beyond the linear approximation to show this. The reason has to do with our use of the geodesic equation to get the motion of a test body. As mentioned at the end of section 4.3, the "geodesic hypothesis" follows as a consequence of the condition $\nabla^a T_{ab} = 0$ which, in turn, follows as a consequence of Einstein's equation. However, in the linear approximation, Einstein's equation (4.4.7) or (4.4.12) actually implies the condition $\partial^a T_{ab} = 0$. (This is reasonable since in the linear approximation, T_{ab} is already "small," so deviations of the derivative operator from the flat derivative operator ∂_a contribute only to higher order.) But the condition $\partial^a T_{ab} = 0$ implies that test bodies move on geodesics of the *flat* metric η_{ab}; i.e., if one stays consistently within the linear approximation, one predicts that test bodies are unaffected by gravity. Thus, in obtaining equation (4.4.21) we actually have gone beyond the linear approximation. This, of course, does not invalidate our discussion. However, it illustrates the difficulties which occur when one tries to derive the equations of motion of bodies from Einstein's equation via a perturbation expansion in the departure from flatness. In order to obtain a good approximation to a solution to a given order, one must use some aspects of the higher order equations.

4.4b Gravitational Radiation

One of the most important changes which occurs when one goes from Coulomb's theory of electrostatics to Maxwell's theory of electromagnetism is that the electromagnetic field becomes a dynamical entity. Electromagnetic radiation can propagate freely through spacetime. A similar change occurs when one goes from Newtonian gravitation to general relativity: Gravitational radiation exists; i.e., ripples in the curvature of spacetime can propagate through spacetime. In the linear approximation the propagation of gravitational radiation is governed by the source-free, linearized Einstein equation (see eqs. [4.4.11] and [4.4.12] above),

$$\partial^a \overline{\gamma}_{ab} = 0 \quad , \tag{4.4.25}$$

$$\partial^c \partial_c \overline{\gamma}_{ab} = 0 \quad . \tag{4.4.26}$$

In obtaining these equations at the beginning of this section, the gauge choice (4.4.25) was made. However, there remains the freedom to make further gauge transformations $\gamma_{ab} \to \gamma_{ab} + \partial_a \xi_b + \partial_b \xi_a$ provided that

$$\partial^b \partial_b \xi^a = 0 \quad , \tag{4.4.27}$$

as such transformations leave equation (4.4.25) unchanged. This is closely analogous to the fact that in electromagnetism the Lorentz gauge condition does not uniquely fix the vector potential A_a; we have the restricted gauge freedom $A_a \to A_a + \partial_a \chi$ with

$$\partial^a \partial_a \chi = 0 \quad . \tag{4.4.28}$$

When treating electromagnetic radiation it is convenient to use the remaining gauge freedom to set the component A_0 in some global inertial coordinate system equal to zero in a source free region ($j_a = 0$). This gauge condition, called the *Coulomb* or *radiation* gauge, can be achieved as follows. On a constant time surface $t = t_0$ of our global inertial coordinate system we solve

$$\nabla^2 \chi = -\vec{\nabla} \cdot \vec{A} \quad . \tag{4.4.29}$$

We define χ throughout spacetime to be the solution of equation (4.4.28) whose initial value on the $t = t_0$ surface is given by equation (4.4.29), and whose initial time derivative is $\partial \chi / \partial t = -A_0$. (That a unique solution of eq. [4.4.28] exists for arbitrarily specified initial values of χ and $\partial \chi / \partial t$ follows from the results of section 10.1 below.) Then the function f defined by

$$f = A_0 + \partial \chi / \partial t \tag{4.4.30}$$

will satisfy

$$\partial^a \partial_a f = -4\pi j_0 \tag{4.4.31}$$

by equations (4.2.32) and (4.4.28). Furthermore, on the initial surface $t = t_0$ we have

$$f = 0 \quad , \tag{4.4.32}$$

$$\frac{\partial f}{\partial t} = \frac{\partial A_0}{\partial t} + \frac{\partial^2 \chi}{\partial t^2} = \vec{\nabla} \cdot \vec{A} + \nabla^2 \chi = 0 \quad . \tag{4.4.33}$$

Hence, if no sources are present in the region under consideration (or, more precisely, if for each point p we have $j_0 = 0$ on the light cone of p between it and the $t = t_0$ surface), the unique solution of equation (4.4.31) with initial data (4.4.32) and (4.4.33) is $f = 0$, and the gauge transformation $A_a \to A_a + \partial_a \chi$ achieves the desired condition $A_0 = 0$ while maintaining the Lorentz gauge condition.

In a very similar manner, in the case of linearized gravity we can use the restricted gauge freedom, equation (4.4.27), to achieve the radiation gauge $\gamma = 0$, $\gamma_{0\mu} = 0$ for $\mu = 1, 2, 3$ in a source-free region ($T_{ab} = 0$). As an extra bonus, we also obtain $\gamma_{00} = 0$ if no sources are present throughout spacetime (i.e., not just in our region) and good behavior at infinity is assumed. To achieve the radiation gauge, we solve on the initial surface $t = t_0$ the equations

$$2\left(-\frac{\partial \xi_0}{\partial t} + \vec{\nabla} \cdot \vec{\xi}\right) = -\gamma \quad , \tag{4.4.34a}$$

$$2[-\nabla^2 \xi_0 + \vec{\nabla} \cdot (\partial \vec{\xi} / \partial t)] = -\,\partial \gamma / \partial t \quad , \tag{4.4.34b}$$

$$\frac{\partial \xi_\mu}{\partial t} + \frac{\partial \xi_0}{\partial x^\mu} = -\gamma_{0\mu} \quad (\mu = 1, 2, 3) \quad , \tag{4.4.34c}$$

$$\nabla^2 \xi_\mu + \frac{\partial}{\partial x^\mu}\left(\frac{\partial \xi_0}{\partial t}\right) = -\frac{\partial \gamma_{0\mu}}{\partial t} \quad (\mu = 1, 2, 3) \quad , \tag{4.4.34d}$$

to obtain the initial values of ξ_0, ξ_1, ξ_2, ξ_3, and their first time derivatives. Then we define ξ^a to be the solution of equation (4.4.27) with these initial data. By the same argument as used in the electromagnetic case, the gauge transformation generated by ξ^a will achieve $\gamma = 0$ and $\gamma_{0\mu} = 0$ ($\mu = 1, 2, 3$) in a source-free region while preserving the gauge condition, equation (4.4.25).

Our bonus, $\gamma_{00} = 0$, comes about as follows. Since $\gamma = 0$, we have $\gamma_{ab} = \bar{\gamma}_{ab}$. Since $\gamma_{0\mu} = 0$ for $\mu = 1, 2, 3$, the gauge condition, equation (4.4.25), yields

$$\frac{\partial \gamma_{00}}{\partial t} = 0 \quad . \tag{4.4.35}$$

The linearized Einstein equation (4.4.12) then yields

$$\nabla^2 \gamma_{00} = -16\pi T_{00} \quad . \tag{4.4.36}$$

But if $T_{00} = 0$ throughout the spacetime, the only solution of equation (4.4.36) which is well behaved at infinity is $\gamma_{00} =$ constant. A further gauge transformation then achieves $\gamma_{00} = 0$ without violating any of the previous conditions.

We employ this radiation gauge to seek solutions of the source-free linearized Einstein equation. Plane waves,

$$\gamma_{ab} = H_{ab} \exp\left(i \sum_{\mu=0}^{3} k_\mu x^\mu\right) \quad , \tag{4.4.37}$$

where H_{ab} is a constant tensor field, will satisfy equation (4.4.26) if and only if

$$\sum_\mu k^\mu k_\mu = \sum_{\mu, \nu} \eta^{\mu\nu} k_\mu k_\nu = 0 \quad . \tag{4.4.38}$$

The radiation gauge conditions require (for $\nu = 0, 1, 2, 3$)

$$\sum_{\mu=0}^{3} k^\mu H_{\mu\nu} = 0 \quad , \tag{4.4.39a}$$

$$H_{0\nu} = 0 \quad , \tag{4.4.39b}$$

$$\sum_{\mu=0}^{3} H^{\mu}{}_{\mu} = 0 \quad . \tag{4.4.39c}$$

Since equations (4.4.39a) and (4.4.39b) both imply $\sum_{\nu} H_{0\nu} k^{\nu} = 0$, only eight of these nine equations are independent. Since there are 10 independent components, $H_{\mu\nu}$, this leaves two linearly independent solutions for H_{ab}. These two solutions describe the two independent polarization states of plane gravitational waves. An arbitrary well behaved solution of the vacuum linearized Einstein equation, i.e., an arbitrary packet of gravitational radiation, can be expressed as a superposition of these plane wave solutions.

How could we detect the presence of gravitational radiation? The most straightforward way is to study the relative acceleration of two masses, i.e., to measure the gravitational tidal force. For two nearby freely falling bodies, this acceleration is governed by the geodesic deviation equation (3.3.18). In our case, if the two bodies are nearly "at rest" in a global inertial coordinate system of η_{ab}, we have

$$\frac{d^2 X^{\mu}}{dt^2} \approx \sum_{\nu} R_{\nu 00}{}^{\mu} X^{\nu} \quad , \tag{4.4.40}$$

where X^a is the deviation vector. In the radiation gauge (assuming $\gamma_{00} = 0$), we obtain from equation (3.4.4) a very simple expression for the relevant components of the linearized Riemann tensor,

$$R_{\nu 0 0 \mu} = \frac{1}{2} \frac{\partial^2 \gamma_{\mu\nu}}{\partial t^2} \quad \text{(radiation gauge).} \tag{4.4.41}$$

(Incidentally, this formula shows that the plane wave solutions we obtained above are physically meaningful—that is, that they cannot be eliminated by further gauge transformations—since they produce nonzero curvature.) Thus, in principle, one could detect gravitational radiation by accurately tracking (say, with laser beams) the separation of two masses suspended freely from supports; such a detection scheme may become practical in the near future. Alternatively, if the masses are not in free motion but are connected by a solid piece of material, the gravitational tidal forces will stress the material. A solid bar of matter would be set into oscillation by these periodic stresses, and this oscillation could be detectable if the frequency of the gravitational radiation is near the resonant frequency of the bar. This scheme for detecting gravitational radiation was pioneered by Joseph Weber and is being pursued by a number of research groups. (The details of this and other schemes for detecting gravitational waves are reviewed by Douglass and Braginsky 1979.) The extreme sensitivity required for the detection of gravitational radiation should be stressed. For physically reasonable astrophysical sources of gravitational waves, one does not expect the radiation gauge components $\gamma_{\mu\nu}$ in the relevant frequency range to have a magnitude greater than about 10^{-17} (see, e.g., Thorne 1978). According to equations (4.4.40) and (4.4.41), this means that the fractional relative displacement $\Delta X/X$ of the two free masses should not exceed 10^{-17}, i.e., free masses separated by 1 meter will be displaced by no more than about 1/100 of a nuclear diameter by a gravitational wave! The stresses on a solid bar are correspondingly small. Nevertheless, many researchers believe that an unambiguous detection of gravitational radiation will be made in the near future.

How are gravitational waves produced? The most likely sources of (relatively) strong bursts of gravitational radiation arise from collapse phenomena (see chapter 12) where gravity is not weak and the linear approximation cannot be used. In such cases, we must solve the full nonlinear Einstein equation, and because of the difficulty of this task, our knowledge of these processes is rudimentary. It is instructive, therefore, to study the radiation generation problem in the linear approximation where the complete solution is readily obtained. Since each component of γ_{ab} satisfies the ordinary, inhomogeneous scalar wave equation (4.4.12), the solution is given in terms of the source by the same retarded Green's function as used for a scalar field and in electromagnetism, namely,

$$\bar{\gamma}_{\mu\nu}(x) = 4\int_{\Lambda} \frac{T_{\mu\nu}(x')}{|\vec{x} - \vec{x}'|} dS(x') \quad , \tag{4.4.42}$$

where Λ denotes the past light cone of the point x and the volume element on the light cone is $dS = r^2 dr\, d\Omega$. The gauge condition (4.4.11) on $\bar{\gamma}_{ab}$ will be satisfied by virtue of the linearized conservation equation $\partial^a T_{ab} = 0$, so equation (4.4.42) gives the solution for the gravitational effects produced by sources in the linear approximation. (The radiation gauge conditions, of course, are not imposed since sources are present.)

It is interesting to evaluate our solution in the slow motion limit where the typical source velocities are much smaller than the speed of light. (More precisely, we consider the limit where the spatial extent of the source is much smaller than the typical wavelengths of the emitted radiation.) In electromagnetism this limit is known as the dipole approximation, since in that case the dominant radiation is generated by the changing dipole moment of the source.

To analyze this limit in the gravitational case, we Fourier transform all quantities in the time variable, t, of our global inertial coordinate system of η_{ab}, leaving the space variables untransformed. We define

$$\hat{\bar{\gamma}}_{\mu\nu}(\omega, \vec{x}) = \frac{1}{\sqrt{2\pi}} \int_{-\infty}^{\infty} \bar{\gamma}_{\mu\nu}(t, \vec{x}) e^{i\omega t} dt \quad , \tag{4.4.43}$$

and we similarly Fourier transform $T_{\mu\nu}$. From equation (4.4.42) it follows that

$$\hat{\bar{\gamma}}_{\mu\nu}(\omega, \vec{x}) = 4 \int \frac{\hat{T}_{\mu\nu}(\omega, \vec{x}')}{|\vec{x} - \vec{x}'|} \exp(i\omega|\vec{x} - \vec{x}'|) d^3x' \quad , \tag{4.4.44}$$

where the "extra" factor of $\exp(i\omega|\vec{x} - \vec{x}'|)$ arises from the fact that the original integral (4.4.42) was over the past light cone. We need only solve for the space-space components of $\hat{\bar{\gamma}}_{\mu\nu}$ since the components $\hat{\bar{\gamma}}_{0\mu}$ are readily obtained in terms of them from the gauge condition (4.4.11) which yields

$$-i\omega\hat{\bar{\gamma}}_{0\mu} = \sum_{\nu=1}^{3} \frac{\partial \hat{\bar{\gamma}}_{\nu\mu}}{\partial x^{\nu}} \quad . \tag{4.4.45}$$

Since we are interested in calculating the radiation, it suffices to obtain our solution in the "far zone," $R >> 1/\omega$, where R denotes the distance from the source. For the limit in which we are interested, the frequencies of interest are sufficiently small that the factor $\exp(i\omega|\vec{x} - \vec{x}'|)$ varies negligibly over the source, so we may replace $\exp((i\omega|\vec{x} - \vec{x}'|)/|\vec{x} - \vec{x}'|)$ by $\exp(i\omega R)/R$ and pull it out of the integral. We evaluate the remaining integral of the space-space components of $\hat{T}_{ab}$ as follows:

$$\begin{aligned}
\int \hat{T}^{\mu\nu} d^3x &= \sum_{\alpha=1}^{3} \left\{ \int \frac{\partial}{\partial x^{\alpha}} (\hat{T}^{\alpha\nu} x^{\mu}) - \int \frac{\partial \hat{T}^{\alpha\nu}}{\partial x^{\alpha}} x^{\mu} \right\} \\
&= -i\omega \int \hat{T}^{0\nu} x^{\mu} \\
&= -\frac{i\omega}{2} \int (\hat{T}^{0\nu} x^{\mu} + \hat{T}^{0\mu} x^{\nu}) \\
&= -\frac{i\omega}{2} \sum_{\beta=1}^{3} \left\{ \int \frac{\partial}{\partial x^{\beta}} (\hat{T}^{0\beta} x^{\mu} x^{\nu}) - \int \frac{\partial \hat{T}^{0\beta}}{\partial x^{\beta}} x^{\mu} x^{\nu} \right\} \\
&= -\frac{\omega^2}{2} \int \hat{T}^{00} x^{\mu} x^{\nu} d^3x \quad ,
\end{aligned} \tag{4.4.46}$$

where in the second and fifth lines we used Gauss's law to get rid of the total divergence and also used the conservation of T_{ab} to express the divergence of spatial components in terms of the time derivative of time components. Thus, we obtain our far-zone solution,

$$\hat{\bar{\gamma}}_{\mu\nu}(\omega, \vec{x}) = -\frac{2\omega^2}{3} \frac{e^{i\omega R}}{R} \hat{q}_{\mu\nu}(\omega) \quad (\mu, \nu = 1, 2, 3) \quad , \tag{4.4.47}$$

where $\hat{q}_{\mu\nu}$ is the Fourier transform of the quadrupole moment tensor,

$$q_{\mu\nu} = 3 \int T^{00} x^{\mu} x^{\nu} d^3x \quad . \tag{4.4.48}$$

The inverse Fourier transform of equation (4.4.47) yields

$$\bar{\gamma}_{\mu\nu}(t, \vec{x}) = \frac{2}{3R} \frac{d^2 q_{\mu\nu}}{dt^2} \bigg|_{\text{ret}} \quad (\mu, \nu = 1, 2, 3) \quad , \tag{4.4.49}$$

where the derivative is evaluated at the retarded time $t' = t - R$. Thus, the dominant gravitational radiation in the slow motion approximation arises from the time rate of change of the quadrupole moment of the source. The absence of dipole radiation can be understood, physically, as resulting from conservation of momentum, which does not permit a time-varying mass dipole moment. On account of the absence of dipole radiation, the emission of gravitational radiation in the slow motion limit is smaller than the radiation in comparable situations in electromagnetism.

The issue of energy in general relativity is a rather delicate one. In general relativity there is no known meaningful notion of local energy density of the gravitational field. The basic reason for this is closely related to the fact that the spacetime metric, g_{ab}, describes both the background spacetime structure and the dynamical aspects of the gravitational field, but no natural way is known to decompose it into its "background" and "dynamical" parts. Since one would expect to attribute energy to the dynamical aspect of gravity but not to the background spacetime structure, it seems unlikely that a notion of local energy density could be obtained without a corresponding decomposition of the spacetime metric. However, for an isolated system, the *total* energy can be defined by examining the gravitational field at large distances from the system. In addition, for an isolated system the flux of energy carried away from the system by gravitational radiation also is well defined.

We shall postpone a full discussion of energy in general relativity until chapter 11. However, for small deviations from flat spacetime, we would expect that—in analogy with the scalar and electromagnetic field (see eqs. [4.2.20] and [4.2.27]—the total energy and energy flux of the gravitational field will be quadratic in the field γ_{ab}. A formula for this energy and energy flux is suggested by the following considerations. The linearized vacuum Einstein equation

$$G^{(1)}_{ab}\,[\gamma_{cd}] = 0 \tag{4.4.50}$$

states that the Einstein tensor for the metric $\eta_{ab} + \gamma_{ab}$ vanishes to first order in γ_{ab}. However, to second order in γ_{ab}, the vacuum Einstein equation will, in general, fail to be satisfied. Indeed, the terms in the Ricci tensor quadratic in γ_{ab} are (problem 4),

$$\begin{aligned} R^{(2)}_{ab} = {} & \frac{1}{2}\gamma^{cd}\partial_a\partial_b\gamma_{cd} - \gamma^{cd}\partial_c\partial_{(a}\gamma_{b)d} \\ & + \frac{1}{4}(\partial_a\gamma_{cd})\partial_b\gamma^{cd} + (\partial^d\gamma^c{}_b)\partial_{[d}\gamma_{c]a} \\ & + \frac{1}{2}\partial_d(\gamma^{dc}\partial_c\gamma_{ab}) - \frac{1}{4}(\partial^c\gamma)\partial_c\gamma_{ab} \\ & - (\partial_d\gamma^{cd} - \frac{1}{2}\partial^c\gamma)\partial_{(a}\gamma_{b)c} \quad . \end{aligned} \tag{4.4.51}$$

Thus, in order to maintain a solution of the vacuum Einstein equation to second order, we must correct γ_{ab} by adding to it the term $\gamma^{(2)}_{ab}$, where $\gamma^{(2)}_{ab}$ satisfies

$$G^{(1)}_{ab}[\gamma^{(2)}_{cd}] + G^{(2)}_{ab}[\gamma_{cd}] = 0 \quad , \tag{4.4.52}$$

where (in the case $R^{(1)}_{ab} = 0$) we have $G^{(2)}_{ab} = R^{(2)}_{ab} - \frac{1}{2}\eta_{ab}R^{(2)}$. We may write equation (4.4.52) in the form

$$G^{(1)}_{ab}[\gamma^{(2)}_{cd}] = 8\pi t_{ab} \quad , \tag{4.4.53}$$

where

$$t_{ab} = -\frac{1}{8\pi}G^{(2)}_{ab}[\gamma_{cd}] \quad . \tag{4.4.54}$$

Thus, in second order, γ_{ab} causes the same correction to the spacetime metric as would be produced by ordinary matter with stress energy tensor t_{ab}. Furthermore, t_{ab} is symmetric and is conserved, $\partial^a t_{ab} = 0$, assuming, of course, that γ_{ab} satisfies the vacuum linearized Einstein equation (4.4.50). This suggests that we should view t_{ab} as the effective stress-energy tensor of the gravitational field, valid to second order in deviation from flatness. However, this interpretation cannot be taken too literally. First, the local construction of t_{ab} quadratically from γ_{ab} as well as its symmetry and conservation properties will not be affected if we add to it a tensor of the form $\partial^c \partial^d U_{acbd}$, where U_{acbd} is locally constructed from γ_{ab}, is quadratic in γ_{ab}, and satisfies the tensor symmetries $U_{acbd} = U_{[ac]bd} = U_{ac[bd]} = U_{bdac}$. (Indeed, the terms quadratic in γ_{ab} of the Landau-Lifshitz "pseudotensor" [Landau and Lifshitz 1962] differ from our expression for t_{ab} by such a term.) Furthermore, t_{ab} is not even gauge invariant; i.e., if we replace γ_{ab} by $\gamma_{ab} + 2\partial_{(a}\xi_{b)}$, then t_{ab} does *not* remain unchanged. This reflects the above mentioned fact that there is no meaningful notion of the local stress-energy of the gravitational field in general relativity. However, the *total* energy associated with γ_{ab},

$$E = \int_\Sigma t_{00} d^3x \tag{4.4.55}$$

(where the integral is taken over the spacelike hyperplane Σ depicted in Fig. 4.2) *is* gauge invariant in the following sense: Suppose the perturbed spacetime metric

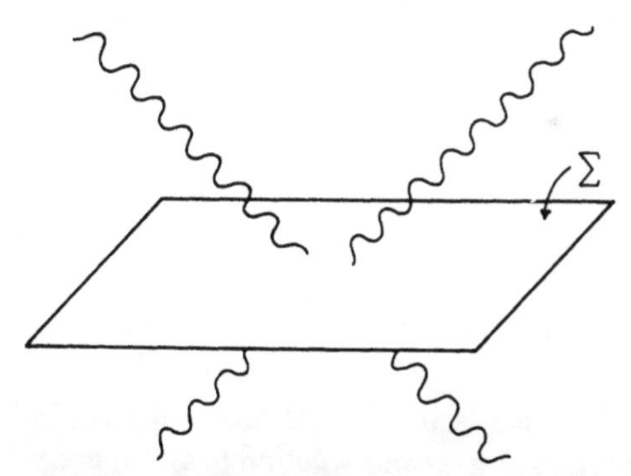

Fig. 4.2. A spacelike hyperplane, Σ. The total energy contained in gravitational radiation is obtained by integrating t_{00} over Σ.

$\eta_{ab} + \gamma_{ab}$ is asymptotically flat in the sense that the inertial components of γ_{ab} and its derivatives go to zero as $r \to \infty$ as: $\gamma_{\mu\nu} = O(1/r)$, $\partial_\rho \gamma_{\mu\nu} = O(1/r^2)$, and $\partial_\gamma \partial_\rho \gamma_{\mu\nu} = O(1/r^3)$. (Note that these conditions ensure the convergence of the integral [4.4.55] which defines E.) Then for any gauge transformation, ξ^a, which preserves these asymptotic conditions, the value of E is unchanged, $E[\gamma_{ab}] = E[\gamma_{ab} + 2\partial_{(a}\xi_{b)}]$. Rather lengthy calculations are needed to demonstrate this gauge invariance of E directly, but a simple proof of restricted gauge invariance is outlined in problem 7. Furthermore, it is not difficult to verify that E is unchanged if a term $\partial^c \partial^d U_{acbd}$ with the above properties is added to t_{ab}, since the volume integral of this term can be converted by Gauss's law to a surface integral which vanishes on account of the asymptotic conditions.

Similarly, although the local flux of energy $-t^a{}_0$ is not gauge invariant, if the spacetime is initially time independent, goes through a time-dependent phase, and becomes time independent again, then the total radiated energy

$$\Delta E = -\int_S t_{a0} dS^a \tag{4.4.56}$$

is gauge independent, where the integral is taken over the three-dimensional timelike surface S depicted in Figure 4.3. Here the limit as $r \to \infty$ for that surface is understood, and the above conditions of asymptotic flatness on γ_{ab} are imposed as $r \to \infty$ along the asymptotically null surfaces Σ_1 and Σ_2 in the stationary regimes. [Note that in the time-dependent regime these conditions would not be appropriate since according to our solution (4.4.42) we expect $\partial_\rho \gamma_{\mu\nu} = O(1/r)$. In chapter 11 we will make precise the taking of limits as $r \to \infty$ along null surfaces by introducing the notion of null infinity.]

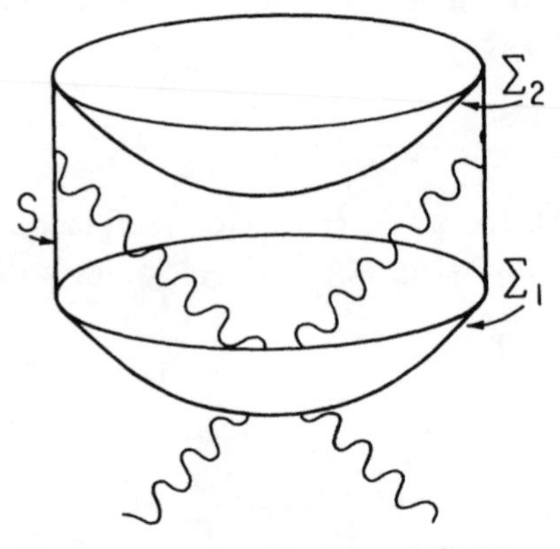

Fig. 4.3. Asymptotically null hypersurfaces Σ_1 and Σ_2. The gravitational radiation "registers" on Σ_1 but does not "register" on Σ_2. The energy, ΔE, carried off to infinity by gravitational radiation is given by an integral over the timelike hypersurface S.

Using equation (4.4.54) and equation (4.4.56), we can calculate the energy carried away by gravitational radiation for our solution (4.4.49) corresponding to the metric perturbation produced by a slowly varying source. A lengthy calculation (where many terms which integrate to zero are dropped) yields the final result,

$$\Delta E = \int P \, dt \quad , \tag{4.4.57}$$

where

$$P = \frac{1}{45} \sum_{\mu,\nu=1}^{3} \left(\left. \frac{d^3 Q_{\mu\nu}}{dt^3} \right|_{\text{ret}} \right)^2 \quad , \tag{4.4.58}$$

and $Q_{\mu\nu}$ is the trace-free quadrupole moment tensor

$$Q_{\mu\nu} = q_{\mu\nu} - \frac{1}{3} \delta_{\mu\nu} q \quad . \tag{4.4.59}$$

This formula shows that the energy which is carried away by gravitational radiation in "ordinary laboratory processes" is extremely small. For example, according to equation (4.4.58), the gravitational energy flux from a rod of mass M and length L which spins about its center at frequency Ω (so that T_{ab} oscillates with frequency 2Ω) is

$$P_{\text{rod}} = \frac{2G}{45c^5} M^2 L^4 \Omega^6 \quad , \tag{4.4.60}$$

where we have put back the G's and c's in our formula to convert back to "nongeometrized" units. Thus, a 1 kilogram rod, of length 1 meter, spinning at an angular velocity of 1 radian per second radiates energy in the form of gravitational waves at the remarkably small rate of about 10^{-47} erg s^{-1}. Even if we scale the rod up to astronomical dimensions, the gravitational energy flux remains small. It is only in phenomena involving strong gravitational fields—such as occurs in gravitational collapse—that one expects large amounts of energy to be radiated away.

Finally, we comment on the validity of our solutions for the gravitational radiation, equations (4.4.42) and (4.4.49), and energy flux, equation (4.4.58), from self-gravitating sources. As should be clear from the above derivation, our solution, equation (4.4.42), is valid for the gravitational radiation from sources such as spinning rods or masses connected by springs, provided that gravity is sufficiently weak that the linear approximation is valid. Equations (4.4.49) and (4.4.58) hold if, in addition, the source velocities are small. However, our derivation is *not* valid, as it stands, for the radiation from sources where self-gravitation is important—e.g., a nearly Newtonian binary star system—even though gravity may be weak in the sense that the inertial components $\gamma_{\mu\nu}$ are much less than 1 and the source motion may be slow. The reason is that, as already mentioned at the end of section 4.4*a*, to obtain the Newtonian limit consistently one must go beyond the linear approximation. In the linear approximation, two stars would not orbit each other but would move on geodesics of the flat metric, i.e., straight lines. Any assumption to the contrary would be inconsistent with $\partial^a T_{ab} = 0$, which follows from the linearized Einstein equation. Even though $\gamma_{\mu\nu} << 1$, higher order terms in γ_{ab} are *not* negligible compared with stress terms in T_{ab}. However, formulas for the radiation from a self-gravitating, nearly Newtonian system can be obtained by restoring all the higher order terms in γ_{ab} into the Einstein field equation. Nearly Newtonian motion of the matter source then is no longer inconsistent. The nonlinear terms in γ_{ab} can be brought to the right-hand side of the equation and viewed as an effective gravitational stress-energy tensor $\tilde{t}_{ab}$. (Here $\tilde{t}_{ab}$ includes all the nonlinear terms in G_{ab}, not just the quadratic terms, as for t_{ab} defined above.) Our solution, equation (4.4.42), is then valid provided we replace T_{ab} by $(T_{ab} + \tilde{t}_{ab})$, although it is no longer really a "solution" since $\tilde{t}_{ab}$ depends on γ_{ab}. In the slow motion approximation, we again get equations (4.4.49) and (4.4.58), where t_{00} is now included in the definition of the quadrupole moment. However, in nearly Newtonian situations, we expect to have $T_{00} >> t_{00}$, so the gravitational contribution to the quadrupole moment should be negligible. Thus, equations (4.4.49) and (4.4.58) should be valid without modification for a nearly Newtonian system. However, a derivation of these equations with the same level of clarity and rigor as in the non-self-gravitating case has not been given.

The above predictions of energy loss by systems due to emission of gravitational radiation have now been confirmed by observations. If two stars in orbit around each other lose energy, their orbital radius will decrease and their orbital frequency will correspondingly increase. However, since the energy loss by gravitational radiation from such a system is extremely small (problem 9), any possibility for observing this speedup caused by emission of gravitational radiation requires (i) an extremely close binary orbit (to make the general relativistic effects as large as possible) and (ii) an ability to measure changes in the period to extremely high accuracy. Remarkably, a binary system satisfying both of these properties was discovered by Hulse and Taylor (1975). This system consists of two compact bodies in an elliptical orbit with a maximum separation of only $\sim 10^{11}$ cm (i.e., approximately one solar radius). Most importantly, one of the bodies is a pulsar (see the end of section 6.2 above) and emits radio pulses in a regular, clocklike fashion. Since the arrival time of the pulsar signals can be measured to very high precision, the time delays or advances caused by changes in the pulsar position and velocity are very well determined, and the orbital parameters of the binary system can be determined to correspondingly high accuracy. Furthermore, the accuracy continues to improve the longer the system is observed. Recently, the accuracy of the determination of the orbital period has been high enough to observe a speedup in the orbital frequency (Taylor and McCulloch 1980). The magnitude of this speedup is in excellent agreement with that associated with the energy loss due to gravitational radiation predicted by equation (4.4.58). Thus, unless some other cause is producing an orbital period change of exactly the same magnitude as that predicted by general relativity, it appears that the effects of energy loss by gravitational radiation have been observed in this system.

Problems

1. Show that Maxwell's equation (4.3.12) implies strict charge conservation, $\nabla_a j^a = 0$.

2. *a*) Let $\boldsymbol{\alpha}$ be a p-form on an n-dimensional oriented manifold with metric g_{ab}, i.e., $\alpha_{a_1 \cdots a_p}$ is a totally antisymmetric tensor field (see appendix B). We define the *dual*, ${}^*\boldsymbol{\alpha}$, of $\boldsymbol{\alpha}$ by

$$ {}^*\alpha_{b_1 \cdots b_{n-p}} = \frac{1}{p!} \alpha^{a_1 \cdots a_p} \epsilon_{a_1 \cdots a_p b_1 \cdots b_{n-p}} \quad , $$

where $\epsilon_{a_1 \cdots a_n}$ is the natural volume element on M, i.e., the totally antisymmetric tensor field determined up to sign by equation (B.2.9). Show that ${}^{**}\boldsymbol{\alpha} = (-1)^{s+p(n-p)}\boldsymbol{\alpha}$, where s is the number of minuses occurring in the signature of g_{ab}.

b) Show that in differential forms notation (see appendix B), Maxwell's equations (4.3.12) and (4.3.13) can be written as

$$d{}^*\boldsymbol{F} = 4\pi {}^*j \quad ,$$

$$d\boldsymbol{F} = 0 \quad .$$

Note that if we apply Stokes's theorem (see appendix B) to the first equation, we obtain $\int_\Sigma {}^*j = (1/4\pi)\int_S {}^*F$, where Σ is a three-dimensional hypersurface with two-dimensional boundary S. But $-\int_\Sigma {}^*j = -\int_\Sigma j^a t_a d\Sigma$ is just the total electric charge e in the volume Σ, where t^a is the unit normal to Σ, and $-\int_S {}^*F = \int_S E^a n_a dA$ is just the integral of the normal component of $E_a = F_{ab}t^b$ on S. Thus, Gauss's law of electromagnetism continues to hold in curved spacetime.

c) Define for each $\beta \in [0, 2\pi]$ the tensor field $\tilde{F}_{ab} = F_{ab}\cos\beta + {}^*F_{ab}\sin\beta$. We call $\tilde{F}_{ab}$ a *duality rotation* of F_{ab} by "angle" β. It follows immediately from part (*b*) that if F_{ab} satisfies the source-free Maxwell's equations ($j^a = 0$), then so does $\tilde{F}_{ab}$. Show that the stress-energy, T_{ab}, of the solution $\tilde{F}_{ab}$ is the same as that of F_{ab}.

3. *a*) Derive equation (4.4.24).

b) Show that the "gravitational electric and magnetic fields" $\vec{E}$ and $\vec{B}$ inside a spherical shell of mass M and radius R (with $M << R$) slowly rotating with angular velocity $\vec{\omega}$ are

$$\vec{E} = 0 \quad , \quad \vec{B} = \frac{2M}{3R}\vec{\omega} \quad .$$

c) An observer at rest at the center of the shell of part (*b*) parallelly propagates along his (geodesic) world line a vector S^a with $S^a u_a = 0$, where u^a is the tangent to his world line. Show that the inertial components, $\vec{S}$, precess according to $d\vec{S}/dt = \vec{\Omega} \times \vec{S}$, where $\vec{\Omega} = 2\vec{B} = \frac{4}{3}(M/R)\vec{\omega}$. This effect, first analyzed by Thirring and Lense (1918) and discussed further by Brill and Cohen (1966), may be interpreted as a "dragging of inertial frames" caused by the rotating shell. At the center of the shell the local standard of "nonrotating," defined by parallel propagation along a geodesic, is changed from what it would be without the shell, in a manner in accord with Mach's principle.

4. Starting with equation (3.4.5) for R_{ab}, derive the formula, equation (4.4.51), for $R^{(2)}_{ab}$ by substituting $\eta_{ab} + \gamma_{ab}$ for g_{ab} and keeping precisely the terms quadratic in γ_{ab}.

5. Let T_{ab} be a symmetric, conserved tensor field (i.e., $T_{ab} = T_{ba}$, $\partial^a T_{ab} = 0$) in Minkowski spacetime. Show that there exists a tensor field U_{acbd} with the symmetries $U_{acbd} = U_{[ac]bd} = U_{ac[bd]} = U_{bdac}$ such that $T_{ab} = \partial^c\partial^d U_{acbd}$. (Hint: For any vector field v^a in Minkowski spacetime satisfying $\partial_a v^a = 0$ there exists a tensor field $s^{ab} = -s^{ba}$ such that $v^a = \partial_b s^{ab}$. [This follows from applying the converse of the Poincaré lemma (see the end of section B.1 in appendix B) to the 3-form $\epsilon_{abcd}v^d$.] Use this fact to show that $T_{ab} = \partial^c W_{cab}$ where $W_{cab} = W_{[ca]b}$. Then use the fact that $\partial^c W_{c[ab]} = 0$ to derive the desired result.)

6. As discussed in the text, in general relativity no meaningful expression is known for the local stress-energy of the gravitational field. However, a four-index tensor T_{abcd} can be constructed out of the curvature in a manner closely analogous to the way in which the stress tensor of the electromagnetic field is constructed out of F_{ab} (eq. [4.2.27]). We define the *Bel-Robinson tensor* in terms of the Weyl tensor by

$$T_{abcd} = C_{aecf}C_b{}^e{}_d{}^f + \frac{1}{4}\epsilon_{ae}{}^{hi}\epsilon_b{}^{ej}{}_k C_{hicf}C_j{}^k{}_d{}^f$$

$$= C_{aecf}C_b{}^e{}_d{}^f - \frac{3}{2}g_{a[b}C_{jk]cf}C^{jk}{}_d{}^f \quad ,$$

where ϵ_{abcd} is defined in appendix B and equation (B.2.13) was used. It follows that $T_{abcd} = T_{(abcd)}$. (This is established most easily from the spinor decomposition of the Weyl tensor given in chapter 13.)

a) Show that $T^a{}_{acd} = 0$.

b) Using the Bianchi identity (3.2.16), show that in vacuum, $R_{ab} = 0$, we have $\nabla^a T_{abcd} = 0$.

7. a) Show that the total energy E, equation (4.4.55), is time independent, i.e., the value of E is unchanged if the integral is performed over a time translate, Σ', of Σ.

b) Let ξ_a be a gauge transformation which vanishes outside a bounded region of space. Show that $E[\gamma_{ab}] = E[\gamma_{ab} + 2\partial_{(a}\xi_{b)}]$ by comparing them with $E[\gamma_{ab} + 2\partial(_{(a}\xi'_{b)}]$ where ξ'_a is a new gauge transformation which agrees with ξ_a in a neighborhood of the hyperplane Σ but vanishes in a neighborhood of another hyperplane Σ'.

8. Two point masses of mass M are attached to the ends of a spring of spring constant K. The spring is set into oscillation. In the quadrupole approximation, equation (4.4.58), what fraction of the energy of oscillation of the spring is radiated away during one cycle of oscillation?

9. A binary star system consists of two stars of mass M and of negligible size in a nearly Newtonian circular orbit of radius R around each other. Assuming the validity of equation (4.4.58) for this system, calculate the rate of increase of the orbital frequency due to emission of gravitational radiation.

FIVE

HOMOGENEOUS, ISOTROPIC COSMOLOGY

The theory of general relativity was formulated in the previous chapter. The essence of the theory is contained in the statement given near the end of section 4.3: Spacetime is a four-dimensional manifold on which is defined a metric, g_{ab}, of Lorentz signature. This metric is related to the matter distribution in spacetime by Einstein's equation, $G_{ab} = 8\pi T_{ab}$.

One of the most vital questions raised by the theory is: Which solution of Einstein's equation describes the spacetime we observe, i.e., which solution corresponds to our universe, or, at least, an idealized model of our universe? In order to answer this question, we first must give sufficient input via observational data and assumptions about the nature of our universe. Armed with this information, we may then solve Einstein's equation to make predictions concerning the dynamical evolution of the universe.

In this chapter, we will investigate the structure of our universe as predicted by general relativity under the assumption that the universe is homogeneous and isotropic. A precise, mathematical formulation of this assumption is given in section 5.1. The dynamical predictions of general relativity are derived in section 5.2. We discuss two important features of the homogeneous, isotropic cosmological models in section 5.3: the cosmological redshift and particle horizons. Finally, we give in section 5.4 a brief account of the history of our universe.

5.1 Homogeneity and Isotropy

In the subject of cosmology, it is very difficult to prove theories by appealing only to observational data. We have direct contact in our lifetime or even in the lifetime of human civilization with only a negligibly small spacetime region of our universe. While our telescopes can observe objects remarkably far away by ordinary human scales, it should be recognized that in cosmic terms they report information about only a portion of our past light cone. Thus, a good deal of our input in the subject of cosmology arises from our philosophical prejudices. Observational data may confirm these prejudices, but in general they cannot be expected to definitively prove that they are correct. Nevertheless, the cosmological models considered in this chapter have provided a remarkably successful account of the nature of the universe.

Since the time of Copernicus, it has generally been assumed that we do not occupy a privileged position in our universe; that if we were located in a different region of our universe, the basic characteristics of our surroundings would appear the same. Similarly, it is natural to assume that the universe is isotropic, that is, that there are no preferred directions in space; that observations on sufficiently large scales should yield results which do not depend on which direction we look. These philosophical prejudices of homogeneity and isotropy have received strong confirmation from modern observations. While observations of the distribution of galaxies in our universe show clustering of galaxies on a wide range of distance scales and recent observations have shown large regions devoid of galaxies (Kirshner *et al.* 1981), on the largest scales the galaxy distribution appears to be homogeneous and isotropic. Counts of radio sources and the isotropy of the X-ray and γ-ray background radiation also support the hypothesis of homogeneity and isotropy of the universe on large scales. Even stronger observational evidence for the homogeneity and isotropy of our universe comes from the discovery of thermal radiation at about 3 K filling our universe, which has been measured to be isotropic to a very high precision. As discussed in section 5.4, this radiation is believed to have a cosmological origin, and it would be very difficult to explain its existence and its isotropy if the hypothesis of the homogeneity and isotropy of the universe were not valid to a very good approximation on large distance scales.

Thus, for the remainder of this chapter, we shall proceed under the assumption that the universe is homogeneous and isotropic. Our first task is to formulate precisely the mathematical meaning of this assumption. Loosely speaking, homogeneity means that at any given "instant of time" each point of "space" should "look like" any other point. A precise formulation can be given as follows: A spacetime is said to be (spatially) *homogeneous* if there exists a one-parameter family of spacelike hypersurfaces Σ_t foliating the spacetime (see Fig. 5.1) such that for each t and for any points $p, q \in \Sigma_t$ there exists an isometry of the spacetime metric, g_{ab}, which takes p into q. (See appendix C for the definition of an isometry.)

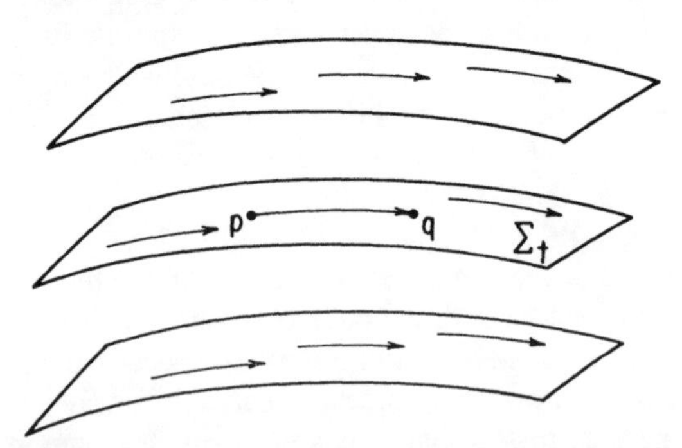

Fig. 5.1. The hypersurfaces of spatial homogeneity in spacetime. By definition of homogeneity, for each t and each $p, q \in \Sigma_t$ there exists an isometry of the space-time which takes p into q.

With regard to isotropy, it first should be pointed out that, in general, at each point, at most one observer can see the universe as isotropic. For example, if ordinary matter fills the universe, any observer in motion relative to the matter must see an anisotropic velocity distribution of the matter. With this fact in mind, a precise formulation of the notion of isotropy can be given as follows: A spacetime is said to be (spatially) *isotropic* at each point if there exists a congruence of timelike curves (i.e., observers), with tangents denoted u^a, filling the spacetime (see Fig. 5.2) and satisfying the following property. Given any point p and any two unit "spatial" tangent vectors s_1^a, $s_2^a \in V_p$ (i.e, vectors at p orthogonal to u^a), there exists an isometry of g_{ab} which leaves p and u^a at p fixed but rotates s_1^a into s_2^a. Thus, in an isotropic universe it is impossible to construct a geometrically preferred tangent vector orthogonal to u^a.

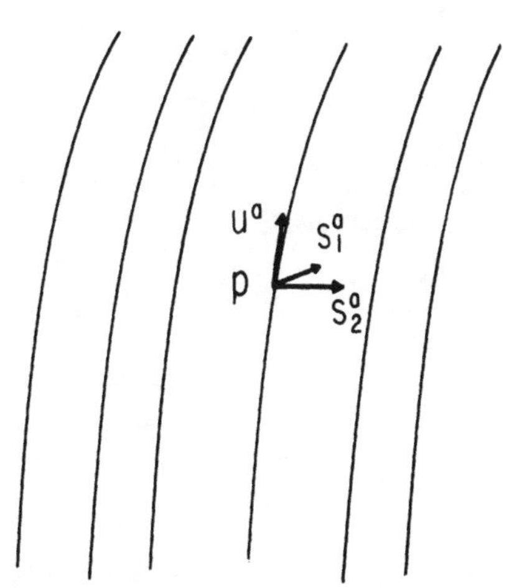

Fig. 5.2. The world lines of isotropic observers in spacetime. By definition of isotropy, for any two vectors s_1^a, s_2^a at p which are orthogonal to u^a, there exists an isometry of the spacetime which leaves p fixed and rotates s_1^a into s_2^a.

It is not difficult to see that in the case of a homogeneous and isotropic spacetime, the surfaces Σ_t of homogeneity must be orthogonal to the tangents, u^a, to the world lines of the isotropic observers. If not, then assuming that the isotropic observers and the family of homogeneous surfaces Σ_t are unique, the failure of the tangent subspace orthogonal to u^a to coincide with the tangent space of Σ_t would enable us to construct a geometrically preferred spatial vector, in violation of isotropy. (If the isotropic observers or family of homogeneous surfaces are not unique, as in special cases such as flat spacetime, it is still possible to show the existence of a family of isotropic observers orthogonal to a family of homogeneous surfaces.) Now, the spacetime metric, g_{ab}, induces a Riemannian metric, $h_{ab}(t)$, on each Σ_t by restricting the action of g_{ab} at each $p \in \Sigma_t$ to vectors tangent to Σ_t. The induced spatial geometry of the surfaces Σ_t is greatly restricted by the following requirements: (i) Because of homogeneity, there must be isometries of h_{ab} which carry any $p \in \Sigma_t$ into any $q \in \Sigma_t$. (ii) Because of isotropy, it must be impossible to construct any geometrically preferred vectors on Σ_t.

We shall now show that the second requirement implied by isotropy is particularly restrictive. Consider the Riemann tensor ${}^{(3)}R_{abc}{}^d$ constructed from h_{ab} on Σ_t. If we raise the third index with h^{ab}, we may view ${}^{(3)}R_{ab}{}^{cd}$ at point p as a linear map, L, of the vector space W of two-forms [i.e., antisymmetric tensors of rank (0,2)] at p into itself $L: W \to W$. By equation (3.2.20), L is symmetric, i.e., it is a self-adjoint map (with the natural, positive definite inner product on W determined by h_{ab}). Therefore, W has an orthonormal basis of eigenvectors of L. If the eigenvalues of these eigenvectors were distinct, then we would be able to give a geometrical prescription for picking out a preferred two-form at p and, consequently, a preferred vector at p. Hence, in order not to violate isotropy, all the eigenvalues of L must be equal. This means that L is a multiple of the identity operator,

$$L = K\,I \quad , \tag{5.1.1}$$

that is,

$$^{(3)}R_{ab}{}^{cd} = K\,\delta^c{}_{[a}\delta^d{}_{b]} \quad . \tag{5.1.2}$$

Thus, lowering the indices, we have

$$^{(3)}R_{abcd} = K\,h_{c[a}h_{b]d} \quad . \tag{5.1.3}$$

The requirement (i) of homogeneity implies that K must be a constant, i.e., it cannot vary from point to point of Σ_t. Actually, it is an interesting fact that the requirement (ii) of isotropy at each point also implies the constancy of K. To prove this, we substitute equation (5.1.3) into the Bianchi identity (3.2.16) to obtain,

$$0 = D_{[e}{}^{(3)}R_{ab]cd} = (D_{[e}K)h_{|c|a}h_{b]d} \quad , \tag{5.1.4}$$

where D_a denotes the derivative operator on Σ_t associated with h_{ab}. (We use the notation D_a rather than ∇_a in order to avoid confusion with the derivative operator on the four-dimensional spacetime associated with g_{ab}.) On a manifold of dimension three or greater, the right side of equation (5.1.4) will vanish if and only if $D_e K = 0$, i.e., K is constant. Thus, we can actually dispense with the assumption of homogeneity in discussing the geometry of Σ_t.

A space where equation (5.1.3) is satisfied (with K = constant) is called a *space of constant curvature*. It can be shown (Eisenhart 1949) that any two spaces of constant curvature of the same dimension and metric signature which have equal values of K must be (locally) isometric. Thus, our task of determining the possible spatial geometries of Σ_t will be completed if we enumerate spaces of constant curvature encompassing all values of K. This is easily done. All positive values of K are attained by the 3-spheres, defined as the surfaces in four-dimensional flat Euclidean space $\mathbb{R}^4$ whose Cartesian coordinates satisfy

$$x^2 + y^2 + z^2 + w^2 = R^2 \quad . \tag{5.1.5}$$

In spherical coordinates, the metric of the unit 3-sphere is

$$ds^2 = d\psi^2 + \sin^2\psi(d\theta^2 + \sin^2\theta\, d\phi^2) \quad . \tag{5.1.6}$$

The value $K = 0$ is attained by ordinary three-dimensional flat space. In Cartesian coordinates, this metric is

$$ds^2 = dx^2 + dy^2 + dz^2 \quad . \tag{5.1.7}$$

Finally, all negative values of K are attained by the three-dimensional hyperboloids, defined as the surfaces in a four-dimensional flat Lorentz signature space (i.e., Minkowski spacetime) whose global inertial coordinates satisfy

$$t^2 - x^2 - y^2 - z^2 = R^2 \quad . \tag{5.1.8}$$

In hyperbolic coordinates, the metric of the unit hyperboloid is

$$ds^2 = d\psi^2 + \sinh^2\psi(d\theta^2 + \sin^2\theta \, d\phi^2) \quad . \tag{5.1.9}$$

The new possibilities for the global spatial structure of our universe should be stressed. In prerelativity physics, as well as in special relativity, it was assumed that space had the flat structure given by the possibility $K = 0$ above. But even under the very restrictive assumptions of homogeneity and isotropy, the framework of general relativity admits two other distinct possibilities. The possibility of a 3-sphere spatial geometry is particularly interesting, as it is a compact manifold (see appendix A) and thus describes a universe which is finite but has no boundary. Such a universe is called "closed," while the universes with noncompact spatial sections such as those given by flat and hyperboloid geometries are called "open." (One could construct closed universes with flat or hyperboloid geometries by making topological identifications, but it does not appear to be natural to do so.) Thus, an intriguing question raised by general relativity is whether our universe is closed or open. We shall discuss the evidence on this issue at the end of section 5.4.

Since the isotropic observers are orthogonal to the homogeneous surfaces, we may express the four-dimensional spacetime metric g_{ab} as

$$g_{ab} = -u_a u_b + h_{ab}(t) \quad , \tag{5.1.10}$$

where for each t, $h_{ab}(t)$ is the metric of either (a) a sphere, (b) flat Euclidean space, or (c) a hyperboloid, on Σ_t. We can choose convenient coordinates on the four-dimensional spacetime as follows. We choose, respectively, either (a) spherical coordinates, (b) Cartesian coordinates, or (c) hyperbolic coordinates on one of the homogeneous hypersurfaces. We then "carry" these coordinates to each of the other homogeneous hypersurfaces by means of our isotropic observers; i.e., we assign a fixed spatial coordinate label to each observer. Finally, we label each hypersurface by the proper time, τ, of a clock carried by any of the isotropic observers. (By homogeneity, all the isotropic observers must agree on the time difference between any two hypersurfaces.) Thus, τ and our spatial coordinates label each event in the universe.

Expressed in these coordinates, the spacetime metric takes the form

$$ds^2 = -d\tau^2 + a^2(\tau) \begin{cases} d\psi^2 + \sin^2\psi(d\theta^2 + \sin^2\theta \, d\phi^2) \\ dx^2 + dy^2 + dz^2 \\ d\psi^2 + \sinh^2\psi(d\theta^2 + \sin^2\theta \, d\phi^2) \end{cases} \tag{5.1.11}$$

where the three possibilities in the bracket correspond to the three possible spatial geometries. [The metric for the spatially flat case could be made to look more similar to the other cases by writing it in spherical coordinates as $d\psi^2 + \psi^2(d\theta^2 + \sin^2\theta \, d\phi^2)$.] The general form of the metric, equation (5.1.11) is called a *Robertson-Walker* cosmological model. Thus, our assumptions of homogeneity and isotropy alone have determined the spacetime metric up to the three discrete possibilities of spatial geometry and the arbitrary positive function $a(\tau)$. To determine the spatial geometry and $a(\tau)$, we turn to Einstein's equation.

5.2 Dynamics of a Homogeneous, Isotropic Universe

Our aim now is to substitute the spacetime metric, equation (5.1.11), into Einstein's equation (4.3.21) to obtain predictions for the dynamical evolution of the universe. The first step is to describe the matter content of the universe in terms of its stress-energy, T_{ab}, which enters the right-hand side of Einstein's equation. Most of the mass-energy in the present universe is believed to be found in ordinary matter, concentrated in galaxies, although, as discussed at the end of this chapter, there are sufficient uncertainties and discrepancies in the determinations of mass that even this statement is not completely certain. On the cosmic scales with which we are dealing, each of the galaxies can be idealized as a "grain of dust." The random velocities of the galaxies are small, so the "pressure" of this dust of galaxies is negligible. By isotropy, the world lines of the galaxies must coincide with those of the isotropic observers. (If they did not, the relative motion of the galaxies and observers could be used to define a preferred spatial direction.) Thus, to a good approximation, the stress-energy tensor of matter in the present universe takes the form

$$T_{ab} = \rho u_a u_b \quad , \tag{5.2.1}$$

where ρ is the (average) mass density of matter. However, other forms of mass-energy are also present in the universe. As already briefly mentioned above, a thermal distribution of radiation at a temperature of about 3 K fills the universe. This radiation can also be described by a perfect fluid stress-energy tensor, but its pressure is nonzero; indeed, for massless thermal radiation, we have $P = \rho/3$. The contribution of this radiation to the stress-energy of the present universe is negligible, but, as will be discussed further in this section and in section 5.4, this radiation is predicted to make the dominant contribution to T_{ab} in the early universe. Thus, in treating Einstein's equation, we shall take T_{ab} to be of the general perfect fluid form,

$$T_{ab} = \rho u_a u_b + P(g_{ab} + u_a u_b) \quad . \tag{5.2.2}$$

There is no loss of generality in restricting consideration to T_{ab} of this form, as it is actually the most general form T_{ab} can take consistent with homogeneity and isotropy.

We now have the task of computing G_{ab} from the metric, equation (5.1.11), and equating it with $8\pi T_{ab}$, equation (5.2.2). *A priori*, we will get 10 equations corresponding to the 10 independent components of a symmetric two-index tensor. However, it is not difficult to see that on account of the spacetime symmetries, there will be only two independent equations in this case. Namely, the vector $G^{ab}u_b$ (as well as $T^{ab}u_b$) cannot have a spatial component, or isotropy would be violated. Thus, the "time-space" components of Einstein's equation are identically zero. Similarly, if we project both indices of G_{ab} into the homogeneous hypersurface and raise an index with the spatial metric, the same type of argument which led to equation (5.1.1) tells us that the resulting tensor must be a multiple of the identity operator. Thus, the off-diagonal "space-space" components of Einstein's equation must vanish, and the diagonal "space-space" components yield the same equations. Hence, the independent components of Einstein's equation are simply

$$G_{\tau\tau} = 8\pi T_{\tau\tau} = 8\pi\rho \quad , \tag{5.2.3}$$

$$G_{**} = 8\pi T_{**} = 8\pi P \quad , \tag{5.2.4}$$

where $G_{\tau\tau} = G_{ab}u^a u^b$ and $G_{**} = G_{ab}s^a s^b$, where s^a is any unit vector tangent to the homogeneous hypersurfaces.

We now have only the mechanical task of computing $G_{\tau\tau}$ and G_{**} in terms of $a(\tau)$. We shall do this explicitly for the case of flat spatial geometry, i.e.,

$$ds^2 = -d\tau^2 + a^2(\tau)(dx^2 + dy^2 + dz^2) \tag{5.2.5}$$

using the coordinate basis method. By equation (3.1.30), the nonvanishing components of the Christoffel symbol are merely

$$\Gamma^\tau{}_{xx} = \Gamma^\tau{}_{yy} = \Gamma^\tau{}_{zz} = a\dot{a} \quad , \tag{5.2.6}$$

$$\Gamma^x{}_{x\tau} = \Gamma^x{}_{\tau x} = \Gamma^y{}_{y\tau} = \Gamma^y{}_{\tau y} = \Gamma^z{}_{z\tau} = \Gamma^z{}_{\tau z} = \dot{a}/a \quad , \tag{5.2.7}$$

where $\dot{a} = da/d\tau$. Hence, by equation (3.4.5) the independent Ricci tensor components are calculated to be

$$R_{\tau\tau} = -3\ddot{a}/a \quad , \tag{5.2.8}$$

$$R_{**} = a^{-2}R_{xx} = \frac{\ddot{a}}{a} + 2\frac{\dot{a}^2}{a^2} \quad . \tag{5.2.9}$$

Since we have

$$R = -R_{\tau\tau} + 3R_{**} = 6\left(\frac{\ddot{a}}{a} + \frac{\dot{a}^2}{a^2}\right) \quad , \tag{5.2.10}$$

we thus obtain

$$G_{\tau\tau} = R_{\tau\tau} + \tfrac{1}{2}R = 3\dot{a}^2/a^2 = 8\pi\rho \quad , \tag{5.2.11}$$

$$G_{**} = R_{**} - \tfrac{1}{2}R = -2\frac{\ddot{a}}{a} - \frac{\dot{a}^2}{a^2} = 8\pi P \quad . \tag{5.2.12}$$

Using the first equation, we may rewrite the second equation as

$$3\ddot{a}/a = -4\pi(\rho + 3P) \quad . \tag{5.2.13}$$

Repeating the calculation for the cases of spherical and hyperboloid geometries (problem 2), we obtain the general evolution equations for homogeneous, isotropic cosmology:

$$3\dot{a}^2/a^2 = 8\pi\rho - 3k/a^2 \quad , \tag{5.2.14}$$

$$3\ddot{a}/a = -4\pi(\rho + 3P) \quad , \tag{5.2.15}$$

where $k = +1$ for the 3-sphere, $k = 0$ for flat space, and $k = -1$ for the hyperboloid. We will present the exact solutions of these equations for the cases of dust ($P = 0$) and radiation ($P = \rho/3$) below in Table 5.1, but first we shall examine some important qualitative properties of the solutions.

The first striking result is that the universe cannot be static, provided only that $\rho > 0$ and $P \geqq 0$. This conclusion follows immediately from equation (5.2.15), which tells us that $\ddot{a} < 0$. Thus, the universe must always either be expanding ($\dot{a} > 0$) or contracting ($\dot{a} < 0$) (with the possible exception of an instant of time when expansion changes over to contraction). Note the nature of this expansion or contraction: The distance scale between all isotropic observers (in particular, between galaxies) changes with time, but there is no preferred center of expansion or contraction. Indeed, if the distance (measured in the homogeneous surface) between two isotropic observers at time τ is R, the rate of change of R is

$$v \equiv \frac{dR}{d\tau} = \frac{R}{a}\frac{da}{d\tau} = HR \quad , \tag{5.2.16}$$

where $H(\tau) = \dot{a}/a$ is called *Hubble's constant*. (Note, however, that the value of H changes with time.) Equation (5.2.16) is known as *Hubble's law*. Note that v can be greater than the speed of light if R is large enough. This does not contradict the fundamental tenet of special and general relativity that "nothing can travel faster than the speed of light," since this tenet refers to the locally measured relative velocity of two objects at the same spacetime event, not a globally defined velocity between distant objects.

The expansion of the universe in accordance with equation (5.2.16) has been confirmed by the observation of the redshifts of distant galaxies, as will be explained

Table 5.1

DUST AND RADIATION FILLED ROBERTSON-WALKER COSMOLOGIES

	TYPE OF MATTER	
	"Dust"	Radiation
SPATIAL GEOMETRY	$P = 0$	$P = \frac{1}{3}\rho$
3-sphere, $k = +1$	$a = \frac{1}{2}C(1 - \cos\eta)$ $\tau = \frac{1}{2}C(\eta - \sin\eta)$	$a = \sqrt{C'}[1 - (1 - \tau/\sqrt{C'})^2]^{1/2}$
Flat, $k = 0$	$a = (9C/4)^{1/3}\,\tau^{2/3}$	$a = (4C')^{1/4}\,\tau^{1/2}$
Hyperboloid, $k = -1$	$a = \frac{1}{2}C(\cosh\eta - 1)$ $\tau = \frac{1}{2}C(\sinh\eta - \eta)$	$a = \sqrt{C'}\,[(1 + \tau/\sqrt{C'})^2 - 1]^{1/2}$

in more detail later in this chapter. The confirmation of this striking prediction of general relativity is a dramatic success of the theory. Unfortunately, the historical development of events clouded this success. Einstein was sufficiently unhappy with the prediction of a dynamic universe that he proposed a modification of his equation, the addition of a new term, as follows:

$$G_{ab} + \Lambda g_{ab} = 8\pi T_{ab} \quad , \tag{5.2.17}$$

where Λ is a new fundamental constant of nature, called the *cosmological constant*. (It can be shown [Lovelock 1972] that a linear combination of G_{ab} and g_{ab} is the most general two-index symmetric tensor which is divergence free and can be constructed locally from the metric and its derivatives up to second order, so eq. [5.2.17] gives the most general modification which does not grossly alter the basic properties of Einstein's equation. If $\Lambda \neq 0$, one does not obtain Newtonian theory in the slow motion, weak field limit; but if Λ is small enough, the deviations from Newtonian theory would not be noticed.) With this additional one-parameter degree of freedom, static solutions exist, though they require exact adjustment of the parameters and are unstable, much like a pencil standing on its point (see problem 3). Thus, Einstein was able to modify the theory to yield static solutions. After Hubble's redshift observations in 1929 demonstrated the expansion of the universe, the original motivation for the introduction of Λ was lost. Nevertheless, Λ has been reintroduced on numerous occasions when discrepancies have arisen between theory and observations, only to be abandoned again when these discrepancies have been resolved. In the following, we shall assume that $\Lambda = 0$.

Given that the universe is expanding, $\dot{a} > 0$, we know from equation (5.2.15) that $\ddot{a} < 0$, so the universe must have been expanding at a faster and faster rate as one goes backward in time. If the universe had always expanded at its present rate, then at the time $T = a/\dot{a} = H^{-1}$ ago, we would have had $a = 0$. Since its expansion actually was faster, the time at which a was zero was even closer to the present. Thus, under the assumption of homogeneity and isotropy, general relativity makes the striking prediction that at a time less than H^{-1} ago, the universe was in a singular state: The distance between all "points of space" was zero; the density of matter and the curvature of spacetime was infinite. This singular state of the universe is referred to as the *big bang*.

Note that the nature of this singularity is that resulting from a homogeneous contraction of space down to "zero size." The big bang does not represent an explosion of matter concentrated at a point of a preexisting, nonsingular spacetime, as it is sometimes depicted and as its name may suggest. Since spacetime structure itself is singular at the big bang, it does not make sense, either physically or mathematically, to ask about the state of the universe "before" the big bang; there is no natural way to extend the spacetime manifold and metric beyond the big bang singularity. Thus, general relativity leads to the viewpoint that the universe began at the big bang. For many years it was generally believed that the prediction of a singular origin of the universe was due merely to the assumptions of exact homogeneity and isotropy, that if these assumptions were relaxed one would get a nonsingular "bounce" at small a rather than a singularity. However, the singularity theorems of general relativity (see chapter 9) show that singularities are generic features of cosmological solutions; they have ruled out the possibility of "bounce" models close to the homogeneous, isotropic models. Of course, at the extreme conditions very near the big bang singularity one expects that quantum effects will become important, and the predictions of classical general relativity are expected to break down (see chapter 14).

Before discussing the qualitative predictions of general relativity for the future evolution of the universe, it is useful to obtain an equation for the evolution of the mass density. By multiplying equation (5.2.14) by a^2, differentiating it with respect to τ, and then eliminating $\ddot{a}$ via equation (5.2.15) (or, directly from eq. [4.3.7]), we obtain

$$\dot{\rho} + 3(\rho + P)\dot{a}/a = 0 \quad . \tag{5.2.18}$$

Thus, for dust ($P = 0$) we find

$$\rho a^3 = \text{constant} \quad , \tag{5.2.19}$$

which expresses conservation of rest mass, while for radiation ($P = \rho/3$) we find

$$\rho a^4 = \text{constant} \quad . \tag{5.2.20}$$

In this case the energy density decreases more rapidly as a increases than by the volume factor a^3, since the radiation in each volume element does work on its surroundings as the universe expands. (Alternatively, in terms of photons, the photon number density decreases as a^{-3}, but each photon loses energy as a^{-1} because of redshift [see section 5.3].) Comparison of equations (5.2.19) and (5.2.20) shows that although the radiation content of the present universe may be negligible, its contribution to the total mass density far enough into the past ($a \to 0$) should dominate over that of ordinary matter.

The qualitative features of the future evolution of our universe can now be seen. If $k = 0$ or -1, equation (5.2.14) shows that $\dot{a}$ never can become zero. Thus, if the universe is presently expanding, it must continue to expand forever. Indeed, for any matter with $P \geqq 0$, ρ must decrease as a increases at least as rapidly as a^{-3}, the value for dust. Thus $\rho a^2 \to 0$ as $a \to \infty$. Hence, if $k = 0$, the "expansion velocity" $\dot{a}$ asymptotically approaches zero as $\tau \to \infty$, while if $k = -1$ we have $\dot{a} \to 1$ as $\tau \to \infty$.

However, if $k = +1$, the universe cannot expand forever. The first term on the right-hand side of equation (5.2.14) decreases with a more rapidly than the second term, and thus, since the left-hand side must be positive, there is a critical value, a_c, such that $a \leqq a_c$. Furthermore, a cannot asymptotically approach a_c as $\tau \to \infty$ because the magnitude of $\ddot{a}$ is bounded from below on account of equation (5.2.15). Thus, if $k = +1$, then at a finite time after the big bang origin of the universe, the universe will achieve a maximum size a_c and then will begin to recontract. The same argument as given above for the occurrence of a big bang origin of the universe now shows that a finite time after recontraction begins, a "big crunch" end of the universe will occur. Thus, the dynamical equations of general relativity show that the spatially closed 3-sphere universe will exist for only a finite span of time.

Let us now turn our attention to solving equations (5.2.14) and (5.2.15) exactly for the cases of dust and radiation. The most efficient procedure for doing this is to eliminate ρ using equation (5.2.19) or, respectively, equation (5.2.20), and substitute into equation (5.2.14). We obtain, for dust,

$$\dot{a}^2 - C/a + k = 0 \quad , \tag{5.2.21}$$

where $C = 8\pi\rho a^3/3$ is constant; and for radiation,

$$\dot{a}^2 - C'/a^2 + k = 0 \quad , \tag{5.2.22}$$

where $C' = 8\pi\rho a^4/3$. Given equation (5.2.19) (or eq. [5.2.20]), equation (5.2.15) is redundant, so the first order ordinary differential equation (5.2.21) (or, respectively, [5.2.22]) is all we need solve. The solutions for $a(\tau)$ are readily obtained by elementary methods. These solutions for the six cases of interest are tabulated in Table 5.1. Graphs of $a(\tau)$ versus τ are displayed in Figures 5.3 and 5.4. The solution for the dust-filled universe with 3-sphere geometry was first given by Friedmann (1922) and is called the *Friedmann cosmology,* although in some references all the solutions in Table 5.1 are referred to as Friedmann solutions.

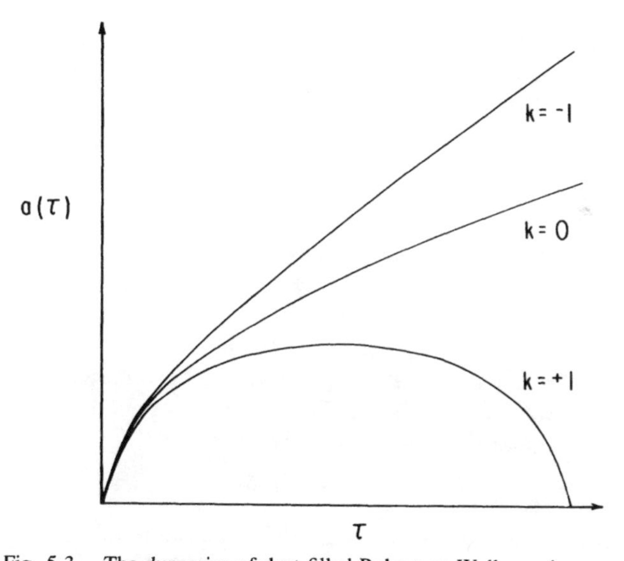

Fig. 5.3. The dynamics of dust-filled Robertson-Walker universes.

5.3 The Cosmological Redshift; Horizons

5.3a Redshift

We have mentioned above that the most direct observational evidence for the expansion of the universe comes from the redshift of the spectral lines of distant galaxies. In this section we shall obtain the redshift formula for a general Robertson-Walker cosmological model, equation (5.1.11). Suppose that at event P_1 at time τ_1

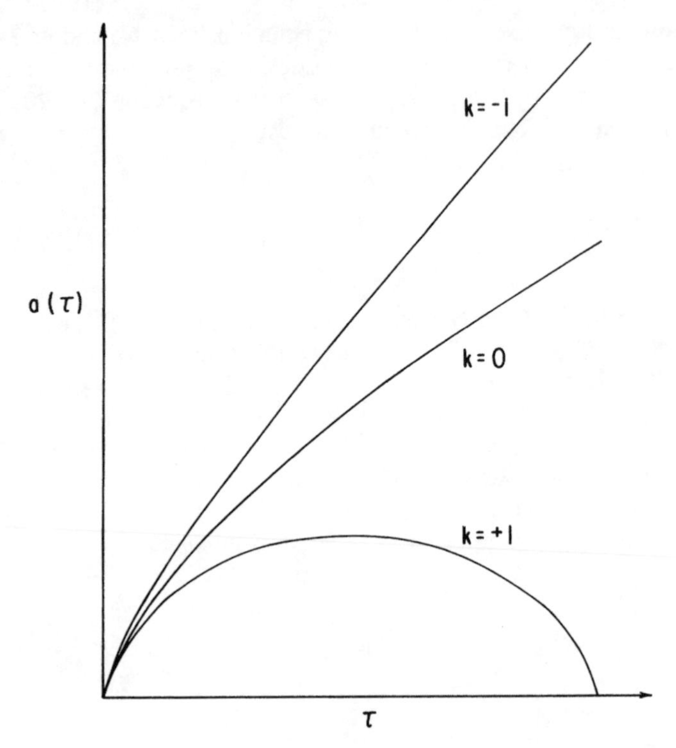

Fig. 5.4. The dynamics of radiation-filled Robertson-Walker universes.

an isotropic observer emits a photon of frequency ω_1. Suppose this photon is observed by another isotropic observer at event P_2 at time τ_2 as illustrated in Figure 5.5. We wish to find the frequency, ω_2, which this second observer will measure.

The solution of all redshift problems in special and general relativity is governed by the following two facts: (1) In the geometric optics approximation, light travels on null geodesics (see section 4.3); (2) The frequency of a light signal of wave vector k^a measured by an observer with 4-velocity u^a is

$$\omega = -k_a u^a \tag{5.3.1}$$

(see eq. [4.2.38]). Thus, we always can find the observed frequency by calculating the null geodesic determined by the initial value of k^a at the emission point and then calculating the right side of equation (5.3.1) at the observation point.

However, when symmetries are present, we often can shortcut this procedure by using the following fact proven in section C.3 of appendix C. Let ξ^a be a Killing vector field, i.e., a vector field which generates a one-parameter group of isometries, as discussed in appendix C. Let t^a be the tangent to a geodesic curve. Then $t^a \xi_a$ is constant along the geodesic. In this case, it is not difficult to calculate the redshift directly without appealing to symmetry arguments (see problem 4), but we shall calculate the redshift using these arguments because they give more insight into why the simple final result, equation (5.3.6), is obtained.

The first step is to notice that for all three choices of spatial geometry, we can find a spacetime Killing vector field ξ^a which points in the direction of the projection of k^a into Σ_1 at P_1 *and* points in the direction of the projection of k^a into Σ_2 at P_2. For example, in the case of flat spatial geometry, without loss of generality, we may assume that the projection of k^a into Σ_1 at P_1 is in the $(\partial/\partial x)^a$ direction. Then $k^a(\partial/\partial y)_a = k^a(\partial/\partial z)_a = 0$ initially, and, since $(\partial/\partial y)^a$ and $(\partial/\partial z)^a$ are Killing vector fields, these inner products also vanish at P_2. Thus, the projection of k^a into Σ_2 at P_2 also points in the $(\partial/\partial x)^a$ direction, and $\xi^a = (\partial/\partial x)^a$ is the required Killing vector field. Similar arguments establish the existence of ξ^a in the spherical and hyperboloid cases. Furthermore, in all cases the length of ξ^a at P_2 varies from its length at P_1 in proportion to the change in the length scale factor a of the universe in going from Σ_1 to Σ_2, i.e.,

$$\frac{(\xi^a\xi_a)^{1/2}|_{P_1}}{(\xi^a\xi_a)^{1/2}|_{P_2}} = \frac{a(\tau_1)}{a(\tau_2)} \quad . \tag{5.3.2}$$

To find the redshift, we note that since k^a is null, at any point its projection onto u^a must have the same magnitude as its projection into Σ, so at P_1

$$k_a u_1^a = -k_a[\xi^a/(\xi^b\xi_b)^{1/2}]|_{P_1} \quad . \tag{5.3.3}$$

Thus, we have

$$\omega_1 = [(k_a\xi^a)/(\xi^b\xi_b)^{1/2}]|_{P_1} \quad . \tag{5.3.4}$$

Similarly, we have

$$\omega_2 = [(k_a\xi^a)/(\xi^b\xi_b)^{1/2}]_{P_2} \quad . \tag{5.3.5}$$

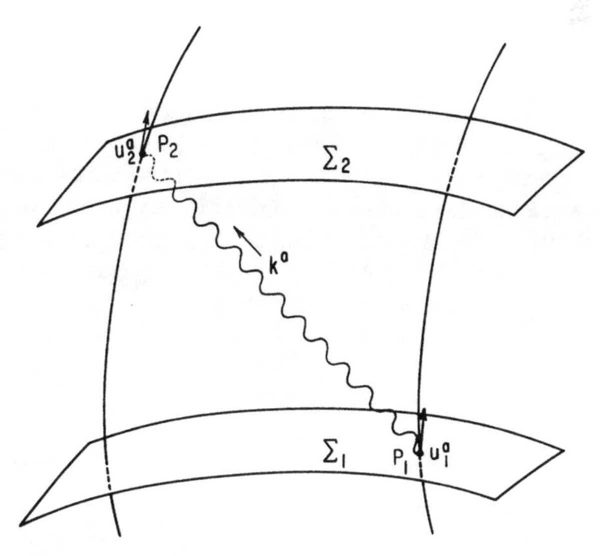

Fig. 5.5. A spacetime diagram showing the emission of a light signal at event P_1 and its reception at event P_2.

But we have $(k_a \xi^a)|_{P_1} = (k_a \xi^a)|_{P_2}$ by the above result on the inner product of Killing vector fields and geodesic tangent vectors. Thus, we find

$$\frac{\omega_2}{\omega_1} = \frac{(\xi^b \xi_b)^{1/2}|_{P_1}}{(\xi^b \xi_b)^{1/2}|_{P_2}} = \frac{a(\tau_1)}{a(\tau_2)} \quad , \tag{5.3.6}$$

where we have used equation (5.3.2). This result has the simple interpretation that as the universe expands, the wavelength of each photon increases in proportion to the amount of expansion.

The redshift factor, z, is given by

$$z \equiv \frac{\lambda_2 - \lambda_1}{\lambda_1} = \frac{\omega_1}{\omega_2} - 1 = \frac{a(\tau_2)}{a(\tau_1)} - 1 \quad . \tag{5.3.7}$$

For light emitted by nearby galaxies, we have $\tau_2 - \tau_1 \approx R$, where R is the present proper distance to the galaxy. Furthermore, for nearby galaxies we have

$$a(\tau_2) \approx a(\tau_1) + (\tau_2 - \tau_1)\dot{a} \quad . \tag{5.3.8}$$

Thus, we find

$$z \approx \frac{\dot{a}}{a} R = HR \quad , \tag{5.3.9}$$

which is the linear redshift-distance relationship discovered by Hubble. The redshifts of distant galaxies will deviate from this linear law, depending on exactly how $a(\tau)$ varies with τ.

5.3b Particle Horizons

The following question arises in the study of cosmological models in general relativity: In principle, how much of our universe can be observed at a given event P? More precisely, in the particular case of the Robertson-Walker cosmological models we may ask which isotropic observers (i.e., galaxies) could have sent a signal which reaches a given isotropic observer at (or before) event P. The boundary between the world lines that can be seen at P and those that cannot is called the *particle horizon* at P. Since the universe "shrinks to zero size" as one approaches the big bang singularity, one might expect that all isotropic observers can communicate with each other by sending signals to each other very early in the history of the universe when they were very close to each other. However, we shall show now that this is not the case for Robertson-Walker models which expand sufficiently rapidly from an initial big bang singularity. Thus, we shall demonstrate the existence of nontrivial particle horizons in a class of Robertson-Walker models which includes all the solutions of Table 5.1.

This demonstration is most easily made in the case of flat spatial geometry,

$$ds^2 = -d\tau^2 + a^2(\tau)(dx^2 + dy^2 + dz^2) \quad , \tag{5.3.10}$$

and we will focus our attention on that case. By making the coordinate transformation $\tau \rightarrow t$ defined by

$$t = \int \frac{d\tau}{a(\tau)} \tag{5.3.11}$$

we can reexpress the metric, equation (5.3.10), as

$$ds^2 = a^2(t)(-dt^2 + dx^2 + dy^2 + dz^2) \quad . \tag{5.3.12}$$

Written in this form, it becomes manifest that this metric is merely a multiple of the metric of the flat Minkowski spacetime metric. Such a metric is called *conformally flat*. The relevance of this remark arises from the fact that a vector will be timelike, null, or spacelike in the metric of equation (5.3.12) if and only if it has the same property with respect to the flat metric

$$ds^2 = -dt^2 + dx^2 + dy^2 + dz^2 \quad . \tag{5.3.13}$$

Thus, it is possible to send a signal between two events (i.e., join the two events by a timelike or null curve) in the metric of equation (5.3.12) if and only if this can be done in the flat metric, equation (5.3.13). With this in mind, it is not difficult to see that an observer at an event P will be able to receive a signal from all other isotropic observers if and only if the integral, equation (5.3.11), which defines t, diverges as one approaches the big bang singularity, $\tau \rightarrow 0$. Namely, if this integral diverges—which will be the case if $a(\tau) \leqq \alpha\tau$ for some constant α as $\tau \rightarrow 0$—then the Robertson-Walker model will be conformally related to all of Minkowski spacetime (i.e., t will range down to $-\infty$) and thus there will be no particle horizon. On the other hand, if the integral converges, the Robertson-Walker model will be conformally related only to the portion of Minkowski spacetime above a t = constant surface, and particle horizons will exist, as illustrated in Figure 5.6. As seen from Table 5.1, for $k = 0$, even in the case of dust we have $a(\tau) \propto \tau^{2/3}$. Since $a(\tau)$ will

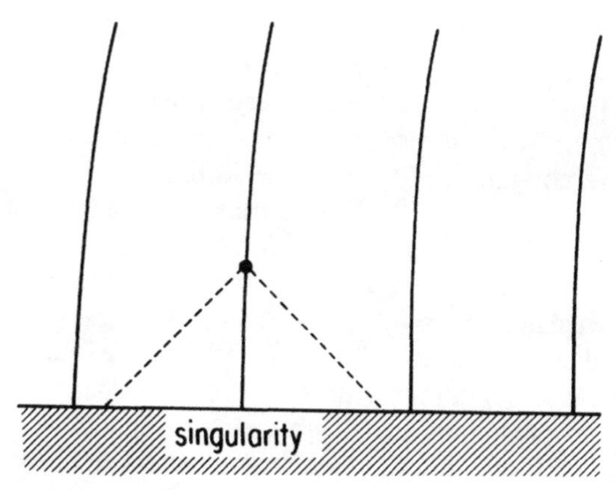

Fig. 5.6. The causal structure of the Robertson-Walker solutions near the big-bang singularity. Particle horizons exist since an observer cannot "see" all other isotropic observers in the universe.

be larger if $P > 0$, for all spatially flat Robertson-Walker solutions of Einstein's equation the integral, equation (5.3.11), will converge as $\tau \to 0$ and particle horizons do indeed occur.

For hyperboloid and spherical geometries, as $\tau \to 0$ the behavior of $a(\tau)$ goes over to the flat case, since the term involving k in equation (5.2.14) becomes negliglible. A similar analysis shows that particle horizons of the same nature as in the flat case also exist in all of these solutions. In the case of the spherical geometry, the spatial extent of the universe is finite, and one may ask if the particle horizons eventually cease to exist or whether they are still present when the universe recollapses to the "big crunch" singularity (problem 5). The answer is that for the dust filled spherical universe of Table 5.1, the particle horizon ceases to exist at the moment of maximum expansion, i.e., a light signal emitted at the big bang would travel exactly halfway around the universe by the moment of maximum expansion, so by looking in all directions at this time an observer could receive signals from all other isotropic observers. However, for the radiation filled spherical universe, a light signal would travel exactly halfway around the universe in the entire history of the universe, so the particle horizons remain present until the "big crunch."

The existence of particle horizons in the Robertson-Walker cosmological models leads to the following interesting issue. From the cosmic microwave background (discussed briefly in the next section) we have good reason to believe that the present universe is homogeneous and isotropic to a very high degree of precision. Now, many ordinary systems, such as a gas confined by a box, often are found in extremely homogeneous and isotropic states. However, the usual explanation of why such systems are in such homogeneous and isotropic states is that they have had an opportunity to self-interact and thermalize. Thus, for example, even if a gas in a box were initially, say, in an inhomogeneous state, these inhomogeneities would quickly "wash out" on a time scale of the order of the transit time across the box. However, this type of explanation cannot possibly apply to a universe with particle horizons, since different portions of the universe cannot even send signals to each other, far less interact sufficiently to thermalize each other. Thus, in order to explain the homogeneity and isotropy of the present universe, one must postulate that either (*a*) the universe was "born" in an extremely homogeneous, isotropic state or (*b*) the very early universe differed significantly from the Robertson-Walker models so that no horizons were present; the inhomogeneities and anisotropy then "damped out"—perhaps due to effects such as viscosity of matter or the back-reaction of quantum particle creation—and the universe approached the Robertson-Walker model. The first "explanation" may appear rather unnatural (see, however, Penrose 1979). The second explanation has been investigated extensively with regard to damping of anisotropy, beginning with the work of Misner (1969), but has not yet proven successful in presenting a plausible picture of evolution from a "chaotic" initial state to a Robertson-Walker model. It also suffers from the potentially serious difficulty that inhomogeneities (which have not been investigated extensively) on a sufficiently large scale generally would be expected to grow rather than damp out in a self-gravitating system. Recently, however, it has been suggested that the very early universe may have undergone an "inflationary phase" (discussed briefly in the next section) resulting in an enormous enlargement of the particle horizon in the Robertson-Walker models and offering a possible explanation of how an initially "chaotic" universe could evolve to one which has large homogeneous and isotropic regions. However, it should be kept in mind that we do not, at present, have a quantum theory of gravity (see chapter 14) and such a theory may play a large role in accounting for the initial state of our universe.

5.4 The Evolution of Our Universe

In this section we shall briefly outline the history of our universe from the big bang to the present. The picture we shall present is the "standard" one, which assumes that the universe is well described throughout its history by a Robertson-Walker solution and which makes further assumptions concerning the matter content of the universe. While this picture has received strong supporting evidence from the existence of the cosmic microwave background and its explanation of the cosmic abundance of helium, it should be kept in mind that none of its assumptions is unchallengeable. Many of the calculations and much of the observational evidence which go into constructing this history of the universe are reviewed by Peebles (1971) and Weinberg (1972), and we refer the reader to these references for more quantitative details.

A good deal of the nature of the early universe can be understood from the fact that the decrease of the scale factor a as one goes back toward the past has the same local effect on the matter as if the matter were placed in a box whose walls contract at the same rate. Thus, as would happen in a contracting box, the contribution of radiation compared with ordinary matter (i.e., baryons) increases in the past, as was already noted in section 5.2. The energy density of the cosmic microwave background in the present universe is estimated to be about 1,000 times smaller than the mass density contribution of matter, although this number is uncertain by as much as a factor of 10 because of the difficulties involved in determining the present density of ordinary matter. Thus, assuming that this radiation continues to exist as one goes back into the past (e.g., that it was not, say, emitted by galaxies), then according to equations (5.2.19) and (5.2.20), when the scale factor a was more than 1,000 times smaller than its present value, this radiation should have been the dominant contribution to the energy density of the universe. Thus, one would expect the radiation filled model of the universe to be a good approximation for the dynamics of the universe before this stage, while the dust filled model should be a good approximation afterwards.

Just as the temperature of a box of gas increases as one compresses it, one would expect the matter and radiation in the universe to get hotter as a decreases, and to become infinitely hot as one approaches the big bang, $a \to 0$.[1] If the early universe

1. It would also be self-consistent to assume that no radiation was present in the very early universe and that only cold matter, such as baryons, was present. However, a "cold big bang" model would have the major tasks of accounting for the present existence of the cosmic microwave background and the correct helium abundance of the universe, both of which are naturally explained by the standard "hot big bang" model described here.

was radiation dominated, as we expect it to be, then for all models ($k = 0, \pm 1$) the dependence of a and ρ on τ for small τ goes over to the $k = 0$ solution,

$$a(\tau) = (4C')^{1/4}\tau^{1/2} \quad , \tag{5.4.1}$$

$$\rho = \frac{3}{32\pi G\tau^2} \quad , \tag{5.4.2}$$

where here, and throughout this section, we shall restore the G's and c's in our equations. If the radiation is thermally distributed, the mass density ρ is given by the following expression, derived from the quantum statistical mechanics of massless particles:

$$\rho = \sum_{i=1}^{n} \alpha_i g_i \frac{\pi^2}{30\hbar^3 c^5} (kT)^4 \quad , \tag{5.4.3}$$

where n is the number of species of radiation, g_i is the spin degeneracy factor, and α_i takes the value 1 for bosons and 7/8 for fermions. Massive particles whose mass is much less than kT effectively act like zero mass particles and should be included in equation (5.4.3) as a "species of radiation." Equations (5.4.1), (5.4.2), and (5.4.3) imply that $T \propto \rho^{1/4} \propto a^{-1}$, as expected from the redshift relation, equation (5.3.6).

An important issue is whether the interactions of matter and radiation in the early universe proceed on a rapid enough time scale for thermalization to occur locally (i.e., within the particle horizon). If they do not, then the self-consistency of assuming that the matter is thermally distributed is questionable, and the predicted evolution of the matter and radiation may depend on the details of the assumed initial distribution. If they do, then we have the relatively simple task of evolving a thermal distribution of matter until such time as equilibrium is no longer maintained. The expansion time scale, t_E, of the universe, i.e., the time over which a significant change in the scale factor a occurs, is

$$t_E \sim a/\dot{a} = 2\tau \tag{5.4.4}$$

by equation (5.4.1). On the other hand, the time scale for interactions is

$$t_I \sim \frac{1}{n\sigma c} \propto \frac{a^3}{\sigma} \propto \tau^{3/2}/\sigma(T) \quad , \tag{5.4.5}$$

where the number of interacting particles is assumed to be conserved, so that their number density, n, scales as a^{-3}, and the possible energy dependence of the interaction cross section σ is explicitly indicated by writing σ as a function of temperature. Comparison of equations (5.4.4) and (5.4.5) shows that unless σ falls off rapidly at high energies, at sufficiently early times we will have $t_I << t_E$, so thermalization can be achieved. (Actually, it is possible that at very high energies the interactions of particle physics become "asymptotically free," and σ does fall off sufficiently rapidly to make $t_I > t_E$ as $\tau \rightarrow 0$. However, even if this occurs, for energies smaller than about 10^{15} GeV we should recover $t_I > t_E$, and thermalization still should be achieved at a very early time.) On the other hand, as the universe evolves, we eventually find $t_I > t_E$, and the matter distribution will drop out of thermal equilibrium.

Thus, the above considerations lead to the following basic picture of the evolution of our universe. The universe began as a hot ($T \to \infty$), dense ($\rho \to \infty$) "soup" of matter and radiation in thermal equilibrium. However, as the universe evolved, thermal equilibrium was not maintained, as t_I became greater than t_E. The energy content of the early universe was dominated by radiation. However, by the time a reached about $1/1000$ of its present value, the ordinary matter contribution dominated the energy content of the universe, and the dynamics of the universe became that of a dust filled Robertson-Walker model. We shall now proceed to fill in some of the important details of this evolutionary history.

During the first 10^{-43} seconds of the evolutionary history of the universe predicted by classical general relativity, the magnitude of the curvature of spacetime was greater than the scale given by the Planck length $(G\hbar/c^3)^{1/2} \approx 10^{-33}$ cm. On dimensional grounds, we expect that the quantum effects of the gravitational field will be very important in this era (see chapter 14), and thus the predictions of classical general relativity cannot be taken seriously. After this time, we may expect classical general relativity to be valid, but at the extreme conditions found in the very early universe (at $\tau = 10^{-43}$ s, we have $\rho \sim 10^{92}$ g cm^{-3}!) it hardly needs to be emphasized that all statements about the behavior of matter during this epoch are speculative.

There are two interesting and important effects that may have occurred in the very early universe, only several orders of magnitude later than the Planck time. The first concerns the dynamics of the very early universe. Some quantum field theory models which attempt to unify the strong and electro-weak interactions predict that at very high temperatures, the thermal equilibrium state of the quantum field will undergo a phase transition. In these models, if "supercooling" occurs there may be an important "vacuum" contribution, of the form $-\lambda g_{ab}$ (where λ is a large, positive constant) to the stress-energy tensor, T_{ab}, of the field. Thus, the very early universe may have gone through a phase where the dynamics was the same as would occur in an empty universe with a large, positive cosmological constant. Hence, if these models are correct, then, as first suggested by Guth (1981), there may be an "inflationary" regime in the very early universe where the universe is approximately a de Sitter solution [see problem 3(b)] and expands very rapidly. If this occurs, then as mentioned at the end of section 5.3, the particle horizons in the present universe could be much larger than would be calculated by extrapolating the solutions of Table 5.1 all the way back to the "big bang." For further discussion of inflationary cosmological models, we refer the reader to Gibbons, Hawking, and Siklos (1983).

The second effect concerns the production of baryons. There is strong reason to believe (Steigman 1976) that the matter content of the universe consists of baryons as opposed to antibaryons; that one does not have matter-antimatter symmetry. It is possible that our universe was simply born with an excess of baryons over antibaryons, and, indeed this must be so if baryon number is strictly conserved. However, it is also possible that the universe began in a matter-antimatter symmetric state and that the baryon excess was manufactured in the very early universe. In order for this to happen, it is necessary that the high energy particle interactions occurring in the very early universe satisfy the following properties: (1) Clearly, they must fail to conserve baryon number. (2) They must fail to preserve charge conjugation, C, and the composition of charge conjugation with parity, CP. (If either of these symmetries are preserved, equal numbers of baryons and antibaryons will be produced.) (3) They must result in departures from thermal equilibrium. (This is because particles and antiparticles have equal masses, so in thermal equilibrium they will occur in equal numbers.) Nonequilibrium phenomena could be produced naturally by the existence of a massive particle whose decay lifetime is greater than the expansion time, t_E, at the time when the temperature of the universe drops below the mass of the particle (so that production of the particle is essentially shut off). Remarkably, the "grand unified" theories of the strong, electromagnetic, and weak interactions currently predict all three of these properties and thus may provide an explanation of the matter-antimatter asymmetry in our universe.

We pick up our account of the history of the early universe at time $\tau = 1$ second, when the density is $\rho \approx 5 \times 10^5$ g cm^{-3}, and the temperature is $T \approx 10^{10}$ K. Although these conditions are extreme by ordinary standards, we are now in a low enough energy and density regime to be able to make solid predictions. At this time the matter in the universe consists almost entirely of neutrinos, photons, electrons, positrons, neutrons, and protons in thermal equilibrium; the temperature is low enough that the equilibrium abundance of the more massive elementary particles is negligible. By about this stage, the interactions of the neutrinos have become sufficiently weak that they decouple from the rest of the matter. For the remainder of the history of universe, they merely passively get redshifted to lower energy. Since a redshifted thermal spectrum ($\omega \to \omega/a$) is simply a thermal spectrum at a lower temperature ($T \to T/a$), the present universe should be filled with a blackbody distribution of neutrinos at temperature $T \approx 2$ K. However, the detection of these neutrinos would be a task very far beyond presently available sensitivities.

As the universe continues to cool, the rates of reactions which convert protons to neutrons and vice versa quickly drop to much lower than the expansion rate of the universe. Consequently, the neutron to proton ratio "freezes out" at $\tau \sim 1.5$ seconds at roughly the value $1/6$. (The protons are more abundant than the neutrons because they are over 1 MeV lighter and thus are more prevalent in thermal equilibrium before the "freezeout" occurs.) Of course, the term "freezeout" should not be taken too literally, since the "turnoff" of the interactions does not occur instantaneously and, furthermore, the neutron-proton ratio continues to decrease slowly with time due to the decay of the neutrons.

At $\tau = 4$ seconds, we have $\rho = 3 \times 10^4$ g cm^{-3}, and $T \approx 5 \times 10^9$ K $\approx$ 0.5 MeV, which is approximately the mass of electrons and positrons. At this stage the equilibrium population of electrons and positrons decreases rapidly; the production rate drops below the annihilation rate, and shortly after this time all the positrons will have annihilated, leaving a relatively small population of residual electrons. Essentially all the energy of the electron-positron pairs is transferred to the photons, heating them to a temperature ~ 1.4 times higher than the temperature of the neutrinos.

When the temperature drops to about 10^9 K at $\tau \approx 3$ minutes, nucleosynthesis begins rather abruptly, producing ^{4}He nuclei. Actually, in thermal equilibrium the abundance of ^{4}He at these baryon densities would be appreciable at even higher temperatures ($\sim 3.5 \times 10^9$ K), but very little nucleosynthesis occurs before $\tau = 3$ minutes because deuterium (^{2}H) plays a key role in the nuclear reactions which build up helium, but the equilibrium abundance of ^{2}H is very low until the temperature drops to 10^9 K. Almost no nucleosynthesis of elements beyond ^{4}He occurs because of the large Coulomb barriers and the lack of stable nuclei with atomic weights 5 and 8. Within a time span of a few minutes, essentially all of the neutrons present at the "freezeout" time which did not decay are converted to ^{4}He, resulting in an abundance of ^{4}He of about 25% by mass, with much smaller abundances of ^{2}H, ^{3}He, and ^{7}Li also produced, but negligible amounts of other elements. The percentage of ^{4}He is not very sensitive to the assumed value of baryon density, since it is governed mainly by the neutron-proton ratio at "freezeout," but the abundances of the other elements, particularly ^{2}H, are very sensitive to the baryon density. A relatively low baryon density can produce a ^{2}H abundance of over 5×10^{-4} by mass, whereas a high baryon density—specifically, a density high enough to "close the universe," as discussed below—increases the efficiency of the chain of reactions producing ^{4}He, and results in a ^{2}H abundance many orders of magnitude lower.

It is difficult to observe the cosmic abundance of helium, but the abundance of 25% seems to be in agreement with observations. The presence of this amount of helium in the universe cannot readily be accounted for by other processes—in particular, nucleosynthesis in stars is estimated to produce an abundance of helium of only a few percent—and thus the prediction of helium production via "big bang nucleosynthesis" must be viewed as a major success of the theory.

After the period of nucleosynthesis, the universe, of course, continued its expansion and cooling. The next cosmic occurrence of major importance took place when the temperature had dropped to about 4000 K. This occurred at $\tau \sim 4 \times 10^5$ years if the universe was still radiation dominated, but it would have occurred somewhat earlier if the universe had already become matter dominated. At this temperature and below it, the free electrons and protons combined to form neutral hydrogen. Indeed, by the time the temperature had dropped to 2000 K, the fraction of ionized hydrogen was only $\sim 10^{-4}$. As a result of this occurrence—called *recombination*, although, of course, the electrons and protons had never combined earlier—the interaction between the matter and radiation dropped precipitously, since the scattering cross section of photons off free charged particles is much greater than off neutral H and He. Indeed, the photons effectively decoupled completely from the matter after recombination, and merely cooled with the expansion of the universe during the remainder of its evolution. Thus, the present universe should be filled with this low temperature blackbody radiation originating from the big bang, whose photons last interacted with matter at the recombination time.

Exactly such a radiation background at temperature $T \approx 2.7$ K (corresponding to wavelengths mainly in the microwave regime) was discovered by Penzias and Wilson (1965). The existence of this radiation would be difficult to account for in any other way, and thus it provides a major confirmation of the above picture of the evolution of our universe. Furthermore, the radiation has been measured to be isotropic to a high degree of precision. (A "dipole type" anisotropy of $\sim 0.1\%$ has been detected [Smoot *et al.* 1977], but this presumably is due to motion of the Earth with respect to the "preferred rest frame" of the Robertson-Walker cosmology.) This provides very strong evidence that the universe was very nearly homogeneous and isotropic at least down to the recombination time.

The decoupling of matter and radiation in this era also had a major effect on the growth of gravitational perturbations, leading to the formation of galaxies. Just before recombination, the pressure provided by the radiation inhibited the growth of gravitational perturbations involving masses smaller than about $10^{17}\, M_\odot$, which is much larger than typical galactic masses ($\sim 10^{11}\, M_\odot$). However, after recombination, the radiation pressure had no effect on the matter, and gravitational instabilities could occur on all mass scales $\gtrsim 10^5\, M_\odot$. Thus, irregularities in the distribution of matter began to grow after recombination, resulting in the formation of galaxies, star clusters, and stars. However, the details of this process are not very well understood at present (see Peebles 1980 for further discussion).

At some time between $\tau \sim 10^3$ years and $\tau \sim 10^7$ years, ordinary matter became the dominant form of energy in the universe. (The exact time at which this occurred is uncertain because there is substantial uncertainty in the matter density of the present universe.) The dynamics of the universe was transformed from a "radiation filled" to a "dust filled" solution. Finally, at $\tau \sim$ 10–20 billion years the universe reached its present state.

Thus, general relativity, together with the assumption of homogeneity and isotropy and assumptions about the matter content of the universe, has produced a remarkably successful picture of the history of our universe. Among its most notable achievements are its explanations of the cosmic abundance of helium and the existence of the cosmic microwave radiation.

It is interesting that if one accepts this picture, some nontrivial constraints are placed on (1) the masses of stable, weakly interacting, elementary particles and (2) the number of species of massless particles which are in thermal equilibrium in the early universe. To explain the first constraint, suppose that the electron or muon neutrino has a mass, m, or that some massive stable particle exists which participates only in the weak interactions. Then the behavior of this particle in the early universe would be the same as that of the massless neutrino. However, in the present universe, instead of contributing energy $\sim 10^{-4}$ eV per particle (corresponding to $T \sim 2$ K), it would contribute energy m per particle. If the particles are very massive, their population will be "frozen in" at a very low value when they drop out of thermal equilibrium in the early universe and they will not contribute much to the present energy density of the universe. But if their mass lies within the range $10\ \text{GeV} \gtrsim m \gtrsim 100\ \text{eV}$, these particles would become by far the dominant contributors of mass-energy in the universe, and they would produce dynamical effects on the expansion rate of the universe which are incompatible with observations. Thus, our cosmological theory and observations place some stringent limits on the possible masses of stable, weakly interacting particles.

The existence of other species of massless particles (e.g., neutrinos) in thermal equilibrium in the early universe would not significantly affect the present mass density of the universe, since their energy would get redshifted away as the universe expands. However, they would have an important dynamical effect upon the early universe. This is because, according to equation (5.4.3), they would affect the relation between ρ and T, making T smaller for a given ρ. Since for small τ the relation between ρ and τ always is given by equation (5.4.2) for a radiation dominated universe, we see that $T(\tau)$ will be decreased, i.e., a given temperature will occur at an earlier epoch, when the expansion rate is higher. The most important consequence of this effective speedup of the expansion rate of the universe is that the neutron to proton ratio will "freeze out" at a higher temperature, yielding a higher percentage of neutrons. Consequently, more helium will be manufactured during the era of nucleosynthesis.[2] Assuming that the baryon density in the early universe corresponds at least to the present mass density observed in small groups of galaxies (see below), it is found that an unacceptably high abundance of helium will be manufactured if there are more than four species of neutrinos. Since the electron and muon neutrinos are known to exist and the existence of a neutrino associated with the recently discovered τ-lepton seems likely, this cosmological limit appears to leave little room for the existence of other species of massless particles in thermal equilibrium in the early universe.

What about the future evolution of our universe? The most important issue with regard to future evolution is whether our universe is "open" or "closed," i.e., does it correspond to the cases $k = 0, -1$, or the case $k = +1$? As discussed in section 5.2, if the universe is open it will expand forever, while if it is closed it will eventually recontract. We may express the basic equations (5.2.14) and (5.2.15) governing the dynamics of the present universe in terms of Hubble's constant, $H = \dot{a}/a$, and the *deceleration parameter, q*, defined by

$$q = -\ddot{a}\, a/(\dot{a})^2 \quad . \tag{5.4.6}$$

Since $P \approx 0$ in the present universe, we have

$$H^2 = 8\pi G\rho/3 - kc^2/a^2 \quad , \tag{5.4.7}$$

$$q = \frac{4\pi G\rho}{3H^2} \quad . \tag{5.4.8}$$

Defining Ω by

$$\Omega = 8\pi G\rho/3H^2 \quad , \tag{5.4.9}$$

we see that $q = \Omega/2$, and the universe is closed ($k = 1$) if and only if $\Omega > 1$, i.e., $\rho > \rho_c \equiv 3H^2/8\pi G$.

There are, at present, four basic, independent pieces of observational evidence, which, in conjunction with the above equations, yield indications of whether the universe is open or closed: (i) the redshift–apparent magnitude relation for distant

2. However, if the expansion rate is tremendously speeded up, there will not even be enough time for the nucleosynthetic reactions and the helium synthesis will decrease.

objects, (ii) the mass density of the present universe, (iii) the age of the universe, and (iv) the cosmic abundance of deuterium. We shall briefly discuss each of these in turn.

As derived at the end of section 5.3*a*, for sufficiently nearby objects we have a linear relation between redshift, z, and present proper distance, R, equation (5.3.9). Also, for sufficiently nearby objects, the apparent luminosity of an object is proportional to its intrinsic luminosity divided by R^2. (For very distant objects, one must correct this simple relationship because of the expansion of the universe and the curvature of space.) Thus, if we consider objects of a fixed intrinsic luminosity, for sufficiently nearby objects, one will obtain a linear relation between $\ln z$ and ln [apparent luminosity] = apparent magnitude. We will not discuss here the difficult tasks of (*a*) how to find "standard candles," i.e., objects of fixed intrinsic luminosity, and (*b*) how to calibrate their intrinsic luminosity (or, equivalently, their distance) by a chain of arguments whereby nearby standard candles are used to calibrate more distant standard candles (see e.g., Weinberg 1972). The relation between $\ln z$ and apparent magnitude is, indeed, found to be linear, corresponding to a value of H given by

$$H \sim 50 \text{ km s}^{-1} \text{ Mpc}^{-1} \quad , \tag{5.4.10}$$

where 1 megaparsec (Mpc) $\approx 3 \times 10^{19}$ km. Because of uncertainties in the absolute distance determinations, this value of H quite possibly could be in error by as much as a factor of 2 and, indeed, some recent determinations have given the value $H \sim 100 \text{ km s}^{-1} \text{ Mpc}^{-1}$. For sufficiently distant objects the relation between apparent magnitude and $\ln z$ will depart from linearity in a way that depends on the dynamical evolution of $a(\tau)$ and, to a lesser extent, on the spatial geometry. In principle, by examining these departures from linearity, we can determine q, which, in turn, will tell us if the universe is open ($q \leqq 1/2$) or closed ($q > 1/2$). However, in practice, these departures begin to occur only for objects at such large distances ($z \gtrsim 0.2$, corresponding to $R \gtrsim 10^3$ Mpc) that the light reaching us was emitted at a significantly earlier epoch. Thus, there is reason to doubt whether our most distant standard candles—the brightest galaxies in clusters of galaxies—are really "standard" on account of evolutionary effects; these galaxies may well have been significantly brighter (or dimmer) in the past. Until these evolutionary effects are better understood, it will not be possible to draw definitive conclusions as to whether the universe is open or closed by this approach.

The parameter Ω can be determined directly by measuring the mass density, ρ, of the present universe and Hubble's constant H. As mentioned above, the uncertainty in calibration of distances leads to considerable uncertainty in H. Fortunately, however, the distance scale calibration also enters into the determination of ρ in such a way that the ratio $\Omega \propto \rho/H^2$ is unaffected by these uncertainties, and thus Ω may be determined much more precisely than either ρ or H. In the present universe, ordinary matter appears to make the dominant contribution to ρ; the energy density of the cosmic microwave background is $\sim 10^{-3}$ times the estimated matter density quoted at the end of this paragraph. It appears that essentially all the matter in the universe is concentrated in galaxies, since most possibilities for intergalactic matter are excluded on observational or theoretical grounds. The masses of spiral galaxies sufficiently nearby ($\lesssim$15 Mpc) can be estimated from their rotational velocities. The mass of a cluster of galaxies can be estimated from the random velocities of the individual galaxies, assuming that the cluster is gravitationally bound. When these masses are compared, some interesting discrepancies arise: The masses of galaxies in binary systems and small clusters as determined by the second method appears to be higher (by a factor of about 3) than the masses of galaxies determined by the first method. Furthermore, the masses of galaxies in large clusters appears to be larger (by a factor of at least 2 and possibly much larger) than the masses obtained for small clusters. A possible explanation of the first discrepancy is that galaxies possess a massive halo of underluminous material; this extra mass would not affect the galactic rotation curve and would cause the first method (which estimates only the mass of the luminous inner portion of the galaxy) to underestimate the total mass. The cause of the second discrepancy is less clear, but it might be explained by the presence of intercluster material or simply by the fact that galaxies in large clusters might tend to be significantly more massive than galaxies in small clusters. If we take the typical mass of galaxies to be that determined by the second method for small clusters, we obtain $\Omega \sim 0.04$, corresponding to $\rho \sim 2 \times 10^{-31}$ g cm^{-3} if the value of H given in equation (5.4.10) is used. This provides evidence in favor of an open universe, but many possibilities for hidden mass remain, so this evidence is not conclusive.

Since our universe is presently matter dominated and should have been so for almost all of its history, from our solution of Table 5.1 we find that if $k = 0$ the age of the universe is related to Hubble's constant by

$$\tau_c = \frac{2}{3H} \sim 1.3 \times 10^{10} \text{ years} \quad , \tag{5.4.11}$$

using the value of H given in equation (5.4.10). If $k = -1$, τ will be larger than τ_c, while if $k = +1$, it will be smaller. Thus, comparison of the actual age of the universe with the critical age, τ_c, will tell us whether the universe is open or closed. Recall, however, that there is considerable uncertainty in the value of H, and thus in the value of τ_c. The age of the universe can be estimated from the age of the oldest objects we can find: globular star clusters (whose age can be estimated from the theory of stellar evolution) and elements manufactured by the first stars (whose age can be determined by radioactive dating if the element is radioactive). Study of these objects yields an age of the universe in the range of 10 to 20 billion years. The rough equality of this age with τ_c provides further strong confirmation of the "big bang" origin of the universe. However, the uncertainties in both are too great to allow us to conclude that the universe is open or that it is closed.

A final clue as to whether the universe is open or closed is provided by the cosmic abundance of deuterium. As mentioned above, if the density of baryons in the early universe was sufficient to give a baryon density in the present universe greater than the critical density $\rho_c = 3H^2/8\pi G$, then very little deuterium should remain after the era of nucleosynthesis. However, measurements of the cosmic abundance of deuterium have yielded mass fractions of $\sim 2 \times 10^{-5}$, corresponding to a present value of Ω of ~ 0.1. This provides evidence that the universe is open, but significant loopholes remain, such as the possibility that deuterium was manufactured by some other process, or that most of the present mass density of the universe is in non-baryonic matter (e.g., a neutrino species with mass $\sim$100 eV).

In summary, many important issues in cosmology remain to be resolved. But it is already clear that general relativity has provided us with a remarkably successful picture of the spacetime structure of our universe.

Problems

1. Show that the Robertson-Walker metric, equation (5.1.11), can be expressed in the form

$$ds^2 = -d\tau^2 + a^2(\tau)\left[\frac{dr^2}{1 - kr^2} + r^2(d\theta^2 + \sin^2\theta\, d\phi^2)\right] \quad .$$

What portion of the 3-sphere ($k = +1$) is covered by these coordinates?

2. Derive Einstein's equations, (5.2.14) and (5.2.15), for the 3-sphere ($k = +1$) and hyperboloid ($k = -1$) cases.

3. *a*) Consider the modified Einstein's equation (5.2.17) with cosmological constant Λ. Write out the analogs of equations (5.2.14) and (5.2.15) with the Λ-terms. Show that static solutions of these equations are possible if and only if $k = +1$ (3-sphere) and $\Lambda > 0$. (These solutions are called *Einstein static universes*.) For a dust filled Einstein universe ($P = 0$), relate the "radius of the universe" a to the density ρ. Examine small perturbations from this "equilibrium" value of a, and show that the Einstein static universe is unstable.

b) Consider the modified Einstein equation with $\Lambda > 0$ and $T_{ab} = 0$. Obtain the spatially homogeneous, isotropic solution in the case $k = 0$. (The resulting spacetime actually is spacetime homogeneous and isotropic and is known as the *de Sitter spacetime*. The solutions with $k = \pm 1$ correspond to different choices of spacelike hypersurfaces in this spacetime. See Hawking and Ellis 1973 for further details.)

4. Derive the cosmological redshift formula (5.3.6) by the following argument:

a) Show that $\nabla_a u_b = (\dot{a}/a)h_{ab}$, where h_{ab} is defined by equation (5.1.10) and $\dot{a} = da/d\tau$.

b) Show that along any null geodesic we have $d\omega/d\lambda = -k^a k^b \nabla_a u_b = -(\dot{a}/a)\omega^2$, where λ is an affine parameter along the geodesic.

c) Show that the result of (*b*) yields equation (5.3.6).

5. Consider a radial ($d\theta/d\lambda = d\phi/d\lambda = 0$) null geodesic propagating in a Robertson-Walker cosmology, equation (5.1.11).

a) Show that for all three spatial geometries the change in the coordinate ψ of the ray between times τ_1 and τ_2 is $\Delta\psi = \int_{\tau_1}^{\tau_2} d\tau/a(\tau)$. [Here, in the flat case ($k = 0$), ψ is defined to be the ordinary radial coordinate, $\psi = (x^2 + y^2 + z^2)^{1/2}$.]

b) Show that in the dust filled spherical model, a light ray emitted at the big bang travels precisely all the way around the universe by the time of the "big crunch."

c) Show that in the radiation filled spherical model, a light ray emitted at the big bang travels precisely halfway around the universe by the time of the "big crunch."

SIX

THE SCHWARZSCHILD SOLUTION

As described in the previous chapter, the theory of general relativity has made a number of strikingly successful predictions concerning the spacetime structure of our universe. However, cosmological observations presently are not good enough to provide stringent quantitative tests of general relativity. Such quantitative tests are provided by the gravitational fields occurring in our solar system, where precise measurements can be made. Thus, it is of great interest to determine the solution of Einstein's equation corresponding to the exterior gravitational field of a static, spherically symmetric body (such as are our Sun and many other bodies, to an excellent approximation). This problem was solved by Karl Schwarzschild (1916*a*), only a few months after Einstein published his vacuum field equations. The Schwarzschild solution without question remains one of the most important known exact solutions of Einstein's equation.

As was previously discussed in section 4.4*a*, in the slow motion, weak field limit, the predicitons of general relativity reduce to those of Newtonian theory. However, the Schwarzschild solution, which describes the exact exterior field of a spherical body, predicts tiny departures from Newtonian theory for the motion of planets in our solar system, and, in addition, predicts the "bending of light," the gravitational redshift of light, and "time delay" effects. These four predictions have been accurately confirmed by precise measurements. Indeed, except for the binary pulsar measurements (see the end of chapter 4), the predictions of the Schwarzschild solution in the weak field regime of our solar system are the only predictions of general relativity to have been tested in a quantitatively precise manner.

But the Schwarzschild solution has provided us with a great deal more than the ability to predict tiny effects occurring in our solar system. As will be discussed further in section 6.2, sufficiently massive bodies are unable to support themselves against complete gravitational collapse. After the collapse of a spherical body has occurred, the entire spacetime geometry will be described by the Schwarzschild solution, and thus the Schwarzschild solution tells us a great deal about the strong field behavior of general relativity. As will be seen in section 6.4, the vacuum Schwarzschild solution describing the end-product of gravitational collapse contains a spacetime singularity which is hidden within a black hole.

We shall derive the Schwarzschild solution in section 6.1, using the tetrad method of section 3.4*b*. In section 6.2 we consider the interior matter sources of the exterior vacuum Schwarzschild solution and thereby derive the relativistic equations of stellar structure. The timelike and null geodesics of the Schwarzschild metric are studied in section 6.3 and are used there to make predictions for the four tests of general relativity: the gravitational redshift, the precession of Mercury's orbit, the bending of light, and the "time delay" of light. Finally, in section 6.4 we examine the strong field regime of the vacuum Schwarzschild solution.

6.1 Derivation of the Schwarzschild Solution

We seek all solutions of Einstein's equation which describe the exterior gravitational field of a static, spherically symmetric body. Thus, we wish to find all four-dimensional Lorentz signature metrics whose Ricci tensor vanishes and which are static and possess spherical symmetry. Our first task is to define more precisely the meaning of the terms "static" and "spherically symmetric" and to choose a convenient coordinate system for analyzing this class of spacetimes. It will be assumed in the discussion below that the reader has read (or will refer to) appendix C.

A spacetime is said to be *stationary* if there exists a one-parameter group of isometries, ϕ_t, whose orbits are timelike curves. This groups of isometries expresses the "time translation symmetry" of the spacetime. Equivalently, a stationary spacetime is one which possesses a timelike Killing vector field, ξ^a. A spacetime is said to be *static* if it is stationary and if, in addition, there exists a (spacelike) hypersurface Σ which is orthogonal to the orbits of the isometry. By Frobenius's theorem (see appendix B) this is equivalent to the requirement that the timelike Killing vector field ξ^a satisfy

$$\xi_{[a} \nabla_b \xi_{c]} = 0 \quad . \tag{6.1.1}$$

The meaning of this extra condition of hypersurface orthogonality perhaps can be seen best by introducing convenient coordinates for static spacetimes as follows. If $\xi^a \neq 0$ everywhere on Σ, then in a neighborhood of Σ, every point will lie on a unique orbit of ξ^a which passes through Σ. Assuming $\xi^a \neq 0$, we choose arbitrary coordinates $\{x^\mu\}$ on Σ and label each point p in this neighborhood of Σ by the parameter, t, of the orbit which starts from Σ and ends at p, and the coordinates, x^1, x^2, x^3, of the orbit at Σ. Since this coordinate system employs the Killing parameter, t, as one of the coordinates, the metric components in this coordinate basis will be independent of t. Furthermore, since the surface Σ_t—defined as the set of points whose "time coordinate" has the value t—is the image of Σ under the isometry ϕ_t, it follows that each Σ_t is also orthogonal to ξ^a. Thus, in these coordinates, the metric components take the form

$$ds^2 = -V^2(x^1, x^2, x^3)\, dt^2 + \sum_{\mu, \nu=1}^{3} h_{\mu\nu}(x^1, x^2, x^3) dx^\mu dx^\nu \quad , \tag{6.1.2}$$

where $V^2 = -\xi_a \xi^a$, and the absence of $dt\, dx^\mu$ cross terms expresses the orthogonality of ξ^a with Σ. A stationary but nonstatic metric unavoidably must have $dt\, dx^\mu$ cross terms in any coordinate system which uses the Killing parameter as one of the coordinates.

From the explicit form of a static metric, equation (6.1.2), it can be seen that the diffeomorphism defined by $t \to -t$ (i.e., the map which takes each point on each Σ_t to the point with the same spatial coordinates on Σ_{-t}), is an isometry. Thus, in addition to the "time translation" symmetry, $t \to t +$ constant, possessed by all stationary spacetimes, the static spacetimes also possess a "time reflection" symmetry. Physically, fields which are time translation invariant can fail to be time reflection invariant when rotational motion is involved, since the time reflection will change the direction of rotation and thus will not restore one to the original configuration. Thus, for example, a rotating fluid ball may have a time-independent matter and velocity distribution, but does not possess a time reflection symmetry. In the case considered here, the failure of equation (6.1.1) to hold implies that neighboring orbits of ξ^a "twist" around each other. This helps explain why equation (6.1.1) is the necessary and sufficient condition for the existence of a time reflection symmetry.

A spacetime is said to be *spherically symmetric* if its isometry group contains a subgroup isomorphic to the group SO(3), and the orbits of this subgroup (i.e., the collection of points resulting from the action of the subgroup on a given point) are two-dimensional spheres. The SO(3) isometries may then be interpreted physically as rotations, and thus a spherically symmetric spacetime is one whose metric remains invariant under rotations. The spacetime metric induces a metric on each orbit 2-sphere, which, because of the rotational symmetry, must be a multiple of the metric of a unit 2-sphere, and thus can be completely characterized by the total area, A, of the 2-sphere. In spherically symmetric spacetimes it is convenient to introduce the function r, defined by

$$r = (A/4\pi)^{1/2} \quad . \tag{6.1.3}$$

Thus, in spherical coordinates (θ, ϕ), the metric on each orbit 2-sphere takes the form

$$ds^2 = r^2(d\theta^2 + \sin^2\theta \, d\phi^2) \quad . \tag{6.1.4}$$

In flat, three-dimensional Euclidean space, r would also be the value of the radius of the sphere, i.e., the distance from the surface of the sphere to its center. However, in a curved space, a sphere need not have a center (e.g., the manifold structure could be, say, $\mathbb{R} \times S^2$, rather than $\mathbb{R}^3$); and even if it does, r need not bear any relation to the distance to the center. Nevertheless, we shall refer to r as the "radial coordinate" of the sphere.

If a spacetime is both static and spherically symmetric, and if the static Killing field ξ^a is unique (as we shall assume), then ξ^a must be orthogonal to the orbit 2-spheres. Namely, if ξ^a is unique, it must be invariant under all the rotational isometries. However, this requires its projection onto any orbit sphere to vanish, since a nonvanishing vector field on a sphere cannot be invariant under all rotations. Thus the orbit spheres lie wholly within the hypersurfaces, Σ_t, orthogonal to ξ^a. Convenient coordinates on the spacetime may be chosen as follows: We select a sphere on $\Sigma = \Sigma_0$ and choose spherical coordinates (θ, ϕ) on it. We "carry" these spherical coordinates to the other spheres of Σ by means of geodesics orthogonal to the two-sphere (similar to the manner in which we used the isotropic observers to "carry" coordinate labels to other hypersurfaces in section 5.1). Provided that $\nabla_a r \neq 0$, we choose (r, θ, ϕ) as coordinates in Σ_t and, finally, we choose (t, r, θ, ϕ) as coordinates for the spacetime according to the prescription described above equation (6.1.2). In these coordinates, the metric of an arbitrary static, spherically symmetric spacetime[1] takes the simple form

$$ds^2 = -f(r)dt^2 + h(r)dr^2 + r^2(d\theta^2 + \sin^2\theta \, d\phi^2) \quad . \tag{6.1.5}$$

It should be kept in mind, however, that, in addition to the well understood breakdown of the spherical coordinates at the north and south poles of the spheres, this coordinate system breaks down at points where $\xi^a = 0$ or $\nabla_a r = 0$ (or, more generally, where ξ^a and $\nabla^a r$ become collinear). As we shall see in section 6.4, this occurs in the strong field region of the Schwarzschild solution.

Most of the work required to obtain the static, spherically symmetric vacuum solutions of Einstein's equation has now been completed, as we have reduced the general problem of determining 10 unknown functions (the metric components, $g_{\mu\nu}$) of four variables (the four coordinates) to determining two functions (f and h) of one variable (r). The remaining tasks are to compute the Ricci tensor, R_{ab}, of the metric, equation (6.1.5), and solve the equation $R_{ab} = 0$ for f and h. We shall calculate R_{ab} using the second tetrad approach of section 3.4*b*.

A convenient orthonormal basis for the metric of equation (6.1.5) is

$$(e_0)_a = f^{1/2}(dt)_a \quad , \tag{6.1.6a}$$

$$(e_1)_a = h^{1/2}(dr)_a \quad , \tag{6.1.6b}$$

$$(e_2)_a = r(d\theta)_a \quad , \tag{6.1.6c}$$

$$(e_3)_a = r \sin\theta(d\phi)_a \quad . \tag{6.1.6d}$$

Using the ordinary derivative, ∂_a, of our coordinate system, we find

$$\partial_{[a}(e_0)_{b]} = \frac{1}{2} f^{-1/2} f'(dr)_{[a}(dt)_{b]} \quad , \tag{6.1.7}$$

$$\partial_{[a}(e_1)_{b]} = 0 \quad , \tag{6.1.8}$$

$$\partial_{[a}(e_2)_{b]} = (dr)_{[a}(d\theta)_{b]} \quad , \tag{6.1.9}$$

$$\partial_{[a}(e_3)_{b]} = \sin\theta(dr)_{[a}(d\phi)_{b]} + r\cos\theta(d\theta)_{[a}(d\phi)_{b]} \quad , \tag{6.1.10}$$

where $f' = df/dr$. Thus, according to equation (3.4.25) (or equivalently, eq. [3.4.27]), we must solve the following equations for the connection one-forms $\omega_{a\mu\nu} = -\omega_{a\nu\mu}$:

1. It is worth pointing out that every stationary, spherically symmetric spacetime must be static. This is because, as in the static case, the stationary Killing field ξ^a must be orthogonal to the orbit spheres and, in addition, $\xi^a \nabla_a r = 0$ since r is a geometrical quantity. Hence, ξ^a must be orthogonal to the hypersurface, Σ, generated by the integral curves of $\nabla^a r$ starting from an orbit sphere. (If $\nabla_a r = 0$, this argument breaks down, but the conclusion of hypersurface orthogonality of ξ^a remains valid.)

$$\frac{1}{2}f^{-1/2}f'(dr)_{[a}(dt)_{b]} = h^{1/2}(dr)_{[a}\omega_{b]01} + r(d\theta)_{[a}\omega_{b]02} + r\sin\theta(d\phi)_{[a}\omega_{b]03} \quad , \tag{6.1.11}$$

$$0 = f^{1/2}(dt)_{[a}\omega_{b]01} + r(d\theta)_{[a}\omega_{b]12} + r\sin\theta(d\phi)_{[a}\omega_{b]13} \quad , \tag{6.1.12}$$

$$(dr)_{[a}(d\theta)_{b]} = -f^{1/2}(dt)_{[a}\omega_{b]20} + h^{1/2}(dr)_{[a}\omega_{b]21} + r\sin\theta(d\phi)_{[a}\omega_{b]23} \quad , \tag{6.1.13}$$

$$\sin\theta(dr)_{[a}(d\phi)_{b]} + r\cos\theta(d\theta)_{[a}(d\phi)_{b]} = -f^{1/2}(dt)_{[a}\omega_{b]30} + h^{1/2}(dr)_{[a}\omega_{b]31} + r(d\theta)_{[a}\omega_{b]32} \quad . \tag{6.1.14}$$

A plausible guess at solving equation (6.1.11) is

$$\omega_{b02} = \omega_{b03} = 0 \quad , \tag{6.1.15}$$

$$\omega_{b01} = \frac{1}{2}\frac{f'}{(fh)^{1/2}}(dt)_b + \alpha_1(dr)_b \quad , \tag{6.1.16}$$

where the function α_1 is undetermined by equation (6.1.11). Substitution of this trial solution in equation (6.1.12) then requires $\alpha_1 = 0$. From equation (6.1.12), we might also guess $\omega_{b12} = \omega_{b13} = 0$, but this leads to inconsistency later, so we merely conclude that

$$\omega_{b12} = \alpha_2(d\theta)_b + \alpha_3(d\phi)_b \quad , \tag{6.1.17}$$

$$\omega_{b13} = \alpha_4(d\phi)_b + \frac{\alpha_3}{\sin\theta}(d\theta)_b \quad . \tag{6.1.18}$$

Substituting our trial solution in equation (6.1.13), we find

$$\alpha_2 = -h^{-1/2} \quad , \tag{6.1.19}$$

$$\omega_{b23} = -\frac{h^{1/2}}{r\sin\theta}\alpha_3(dr)_b + \alpha_5(d\phi)_b \quad . \tag{6.1.20}$$

Finally, substitution into equation (6.1.14) yields

$$\alpha_3 = 0 \quad , \tag{6.1.21}$$

$$\alpha_4 = -h^{-1/2}\sin\theta \quad , \tag{6.1.22}$$

$$\alpha_5 = -\cos\theta \quad . \tag{6.1.23}$$

Since we have found no inconsistency, this means that our trial solution (generated by the initial guess $\omega_{b02} = \omega_{b03} = 0$) is, in fact, *the* solution. Thus, we have found

$$\omega_{b02} = \omega_{b03} = 0 \quad , \tag{6.1.24}$$

$$\omega_{b01} = \frac{f'}{2(fh)^{1/2}}(dt)_b \quad , \tag{6.1.25}$$

$$\omega_{b12} = -h^{-1/2}(d\theta)_b \quad , \tag{6.1.26}$$

$$\omega_{b13} = -h^{-1/2}\sin\theta(d\phi)_b \quad , \tag{6.1.27}$$

$$\omega_{b23} = -\cos\theta(d\phi)_b \quad . \tag{6.1.28}$$

From equation (3.4.20) (or, equivalently, eq. [3.4.28]) we obtain the Riemann tensor with remarkably little total computation compared with other approaches (see problem 2),

$$R_{ab01} = -R_{ab10} = \frac{d}{dr}[(fh)^{-1/2}f'](dr)_{[a}(dt)_{b]} \quad , \tag{6.1.29}$$

$$R_{ab02} = -R_{ab20} = f^{-1/2}h^{-1}f'(d\theta)_{[a}(dt)_{b]} \quad , \tag{6.1.30}$$

$$R_{ab03} = -R_{ab30} = f^{-1/2}h^{-1}f' \sin\theta(d\phi)_{[a}(dt)_{b]} \quad , \tag{6.1.31}$$

$$R_{ab12} = -R_{ab21} = h^{-3/2}h'(dr)_{[a}(d\theta)_{b]} \quad , \tag{6.1.32}$$

$$R_{ab13} = -R_{ab31} = \sin\theta\, h^{-3/2}h'(dr)_{[a}(d\phi)_{b]} \quad , \tag{6.1.33}$$

$$R_{ab23} = -R_{ab32} = 2(1 - h^{-1}) \sin\theta(d\theta)_{[a}(d\phi)_{b]} \quad . \tag{6.1.34}$$

The Ricci tensor is easily computed from the Riemann tensor via equation (3.4.22). Equating it to zero, we obtain the vacuum Einstein equation for a static, spherically symmetric spacetime,

$$\begin{aligned} 0 = R_{00} &= R_{010}{}^{1} + R_{020}{}^{2} + R_{030}{}^{3} \\ &= \frac{1}{2}(fh)^{-1/2}\frac{d}{dr}[(fh)^{-1/2}f'] + (rfh)^{-1}f' \quad , \end{aligned} \tag{6.1.35}$$

$$0 = R_{11} = -\frac{1}{2}(fh)^{-1/2}\frac{d}{dr}[(fh)^{-1/2}f'] + (rh^2)^{-1}h' \quad , \tag{6.1.36}$$

$$0 = R_{22} = -\frac{1}{2}(rfh)^{-1}f' + \frac{1}{2}(rh^2)^{-1}h' + r^{-2}(1 - h^{-1}) \quad , \tag{6.1.37}$$

where $R_{\mu\nu} \equiv R_{ab}(e_\mu)^a(e_\nu)^b$. We also find that $R_{33} = R_{22}$ and that the off-diagonal components of $R_{\mu\nu}$ vanish identically, as could be predicted from symmetry arguments similar to those used in section 5.2.

Adding equations (6.1.35) and (6.1.36), we obtain

$$f'/f + h'/h = 0 \quad , \tag{6.1.38}$$

which implies

$$f = Kh^{-1} \quad , \tag{6.1.39}$$

where K is a constant. By re-scaling the time coordinate, $t \to K^{1/2}t$, we may set $K = 1$. Equation (6.1.37) now yields

$$-f' + \frac{1-f}{r} = 0 \quad , \tag{6.1.40}$$

i.e.,

$$\frac{d}{dr}(rf) = 1 \quad , \tag{6.1.41}$$

which implies

$$f = 1 + C/r \quad , \tag{6.1.42}$$

where C is a constant. Equation (6.1.42) together with equation (6.1.39) (with $K = 1$) solves equations (6.1.35)–(6.1.37), and thus we have found the general solution, first discovered by Schwarzschild, of the vacuum Einstein equation for static, spherically symmetric spacetimes,

$$ds^2 = -\left(1 + \frac{C}{r}\right)dt^2 + \left(1 + \frac{C}{r}\right)^{-1} dr^2 + r^2 d\Omega^2 \quad , \tag{6.1.43}$$

where $d\Omega^2$ is shorthand for $(d\theta^2 + \sin^2\theta\, d\phi^2)$.

Perhaps the first point to notice about the Schwarzschild solution is that it is asymptotically flat,[2] i.e., as $r \to \infty$, the metric components approach those of Minkowski spacetime in spherical coordinates. This allows us to interpret the Schwarzschild metric as the exterior gravitational field of an isolated body. We may interpret the constant C by comparing the behavior of a test body in the weak field regime $(r \to \infty)$ with the behavior of a test body in the Newtonian theory of gravity. Examination of the geodesics of the Schwarzschild metric (see section 6.3)—or, equivalently, direct comparison of the Schwarzschild metric with the Newtonian metric discussed in section 4.4*a* (which, however, is not expressed in a convenient gauge to facilitate the comparison)—shows that for large r, the behavior of a test body in the Schwarzschild solution with parameter C agrees with the behavior of a test body in a Newtonian gravitational field of mass $M = -C/2$ (i.e., restoring the G's and c's, $GM/c^2 = -C/2$). Thus, we interpret $-C/2$ as the total mass[3] of the Schwarzschild field, and we may write the Schwarzschild metric in its final form,

$$ds^2 = -\left(1 - \frac{2M}{r}\right)dt^2 + \left(1 - \frac{2M}{r}\right)^{-1} dr^2 + r^2 d\Omega^2 \quad . \tag{6.1.44}$$

A striking feature of the Schwarzschild solution is that the metric components become singular in the strong field regime at both $r = 2M$ and $r = 0$. This singular behavior of the components could be due to either (i) a breakdown of the coordinates used to obtain the general form of the metric, equation (6.1.5), because $\xi^a = 0$ or $\nabla_a r = 0$ (or ξ^a and $\nabla^a r$ become collinear) or (ii) a true singularity of the spacetime structure. We shall see in section 6.4 that the "singularity" at $r = 2M$ is caused merely by a breakdown of the coordinates, while the singularity at $r = 0$ is a true, physical singularity. However, we note here that the "singularity" at $r = 2M$ occurs at a numerical value of the radial coordinate given by

$$r_S = \frac{2GM}{c^2} \approx 3\left(\frac{M}{M_\odot}\right) \text{km} \quad , \tag{6.1.45}$$

2. A precise, general notion of asymptotic flatness will be given in chapter 11.
3. A general definition of total mass for asymptotically flat solutions (based on this idea of examining the distant gravitational field) will be given in chapter 11.

where $M_\odot \approx 2 \times 10^{33}$ g is the mass of the Sun. Thus, for an "ordinary body," such as the Sun or the Earth, the Schwarzschild radius, r_S, is well inside the radius of the body where, of course, the vacuum Schwarzschild solution is no longer valid. Indeed, the coordinate and physical singularities respectively at $r = r_S$ and $r = 0$ are relevant only for bodies which have undergone complete gravitational collapse.

Finally, we mention that the vacuum Einstein equation also can be solved for a general spherically symmetric spacetime, without the assumption of staticity. As was first shown by Birkhoff (1923), the Schwarzschild solution remains the only solution of this more general system of equations (see, e.g., Hawking and Ellis 1973 for a proof). Thus, all spherically symmetric spacetimes with $R_{ab} = 0$ are static. This result, known as *Birkhoff's theorem,* is closely analogous to the fact that the Coulomb solution is the only spherically symmetric solution of Maxwell's equations in vacuum. It can be interpreted as saying that in gravity, as in electromagnetism, there exists no monopole (i.e., spherically symmetric) radiation.

6.2 Interior Solutions

We turn our attention, now, to the static, spherically symmetric solutions of Einstein's equation with a perfect fluid stress-energy tensor,

$$T_{ab} = \rho u_a u_b + P(g_{ab} + u_a u_b) \quad . \tag{6.2.1}$$

In order to be compatible with the static symmetry of the spacetime the fluid 4-velocity, u^a, must point in the same direction as the static Killing vector field, ξ^a, i.e.,

$$u^a = -(e_0)^a = -f^{1/2}(dt)^a \quad , \tag{6.2.2}$$

where f is the function appearing in the general metric form equation (6.1.5). The solutions we seek describe the possible interior fluid sources of the exterior Schwarzschild metric, and thus our investigation will yield the equations of structure for static, fluid objects, such as stars.

Einstein's equation with a fluid present is obtained simply by adding the appropriate stress-energy terms to equations (6.1.35)–(6.1.37). We shall take the three independent equations in the form

$$\begin{aligned} 8\pi T_{00} = 8\pi\rho = G_{00} &= R_{00} + \frac{1}{2}(R_0{}^0 + R_1{}^1 + R_2{}^2 + R_3{}^3) \\ &= (rh^2)^{-1}h' + r^{-2}(1 - h^{-1}) \quad , \end{aligned} \tag{6.2.3}$$

$$\begin{aligned} 8\pi T_{11} = 8\pi P = G_{11} &= R_{11} - \frac{1}{2}(R_0{}^0 + R_1{}^1 + R_2{}^2 + R_3{}^3) \\ &= (rfh)^{-1}f' - r^{-2}(1 - h^{-1}) \quad , \end{aligned} \tag{6.2.4}$$

$$\begin{aligned} 8\pi T_{22} = 8\pi P = G_{22} &= \frac{1}{2}(fh)^{-1/2}\frac{d}{dr}[(fh)^{-1/2}f'] \\ &+ \frac{1}{2}(rfh)^{-1}f' - \frac{1}{2}(rh^2)^{-1}h' \quad . \end{aligned} \tag{6.2.5}$$

Equation (6.2.3) involves only h. (This could have been predicted from the general analysis of the structure of Einstein's equation given in chapter 10, which shows that the equation $G_{00} = 8\pi T_{00}$ is an "initial value constraint" which, in the static case, involves only the geometry of the spacelike hypersurface Σ orthogonal to ξ^a and thus cannot involve f.) It can be rewritten in the form

$$\frac{1}{r^2}\frac{d}{dr}[r(1 - h^{-1})] = 8\pi\rho \quad , \tag{6.2.6}$$

from which it follows immediately that the solution for h is

$$h(r) = \left[1 - \frac{2m(r)}{r}\right]^{-1} \quad , \tag{6.2.7}$$

where

$$m(r) = 4\pi \int_0^r \rho(r')r'^2 dr' + a \quad , \tag{6.2.8}$$

where a is a constant. Smoothness of the metric on Σ at $r = 0$ requires that as $r \to 0$ the area of spheres approach 4π times the square of their proper radius, i.e., that $h(r) \to 1$ as $r \to 0$. Thus, in order to avoid a "conical singularity" in the metric at $r = 0$, we must set $a = 0$. Since Σ must be spacelike for a static configuration, we see that a necessary condition for staticity is $h \geq 0$, i.e.,

$$r \geqq 2m(r) \quad . \tag{6.2.9}$$

If $\rho = 0$ for $r > R$, our solution for h, equation (6.2.7), joins on to the vacuum Schwarzschild solution with total mass

$$M = m(R) = 4\pi \int_0^R \rho(r)r^2 dr \quad . \tag{6.2.10}$$

Equation (6.2.10) is formally identical to the expression for total mass in Newtonian gravity. Note, however, that this formal analogy is misleading because the proper volume element on Σ (see appendix B) is $\sqrt{^{(3)}g}\, d^3x = h^{1/2}r^2 \sin\theta\, dr d\theta d\phi$ so that the total *proper mass* is

$$M_p = 4\pi \int_0^R \rho(r)r^2\left[1 - \frac{2m(r)}{r}\right]^{-1/2} dr \quad . \tag{6.2.11}$$

The difference between M and M_p can be interpreted as the gravitational binding energy, E_B, of the configuration,

$$E_B = M_p - M \quad , \tag{6.2.12}$$

which is always positive since $M_p > M$.

If we write

$$f = e^{2\phi} \quad , \tag{6.2.13}$$

equation (6.2.4) becomes

$$\frac{d\phi}{dr} = \frac{m(r) + 4\pi r^3 P}{r[r - 2m(r)]} \quad . \tag{6.2.14}$$

In the Newtonian limit, we have $r^3 P << m(r)$ and $m(r) << r$, so equation (6.2.14) reduces to

$$\frac{d\phi}{dr} \approx \frac{m(r)}{r^2} \quad , \tag{6.2.15}$$

which is simply the spherically symmetric version of Poisson's equation for the Newtonian gravitational potential. Thus, in the static spherically symmetric case we may view $\phi = \frac{1}{2} \ln f$ as the general relativistic analog of the Newtonian potential. For nonstationary configurations, however, there is no known analog in general relativity of the Newtonian potential.

If we substitute our results, equations (6.2.7) and (6.2.14), into our final equation (6.2.5), it is apparent that we will obtain an equation for dP/dr. However, the rather messy algebra required for doing this can be circumvented by noting that

$$\begin{aligned}(e_\mu)_b 8\pi \nabla_a T^{ab} &= (e_\mu)_b \nabla_a (8\pi T^{ab} - G^{ab}) \\ &= \nabla_a[(e_\mu)_b(8\pi T^{ab} - G^{ab})] - \omega_{ab\mu}(8\pi T^{ab} - G^{ab}) \quad ,\end{aligned} \tag{6.2.16}$$

where $\omega_{ab\mu} = \nabla_a(e_\mu)_b$. Setting $\mu = 1$, we find the first term on the right-hand side of equation (6.2.16) vanishes if equation (6.2.4) is satisfied. Thus, since $\omega_{221} \neq 0$, we see that (given eqs. [6.2.3] and [6.2.4]) the vanishing of $(8\pi T^{22} - G^{22})$ is equivalent to

$$(e_1)_b \nabla_a T^{ab} = 0 \quad , \tag{6.2.17}$$

and hence we may replace equation (6.2.5) by equation (6.2.17). We have already calculated this component of $\nabla_a T^{ab}$ for a perfect fluid in equation (4.3.8), and thus, without further work, we obtain [using $(e_1)_b = h^{1/2}(dr)_b = h^{-1/2}(\partial/\partial r)_b$],

$$h^{-1/2}\frac{dP}{dr} = -(P + \rho)(e_1)_b u^a \nabla_a u^b = -h^{-1/2}(P + \rho)\frac{d\phi}{dr} \quad , \tag{6.2.18}$$

where equations (6.2.2) and (6.2.13) were used for the last equality. Using equation (6.2.14), we may eliminate $d\phi/dr$ to obtain our final result,

$$\frac{dP}{dr} = -(P + \rho)\frac{m(r) + 4\pi r^3 P}{r[r - 2m(r)]} \quad . \tag{6.2.19}$$

Equation (6.2.19) is known as the *Tolman-Oppenheimer-Volkoff equation* of hydrostatic equilibrium. In the Newtonian limit ($P << \rho$, $m(r) << r$) it reduces to the Newtonian hydrostatic equilibrium equation.[4]

$$\frac{dP}{dr} \approx -\frac{\rho m(r)}{r^2} \quad . \tag{6.2.20}$$

4. It is interesting to note that although general relativity has little effect on the equilibrium configurations of stars with $P << \rho$ and $m(r) << r$, it still can have a significant influence on the stability of stars with equation of state $P = C\rho^\lambda$ with λ near the critical value of 4/3; see Chandrasekhar (1964).

In summary, we have found that the spacetime geometry inside a static, spherical fluid star is

$$ds^2 = -e^{2\phi}dt^2 + \left(1 - \frac{2m(r)}{r}\right)^{-1} dr^2 + r^2 d\Omega^2 \quad , \tag{6.2.21}$$

where

$$m(r) = 4\pi \int_0^r \rho(r')r'^2 dr' \tag{6.2.22}$$

and ϕ is determined from equation (6.2.14). The necessary and sufficient condition for equilibrium is that equation (6.2.19) be satisfied.

Thus, for fluid matter with a given equation of state, $P = P(\rho)$, equilibrium configurations can be determined as follows: We arbitrarily prescribe a central density ρ_c, and hence a central pressure $P_c = P(\rho_c)$. Then we integrate equations (6.2.19) and (6.2.22) outward until we reach the surface of the star, $P = \rho = 0$, at which point we join the solution onto the vacuum Schwarzschild solution, equation (6.1.44). Finally, we solve for ϕ by integrating equation (6.2.14), subject to the boundary condition $\phi \rightarrow 0$ as $r \rightarrow \infty$ (or, equivalently, by requiring ϕ to match onto the Schwarzschild value at the surface of the star). This procedure differs from that used in Newtonian theory only in that equation (6.2.19) has replaced equation (6.2.20), and equation (6.2.14) has replaced equation (6.2.15).

The most important difference between equilibrium configurations in general relativity and Newtonian gravity can be seen from the fact that (assuming $P \geq 0$) for a given density profile $\rho(r) \geqq 0$, the right-hand side of the relativistic hydrostatic equilibrium equation (6.2.19) is always larger in magnitude than the right-hand side of the Newtonian equation (6.2.20). This means that for a given $\rho(r)$, the central pressure, P_c, required for equilibrium is always higher in general relativity than in Newtonian theory; i.e., it is harder to maintain equilibrium in general relativity. This is dramatically illustrated by consideration of uniform density configurations, corresponding to an incompressible fluid of density ρ_0,

$$\rho(r) = \begin{cases} \rho_0 & (r \leqq R) \\ 0 & (r > R) \end{cases} \quad , \tag{6.2.23}$$

and hence (in both general relativity and Newtonian theory)

$$m(r) = \frac{4}{3}\pi r^3 \rho_0 \quad (r \leqq R) \quad . \tag{6.2.24}$$

The Newtonian equation of hydrostatic equilibrium (6.2.20) is easily integrated to yield (for $r \leqq R$),

$$P(r) = \frac{2}{3}\pi\rho_0^2(R^2 - r^2) \quad , \tag{6.2.25}$$

where the boundary condition $P(R) = 0$ has been imposed. Thus, the central pressure of a Newtonian uniform density star is

$$P_c = \frac{2}{3}\pi\rho_0^2 R^2 = \left(\frac{\pi}{6}\right)^{1/3} M^{2/3}\rho_0^{4/3} \quad , \tag{6.2.26}$$

which is finite for all values of ρ_0 and R, i.e., equilibrium can be achieved with sufficiently large pressures for all ρ_0 and R. On the other hand, the general relativistic equation of hydrostatic equilibrium (6.2.19) also can be integrated exactly, as was first done by Schwarzschild (1916*b*), yielding

$$P(r) = \rho_0\left[\frac{(1 - 2M/R)^{1/2} - (1 - 2Mr^2/R^3)^{1/2}}{(1 - 2Mr^2/R^3)^{1/2} - 3(1 - 2M/R)^{1/2}}\right] \quad . \tag{6.2.27}$$

Thus, the central pressure required for equilibrium of a uniform density star in general relativity is

$$P_c = \rho_0\left[\frac{1 - (1 - 2M/R)^{1/2}}{3(1 - 2M/R)^{1/2} - 1}\right] \quad . \tag{6.2.28}$$

For $R >> M$, equation (6.2.28) reduces to the Newtonian value, equation (6.2.26). However, now P_c becomes infinite when

$$3(1 - 2M/R)^{1/2} = 1 \quad , \tag{6.2.29}$$

i.e., when

$$R = \frac{9}{4}M \quad . \tag{6.2.30}$$

Thus, in general relativity, uniform density stars with $M > 4R/9$ simply cannot exist. Another way of stating this result is that the maximum possible mass of a star of uniform density ρ_0 is

$$M_{\max} = \frac{4}{9(3\pi)^{1/2}}\rho_0^{-1/2} \quad . \tag{6.2.31}$$

The existence of upper mass limits in general relativity is not an artifact of having restricted consideration to stars of uniform density. In fact, if we assume[5] only that the density, $\rho(r)$, is nonnegative and is a monotone decreasing function of r, $d\rho/dr \leqq 0$, then we can derive the following two types of upper mass limits for static, spherical stars in general relativity: (i) For stars of fixed radius R, the maximum possible mass is given by the uniform density value, $M_{\max} = 4R/9$. (ii) For a fixed equation of state below a density ρ_0 (which is physically realistic at low densities), an upper mass limit exists independent of the equation of state at higher densities than ρ_0. (The value of this upper mass limit depends, of course, on the value of ρ_0 and the equation of state assumed to hold below ρ_0.) We now shall derive the

5. In fact, the assumption that $\rho \geqq 0$ follows from the monotone decrease assumption, $d\rho/dr \leqq 0$, since the interior solution must eventually match onto the exterior Schwarzschild solution with $\rho = 0$. Conversely, the assumption that $d\rho/dr \leqq 0$ follows from $\rho \geqq 0$ if we assume, in addition, that the fluid has an equation of state, $P = P(\rho)$, with $P \geqq 0$ and $dP/d\rho \geqq 0$, since equation (6.2.19) implies $dP/dr \leqq 0$.

upper mass limit (i), and then indicate how the upper mass limit (ii) is obtained.

The existence of an upper mass limit at fixed star radius R already follows from the condition $h \geqq 0$, which, as seen from equation (6.2.9), implies that a necessary condition for staticity is

$$M \leqq R/2 \quad . \tag{6.2.32}$$

We can sharpen this limit to $M \leqq 4R/9$ using the condition $f \geqq 0$ which states that the Killing field ξ^a is timelike everywhere. To do so—assuming only that $\rho \geqq 0$ and $d\rho/dr \leqq 0$ but without any assumptions whatsoever about P—we must examine the two independent equations which do not involve P from the basic set (6.2.3)–(6.2.5). We already have solved equation (6.2.3) for h, so the remaining equation is obtained by taking the difference of equations (6.2.4) and (6.2.5),

$$0 = G_{11} - G_{22} = \frac{1}{2}(rfh)^{-1}f' - r^{-2}(1 - h^{-1}) + \frac{1}{2}(rh^2)^{-1}h' \\ - \frac{1}{2}(fh)^{-1/2}\frac{d}{dr}[(fh)^{-1/2}f'] \quad . \tag{6.2.33}$$

Substituting our solution for h, equation (6.2.7), in the second and third terms, and performing a few algebraic manipulations, we find

$$\frac{d}{dr}\left[r^{-1}h^{-1/2}\frac{df^{1/2}}{dr}\right] = (fh)^{1/2}\frac{d}{dr}\left[\frac{m(r)}{r^3}\right] \quad . \tag{6.2.34}$$

Since $d\rho/dr \leqq 0$, the average density, which is proportional to $m(r)/r^3$, also must decrease monotonically with r, so we have

$$\frac{d}{dr}\left[r^{-1}h^{-1/2}\frac{df^{1/2}}{dr}\right] \leqq 0 \quad . \tag{6.2.35}$$

Integrating this inequality *inward* from the surface, R, of the star to radius r, we obtain

$$\frac{1}{rh^{1/2}(r)}\frac{df^{1/2}}{dr}(r) \geqq \frac{1}{Rh^{1/2}(R)}\frac{df^{1/2}}{dr}(R) \\ = \frac{(1 - 2M/R)^{1/2}}{R}\frac{d}{dr}\left(1 - \frac{2M}{r}\right)^{1/2}\bigg|_{r=R} = M/R^3 \quad , \tag{6.2.36}$$

where we have used the fact that our interior solution must join on smoothly[6] to the Schwarzschild solution, equation (6.1.44), and hence we have calculated the derivative of $f^{1/2}$ at $r = R$ using the exterior Schwarzschild metric. Multiplying equation (6.2.36) by $rh^{1/2}$ and integrating inward again from R to 0, we obtain

$$f^{1/2}(0) \leqq (1 - 2M/R)^{1/2} - \frac{M}{R^3}\int_0^R \left[1 - \frac{2m(r)}{r}\right]^{-1/2} r\,dr \quad , \tag{6.2.37}$$

6. As long as ρ and P are continuous at $r = R$, equations (6.2.3)–(6.2.5) imply that h and f are at least C^1 there.

where again we have used $f^{1/2}(R) = (1 - 2M/R)^{1/2}$ and we have also used the explicit solution, equation (6.2.7), for h. Now, the condition $d\rho/dr \leqq 0$ implies that $m(r)$ cannot be smaller than the value it would have for a uniform density star ($d\rho/dr = 0$),

$$m(r) \geqq Mr^3/R^3 \quad . \tag{6.2.38}$$

Hence, the second term on the right-hand side of equation (6.2.37) will be smallest in magnitude when equality holds in equation (6.2.38). Thus, we obtain

$$\begin{aligned} f^{1/2}(0) &\leqq (1 - 2M/R)^{1/2} - \frac{M}{R^3}\int_0^R \left(1 - \frac{2Mr^2}{R^3}\right)^{-1/2} rdr \\ &= \frac{3}{2}(1 - 2M/R)^{1/2} - \frac{1}{2} \quad . \end{aligned} \tag{6.2.39}$$

Thus, the condition $f^{1/2}(0) \geqq 0$ implies that a necessary condition for staticity is

$$(1 - 2M/R)^{1/2} \geqq \frac{1}{3} \quad , \tag{6.2.40}$$

i.e.,

$$M \leqq 4R/9 \quad , \tag{6.2.41}$$

which is the desired result. Again, we emphasize that absolutely no assumptions concerning the pressure, P, of the star entered into the derivation of equation (6.2.41).

Given the existence of an upper mass limit at fixed radius, the existence of an upper mass limit for a given equation of state below density ρ_0 (assuming this equation of state is physically reasonable and that $d\rho/dr \leqq 0$) should not be surprising. If the equation of state does not become very "stiff" (i.e., large P) at low densities (which is what we mean by "physically reasonable"), an upper mass limit will exist for stars whose density is everywhere less than ρ_0. Since $d\rho/dr \leqq 0$, stars whose density fails to be less than ρ_0 must consist of a "core" of mass m_0 and radius r_0 where $\rho \geqq \rho_0$, surrounded by an "envelope" where $\rho < \rho_0$. Given the equation of state for $\rho < \rho_0$, the total mass, M, is determined by the parameters m_0 and r_0. The function $M(m_0, r_0)$ will be continuous in m_0 and r_0. However, since the core density is at least ρ_0 everywhere, we have the lower mass limit for the core,

$$m_0 \geqq \frac{4}{3}\pi r_0^3 \rho_0 \quad . \tag{6.2.42}$$

On the other hand, from the above upper total mass limit, equation (6.2.41), we would expect an upper limit on the core mass m_0 for a given core radius r_0. Indeed, the same derivation as led to equation (6.2.41)—except that we use the core radius, r_0, rather than the surface of the star as the boundary, and we use equation (6.2.14) rather than matching to the Schwarzschild solution to evaluate $df^{1/2}/dr$ at r_0—yields the more stringent limit,

$$m_0 \leqq \frac{2}{9} r_0[1 - 6\pi r_0^2 P_0 + (1 + 6\pi r_0^2 P_0)^{1/2}] \quad , \tag{6.2.43}$$

where $P_0 = P(\rho_0)$ is the pressure at the core-envelope boundary. But equations (6.2.42) and (6.2.43) restrict m_0 and r_0 to a compact region of the m_0-r_0 plane. Thus, $M(m_0, r_0)$ is a continuous function defined on a compact set. Therefore, M is bounded!

It should be pointed out that upper mass limits for a given equation of state at *all* densities occur in Newtonian theory also. The important difference which occurs in general relativity is that one obtains a limit which is independent of the equation of state at sufficiently high densities. Since there is unavoidable uncertainty in our knowledge of the physically correct equation of state at very high densities, this independence is essential to a derivation of a firm upper mass limit.

For cold matter, the dominant contribution to the pressure at densities much less than nuclear density ($\sim 10^{14}$ g cm^{-3}) arises from electron degeneracy pressure. At "low" densities ($n \ll m_e{}^3c^3/\hbar^3 \sim 10^{31}$ cm^{-3}, where n is the number density of electrons and m_e is the mass of the electron) this source provides a pressure at $T = 0$ of

$$P = \frac{\hbar^2(3\pi^2)^{2/3}}{5m_e} n^{5/3} \quad , \tag{6.2.44}$$

whereas at high densities ($n \gg m_e{}^3c^3/\hbar^3$) we have

$$P = \frac{\hbar c(3\pi^2)^{1/3}}{4} n^{4/3} \quad . \tag{6.2.45}$$

At densities approaching nuclear density, cold matter will be mainly in the form of free neutrons, and a similar degeneracy pressure due to neutrons becomes important. The precise form obtained for the equation of state depends, of course, on assumptions concerning the form of matter at the density under consideration and the microscopic forces acting between the fundamental constituents. While these factors are well understood at low densities, significant uncertainties arise at nuclear density and above.

The mass, M, and radius, R, of equilibrium configurations of cold matter for central densities between $\sim 10^5$ g cm^{-3} and $\sim 10^{17}$ g cm^{-3} are shown in Figure 6.1. Near and above nuclear density, Figure 6.1 is only a sketch of the qualitative features found for physically reasonable equations of state, since the precise values of M and R in this region depend upon many assumptions. (For a tabulation of exact values in the nuclear density regime for a wide variety of equations of state, see Arnett and Bowers 1977.) Even well below nuclear density the predicted equilibrium configurations depend somewhat upon assumptions involving the composition of the star. Cold bodies on the segment AB of the curve comprise stable equilibrium configurations supported by electron degeneracy pressure. These bodies are known as *white dwarfs*. As first shown by Chandrasekhar (see Chandrasekhar 1939), the maximum mass of a white dwarf is given by[7]

7. We can see why an upper mass limit of this magnitude should exist by comparing the central pressure needed to hold up a uniform density Newtonian star, equation (6.2.26), with the high density limit of the electron degeneracy pressure, equation (6.2.45). Taking $\rho = \mu_N m_N n$, where m_N denotes the

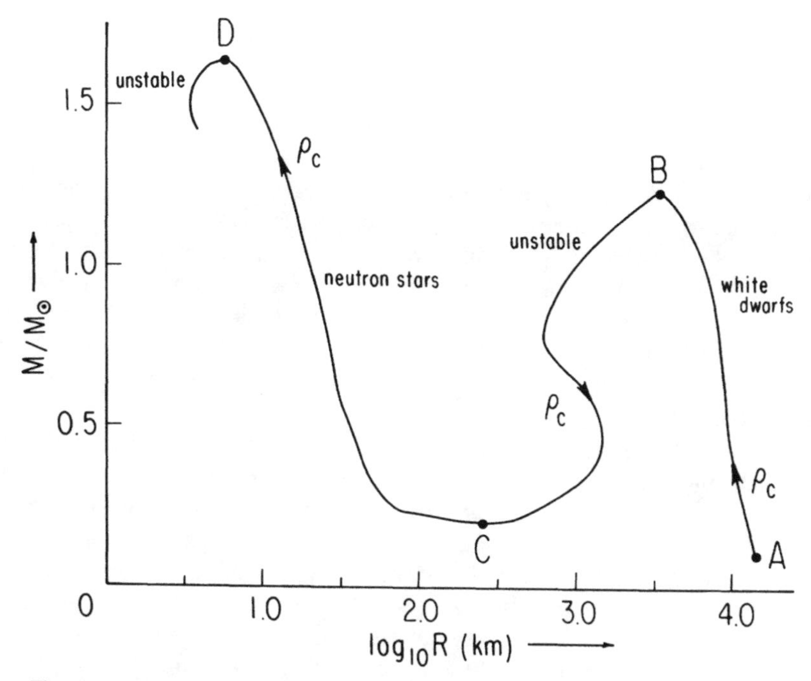

Fig. 6.1. The equilibrium configurations of cold matter. Given an equation of state, $P = P(\rho)$, an equilibrium configuration is uniquely determined by the value of the central density ρ_C. The masses and radii of these configurations are shown here for values of ρ_C ranging from $\sim 10^5$ g cm^{-3} at point A to $\sim 10^{17}$ g cm^{-3} beyond point D. In the white dwarf regime, the values of M and R depend somewhat upon the assumed composition of the star. In the neutron star regime, the values of M and R depend greatly upon assumptions about the state of matter as well as assumptions about the interactions between the fundamental constituents of matter. In this latter regime, the figure is only a rough sketch of the qualitative features found for most equations of state.

$$M_C \approx 1.4\left(\frac{2}{\mu_N}\right)^2 M_\odot \quad , \tag{6.2.46}$$

where μ_N is the number of nucleons per electron and $M_\odot$ denotes the mass of the Sun ($\sim 2 \times 10^{33}$ g). In Newtonian theory with a fixed value of μ_N, the mass monotonically increases with central density, approaching the limiting mass (6.2.46) for a configuration with $\rho_c \to \infty$ at radius $R \to 0$. In general relativity with a fixed value of μ_N, there is a "turnover" in the curve (i.e., the mass begins to decrease with ρ_c) at a finite value of ρ_c (Chandrasekhar and Tooper 1964), although the value of M_C is not significantly changed. However, the turnover at point B shown in Figure 6.1

---

mass of a nucleon, equating the two pressures, and restoring the G's and c's, we obtain the order of magnitude estimate

$$M_C \sim \left(\frac{\hbar c}{G}\right)^{3/2} \frac{1}{\mu_N{}^2 m_N{}^2} \sim \frac{M_\odot}{\mu_N{}^2}.$$

is due to the increase in μ_N at high density caused by the conversion of protons and electrons to neutrons. General arguments (see Sorkin 1981) show that an instability sets in at the "turnover" point B of the curve, making the configurations on the segment BC unstable. However, configurations on the segment CD are stable. Such bodies, called *neutron stars,* are composed mainly of neutrons and are supported largely by neutron degeneracy pressure. Since the density of matter in a neutron star is comparable to that of atomic nuclei, a neutron star is much like a huge, highly neutron-enriched nucleus (with atomic weight $\sim 10^{57}$) except that it is bound together by gravitation rather than by nuclear forces. On accout of uncertainties in the equation of state, the value of the maximum possible mass, $M(D)$, of a neutron star is considerably more uncertain than the white dwarf limit (6.2.46). However, most physically reasonable equations of state yield $M(D)$ below 2 $M_\odot$. Beyond point D on the curve, the configurations again become unstable.

If one applies the above upper mass limit argument to the equation of state for cold matter, choosing ρ_0 of the order of nuclear densities (so that there are few uncertainties in the equation of state below ρ_0), a firm upper mass limit of $\sim 5\ M_\odot$ is obtained (see Hartle 1978). If the general features of the equation of state are trusted somewhat beyond nuclear density, the above limit of $\sim 2\ M_\odot$ appears to be a realistic upper limit to the maximum possible mass of a spherical body composed of cold matter. We should emphasize, however, that our analysis applies only to the spherical case, and thus, for example, these upper mass limits need not hold for rapidly rotating bodies.

The existence of an upper mass limit for cold matter has extremely important consequences for the ultimate fate of stars. Ordinary stars are supported against collapse under their own weight by ideal gas pressure resulting from high temperature. This pressure is much higher than the pressure that could be produced by cold matter at comparable densities, and the above mentioned upper mass limits for cold stars do not apply. However, a hot star radiates energy from its surface, and unless this energy is replenished, hydrostatic equilibrium cannot be maintained. If stars had no source of energy, they would contract—very slowly at first, but eventually on dynamic time scales when instabilities set in—until such high densities were reached that the "cold matter" pressure would dominate the thermal contribution to the pressure. At that point, if the mass of the star were sufficiently small that stable equilibrium could be achieved with this cold matter pressure, the star would simply cool down, remaining in static equilibrium forever. On the other hand, if the mass of the star were greater than the cold matter upper mass limit, equilibrium never could be achieved, and the star would have to undergo complete gravitational collapse.

In fact, stars, of course, do have an internal source of energy: nuclear reactions. Consequently, they are able to stave off—typically for billions of years, though for much less time for very massive stars—their ultimate fate of either support by cold matter pressure or complete collapse. However, this does not alter the fundamental conclusion that stars with mass greater than the cold matter upper mass limit must eventually undergo complete gravitational collapse, unless, of course, they shed enough mass during the course of stellar evolution to fall below this limit.

When a star forms by condensation of a gas cloud, it contracts and heats up until the central temperature and density are sufficiently high that nuclear reactions converting hydrogen to helium occur there. The collapse of the star is then halted, and a long lasting equilibrium is maintained until finally a large core of helium is built up. If the star is sufficiently massive, contraction of this core then will take place until helium reactions begin to occur, resulting in the formation of heavier elements. This process may then repeat itself until a large core of the most stable nuclei, nickel and iron, is synthesized.

The crucial issue governing how far along this evolutionary sequence a star proceeds is whether the electron degeneracy pressure at any stage becomes sufficient to support the star. If the mass of the star is greater than M_C, equation (6.2.46), electron degeneracy pressure cannot support the star, and the large nickel and iron core will be built up (unless instabilities caused by explosive nuclear burning occur prior to this stage and literally blow the star apart). However, if the mass of the star is less than M_C, the contraction will be permanently halted at a prior stage. No further nuclear reactions will occur, and the star will simply cool down forever in a stable, white dwarf configuration.

If M is greater than M_C, then after a core of nickel and iron of mass $\sim M_C$ is built up, this core simply will not be able to support itself: Electron degeneracy pressure is not sufficient, and no further energy generating nuclear reactions can occur. Hence, the core will undergo gravitational collapse. By the time the density of the collapsing core has reached nuclear densities, neutron degeneracy pressure and nuclear forces provide a significant cold matter pressure. If the total mass of the collapsing part of the star is below the upper mass limit for cold matter ($\sim 2\, M_\odot$), the collapse may be halted, resulting in a neutron star. Confirmation of the existence of neutron stars has been provided by the discovery of pulsars—the astronomically observed objects which emit exactly reproduced signals with periods of fractions of a second. The only viable theoretical explanation of pulsars is that they are neutron stars which possess a "hot spot" rotating at the signal period which becomes oriented toward us during each revolution.

When the collapse of the core is halted or slowed down at nuclear densities, a shock wave will be produced and will propagate outward into the envelope of the star. It appears likely that this shock wave is responsible, in many cases, for blowing off the outer envelope of the star with an enormous release of energy, thus producing a *supernova*. The details of exactly how this occurs, are, however, not well understood at present. Nevertheless, the discovery of pulsars at the sites of the Crab and Vela supernova remnants has provided strong observational confirmation of the picture that supernovae are produced in conjunction with the collapse of the core of a star at the endpoint of stellar evolution.

Thus, in summary, if the mass of a star is sufficiently small, it can attain final equilibrium as a white dwarf or a neutron star. However, if the mass of the collapsing portion of a star is greater than the cold matter upper mass limit, equilibrium can never be achieved, and complete gravitational collapse will occur. As discussed at the end of section 6.4, the endpoint of such a collapse in the spherical case will be a Schwarzschild black hole.

6.3 Geodesics of Schwarzschild: Gravitational Redshift, Perihelion Precession, Bending of Light, and Time Delay of Radar Signals

In this section, we will analyze the behavior of test bodies and light rays in the exterior region ($r > 2M$) of the Schwarzschild solution by solving for the timelike and null geodesics of the Schwarzschild geometry. Our analysis of the geodesics in the strong field regime (r near $2M$) is physically relevant, of course, only for highly condensed stars or totally collapsed objects, but our results in the weak field regime ($r >> M$) apply to the exterior field of "ordinary bodies" such as the Sun.

It would involve a fair amount of labor to solve directly the geodesic equation in the form (3.3.5). Fortunately, almost all of this labor can be avoided by taking advantage of the symmetries of the Schwarzschild solution using proposition C.3.1 of appendix C: The inner product, $u^a \xi_a$, of a Killing field ξ^a with a geodesic tangent u^a is constant along the geodesic. As we shall see below, these constants of the motion enable us to reduce the problem of finding the geodesics to the problem of one-dimensional motion of a particle in an effective potential.

Proposition C.3.1 immediately allows us to derive a formula for the change between emitted and observed frequency of light signals sent between two static observers, i.e., for the gravitational redshift. (A similar derivation of the cosmological redshift in an expanding homogeneous and isotropic universe was given above in section 5.3a.) Consider two static observers (i.e., observers whose 4-velocity is tangent to the static Killing field ξ^a) O_1 and O_2 whose 4-velocities are u_1^a and u_2^a. Suppose O_1 emits a light signal at event P_1 which is received by O_2 at event P_2, as illustrated in Figure 6.2. As discussed in section 4.3., in the geometrical optics

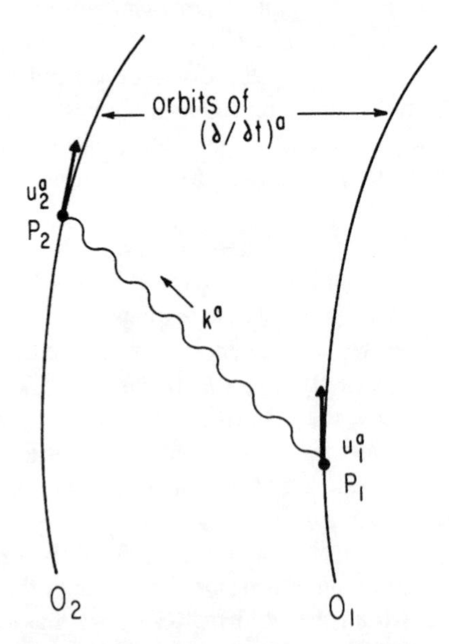

Fig. 6.2. A spacetime diagram showing a light signal sent from static observer O_1 to static observer O_2.

approximation this light signal travels on a null geodesic, whose tangent we denote by k^a. The frequency of emission is

$$\omega_1 = -(k_a u_1^a)\,|_{P_1} \quad , \tag{6.3.1}$$

while the frequency measured by the observer receiving the signal is

$$\omega_2 = -(k_a u_2^a)\,|_{P_2} \quad . \tag{6.3.2}$$

However, since u_1^a and u_2^a both are unit vectors which point in the direction of the timelike Killing field ξ^a, we have

$$u_1^a = [\xi^a/(-\xi^b\xi_b)^{1/2}]\,|_{P_1} \quad , \tag{6.3.3}$$

$$u_2^a = [\xi^a/(-\xi^b\xi_b)^{1/2}]\,|_{P_2} \quad . \tag{6.3.4}$$

By proposition C.3.1 we have $(k_a\xi^a)\,|_{P_1} = (k_a\xi^a)\,|_{P_2}$, so we obtain

$$\frac{\omega_1}{\omega_2} = \frac{(-\xi^b\xi_b)^{1/2}\,|_{P_2}}{(-\xi^b\xi_b)^{1/2}\,|_{P_1}} = \frac{(1 - 2M/r_2)^{1/2}}{(1 - 2M/r_1)^{1/2}} \quad , \tag{6.3.5}$$

where the explicit form of $\xi^b\xi_b = g_{tt} = -(1 - 2M/r)$ for Schwarzschild spacetime has been used and r_1 and r_2 are, respectively, the radial coordinates of O_1 and O_2. Note, however, that the above derivation is valid in an arbitrary stationary spacetime.

Equation (6.3.5) shows that for $r_2 > r_1$ (i.e., for the emitter closer to the center of gravitational attraction than the receiver), we have $\omega_2 < \omega_1$, i.e., the frequency of the light will be decreased ("shifted toward the red"). Physically, this makes sense because according to quantum theory the energy of a photon is proportional to its frequency, $E = \hbar\omega$, and we would expect the photon energy to be degraded as it "climbs out of the gravitational potential well." Indeed, in the limit where M is much less than r_1 and r_2, as is the case outside the surface of ordinary bodies, equation (6.3.5) becomes

$$\frac{\Delta\omega}{\omega} \approx -\frac{GM}{c^2 r_1} + \frac{GM}{c^2 r_2} \quad , \tag{6.3.6}$$

where we have restored the G's and c's in this equation. This can be interpreted as saying that the change in the locally measured energy of a photon, $\hbar\Delta\omega$, equals the change in its Newtonian gravitational potential energy

$$(\hbar\omega/c^2)(-GM/r_1 + GM/r_2) \quad .$$

The gravitational redshift has been measured and found to be in agreement with the prediction of general relativity to within the 1% experimental uncertainty by a laboratory experiment devised by Pound and Rebka (1960) which uses the Mössbauer effect to measure the tiny frequency shift of photons on the surface of the Earth which fall down a tower. The gravitational redshift of spectral lines from the Sun and some white dwarf stars has also been observed, but the accuracy of these measurements for testing equation (6.3.5) is not as high due to convective motions (which broaden the spectral lines) in the case of the Sun and a number of other uncertainties in the case of the white dwarfs. More recently, the gravitational redshift has been confirmed to about 0.01% accuracy by the tracking of hydrogen masers launched by rocket (Vessot and Levine 1979; Vessot *et al.* 1980). Confirmation of the existence of a similar "gravitational time dilation" effect also has been achieved by comparison of atomic clocks flown in airplanes with clocks on the ground (Hafele and Keating 1972; Alley 1979).

It is worth pointing out that, as proven in section 6.2, the maximum value of M/R at the surface of a static spherical body (with $d\rho/dr \leqq 0$ everywhere inside the body) is 4/9. Thus, by equation (6.3.5), the maximum redshift of light emitted from the surface of a static star is

$$\left.\frac{\omega_1}{\omega_2}\right|_{\max} = \frac{\omega(r = 9M/4)}{\omega(r = \infty)} = 3 \quad ; \tag{6.3.7}$$

i.e., in terms of the redshift factor z defined in chapter 5, we have

$$z_{\max} = \left.\frac{\omega_1}{\omega_2}\right|_{\max} - 1 = 2 \quad . \tag{6.3.8}$$

This rules out the possibility that observed redshifts of greater than 2 (as are commonly found for quasars) could arise solely from the gravitational redshift of light emitted from the surface of a static, spherical body.

In order to learn more about the behavior of light rays and test bodies in the exterior gravitational field of a spherical body, we need to solve for the timelike and null geodesics. First, we note that because of the parity reflection symmetry, $\theta \to \pi - \theta$, of the Schwarzschild metric, if the initial position and tangent vector of a geodesic lie in the "equatorial plane" $\theta = \pi/2$, then the entire geodesic must lie in this "plane." Since every geodesic can be brought to an initially (and hence everywhere) equatorial geodesic by a rotational isometry, this means that without loss of generality we may restrict attention to study of the equatorial geodesics, and we shall do so.

The coordinate basis components of the tangent u^a to a curve parameterized by τ are (see eq. [2.2.12])

$$u^\mu = \frac{dx^\mu}{d\tau} \equiv \dot{x}^\mu \quad . \tag{6.3.9}$$

For timelike geodesics, we choose τ to be the proper time; and for null geodesics, we choose τ to be an affine parameter. Thus for these cases we have (recalling that $\theta = \pi/2$)

$$-\kappa = g_{ab}u^a u^b = -(1 - 2M/r)\dot{t}^2 + (1 - 2M/r)^{-1}\dot{r}^2 + r^2\dot{\phi}^2 \quad , \tag{6.3.10}$$

where

$$\kappa = \begin{cases} 1 & \text{(timelike geodesics)} \\ 0 & \text{(null geodesics)} \end{cases} \quad . \tag{6.3.11}$$

In the derivation of the gravitational redshift, we already used the fact that the quantity

$$E = -g_{ab}\xi^a u^b = (1 - 2M/r)\dot{t} \qquad (6.3.12)$$

is a constant of the motion, where $\xi^a = (\partial/\partial t)^a$ denotes the static Killing field. In the case of timelike geodesics, at large distances from the center of attraction ($r >> M$) where the metric becomes flat and the norm of ξ^a becomes -1, E reduces to the special relativistic formula for the total energy per unit rest mass of a particle as measured by a static observer. In general, we may interpret E for timelike geodesics as representing the total energy (including gravitational potential energy) per unit rest mass of a particle following the geodesic in question, relative to a static observer at infinity, since it is the energy that would be required of such an observer in order to put a unit rest mass particle in the given orbit. Similarly, in the null case, $\hbar E$ represents the total energy of a photon.

The rotational Killing field $\psi^a = (\partial/\partial\phi)^a$ also yields a constant of the motion, L, via proposition C.3.1,

$$L = g_{ab}\psi^a u^b = r^2\dot{\phi} \quad . \qquad (6.3.13)$$

We may interpret L as the angular momentum per unit rest mass of a particle in the timelike case, and we may interpret $\hbar L$ as the angular momentum of a photon in the null case. In the Newtonian limit, equation (6.3.13) simply expresses Kepler's second law: equal areas are swept out in equal times. In general relativity, the spatial geometry is not Euclidean and we cannot interpret equation (6.3.13) in terms of "areas swept out," but it is interesting that the simple form of equation (6.3.13) (with the appropriate interpretations of r and $\dot{\phi} = d\phi/d\tau$) remains exactly valid in the strong field regime.

Substituting equations (6.3.12) and (6.3.13) in equation (6.3.10) and rearranging terms, we obtain our final equation for geodesics with remarkably little labor,

$$\frac{1}{2}\dot{r}^2 + \frac{1}{2}\left(1 - \frac{2M}{r}\right)\left(\frac{L^2}{r^2} + \kappa\right) = \frac{1}{2}E^2 \quad . \qquad (6.3.14)$$

This equation shows that the radial motion of a geodesic is the same as that of a unit mass particle of energy $E^2/2$ in ordinary one-dimensional, nonrelativistic mechanics moving in the effective potential,

$$V = \frac{1}{2}\kappa - \kappa\frac{M}{r} + \frac{L^2}{2r^2} - \frac{ML^2}{r^3} \quad . \qquad (6.3.15)$$

Once the radial motion is determined using this effective potential, the angular motion and time coordinate change are easily found from equations (6.3.13) and (6.3.12). The crucial new feature provided by general relativity is that in equation (6.3.15) in addition to the "Newtonian term," $-\kappa M/r$, and the "centrifugal barrier term," $L^2/2r^2$, we have the new term, $-ML^2/r^3$, which dominates over the centrifugal barrier term at small r.

We consider, first, the timelike geodesics, $\kappa = 1$. The extrema of the effective potential V are given by

$$0 = \frac{\partial V}{\partial r} = r^{-4}[Mr^2 - L^2 r + 3ML^2] \quad . \tag{6.3.16}$$

Equation (6.3.16) has the roots

$$R_\pm = \frac{L^2 \pm (L^4 - 12L^2M^2)^{1/2}}{2M} \quad . \tag{6.3.17}$$

Thus, if $L^2 < 12M^2$, there are no extrema of V, as illustrated in Figure 6.3. A particle heading toward the center of attraction ($\dot{r} \leqq 0$) with $L^2 < 12M^2$ will fall directly to the $r = 2M$ surface and, indeed, will continue its fall into the spacetime singularity at $r = 0$ (see section 6.4).

For $L^2 > 12M^2$, it is easy to check that the extremum R_+ is a minimum of V, while R_- is a maximum, as illustrated in Figure 6.4. Thus, stable circular orbits ($\dot{r} = 0$) exist at the radius $r = R_+$, and unstable circular orbits exist at $r = R_-$. For $L >> M$, the formula for R_+ becomes

$$R_+ \approx L^2/M \quad , \tag{6.3.18}$$

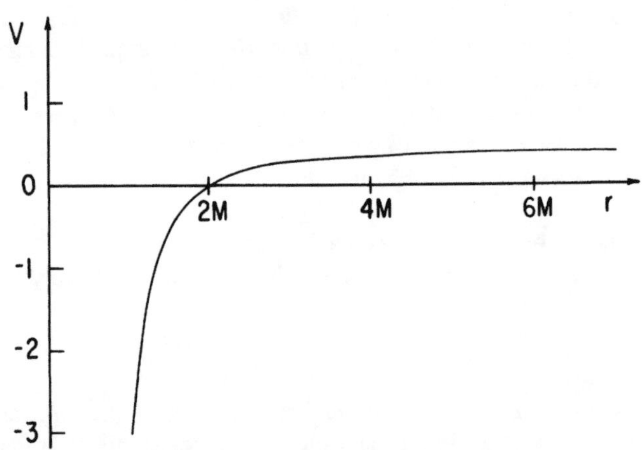

Fig. 6.3. The effective potential, V, for timelike geodesics in the case $L^2 = 6M^2$.

which is just the Newtonian formula for the radius of a circular orbit of a particle of angular momentum per mass L orbiting a spherical body of mass M. This justifies the interpretation of the constant C in our derivation of the Schwarzschild solution, equation (6.1.43), as $-2M$. Note that according to equation (6.3.17), R_+ is restricted to the range

$$R_+ > 6M \quad . \tag{6.3.19}$$

Thus, in general relativity, no stable circular orbits exist at radii smaller than $6M$. Furthermore, the unstable circular orbits are restricted to the range

$$3M < R_- < 6M \quad . \tag{6.3.20}$$

Thus, no circular orbits at all exist at radii less than $3M$.

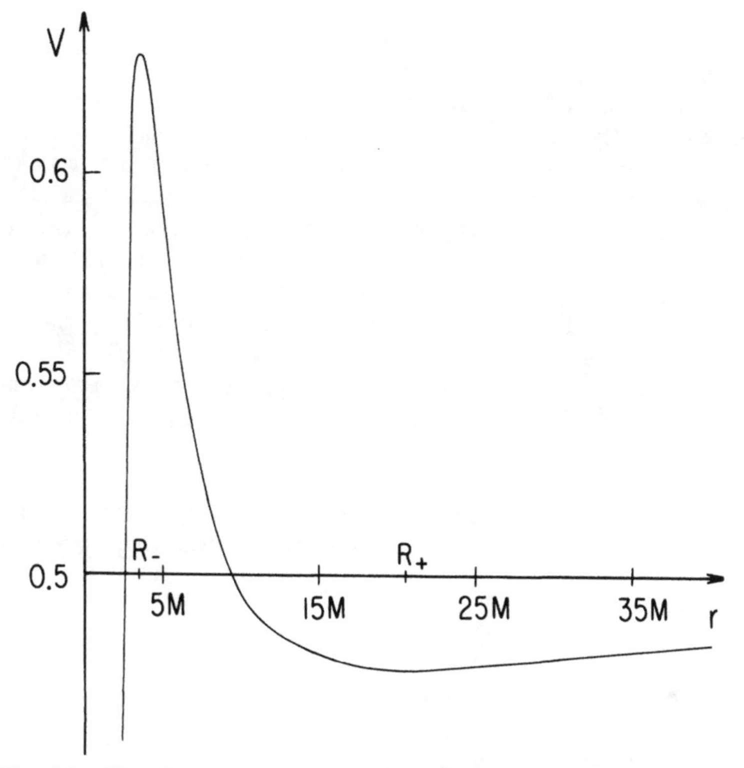

Fig. 6.4. The effective potential, V, for timelike geodesics in the case $L^2 = 24M^2$. (Note that the scale of V is greatly expanded as compared with Fig. 6.3.)

The energy of an ordinary particle in one-dimensional motion which sits at the minimum or maximum of a potential V is, of course, just the value of V at that point. Thus, from our mathematical analogy, equation (6.3.14), we see that the true energy per unit rest mass, E, of a particle in a circular orbit of radius R is given by

$$\frac{1}{2}E^2(R) = V(R) = \frac{1}{2}\frac{(R - 2M)^2}{R(R - 3M)} \quad , \tag{6.3.21}$$

so that

$$E(R) = \frac{R - 2M}{R^{1/2}(R - 3M)^{1/2}} \quad , \tag{6.3.22}$$

where equation (6.3.16) was used to eliminate L^2 in terms of R ($= R_+$ or R_-). Note that if $R \leqq 4M$, we have $E \geqq 1$ and, indeed $E \to \infty$ as $R \to 3M$. Thus, particles in the unstable circular orbits between $3M$ and $4M$ would escape to infinity if perturbed radially outward.

The binding energy, E_B, per unit rest mass of the last stable circular orbit at $R = 6M$ is

$$E_B = 1 - E = 1 - (8/9)^{1/2} \approx 0.06. \tag{6.3.23}$$

Now, a particle orbiting in the Schwarzschild geometry will emit gravitational radiation, as discussed for weak fields in section 4.4*b*. Because of radiation reaction, it will deviate slightly from geodesic motion. A particle initially in a circular orbit with $R >> M$ (and thus with $E \approx 1$) should slowly spiral in to smaller radii as it loses energy by gravitational radiation, remaining in a nearly circular orbit until it reaches the orbital radius $R = 6M$. At that point, the orbit becomes unstable and the particle should rapidly fall to $r = 0$. According to equation (6.3.23), about 6% of the original mass-energy of the particle will be radiated away during the time in which the particle spirals to $R = 6M$. (For the Kerr metric with $a = M$, the corresponding fraction of energy radiated is 42%; see chapter 12.) This illustrates that although the emission gravitational radiation was found to be very weak in the examples of section 4.4*b*, large amounts of energy can be converted to gravitational radiation in astrophysically plausible processes.

If a particle is displaced slightly from the "equilibrium" radius R_+ of a stable circular orbit, the particle will oscillate in radius about R_+. For sufficiently small displacements, it will execute simple harmonic motion with frequency ω_r given by

$$\omega_r{}^2 = k_{\text{eff}} = \left.\frac{d^2V}{dr^2}\right|_{R_+} = \frac{M(R_+ - 6M)}{R_+^3(R_+ - 3M)} \quad , \tag{6.3.24}$$

where equation (6.3.16) was used to eliminate L^2, and we should emphasize that the "time" implicit in ω_r is the proper time τ measured by the particle as opposed to the coordinate time t of the Schwarzschild geometry. On the other hand, for a circular orbit, the angular frequency, $\omega_\phi = \dot{\phi}$, is given by equation (6.3.13),

$$\omega_\phi{}^2 = \frac{L^2}{R_+^4} = \frac{M}{R_+^2(R_+ - 3M)} \quad , \tag{6.3.25}$$

where, again, equation (6.3.16) was used to eliminate L^2. In the limit of Newtonian orbits, $R_+ >> M$, we have $\omega_r \approx \omega_\phi$. If $\omega_r = \omega_\phi$, then the particle would return to a given value of r exactly one orbital period later; i.e., the orbit will close. Indeed, in the Newtonian theory of gravity, all bound orbits—not just the nearly circular orbits—are closed ellipses. The failure of ω_r to equal ω_ϕ in general relativity means that the orbits do not close; rather there is a precession of the angle at which the maximum and minimum values of r are achieved. For nearly circular orbits, this precession rate is given by

$$\omega_p = \omega_\phi - \omega_r = -[(1 - 6M/R_+)^{1/2} - 1]\omega_\phi \quad . \tag{6.3.26}$$

In the limit $R_+ >> M$, we have, to lowest nonvanishing order,

$$\omega_p \approx \frac{3M^{3/2}}{R_+^{5/2}} = \frac{3(GM)^{3/2}}{c^2R_+^{5/2}} \quad , \tag{6.3.27}$$

where we have restored the G's and c's in the final formula. We have considered only nearly circular orbits here, but a more general analysis (see, e.g., Weinberg 1972) shows that to lowest nonvanishing order the precession of an arbitrary elliptical orbit is given by

$$\omega_p \approx \frac{3(GM)^{3/2}}{c^2(1 - e^2)a^{5/2}} \quad , \tag{6.3.28}$$

where a is the semimajor axis of the ellipse and e is its eccentricity.

For the orbit of the planet Mercury, general relativity predicts a precession rate of 43 seconds of arc per century. Precisely this residual precession rate had been observed (after taking into account known effects such as perturbations of the planet Venus) prior to the formulation of general relativity and had been an unexplained mystery. The explanation of the precession of Mercury's orbit by general relativity was one of the most dramatic early successes of the theory. Although the general relativistic precession of Mercury's orbit is extremely small, the similar precession observed in the orbit of the binary pulsar mentioned at the end of chapter 4 is about 4° per year. This result has been used to estimate the masses of the two bodies in this system.

We turn, now, to consideration of the null geodesics. Setting $\kappa = 0$ in equation (6.3.14), we find the effective potential for null geodesics to be simply

$$V = \frac{L^2}{2r^3}(r - 2M) \quad . \tag{6.3.29}$$

Thus, the shape of V is independent of L and, as illustrated in Figure 6.5, the only extremum of V is a maximum occurring at $r = 3M$. Thus, in general relativity, unstable circular orbits of photons exist at radius $3M$, so that, physically, gravity has a very significant effect on the propagation of light rays in the strong field regime.

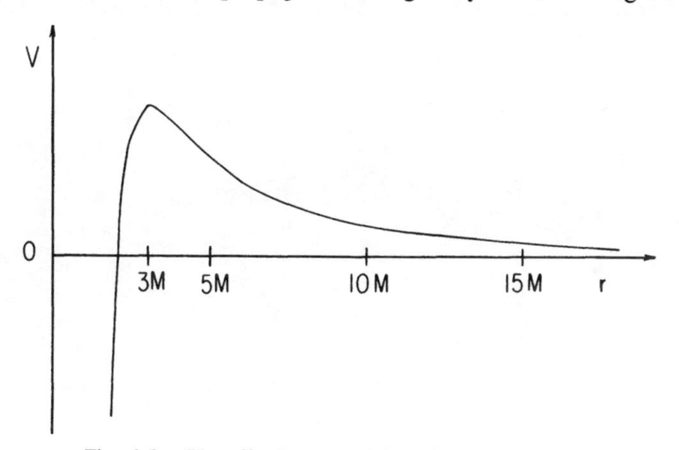

Fig. 6.5. The effective potential, V, for null geodesics.

The minimum energy, E, required to surmount the top of the potential barrier is given by (see eq. [6.3.21] above)

$$\frac{1}{2}E^2 = V(R = 3M) = L^2M/[2(3M)^3] \quad , \tag{6.3.30}$$

i.e.,

$$L^2/E^2 = 27M^2 \quad . \tag{6.3.31}$$

Now, for a light ray propagating in flat spacetime, it follows directly from the definitions of L and E that L/E is the impact parameter of the light ray, i.e., the distance of closest approach to the origin $r = 0$. Since the Schwarzschild geometry is asymptotically flat, for a light ray initially in the asymptotically flat region ($r >> M$), L/E will represent the *apparent impact parameter,*

$$b \equiv L/E \quad , \tag{6.3.32}$$

of the light ray even though it no longer represents the "distance of closest approach." Thus, the Schwarzschild geometry will capture any photon sent toward it with an apparent impact parameter smaller than the critical value, b_c, given by

$$b_c = 3^{3/2}M \quad . \tag{6.3.33}$$

Hence, the capture cross section, σ, for photons in the Schwarzschild geometry is

$$\sigma = \pi b_c{}^2 = 27\pi M^2 \quad . \qquad 6.3.34)$$

To analyze the "light bending" effects of the Schwarzschild geometry on the light rays which are not captured, it is convenient to derive an equation for the spatial orbit of the light ray by solving equation (6.3.14) for $\dot{r}$ and then dividing $\dot{\phi}$, equation (6.3.13), by $\dot{r}$. We obtain

$$\frac{d\phi}{dr} = \frac{L}{r^2}\left[E^2 - \frac{L^2}{r^3}(r - 2M)\right]^{-1/2} \quad . \tag{6.3.35}$$

We wish to find the change, $\Delta\phi = \phi_{+\infty} - \phi_{-\infty}$, in the angular coordinate ϕ of a light ray in the Schwarzschild geometry traversing a path as illustrated in Figure 6.6. In

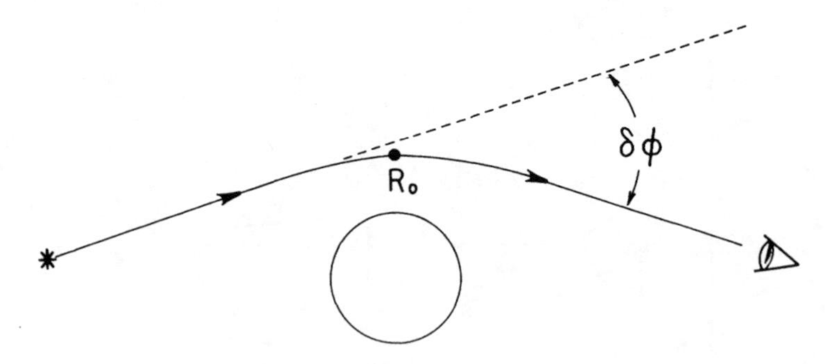

Fig. 6.6. Diagram illustrating the "bending of light" effect.

order not to be captured, the impact parameter, b, must be greater than the critical value, equation (6.3.33). In that case, the orbit of the light ray will have a "turning point" at the largest radius, R_0, for which $V(R_0) = E^2/2$, i.e., at the largest root of

$$R_0^3 - b^2(R_0 - 2M) = 0 \quad , \tag{6.3.36}$$

which is

$$R_0 = \frac{2b}{\sqrt{3}} \cos\left[\frac{1}{3}\cos^{-1}\left(-\frac{3^{3/2}M}{b}\right)\right] \quad . \tag{6.3.37}$$

By symmetry, the contributions to $\Delta\phi$ made by the parts of the path prior to the turning point and after the turning point will be equal. Hence, by equation (6.3.35), the total change in the angular coordinate, $\Delta\phi$, is given by

$$\Delta\phi = 2\int_{R_0}^{\infty} \frac{dr}{[r^4 b^{-2} - r(r-2M)]^{1/2}} \quad . \tag{6.3.38}$$

It is convenient to make the change of variables, $u = 1/r$, in terms of which equation (6.3.38) becomes

$$\Delta\phi = 2\int_0^{1/R_0} \frac{du}{(b^{-2} - u^2 + 2Mu^3)^{1/2}} \quad . \tag{6.3.39}$$

In the case of flat spacetime, $M = 0$, we have $R_0 = b$, and equation (6.3.39) yields simply

$$\Delta\phi|_{M=0} = 2\sin^{-1}(b/R_0) = \pi \quad , \tag{6.3.40}$$

as obviously must be the case since the trajectory is a straight line. When $M \neq 0$, according to equation (6.3.39) $\Delta\phi$ will not equal π; i.e., there will be a nonzero deflection of the light ray, which we may interpret physically as being due to the gravitational attraction of the Schwarzschild geometry. We wish to calculate the contribution to the "light bending" valid to first order in M. Actually, it is rather tricky to calculate this contribution by varying M keeping b fixed in equation (6.3.39) because of the M dependence of the integration limit $1/R_0$ and the singular behavior of the integrand at this limit. However, we can circumvent these difficulties by working with M and R_0 as the independent variables. In other words, as we vary M, we compare $\Delta\phi$ for light rays which have the same Schwarzschild radial coordinate, R_0, at the point of closest approach rather than, say, light rays with the same apparent impact parameter, b. (It is not difficult to see that to first order in M, it makes no difference whether we keep b or R_0 fixed, but to higher orders in M it *does* matter whether we compare light rays with the same value of b or the same value of R_0.) Eliminating b via equation (6.3.36), we obtain

$$\Delta\phi = 2\int_0^{1/R_0} \frac{du}{(R_0^{-2} - 2MR_0^{-3} - u^2 + 2Mu^3)^{1/2}} \quad . \tag{6.3.41}$$

Differentiating with respect to M at fixed R_0 and evaluating the result at $M = 0$, we obtain

$$\begin{aligned} \left.\frac{\partial(\Delta\phi)}{\partial M}\right|_{M=0} &= 2\int_0^{1/R_0} \left.\frac{(R_0^{-3} - u^3)du}{(R_0^{-2} - 2MR_0^{-3} - u^2 + 2Mu^3)^{3/2}}\right|_{M=0} \\ &= 2\int_0^{1/b} \frac{(b^{-3} - u^3)}{(b^{-2} - u^2)^{3/2}}\, du \\ &= 4b^{-1} \quad . \end{aligned} \tag{6.3.42}$$

Thus, to first order in M, the deflection of light is given by

$$\delta\phi = \Delta\phi - \pi \approx M \frac{\partial(\Delta\phi)}{\partial M}\bigg|_{M=0} = \frac{4M}{b}$$
$$= \frac{4GM}{bc^2} , \tag{6.3.43}$$

where we have reinserted the G's and c's in the last step.

For a light ray which grazes the surface of the Sun, equation (6.3.43) predicts a deflection of 1.75 seconds of arc. This "bending" of starlight passing near the Sun has been observed during solar eclipses beginning with the 1919 expedition led by Eddington (Dyson, Eddington, and Davidson 1920), thus confirming this important prediction of general relativity, but because of numerous difficulties the accuracy of these measurements has only been about 10%. However, the bending of radio waves emitted from a quasar as it approaches eclipse by the Sun has been measured to about 1% accuracy (Fomalont and Sramek 1976) and has been found to agree with equation (6.3.43).

A further measurable effect concerning the behavior of null geodesics in the Schwarzschild geometry is the "time delay" of radar signals emitted from Earth. To analyze this effect, we divide $\dot{t}$ (as given by eq. [6.3.12]) by $\dot{r}$ (as determined from eq. [6.3.14]), thereby obtaining

$$\frac{dt}{dr} = \left(1 - \frac{2M}{r}\right)^{-1}\left[1 - \left(1 - \frac{2M}{r}\right)\frac{b^2}{r^2}\right]^{-1/2} . \tag{6.3.44}$$

Integration of this equation over the trajectory of a null geodesic yields the total change, Δt, in Schwarzschild time coordinate along the trajectory. Consider, now, the following situation. A radar signal is emitted from Earth, located at Schwarzschild radial coordinate R_E. The signal passes near the Sun, with radius of closest approach R_0, and then is reflected off a planet, located at R_p. The signal then retraces its trajectory and returns to Earth, as illustrated in Figure 6.7. (The motion of Earth and the planet during the intervening time is neglected.) How much time, $\Delta\tau$, elapses on the clock of an observer on Earth between emission and reception of the signal?

We wish to calculate $\Delta\tau$ to first order in M. By integrating equation (6.3.44) and then differentiating with respect to M (holding R_0 fixed) we obtain (problem 5), in close analogy to the derivation of the light bending effect,

$$\Delta t = \frac{2}{c}[(R_E^2 - R_0^2)^{1/2} + (R_p^2 - R_0^2)^{1/2}] + \frac{2GM}{c^3}\left\{2 \ln\left[\frac{R_E + (R_E^2 - R_0^2)^{1/2}}{R_0}\right]\right.$$
$$\left. + 2 \ln\left[\frac{R_p + (R_p^2 - R_0^2)^{1/2}}{R_0}\right] + \left(\frac{R_E - R_0}{R_E + R_0}\right)^{1/2} + \left(\frac{R_p - R_0}{R_p + R_0}\right)^{1/2}\right\} . \tag{6.3.45}$$

It should be noted that we have formulated the question—purely for mathematical convenience—so that as we vary M, we compare null geodesics with the same R_0, rather than, say, null geodesics with the same b or the same total change, $\Delta\phi$, in angle in passing between R_E and R_p. Unlike the light bending case, here the first order terms in M depend sensitively on what parameters of the null geodesics are held fixed as M is varied.

The proper time, $\Delta\tau$, that elapses on Earth is related to the change in coordinate time, Δt, by

$$\Delta\tau = (1 - 2M/R_E)^{1/2}\,\Delta t \quad . \tag{6.3.46}$$

Thus, to first order in M, the reflected radar signal will be measured to arrive back on Earth at a time after emission given by

$$\Delta\tau = -\frac{2GM}{c^3 R_E}[(R_E^2 - R_0^2)^{1/2} + (R_p^2 + R_0^2)^{1/2}] + \Delta t \quad , \tag{6.3.47}$$

where Δt is given by equation (6.3.45).

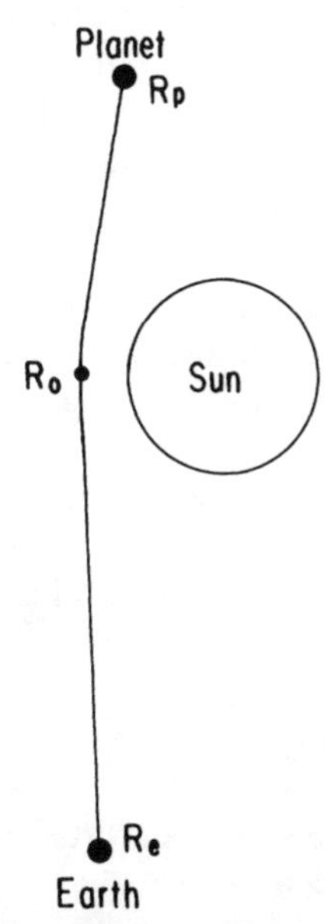

Fig. 6.7. Diagram illustrating the "time delay of light" effect.

Actually, equation (6.3.47) by itself is not very useful for a direct test of the general relativistic time delay effect, since the parameters R_E, R_p, and R_0 appearing in the equation are not known to the necessary precision. Therefore, the procedure used to test the time delay prediction is to write down a formula for how $\Delta\tau$ is expected to vary with time (due to the motion of the Earth and planets), treating all parameters (in particular, the orbital parameters of the planets) as unknowns. A "best fit" to the observed data then is used to determine all the parameters. When this is done, the agreement between the theoretical formula and observation is excellent. On the other hand, a slight modification of the general relativistic contribution—in particular, the change that would be produced in the formula by altering g_{rr} to $(1 - 2\gamma M/r)^{-1}$ where γ differs from unity by only 0.2%—produces a formula which cannot be satisfactorily fit by the observed data. (These highest precision data actually have come from the tracking of spacecraft which emit signals rather than reflection of radar off of planets [Reasenberg *et al.* 1979].) Thus, this prediction of general relativity has been confirmed to high precision.

In summary, the analysis of timelike and null geodesics in the Schwarzschild geometry leads to a number of important predictions which have been tested by solar system observations: the gravitational redshift, the precession of planetary orbits, the bending of light, and the time delay of radar signals. These predictions provide stringent quantitative tests of general relativity, and it is gratifying that general relativity thus far has passed these tests very successfully.

6.4 The Kruskal Extension

We turn our attention now to the analysis of the singularities at $r = 2M$ and $r = 0$ in the coordinate basis components of the metric of the Schwarzschild solution. As discussed in section 6.2, for any static equilibrium configuration the region $r \leqq 2M$ will be within the matter-filled interior, so analysis of the singularities at $r = 2M$ and $r = 0$ for the vacuum Schwarzschild solution is irrelevant to the study of the gravitational field of a static star, as was first pointed out by Schwarzschild. However, as mentioned at the end of section 6.2, sufficiently massive bodies will undergo complete gravitational collapse, and the region $r \leqq 2M$ of the vacuum Schwarzschild solution is very relevant to the description of the endpoint of this collapse.

Whenever the metric components in a coordinate basis are badly behaved for certain values of the coordinates, there are two possible causes: (1) The spacetime geometry is, in fact, singular; or (2) the spacetime geometry is nonsingular, but the coordinates fail to properly cover a region of spacetime. In general, it is not an easy task to determine in a given situation which of these two possibilities holds. (Indeed, it is nontrivial even to formulate a precise general definition of the notion of "singularity in the spacetime geometry." We will postpone a full discussion of this issue until chapter 9.) Normally, possibility (1) would be demonstrated by calculating a curvature scalar, such as $R_{abcd}R^{abcd}$, and showing that it blows up "at the singularity of the metric components" and furthermore, that this singularity lies at a finite affine parameter along some geodesic (so that the "singularity" is not "at infinity," in which case it is not really a singularity). However, as discussed in chapter 9, one may have spacetime singularities where no curvature scalars blow up. Normally, possibility (2) would be demonstrated by explicitly displaying an extension of the "nonsingular region" of the original metric, i.e., a (nonsingular) spacetime $(\tilde{M}, \tilde{g}_{ab})$ which includes the original spacetime (M, g_{ab}) as a proper subset. Normally, this would be accomplished by finding a coordinate transformation which eliminates the singularities of the original metric components.

In the case of the Schwarzschild metric, we already pointed out that the Schwarzschild coordinates will be badly behaved where the timelike Killing field ξ^a becomes collinear with $\nabla^a r$. We shall see below that this occurs at $r = 2M$, and that the "singularity" at $r = 2M$ is merely a coordinate singularity. On the other hand, the singularity at $r = 0$ is a true singularity in the spacetime geometry, as can be demonstrated by calculating via equations (6.1.29)–(6.1.34) the curvature invariant $R_{abcd}R^{abcd}$.

We begin, however, by considering two simple examples which help illuminate the nature of the problem. As a first example, consider the two-dimensional metric

$$ds^2 = -\frac{1}{t^4}\,dt^2 + dx^2 \tag{6.4.1}$$

defined over the coordinate range $-\infty < x < \infty$, $0 < t < \infty$. This metric appears to have a singularity at $t = 0$. However, the true nature of the spacetime geometry can be seen by making the coordinate transformation $t \to t' = 1/t$. In the new coordinates, the metric is seen to be simply the flat metric,

$$ds^2 = -(dt')^2 + dx^2 \quad , \tag{6.4.2}$$

and the original spacetime is seen to be the portion $t' > 0$ of Minkowski spacetime. Thus, the apparent singularity at $t = 0$ of the original metric, equation (6.4.1), really represents $t' \to \infty$ in Minkowski spacetime and is not a singularity at all but merely corresponds to the covering of an infinite region of spacetime with a finite range of a coordinate. The spacetime geometry is geodesically complete as $t \to 0$ ($t' \to \infty$), i.e., all the geodesics approaching $t = 0$ extend to arbitrarily large values of their affine parameter. On the other hand, the spacetime of equation (6.4.1) is *not* geodesically complete as $t \to \infty$ ($t' \to 0$). However, we can extend the original spacetime "beyond $t = \infty$" by adding on the portion $t' \leqq 0$ of Minkowski spacetime. This example provides an excellent illustration of how one can be misled by interpreting coordinate labels such as t as physically meaningful quantities. It also illustrates how an appropriate coordinate transformation can help one analyze what is really going on in the spacetime.

Our second example will be seen to be closely analogous to the Schwarzschild solution, and for this reason we will analyze it in detail. We consider the *Rindler spacetime*,

$$ds^2 = -x^2 dt^2 + dx^2 \tag{6.4.3}$$

with coordinate ranges $-\infty < t < \infty$, $0 < x < \infty$. This metric appears to have a singularity at $x = 0$. (The determinant of $g_{\mu\nu}$ vanishes at $x = 0$, so the inverse metric components $g^{\mu\nu}$ are singular at $x = 0$.) Geodesics terminate with finite length at $x = 0$, but calculation of curvature scalars shows no bad behavior[8] as $x \to 0$, suggesting that the singularity may be simply a coordinate singularity. This metric is not simple enough for trial and error guesses at coordinate transformations to have much chance of success, so we need a more systematic approach. In general, the best procedure is to introduce new coordinates which are linked closely to the spacetime geometry. This can be achieved, for example, by introducing a family of geodesics which "head toward the singularity" and using the affine parameter along the geodesics as one of the coordinates. In general, there is no foolproof method for elimination of coordinate singularities; for example, for coordinates based on a family of geode-

8. In fact, the curvature of the Rindler metric vanishes, showing immediately that the Rindler spacetime must be simply a portion of two-dimensional Minkowski spacetime. However, in order to pursue the analogy with the Schwarzschild metric, we prefer not to make use of this fact in our analysis.

sics, new coordinate singularities will be produced whenever the geodesics cross. However, in two-dimensional spacetimes, there does exist an essentially foolproof method for analyzing coordinate singularities. This is because in two dimensions, the null geodesics divide up (at least locally) into two classes—"ingoing" and "outgoing"—and within each class two distinct null geodesics cannot cross, since their tangents would have to coincide at their intersection point, thus implying that the geodesics coincide everywhere. This suggests that we introduce null coordinates, i.e., coordinates such that the first is constant along each "outgoing" geodesic and the second is constant along each "ingoing" geodesic. In this way, our coordinate grid will be based on the geometrical (and well behaved) "grid" of null geodesics. The only coordinate singularities which can result from using null coordinates in two-dimensional spacetimes arise from bad parameterization of the geodesics, and this can be investigated and corrected by comparing the coordinate parameterization with an affine parameterization.

The null geodesics of Rindler spacetime are easily found from the null condition,

$$0 = g_{ab}k^a k^b = -x^2 \dot{t}^2 + \dot{x}^2 \quad , \tag{6.4.4}$$

where k^a is the geodesic tangent and the dot denotes derivative with respect to affine parameter. Equation (6.4.4) implies that

$$(dt/dx)^2 = 1/x^2 \quad , \tag{6.4.5}$$

so that along each geodesic, we have

$$t = \pm \ln x + \text{constant} \quad , \tag{6.4.6}$$

where the plus sign refers to the "outgoing" geodesics, and the minus sign refers to "ingoing" geodesics. Thus, we may define null coordinates (u, v) by

$$u = t - \ln x \quad , \tag{6.4.7}$$

$$v = t + \ln x \quad . \tag{6.4.8}$$

In the coordinates (u, v), the metric components are simply

$$ds^2 = -e^{v-u}\, du\, dv \quad . \tag{6.4.9}$$

In making the coordinate transformation, equations (6.4.7) and (6.4.8), we have not yet achieved our goal of analyzing the singularity at $x = 0$, since the coordinate ranges $-\infty < u < \infty$ and $-\infty < v < \infty$ still correspond only to the region $x > 0$ of the original Rindler spacetime. However, we are now in a position to reparameterize the null geodesics by new coordinates $U = U(u)$, $V = V(v)$, which will show how to extend the spacetime beyond "$x = 0$" (or, beyond "$u = \infty$" and "$v = -\infty$"). The form of the metric, equation (6.4.9) is sufficiently simple that we could easily guess this transformation, but in order to be more systematic we calculate the affine parameter along the null geodesics. This is most easily done by using the fact that the "time translation vector" $(\partial/\partial t)^a$ of the original Rindler metric, equation (6.4.3), is a Killing field, and thus

$$E = -g_{ab}k^a(\partial/\partial t)^b = x^2\, dt/d\lambda \tag{6.4.10}$$

is a constant of the motion, where λ is the affine parameter. Thus, for the outgoing null geodesics, substituting for x and t from equations (6.4.7) and (6.4.8) and setting u = constant, we find

$$\lambda = \frac{1}{2E}\int e^{v-u}dv = C + (e^{-u}/2E)e^{v} \quad , \tag{6.4.11}$$

where C is a constant. Thus, $\lambda_{\text{out}} = e^{v}$ is an affine parameter along the outgoing geodesics. A similar calculation shows that $\lambda_{\text{in}} = -e^{-u}$ is an affine parameter along the ingoing geodesics. (Note that the finite ranges of λ_{out} and λ_{in} show that all the null geodesics of the original Rindler spacetime are incomplete.) This suggests that we make the coordinate transformation $U = -e^{-u}$, $V = e^{v}$, which puts the metric in the extremely simple form,

$$ds^2 = -dU\, dV \quad . \tag{6.4.12}$$

The original Rindler spacetime corresponds to the coordinate ranges $U < 0$, $V > 0$. However, there is no longer any singularity in the metric components at $U = 0$ or $V = 0$, so we may now extend the spacetime by allowing the ranges of U and V to be unrestricted, $-\infty < U < \infty$, $-\infty < V < \infty$. The final coordinate transformation $T = (U + V)/2$, $X = (V - U)/2$, converts the metric to the immediately recognizable form,

$$ds^2 = -dT^2 + dX^2 \quad , \tag{6.4.13}$$

showing that our extended spacetime is just Minkowski spacetime!

The original coordinates (t, x) are given in terms of the final Minkowski coordinates (T, X) by

$$x = (X^2 - T^2)^{1/2} \quad , \tag{6.4.14}$$

$$t = \tanh^{-1}(T/X) \quad . \tag{6.4.15}$$

From equations (6.4.14) and (6.4.15), it can be seen that Rindler spacetime is simply the wedge $X > |T|$ of Minkowski spacetime, i.e., region I of Figure 6.8. Exam-

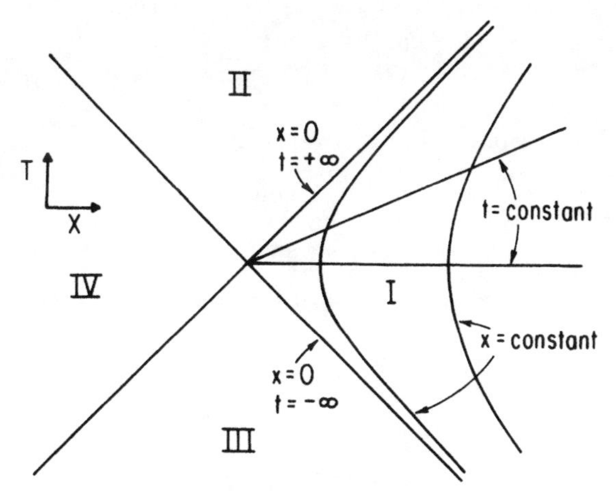

Fig. 6.8. Rindler spacetime, displayed as the "wedge," I, of two-dimensional Minkowski spacetime.

ination of equations (6.4.14) and (6.4.15) or of Figure 6.8 shows the nature of the coordinate singularity: The null lines $X = \pm T$ are mislabeled by the original coordinates as $x = 0$, $t = \pm\infty$. Our transformation to the null geodesic coordinates (U, V) allowed us to "break through" the coordinate barrier at $X = |T|$ and extend the spacetime to all of Minkowski spacetime.

Note that the "time translation symmetry" of the Rindler metric, equation (6.4.3), really corresponds to the "boost symmetry" of Minkowski spacetime. The observers at constant x undergo the uniform acceleration $a = 1/x$, which diverges as $x \to 0$. It is easy to check that static observers in the Schwarzschild spacetime must undergo a proper acceleration (in order to "stand still" in the "gravitational field") given by $a = (1 - 2M/r)^{-1/2}M/r^2$, which diverges as $r \to 2M$. Thus, the behavior of the Schwarzschild time coordinate as $r \to 2M$ is analogous to the behavior of the Rindler time coordinate as $x \to 0$. (Indeed, the mathematical analogy between the two metrics could be made more manifest by introducing a new space coordinate $y = x^2$ to put the Rindler metric in the form $ds^2 = -y\,dt^2 + 4y^{-1}\,dy^2$.) Therefore, it should not be surprising that the coordinate singularity of the Schwarzschild spacetime at $r = 2M$ is closely analogous to that of the Rindler spacetime, as we now shall show.

The Schwarzschild spacetime is, of course, four-dimensional, but, because of the spherical symmetry, only the two-dimensional "r-t part" of the metric is of importance for analyzing the nature of the singularity at $r = 2M$. Hence, we shall study the two-dimensional metric

$$ds^2 = -(1 - 2M/r)dt^2 + (1 - 2M/r)^{-1}\,dr^2 \quad . \tag{6.4.16}$$

We apply the two-dimensional method described above, using the outgoing and ingoing radial null geodesics of the Schwarzschild spacetime. The null condition, analogous to equation (6.4.4), is

$$0 = g_{ab}k^a k^b = -(1 - 2M/r)\dot{t}^2 + (1 - 2M/r)^{-1}\,\dot{r}^2 \quad , \tag{6.4.17}$$

which implies

$$\left(\frac{dt}{dr}\right)^2 = \left(\frac{r}{r - 2M}\right)^2 \quad . \tag{6.4.18}$$

Thus, the radial null geodesics of Schwarzschild satisfy

$$t = \pm r_* + \text{constant} \quad , \tag{6.4.19}$$

where the "Regge-Wheeler tortoise coordinate" r_* is defined by

$$r_* = r + 2M\ln(r/2M - 1) \tag{6.4.20}$$

so that $dr_*/dr = (1 - 2M/r)^{-1}$. We define the null coordinates u, v by

$$u = t - r_* \quad , \tag{6.4.21}$$

$$v = t + r_* \quad . \tag{6.4.22}$$

In these coordinates,[9] the metric (6.4.16) takes the form

$$ds^2 = -(1 - 2M/r)\, du\, dv \quad , \tag{6.4.23}$$

where r is now viewed as a function of u and v, defined implicitly by

$$r + 2M \ln\left(\frac{r}{2M} - 1\right) = r_* = (v - u)/2 \quad . \tag{6.4.24}$$

Using equation (6.4.24), we can rewrite equation (6.4.23) as

$$ds^2 = -\frac{2Me^{-r/2M}}{r} e^{(v-u)/4M}\, du\, dv \quad , \tag{6.4.25}$$

where we have factored the metric components into a piece, $e^{-r/2M}/r$, which is nonsingular as $r \to 2M$ (i.e., as $u \to \infty$ or $v \to -\infty$) times a piece with simple u and v dependence. Comparison with the Rindler case, equation (6.4.9) (or calculation of the affine parameter along the null geodesics) suggests that we define new coordinates U and V by

$$U = -e^{-u/4M} \quad , \tag{6.4.26}$$

$$V = e^{v/4M} \quad , \tag{6.4.27}$$

in terms of which the metric becomes

$$ds^2 = -\frac{32M^3 e^{-r/2M}}{r} dU\, dV \quad . \tag{6.4.28}$$

There is now no longer a singularity at $r = 2M$ (i.e., at $U = 0$ or $V = 0$), and thus we can extend the Schwarzschild solution by allowing U and V to take on all values compatible with $r > 0$. The singularity which persists at $r = 0$ is physical—as mentioned above, the curvature scalar $R_{abcd}R^{abcd}$ blows up there—so it cannot be eliminated by a further coordinate transformation.

If we make the final transformation $T = (U + V)/2$, $X = (V - U)/2$, the full Schwarzschild metric takes the final form given by Kruskal (1960) (see also Szekeres 1960),

$$ds^2 = \frac{32M^3 e^{-r/2M}}{r}(-dT^2 + dX^2) + r^2(d\theta^2 + \sin^2\theta\, d\phi^2) \quad . \tag{6.4.29}$$

The relation between the old coordinates (t, r) and the new coordinates (T, X) is given by

$$\left(\frac{r}{2M} - 1\right)e^{r/2M} = X^2 - T^2 \quad , \tag{6.4.30}$$

$$\frac{t}{2M} = \ln\left(\frac{T + X}{X - T}\right) = 2\tanh^{-1}(T/X) \quad , \tag{6.4.31}$$

9. The "hybrid" coordinates (u, r) or (v, r) are known as *Eddington-Finkelstein coordinates* (Eddington 1924; Finkelstein 1958).

and in equation (6.4.29), r is to be viewed as the function of X and T defined by equation (6.4.30). The allowed range of the coordinates X and T is given by the condition $r > 0$, which yields

$$X^2 - T^2 > -1 \quad . \tag{6.4.32}$$

A spacetime diagram for the Kruskal extension is shown in Figure 6.9. The causal structure of the extended Schwarzschild spacetime is easily seen from the diagram since, by construction, the radial null geodesics are 45° lines in Kruskal coordinates. The Kruskal extension is remarkably similar to the extension of the Rindler spacetime, the major differences being (1) the Schwarzschild spacetime is four-dimensional, so each point in Figure 6.9 really represents a two-dimensional sphere of radius r and (2) there are physical singularities in the extended region at $X = \pm(T^2 - 1)^{1/2}$, as shown. Note that the singularities at "$r = 0$" have a spacelike character and exist in the future of region II and the past of region III, rather than corresponding to "a timelike line at the origin of coordinates" as a naive interpretation of the Schwarzschild coordinates (t, r) might have suggested. The bad behavior of the coordinates (t, r) can also be seen in the "Kruskal diagram" 6.9. From equation (6.4.30) we see that $\nabla_a r = 0$ at $X = T = 0$, and it is not difficult to verify that the static Killing field ξ^a vanishes there also. Note also that $\nabla_a r$ and ξ^a become collinear along the null lines $X = \pm T$. The vanishing of ξ^a at $X = T = 0$ leads to a mislabeling of the lines $X = \pm T$ as "$t = \pm\infty$" analogous to the behavior of the t-coordinate in Rindler spacetime. Note, however, that although the Kruskal coordinates are very convenient for analyzing the "strong field" region of the Schwarzschild geometry, they are not convenient for analyzing the asymptotically flat region, $r \to \infty$.

The extended Schwarzschild spacetime has a rather surprising structure. The region I of Figure 6.9 corresponds to the original region $r > 2M$ of the Schwarzschild solution, and can be interpreted physically as representing the exterior grav-

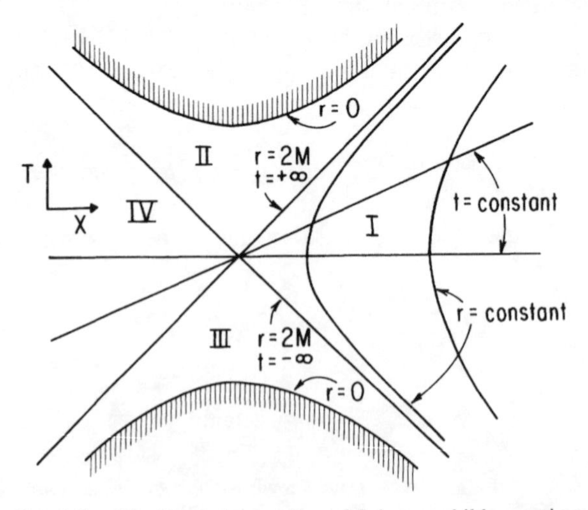

Fig. 6.9. The Kruskal extension of Schwarzschild spacetime.

itational field of a spherical body. However, a radially infalling observer in region I will cross the null line $X = T$ and enter region II. Once this observer has entered region II, he can never escape from it. Within a finite proper time (see problem 6) he will unavoidably fall into the singularity at $X = (T^2 - 1)^{1/2}$, and, indeed, any light signal he sends from region II will remain in region II and also will fall into the singularity. For this reason, region II is referred to as a *black hole*. (A general definition of the notion of a black hole will be given in chapter 12.) Region III has exactly the "time reversed" properties of region II, and is referred to as a *white hole*. Any observer present in region III must have originated in the spacetime singularity at $X = -(T^2 - 1)^{1/2}$ and, within a finite time, must leave region III. Finally region IV has properties identical to the original region I. It represents another asymptotically flat region of spacetime which lies "inside" the "radius" $r = 2M$. How this can occur is best illustrated by examining the geometry of the spacelike surface $T = 0$, which is shown in Figure 6.10. Note, however, that an observer in region I cannot communicate with any observer in region IV; as seen from Figure 6.9, a light signal sent from region I toward region IV will instead go into the black hole and be "swallowed up" by the spacetime singularity.

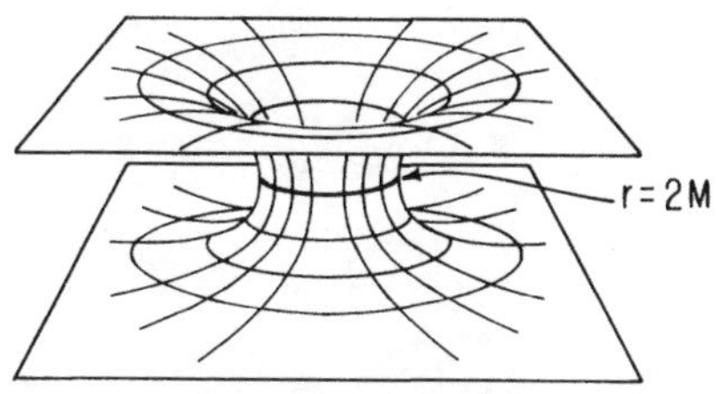

Fig. 6.10. The spatial geometry of the hypersurface $t = 0$ of the Schwarzschild spacetime, shown as it would look if it were embedded in flat space. (One dimension is suppressed, i.e., the topology of the hypersurface is $\mathbb{R} \times S^2$, not $\mathbb{R} \times S^1$, so each circle—such as the one shown at $r = 2M$—really represents a two-sphere.) The portion of the surface lying above the "throat" at $r = 2M$ in the figure corresponds to the portion lying in region I of Figure 6.9; the portion below $r = 2M$ lies in region IV of Figure 6.9.

How much of this picture of the extended Schwarzschild solution should we take seriously? The extended Schwarzschild solution is, of course, a perfectly valid solution of the vacuum Einstein equation, and as such, represents a possible structure for spacetime in general relativity. However, in order to "produce" the fully extended Schwarzschild solution, we must "start" with two asymptotically flat regions of spacetime together with an initial singularity in the region III which connects them. There is no reason to believe that the initial configuration of any region of our universe corresponds to these initial conditions, so there is no reason to believe that any region of our universe corresponds to the fully extended Schwarzschild solution (although, of course, this possibility cannot easily be ruled out). However, as discussed in section 6.2, sufficiently massive bodies will undergo complete gravitational collapse. The interior metric of these bodies will not be the Schwarzschild metric since $T_{ab} \neq 0$ there, but at all stages of the collapse the metric outside a spherical body will be the Schwarzschild metric, since, as mentioned at the end of section 6.1, it is the only spherically symmetric vacuum solution of Einstein's equation. Therefore, the spacetime geometry corresponding to the gravitational collapse of a spherical body will be as shown in Figure 6.11. All of regions III and IV (as well as parts of regions I and II) will be "covered up" by the matter and thus replaced by a "normal" spacetime region. However, a spacetime region corresponding to region II of the extended vacuum Schwarzschild spacetime will be produced when the radial coordinate of the collapsing body becomes less than $2M$. Another representation of the spacetime geometry resulting from spherical collapse is shown in Figure 6.12.

Thus, regions III and IV of the extended Schwarzschild solution are probably unphysical, but region II is of great physical importance: *The complete gravitational collapse of a spherical body always produces a Schwarzschild black hole region of spacetime.*

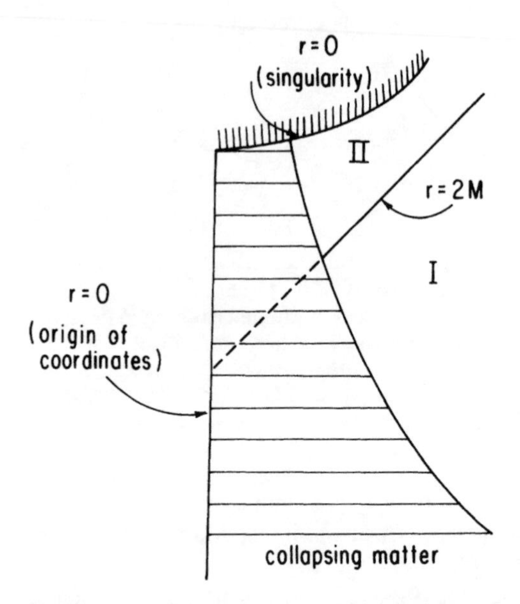

Fig. 6.11. The spacetime resulting from the complete gravitational collapse of a spherical body. All of regions III and IV of the extended Schwarzschild spacetime (Fig. 6.9) are "covered up" by the collapsing matter. However, (part of) the black hole region II is produced.

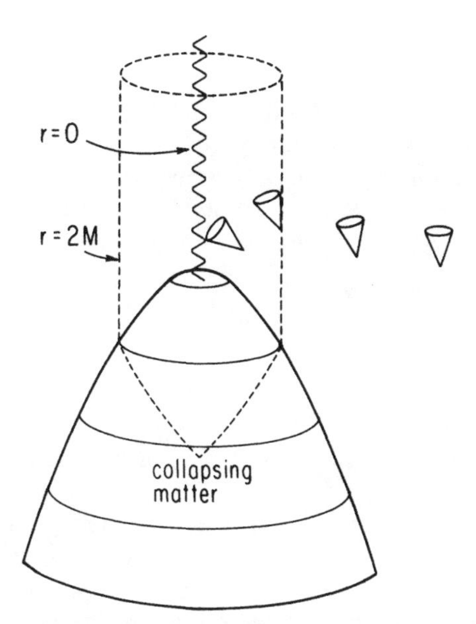

Fig. 6.12. Another representation of the spacetime of Figure 6.11. Here, one of the two suppressed spatial dimensions is restored, so each of the circles shown on the collapsing body corresponds to the 2-sphere surface of the body at an instant of time. However, the light cones no longer are represented by 45° lines. Indeed, the spacelike nature of the singularity and the inevitable capture by the singularity of any particle or light ray in the region $r < 2M$ is illustrated here by the "tipping over" of the future light cones in the strong field region.

Problems

1. Let M be a three-dimensional manifold possessing a spherically symmetric Riemannian metric with $\nabla_a r \neq 0$, where r is defined by equation (6.1.3).

a) Show that a new "isotropic" radial coordinate $\bar{r}$ can be introduced so that the metric takes the form $ds^2 = H(\bar{r})[d\bar{r}^2 + \bar{r}^2\, d\Omega^2]$. (This shows that every spherically symmetric three-dimensional space is conformally flat.)

b) Show that in isotropic coordinates the Schwarzschild metric is

$$ds^2 = -\frac{(1 - M/2\bar{r})^2}{(1 + M/2\bar{r})^2}dt^2 + \left(1 + \frac{M}{2\bar{r}}\right)^4[d\bar{r}^2 + \bar{r}^2\, d\Omega^2].$$

2. Calculate the Ricci tensor, R_{ab}, for a static, spherically symmetric spacetime, equation (6.1.5), using the coordinate component method of section 3.4*a*. Compare the amount of labor involved with that of the tetrad approach given in the text.

3. Consider the source-free ($j^a = 0$) Maxwell's equations (4.3.12) and (4.3.13) in a static, spherically symmetric spacetime, equation (6.1.5).

a) Argue that the general form of a Maxwell tensor which shares the static and spherical symmetries of the spacetime is $F_{ab} = 2A(r)(e_0)_{[a}(e_1)_{b]} + 2B(r)(e_2)_{[a}(e_3)_{b]}$, where the $(e_\mu)_a$ are defined by equation (6.1.6).

b) Show that if $B(r) = 0$, the general solution of Maxwell's equations with the form of part (*a*) is $A(r) = -q/r^2$, where q may be interpreted as the total charge. [The solution obtained with $B(r) \neq 0$ is a "duality rotation" of this solution, representing the field of a magnetic monopole.]

c) Write down and solve Einstein's equation, $G_{ab} = 8\pi T_{ab}$, with electromagnetic stress-energy tensor corresponding to the solution of part (*b*). Show that the general solution is the *Reissner-Nordstrom metric,*

$$ds^2 = -\left(1 - \frac{2M}{r} + \frac{q^2}{r^2}\right)dt^2 + \left(1 - \frac{2M}{r} + \frac{q^2}{r^2}\right)^{-1} dr^2 + r^2 d\Omega^2 \quad .$$

4. Let (M, g_{ab}) be a stationary spacetime with timelike Killing field ξ^a. Let $V^2 = -\xi^a \xi_a$.

a) Show that the acceleration $a^b = u^a \nabla_a u^b$ of a stationary observer is given by $a^b = \nabla^b \ln V$.

b) Suppose in addition that (M, g_{ab}) is asymptotically flat, i.e., that there exist coordinates t, x, y, z [with $\xi^a = (\partial/\partial t)^a$] such that the components of g_{ab} approach diag$(-1, 1, 1, 1)$ as $r \rightarrow \infty$, where $r = (x^2 + y^2 + z^2)^{1/2}$. (See chapter 11 for further discussion of asymptotic flatness.) As in the case of the Schwarzschild metric, the "energy as measured at infinity" of a particle of mass m and 4-velocity u^a is $E = -m\xi^a u_a$. Suppose a particle of mass m is held stationary by a (massless) string, with the other end of the string held by a stationary observer at large r. Let F denote the force exerted by the string on the particle. According to part (*a*) we have $F = mV^{-1}[\nabla_a V \nabla^a V]^{1/2}$. Use conservation-of-energy arguments to show that the force exerted by the observer at infinity on the other end of the string is $F_\infty = VF$. Thus, the magnitude of the force exerted at infinity differs from the force exerted locally by the redshift factor.

5. Derive the formula, equation (6.3.45), for the general relativistic time delay.

6. Show that any particle (not necessarily in geodesic motion) in region II ($r < 2M$) of the extended Schwarzschild spacetime, Figure 6.9, must decrease its radial coordinate at a rate given by $|dr/d\tau| \geqq [2M/r - 1]^{1/2}$. Hence, show that the maximum lifetime of any observer in region II is $\tau = \pi M$ [$\sim 10^{-5}(M/M_\odot)$ s], i.e., any observer in region II will be pulled into the singularity at $r = 0$ within this proper time. Show that this maximum time is approached by freely falling (i.e., geodesic) motion from $r = 2M$ with $E \rightarrow 0$.

PART II

ADVANCED TOPICS

SEVEN

METHODS FOR SOLVING EINSTEIN'S EQUATION

The theory of general relativity presented in chapter 4 provides a complete classical description of spacetime structure and gravitation. All physically possible spacetimes correspond to solutions of Einstein's equation (4.3.21). Thus, by solving Einstein's equation for spacetimes of physical interest, we can obtain complete predictions for the phenomena of interest. Unfortunately, Einstein's equation translates into a complicated coupled system of nonlinear partial differential equations and, as with most nonlinear partial differential equations, there exist no general methods for obtaining all solutions. Indeed, aside from the Robertson-Walker and Schwarzschild solutions (discussed in detail in chapters 5 and 6), the Kerr solution (see chapter 12), and some other solutions mentioned in the present chapter, very few solutions of physical interest have been found.

In this chapter, we shall discuss some of the methods which have been employed to obtain physically relevant solutions. Our discussion of successful methods is far from exhaustive. Furthermore, we shall make no attempt to enumerate all the solutions which have been obtained or discuss their properties. Rather, our purpose here is to present some of the most important techniques for extracting solutions.

In section 7.1 we analyze solutions with stationary and axisymmetric symmetry and show how Einstein's equation in vacuum can be reduced to a system of two coupled equations for two unknown functions together with a further quadrature. Unfortunately, even these equations are difficult to solve directly, except in the static case, where the complete solution can be found. In section 7.2 we analyze spatially homogeneous (but not isotropic) cosmological models and show how Einstein's equation can be reduced to solving a coupled system of ordinary differential equations. (Such a system always can be solved numerically, if not analytically, so there is no obstacle to obtaining all homogeneous cosmological solutions.) In section 7.3 we give a brief description of the procedure for obtaining solutions by assuming special algebraic properties of the Weyl tensor as a simplifying hypothesis. This technique has produced a wealth of solutions, though only a few of direct physical interest. In section 7.4 we outline a procedure for constructing new solutions from a given solution with a Killing vector field. This technique has proven particularly useful for generating stationary, axisymmetric solutions. Finally, in section 7.5 we derive the equations governing small perturbations from a known exact solution.

Although it may not be possible to find nearby exact solutions, a great deal of useful information often can be obtained from perturbation theory.

7.1 Stationary, Axisymmetric Solutions

As already mentioned in section 6.1, a spacetime is said to be *stationary* if there exists a one-parameter group of isometries σ_t whose orbits are timelike curves. Thus, every stationary spacetime possesses a timelike Killing vector field ξ^a (see appendix C). (Conversely, every spacetime with a timelike Killing vector field whose orbits are complete is stationary.) Similarly, we call a spacetime *axisymmetric* if there exists a one-parameter group of isometries χ_ϕ whose orbits are closed spacelike curves, which implies the existence of a spacelike Killing field ψ^a whose integral curves are closed. We call a spacetime stationary and axisymmetric if it possesses both these symmetries and if, in addition, the actions of σ_t and χ_ϕ commute:

$$\sigma_t \circ \chi_\phi = \chi_\phi \circ \sigma_t \quad , \tag{7.1.1}$$

i.e., the rotations commute with the time translations. This is easily seen to be equivalent to the condition that the Killing vector fields ξ^a and ψ^a commute,

$$[\xi, \psi] = 0 \quad , \tag{7.1.2}$$

Stationary, axisymmetric spacetimes are of considerable interest in general relativity since they describe equilibrium configurations of axisymmetric, rotating bodies. Thus, in particular, in order to generalize the results of section 6.2 to account for rotation, we must be able to solve Einstein's equation with perfect fluid source in the presence of these symmetries. Stationary, axisymmetric vacuum solutions are also of great interest, since they describe the exterior gravitational field of rotating bodies and would be needed to match onto the interior solutions.

The commutativity of ξ^a and ψ^a implies that we can choose coordinates ($x^0 = t$, $x^1 = \phi$, x^2, x^3) so that both $\xi^a = (\partial/\partial t)^a$ and $\psi^a = (\partial/\partial\phi)^a$ are coordinate vector fields. As discussed in appendix C, the metric components in such a coordinate system will be independent of t and ϕ, so the metric will take the form

$$ds^2 = \sum_{\mu,\nu} g_{\mu\nu}(x^2, x^3) dx^\mu dx^\nu \quad . \tag{7.1.3}$$

Thus, we must solve for 10 unknown functions, $g_{\mu\nu}$, of two variables. We shall show, now, that by a further careful selection of coordinate system, a weak further assumption (see hypothesis i of theorem 7.1.1 below) and some use of Einstein's equation, we can reduce the metric to a form involving only three functions of two variables (see eq. [7.1.22] below), and reduce the problem of solving Einstein's equation to that of solving two equations for two unknown functions (see eqs. [7.1.24] and [7.1.25] below) together with a quadradure for a third function (see eqs. [7.1.26] and [7.1.27]).

The first crucial simplification arises from the fact that under the hypotheses of the following theorem, the two-dimensional subspaces of the tangent space at each point which are spanned by the vectors orthogonal to ξ^a and ψ^a are integrable, i.e., tangent to two-dimensional surfaces. We shall state and prove the theorem first and then discuss its implications for simplifying the general form of stationary, axisymmetric metrics.

THEOREM 7.1.1. *Let ξ^a and ψ^a be two commuting Killing fields such that* (i) $\xi_{[a}\psi_b\nabla_c\xi_{d]}$ *and* $\xi_{[a}\psi_b\nabla_c\psi_{d]}$ *each vanishes at at least one point of the spacetime (which, in particular, will be true if either ξ^a or ψ^a vanishes at one point) and* (ii) $\xi^a R_a{}^{[b}\xi^c\psi^{d]} = \psi^a R_a{}^{[b}\xi^c\psi^{d]} = 0$. *Then the 2-planes orthogonal to ξ^a and ψ^a are integrable.*

Proof. This theorem is a direct application of Frobenius's theorem (see appendix B). We want to know when the planes orthogonal to the 1-forms ξ_a and ψ_a are integrable. According to the dual formulation of Frobenius's theorem (theorem B.3.2) the necessary and sufficient conditions for integrability are

$$\nabla_{[a}\xi_{b]} = (\mu^1)_{[a}\xi_{b]} + (\mu^2)_{[a}\psi_{b]} \quad , \tag{7.1.4}$$

$$\nabla_{[a}\psi_{b]} = (\sigma^1)_{[a}\xi_{b]} + (\sigma^2)_{[a}\psi_{b]} \quad , \tag{7.1.5}$$

where $(\mu^1)_a$, $(\mu^2)_a$, $(\sigma^1)_a$, $(\sigma^2)_a$ are arbitrary 1-forms. We shall establish the theorem by proving that equation (7.1.4) holds; the proof that equation (7.1.5) also holds follows in an identical manner. Equation (7.1.4) is equivalent to

$$\psi_{[a}\xi_b\nabla_c\xi_{d]} = 0 \quad , \tag{7.1.6}$$

which, in turn, is equivalent to

$$\epsilon_{abcd}\psi^a\xi^b\nabla^c\xi^d = 0 \quad , \tag{7.1.7}$$

where ϵ_{abcd} is the volume element associated with the metric (see appendix B). We define the *twist*, ω_a, of ξ^a by

$$\omega_a = \epsilon_{abcd}\xi^b\nabla^c\xi^d \quad . \tag{7.1.8}$$

(According to eq. [B.3.6], ω_a measures the failure of ξ^a to be hypersurface orthogonal.) Our task is to show that $\psi^a\omega_a = 0$.

By hypothesis (i), we know $\psi^a\omega_a$ vanishes at at least one point, so it will vanish everywhere if and only if its derivative is identically zero. We have

$$\begin{aligned}\nabla_b(\psi^a\omega_a) &= \psi^a\nabla_b\omega_a + \omega_a\nabla_b\psi^a \\ &= \psi^a\nabla_a\omega_b + \omega_a\nabla_b\psi^a + 2\psi^a\nabla_{[b}\omega_{a]} \\ &= £_\psi\omega_b + 2\psi^a\nabla_{[b}\omega_{a]} \quad ,\end{aligned} \tag{7.1.9}$$

where the formula (C.2.13) for the Lie derivative was used. However, the group of diffeomorphisms χ_ϕ generated by ψ^a leaves ξ^a invariant (since ψ^a and ξ^a commute) and leaves the metric invariant (since ψ^a is a Killing field). Hence, χ_ϕ must leave invariant any tensor field that can be constructed uniquely out of ξ^a and the metric.

Since ω_b is such a tensor field, we must have[1]

$$\pounds_\psi \omega_b = 0 \quad . \tag{7.1.10}$$

Thus, to complete the proof, we need only compute $\nabla_{[b}\omega_{a]}$. We have

$$\begin{aligned} \epsilon^{abcd}\nabla_c\omega_d &= \epsilon^{abcd}\epsilon_{defg}\nabla_c(\xi^e\nabla^f\xi^g) \\ &= 6\nabla_c(\xi^{[c}\nabla^a\xi^{b]}) \quad , \end{aligned} \tag{7.1.11}$$

where equations (B.2.11) and (B.2.13) were used. Now, using Killing's equation (C.3.1), we have

$$\xi^{[c}\nabla^a\xi^{b]} = \frac{1}{3}\{\xi^c\nabla^a\xi^b + \xi^b\nabla^c\xi^a + \xi^a\nabla^b\xi^c\} \quad . \tag{7.1.12}$$

However, we find

$$\begin{aligned} \nabla_c(\xi^c\nabla^a\xi^b) &= (\nabla_c\xi^c)\nabla^a\xi^b + \xi^c\nabla_c\nabla^a\xi^b \\ &= -\xi^c R^{ab}{}_{cd}\xi^d \\ &= 0 \quad . \end{aligned} \tag{7.1.13}$$

Here the trace of Killing's equation was used to eliminate the first term, while equation (C.3.6) was used to express the second term in terms of the Riemann tensor. The antisymmetry of the Riemann tensor in its last two indices was used to get the final conclusion. On the other hand, we have

$$\begin{aligned} \nabla_c(\xi^b\nabla^c\xi^a + \xi^a\nabla^b\xi^c) &= (\nabla_c\xi^b)(\nabla^c\xi^a) + (\nabla_c\xi^a)(\nabla^b\xi^c) + \xi^b\nabla_c\nabla^c\xi^a + \xi^a\nabla_c\nabla^b\xi^c \\ &= 2\xi^{[b}\nabla_c\nabla^{|c|}\xi^{a]} \\ &= -2\xi^{[b}R^{a]}{}_c\xi^c \quad , \end{aligned} \tag{7.1.14}$$

where Killing's equation was used to get the second line and equation (C.3.9) was used to get the third line. Thus, we find

$$\begin{aligned} \nabla_{[a}\omega_{b]} &= -\frac{1}{4}\epsilon_{abcd}\epsilon^{cdef}\nabla_e\omega_f \\ &= -\epsilon_{abcd}\xi^c R^d{}_e\xi^e \quad , \end{aligned} \tag{7.1.15}$$

and, finally,

$$\begin{aligned} \nabla_b(\psi^a\omega_a) &= 2\psi^a\nabla_{[b}\omega_{a]} \\ &= -2\epsilon_{bacd}\psi^a\xi^c R^d{}_e\xi^e \\ &= 0 \end{aligned} \tag{7.1.16}$$

by hypothesis (ii). This completes the proof. □

1. Another way of seeing this is that in a coordinate system adapted to both ξ^a and ψ^a, any tensor $T^{a\ldots b}{}_{c\ldots d}$ whose components $T^{\alpha\ldots\beta}{}_{\rho\ldots\sigma}$ are constructed out of the components ξ^α and $g_{\mu\nu}$ and their coordinate derivatives must satisfy $\partial(T^{\alpha\cdots\beta}{}_{\rho\ldots\sigma})/\partial\phi = 0$, which proves that $\pounds_\psi T^{a\cdots b}{}_{c\cdots d} = 0$ accord-

The hypotheses of theorem 7.1.1 will be satisfied in a wide range of stationary, axisymmetric spacetimes of physical interest. In particular, if the spacetime is asymptotically flat, there must be a "rotation axis" on which ψ^a vanishes, so hypothesis (i) will be satisfied. For a vacuum spacetime, we have $R_{ab} = 0$, so hypothesis (ii) is trivially satisfied. In addition, hypothesis (ii) will also be satisfied when the matter stress-energy $T_{ab} = (8\pi)^{-1}(R_{ab} - \frac{1}{2}g_{ab}R)$ has the form of a perfect fluid with 4-velocity in the plane spanned by ξ^a and ψ^a (i.e., circular flow) or if T_{ab} is the stress tensor of a stationary, axisymmetric electromagnetic field (Carter 1969).

For spacetimes which satisfy the hypotheses of theorem 7.1.1, we may choose coordinates x^2, x^3 in one of the orthogonal two-surfaces and "carry" these coordinates to the rest of the spacetime along the integral curves of ξ^a and ψ^a. In the coordinates (t, ϕ, x^2, x^3), the metric components take the form

$$g_{\mu\nu} = \begin{pmatrix} -V & W & 0 & 0 \\ & X & 0 & 0 \\ & & g_{22} & g_{23} \\ \text{(sym.)} & & & g_{33} \end{pmatrix} \quad , \tag{7.1.17}$$

where $V = -g_{00} = -\xi_a \xi^a$, $W = g_{01} = \xi^a \psi_a$, $X = g_{11} = \psi^a \psi_a$, and the 2×2 block of zeros expresses the orthogonality of $\partial/\partial x^2$ and $\partial/\partial x^3$ with $\partial/\partial t$ and $\partial/\partial\phi$. (A physical interpretation of W is established in problem 3.) Thus, theorem 7.1.1 allows us to reduce the number of nonvanishing metric components to six. Without theorem 7.1.1, only two metric components in the 2×2 block could be set to zero by use of the coordinate freedom available in a coordinate system adapted to ξ^a and ψ^a.

We still have not specified how the two-surface coordinates x^2 and x^3 are to be chosen, and, as we shall see, significant further simplification can be achieved by a judicious choice of these coordinates. We define the scalar function ρ by

$$\rho^2 = VX + W^2 \quad ; \tag{7.1.18}$$

i.e., ρ^2 is minus the determinant of the $t - \phi$ part of the metric. Assuming $\nabla_a \rho \neq 0$, we choose ρ as one of the coordinates, x^2, of the two-surface. We choose the other coordinate, $z = x^3$, so that $\nabla_a z$ is orthogonal to $\nabla_a \rho$. [This is accomplished by setting $z =$ constant along the integral curves of $\nabla^a \rho$, which uniquely determines z up to the transformations $z \to z' = f(z)$.] In the coordinates t, ϕ, ρ, z the metric takes the form

$$ds^2 = -V(dt - w\, d\phi)^2 + V^{-1}\rho^2 d\phi^2 + \Omega^2(d\rho^2 + \Lambda\, dz^2) \quad , \tag{7.1.19}$$

where $w = W/V$. Thus, we have reduced the unknown metric components to four functions, V, w, Ω, Λ of two variables ρ, z. Equation (7.1.19) is the general form of a stationary, axisymmetric spacetime satisfying the hypotheses of theorem 7.1.1.

---

ing to equation (C.2.4). That $\pounds_\psi \omega_b = 0$ holds, of course, also could be proven directly using the formulas of appendix C (problem 1).

The form of the metric can be simplified further for vacuum spacetimes, $R_{ab} = 0$. The components of R_{ab} in the plane spanned by ξ^a and ψ^a can be computed from equation (C.3.9) or, alternatively, from the general coordinate basis formulas of section 3.4*a*. The equation

$$0 = R^t{}_t + R^\phi{}_\phi = (\nabla_a t) R^a{}_b \xi^b + (\nabla_a \phi) R^a{}_b \psi^b \tag{7.1.20}$$

yields

$$D^a D_a \rho = 0 \quad , \tag{7.1.21}$$

where D_a is the covariant derivative operator on the two-dimensional surface spanned by ρ and z with the induced metric $ds^2 = \Omega^2(d\rho^2 + \Lambda\, dz^2)$. Thus, ρ is harmonic, i.e., it satisfies the two-dimensional Laplace equation in the two-surfaces. This has two important consequences: (1) If $\rho \neq$ constant, it can be shown that $\nabla_a \rho$ can vanish only at isolated points. Since our coordinates ρ, z will be well behaved except where $\nabla_a \rho = 0$, this shows that our coordinate system can break down only at isolated points. In fact, in many situations it is possible to show that $\nabla_a \rho \neq 0$ everywhere, so the coordinates ρ, z are globally well behaved (Carter 1973). (2) It follows directly from equation (7.1.21) that Λ is a function of z alone. Thus we may use the remaining coordinate freedom to transform z via $z \to \int \Lambda^{1/2} dz$ and thereby set $\Lambda = 1$. (Note that z then becomes the harmonic function conjugate to ρ; see problem 2 of chapter 3.) Except for the trivial constant re-scaling or shift of origin of the coordinates t, ϕ, z, we have now completely specified our coordinate system. In these coordinates the metric takes the remarkably simple form (Papapetrou 1953, 1966)

$$ds^2 = -V(dt - w\, d\phi)^2 + V^{-1}[\rho^2 d\phi^2 + e^{2\gamma}(d\rho^2 + dz^2)] \quad , \tag{7.1.22}$$

where $\gamma = \frac{1}{2}\ln(V\Omega^2)$. Note that in flat spacetime ($V = 1$, $w = \gamma = 0$), the coordinates (ϕ, ρ, z) are ordinary cylindrical coordinates.

In deriving equation (7.1.22), we have used Einstein's equation thus far only in hypothesis (ii) of theorem 7.1.1 and in equation (7.1.20). The remaining components of the vacuum Einstein's equation $R_{ab} = 0$ can be computed in a straightforward manner using the methods of section 3.4*a*. We shall not present the details here but will merely quote the final results. We obtain four independent equations for V, w, and γ. The first two involve only V and w and are most conveniently formulated by defining an (unphysical) flat three-dimensional metric,

$$d\tilde{s}^2 = \rho^2 d\phi^2 + d\rho^2 + dz^2 \quad , \tag{7.1.23}$$

and expressing the equations in terms of the flat derivative operator $\vec{\nabla}$ associated with this metric. (We use the usual vector notation $\vec{\nabla}$ rather than the index notation to avoid confusion with the derivative operator associated with the true spacetime metric.) In this way, the first two equations may be viewed as equations for axisymmetric scalar fields V, w in the unphysical three-dimensional flat space of equation (7.1.23). The equations are:

$$\vec{\nabla} \cdot \{V^{-1}\vec{\nabla} V + \rho^{-2} V^2 w \vec{\nabla} w\} = 0 \quad , \tag{7.1.24}$$

$$\vec{\nabla} \cdot \{\rho^{-2} V^2 \vec{\nabla} w\} = 0 \quad , \tag{7.1.25}$$

$$\frac{\partial \gamma}{\partial \rho} = \frac{1}{4}\rho V^{-2}\left[\left(\frac{\partial V}{\partial \rho}\right)^2 - \left(\frac{\partial V}{\partial z}\right)^2\right] - \frac{1}{4}\rho^{-1}V^2\left[\left(\frac{\partial w}{\partial \rho}\right)^2 - \left(\frac{\partial w}{\partial z}\right)^2\right] \quad , \tag{7.1.26}$$

$$\frac{\partial \gamma}{\partial z} = \frac{1}{2}\rho V^{-2}\frac{\partial V}{\partial \rho}\frac{\partial V}{\partial z} - \frac{1}{2}\rho^{-1}V^2\frac{\partial w}{\partial \rho}\frac{\partial w}{\partial z} \quad . \tag{7.1.27}$$

The last two equations are in danger of overdetermining γ. However, the integrability condition $\partial^2\gamma/\partial z\partial\rho = \partial^2\gamma/\partial\rho\partial z$ is identically satisfied by virtue of equations (7.1.24) and (7.1.25). Hence, given a solution of equations (7.1.24) and (7.1.25), a solution of equations (7.1.26) and (7.1.27) always exists and is unique up to the addition of a constant. This solution can be obtained by performing a line integral of the right sides of equations (7.1.26) and (7.1.27), i.e., explicitly,

$$\gamma(p) - \gamma(q) = \int_q^p \left(\frac{\partial \gamma}{\partial \rho}\,d\rho + \frac{\partial \gamma}{\partial z}\,dz\right) \quad . \tag{7.1.28}$$

Thus, apart from the computation required to find γ explicitly, the problem of solving for all stationary, axisymmetric vacuum solutions of Einstein's equation has been reduced to solving equations (7.1.24) and (7.1.25) for the two axisymmetric functions V and w in ordinary three-dimensional Euclidean space. This is a remarkable simplification of the original problem of solving the full set of Einstein's equation for 10 unknown functions $g_{\mu\nu}$. Nevertheless, these equations are still sufficiently difficult to solve that—with the exception of static solutions discussed below—almost no solutions have been found by direct attack on equations (7.1.24) and (7.1.25). (In cases where solutions have been obtained, it often has proven useful to work with coordinates other than ρ and z [Zipoy 1966; Chandrasekhar 1983].) The equations can be reformulated through the introduction of potentials (Ernst 1968), and a few solutions of interest have been found by direct study of these modified equations (Tomimatsu and Sato 1972, 1973). However, recent progress in methods for generating solutions (discussed briefly in section 7.4 below) has produced algorithms for obtaining *all* the asymptotically flat stationary, axisymmetric solutions, although the computations required in this procedure remain formidable.

The major exception where direct attack has been completely successful is the case of static, axisymmetric spacetimes (with the axisymmetric Killing field ψ^a lying in the hypersurfaces orthogonal to the static Killing field). In that case, we have $w = 0$, so equation (7.1.25) is trivially satisfied, and if we define $\chi = \ln V$, equation (7.1.24) reduces to simply

$$\nabla^2 \chi = 0 \quad , \tag{7.1.29}$$

i.e., χ is an axisymmetric solution of the ordinary Laplace equation in three-dimensional flat space. Since all such solutions of this equation are explicitly known, all static, axisymmetric solutions of Einstein's equation can be obtained. This analysis of the static, axisymmetric spacetimes was first carried out by Weyl (1917), and the solutions are often referred to as the Weyl solutions. It should be pointed out that the properties of the solution of equation (7.1.29) do not translate in a simple way into properties of the spacetime metric it generates. Specifically, the monopole solution of Laplace's equation does *not* generate a spherically symmetric spacetime. To obtain the spherically symmetric Schwarzschild solution by this procedure, one must choose X to be the potential of a finite rod on the z-axis (centered on the origin) with constant mass per unit length.

7.2 Spatially Homogeneous Cosmologies

In chapter 5 we studied in detail the solutions of Einstein's equation which are spatially homogeneous and isotropic. These models successfully account for many of the observed properties of our universe. However, our universe certainly is not *exactly* homogeneous and isotropic, and one would like to have a better understanding of the possible dynamical behavior of the universe in the absence of these symmetries. We already showed in chapter 5 that spatial isotropy at each point implies spatial homogeneity, so the simplest first step toward generalizing the Robertson-Walker models would be to obtain the spatially homogeneous but anisotropic solutions. We would expect this to be a tractable problem, since on account of the spatial symmetry, only the time variations should be nontrivial. Thus, Einstein's equation should reduce to a system of ordinary differential equations. The main purpose of this section is to demonstrate that this is indeed the case by outlining the derivation of equations (7.2.41)–(7.2.43) below. The solutions to these equations provide interesting models for the possible behavior of the universe near an initial singularity (Collins and Ellis 1979).

We already defined a spatially homogeneous spacetime at the beginning of chapter 5 as one which possesses a group of isometries whose orbits are spacelike hypersurfaces which foliate (i.e., pass through every point of) the spacetime. Our first task is to define more precisely the nature of the group of isometries we seek and to develop some important properties of such groups. The notion of a Lie group which we are about to define generalizes to "*m* parameters" the notion of a 1-parameter group of transformations discussed in section 2.2.

First, we remind the reader that a group $\mathscr{G}$ is simply a set together with a map $\mathscr{G} \times \mathscr{G} \rightarrow \mathscr{G}$ (called "multiplication") and a preferred element e (called the identity element) such that: (1) the multiplication law is associative, $g_1(g_2g_3) = (g_1g_2)g_3$; (2) for all $g \in \mathscr{G}$, we have $eg = ge = g$; and (3) for every $g \in \mathscr{G}$ there is an element of $\mathscr{G}$—denoted g^{-1} and called the inverse of g—such that $gg^{-1} = g^{-1}g = e$.

Many sets consisting of a finite number of elements provide examples of groups. A good example of a group with infinitely many elements is the collection of diffeomorphisms of a manifold M with "multiplication" given by composition, $\psi\phi = \psi \circ \phi$, and e taken as the identity map, $e(p) = p$ for all $p \in M$. Similarly, the collection of isometries of a manifold M with metric g_{ab} also forms a group (which is a subgroup of the group of diffeomorphisms), since the composition of two isometries is an isometry, and the inverse of any isometry is an isometry.

A *Lie group*, G, of dimension m is a group which is also an m-dimensional manifold such that the inverse map $i(g) = g^{-1}$ and the multiplication map $f(g_1, g_2) = g_1g_2$ are smooth (C^∞). Thus, the elements of a Lie group can be characterized locally by m parameters, and the multiplication and inverse operations depend smoothly on these parameters. The group of diffeomorphisms of a manifold M does not yield a (finite dimensional) Lie group, since this group is "too big" to be characterized by m parameters. However, it is possible to show that the group of isometries always yields a (possibly zero-dimensional and possibly disconnected) Lie group. In fact, an argument given in section C.3 of appendix C shows that on a manifold, M, of dimension n, the Lie group of isometries cannot have dimension greater than $n(n+1)/2$. We establish, now, a number of properties of Lie groups which will be used below in our analysis of homogeneous cosmologies.

Let G be a Lie group of dimension m. It follows from the smoothness of i and f that for each $h \in G$ the map

$$\psi_h(g) = hg \quad , \tag{7.2.1}$$

called *left translation* by h, is a diffeomorphism. Let ψ_h^* denote the map on tensors induced by ψ_h (see appendix C). If a vector field v^a on G satisfies

$$\psi_h^* v^a = v^a \tag{7.2.2}$$

for all $h \in G$, then v^a is called *left invariant*. More generally, a tensor field invariant under ψ_h^* for all $h \in G$ is said to be left invariant. It is easy to see that the left invariant vector fields form a vector space since sums and scalar multiples of left invariant vector fields are also left invariant. Clearly, a left invariant vector field is determined by its value at V_e, the tangent space at the identity element, e, since if v^a is left invariant, we have

$$v^a|_h = \psi_h^*[v^a|_e] \quad . \tag{7.2.3}$$

Conversely, if we define a vector field in terms of its value at e by equation (7.2.3), we produce a left invariant vector field. Thus, the left invariant vector fields on G form an m-dimensional vector space.

Now, if ϕ is any diffeomorphism on any manifold and v^a and w^a are any two vector fields, we have

$$\phi^*([v,w]) = [\phi^*(v),\phi^*(w)] \quad , \tag{7.2.4}$$

where [,] denotes the commutator, defined in section 2.2 above. Thus, if v^a and w^a are left invariant vector fields on a Lie group, the commutator $[v, w]^a$ also will be a left invariant vector field. Since the commutator depends linearly on v^a and w^a, this implies that there exists a left invariant tensor field $c^a{}_{bc}$ such that

$$[v,w]^a = c^a{}_{bc}v^b w^c \quad . \tag{7.2.5}$$

The tensor field $c^a{}_{bc}$ is called the *structure constant tensor* of the Lie group. It follows immediately from its definition that

$$c^a{}_{bc} = -c^a{}_{cb} \quad . \tag{7.2.6}$$

Furthermore, the Jacobi identity for commutators,

$$[u,[v,w]] + [v,[w,u]] + [w,[u,v]] = 0 \quad , \tag{7.2.7}$$

implies that

$$c^e{}_{d[a}c^d{}_{bc]} = 0 \quad . \tag{7.2.8}$$

A finite dimensional vector space with a tensor $c^a{}_{bc}$ of type $(1,2)$ satisfying equations (7.2.6) and (7.2.8) is called a *Lie algebra*. We have shown above that the left invariant vector fields of a Lie group comprise a Lie algebra. Thus, every Lie group gives rise to a Lie algebra. Conversely, it is possible to show that every Lie algebra gives rise to a Lie group in a unique way up to global topological structure. More precisely, given any Lie algebra, there exists a unique connected, simply connected Lie group whose Lie algebra coincides with the given Lie algrebra.[2] This fact tremendously simplifies the analysis of Lie groups, since Lie algebras are much simpler objects to work with than Lie groups.

Let α_a be a left invariant dual vector field. If v^a is a left invariant vector field, then $\alpha_a v^a$ is constant, so

$$0 = \nabla_b(\alpha_a v^a) = v^a \nabla_b \alpha_a + \alpha_a \nabla_b v^a \quad , \tag{7.2.9}$$

where ∇_a is any derivative operator on G. If v^a and w^a are both left invariant vector fields, we have

$$\begin{aligned} 2v^a w^b \nabla_{[a}\alpha_{b]} &= (v^a w^b - v^b w^a)\nabla_a \alpha_b \\ &= -v^a \alpha_b \nabla_a w^b + \alpha_b w^a \nabla_a v^b \\ &= -\alpha_b [v, w]^b \\ &= -\alpha_b c^b{}_{cd} v^c w^d \quad . \end{aligned} \tag{7.2.10}$$

Thus, we find that every left invariant dual vector field α_b satisfies

$$2\nabla_{[a}\alpha_{b]} = -\alpha_c c^c{}_{ab} \quad . \tag{7.2.11}$$

The *right translation* map χ_h, defined by $\chi_h(g) = gh$, also is a diffeomorphism. We define right invariant tensor fields analogously to left invariant tensor fields. As discussed in section 2.2, any vector field generates a one-parameter group of diffeomorphisms. Let x^a be a right invariant vector field, and let ϕ_t be its one-parameter group of diffeomorphisms. Since x^a is right invariant, we have

$$\chi_h \circ \phi_t = \phi_t \circ \chi_h \tag{7.2.12}$$

for all $h \in G$. Consequently, defining $h(t) = \phi_t(e)$, we have, for all $g \in G$,

$$\phi_t(g) = \phi_t \circ \chi_g(e) = \chi_g \circ \phi_t(e) = \chi_g[h(t)] = h(t)g = \psi_{h(t)}(g) \quad . \tag{7.2.13}$$

Thus, we find

$$\phi_t = \psi_{h(t)} \quad ; \tag{7.2.14}$$

i.e., the *right* invariant vector fields are infinitesimal generators of *left* translations.

2. Existence of a Lie group associated with a Lie algebra follows from Ado's theorem (see, e.g., Jacobson 1962). Existence of a simply connected Lie group follows from the universal covering group construction described in chapter 13. Uniqueness is proven, e.g., in Warner (1971).

This immediately implies that if x^a is a right invariant vector field and v^a is a left invariant vector field, we have

$$0 = \pounds_x v^a = [x, v]^a \quad . \tag{7.2.15}$$

Similarly, if x^a is a right invariant vector field and α_a is a left invariant dual vector field, we have

$$0 = \pounds_x \alpha_b = x^a \nabla_a \alpha_b + \alpha_a \nabla_b x^a \tag{7.2.16}$$

where ∇_a is any derivative operator. Consequently, if x^a and y^a are both right invariant but α_b is left invariant, we find, by a calculation similar to that of equation (7.2.10),

$$\begin{aligned} 2x^a y^b \nabla_{[a}\alpha_{b]} &= y^b x^a \nabla_a \alpha_b - x^b y^a \nabla_a \alpha_b \\ &= -y^b \alpha_a \nabla_b x^a + x^b \alpha_a \nabla_b y^a \\ &= \alpha_a [x, y]^a \quad . \end{aligned} \tag{7.2.17}$$

On the other hand, by equation (7.2.11), we have

$$2x^a y^b \nabla_{[a}\alpha_{b]} = x^a y^b (-\alpha_c c^c{}_{ab}) \quad . \tag{7.2.18}$$

Thus, we find

$$[x, y]^c = -c^c{}_{ab} x^a y^b \quad ; \tag{7.2.19}$$

i.e., except for a change in sign, the right invariant vector fields satisfy the same commutation relations as the left invariant vector fields. Note that equation (7.2.19) shows that $c^c{}_{ab}$ is also right invariant.[3]

Let us apply the above facts about Lie groups to homogeneous cosmological models. By definition, in a spatially homogeneous spacetime (M, g_{ab}), there exists a family of spacelike hypersurfaces Σ_t such that for any two points $p, q \in \Sigma_t$ there exists an element $g: M \to M$ of the Lie group, G, of isometries such that $g(p) = q$. (G is said to act *transitively* on each Σ_t.) We will restrict attention to the case where for all Σ_t and for all $p, q \in \Sigma_t$ there is a *unique* element $g \in G$ such that $g(p) = q$, in which case G is said to act *simply transitively* on each Σ_t. This implies $\dim G = \dim \Sigma_t = 3$. In fact, almost no loss of generality results from restricting consideration to a simply transitive action because it turns out (see, e.g. MacCallum 1979) that the only case where G does not act simply transitively or does not possess a subgroup with simply transitive action[4] is the group $SO(3) \times \mathbb{R}$ acting on the "cylinder" $S^2 \times \mathbb{R}$ [with the orbits of SO(3) being the two-dimensional spheres]; spatially homogeneous models with this isometry group (called Kantowski-Sachs models) can be treated separately by similar techniques (see, e.g., Ryan and Shepley 1975).

3. The tensor $c^c{}_{da} c^d{}_{cb}$ is symmetric in a and b. For semisimple Lie groups it also is nondegenerate. Thus, semisimple Lie groups possess a natural bi-invariant (i.e., left *and* right invariant) metric $g_{ab} = c^c{}_{da} c^d{}_{cb}$.

4. In particular, the isometry groups of the homogeneous, isotropic models of chapter 5 all possess subgroups which have a simply transitive action on the homogeneous surfaces.

The advantage of considering simply transitive action is that if we (arbitrarily) choose a point $p \in \Sigma_t$, we can put the elements of G into correspondence with the points of Σ_t by the association $g \to g(p)$. (A simply transitive action is needed to ensure that this correspondence is one-to-one.) Under this identification of G and Σ_t, the action on Σ_t of the isometry g corresponds to left multiplication by g on G. Thus, the tensor fields on Σ_t which are preserved under the isometries—in particular, the spatial metric, h_{ab}, on Σ_t—correspond precisely to the left invariant tensor fields on G. In particular, this means that the vector fields on Σ_t which are preserved under the isometries satisfy the commutation relations (7.2.5), and, similarly, the invariant dual vectors satisfy equation (7.2.11). Furthermore, the infinitesimal generators of the isometries on Σ_t—i.e., the Killing vector fields of Σ_t—correspond to the right invariant vector fields of G. Thus, the Killing vector fields of Σ_t satisfy the commutation relations (7.2.19).

Our next task is to put the metric of a spatially homogeneous spacetime with simply transitive group action into a useful canonical form. Consider a single homogeneous hypersurface Σ_0. Ideally, one would like to choose spatial coordinates adapted to the Killing vector fields so that the coordinate components of the spacetime metric would be independent of the spatial coordinates. However, except for the simple case of the translation group, $G = \mathbb{R}^3$, this cannot be done, since coordinate vector fields must commute, but according to equation (7.2.19) the Killing vector fields do not commute unless $c^c{}_{ab} = 0$. Thus, in general, only one Killing vector field can be employed as a coordinate vector field, and this is not particularly useful. Thus, instead, we choose a basis of dual vector fields $(\sigma^1)_a$, $(\sigma^2)_a$, $(\sigma^3)_a$ which are preserved under the isometries (which we can do by choosing an arbitrary basis at $p \in \Sigma_0$ and defining the basis elsewhere by "left translation"). Since the spatial metric, h_{ab}, on Σ_0 (obtained by restriction of the spacetime metric g_{ab} to vectors tangent to Σ_0) is left invariant, we have

$$h_{ab} = \sum_{\alpha,\beta=1}^{3} h_{\alpha\beta}(\sigma^\alpha)_a(\sigma^\beta)_b \quad , \tag{7.2.20}$$

where the components, $h_{\alpha\beta}$, are constant on Σ_0, i.e., independent of spatial position.

Let $p \in \Sigma_0$, let t^a denote the unit normal to Σ_0 at p, and let γ be the geodesic determined by (p, t^a). Then γ will be orthogonal to all the spatial hypersurfaces it intersects because, by proposition C.3.1, the tangent to γ must always remain orthogonal to all the spatial Killing vector fields since it is initially orthogonal to them. We label the other spatially homogeneous hypersurfaces by the proper time t of the intersection of the geodesic γ with the hypersurface. Then, the vector field $t^a = -\nabla^a t$ will be everywhere orthogonal to each Σ_t (since t is constant on Σ_t), and the integral curves of t^a are all geodesics (with $t^a t_a = -1$), since this is true along γ by construction and hence is true everywhere on each Σ_t by spatial homogeneity. We define the dual vector fields $(\sigma^\alpha)_a$ throughout the spacetime in terms of their values on the initial surface Σ_0 by "Lie transport" along t^a, i.e., by setting

$$\pounds_t(\sigma^\alpha)_a = 0 \quad , \tag{7.2.21}$$

or, equivalently, by defining $(\sigma^\alpha)_a$ on the hypersurface Σ_t by

$$(\sigma^\alpha)_a(t) = \phi_t^*[\sigma^\alpha(0)]_a \quad , \tag{7.2.22}$$

where ϕ_t denotes the one-parameter group of diffeomorphisms generated by t^a. It follows directly from equation (7.2.21) that $(\sigma^\alpha)_a t^a = 0$ everywhere.

Since ϕ_t is constructed in an entirely geometrical manner, it is not difficult to see that it must commute with any isometry g,

$$\phi_t \circ g = g \circ \phi_t \quad . \tag{7.2.23}$$

It follows immediately from this fact that on each Σ_t (i.e., not only on Σ_0), the basis vectors $(\sigma^\alpha)_a$ are invariant under the spatial isometries. Note also that

$$\begin{aligned} 2t^a \nabla_{[a}(\sigma^\alpha)_{b]} &= t^a \nabla_a (\sigma^\alpha)_b - t^a \nabla_b (\sigma^\alpha)_a \\ &= t^a \nabla_a (\sigma^\alpha)_b + (\sigma^\alpha)_a \nabla_b t^a \\ &= \pounds_t (\sigma^\alpha)_b \\ &= 0 \quad . \end{aligned} \tag{7.2.24}$$

Thus, $\nabla_{[a}(\sigma^\alpha)_{b]}$ has no component perpendicular to Σ_t. Since the part of $\nabla_{[a}(\sigma^\alpha)_{b]}$ projected into Σ_t satisfies equation (7.2.11) since $(\sigma^\alpha)_a$ is invariant under the spatial isometries, we have

$$2\nabla_{[a}(\sigma^\alpha)_{b]} = -c^c{}_{ab}(\sigma^\alpha)_c \quad . \tag{7.2.25}$$

The conclusion of all of the above discussion is the following: For a spatially homogeneous spacetime on which the spatial isometry group G acts in a simply transitive manner, the manifold structure is $M = \mathbb{R} \times G$. By defining the function t and the left invariant dual vector fields $(\sigma^1)_a$, $(\sigma^2)_a$, and $(\sigma^3)_a$ on M in the manner described above, we can express the spacetime metric g_{ab} in the form

$$g_{ab} = -\nabla_a t \nabla_b t + \sum_{\alpha,\beta=1}^{3} h_{\alpha\beta}(t)(\sigma^\alpha)_a(\sigma^\beta)_b \quad , \tag{7.2.26}$$

where the vector fields $(\sigma^\alpha)_a$ satisfy

$$\nabla_{[a}(\sigma^\alpha)_{b]} = -\frac{1}{2} c^c{}_{ab}(\sigma^\alpha)_c \quad . \tag{7.2.27}$$

Thus, to construct spatially homogeneous cosmological models, we simply choose a three-dimensional Lie group G, choose a basis of left invariant dual vector fields on G, and choose the functions $h_{\alpha\beta}(t)$ [or, equivalently, choose a time-dependent left invariant metric $h_{ab}(t)$ on G]. We then define the spacetime metric on $\mathbb{R} \times G$ by equation (7.2.26). All homogeneous cosmologies with simply transitive action can be constructed in this manner.

Thus, apart from the exceptional case mentioned above where the Lie group action is not simply transitive, the program for obtaining all general relativistic spatially homogeneous cosmologies will be completed by (1) obtaining all three-dimensional

Lie groups G and (2) writing down and solving Einstein's equation for the metric (7.2.26).

The first task was accomplished by Bianchi (1897), who obtained all three-dimensional Lie algebras and classified them into nine types. (As mentioned above, these Lie algebras are in one-to-one correspondence with all connected, simply connected three-dimensional Lie groups.) We shall derive here a slightly modified version of the Bianchi classification (see Ellis and MacCallum 1969). We seek all possible tensors $c^c{}_{ab}$ on a three-dimensional vector space V satisfying $c^c{}_{ab} = c^c{}_{[ab]}$ and the Jacobi identity, equation (7.2.8). Let ϵ_{abc} be a fixed 3-form on V, i.e., a totally antisymmetric tensor of type (0, 3). Given $c^c{}_{ab}$, we define A_a and M^{ab} by

$$A_a = c^b{}_{ba} \quad , \tag{7.2.28}$$

$$M^{ab} = \frac{1}{2}\,\epsilon^{acd}(c^b{}_{cd} - \delta^b{}_c A_d) \quad , \tag{7.2.29}$$

where ϵ^{abc} is the unique totally antisymmetric tensor of type (3, 0) satisfying $\epsilon^{abc}\epsilon_{abc} = 3! = 6$. Contracting equation (7.2.29) with ϵ_{aef} and using equation (B.2.13), we obtain

$$\epsilon_{aef} M^{ab} = c^b{}_{ef} - \delta^b{}_{[e} A_{f]} \quad . \tag{7.2.30}$$

Contracting this equation over e and b and using the definition (7.2.28) of A_a, we find that M^{ab} is symmetric,

$$M^{[ab]} = 0 \quad . \tag{7.2.31}$$

Thus, we have shown that on a three-dimensional vector space, any $c^c{}_{ab}$ satisfying $c^c{}_{ab} = c^c{}_{[ab]}$ can be written in the form

$$c^c{}_{ab} = M^{cd}\epsilon_{dab} + \delta^c{}_{[a} A_{b]} \quad , \tag{7.2.32}$$

where $M^{cd} = M^{dc}$. Substitution of this expression into the Jacobi identity (7.2.8) yields the remarkably simple result,

$$M^{ab} A_b = 0 \quad . \tag{7.2.33}$$

Thus, a three-dimensional Lie algebra is determined by a dual vector A_a and a symmetric tensor M^{ab} satisying equation (7.2.33).

If $A_a = 0$, then equation (7.2.33) is trivially satisfied. In this case (referred to as "class A"), the resulting Lie algebras are classified (i.e., uniquely determined up to isomorphisms) by the rank and signature (up to overall sign) of M^{ab}. Hence, there exist precisely six distinct Lie algebras in this case: (i) $M^{ab} = 0$, (ii) rank $(M^{ab}) = 1$, (iii) rank $(M^{ab}) = 2$, signature $+\ -$, (iv) rank $(M^{ab}) = 2$, signature $+\ +$, (v) rank $(M^{ab}) = 3$, signature $+\ +\ -$, (vi) rank $(M^{ab}) = 3$, signature $+\ +\ +$. If $A_b \neq 0$ (referred to as "class B"), then by equation (7.2.33) the rank of M^{ab} cannot be greater than two. Hence, in this case there exist four possibilities for the rank and signature (up to overall sign) of M^{ab}. However, in the two cases where the rank of M^{ab} is equal to two, a scalar α is determined by the formula

$$A_e A_f = \alpha M^{ac} M^{bd} \epsilon_{cde}\epsilon_{abf} \quad , \tag{7.2.34}$$

since the tensors appearing on both sides of this equation are nonvanishing and must be proportional to each other. As a result, in class B there exist two one-parameter families of Lie algebras classified by the signature of the rank-two tensor M^{ab} and the value of α, as well as two further Lie algebras corresponding to the cases where the rank of M^{ab} is zero (i.e., $M^{ab} = 0$) or one. Tables giving explicit formulas for the components of $c^c{}_{ab}$ in conveniently chosen bases for all the above Lie algebras can be found in Taub (1951) and Ryan and Shepley (1975).

To write down Einstein's equation, we need to compute the curvature of the metric (7.2.26). The most straightforward procedure to do so is to define a left invariant orthonormal basis $(e_\mu)_a$ (which, however, will not satisfy [7.2.21]) by

$$(e_0)_a = \nabla_a t \quad , \tag{7.2.35}$$

$$(e_\alpha)_a = \sum_{\beta=1}^{3} B_{\alpha\beta}(t)(\sigma^\beta)_a \quad (\alpha = 1, 2, 3) \quad , \tag{7.2.36}$$

where, in order for the $(e_\mu)_a$ to be orthonormal, $B_{\alpha\beta}$ must satisfy

$$\delta_{\alpha\beta} = \sum_{\gamma,\delta} B_{\alpha\gamma} B_{\beta\delta} h^{\gamma\delta} \quad (\alpha, \beta = 1, 2, 3) \quad , \tag{7.2.37}$$

where $h^{\gamma\delta}$ are the components of the inverse metric, h^{ab}, in the basis dual to $(\sigma^\alpha)_a$. (There is, of course, considerable freedom involved in the choice of $B_{\alpha\beta}$; additional conditions, such as $B_{\alpha\beta} = B_{\beta\alpha}$ [Ryan and Shepley 1975] may be imposed in order to uniquely specify $B_{\alpha\beta}$.) For this orthonormal basis, we have

$$\nabla_{[a}(e_0)_{b]} = \nabla_{[a}\nabla_{b]}t = 0 \quad ; \tag{7.2.38}$$

and, for $\alpha = 1, 2, 3$,

$$\begin{aligned} \nabla_{[a}(e_\alpha)_{b]} &= \sum_{\beta=1}^{3} \left\{ \frac{dB_{\alpha\beta}}{dt} \nabla_{[a}t(\sigma^\beta)_{b]} - \frac{1}{2} B_{\alpha\beta} c^c{}_{ab}(\sigma^\beta)_c \right\} \\ &= \sum_{\gamma=1}^{3} A_\alpha{}^\gamma \nabla_{[a}t(e_\gamma)_{b]} - \frac{1}{2} c^c{}_{ab}(e_\alpha)_c \quad , \end{aligned} \tag{7.2.39}$$

where $A_\alpha{}^\gamma = \sum_\beta (dB_{\alpha\beta}/dt)(B^{-1})^{\beta\gamma}$ and equation (7.2.27) was used. Equations (7.2.38) and (7.2.39) determine the connection one-forms via equation (3.4.24), and, thus, the curvature can be calculated by the tetrad method of section 3.4*b*. This procedure will yield a formula for the Riemann and Ricci curvature tensors in terms of the matrix $B_{\alpha\beta}$ and the structure constant tensor $c^c{}_{ab}$, and such formulas for the Ricci curvature are given by Ryan and Shepley (1975). However, using equation (7.2.37), we can eliminate $B_{\alpha\beta}$ in terms of $h_{\alpha\beta}$ (as must be possible since the choices of Lie group and $h_{\alpha\beta}$ completely determine the metric [7.2.26] and thus the curvature must be expressible in terms of them). We then can write the component formulas thus obtained as tensor equations for the curvature in terms of the metric, $h_{ab}(t)$, on Σ_t and the structure constant tensor $c^c{}_{ab}$. We shall not present the details of these rather lengthy calculations here but merely present the final result. Defining K_{ab} by

$$K_{ab} = \frac{1}{2} \pounds_t h_{ab} \tag{7.2.40}$$

[so that, as discussed more generally in chapters 9 and 10, $K_{ab}(t)$ is the extrinsic curvature of Σ_t], we obtain the following formulas for the components $G_{ab}(e_0)^a(e_0)^b$, $G_{bc}h^b{}_a(e_0)^c$ and $h^c{}_ah^d{}_bR_{cd}$ of the Einstein and Ricci curvature tensors:

$$2G_{ab}(e_0)^a(e_0)^b = (K^a{}_a)^2 - K_{ab}K^{ab} - c^a{}_{ab}c^c{}_c{}^b - \frac{1}{2}c^a{}_{cb}c^c{}_a{}^b - \frac{1}{4}c_{abc}c^{abc} \quad , \tag{7.2.41}$$

$$G_{bc}h^b{}_a(e_0)^c = R_{bc}h^b{}_a(e_0)^c = K^b{}_c c^c{}_{ba} + K^b{}_a c^c{}_{bc} \quad , \tag{7.2.42}$$

$$h^c{}_ah^d{}_bR_{cd} = \pounds_t K_{ab} + K^c{}_c K_{ab} - 2K_{ac}K^c{}_b + \frac{1}{4}c_{acd}c_b{}^{cd}$$
$$- c^c{}_{cd}c_{(ab)}{}^d - c_{cda}c^{(cd)}{}_b \quad . \tag{7.2.43}$$

In the right-hand sides of equations (7.2.41)–(7.2.43) all indices are lowered and raised by the spatial metric, h_{ab}, and its inverse, h^{ab}.

Now, the vacuum Einstein equation, $R_{ab} = 0$, is equivalent to the vanishing of the left-hand sides of equations (7.2.41)–(7.2.43). (The non-vacuum Einstein equation, of course, is obtained by setting the left-hand sides of equations [7.2.41]–[7.2.43] equal to appropriate stress-energy terms.) Using the definition of K_{ab}, equation (7.2.40), it may be verified that if equations (7.2.41) and (7.2.42) with the left-hand sides set equal to zero hold "initially," i.e., on the hypersurface Σ_0, then they hold everywhere if equation (7.2.43) is satisfied (with the left-hand side set equal to zero); in other words, the time derivatives of equations (7.2.41) and (7.2.42) vanish automatically by virtue of equations (7.2.40) and (7.2.43). The reason this occurs can be traced directly to the Bianchi identity, $\nabla_a G^{ab} = 0$, and will be discussed more generally in chapter 10. Thus, equations (7.2.41) and (7.2.42) act as constraint equations on the initial values of K_{ab} and h_{ab}. (The right-hand sides of eqs. [7.2.41]–[7.2.43], of course, contain h_{ab} implicitly in the index raisings and lowerings.) On the other hand, equations (7.2.40) and (7.2.43) act as evolution equations for h_{ab} and K_{ab}. If we take the components of equations (7.2.40) and (7.2.43) in our original basis $(\sigma^\alpha)_a$, we obtain directly a system of ordinary differential equations which express $dh_{\alpha\beta}/dt$ and $dK_{\alpha\beta}/dt$ as functions of $h_{\alpha\beta}$, $K_{\alpha\beta}$, and the group structure constants $c^\gamma{}_{\alpha\beta}$. From the theory of ordinary differential equations, it is well known that a unique solution of equations (7.2.40) and (7.2.43) always exists for given initial values of h_{ab} and K_{ab}.

Thus, to obtain a spatially homogeneous solution of the vacuum Einstein equation, we need only solve the algebraic constraint equations, (7.2.41) and (7.2.42), for the initial data h_{ab}, K_{ab} at $t = 0$. A unique solution then is generated by the evolution equations (7.2.40) and (7.2.43). Although in most cases it may not be possible to integrate these equations analytically, numerical integration of such a system of ordinary differential equations always can be carried out in a straightforward manner. Similar conclusions hold for the non-vacuum Einstein equation with appropriate matter sources.

As an example of the above procedure, we shall determine all the spatially homogeneous vacuum solutions for the simplest case of the Lie group $G = \mathbb{R}^3$ with group "multiplication" given by addition. (This case is called "type I" in the Bianchi classification of Lie groups.) Since the group multiplication is commutative, the vector field generators of left translations must commute, so we have $c^c{}_{ab} = 0$ by equation (7.2.19). Hence any basis of left invariant vector fields forms a coordinate basis, and for any choice of $(\sigma^\alpha)_a$ constructed in the manner described above, we can find coordinates x, y, z on $\mathbb{R}^3$ so that $(\sigma^1)_a = (dx)_a$, $(\sigma^2)_a = (dy)_a$, $(\sigma^3)_a = (dz)_a$. (The simple case $c^c{}_{ab} = 0$, of course, is the only case for which this can be done.) Without loss of generality, we may choose $(\sigma^\alpha)_a$ to be orthonormal on the initial surface Σ_0, and we may rotate this basis to diagonalize $K_{\alpha\beta}$ on Σ_0. The evolution equations (7.2.40) and (7.2.43) with $c^c{}_{ab} = 0$ then imply that in the basis $(\sigma^\alpha)_a$ (which is defined off the initial surface by eq. [7.2.21]) both $h_{\alpha\beta}$ and $K_{\alpha\beta}$ must remain diagonal for all time. Thus, we have

$$K_{\alpha\beta} = \text{diag}\,(k_1, k_2, k_3) \quad , \tag{7.2.44}$$

$$h_{\alpha\beta} = \text{diag}\,(f_1, f_2, f_3) \quad , \tag{7.2.45}$$

$$h^{\alpha\beta} = \text{diag}\,(f_1^{-1}, f_2^{-1}, f_3^{-1}) \quad . \tag{7.2.46}$$

By construction, on the initial surface, we have $f_\alpha = 1$. The initial value constraint (7.2.42) is trivially satisfied, so the only requirement on the initial values of k_α comes from equation (7.2.41) which yields (using $f_\alpha = 1$),

$$(k_1 + k_2 + k_3)^2 = k_1^2 + k_2^2 + k_3^2 \quad . \tag{7.2.47}$$

The evolution equations (7.2.40) and (7.2.43) yield

$$k_\alpha = \frac{1}{2}\frac{df_\alpha}{dt} \quad , \tag{7.2.48}$$

$$\frac{dk_\alpha}{dt} = -\left(\sum_\beta f_\beta^{-1} k_\beta\right) k_\alpha + 2 f_\alpha^{-1} k_\alpha^2 \quad . \tag{7.2.49}$$

Hence, we find

$$\begin{aligned}\frac{d}{dt}(f_\alpha^{-1} k_\alpha) &= f_\alpha^{-1}\left\{-\left(\sum_\beta f_\beta^{-1} k_\beta\right) k_\alpha + 2 f_\alpha^{-1} k_\alpha^2\right\} - k_\alpha f_\alpha^{-2}(2k_\alpha) \\ &= \left(\sum_\beta f_\beta^{-1} k_\beta\right) f_\alpha^{-1} k_\alpha \quad .\end{aligned} \tag{7.2.50}$$

Summing equation (7.2.50) over α, we obtain

$$dK/dt = -K^2 \quad , \tag{7.2.51}$$

where $K = \sum_\alpha f_\alpha^{-1} k_\alpha = K^a{}_a$. If we exclude the trivial solution $K = 0$ (which leads uniquely to the Minkowski solution), the general solution of equation (7.2.51) is

$$K = 1/t \quad , \tag{7.2.52}$$

where the constant of integration has been absorbed in the definition of the zero of t. Substitution of our solution (7.2.52) into equation (7.2.50) yields

$$\frac{d}{dt}(f_\alpha^{-1} k_\alpha) = -\frac{1}{t}(f_\alpha^{-1} k_\alpha) \quad , \tag{7.2.53}$$

which has the general solution

$$f_\alpha^{-1} k_\alpha = p_\alpha / t \quad , \tag{7.2.54}$$

where p_α is a constant. Hence, by equation (7.2.48), we have

$$\frac{1}{2} f_\alpha^{-1} \frac{df_\alpha}{dt} = \frac{p_\alpha}{t} \quad , \tag{7.2.55}$$

which has the general solution

$$f_\alpha = C_\alpha t^{2p_\alpha} \tag{7.2.56}$$

where C_α is a constant. Equation (7.2.48) (or eq. [7.2.54]) then yields

$$k_\alpha = C_\alpha p_\alpha t^{2p_\alpha - 1} \quad . \tag{7.2.57}$$

Substitution of equations (7.2.56) and (7.2.57) in equation (7.2.52) yields

$$\sum_\alpha p_\alpha = 1 \quad . \tag{7.2.58}$$

Equations (7.2.56) and (7.2.57), subject to the condition (7.2.58), are the general solutions (aside from Minkowski spacetime) of the evolution equations. Taking $t = 1$ as our initial surface on which to impose the requirement $f_\alpha = 1$ and k_α subject to the constraint (7.2.47), we obtain

$$C_\alpha = 1 \quad , \tag{7.2.59}$$

$$\left(\sum_\alpha p_\alpha \right)^2 = \sum_\alpha p_\alpha{}^2 \quad . \tag{7.2.60}$$

Thus, the general spatially homogeneous vacuum solution (aside from Minkowski spacetime) of Einstein's equation with simply transitive Lie group $G = \mathbb{R}^3$ is

$$ds^2 = -dt^2 + t^{2p_1} dx^2 + t^{2p_2} dy^2 + t^{2p_3} dz^2 \quad , \tag{7.2.61}$$

where p_1, p_2, p_3 are constants subject to equations (7.2.58) and (7.2.60). Equation (7.2.61) is known as the Kasner solution (Kasner 1925).

One solution of equations (7.2.58) and (7.2.60) is $p_1 = 1$, $p_2 = p_3 = 0$. The Kasner solution with this choice of p_α can be recognized to be simply the Rindler spacetime of section 6.4 (crossed with $\mathbb{R}^2$), with t and x interchanged. Thus, in this case, the apparent singularity at $t = 0$ is only a coordinate singularity, and the Kasner solution (7.2.61) is just the wedge of Minkowski spacetime labeled as region II in Figure 6.8. However, all solutions other than the trivial solutions where two of the p_α vanish yield nonflat spacetimes with a physical singularity at $t = 0$. Note that with the exception of the trivial solutions, all solutions of equations (7.2.58) and (7.2.60) must have two of the p_α positive and the other negative. Thus, the Kasner solution describes a homogeneous universe which expands in two directions but contracts in the other direction. Interestingly, the analysis of the dynamics of homogeneous vacuum solutions with $G = SU(2)$ (Bianchi type IX) shows that near the initial singularity the evolution can be described as a series of "Kasner epochs" connected by "transition regimes" where the values of p_α change (see Belinskii *et al.* 1970).

7.3 Algebraically Special Solutions

In the previous two sections (as well as in chapters 5 and 6) progress was made toward obtaining physically interesting solutions of Einstein's equation by restricting attention to spacetimes with a high degree of symmetry. In this section, we shall very briefly discuss a different type of simplifying assumption for obtaining solutions, which relates to the algebraic properties of the curvature tensor. Unfortunately, since the character of this simplifying assumption is more mathematical than physical, many of the solutions obtained by this approach do not appear to be of direct physical relevance. Nevertheless, this approach has been one of the most successful in providing us with a large class of exact solutions, and some physically very important solutions, such as the Kerr metric (see chapter 12), have been found in this way.

In the same manner as for the spatial metrics considered at the beginning of section 5.1, at each point in spacetime we may view the Riemann tensor $R_{ab}{}^{cd}$ as a linear map from the six-dimensional space of antisymmetric tensors of type $(0, 2)$ (i.e., two-forms) into itself. This map is self-adjoint since $R_{abcd} = R_{cdab}$. However, it should be noted that unlike the case of a Riemannian metric considered in section 5.1, the inner product induced on the two-forms by the spacetime metric g_{ab} is not positive definite, so the familiar theorem that a self-adjoint linear map has a complete orthonormal basis of eigenvectors (which holds in any finite dimensional vector space with a positive definite inner product) does not apply here.

The analysis of the structure of the Riemann tensor as a linear map was first carried out by Petrov (1954, 1969). We shall not present the details of the analysis of the curvature tensor here but turn immediately to an important conclusion of the analysis: In general, there exist precisely four distinct null directions (i.e., null vectors k^a, defined up to scaling $k^a \to \lambda k^a$) which satisfy the relation

$$k^b k^c k_{[e} C_{a]bc[d} k_{f]} = 0 \quad , \tag{7.3.1}$$

where C_{abcd} is the Weyl tensor, defined by equation (3.2.28). The null directions which satisfy equation (7.3.1) are called *principal null directions*. Thus, every tensor satisfying the algebraic conditions satisfied by the Weyl tensor over a four-dimensional vector space with a metric of Lorentz signature possesses, in general, four principal null directions. A proof of this result by tensor methods requires a considerable amount of analysis. However, a simple proof of this result can be obtained by spinor methods, and we will present this proof in chapter 13.

Although, in general, the Weyl tensor possesses four distinct principal null directions, it is possible for some of these null directions to coincide—in which case they satisfy stronger relations than equation (7.3.1)—resulting in fewer than four principal null directions. Spacetimes for which fewer than four distinct principal null directions exist at each point are called *algebraically special spacetimes*. The different types of algebraically special spacetimes are classified in Table 7.1.

In algebraically special spacetimes, we can take advantage of the conditions satisfied by the repeated principal null direction by choosing a null tetrad (see the

Table 7.1
ALGEBRAIC CLASSIFICATION OF SPACETIMES

Type	Description	Condition Satisfied by (repeated) Principal Null Direction k^a
I	Algebraically general; four distinct principal null directions	$k^b k^c k_{[e} C_{a]bc[d} k_{f]} = 0$
II	One pair of principal null directions coincides	$k^b k^c C_{abc[d} k_{e]} = 0$
II–II [D]	Two pairs of principal null directions coincide	$k^b k^c C_{abc[d} k_{e]} = 0$ (two solutions)
III	Three principal null directions coincide	$k^c C_{abc[d} k_{e]} = 0$
IV [N]	All four principal null directions coincide	$k^c C_{abcd} = 0$

discussion of the Newman-Penrose formalism at the end of section 3.4) with one of the null vectors aligned with the repeated principal null direction. Certain tetrad components of the Weyl tensor then will vanish. This yields extra equations on the connection components (i.e., the spin coefficients in the Newman-Penrose formalism) in addition to the conditions imposed by Einstein's equation. Using these additional conditions, it has proven possible to integrate Einstein's equation exactly in many cases.

We shall not attempt to present here any of the details of how algebraically special solutions can be found. For an excellent illustration of this approach, we refer the reader to the paper of Kinnersley (1969) who explicitly found all type II–II solutions of the vacuum Einstein equation by direct integration of the Newman-Penrose equations. A survey of the algebraically special solutions known by the late 1970s can be found in the book of Kramer et al. (1980).

7.4 Methods for Generating Solutions

It often happens when one is trying to solve an equation that an algorithm will exist for constructing new solutions from a given solution. For example, for Laplace's equation in ordinary electrostatics one can construct new solutions from a given solution by the "method of inversion" (see, e.g., Jackson 1962).[5] Although such prescriptions often can be applied readily, they usually suffer from the serious drawback that the solutions they generate may be of no physical interest, or at least, not applicable to the problem that one desires to solve.

5. The method of inversion works because sphere inversion maps the Euclidean metric into a multiple of itself (i.e., it is a "conformal isometry") and Laplace's equation has simple conformal transformation properties (see appendix D).

For the vacuum Einstein equation, $R_{ab} = 0$, it turns out that if one is given a solution with a Killing vector field ξ^a there exists an algorithm for constructing a one-parameter family of solutions, for which ξ^a remains a Killing field. Algorithms of this sort have been given by Ehlers (1957) and others. We shall present the algorithm in the more general form given by Geroch (1971). To define it, we need to use the fact that as proven in section 7.1 (see eq. [7.1.15]) in a vacuum spacetime, the twist ω_a, of a Killing field ξ^a, defined by

$$\omega_a = \epsilon_{abcd} \xi^b \nabla^c \xi^d \quad , \tag{7.4.1}$$

satisfies

$$\nabla_{[a} \omega_{b]} = 0 \quad . \tag{7.4.2}$$

Consequently, as remarked at the end of section B.1 of appendix B, locally there exists a scalar function, ω, such that

$$\omega_a = \nabla_a \omega \tag{7.4.3}$$

(ω is called the *scalar twist* of ξ^a). Furthermore, using equation (B.2.13), we find

$$\epsilon^{abef} \nabla_e [\epsilon_{abcd} \nabla^c \xi^d] = -4 \nabla_e \nabla^e \xi^f \quad . \tag{7.4.4}$$

Thus, using equation (C.3.9) and the vacuum Einstein equation $R_{ab} = 0$, we find

$$\nabla_{[e} \{\epsilon_{ab]cd} \nabla^c \xi^d\} = 0 \quad , \tag{7.4.5}$$

which implies the (local) existence of a 1-form field α_b such that

$$\nabla_{[a} \alpha_{b]} = \frac{1}{2} \epsilon_{abcd} \nabla^c \xi^d \quad . \tag{7.4.6}$$

By adding a gradient to α_b, we can adjust it so that

$$\xi^a \alpha_a = \omega \quad . \tag{7.4.7}$$

Finally, similar calculations (problem 5) show that

$$\nabla_{[e} \{2\lambda \nabla_a \xi_{b]} + \omega \epsilon_{ab]cd} \nabla^c \xi^d\} = 0 \quad , \tag{7.4.8}$$

where

$$\lambda = \xi^c \xi_c \quad , \tag{7.4.9}$$

and thus, locally, there exists a 1-form β_b such that

$$\nabla_{[a} \beta_{b]} = 2\lambda \nabla_a \xi_b + \omega \epsilon_{abcd} \nabla^c \xi^d \quad , \tag{7.4.10}$$

where β_b can be adjusted to satisfy

$$\xi^a \beta_a = \omega^2 + \lambda^2 - 1 \quad . \tag{7.4.11}$$

The algorithm for generating new solutions from a given solution now may be stated as follows: Let g_{ab} be a solution of the vacuum Einstein equation $R_{ab} = 0$ with

Killing field ξ^a. Define ω, α_a, and β_a as above. For each $\theta \in [0, \pi]$, define $\tilde{g}_{ab}(\theta)$ by

$$\tilde{g}_{ab}(\theta) = \sigma g_{ab} + 2 \sin \theta \, \xi_{(a} \gamma_{b)} + \frac{\lambda \sin^2 \theta}{\sigma} \gamma_a \gamma_b \quad , \tag{7.4.12}$$

where $\sigma(\theta)$ and $\gamma_a(\theta)$ are defined by

$$\sigma = (\cos \theta - \omega \sin \theta)^2 + \lambda^2 \sin^2 \theta \quad , \tag{7.4.13}$$

$$\gamma_a = 2\alpha_a \cos \theta - \beta_a \sin \theta \quad . \tag{7.4.14}$$

Then for all θ, $\tilde{g}_{ab}(\theta)$ is a solution of the vacuum Einstein equation. Furthermore, ξ^a remains a Killing field of $\tilde{g}_{ab}(\theta)$. Note that for $\theta = \pi$, $\tilde{g}_{ab}$ reduces to g_{ab}; however, otherwise $\tilde{g}_{ab}(\theta)$ is, in general, a distinct solution. Note also that even if g_{ab} is nonsingular, $\tilde{g}_{ab}$ may develop singularities on account of the bad behavior of ω, α_a, or β_a resulting from the possibility that one cannot define them globally.

The reader may verify by direct computation that equation (7.4.12) indeed does define a solution of the vacuum Einstein equation. However, for a derivation of equation (7.4.12) which gives more insight into why this algorithm works, we refer the reader to Geroch (1971).

Unfortunately, the algorithm (7.4.12) does not reproduce most physically interesting features of the original metric. In particular, $\tilde{g}_{ab}$ will not, in general, be asymptotically flat even if g_{ab} satisfies this property. Furthermore, equation (7.4.12) produces only a one-parameter family of solutions. If one reapplies the algorithm to $\tilde{g}_{ab}(\theta)$, one does not generate any new solutions.

However, if g_{ab} has two commuting Killing vector fields, ξ^a and ψ^a (as is the case in the stationary axisymmetric spacetimes considered in section 7.1), it can be shown (Geroch 1972*a*) that ψ^a also remains a Killing vector field of the metric $\tilde{g}_{ab}(\theta)$ generated by the algorithm (7.4.12) using the Killing field ξ^a. Hence, we may apply the algorithm to $\tilde{g}_{ab}(\theta)$ using the Killing vector ψ^a to generate a two-parameter family of solutions $\tilde{g}_{ab}(\theta, \theta')$ from our original solution g_{ab}. We then may reapply the transformation (7.4.12) to $\tilde{g}_{ab}(\theta, \theta')$. Geroch (1972*a*) has shown that this procedure does *not*, in general, reproduce previous solutions, and thus a three-parameter family of solutions is generated. Indeed, an infinite dimensional group of transformations is generated by repeated application of the transformation (7.2.12). Recently, it has been proven (Hauser and Ernst 1981) that—as conjectured by Geroch (1972*a*)—*all* asymptotically flat, stationary, axisymmetric vacuum solutions of Einstein's equation can be generated from Minkowski spacetime by an element of this group. Furthermore, it is known how to generate such asymptotically flat solutions with desired values for all multipole moments (Hoenselaers, Kinnersley, and Xanthopoulos 1979; Xanthopoulos 1981). Unfortunately, the algebraic computations required to obtain these solutions explicitly remain formidable, so in fact very few solutions are presently known in explicit form. Nevertheless, this method for generating solutions is unquestionably one of the most important techniques for solving Einstein's equation.

7.5 Perturbations

As emphasized at the beginning of this chapter, relatively few physically interesting exact solutions of Einstein's equation are known. The known solutions, such as the Robertson-Walker solutions (chapter 5) and the Schwarzschild solution (chapter 6), may tell us a great deal about cosmology and the gravitational fields of isolated bodies, but they leave many questions unanswered. For example, we might wish to know how a small inhomogeneity in the matter distribution in a nearly Robertson-Walker universe develops with time, or how a nearly spherical star behaves if displaced slightly from equilibrium, or what happens if a small amount of gravitational radiation is incident on a Schwarzschild black hole. It appears hopeless to attempt to find exact solutions describing these processes. However, if the deviation from a known exact solution is small, it makes sense to look for an approximate solution by writing $g_{ab} = {}^0g_{ab} + \gamma_{ab}$ (where ${}^0g_{ab}$ is the known exact solution) and "linearizing" Einstein's equation in γ_{ab}. We already followed this procedure in section 4.4 for the case where ${}^0g_{ab}$ was the Minkowski metric, η_{ab}. In this section, we shall systematize the general procedure for deriving linearized equations, and for Einstein's vacuum equation we shall obtain explicitly the vacuum perturbation equations off an arbitrary exact solution.

Consider an equation

$$\mathscr{E}(g) = 0 \tag{7.5.1}$$

for an unknown function g (which, more generally, may be a collection of functions or tensor fields, etc.). In the case of interest, g is the spacetime metric possibly together with variables describing the matter distribution, and $\mathscr{E}$ is Einstein's equation. Suppose an exact solution, ${}^0g$, is known and suppose we are interested in studying situations where the deviation from ${}^0g$ is small. What we would really like to have is a one-parameter (or multiparameter) family $g(\lambda)$ of exact solutions,

$$\mathscr{E}[g(\lambda)] = 0 \quad , \tag{7.5.2}$$

where λ measures the size of perturbation in the sense that (i) $g(\lambda)$ depends differentiably on λ, and (ii) $g(0) = {}^0g$. Thus, small λ corresponds to small deviations from ${}^0g$, and a knowledge of $g(\lambda)$ for small λ would give us an exact perturbed solution. However, equation (7.5.2) may be too difficult to solve. Nevertheless, we can derive a much simpler equation from equation (7.5.2) by differentiating it with respect to λ and setting λ equal to zero,

$$\left.\frac{d}{d\lambda}[\mathscr{E}(g(\lambda))]\right|_{\lambda=0} = 0 \quad . \tag{7.5.3}$$

Equation (7.5.3) is a linear equation for

$$\gamma = \left.\frac{dg}{d\lambda}\right|_{\lambda=0} \quad ; \tag{7.5.4}$$

i.e., it can be expressed in the form

$$\mathcal{L}(\gamma) = 0 \quad , \tag{7.5.5}$$

where $\mathcal{L}$ is a linear operator. (Eq. [7.5.5] is referred to as the "linearization" of eq. [7.5.1] about ${}^0g$.) Since linear equations are generally much easier to solve than nonlinear equations, it may be feasible to solve equation (7.5.5) even if equation (7.5.1) is intractable. If we can solve equation (7.5.5), then ${}^0g + \lambda\gamma$ should yield a good approximation to $g(\lambda)$ for sufficiently small λ, and issues of physical interest thus can be investigated.

The above procedure provides a powerful tool for obtaining approximate solutions. However, two important points should be kept in mind when employing perturbation techniques: (1) It is, in general, very difficult to estimate the error involved in replacing $g(\lambda)$ by ${}^0g + \lambda\gamma$, i.e., to determine how small λ must be in order that the approximate solution have sufficient accuracy. (2) As derived above, existence of a one-parameter family of solutions $g(\lambda)$ implies the existence of a solution of equation (7.5.5), where $\gamma = (dg/d\lambda)|_{\lambda=0}$. However, the existence of a solution γ of equation (7.5.5) does not necessarily imply the existence of a corresponding one-parameter family of solutions, $g(\lambda)$, of (7.5.2), i.e., there may be spurious solutions of equation (7.5.5). Thus, the issue of "linearization stability," i.e., the existence of exact solutions corresponding to a solution of the linearized equations, must be investigated before a perturbation analysis can be applied with complete reliability.

Let us now derive the linearized vacuum Einstein equation (7.5.5) for a metric perturbation γ_{ab} of an exact solution ${}^0g_{ab}$ of Einstein's equation in vacuum,

$$R_{ab} = 0 \quad . \tag{7.5.6}$$

To do so, we need to compute the Ricci tensor $R_{ab}(\lambda)$ for the metric $g_{ab}(\lambda)$ in a useful form, specifically in terms of quantities related to the "background metric" ${}^0g_{ab}$. Differentiation of this expression with respect to λ at $\lambda = 0$ then will yield the equation we seek.

Let ${}^\lambda\nabla_a$ denote the derivative operator associated with $g_{ab}(\lambda)$, and let ${}^0\nabla_a$ denote the derivative operator associated with ${}^0g_{ab}$. According to the general analysis of section 3.1, the difference between ${}^\lambda\nabla_a$ and ${}^0\nabla_a$ is determined by a tensor field $C^c{}_{ab}(\lambda)$, which, according to equation (3.1.28), is given by

$$C^c{}_{ab}(\lambda) = \frac{1}{2} g^{cd}(\lambda)\{{}^0\nabla_a g_{bd}(\lambda) + {}^0\nabla_b g_{ad}(\lambda) - {}^0\nabla_d g_{ab}(\lambda)\} \quad . \tag{7.5.7}$$

The Riemann tensor $R_{abc}{}^d(\lambda)$ associated with $g_{ab}(\lambda)$ can be computed in terms of ${}^0R_{abc}{}^d$ and $C^c{}_{ab}(\lambda)$ by replacing the derivative operator ${}^\lambda\nabla_a$ in the definition of the Riemann tensor, equation (3.2.3), by its expression in terms of ${}^0\nabla_a$ and $C^c{}_{ab}$. Proceeding in the same manner as in the derivation of equation (3.4.3), we find

$$R_{abc}{}^d = {}^0R_{abc}{}^d - 2\,{}^0\nabla_{[a} C^d{}_{b]c} + 2C^e{}_{c[a} C^d{}_{b]e} \quad . \tag{7.5.8}$$

Thus, the Ricci tensor of $g_{ab}(\lambda)$ is given by

$$R_{ac} = -2\,{}^0\nabla_{[a} C^b{}_{b]c} + 2C^e{}_{c[a} C^b{}_{b]e} \quad , \tag{7.5.9}$$

where we have used the fact that ${}^0R_{ac} = 0$ since ${}^0g_{ab}$ is a solution of the vacuum Einstein equation.

Equation (7.5.9) expresses $R_{ac}(\lambda)$ in a convenient form for evaluating its derivative $\dot{R}_{ac}(\lambda) = (dR_{ac}/d\lambda)|_{\lambda=0}$, with respect to λ at $\lambda = 0$. It follows immediately from its definition that $C^c{}_{ab}(\lambda)$ vanishes when $\lambda = 0$, so the term quadratic in $C^c{}_{ab}$ will make no contribution to $\dot{R}_{ac}$. Thus, we obtain

$$\dot{R}_{ac} = -2\,{}^0\nabla_{[a} \dot{C}^b{}_{b]c} \quad , \tag{7.5.10}$$

where

$$\dot{C}^c{}_{ab} = \left.\frac{dC^c{}_{ab}}{d\lambda}\right|_{\lambda=0} \quad . \tag{7.5.11}$$

From equation (7.5.7) we find that

$$\dot{C}^c{}_{ab} = \frac{1}{2}\,{}^0g^{cd}\{{}^0\nabla_a \gamma_{bd} + {}^0\nabla_b \gamma_{ad} - {}^0\nabla_d \gamma_{ab}\} \quad , \tag{7.5.12}$$

where

$$\gamma_{ab} = \left.\frac{dg_{ab}}{d\lambda}\right|_{\lambda=0} \tag{7.5.13}$$

and we have used the fact that ${}^0\nabla_a({}^0g_{bc}) = 0$. Thus, substituting equation (7.5.12) in equation (7.5.10), the linearized Einstein equation for γ_{ab} is found to be

$$0 = \dot{R}_{ac} = -\frac{1}{2}\,{}^0g^{bd}\,{}^0\nabla_a {}^0\nabla_c \gamma_{bd} - \frac{1}{2}\,{}^0g^{bd}\,{}^0\nabla_b {}^0\nabla_d \gamma_{ac} + {}^0g^{bd}\,{}^0\nabla_b {}^0\nabla_{(c} \gamma_{a)d} \quad . \tag{7.5.14}$$

Since all quantities aside from γ_{ab} now refer only to the background metric ${}^0g_{ab}$, we shall in the following drop the superscript zero on the background metric and its derivative operator, and we shall use the background metric g_{ab} and its inverse g^{ab} to raise and lower indices. In this notation, equation (7.5.14) becomes

$$0 = -\frac{1}{2}\nabla_a \nabla_c \gamma - \frac{1}{2}\nabla^b \nabla_b \gamma_{ac} + \nabla^b \nabla_{(c} \gamma_{a)b} \quad , \tag{7.5.15}$$

where $\gamma = \gamma^a{}_a = g^{ab}\gamma_{ab}$. Note that equation (7.5.15) agrees with equation (4.4.4) in the special case of Minkowski spacetime, where $g_{ab} = \eta_{ab}$ and $\nabla_a = \partial_a$.

Equation (7.5.15) can be simplified by a convenient choice of gauge. As shown at the end of section C.2 of appendix C, γ_{ab} and $\gamma_{ab} + 2\nabla_{(a} v_{b)}$ represent the same physical perturbation, where v^a is an arbitrary vector field. As in the special case of perturbations of Minkowski spacetime considered in section 4.4, we can solve the curved spacetime wave equation (see theorem 10.1.2 of chapter 10),

$$\nabla^b \nabla_b v_a + R_a{}^b v_b = -\nabla^b \bar{\gamma}_{ab} \quad , \tag{7.5.16}$$

where

$$\bar{\gamma}_{ab} \equiv \gamma_{ab} - \frac{1}{2} g_{ab} \gamma \quad , \tag{7.5.17}$$

and thereby set

$$\nabla^b \bar{\gamma}_{ab} = 0 \quad . \tag{7.5.18}$$

In this gauge, the trace of equation (7.5.15) yields

$$\nabla_a \nabla^a \gamma = 0 \quad . \tag{7.5.19}$$

Hence, we can use the restricted gauge freedom $\gamma_{ab} \to \gamma_{ab} + 2\nabla_{(a} w_{b)}$ with $\nabla^b \nabla_b w_a + R_a{}^b w_b = 0$ to satisfy the curved space analogs of the initial value conditions (4.4.34a) and (4.4.34b), thereby obtaining $\gamma = 0$ throughout the spacetime by theorem 10.1.2. Thus, for an arbitrary vacuum perturbation γ_{ab} of an arbitrary vacuum solution g_{ab}, we can always choose a *transverse traceless gauge* whereby

$$\nabla^a \gamma_{ab} = 0 \quad , \tag{7.5.20}$$

$$\gamma = 0 \quad . \tag{7.5.21}$$

However, in general no analog of the radiation gauge conditions $\gamma_{0\mu} = 0$ ($\mu = 1, 2, 3$) used in perturbations of flat spacetime can be imposed in addition to (7.5.20) and (7.5.21).

From the properties of the Riemann tensor, we have

$$\begin{aligned} \nabla^b \nabla_{(c} \gamma_{a)b} &= \nabla_{(c} \nabla^b \gamma_{a)b} + R^b{}_{(ca)}{}^d \gamma_{db} + R^b{}_{(c|b|}{}^d \gamma_{a)d} \\ &= \nabla_{(c} \nabla^b \gamma_{a)b} + R^b{}_{ca}{}^d \gamma_{db} \quad , \end{aligned} \tag{7.5.22}$$

where in the second line we have used the fact that $R^b{}_{ca}{}^d \gamma_{db}$ is symmetric in c and a and $R_c{}^d = R^b{}_{cb}{}^d = 0$ since g_{ab} is a vacuum solution. Thus, in the transverse traceless gauge, equations (7.5.20) and (7.5.21), the linearized Einstein equation becomes simply

$$\nabla^b \nabla_b \gamma_{ac} - 2R^b{}_{ac}{}^d \gamma_{bd} = 0 \quad . \tag{7.5.23}$$

This is remarkably similar in form to Maxwell's equation (4.3.15) in the Lorentz gauge.

While the form of equation (7.5.23) is simple, it should be kept in mind that, with the notable exception of a flat background metric, in practice equation (7.5.23) comprises a very complicated system of coupled partial differential equations. Success in perturbation analyses has been achieved only in a few cases where the background metric has a great deal of symmetry or possesses other simplifying properties, and even in these cases success has not usually been achieved by a direct attack on equation (7.5.23) or (7.5.15).

Finally, the issue of linearization stability—i.e., the existence of a one-parameter family of exact solutions corresponding to a solution of the linearized equation—has been studied extensively for the vacuum Einstein equation. If the background spacetime (M, g_{ab}) is "closed"—i.e., if it possesses a compact, spacelike Cauchy hypersurface Σ (see chapter 8)—then the Einstein equation is linearization stable about (M, g_{ab}) if and only if (M, g_{ab}) does not possesses a Killing vector field. If (M, g_{ab}) possesses a Killing field, then it is necessary that γ_{ab} satisfy a second-order integral constraint involving the Killing field in order that a one-parameter family $g_{ab}(\lambda)$ exist. On the other hand, linearization stability is believed to hold for asymptotically flat perturbations of all asymptotically flat background spacetimes (even if Killing fields are present), although this has been proven only for the flat background spacetime. Details of the linearization stability analyses can be found in Fischer and Marsden (1979).

Problems

1. Prove that $\pounds_\psi \omega_a = 0$ (see eq. [7.1.10]) directly, without appealing to the arguments given in the text, i.e., prove first that $\pounds_\psi \epsilon_{abcd} = 0$, and then use only $\pounds_\psi \xi^a = 0$ and the formulas of appendix C.

2. Derive Einstein's equation (7.1.24)–(7.1.27) for the metric (7.1.22).

3. For a stationary, axisymmetric metric of the form (7.1.17), we define the *locally nonrotating observers* to be the family of observers which are "at rest" with respect to the t = constant hypersurfaces, i.e., whose 4-velocity, u^a, is proportional to $\nabla^a t$.
 a) Show that the angular momentum, L, of such observers vanishes, where L is defined by $L = u^a \psi_a$ (see eq. [6.3.13]).
 b) Show that such observers rotate with coordinate angular velocity $d\phi/dt = -W/X$. Since, in general, the metric (7.1.17) represents the exterior field of a stationary, axisymmetric rotating body (or black hole), we may interpret this $d\phi/dt$ as resulting from the "dragging of inertial frames" produced by rotating matter, in accord with Mach's principle.

4. Since any isometry ψ leaves the Weyl tensor invariant, $\psi^* C_{abcd} = C_{abcd}$, it follows that if k^a is a principal null vector, then so is $\psi^* k^a$. Since there exists only a discrete set of principal null vectors at each point, it follows that if ψ_t is a smooth one-parameter group of isometries which leaves $p \in M$ fixed, then ψ_t^* must leave invariant all principal directions at p. Use this fact to prove the following results (without resorting to a calculation of C_{abcd} and its principal null directions):
 a) Every spherically symmetric spacetime is algebraically special. Furthermore, every static, spherically symmetric spacetime, equation (6.1.5), is of algebraic type II–II.
 b) Every Robertson-Walker spacetime, equation (5.1.11), is conformally flat, $C_{abcd} = 0$.

5. Derive equation (7.4.8).

EIGHT

CAUSAL STRUCTURE

The causal structure of spacetime in special relativity already was described briefly in section 1.2. Associated with each event, p, in spacetime is a light cone, as illustrated in Figure 1.2. We assign the label "future" to half of the cone and the label "past" to the other half. The events lying in the interior of the future light cone represent events which can be reached by a material particle starting at p; these comprise the "chronological future" of p. The chronological future of p together with the events lying on the cone itself comprise the "causal future" of p, which physically represents events which, in principle, can be influenced by a signal emitted from p.

In general relativity, the causal structure of spacetime is *locally* of the same qualitative nature as in the flat spacetime of special relativity. However, significant differences can occur globally because of nontrivial topology, spacetime singularities, or the "twisting" of the directions of light cones as one moves from point to point. The purpose of this chapter is to give an account of the definitions and basic results concerning the causal structure of spacetimes in general relativity. These results not only are of interest in their own right but also are a crucial ingredient in the proof of the singularity theorems, which we shall discuss in the next chapter. Further discussion of topics on causal structure can be found in Hawking and Ellis (1973), Penrose (1972), and Geroch (1970*b*); most of the discussion presented here is based on these references.

The definitions of chronological and causal futures in general spacetimes are given in section 8.1 and several properties of these sets are derived. The conditions that express the notion that a spacetime be "causally well behaved" are discussed in section 8.2. Finally, in section 8.3 the notion of domains of dependence and global hyperbolicity are defined and numerous properties of globally hyperbolic spacetimes are derived. All the arguments given in this chapter rely heavily on the machinery of topological spaces outlined in appendix A, and it is assumed that the reader has read (or will refer to) this appendix.

The discussion throughout this chapter will concern arbitrary spacetimes (M, g_{ab}) in the sense that we shall not attempt to impose Einstein's equation on g_{ab}. Some of the examples given here may appear to be artificial in that they are constructed by removing points and/or making topological identifications of Minkowski spacetime. Nevertheless, they provide excellent illustrations of the types of phenomena that can occur in much less artifical spacetime models in general relativity.

8.1 Futures and Pasts: Basic Definitions and Results

Let (M, g_{ab}) be a spacetime. At each event $p \in M$, the tangent space, V_p, is, of course, isomorphic to Minkowski spacetime. We will refer to the light cone passing through the origin of V_p as the *light cone of p*. Thus, we emphasize that the light cone of p is a subset of V_p, not M.[1] As in special relativity, at each $p \in M$ we may designate half of the light cone as "future" and the other half as "past." However, in a non–simply connected[2] spacetime it may not be possible to make a continuous designation of "future" and "past" as p varies over M. An example of a spacetime for which no such continuous designation can be made is shown in Figure 8.1. If a continuous choice can be made, (M, g_{ab}) is said to be *time orientable*. (This property of a spacetime is analogous to, but distinct from, the notion of orientability of a manifold defined in section B.2 of appendix B.) Thus, non–time orientable spacetimes have the physically pathological property that we cannot consistently distinguish between the notions of going "forward in time" as opposed to "backward in time." In the following, we will consider only time orientable spacetimes and will assume that a continuous designation has been made of the "future" and "past" halves of the light cones at each point. A timelike or null vector lying in the "future half" of the light cone will be called *future directed*.

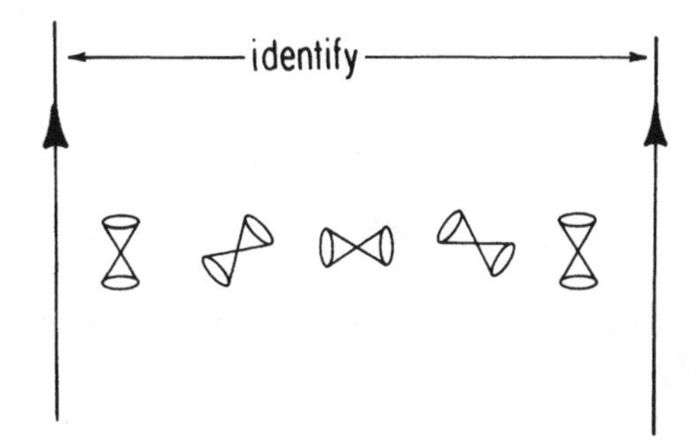

Fig. 8.1. A non-time-orientable spacetime.

An important property satisfied by every time orientable spacetime is expressed in the following lemma:

LEMMA 8.1.1. Let (M, g_{ab}) be time orientable. Then there exists a (highly nonunique) smooth nonvanishing timelike vector field t^a on M.

Proof. Since M is paracompact, we can choose a smooth Riemannian metric k_{ab} on M (see appendix A). At each $p \in M$ there will be a unique future directed timelike vector t^a which minimizes the value of $g_{ab}v^a v^b$ for vectors v^a subject to the condition that $k_{ab}v^a v^b = 1$. This t^a will vary smoothly over M and thus provide the desired vector field. □

1. It should be noted that some authors use the term "light cone" to designate the subset of M generated by null geodesics from p (as, in fact, we did in chapter 1). They use the term "null cone" to designate what we have called the "light cone."

2. See chapter 13 for the definition of a non–simply connected manifold.

Conversely, if a continuous, timelike vector field can be chosen, then (M, g_{ab}) is time orientable.

Let (M, g_{ab}) be a time orientable spacetime. A differentiable curve $\lambda(t)$ is said to be a *future directed timelike curve* if at each $p \in \lambda$ the tangent t^a is a future directed timelike vector. Note that according to this definition, if t^a vanishes at one point, λ is *not* considered to be a timelike curve.[3] Similarly, λ is said to be a *future directed causal curve* if at each $p \in \lambda$, t^a is either a future directed timelike or null vector. Thus, on a causal curve t^a may vanish. Analogous definitions apply to past directed timelike and causal curves.

The *chronological future* of $p \in M$, denoted $I^+(p)$, is defined as the set of events that can be reached by a future directed timelike curve starting from p,

$$I^+(p) = \left\{ q \in M \,\middle|\, \begin{array}{c} \text{There exists a future directed timelike curve} \\ \lambda(t) \text{ with } \lambda(0) = p \text{ and } \lambda(1) = q \end{array} \right\} . \quad (8.1.1)$$

Since one always can perform a sufficiently small deformation of the endpoint of a timelike curve while preserving the timelike nature of the curve, it follows that for all $q \in I^+(p)$ there exists an open neighborhood O of q such that $O \subset I^+(p)$. (A formal proof of this statement can be given using theorem 8.1.2 below.) Thus, $I^+(p)$ always is an open subset of M. Note that in general $p \notin I^+(p)$, but p will be in $I^+(p)$ if there is a closed timelike curve beginning and ending at p.

For any subset $S \subset M$, we define $I^+(S)$ by

$$I^+(S) = \bigcup_{p \in S} I^+(p) \quad . \quad (8.1.2)$$

Since an arbitrary union of open sets is open, it follows that $I^+(S)$ always is an open set. We also have $I^+[I^+(S)] = I^+(S)$ and $I^+(\overline{S}) = I^+(S)$, where $\overline{S}$ denotes the closure of S (see problem 1). Analogous definitions and properties apply to the *chronological pasts* $I^-(p)$ and $I^-(S)$.

The *causal future* of $p \in M$, denoted $J^+(p)$, is defined in the same way as $I^+(p)$ except that "future directed causal curve" replaces "future directed timelike curve" in equation (8.1.1). Thus, we always have $p \in J^+(p)$. In flat spacetime, $J^+(p)$ is a closed set. However, in a general spacetime, $J^+(p)$ can fail to be closed; an example of such a spacetime is given in Figure 8.2. In section 8.3 we shall prove that $J^+(p)$ must be closed in any globally hyperbolic spacetime. Again, we define

$$J^+(S) = \bigcup_{p \in S} J^+(p) \quad (8.1.3)$$

and define the causal pasts, $J^-(p)$ and $J^-(S)$, analogously.[4]

In Minkowski spacetime, $I^+(p)$ consists precisely of the points that can be reached by future directed timelike geodesics starting from p. The boundary $\dot{I}^+(p)$, of $I^+(p)$ is generated by the future directed null geodesics starting from p. In arbitrary

3. The strictly timelike character of t^a is imposed in order that the curve $\lambda(t) = p$ for all t is not considered to be timelike. Note that this makes the notion of a differentiable timelike curve parameterization dependent. However, below we shall define a parameterization-independent notion of a continuous, future-directed timelike curve.

4. In other references the notation $p \ll q$ and $p < q$ is sometimes used, respectively, for $q \in I^+(p)$ and $q \in J^+(p)$.

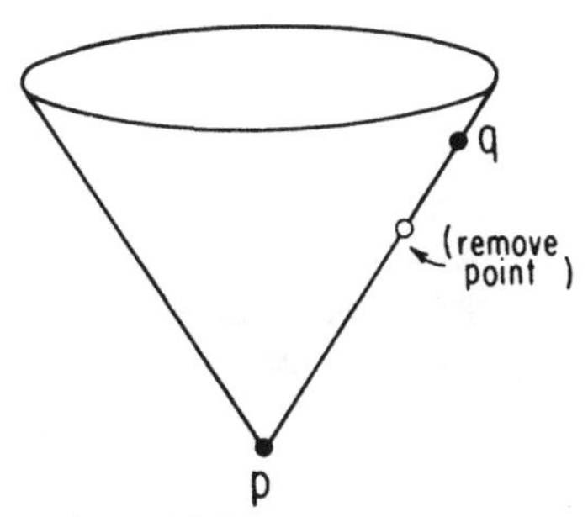

Fig. 8.2. Minkowski spacetime with a point on the future light cone of p removed. In this spacetime, no causal curve connects p and q so $q \notin J^+(p)$ but $q \in \overline{J^+(p)}$. Thus, $J^+(p)$ is not closed.

spacetimes neither of these statements is valid in general, as shown by simple examples obtained by removing points from Minkowski spacetime such as in Figure 8.2. However, *locally* these properties always remain valid, as expressed by the following theorem:

THEOREM 8.1.2. *Let (M, g_{ab}) be an arbitrary spacetime, and let $p \in M$. Then there exists a convex normal neighborhood of p, i.e., an open set U with $p \in U$ such that for all $q,r \in U$ there exists a unique geodesic γ connecting q and r and staying entirely within U. Furthermore, for any such U, $I^+(p)|_U$ consists of all points reached by future directed timelike geodesics starting from p and contained within U, where $I^+(p)|_U$ denotes the chronological future of p in the spacetime (U, g_{ab}). In addition, $\dot{I}^+(p)|_U$ is generated by the future directed null geodesics in U emanating from p.*

Although theorem 8.1.2 may seem obvious intuitively, it is nontrivial to give a formal proof of it. A proof of the first half is given beginning on page 134 of Hicks (1965), while the second half is proven as proposition 4.5.1 of Hawking and Ellis (1973).

If $q \in J^+(p)$ and λ is a causal curve beginning at p and ending at q, we can cover λ by convex normal neighborhoods, and using the compactness of λ (since it is the continuous image of a closed interval), we can extract a finite subcover (see appendix A). If λ failed to be a null geodesic in any such neighborhood, then with the help of theorem 8.1.2 we could deform λ into a timelike curve in that neighborhood and then extend this deformation to the other neighborhoods to obtain a timelike curve from p to q. Thus, we obtain the following corollary of theorem 8.1.2, which will be strengthened by theorem 9.3.8 of chapter 9:

COROLLARY. If $q \in J^+(p) - I^+(p)$, then any causal curve connecting p to q must be a null geodesic.

Using similar arguments, we also see that for any set $S \subset M$, we have $J^+(S) \subset \overline{I^+(S)}$. Since, clearly $I^+(S) \subset J^+(S)$, it follows immediately that $\overline{J^+(S)} = \overline{I^+(S)}$. Similarly, we have $I^+(S) = \text{int}[J^+(S)]$, and hence the boundaries of the chronological and causal futures of a set are always equal, $\dot{I}^+(S) = \dot{J}^+(S)$ (see problem 2).

A subset $S \subset M$ is said to be *achronal* if there do not exist $p, q \in S$ such that $q \in I^+(p)$, i.e., if $I^+(S) \cap S = \emptyset$, where $\emptyset$ denotes the empty set. The next theorem asserts that the boundary of the chronological future of a set always forms a "well behaved," three-dimensional, achronal surface.

THEOREM 8.1.3. *Let (M, g_{ab}) be a time orientable spacetime, and let $S \subset M$. Then $\dot{I}^+(S)$ (if nonempty) is an achronal, three-dimensional, embedded, C^0-submanifold of M.*

Proof. Let $q \in \dot{I}^+(S)$. If $p \in I^+(q)$, then $q \in I^-(p)$ and since $I^-(p)$ is open, an open neighborhood O of q is contained in $I^-(p)$ as shown in Figure 8.3. Since q is on the boundary of $I^+(S)$, we have $O \cap I^+(S) \neq \emptyset$ and thus $p \in I^+[O \cap I^+(S)] \subset I^+(S)$. This proves that $I^+(q) \subset I^+(S)$. Similarly we have $I^-(q) \subset M - I^+(S)$. Now, if $\dot{I}^+(S)$ failed to be achronal, we could find $q, r \in \dot{I}^+(S)$ such that $r \in I^+(q)$ and hence $r \in I^+(S)$. However, this is impossible since $I^+(S)$ is open and therefore $\dot{I}^+(S) \cap I^+(S) = \emptyset$. This proves that $\dot{I}^+(S)$ is achronal. To obtain the manifold structure of $\dot{I}^+(S)$, we introduce Riemannian normal coordinates x^0, x^1, x^2, x^3 at $q \in \dot{I}^+(S)$ and consider a sufficiently small neighborhood of q that $(\partial/\partial x^0)^a$ is everywhere timelike and each of the integral curves of $(\partial/\partial x^0)^a$ enters $I^+(q) \subset I^+(S)$ and $I^-(q) \subset M - I^+(S)$. But this implies that each such integral curve intersects $\dot{I}^+(S)$, and since $\dot{I}^+(S)$ is achronal, it must intersect it at precisely one point. Thus,

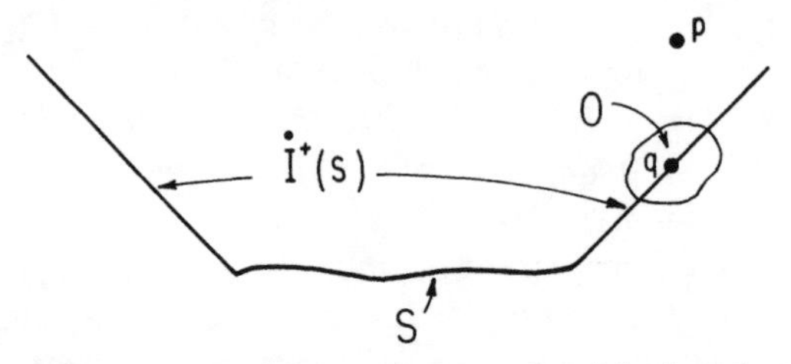

Fig. 8.3. A spacetime diagram showing a set S and the boundary of its future, $\dot{I}^+(S)$ (see theorem 8.1.3).

in each such neighborhood, we get a one-to-one association of points of $\dot{I}^+(S)$ with coordinates (x^1, x^2, x^3) characterizing the integral curve of $(\partial/\partial x^0)^a$. Furthermore, using the achronality of $\dot{I}^+(S)$, the value of x^0 at the intersection point must be a continuous (in fact,[5] C^{1-}) function of the coordinates (x^1, x^2, x^3), and thus the above map from a neighborhood of q in $\dot{I}^+(S)$ into $\mathbb{R}^3$ is a homeomorphism in the induced topology on $\dot{I}^+(S)$. By repeating this construction for all $q \in \dot{I}^+(S)$ we obtain a C^0-compatible family of charts covering $\dot{I}^+(S)$ which makes $\dot{I}^+(S)$ an embedded submanifold. □

Before proceeding further, we need to introduce several notions that will play an important role in many considerations of this chapter and of chapter 9. First, although for many purposes it suffices to consider only differentiable (or even C^∞) curves, when it comes to taking limits (as we shall do below) it is essential to extend consideration to continuous curves. The definitions of a timelike or causal curve can

5. See Hawking and Ellis (1973) for the definition of C^{1-}.

be extended to the continuous case by requiring that, locally, pairs of points on the curve can be joined by a differentiable timelike or, respectively, causal curve. More precisely, a *continuous* curve λ is said to be a *future directed timelike* (or *causal*) *curve* if for each $p \in \lambda$ there exists a convex normal neighborhood U of p such that if $\lambda(t_1)$, $\lambda(t_2) \in U$ with $t_1 < t_2$, then there exists a future directed *differentiable* timelike (or, respectively, causal) curve in U from $\lambda(t_1)$ to $\lambda(t_2)$. The timelike or causal nature of a continuous curve clearly is unchanged by a continuous, one-to-one reparameterization, and in the following we shall consider two curves differing by such a reparameterization to be equivalent.

Next, we define the notion of extendibility of a continuous curve. We need to distinguish clearly between the possibilities that a curve "runs off to infinity" or "runs around and around forever" or "runs into a singularity" as opposed to the possibility that a curve "stops" somewhere simply because one did not define it to go further. This distinction can be made precise via the notion of an endpoint of a curve. Let $\lambda(t)$ be a future directed causal curve. We say that $p \in M$ is a *future endpoint* of λ if for every neighborhood O of p there exists a t_0 such that $\lambda(t) \in O$ for all $t > t_0$. (Thus, by the Hausdorff property of M, λ can have, at most, one future endpoint. Note also that the endpoint need not lie on the curve, i.e., there need not exist a value of t such that $\lambda(t) = p$.) The curve λ is said to be *future inextendible* if it has no future endpoint. Past inextendibility is defined similarly. Note that if λ is a differentiable causal curve with future endpoint p, then it may not be possible to extend λ beyond p as a differentiable causal curve, but λ always can be extended as a continuous causal curve by adjoining a continuous causal curve to λ at p.

An important technical lemma relating to extendibility of causal curves which will be used in section 8.3 is the following.

LEMMA 8.1.4. Let λ be a past inextendible causal curve passing through point p. Then through any $q \in I^+(p)$ there exists a past inextendible timelike curve γ such that $\gamma \in I^+(\lambda)$.

Proof. Without loss of generality we may assume that the curve parameter, t, of λ has the range $[0, \infty)$. Following Geroch (1970*b*), we choose an arbitrary Riemannian metric on M. Using theorem 8.1.2 we can construct a timelike curve $\gamma(t)$ for $t \in [0, 1]$ which starts at q and satisfies the property that $\gamma \subset I^+(\lambda)$ and $d[\gamma(t), \lambda(t)] < C/(1 + t)$, where C is a constant and the distance, d, is the greatest lower bound of the length, measured using the Riemannian metric, of all curves connecting a point $\gamma(t)$ to a point $\lambda(t')$ with $t, t' \in [0, 1]$. By induction, we continue to extend $\gamma(t)$ to a curve defined for $t \in [0, n]$ for all n (i.e., for $t \in [0, \infty)$) preserving these properties. The resulting γ will be past inextendible because any endpoint of γ would also be an endpoint of λ. Thus, γ is the desired curve. $\square$

Another notion that will play a prominent role in many arguments of this chapter and of chapter 9 is the convergence of causal curves. Let $\{\lambda_n\}$ be a sequence of causal curves. A point $p \in M$ is said to be a *convergence point* of $\{\lambda_n\}$ if, given any open neighborhood O of p, there exists an N such that $\lambda_n \cap O \neq \emptyset$ for all $n > N$. A curve λ is said to be a *convergence curve* of $\{\lambda_n\}$ if each $p \in \lambda$ is a convergence point. Similarly, p is said to be a *limit point* of $\{\lambda_n\}$ if every open neighborhood of p

intersects infinitely many λ_n. A curve λ is said to be a *limit curve* of $\{\lambda_n\}$ if there exists a subsequence $\{\lambda_n'\}$ for which λ is a convergence curve. (Thus, if λ is a limit curve, then each $p \in \lambda$ is a limit point. Note, however, that a curve γ such that each $p \in \gamma$ is a limit point need *not* be a limit curve.) The following result plays a crucial role in many of the proofs given below.

LEMMA 8.1.5. Let $\{\lambda_n\}$ be a sequence of future inextendible causal curves which have a limit point p. Then there exists a future inextendible causal curve λ passing through p which is a limit curve of the $\{\lambda_n\}$.

Sketch of proof. We choose a convex normal neighborhood of p and a ball of Riemannian normal coordinate radius R about p contained in this neighborhood. We pick a subsequence of $\{\lambda_n\}$ that converges to p and, using the compactness of the coordinate sphere of radius R, a sub-subsequence which converges to a point on this sphere. Then we examine, in turn, all the coordinate spheres whose radii are rational multiples, between 0 and 1, of R and continue to extract limit points lying on these spheres and subsequences converging to these points. Finally, we take the closure of the set of these limit points and show that it defines a continuous causal limit curve λ. We then go to the endpoint of λ on the sphere of radius R and repeat the procedure. Continuing in this manner, we extend λ indefinitely.[6] Technical details of this proof can be found in Hawking and Ellis (1973). □

It should be noted that in theorem 8.1.5 if each λ_n is a timelike curve, the limit curve λ still may be only a causal curve, i.e., a sequence of timelike curves may converge to a null curve. Similarly, even if all the λ_n are smooth curves, λ may be only continuous.

As a direct application of theorem 8.1.5, we prove, now, a theorem characterizing the nature of boundaries of chronological futures.

THEOREM 8.1.6. *Let C be a closed subset of the spacetime manifold M. Then every point $p \in \dot{I}^+(C)$ with $p \notin C$ lies on a null geodesic λ which lies entirely in $\dot{I}^+(C)$ and either is past inextendible or has a past endpoint on C.*

Proof. Choose a sequence of points $\{q_n\}$ in $I^+(C)$ which converges to p, as illustrated in Figure 8.4. For each q_n let λ_n be a past directed timelike curve connecting q_n to a point in C. Consider, now, the new spacetime manifold $M - C$ obtained by removing the set C. (It is here that we use the assumption that C is closed, for otherwise $M - C$ would not define a manifold.) On $M - C$, each λ_n is past inextendible and p is a limit point of the sequence $\{\lambda_n\}$. Hence, by lemma 8.1.5 there exists a past inextendible causal limit curve λ passing through p. Each point of λ is a limit point of sequences in $I^+(C)$, so $\lambda \subset \overline{I^+(C)}$. On the other hand, if any point of λ were in $I^+(C)$, then by the corollary to theorem 8.1.2 we would have $p \in I^+(C)$, since p could be connected to C by a causal curve which is not a null

6. It should be noted that the above argument shows the existence of an extendible limit curve through p, together with the fact that any extendible limit curve can be further extended as a limit curve. To assert existence of an inextendible limit curve (i.e., a maximal element in the set of limit curves under ordering by inclusion) by this argument we must appeal to Zorn's lemma. However, it appears that a proof of lemma 8.1.5 could be given without invoking Zorn's lemma.

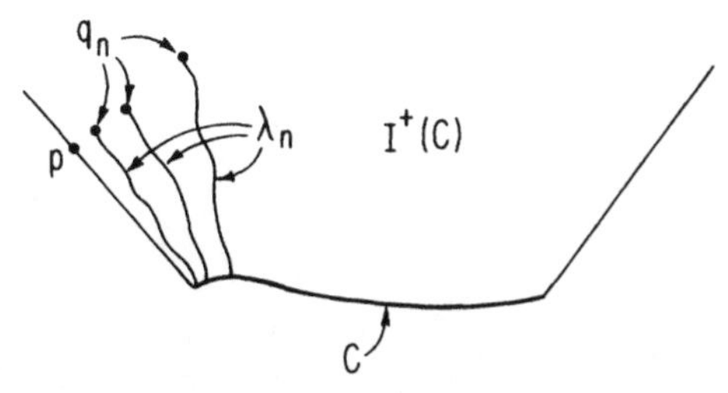

Fig. 8.4. A spacetime diagram showing a sequence of points in $I^+(C)$ converging to $p \in \dot{I}^+(C)$ (see theorem 8.1.6).

geodesic. This contradicts the fact that $p \in \dot{I}^+(C)$. Thus, $\lambda \subset \dot{I}^+(C)$. Furthermore, the achronality of $\dot{I}^+(C)$ (theorem 8.1.3), together with the corollary to theorem 8.1.2, implies that λ must be a null geodesic. Finally, since λ is past inextendible in $M - C$, in M it must either remain past inextendible or have a past endpoint on C. □

An example where λ is past inextendible is provided by point q in Figure 8.2. Note that although λ cannot have a past endpoint except on C, it may have a future endpoint; an example where λ has a future endpoint is given in Figure 8.5.

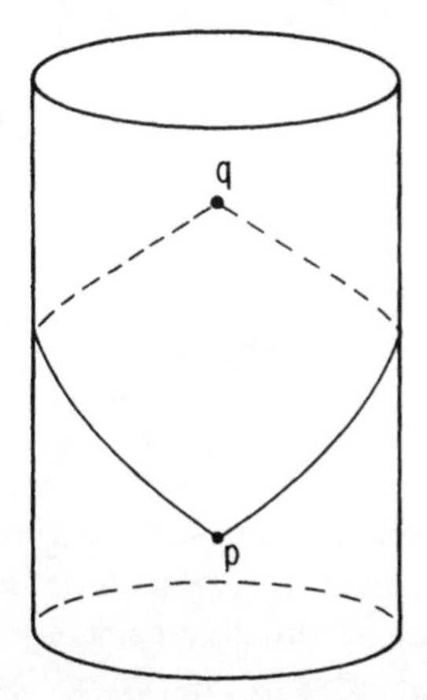

Fig. 8.5. A spacetime diagram of a two-dimensional flat spacetime with topology $\mathbb{R} \times S^1$. The null geodesic generators of $\dot{I}^+(p)$ have q as a future endpoint.

8.2 Causality Conditions

In this section, we shall give a brief discussion of formulations of the notion that a spacetime be "causally well behaved." Although according to theorem 8.1.2 all spacetimes in general relativity *locally* have the same qualitative causal structure as in special relativity, globally very significant differences can occur. For example, we can construct a (flat) spacetime with topology $S^1 \times \mathbb{R}^3$ by identifying the $t = 0$ and $t = 1$ hyperplanes of Minkowski spacetime, as illustrated in Figure 8.6. The integral curves of $(\partial/\partial t)^a$ in this spacetime will be closed timelike curves, and it is not difficult to see that for all $p \in M$ we have $I^+(p) = I^-(p) = M$. Thus, in this spacetime an observer with "free will" should have no difficulty altering past events.

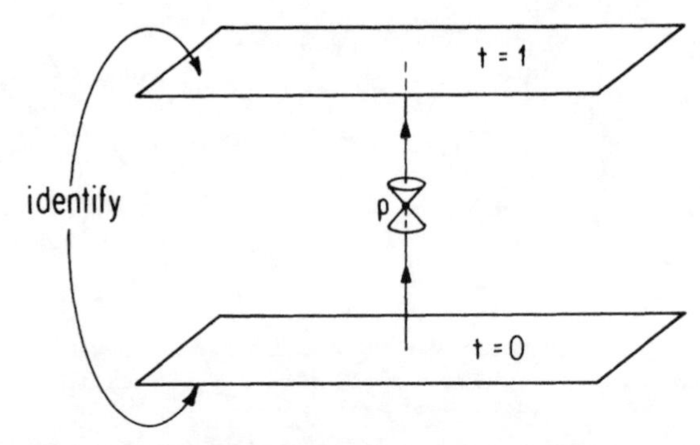

Fig. 8.6. Minkowski spacetime with the hyperplanes $t = 0$ and $t = 1$ identified. A closed timelike curve through point p is shown.

In addition, in spacetimes with nontrivial closed causal curves [i.e., closed causal curves other than the trivial curve $\lambda(t) = p$ for all t], severe consistency conditions may exist on solutions of the equations describing the propagation of physical fields. Note that the existence of spacetimes with closed causal curves cannot be blamed entirely on "artificial" topological identifications such as in Figure 8.6, since examples with topology $\mathbb{R}^4$ can be constructed easily by "twisting" the light cones, as illustrated in Figure 8.7.

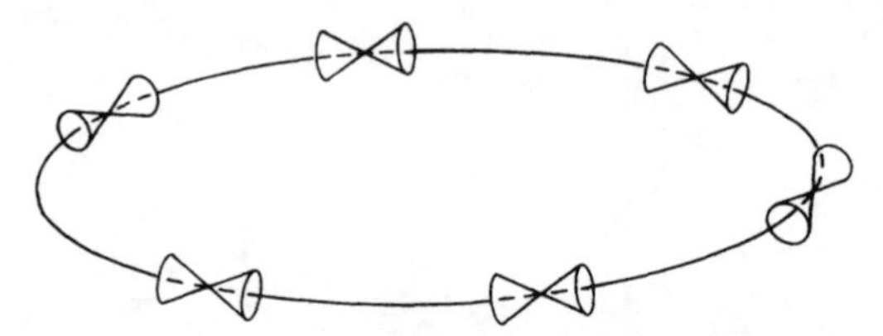

Fig. 8.7. A spacetime with topology $\mathbb{R}^4$ where the light cones "tip over" sufficiently to permit the existence of closed timelike curves.

It is generally believed that spacetimes with nontrivial closed causal curves are not physically realistic. However, even if a spacetime does not possess closed causal curves, it can be "on the verge" of violating causality as illustrated in Figure 8.8. In this example, there exist causal curves which come "arbitrarily close" to intersecting themselves, although none of them actually do. Since an arbitrarily small perturbation of the metric in spacetimes such as this would produce causality violation, these spacetimes also seem physically unreasonable. The mathematical possibility of spacetimes such as Figure 8.8 frequently arises in proofs of theorems concerning the possible global structures of spacetimes, so it is very useful to formulate precise conditions which characterize this type of behavior.

One such characterization is the strong causality condition. A spacetime (M, g_{ab}) is said to be *strongly causal* if for all $p \in M$ and every neighborhood O of p, there exists a neighborhood V of p contained in O such that no causal curve intersects V more than once. Thus, if a spacetime violates strong causality at p, then near p there exist causal curves which come arbitrarily close to intersecting themselves. In such

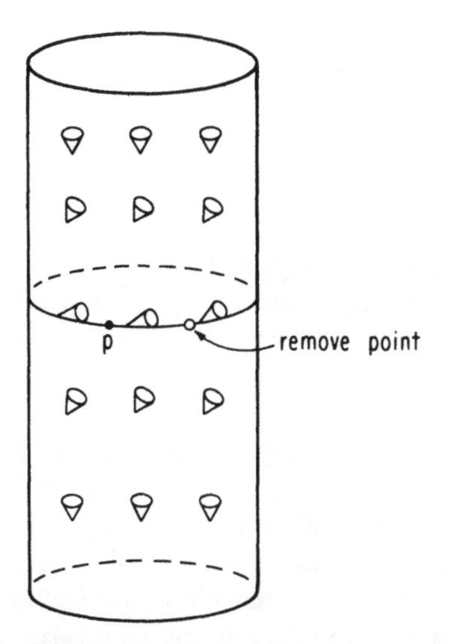

Fig. 8.8. A spacetime which violates strong causality. The light cones "tip over" sufficiently that the curve drawn through p is a null geodesic (which, however, is not a closed curve because of the point removed from the manifold). There exist no closed causal curves in this spacetime, but there are causal curves through p which come "arbitrarily close" to intersecting themselves.

a spacetime closed causal curves could be produced by a small modification of g_{ab} in an arbitrarily small neighborhood of p. In the example of Figure 8.8, strong causality is violated at the event labeled p.

A useful consequence of strong causality is expressed by the following lemma:

LEMMA 8.2.1. Let (M, g_{ab}) be strongly causal and let $K \subset M$ be compact. Then every causal curve λ confined within K must have past and future endpoints in K.

Proof. Without loss of generality, we may assume that the curve parameter, t, of λ runs from $-\infty$ to ∞. Let $\{t_i\}$ be an increasing sequence of numbers which diverges to infinity, and let $p_i = \lambda(t_i)$. Since $\{p_i\}$ is a sequence in K, by theorem A.9 of appendix A it has an accumulation point $p \in K$. Suppose one could find an open neighborhood O of p such that no $t_0 \in \mathbb{R}$ exists for which $\lambda(t) \in O$ for all $t > t_0$. Then the same must hold for every open neighborhood $V \subset O$. This means that λ enters every such V more than once, since infinitely many points of the sequence $\{\lambda(t_i)\}$ enter V but $\lambda(t)$ never remains in V. This contradicts the hypothesis that strong causality holds at p. Thus, p is a future endpoint of λ. Similarly, a past endpoint $q \in K$ of λ also exists. □

Although the imposition of the strong causality condition suffices to rule out the example of Figure 8.8, one can construct more complicated examples where strong causality is satisfied, but a modification of g_{ab} in an arbitrarily small neighborhood of two or more points produces closed causal curves. Thus, strong causality does not fully express the condition that one is not on the verge of producing causality violation. However, this condition is expressed satisfactorily by the stronger notion of stable causality, defined as follows.

Let t^a be a timelike vector at point $p \in M$, and define $\tilde{g}_{ab}$ at p by

$$\tilde{g}_{ab} = g_{ab} - t_a t_b \quad , \tag{8.2.1}$$

where g_{ab} is the spacetime metric. It is easy to see that $\tilde{g}_{ab}$ is also a Lorentz signature metric at p. Furthermore, the light cone of $\tilde{g}_{ab}$ is strictly larger than that of g_{ab}; i.e., every timelike and null vector of g_{ab} is a timelike vector of $\tilde{g}_{ab}$. If a spacetime were "on the verge" of having closed causal curves, then if we "open out" the light cone at each point, we should be able to produce closed timelike curves. Thus, we define a spacetime (M, g_{ab}) to be *stably causal* if there exists a continuous nonvanishing timelike vector field t^a such that the spacetime $(M, \tilde{g}_{ab})$ (where $\tilde{g}_{ab}$ is defined by [8.2.1]) possesses no closed timelike curves. The following theorem shows that stable causality is equivalent to the existence of a "global time function" on the spacetime. This greatly strengthens the above suggestion that the requirement of stable causality should suffice to rule out any causal pathologies.

THEOREM 8.2.2. *A spacetime (M, g_{ab}) is stably causal if and only if there exists a differentiable function f on M such that $\nabla^a f$ is a past directed timelike vector field.*

Proof of "if." Since $\nabla^a f$ is a past directed timelike vector field, along every future directed timelike curve with tangent v^a, we have $g_{ab} v^a \nabla^b f > 0$, and thus $v(f) > 0$. Hence, f strictly increases along every future directed timelike curve. Clearly, then, there can be no closed timelike curves in (M, g_{ab}) since f cannot return to its initial value. Now let $t^a = \nabla^a f$ and define the metric $\tilde{g}_{ab}$ by equation (8.2.1). It is easy to check that the inverse metric to $\tilde{g}_{ab}$ is

$$\tilde{g}^{ab} = g^{ab} + t^a t^b/(1 - t^c t_c) \quad . \tag{8.2.2}$$

Thus, we obtain

$$\tilde{g}^{ab} \nabla_a f \nabla_b f = t_a t^a + (t^a t_a)^2/(1 - t^c t_c) = t_a t^a/(1 - t^c t_c) < 0 \quad , \tag{8.2.3}$$

where all index raisings and lowerings on the right-hand sides of equations (8.2.2) and (8.2.3) are done with g_{ab} and g^{ab}. Hence $\tilde{g}^{ab} \nabla_b f$ is a timelike vector in the metric $\tilde{g}_{ab}$. By the same argument as already given above, it follows that no closed timelike curves exist in the spacetime $(M, \tilde{g}_{ab})$. Thus, (M, g_{ab}) is stably causal.

Sketch of proof of "only if." We are given that (M, g_{ab}) is stably causal and wish to construct a global time function. A promising candidate can be obtained as follows: Using the paracompactness of M, it is possible to show by arguments similar to those discussed at the end of appendix A, that one can always define a continuous volume measure, μ, on M such that the total volume of M is finite, $\mu[M] < \infty$. We define

$$F(p) = \mu[I^-(p)] \quad . \tag{8.2.4}$$

Then F strictly increases along all future directed causal curves (with nonvanishing tangent), and thus is a promising candidate for a global time function. Unfortunately, F need not be continuous. Nevertheless, for stably causal spacetimes we can obtain a continuous function with these properties by averaging F over nearby spacetimes with "opened out" light cones. More precisely, let t^a be a timelike vector field such that $\tilde{g}_{ab}$, defined by equation (8.2.1), has no closed timelike curves. For $0 \leqq \alpha \leqq 1$ we define $(g_\alpha)_{ab}$ by

$$(g_\alpha)_{ab} = g_{ab} - \alpha t_a t_b \tag{8.2.5}$$

and define $F_\alpha(p)$ by

$$F_\alpha(p) = \mu[I_\alpha^-(p)] \quad , \tag{8.2.6}$$

where $I_\alpha^-(p)$ denotes the chronological past of p in the metric $(g_\alpha)_{ab}$. By averaging F_α over α, it can be shown that we produce a continuous function which strictly increases along causal curves. Further "smoothing out" of this function produces a differentiable function with these properties. Details of this proof can be found in Hawking and Ellis (1973). □

As a corollary, we have:

COROLLARY. Stable causality implies strong causality.

Proof. Let f be a global time function on M. Given any $p \in M$ and any open neighborhood O of p, we can choose an open neighborhood $V \subset O$ of p shaped so that the limiting value of f along every future directed causal curve leaving V is greater than the limiting value of f on every future directed causal curve entering V (see Fig. 8.9). Thus, since f increases along every future directed causal curve, no causal curve can enter V twice. □

In conclusion, stable causality appears to be the appropriate notion which expresses the idea that a spacetime is not "on the verge" of displaying bad causal behavior.

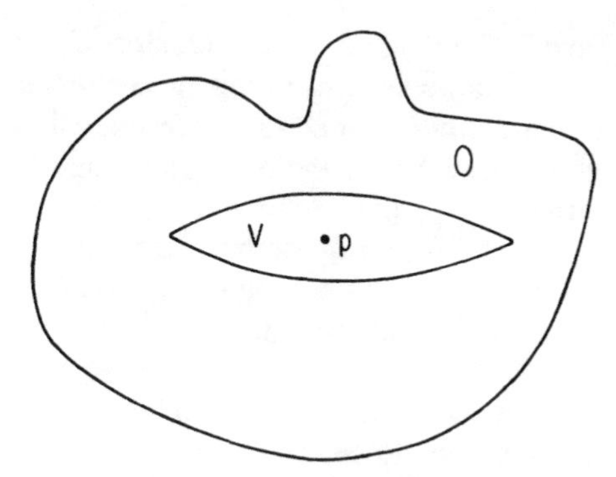

Fig. 8.9. The neighborhood, V, of p used in the proof of the corollary to theorem 8.2.2.

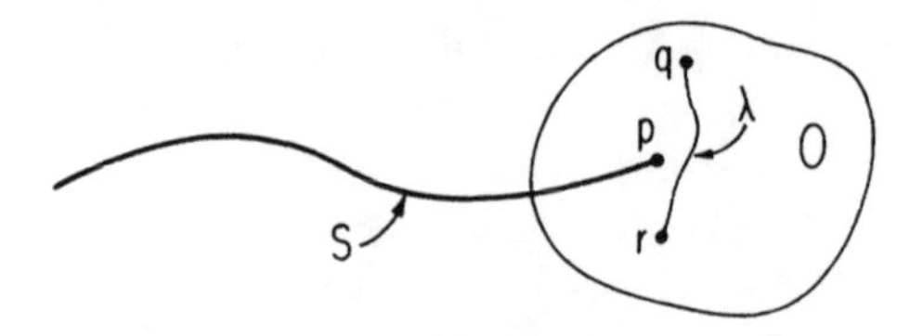

Fig. 8.10. A spacetime diagram illustrating the definition of the edge of a closed, achronal set S.

8.3 Domains of Dependence; Global Hyperbolicity

In the previous sections, attention was focused on the collection of events, $I^+(S)$ or $J^+(S)$, that could be influenced by a set, S, of events. In this section, we shall be concerned with the collection of events which, in the sense described below, are "entirely determined" by a set of events, S, (which, for technical reasons, will be taken to be a closed, achronal set). We also shall explore the properties of spacetimes (or spacetime regions) in which all events are "determined" by an appropriate S.

We begin, however, by pointing out an important property of closed, achronal sets. For S closed and achronal, we define the *edge* of S as the set of points $p \in S$ such that every open neighborhood O of p contains a point $q \in I^+(p)$, a point $r \in I^-(p)$ and a timelike curve λ from r to q which does not intersect S (see Fig. 8.10). We often will be interested in closed, achronal sets without edge. (Such sets are sometimes referred to as *slices*.) A repetition of the proof of theorem 8.1.3 establishes the following result:

THEOREM 8.3.1. *Let S be a (nonempty) closed achronal set with edge $(S) = \emptyset$. Then S is a three-dimensional, embedded, C^0 submanifold of M.*

Now, let S be a closed, achronal set (possibly with edge). We define the *future domain of dependence* of S, denoted $D^+(S)$, by

$$D^+(S) = \left\{ p \in M \middle| \begin{array}{l} \text{Every past inextendible causal curve} \\ \text{through } p \text{ intersects } S \end{array} \right\} . \tag{8.3.1}$$

Note that we always have $S \subset D^+(S) \subset J^+(S)$ and, since S is achronal, we also have $D^+(S) \cap I^-(S) = \emptyset$. Two examples illustrating the nature of $D^+(S)$ are given in Figures 8.11 and 8.12. Our definition of $D^+(S)$ agrees with that of Hawking and Ellis (1973) but differs slightly from that of Penrose (1972) and Geroch (1970*b*) in that they replace "causal curve" by "timelike curve."

The set $D^+(S)$ is of interest because if "nothing can travel faster than light," then any signal sent to $p \in D^+(S)$ must have "registered" on S. Thus, if we are given appropriate information about "initial conditions" on S, we should be able to predict what happens at $p \in D^+(S)$. Conversely, if, say, $p \in I^+(S)$ but $p \notin D^+(S)$, then it should be possible to send a signal to p without influencing S, and a knowledge of conditions on S should not suffice to determine conditions at p. These expectations are confirmed by an analysis of the propagation of solutions to hyperbolic wave equations representing physical fields in curved spacetime (see chapter 10).

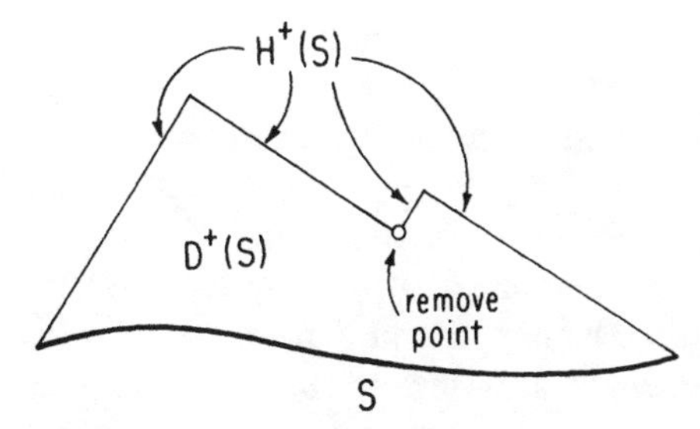

Fig. 8.11. A spacetime diagram showing the future domain of dependence, $D^+(S)$, and Cauchy horizon $H^+(S)$ of a particular closed achronal set S in Minkowski spacetime with a point removed.

The *past domain of dependence* of S, denoted $D^-(S)$, is defined by interchanging "future" and "past" in (8.3.1). Again, from a knowledge of conditions on S, we expect to be able to "retrodict" conditions at all $q \in D^-(S)$. The (full) *domain of dependence* of S, denoted $D(S)$, is defined as simply,

$$D(S) = D^+(S) \cup D^-(S) \quad . \tag{8.3.2}$$

Thus, $D(S)$ represents the complete set of events for which all conditions should be determined by a knowledge of conditions on S.

A closed achronal set Σ for which $D(\Sigma) = M$ is called a *Cauchy surface*. It follows immediately that for any Cauchy surface Σ, we have edge $(\Sigma) = \emptyset$. Thus, by theorem 8.3.1, every Cauchy surface is an embedded C^0 submanifold of M. This justifies our use of the terminology "surface" to describe Σ, and, since Σ is achronal, we may think of Σ as representing an "instant of time" throughout the universe.

A spacetime (M, g_{ab}) which possesses a Cauchy surface Σ is said to be *globally hyperbolic*. (This definition differs significantly from both the original definition of Leray 1952 and the definition used in Hawking and Ellis 1973, but all three definitions are equivalent; see the remark at the end of this chapter.) Thus, in a

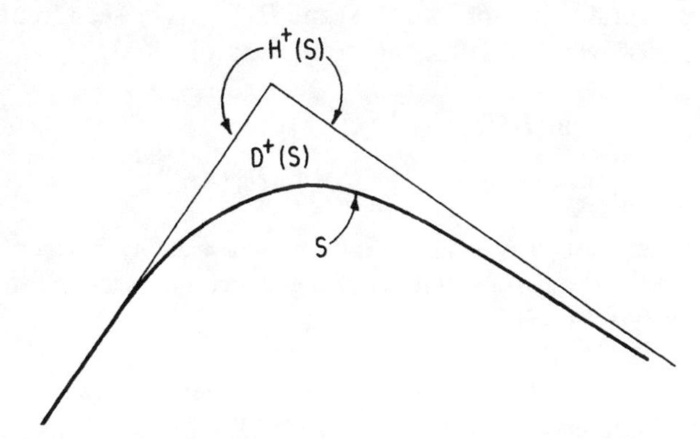

Fig. 8.12. A spacetime diagram showing $D^+(S)$ and $H^+(S)$ for a particular closed achronal set S in Minkowski spacetime. Here S is "asymptotically null" to the right and becomes "exactly null" to the left. This example shows that $H^+(S)$ can intersect S even if edge $(S) = \emptyset$.

globally hyperbolic spacetime, the entire future and past history of the universe can be predicted (or retrodicted) from conditions at the instant of time represented by Σ. Conversely, in a non–globally hyperbolic spacetime we have a breakdown of predictability in the sense that a complete knowledge of conditions at a single "instant of time" can never suffice to determine the entire history of the universe. There are some good reasons for believing that all physically realistic spacetimes must be globally hyperbolic (see chapter 12 and Penrose 1979). However, even if one does not accept these arguments and wishes to consider a non–globally hyperbolic spacetime, one still can apply the theorems proven below on globally hyperbolic spacetimes to any region of the form $\text{int}[D(S)]$ for any closed achronal set S.

The closure of the future domain of dependence of S, $\overline{D^+(S)}$, is characterized by the following property.

PROPOSITION 8.3.2. $p \in \overline{D^+(S)}$ if and only if every past inextendible timelike curve from p intersects S.

Proof. If there exists a past inextendible timelike curve from p which does not intersect S, it is easy to see that the same property must hold for an open neighborhood O of p. In that case, $O \cap D^+(S) = \emptyset$ and $p \notin \overline{D^+(S)}$. Conversely, suppose every past inextendible timelike curve from p intersects S. Then either $p \in S \subset D^+(S) \subset \overline{D^+(S)}$ or $p \in I^+(S)$. If the latter holds, let $q \in I^-(p) \cap I^+(S)$ and suppose a past inextendible causal curve λ from q failed to intersect S. Then either (i) λ remains in $I^+(S)$ or (ii) λ intersects $\dot{I}^+(S)$ at a point $r \notin S$. In either case, we could construct a past inextendible timelike curve from p which does not intersect S. [In case (i), we use lemma 8.1.4 to get a timelike curve $\gamma \subset I^+(\lambda) \subset I^+(S)$; in case (ii) we use the corollary to theorem 8.1.2 to obtain a timelike curve from p to r and then we extend it arbitrarily into the past.] Thus for any $q \in I^-(p) \cap I^+(S)$ we have $q \in D^+(S)$. Since every open neighborhood of $p \in I^+(S)$ intersects $I^-(p) \cap I^+(S)$, we have $p \in \overline{D^+(S)}$. $\square$

Some properties of the interiors of $D^+(S)$ and $D(S)$ can be seen from the following lemma, the proof of which is left as an exercise (problem 3).

LEMMA 8.3.3.
$$\text{int}[D^+(S)] = I^-[D^+(S)] \cap I^+(S) \quad ,$$
$$\text{int}[D(S)] = I^-[D^+(S)] \cap I^+[D^-(S)] \quad .$$

It follows immediately from the definition of a Cauchy surface that if Σ is a Cauchy surface, then every inextendible causal curve intersects Σ. In fact, we have the following stronger result.

PROPOSITION 8.3.4. Let Σ be a Cauchy surface and let λ be an inextendible causal curve. Then λ intersects Σ, $I^+(\Sigma)$, and $I^-(\Sigma)$.

Proof. Suppose λ did not intersect $I^-(\Sigma)$. By lemma 8.1.4 we could find a past inextendible timelike curve $\gamma \subset I^+(\lambda) \subset I^+[\Sigma \cup I^+(\Sigma)] = I^+(\Sigma)$. If we extend γ indefinitely into the future, it still cannot intersect Σ or the achronality of Σ would be violated. Since every inextendible causal curve intersects Σ, no such γ exists. Thus λ must enter $I^-(\Sigma)$. Similarly, λ must enter $I^+(\Sigma)$. □

Let S be a closed achronal set. We define the future Cauchy horizon of S, denoted $H^+(S)$, by

$$H^+(S) = \overline{D^+(S)} - I^-[D^+(S)] \quad . \tag{8.3.3}$$

As will be made more precise later (see the corollary to proposition 8.3.6 below), $H^+(S)$ and the analogously defined $H^-(S)$ measure the failure of S to be a Cauchy surface. Examples of $H^+(S)$ are shown in Figures 8.11 and 8.12. Clearly $H^+(S)$ always is closed since it is the intersection of the two closed sets $\overline{D^+(S)}$ and $M - I^-[D^+(S)]$. Furthermore, we have

$$\begin{aligned} I^-[H^+(S)] &\subset I^-[\overline{D^+(S)}] \\ &= I^-[D^+(S)] \\ &\subset M - H^+(S) \quad . \end{aligned} \tag{8.3.4}$$

Thus, $I^-[H^+(S)] \cap H^+(S) = \emptyset$, so $H^+(S)$ is achronal. Indeed, $H^+(S)$ is a portion of the boundary of the past of the set $D^+(S)$; specifically we have $H^+(S) = [I^+(S) \cup S] \cap \dot{I}^-[D^+(S)]$ (see problem 5).

One of the most important properties of $H^+(S)$ is stated in the following theorem:

THEOREM 8.3.5. *Every point $p \in H^+(S)$ lies on a null geodesic λ contained entirely within $H^+(S)$ which either is past inextendible or has a past endpoint on the edge of S.*

Proof. The basic idea of the proof is similar to that of theorem 8.1.6. Let $p \in H^+(S)$ with $p \notin \text{edge}(S)$. Then either (i) $p \in I^+(S)$ or (ii) $p \in S$ but $p \notin \text{edge}(S)$. We first will show that in either case, a nontrivial past directed null geodesic contained in $H^+(S)$ passes through p.

In case (i), since $p \notin I^-[D^+(S)]$, for every $q \in I^+(p)$ there exists a past inextendible causal curve from q which does not intersect S. Let $\{q_n\}$ be a sequence of points in $I^+(p)$ which converges to p, and let $\{\lambda_n\}$ be a corresponding sequence of such curves as illustrated in Figure 8.13. Since p is a limit point of $\{\lambda_n\}$, by lemma 8.1.5 there exists a past inextendible causal limit curve, λ, of the $\{\lambda_n\}$ which passes through p. Now, suppose λ entered the open set $I^+(S) \cap I^-[D^+(S)] \subset D^+(S)$. Then so would some λ_n for sufficiently large n, which is a contradiction since $\lambda_n \cap D^+(S) = \emptyset$ because each λ_n fails to intersect S. In particular, since $I^-(p) \subset I^-[\overline{D^+(S)}] = I^-[D^+(S)]$, this implies that within $I^+(S)$, λ is a past directed causal curve from p which does not enter $I^-(p)$. Thus, by the corollary to theorem 8.1.2, within $I^+(S)$, λ must be a null geodesic. Furthermore, if a past directed timelike curve from a point in $\lambda \cap I^+(S)$ failed to intersect S, we could construct a past directed timelike curve from p with the same property. Since $p \in \overline{D^+(S)}$, this is impossible because of proposition 8.3.2, and thus by the same proposition we have $[\lambda \cap I^+(S)] \subset \overline{D^+(S)}$. Thus, $\lambda \cap I^+(S) \subset H^+(S)$. Putting all these results together,

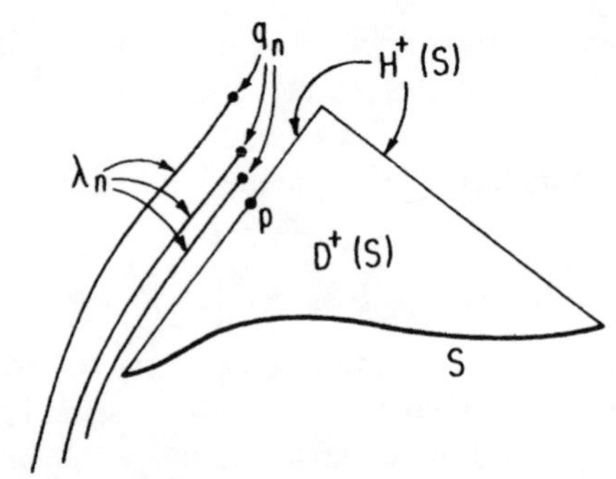

Fig. 8.13. A point $p \in H^+(S) \cap I^+(S)$ with a sequence of points $q_n \in I^+(p)$ converging to p (see theorem 8.3.5).

starting from every $p \in H^+(S) \cap I^+(S)$ we have obtained a nontrivial past directed null geodesic segment λ lying in $H^+(S)$.

In case (ii) where $p \in S$ but $p \notin \text{edge}(S)$, we use the definition of edge(S) to establish existence of an open set O such that no causal curve contained in O from a point $q \in I^+(p) \cap O$ can enter $I^-(S) \cap O$ without intersecting S. The argument establishing existence of a nontrivial past directed null geodesic through p remaining in $H^+(S)$ then proceeds along similar lines.

Finally, suppose a past directed null geodesic λ leaves $H^+(S)$; i.e., suppose the portion of λ contained in $H^+(S)$ has a past endpoint r. Since $H^+(S)$ is closed, we have $r \in H^+(S)$. If $r \notin \text{edge}(S)$, we can find a nontrivial past directed null geodesic segment λ' in $H^+(S)$ starting from r. However, if λ' were not a continuation of λ, by the corollary to theorem 8.1.2 we could find a timelike curve connecting a point of λ to a point of λ'. This would violate the achronality of $H^+(S)$. This proves that λ cannot have a past endpoint except on edge(S). □

The (full) Cauchy horizon of a closed achronal set S is defined by

$$H(S) = H^+(S) \cup H^-(S) \quad , \tag{8.3.5}$$

where $H^-(S)$ is defined by interchanging past and future in the definition of $H^+(S)$. Using lemma 8.3.3, it is not difficult to show that the Cauchy horizon marks the boundary of the domain of dependence of S:

Proposition 8.3.6. $H(S) = \dot{D}(S)$.

We leave the proof of this proposition as an exercise (problem 6). As a direct corollary we have

Corollary. If M is connected, then a nonempty closed achronal set, Σ, is a Cauchy surface for (M, g_{ab}) if and only if $H(\Sigma) = \emptyset$.

Proof. If $\dot{D}(\Sigma) = \emptyset$, we have $\overline{D(\Sigma)} = \text{int}[D(\Sigma)] = D(\Sigma)$, so $D(\Sigma)$ is both open and closed. Thus, since $D(\Sigma) \supset \Sigma \neq \emptyset$ and M is connected, we have $D(\Sigma) = M$. □

This corollary leads to the following useful criterion for Cauchy surfaces.

THEOREM 8.3.7. *If Σ is a closed, achronal, edgeless*[7] *set, then Σ is a Cauchy surface if and only if every inextendible null geodesic intersects Σ and enters $I^+(\Sigma)$ and $I^-(\Sigma)$.*

Proof. The "only if" part is a special case of proposition 8.3.4. To prove the "if" part, it suffices to show that if Σ is not a Cauchy surface, then at least one null geodesic fails to enter $I^+(\Sigma)$ or $I^-(\Sigma)$. But if Σ fails to be a Cauchy surface, then at least one of $H^+(\Sigma)$ or $H^-(\Sigma)$ is nonempty (unless M is disconnected and Σ does not intersect a component of M, in which case the result is trivial). If, say, $H^+(\Sigma) \neq \emptyset$, then since $\text{edge}(\Sigma) = \emptyset$, by theorem 8.3.5 there exists a past inextendible null geodesic which remains forever in $H^+(\Sigma)$ and thus never enters $I^-(\Sigma)$. Clearly, if we extend this geodesic forever into the future, it still cannot enter $I^-(\Sigma)$ or the achronality of Σ would be violated. □

Now, let (M, g_{ab}) be a globally hyperbolic spacetime with Cauchy surface Σ. It is easy to see that no closed timelike curves can exist in M: A closed timelike curve which intersects Σ would violate achronality of Σ, whereas a closed timelike (or even causal) curve which fails to intersect Σ would violate global hyperbolicity, since we could follow this curve "around and around" to define an inextendible causal curve which does not intersect Σ. In fact, we will see later (theorem 8.3.14 below) that (M, g_{ab}) must be stably causal. However, we first show that strong causality holds:

LEMMA 8.3.8. Let (M, g_{ab}) be a globally hyperbolic spacetime. Then (M, g_{ab}) is strongly causal.

Proof. In a globally hyperbolic spacetime with Cauchy surface Σ, we clearly have $M = I^+(\Sigma) \cup \Sigma \cup I^-(\Sigma)$. Suppose strong causality were violated at $p \in I^+(\Sigma)$. It follows directly from the definition of strong causality violation that we could find a convex normal neighborhood U of p contained in $I^+(\Sigma)$ and a nested family of open sets $O_n \subset U$ which converges to p such that for each n we can find a future directed timelike curve λ_n which begins in O_n, leaves U, and ends in O_n. Using lemma 8.1.5, we can find a limit curve λ which passes through p. Although each of the λ_n are extendible, it is clear that λ must either be inextendible or yield a closed causal curve through p, in which case it could be made to be inextendible by going "around and around." Since none of the λ_n can enter $I^-(\Sigma)$ or achronality of Σ would be violated, λ also cannot enter $I^-(\Sigma)$. However, this contradicts proposition 8.3.4. Thus, strong causality cannot be violated at $p \in I^+(\Sigma)$ or, by the same reasoning, at $p \in I^-(\Sigma)$. For the case $p \in \Sigma$, we can choose the $\{O_n\}$ so that any future directed timelike curve starting in O_n must exit O_n in $I^+(\Sigma)$. Again we would find that the limit curve λ could not enter $I^-(\Sigma)$, in contradiction to proposition 8.3.4. □

The remaining results of this section deal with the space of causal curves joining two points in a globally hyperbolic spacetime. The arguments used in the proofs

7. The requirement that Σ be edgeless can be removed from the hypothesis of this theorem. See Geroch (1970*b*).

provide a beautiful illustration of the abstract topological space methods outlined in appendix A.

Let (M, g_{ab}) be a strongly causal spacetime and let $p, q \in M$. We define $C(p, q)$ to be the set of continuous, future directed causal curves from p to q, where curves that differ only by reparameterization are considered to be the same curve. [Of course, $C(p, q)$ will be empty unless $q \in J^+(p)$.] We define a topology, $\mathcal{T}$, on $C(p,q)$, thereby making $(C(p,q), \mathcal{T})$ into a topological space as follows. Let $U \subset M$ be open, and define $O(U) \subset C(p, q)$ by

$$O(U) = \{\lambda \in C(p,q) \,|\, \lambda \subset U\} \quad . \tag{8.3.6}$$

In other words, $O(U)$ consists of all causal curves from p to q which lie entirely within U. We define our topology $\mathcal{T}$ by calling a subset, O, of $C(p,q)$ *open* if it can be expressed as

$$O = \cup\, O(U) \quad , \tag{8.3.7}$$

where each $O(U)$ is of the form (8.3.6).

Since $O(U_1) \cap O(U_2) = O(U_1 \cap U_2)$, it is easy to check that $\mathcal{T}$ does indeed define a topology on $C(p,q)$. Since we restrict consideration to spacetimes in which no closed causal curves exist, it follows that if λ, $\lambda' \in C(p,q)$ are distinct causal curves, the subset λ of M cannot contain or be contained in the subset λ' of M. From the properties of the topology on M, it then follows that the topology $\mathcal{T}$ on $C(p,q)$ is Hausdorff. Furthermore, when no closed causal curves exist, $\mathcal{T}$ is second countable. (See Geroch 1970*b* for a sketch of the proof of this result.) Finally, the notion of convergence defined by $\mathcal{T}$ is the following: $\lambda_n \to \lambda$ if for every open set $U \subset M$ with $\lambda \subset U$, there exists an N such that $\lambda_n \subset U$ for all $n > N$. If strong causality holds in M as we require, one can show that this notion of convergence is equivalent to the notion of convergence of curves defined in section 8.1. Similarly, the notion of a "limit curve," λ, of a sequence $\{\lambda_n\}$ in the sense of section 8.1 coincides with the topological space notion of an accumulation point defined in appendix A.

The key theorem upon which all further results of this section are based is the following:

THEOREM 8.3.9. *Let (M, g_{ab}) be a globally hyperbolic spacetime and let $p, q \in M$. Then $C(p,q)$ is compact.*

Proof. Since the topology on $C(p,q)$ is second countable, by theorem A.9 of appendix A we need only show that every infinite sequence $\{\lambda_n\}$ of points (i.e., curves) in $C(p,q)$ has an accumulation point (i.e., a limit curve, λ) in $C(p,q)$. Consider, first, the case $p, q \in D^-(\Sigma)$, where Σ is a Cauchy surface for (M, g_{ab}), and let $\{\lambda_n\}$ be a sequence in $C(p,q)$ as illustrated in Figure 8.14. If we (temporarily) remove q from M, then $\{\lambda_n\}$ becomes a sequence of future inextendible causal curves starting at p. Hence, by lemma 8.1.5, there exists a future inextendible (in $M - q$) limit curve λ starting at p. Since none of the λ_n enters $I^+(\Sigma)$, neither can λ. If, now, we restore the point q, then in M either (i) λ will remain inextendible or (ii) q will be an endpoint of λ. However, possibility (i) is ruled out by proposition 8.3.4 since λ does not enter $I^+(\Sigma)$. Hence, λ (with its endpoint at q added) provides the desired limit curve.

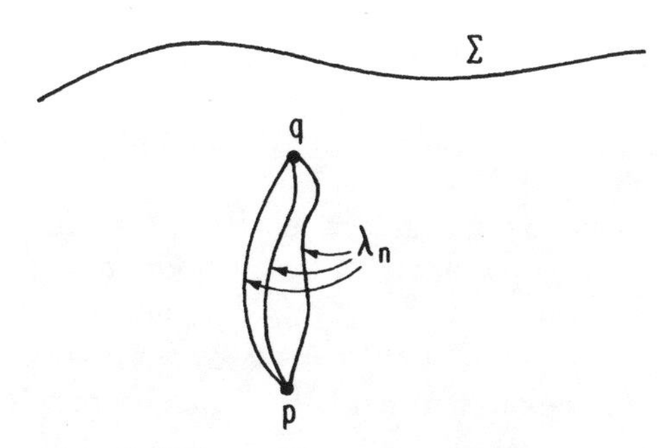

Fig. 8.14. A sequence $\{\lambda_n\}$ of causal curves from p to q in the case $p, q \in D^-(\Sigma)$ (see theorem 8.3.9).

The case $p, q \in D^+(\Sigma)$ obviously follows by the same argument. Thus, the only remaining nontrivial case is $p \in D^-(\Sigma)$, $q \in I^+(\Sigma)$. Given a sequence $\{\lambda_n\}$ in $C(p, q)$, the above argument proves existence of a future directed limit curve λ starting at p which enters $I^+(\Sigma)$. We choose $r \in \lambda \cap I^+(\Sigma)$ and extract a subsequence $\{\lambda'_n\}$ such that every point on the segment of λ between p and r is a convergence point of this subsequence. Now we reverse the procedure and consider the sequence of past inextendible (in $M - p$) causal curves $\{\lambda'_n\}$ starting from q. By the same arguments as given above, we obtain a limit curve λ' from q which enters $I^-(\Sigma)$ and thus must pass through r, since r is a convergence point of $\{\lambda'_n\}$ and if λ' did not extend to r it would have to remain in $I^+(r) \subset I^+(\Sigma)$. Thus, by joining the segment of λ' from r to q with the segment of λ from p to r we get the desired limit curve. □

The compactness of $C(p, q)$ directly yields a corresponding compactness of the subset $J^+(p) \cap J^-(q)$ in the manifold topology:

THEOREM 8.3.10. *Let (M, g_{ab}) be a globally hyperbolic spacetime and let $p, q \in M$. Then $J^+(p) \cap J^-(q)$ is compact.*

Proof. Since paracompactness implies that the manifold topology is second countable (see appendix A), by theorem A.9 in that appendix we need only show that every sequence $\{r_n\}$ of points in $J^+(p) \cap J^-(q)$ has an accumulation point r. To prove this, we let $\{\lambda_n\}$ be a sequence of causal curves from p to q such that λ_n passes through r_n. Since $C(p, q)$ is compact and first countable, by theorem A.9 we find a subsequence $\{\lambda'_n\}$ which converges to a curve $\lambda \in C(p, q)$. Now view λ as a subset of M. Since λ is compact (for it is the continuous image of a closed interval of $\mathbb{R}$), we can find an open neighborhood $U \subset M$ of λ such that $\overline{U}$ is compact. (*Proof.* Cover λ with open sets with compact closure, use compactness of λ to extract a finite subcover, and take the union.) By definition of convergence, there exists an integer N such that for all $n > N$ we have $\lambda'_n \subset U$. Thus, since $r'_n \in \lambda'_n$ and $U \subset \overline{U}$, we have a subsequence $\{r'_n\}$ contained in $\overline{U}$. From the compactness of $\overline{U}$, there exists an accumulation point $r \in \overline{U}$. If $r \notin \lambda$, we would contradict the fact that λ is the limit of $\{\lambda'_n\}$. Thus, we have $r \in \lambda \subset J^+(p) \cap J^-(q)$ and r is the desired accumulation point of $\{r_n\}$. □

From theorem A.2 in appendix A, we obtain the following corollary:

COROLLARY. In a globally hyperbolic spacetime, $J^+(p) \cap J^-(q)$ is closed.

In fact, it is not difficult to see that $J^+(p)$ itself must be closed. Namely, if not we could find a point $r \in \overline{J^+(p)}$ with $r \notin J^+(p)$. Choose $q \in I^+(r)$. Then we would have $r \in \overline{J^+(p) \cap J^-(q)}$ but $r \notin J^+(p) \cap J^-(q)$, which is a contradiction since $J^+(p) \cap J^-(q)$ is closed. One can strengthen these arguments to obtain the following more general result, the proof of which is left as an exercise (problem 7).

THEOREM 8.3.11. *Let (M, g_{ab}) be a globally hyperbolic spacetime and let $K \subset M$ be compact. Then $J^+(K)$ is closed.*

Recall that for any subset S, we have $I^+(S) \subset J^+(S) \subset \overline{I^+(S)}$. Thus, theorem 8.3.11 shows that for a compact set K in a globally hyperbolic spacetime, we have $J^+(K) = \overline{I^+(K)}$. This implies that $\dot{I}^+(K) \subset J^+(K)$, and thus in this case we can strengthen theorem 8.1.6 to conclude that every $p \in \dot{I}^+(K)$ can be joined to K by a past directed null geodesic lying in $\dot{I}^+(K)$. Thus, the phenomenon illustrated in Figure 8.2 cannot occur for compact sets in a globally hyperbolic spacetime.

Let (M, g_{ab}) be a globally hyperbolic spacetime with Cauchy surface Σ. For $q \in D^+(\Sigma)$ we define $C(\Sigma, q)$ as the set of continuous future directed causal curves from Σ to q and we define a topology on $C(\Sigma, q)$ in the same way as for $C(p, q)$. Essentially the same argument as given in theorem 8.3.9 establishes that $C(\Sigma, q)$ also is compact. A repetition of the proof of theorem 8.3.10 then yields the following theorem.

THEOREM 8.3.12. *Let (M, g_{ab}) be a globally hyperbolic spacetime with Cauchy surface Σ and let $q \in D^+(\Sigma)$. Then $J^+(\Sigma) \cap J^-(q)$ is compact.*

Our final results concern the topology and causal structure of globally hyperbolic spacetimes.

PROPOSITION 8.3.13. Suppose Σ and Σ' both are Cauchy surfaces for the spacetime (M, g_{ab}). Then Σ and Σ' are homeomorphic.

Proof. By lemma 8.1.1, there exists a nowhere vanishing timelike vector field t^a on M. Since Σ and Σ' are Cauchy surfaces, each integral curve of t^a must have precisely one intersection point with Σ and Σ'. The map of Σ onto Σ' obtained by associating points which lie on the same integral curve of t^a yields a continuous, one-to-one map with continuous inverse. □

Our final theorem greatly strengthens lemma 8.3.8.

THEOREM 8.3.14. *Let (M, g_{ab}) be a globally hyperbolic spacetime. Then (M, g_{ab}) is* *stably causal. Furthermore, a global time function, f, can be chosen such that each surface of constant f is a Cauchy surface. Thus M can be foliated by Cauchy surfaces and the topology of M is* $\mathbb{R} \times \Sigma$*, where* Σ *denotes any Cauchy surface.*

Sketch of proof. As in theorem 8.2.2, we introduce a volume measure μ on M such that $\mu(M)$ is finite, and we define the function f^- by

$$f^-(p) = \mu[I^-(p)] \quad . \tag{8.3.8}$$

Using the fact that, for all $q \in M$, $J^+(q)$ and $J^-(q)$ are closed, it is possible to show that $f^-(p)$ is continuous (i.e., the "averaging" used in theorem 8.2.2 is not needed); see Hawking and Ellis (1973) or Geroch (1970*b*) for details. Thus, f^- defines a continuous global time function, which can be "smoothed" to yield a differentiable global time function. This proves stable causality. To obtain the stronger conclusions of this theorem, we define f^+ by

$$f^+(p) = \mu[I^+(p)] \quad , \tag{8.3.9}$$

and f by

$$f(p) = f^-(p)/f^+(p) \quad . \tag{8.3.10}$$

Then f is continuous, and each surface of constant f is a closed achronal set. It remains to show that every inextendible causal curve intersects each surface of constant f. This must occur if $f^-(p)$ goes to zero along every past inextendible causal curve, and if, similarly, $f^+(p)$ goes to zero along every future inextendible causal curve, since then f will attain all values in the interval $(0, \infty)$ along every inextendible causal curve. To prove this, we suppose that λ is a past inextendible causal curve starting at q along which f^- does not approach zero in the past. Then there must exist a point p such that $p \in I^-(r)$ for all $r \in \lambda$. Hence λ must be contained within the set $J^-(q) \cap J^+(p)$. However, by theorem 8.3.10 this set is compact, and by lemma 8.3.8 strong causality holds. Thus, by lemma 8.2.1, λ must have a past endpoint, which contradicts the fact that λ is past inextendible. Thus, f^- must approach zero as one goes along λ into the past. Clearly, the analogous property also holds for f^+. □

Remark. Our definition of global hyperbolicity of (M, g_{ab}) is (I) existence of a Cauchy surface. Leray's (1952) original definition was essentially (II) strong causality holds and $C(p, q)$ is compact for all $p, q \in M$. The definition given by Hawking and Ellis (1973) is (III) strong causality holds and $J^+(p) \cap J^-(q)$ is compact. Lemma 8.3.8 and theorem 8.3.9 prove that (I) $\Rightarrow$ (II), while theorem 8.3.10 shows (II) $\Rightarrow$ (III) since the proof used only the compactness of $C(p, q)$. Finally, the proof of theorem 8.3.14 establishes that (III) $\Rightarrow$ (I), since only properties derived from (III) played a role in the construction of Cauchy surfaces. Thus, all three definitions are equivalent.

Problems

1. Prove that for any subset S of the spacetime manifold, M, we have
   a) $I^+[I^+(S)] = I^+(S)$ and
   b) $I^+(\overline{S}) = I^+(S)$, where $\overline{S}$ is the closure of S.

2. Let S be any subset of the spacetime manifold M. Give a proof of the following two claims made in the text:
   a) $J^+(S) \subset \overline{I^+(S)}$ and
   b) $I^+(S) = \text{int}[J^+(S)]$.

3. Prove lemma 8.3.3.

4. Let (M, g_{ab}) be a globally hyperbolic spacetime with Cauchy surface Σ, and let $C \subset \Sigma$ be closed. Show that $J^+(\Sigma) = D^+(C) \cup J^+(\Sigma - C)$.

5. Prove that for any closed, achronal set S, we have

$$H^+(S) = [I^+(S) \cup S] \cap \dot{I}^-[D^+(S)].$$

6. Prove proposition 8.3.6.

7. Prove theorem 8.3.11.

8. *a*) Let (M, g_{ab}) be a globally hyperbolic spacetime with Cauchy surface Σ. Let $C \subset \Sigma$ be closed but not necessarily compact. Prove that $J^+(C)$ is closed.
   b) Find an example of a closed, achronal set S in Minkowski spacetime such that $J^+(S)$ is *not* closed. (Hint: Consider a noncompact subset of a null plane.)

NINE

SINGULARITIES

In chapter 5 we studied the dynamics of a homogeneous, isotropic universe. We found there that if the universe presently is expanding (as it is observed to be), then a finite time ago it must have begun in a singular state, with infinite density and infinite spacetime curvature. In section 6.4 we analyzed the extended Schwarzschild geometry. There we found that at "$r = 0$" an infinite spacetime curvature singularity is present. Thus, a singularity also is predicted in the complete gravitational collapse of a spherical body.

However, the above predictions of singularities in cosmology and gravitational collapse are based solely on the analysis of solutions with a very high degree of symmetry. It certainly is possible that they could give a completely misleading picture of singularity formation. For example, in the Newtonian theory of gravity, if a spherical, nonrotating shell of dust is released from rest, a singularity will be produced at $r = 0$ when all the matter simultaneously reaches the origin. However, if one perturbs the shell away from spherical symmetry or gives it some rotation, then no such singularity will occur. Thus, one might suspect that in general relativity, nonspherical collapse will not, in general, lead to a singularity. Similarly, if one evolves a universe which is not exactly homogeneous and isotropic backward into the past, perhaps instead of a "big bang" singularity, general relativity might predict merely a nonsingular, high density phase which might then reexpand into the past. If this occurred, then in general relativity one would have a viable nonsingular closed universe model with infinitely many cycles of expansion, contraction, and "bounce." Thus, it is of considerable interest to determine if the above predictions of singularities are merely artifacts of the high degree of symmetry of the known exact solutions, or whether singularities are a true, generic feature of solutions describing cosmology and gravitational collapse in general relativity.

The purpose of this chapter is to discuss the singularity theorems of general relativity. These theorems prove that singularities *are* true, generic features of cosmological and collapse solutions. Although they give very little information concerning the detailed nature of these singularities, they show that models such as the nonsingular "bouncing universe" are incompatible with general relativity provided only that certain energy conditions are satisfied by matter and several other conditions hold in the spacetime.

The prediction of singularities undoubtedly represents a breakdown of general relativity in that its classical description of gravitation and matter cannot be expected to remain valid at the extreme conditions expected near a spacetime singularity. Indeed, on dimensional grounds one would expect a classical description of spacetime structure to break down at scales characterized by the Planck length, $l_P = (\hbar G/c^3)^{1/2} \approx 10^{-33}$ cm (see chapter 14). Hence, classical general relativity certainly should break down at or before the stage where it predicts spacetime curvatures of order l_P^{-2}. Although the singularity theorems do not prove that the singularities of classical general relativity must involve unboundedly large curvature, they strongly suggest the occurrence in cosmology and gravitational collapse of conditions in which quantum or other effects which invalidate classical general relativity will play a dominant role.

We begin our discussion of the singularity theorems in section 9.1 by attempting to define the term "singularity." Although it may be easy to recognize the "big bang" of the Robertson-Walker solutions or "$r = 0$" of the Schwarzschild solution as singularities, it is extremely difficult to give a satisfactory general notion of that term. We provide motivation for the notion of timelike and null geodesic incompleteness as a criterion for the presence of a singularity, although even this notion has some unwanted features. It is this criterion which is used in the singularity theorems.

The basic idea of the singularity theorems is as follows. As already shown in section 3.3, geodesics are curves of extremal length. After deriving some useful equations describing geodesic congruences in section 9.2, we obtain criteria in section 9.3 for when a timelike geodesic fails to be a local maximum of "length" (i.e., proper time) between two points or between a point and a three-dimensional surface. We obtain similar criteria for when a null geodesic fails to remain on the boundary of the future of a point or two-dimensional surface. Using an inequality on the Ricci tensor which follows from local positivity properties of the stress-energy tensor—this being the only place in the analysis where Einstein's equation is used—we show that in appropriate circumstances "sufficiently long" timelike geodesics cannot be maximal length curves, and "sufficiently long" null geodesics cannot remain on past or future boundaries. However, in section 9.4 we give global arguments involving compactness of the spaces of causal curves (see section 8.3) to establish existence of timelike curves of maximal length in globally hyperbolic spacetimes. An analogous result (theorem 8.3.11) for null geodesics was already given in chapter 8. The contradiction between these results (even if the spacetime is not assumed to be globally hyperbolic) produces the singularity theorems of section 9.5: In appropriate circumstances "sufficiently long" timelike or null geodesics cannot exist; i.e., one must have timelike or null geodesic incompleteness.

9.1 What Is a Singularity?

Intuitively, a spacetime singularity is a "place" where the curvature "blows up" or other "pathological behavior" of the metric takes place. The difficulty in making this notion into a satisfactory, precise definition of a singularity stems from the above terms placed in quotes.

By far the most serious (and, perhaps, insurmountable) difficulty arises from trying to give meaning to the idea of a singularity as a "place." In all physical theories except general relativity, the manifold and metric structure of spacetime is assumed in advance; we know the "where and when" of all spacetime events, and our task is simply to detemine the values of physical quantities at these events. If a physical quantity is infinite or otherwise undefined at a point in spacetime, we have no difficulty in saying that there is a singularity at that point. Thus, for example, we easily may give precise meaning to the statement that the Coulomb solution of Maxwell's equations in special relativity has a singularity at the events labeled by $r = 0$. However, the situation in general relativity is completely different. Here we are trying to solve for the manifold and metric structure of spacetime itself. Since the notion of an event makes physical sense only when manifold and metric structure are defined around it, the most natural approach in general relativity is to say (as we have been doing) that a spacetime consists of a manifold M and a metric g_{ab} defined *everywhere* on M. Thus, the "big bang" singularity of the Robertson-Walker solution is not considered to be part of the spacetime manifold; it is not a "place" or a "time." Similarly, only the region $r > 0$ is incorporated into the Schwarzschild spacetime; unlike the Coulomb solution in special relativity, the singularity at $r = 0$ is not a "place."

On the basis of the Robertson-Walker and Schwarzschild examples, one might expect that one still could define the notion of a singular boundary of a spacetime. The idea here would be to add points representing the singularity (e.g., the points "$\tau = 0$" in the Robertson-Walker spacetime and "$r = 0$" in the Schwarzschild spacetime) and define a topological space or perhaps even manifold with boundary[1] structure on the resulting collection of points. This would allow one to talk in precise terms of a singularity as "place" even though the metric is not defined there. However, while this could be done "by hand" in a few simple cases like the Robertson-Walker or Schwarzschild spacetimes, severe difficulties arise if one tries to give a meaningful general prescription for defining a singular boundary. In the first place, naive definitions based on coordinate component expressions for the metric have no chance of success as a general prescription since the apparent structure of a singularity can be altered easily by a coordinate transformation, as illustrated by simple examples like those given at the beginning of section 6.4 (where no true singularity was present). Several coordinate independent prescriptions for defining a singular boundary have been proposed, in particular, the "g-boundary" (Geroch 1968*b*) and the "b-boundary" (Schmidt 1971). However, these prescriptions and all others with similar properties produce boundaries with pathological topological properties in simple examples (Johnson 1977; Geroch, Liang, and Wald 1982). Geroch, Kronheimer, and Penrose (1972) have defined the notion of the causal boundary of a spacetime by using equivalence classes of future and past inextendible timelike curves (called TIPs and TIFs, respectively) to define the boundary points. However, some difficulties arise in the identification procedure of TIPs with TIFs and in the definition of a topology on the spacetime with boundary points adjoined.

1. See appendix B for the definition of "manifold with boundary."

Thus, at present, no fully satisfactory general notion of a singular boundary exists.[2] Until a satisfactory definition can be produced, we must abandon the notion of a singularity as a "place."

Of course, our failure to describe a singularity as a "place" in precise mathematical terms does not in any way lessen the obvious fact that singularities exist in, say, the Robertson-Walker and Schwarzschild spacetimes. It simply means that we must find other ways of characterizing a singularity. One approach is to note that the curvature "blows up" in these spacetimes, i.e., it is unbounded as $\tau \to 0$ in the Robertson-Walker spacetimes and as $r \to 0$ in the Schwarzschild spacetime. However, one encounters a number of serious difficulties if one attempts to use the notion of curvature "blowing up" as a general criterion for singularities. In the first place, curvature is described by a tensor field $R_{abc}{}^d$, and if one uses bad behavior of the components of this tensor or its derivatives as a criterion for singularities, one can get into trouble since this bad behavior of components could be due to bad behavior of the coordinate or tetrad basis rather than the curvature. To avoid this problem, one could examine scalars formed out of the curvature, such as R, $R_{ab}R^{ab}$, $R_{abcd}R^{abcd}$, and similar scalars formed by polynomial expressions in derivatives of the curvature tensor. However, even if the value of some curvature scalar is unbounded in a spacetime, the curvature might blow up only "as one goes to infinity," in which case one would not want to say that the spacetime is singular. On the other hand, for "plane gravitational waves," i.e., type IV vacuum solutions of Einstein's equation (see section 7.3), all such polynomial curvature scalars vanish, but the curvature tensor still may be singular! [Analogous behavior occurs for the plane wave solution, $F_{ab} = f(u)n_{[a}\partial_{b]}u$, of Maxwell's equations in special relativity, where $u = t - x$, n^a is a constant vector field orthogonal to $\partial^a u$, and f is an arbitrary function. All field invariants formed by polynomials in F_{ab} and its derivatives vanish identically, but F_{ab} will be singular if $f(u)$ is singular.] Furthermore, in a physically meaningful sense spacetimes may be singular without any bad behavior of the curvature tensor. For example, we can remove the "wedge" of Minkowski spacetime consisting of points with azimuthal coordinate ϕ satisfying $0 < \phi < \phi_0$. Then we can identify the points $\phi = 0$ with the corresponding points at $\phi = \phi_0$. The resulting space can be given the manifold structure of $\mathbb{R}^4$ by redefining the coordinate neighborhoods of the points with $\phi = 0$ (including $r = 0$). The flat metric of the original spacetime at the points with $\phi \neq 0$ (where the manifold structure is unaltered) is naturally defined on our new manifold, and it can be smoothly continued to the points with $\phi = 0$ and $r > 0$. However, it cannot be smoothly continued to $r = 0$. We have a "conical singularity" at $r = 0$ (and, thus, we must exclude $r = 0$ from the spacetime) even though the curvature tensor $R_{abc}{}^d$ vanishes everywhere else! Thus, the general characterization of singularities by the "blowing up" of curvature is unsatisfactory. The characterization of singularities by a detailed enumeration of the possible other types of

2. A manifold with a Riemannian metric defined on it can be made into a metric space by using the greatest lower bound of the length of curves connecting two points as a distance function. For a metric space, the Cauchy completion construction (see, e.g., Royden 1963) gives a fully satisfactory notion of a "singular boundary." However, a Lorentz metric does not naturally give rise to a distance function, so the Cauchy completion construction cannot be applied here.

"pathological behavior" of the spacetime metric also appears to be a hopeless task because of the infinite variety of possible pathological behavior.

How, then, can one characterize singular spacetimes? By far the most satisfactory idea proposed thus far is basically to use the "holes" left behind by the removal of singularities as the criterion for their presence. These "holes" should be detectable by the fact that there will be geodesics which have finite affine length; i.e., more precisely there should exist geodesics which are inextendible in at least one direction but have only a finite range of affine parameter. Such geodesics are said to be *incomplete*. (For timelike and spacelike geodesics, finite affine "length" is equivalent to finite proper time or length so the use of affine parameter simply generalizes the notion of "finite length" to null geodesics.) Thus, we could define a spacetime to be singular if it possesses at least one incomplete geodesic.[3] We then also may classify a singularity represented by an incomplete geodesic according to whether (i) a scalar constructed polynomially from $R_{abc}{}^{d}$ and its covariant derivatives blows up along the geodesic ("scalar curvature singularity") or (ii) no such scalar blows up, but a component of $R_{abc}{}^{d}$ or its covariant derivatives in a parallelly propagated tetrad blows up along the geodesic ("parallelly propagated curvature singularity") or (iii) no such curvature scalar or component blows up ("non–curvature singularity").

One possible objection to this definition is that spacetimes which are otherwise nonsingular but simply have points "artificially" removed would be considered singular. However, we can avoid this possibility by restricting consideration only to *inextendible spacetimes*, i.e., spacetimes which are not isometric to a proper subset of another spacetime. A much more serious objection is that geodesic incompleteness does not always correspond to the intuitive notion that there are "holes" in the spacetime produced by the excision of singularities. If a "hole" is present, then incompleteness should occur for all types of geodesics. However, Figure 9.1 shows an example of a spacetime which is spacelike and null geodesically complete but timelike geodesically incomplete. By "turning this example on its side," we obtain a spacetime which is timelike and null geodesically complete, but spacelike geodesically incomplete. Thus, in this latter spacetime, there is a singularity, but no observer or light ray can ever reach it! Further examples of this nature are given in problem 1. Furthermore, there exist spacetimes which are timelike, null, and spacelike geodesically complete, but possess future inextendible timelike curves of bounded acceleration which have finite proper length (Geroch 1968*c*). Such a spacetime would be considered nonsingular according to the above definition, but given a rocket ship with a sufficiently large but finite amount of fuel, there exists an observer who can end his existence in a finite time.

In fact, much more dramatic examples can be given of the failure of geodesic incompleteness to correspond to the intuitive notion of the excision of singular "holes." In a compact spacetime, every sequence of points has an accumulation point, so in a strong intuitive sense, no "holes" can be present. Yet, compact spacetimes exist which are geodesically incomplete; an example of one due to Misner

3. In the case of a manifold with Riemannian metric, geodesic completeness is equivalent to Cauchy completeness (see n. 2 and see, e.g., Kobayashi and Nomizu 1963 for a proof of equivalence).

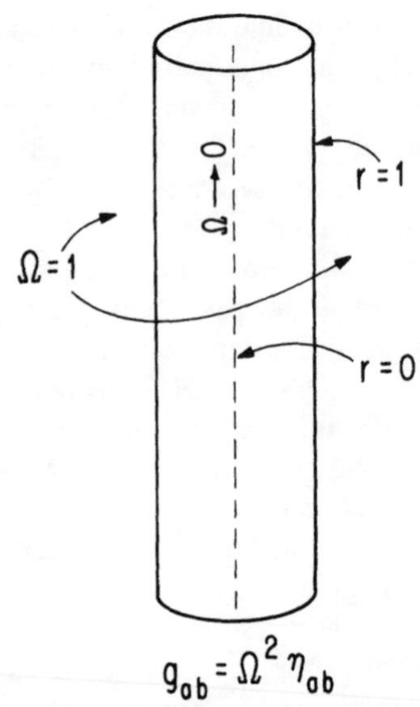

Fig. 9.1. A spacetime diagram of a spacetime $(\mathbb{R}^4, g_{ab})$ conformal to Minkowski spacetime $(\mathbb{R}^4, \eta_{ab})$, i.e., we have $g_{ab} = \Omega^2 \eta_{ab}$ (see appendix D). The conformal factor Ω is chosen to be spherically symmetric with $\Omega = 1$ for $r > 1$. From the spherical symmetry of Ω, it follows that the curve labeled $r = 0$ is a timelike geodesic of g_{ab}. If we choose Ω so that $\Omega(r = 0, t) \to 0$ sufficiently rapidly as $t \to \infty$, this timelike geodesic will be incomplete. However, all spacelike and null geodesics simply "pass through" the world tube of nonzero curvature and thus are complete.

(1963) is given in problem 2. This failure of geodesic incompleteness to correspond properly to the existence of "holes" is, of course, closely related to the difficulty discussed above of defining a singularity as a "place."

Nevertheless, there is a serious physical pathology in any spacetime which is timelike or null geodesically incomplete. In such a spacetime, it is possible for at least one freely falling particle or photon to end its existence within a finite "time" (i.e., affine parameter) or to have begun its existence a finite time ago. Thus, even if one does not have a completely satisfactory general notion of singularities, one would be justified in calling such spacetimes physically singular. It is this property that is proven by the singularity theorems to hold in a wide class of spacetimes. Given the existence of an incomplete timelike or null geodesic, one would like to know more about the character of the singularity, e.g., whether it is a curvature singularity [types (i) or (ii) above] or a non–curvature singularity. Unfortunately, the singularity theorems give virtually no information about the nature of the singularities of which they prove existence.

9.2 Timelike and Null Geodesic Congruences

Let M be a manifold and let $O \subset M$ be open. A *congruence* in O is a family of curves such that through each $p \in O$ there passes precisely one curve in this family. Thus, the tangents to a congruence yield a vector field in O, and, conversely, as discussed in section 2.2, every continuous vector field generates a congruence of curves. The congruence is said to be smooth if the corresponding vector field is smooth. In this section we shall define the expansion, shear, and twist of timelike and null geodesic congruences in a spacetime (M, g_{ab}) and derive equations for their rate of change as one moves along the curves in the congruence.

Consider, first, a smooth congruence of timelike geodesics. Without loss of generality, we may assume that the geodesics are parameterized by proper time τ, so that the vector field, ξ^a, of tangents is normalized to unit length, $\xi^a \xi_a = -1$. Then the tensor field, B_{ab}, defined by

$$B_{ab} = \nabla_b \xi_a \quad , \tag{9.2.1}$$

will be purely "spatial," i.e.,

$$B_{ab}\xi^a = B_{ab}\xi^b = 0 \quad . \tag{9.2.2}$$

The physical interpretation of B_{ab} may be seen from the following considerations. Consider a smooth one-parameter subfamily $\gamma_s(\tau)$ of geodesics in the congruence, and let η^a be the orthogonal deviation vector from γ_0 for this subfamily (see section 3.3). Then η^a represents an infinitesimal spatial displacement from γ_0 to a nearby geodesic in the subfamily. We have

$$\pounds_\xi \eta^a = 0 \quad , \tag{9.2.3}$$

and thus

$$\xi^b \nabla_b \eta^a = \eta^b \nabla_b \xi^a = B^a{}_b \eta^b \quad . \tag{9.2.4}$$

Thus, $B^a{}_b$ measures the failure of η^a to be parallelly transported. An observer on the geodesic γ_0 would find the nearby geodesics surrounding him to be stretched and rotated by the linear map $B^a{}_b$.

We define the "spatial metric" h_{ab} by

$$h_{ab} = g_{ab} + \xi_a \xi_b \quad . \tag{9.2.5}$$

Thus, $h^a{}_b = g^{ac}h_{cb}$ is the projection operator onto the subspace of the tangent space perpendicular to ξ^a. We define the *expansion* θ, *shear* σ_{ab}, and *twist* ω_{ab} of the congruence by

$$\theta = B^{ab}h_{ab} \quad , \tag{9.2.6}$$

$$\sigma_{ab} = B_{(ab)} - \frac{1}{3}\theta h_{ab} \quad , \tag{9.2.7}$$

$$\omega_{ab} = B_{[ab]} \quad . \tag{9.2.8}$$

Thus, B_{ab} is decomposed as

$$B_{ab} = \frac{1}{3}\theta h_{ab} + \sigma_{ab} + \omega_{ab} \quad . \tag{9.2.9}$$

Note that by virtue of equation (9.2.2) and Frobenius's theorem, the congruence is (locally) hypersurface orthogonal if and only if $\omega_{ab} = 0$ (see appendix B).

From equation (9.2.2), we see that σ_{ab} and ω_{ab} are purely spatial, i.e., $\sigma_{ab}\xi^b = \omega_{ab}\xi^b = 0$. From equation (9.2.4), it follows that along any geodesic in the congruence, θ indeed measures the average expansion of the infinitesimally nearby surrounding geodesics; ω_{ab}, being the antisymmetric part of the linear map B_{ab}, measures their rotation; and σ_{ab} measures their shear, i.e., an initial sphere in the tangent space which is Lie transported along ξ^a will distort toward an ellipsoid with principal axes given by the eigenvectors of $\sigma^a{}_b$, with rate given by the eigenvalues of $\sigma^a{}_b$.

It is clear that the geodesic deviation equation (3.3.18) will yield equations for the rate of change of θ, σ_{ab}, and ω_{ab} along each geodesic in the congruence. However, it is just as easy to derive these equations from scratch. We have

$$\begin{aligned} \xi^c \nabla_c B_{ab} &= \xi^c \nabla_c \nabla_b \xi_a = \xi^c \nabla_b \nabla_c \xi_a + R_{cba}{}^d \xi^c \xi_d \\ &= \nabla_b(\xi^c \nabla_c \xi_a) - (\nabla_b \xi^c)(\nabla_c \xi_a) + R_{cba}{}^d \xi^c \xi_d \\ &= -B^c{}_b B_{ac} + R_{cba}{}^d \xi^c \xi_d \quad . \end{aligned} \tag{9.2.10}$$

Taking the trace of equation (9.2.10), we obtain

$$\xi^c \nabla_c \theta = \frac{d\theta}{d\tau} = -\frac{1}{3}\theta^2 - \sigma_{ab}\sigma^{ab} + \omega_{ab}\omega^{ab} - R_{cd}\xi^c \xi^d \quad . \tag{9.2.11}$$

The trace-free, symmetric part of equation (9.2.10) yields

$$\begin{aligned} \xi^c \nabla_c \sigma_{ab} = &-\frac{2}{3}\theta\sigma_{ab} - \sigma_{ac}\sigma^c{}_b - \omega_{ac}\omega^c{}_b + \frac{1}{3}h_{ab}(\sigma_{cd}\sigma^{cd} - \omega_{cd}\omega^{cd}) \\ &+ C_{cbad}\xi^c \xi^d + \frac{1}{2}\tilde{R}_{ab} \quad , \end{aligned} \tag{9.2.12}$$

where $\tilde{R}_{ab}$ is the spatial, trace-free part of R_{ab}, i.e.,

$$\tilde{R}_{ab} = h_{ac}h_{bd}R^{cd} - \frac{1}{3}h_{ab}h_{cd}R^{cd} \quad . \tag{9.2.13}$$

Finally, the antisymmetric part of equation (9.2.10) yields,

$$\xi^c \nabla_c \omega_{ab} = -\frac{2}{3}\theta\omega_{ab} - 2\sigma^c{}_{[b}\omega_{a]c} \quad . \tag{9.2.14}$$

Equation (9.2.11) is known as Raychaudhuri's equation and is the key equation used in the proof of the singularity theorems. We turn our attention, now, to investigate the positivity property of the last term on the right-hand side of this equation. Using Einstein's equation, we may write this term as

$$R_{ab}\xi^a \xi^b = 8\pi\left[T_{ab} - \frac{1}{2}Tg_{ab}\right]\xi^a \xi^b = 8\pi\left[T_{ab}\xi^a \xi^b + \frac{1}{2}T\right] \quad . \tag{9.2.15}$$

Now, $T_{ab}\xi^a \xi^b$ physically represents the energy density of matter as measured by an observer whose 4-velocity is ξ^a. It is generally believed that for all physically reasonable classical matter this energy density is nonnegative, i.e.,

$$T_{ab}\xi^a \xi^b \geqq 0 \tag{9.2.16}$$

for all timelike ξ^a. This assumption is known as the *weak energy condition*. However, it also seems physically reasonable that the stresses of matter will not become so large and negative as to make the right-hand side of equation (9.2.15) negative. This assumption that

$$T_{ab}\xi^a\xi^b \geqq -\frac{1}{2}T \tag{9.2.17}$$

for all unit timelike ξ^a is known as the *strong energy condition*. For completeness, we also mention another energy condition which is believed to hold for physically reasonable matter: For all future directed, timelike ξ^a, $-T^a{}_b\xi^b$ should be a future directed timelike or null vector. Since for an observer with 4-velocity ξ^a the quantity $-T^a{}_b\xi^b$ physically represents the energy-momentum 4-current density of matter as seen by him, this condition, known as the *dominant energy condition*, can be interpreted as saying that the speed of energy flow of matter is always less than the speed of light. Indeed, this interpretation can be made precise: One can prove that if T_{ab} is conserved (i.e., $\nabla^a T_{ab} = 0$), satisfies the dominant energy condition, and vanishes on a closed, achronal set S, then it also vanishes in $D(S)$ (see lemma 4.3.1 of Hawking and Ellis 1973). Note that the dominant energy condition implies the weak energy condition, but otherwise the above three energy conditions are mathematically independent assumptions. In particular, the strong energy condition does not imply the weak energy condition.[4] It is "stronger" only in the sense that it appears to be a stronger physical requirement to assume equation (9.2.17) rather than (9.2.16).

The stress tensor T_{ab} is symmetric in its two indices, but since g_{ab} is not positive definite, the linear map $T^a{}_b$ from vectors into vectors need not be diagonalizable, i.e., it need not have four linearly independent eigenvectors. However, with the exception of a stress-tensor representing a null fluid (specifically, T_{ab} of the form $T_{ab} = \rho l_a l_b + p_1 x_a x_b + p_2 y_a y_b$ where l^a is null and x^a and y^a are orthonormal spacelike vectors orthogonal to l^a), all stress-tensors representing what is believed to be physically reasonable matter are diagonalizable. The eigenvectors of $T^a{}_b$ with different eigenvalues are automatically orthogonal, while eigenvectors with the same eigenvalue can be chosen to be orthogonal, so in the diagonal case, T_{ab} takes the form

$$T_{ab} = \rho t_a t_b + p_1 x_a x_b + p_2 y_a y_b + p_3 z_a z_b \quad , \tag{9.2.18}$$

where $\{t^a, x^a, y^a, z^a\}$ is an orthonormal basis, with t^a timelike. The eigenvalue ρ may be interpreted as the rest energy density of the matter, while the eigenvalues p_1, p_2, p_3 are called the *principal pressures*. For T_{ab} of the form (9.2.18), the above energy conditions are equivalent to the following conditions. The weak energy condition will be satisfied if and only if

$$\rho \geqq 0 \quad \text{and} \quad \rho + p_i \geqq 0 \quad (i = 1, 2, 3) \quad . \tag{9.2.19}$$

4. In some references, the terminology "weak energy condition" is defined to mean $T_{ab}k^a k^b \geqq 0$ for all null k^a. With this definition, the strong energy condition *does* imply the weak energy condition.

The strong energy condition is equivalent to

$$\rho + \sum_{i=1}^{3} p_i \geqq 0 \qquad \text{and} \qquad \rho + p_i \geqq 0 \quad (i = 1, 2, 3) \quad . \tag{9.2.20}$$

Thus, the weak and strong energy conditions will be satisfied provided only that $\rho \geqq 0$ and there do not exist negative pressures (i.e., tensions) comparable in magnitude or larger than ρ. Finally, the dominant energy condition is equivalent to

$$\rho \geqq |p_i| \quad (i = 1, 2, 3) \quad . \tag{9.2.21}$$

Let us return now to Raychaudhuri's equation (9.2.11). As discussed above, if Einstein's equation holds, and the strong energy condition is satisfied by T_{ab}, then the last term of the right-hand side of equation (9.2.11) will be negative. This can be interpreted, physically, as a manifestation of the attractiveness of gravity. If the congruence is hypersurface orthogonal, we have $\omega_{ab} = 0$, so the third term vanishes. The second term, $-\sigma_{ab}\sigma^{ab}$, is manifestly nonpositive. Thus, under these assumptions we find

$$\frac{d\theta}{d\tau} + \frac{1}{3}\theta^2 \leqq 0 \quad , \tag{9.2.22}$$

which implies

$$\frac{d}{d\tau}(\theta^{-1}) \geqq \frac{1}{3} \tag{9.2.23}$$

and hence

$$\theta^{-1}(\tau) \geqq \theta_0^{-1} + \frac{1}{3}\tau \quad , \tag{9.2.24}$$

where θ_0 is the initial value of θ. Suppose, now, that θ_0 is negative, i.e., the congruence is initially converging. Then equation (9.2.24) implies that θ^{-1} must pass through zero, i.e., $\theta \to -\infty$, within a proper time $\tau \leqq 3/|\theta_0|$. Thus, we have proven the following lemma.

LEMMA 9.2.1. Let ξ^a be the tangent field of a hypersurface orthogonal timelike geodesic congruence. Suppose $R_{ab}\xi^a\xi^b \geqq 0$, as will be the case if Einstein's equation holds in the spacetime and the strong energy condition is satisfied by the matter. If the expansion θ takes the negative value θ_0 at any point on a geodesic in the congruence, then θ goes to $-\infty$ along that geodesic within proper time $\tau \leqq 3/|\theta_0|$.

In general, the singularity in θ implied by lemma 9.2.1, of course, represents merely a singularity in the congruence, not a singularity in the structure of spacetime. It simply states that caustics will develop in a congruence if convergence occurs anywhere. The conclusions of lemma 9.2.1 hold for congruences in Minkowski spacetime and in many other singularity free spacetimes. However, in the next three sections we shall show that in certain spacetimes, the conclusions of lemma 9.2.1 in conjunction with properties derived from global arguments prove the existence of spacetime singularities.

We turn our attention, now, to null geodesic congruences. Again, we parameterize the geodesics by an affine parameter λ, but, unlike the timelike case, we now have no natural way of normalizing the tangent field k^a and thereby adjusting the scale of λ on different geodesics. In the timelike case, we restricted consideration to deviation vectors η^a orthogonal to ξ^a. There actually were two independent (though related) reasons for doing so. (1) We have $\xi^a \nabla_a(\xi_b \eta^b) = \xi^a \xi_b \nabla_a \eta^b = \xi_b \pounds_\xi \eta^b + \eta^a \xi_b \nabla_a \xi^b = 0$ provided $\xi_a \xi^a$ is normalized to be constant. Thus, $\xi_a \eta^a$ is constant along each geodesic, and the behavior of the "nonorthogonal" part of η^a is uninteresting. (2) Deviation vectors which differ only by a multiple of ξ^a represent a displacement to the same nearby geodesic. Orthogonality fixes a natural "gauge condition" on η^a.

In the case of a null geodesic congruence, the above reasons for restricting the choice of deviation vector still apply, but now they lead to two independent restrictions. First, for any deviation vector η^a, we again have $k^a \nabla_a(k_b \eta^b) = k_b \pounds_k \eta^b + \eta^a k_b \nabla_a k^b = 0$, so $k_b \eta^b$ does not vary along each geodesic. This implies that an arbitrary deviation vector η^a may be written as the sum of a vector not orthogonal to k^a which is parallelly propagated along the geodesic, plus a vector perpendicular to k^a. (Note, however, that there is no natural, unique way of decomposing η^a in this manner.) Thus, the behavior of the "nonorthogonal" part of η^a again is uninteresting, and we may restrict consideration to deviation vectors satisfying $\eta^a k_a = 0$. Second, deviation vectors which differ only by a multiple of k^a again represent a displacement to the same nearby geodesic. Thus, the physically interesting quantity is really the equivalence class of deviation vectors, where two deviation vectors are considered equivalent if their difference is a multiple of k^a. Since k^a is null and thus is orthogonal to itself, this second restriction is independent of the first restriction, and it reduces the physically interesting class of deviation vectors to a two-dimensional subspace, as we now shall describe with more precision.

Let V_p denote the tangent space at a point $p \in M$. The tangent vectors in V_p which are orthogonal to k^a form a three-dimensional subspace, which we shall denote by $\tilde{V}_p$. We define $\hat{V}_p$ to be the vector space of equivalence classes of vectors in $\tilde{V}_p$, where $x^a, y^a \in \tilde{V}_p$ are defined to be equivalent if there exists a number $c \in \mathbb{R}$ such that $x^a - y^a = ck^a$. Then $\hat{V}_p$ is a two-dimensional vector space, although it cannot be identified as a subspace of V_p in any natural way. As explained above, the deviation vectors of interest live in the vector space $\hat{V}_p$.

A tangent vector $t^a \in V_p$ does not naturally give rise to a tangent vector in $\tilde{V}_p$, because there is no natural way of decomposing it as a sum of a vector in $\tilde{V}_p$ and a vector not lying in $\tilde{V}_p$. However, if $t^a \in \tilde{V}_p$ (i.e., if $t^a k_a = 0$), then t^a naturally gives rise to a vector $\hat{t}^a$ in $\hat{V}_p$ by taking its equivalence class. On the other hand, a dual vector $\mu_a \in V_p^*$ naturally gives rise to a dual vector $\tilde{\mu}_a \in (\tilde{V}_p)^*$ by restricting its action to vectors in $\tilde{V}_p$. However, $\tilde{\mu}_a$ naturally gives rise to a dual vector $\hat{\mu}_a \in (\hat{V}_p)^*$

if and only if $\tilde{\mu}_a k^a = \mu_a k^a = 0$. More generally, a tensor $T^{a_1 \cdots a_k}{}_{b_1 \cdots b_l}$ over V_p naturally gives rise to a tensor $\hat{T}^{a_1 \cdots a_k}{}_{b_1 \cdots b_l}$ over $\hat{V}_p$ if and only if the result of contracting any one of its indices with k_a or k^a and contracting its remaining indices with vectors or dual vectors which give rise to elements of $\hat{V}_p$ or $(\hat{V}_p)^*$ is always zero. For tensors satisfying this property, the operation of taking outer products commutes with the "projection" into tensors over $\hat{V}_p$. Furthermore, for a tensor satisfying the stronger property that contraction of any one of its indices into k^a or k_a gives zero, then projection into $\hat{V}_p$ commutes with contraction.

The spacetime metric g_{ab} satisfies the above property and thus gives rise to a tensor over $\hat{V}_p$, which we shall denote as $\hat{h}_{ab}$. It is not difficult to see that $\hat{h}_{ab}$ is a positive definite (i.e., signature $++$) metric on $\hat{V}_p$. (On the other hand, the tensor $\tilde{h}_{ab}$ over $\tilde{V}_p$ obtained by restricting the action of g_{ab} to vectors in $\tilde{V}_p$ is degenerate, since $\tilde{h}_{ab} k^b = 0$, so $\tilde{h}_{ab}$ is not a metric on $\tilde{V}_p$.) Note that the inverse metric $\hat{h}^{ab}$ is just the "hatted" tensor associated with g^{ab}.

Consider, now, a congruence of null geodesics with tangent field k^a. The tensor field

$$B_{ab} = \nabla_b k_a \tag{9.2.25}$$

also satisfies the above property, and thus gives rise to a "hatted" tensor field $\hat{B}_{ab}$. We can decompose $\hat{B}_{ab}$ as

$$\hat{B}_{ab} = \frac{1}{2} \theta \hat{h}_{ab} + \hat{\sigma}_{ab} + \hat{\omega}_{ab} \tag{9.2.26}$$

with

$$\theta = \hat{h}^{ab} \hat{B}_{ab} \quad , \tag{9.2.27}$$

$$\hat{\sigma}_{ab} = \hat{B}_{(ab)} - \frac{1}{2} \theta \hat{h}_{ab} \quad , \tag{9.2.28}$$

$$\hat{\omega}_{ab} = \hat{B}_{[ab]} \quad , \tag{9.2.29}$$

so that θ, $\hat{\sigma}_{ab}$, and $\hat{\omega}_{ab}$ again have the physical interpretation as, respectively, the expansion, shear, and twist of the congruence. The change in the numerical factor in the term $\frac{1}{2} \theta \hat{h}_{ab}$ (as compared with $\frac{1}{3} \theta h_{ab}$ in the timelike case) arises simply because the relevant vector space is now two-dimensional rather than three-dimensional.

The same derivation as led to equation (9.2.10) in the timelike case now yields

$$k^c \nabla_c B_{ab} + B^c{}_b B_{ac} = R_{cba}{}^d k_d k^c \quad . \tag{9.2.30}$$

"Hatting" this equation, we obtain

$$k^c \nabla_c \hat{B}_{ab} + \hat{B}^c{}_b \hat{B}_{ac} = \widehat{R_{cbad} k^c k^d} \quad . \tag{9.2.31}$$

Finally, substituting our decomposition (9.2.26) for $\hat{B}_{ab}$ and taking the trace, symmetric trace free, and antisymmetric parts of the equation, we obtain, respectively,

$$\frac{d\theta}{d\lambda} = -\frac{1}{2} \theta^2 - \hat{\sigma}_{ab} \hat{\sigma}^{ab} + \hat{\omega}_{ab} \hat{\omega}^{ab} - R_{cd} k^c k^d \quad , \tag{9.2.32}$$

$$k^c \nabla_c \hat{\sigma}_{ab} = -\theta \hat{\sigma}_{ab} + \widehat{C_{cbad} k^c k^d} \quad , \tag{9.2.33}$$

$$k^c \nabla_c \hat{\omega}_{ab} = -\theta \hat{\omega}_{ab} \quad . \tag{9.2.34}$$

Notice that the last term on the right-hand side of equation (9.2.33) vanishes if and only if $C_{cbad} k^c k^d x^a y^b = 0$ for all x^a, y^b orthogonal to k^a. This is equivalent to $k_{[e} C_{b]cd[a} k_{f]} k^c k^d = 0$ which is just the condition that k^a be a principal null vector of the Weyl tensor[5] (see section 7.3).

The nature of equation (9.2.32) for the expansion of null geodesics is very similar to that of Raychaudhuri's equation (9.2.11). The only significant change is that, using Einstein's equation, we now obtain

$$R_{ab} k^a k^b = 8\pi T_{ab} k^a k^b \quad . \tag{9.2.35}$$

Thus, all that is needed to ensure that the last term of equation (9.2.32) is nonpositive is that for all null k^a,

$$T_{ab} k^a k^b \geqq 0 \quad . \tag{9.2.36}$$

If the strong energy condition, equation (9.2.17), holds, then for all timelike ξ^a we have $T_{ab}\xi^a\xi^b - \frac{1}{2}T\xi^a\xi_a \geqq 0$, and by continuity equation (9.2.36) will hold for all null k^a. Similarly, if the weak energy condition (9.2.16) holds, then by continuity equation (9.2.36) also will be satisfied. For a diagonalizable T_{ab}, equation (9.2.18), the necessary and sufficient requirement for satisfying equation (9.2.36) for all null k^a is

$$\rho + p_i \geqq 0 \quad (i = 1, 2, 3) \quad . \tag{9.2.37}$$

Thus, the requirements on T_{ab} needed to ensure that the last term of equation (9.2.32) is nonpositive are weaker than the corresponding requirements in the timelike case.

By the same arguments as led to lemma 9.2.1 we obtain the following result.

LEMMA 9.2.2. Let k^a be the tangent field of a hypersurface orthogonal null geodesic congruence. Suppose $R_{ab}k^a k^b \geq 0$, as will be the case if Einstein's equation holds in the spacetime and the weak or strong energy condition is satisfied by the mattter. If the expansion θ takes the negative value θ_0 at any point on a geodesic in the congruence, then θ goes to $-\infty$ along that geodesic within affine length $\lambda \leqq 2/|\theta_0|$.

9.3 Conjugate Points

Let M be any manifold on which a connection is defined and let γ be a geodesic with tangent v^a. A solution η^a, of the geodesic deviation equation

$$v^a \nabla_a (v^b \nabla_b \eta^c) = -R_{abd}{}^c \eta^b v^a v^d \tag{9.3.1}$$

(see eq. [3.3.18]) is called a *Jacobi field* on γ. A pair of points $p, q \in \gamma$ are said to

5. Thus, it follows immediately from equation (9.2.33) that if the shear, $\hat{\sigma}_{ab}$, of a null geodesic congruence vanishes, then k^a is a principal null vector. In the case of a vacuum spacetime, $R_{ab} = 0$, considerably stronger results hold. If $\hat{\sigma}_{ab} = 0$, then k^a must be a *repeated* principal null vector. Conversely, if k^a is a repeated principal null vector in a vacuum, algebraically special spacetime, then it is tangent to a shear-free, null geodesic congruence. These results are known as the Goldberg-Sachs theorem, and a proof can be found in Newman and Penrose (1962).

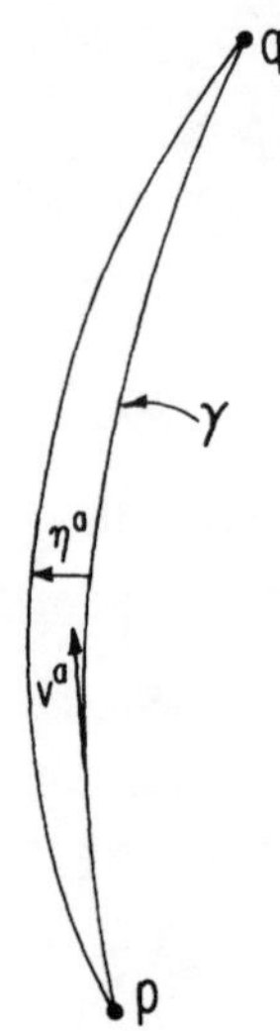

Fig. 9.2. A spacetime diagram illustrating the notion of conjugate points along the geodesic γ.

be *conjugate* if there exists a Jacobi field η^a which is not identically zero but vanishes at both p and q. Thus, roughly speaking, p and q are conjugate if an "infinitesimally nearby" geodesic intersects γ at both p and q as illustrated in Figure 9.2. As a simple example, the north and south poles of the sphere in Riemannian geometry are conjugate points of every "longitudinal geodesic." Note, however, that the definition requires only the existence of a Jacobi field vanishing at p and q; there need not exist an actual geodesic other than γ which passes through p and q. Conversely, the existence of a geodesic other than γ which passes through p and q does not mean that p and q are conjugate or even that some point conjugate to p exists between p and q.

Conjugate points are of interest because, as we shall see below, in spacetimes they characterize the stage at which a timelike geodesic fails to be a local maximum of proper time between two points and a null geodesic fails to remain on the boundary of the future of a point. (In Riemannian geometry, they similarly characterize the stage at which a geodesic fails, locally, to be the minimum length curve connecting points.) We shall consider, first, conjugate points on timelike geodesics. We begin by using the results of the previous section to obtain criteria for the existence of conjugate points.

Let γ be a timelike geodesic with tangent ξ^a and let $p \in \gamma$. Consider the congruence of all timelike geodesics passing through p. (This congruence, of course, is singular at p itself.) Then every Jacobi field which vanishes at p is a deviation vector for this congruence. We shall show that a point $q \in \gamma$ lying to the future of p is conjugate to p if and only if the expansion, θ, of this congruence approaches $-\infty$ at q. For this purpose, it is convenient to introduce an orthonormal basis of spatial vectors e_1^a, e_2^a, e_3^a orthogonal to ξ^a and parallelly propagated along γ. Since the components, η^μ, of the deviation vectors η^a for this congruence satisfy the linear ordinary differential equations,

$$\frac{d^2\eta^\mu}{d\tau^2} = -\sum_{\alpha,\beta,\nu} R_{\alpha\beta\nu}{}^\mu \xi^\alpha \eta^\beta \xi^\nu \quad , \tag{9.3.2}$$

the value of η^μ at time τ must depend linearly on the initial data $\eta^\mu(0)$ and $d\eta^\mu/d\tau(0)$ at p. Since, by construction, $\eta^\mu(0) = 0$ for this congruence, we must have

$$\eta^\mu(\tau) = \sum_{\nu=1}^{3} A^\mu{}_\nu(\tau) \frac{d\eta^\nu}{d\tau}(0) \quad . \tag{9.3.3}$$

Substituting this in equation (9.3.2), we see that the matrix $A^\mu{}_\nu(\tau)$ satisfies the equation

$$\frac{d^2 A^\mu{}_\nu}{d\tau^2} = -\sum_{\alpha,\beta,\sigma} R_{\alpha\beta\sigma}{}^\mu \xi^\alpha \xi^\sigma A^\beta{}_\nu \quad . \tag{9.3.4}$$

Clearly, we also have $A^\mu{}_\nu(0) = 0$ and $dA^\mu{}_\nu/d\tau(0) = \delta^\mu{}_\nu$. Now, q will be conjugate to p if and only if there exists nontrivial initial data [i.e., $d\eta^\mu/d\tau(0) \neq 0$] for which $\eta^\mu = 0$ at q. By equation (9.3.3), this occurs if and only if $\det A^\mu{}_\nu = 0$ at q, so $\det A^\mu{}_\nu = 0$ is the necessary and sufficient condition for a conjugate point to p. Note that between conjugate points we have $\det A^\mu{}_\nu \neq 0$, so the inverse of $A^\mu{}_\nu$ exists.

Clearly, the matrix $A^\mu{}_\nu$ must be related to the tensor field $B_{ab} = \nabla_b \xi_a$ of the congruence. To find this relation, we note that

$$\begin{aligned}
\frac{d\eta^\mu}{d\tau} &= \xi^a \nabla_a \eta^\mu = \xi^a \nabla_a [(e_\mu)_b \eta^b] \\
&= (e_\mu)_b \xi^a \nabla_a \eta^b \\
&= (e_\mu)_b B^b{}_a \eta^a \\
&= \sum_{\alpha=1}^{3} B^\mu{}_\alpha \eta^\alpha \quad .
\end{aligned} \tag{9.3.5}$$

However, using equation (9.3.3), we find

$$\frac{d\eta^\mu}{d\tau} = \sum_\nu \frac{dA^\mu{}_\nu}{d\tau} \frac{d\eta^\nu}{d\tau}(0) \quad , \tag{9.3.6}$$

so we obtain

$$\sum_\nu \frac{dA^\mu{}_\nu}{d\tau} \frac{d\eta^\nu}{d\tau}(0) = \sum_{\alpha,\nu} B^\mu{}_\alpha A^\alpha{}_\nu \frac{d\eta^\nu}{d\tau}(0) \quad . \tag{9.3.7}$$

Thus, in matrix notation, we find

$$dA/d\tau = BA \quad , \tag{9.3.8}$$

i.e.,

$$B = \frac{dA}{d\tau} A^{-1} \quad . \tag{9.3.9}$$

Consequently, we obtain

$$\theta = \mathrm{tr}\, B = \mathrm{tr}\left[\frac{dA}{d\tau}A^{-1}\right] \quad , \tag{9.3.10}$$

where "tr" denotes the trace of the matrix. However, it follows from the formula for the inverse of a matrix that for any nonsingular matrix A, we have [6]

$$\mathrm{tr}\left[\frac{dA}{d\tau}A^{-1}\right] = \frac{1}{\det A}\frac{d}{d\tau}(\det A) \quad , \tag{9.3.11}$$

so that

$$\theta = \frac{d}{d\tau}(\ln|\det A|) \quad . \tag{9.3.12}$$

Since A satisfies the ordinary differential equation (9.3.4), $d(\det A)/d\tau$ cannot become infinite anywhere along γ. Therefore, if $\theta \to -\infty$ at q, it follows from equation (9.3.12) that $\det A \to 0$ at q. Conversely, if $\det A \to 0$ at q, it follows that $\theta \to -\infty$ at q. Thus, as desired, we have proven that a necessary and sufficient condition for q to be conjugate to p is that for the congruence of timelike geodesics emanating from p, we have $\theta \to -\infty$ at q.

The congruence of timelike geodesics passing through p is hypersurface orthogonal. Indeed, as proven in lemma 4.5.2 of Hawking and Ellis (1973), within a convex normal neighborhood of p, the geodesics in this congruence are orthogonal to the surfaces of constant proper time τ along the geodesics and, from equation (9.2.14), it follows that if ω_{ab} vanishes at one time, it must vanish at all times. Thus, we may use lemma 9.2.1 to establish immediately the following result on the existence of conjugate points:

PROPOSITION 9.3.1. Let (M, g_{ab}) be a spacetime satisfying $R_{ab}\xi^a\xi^b \geqq 0$ for all timelike ξ^a. Let γ be a timelike geodesic and let $p \in \gamma$. Suppose the convergence of the congruence of timelike geodesics emanating into the future from p attains the negative value θ_0 at $r \in \gamma$. Then within proper time $\tau \leqq 3/|\theta_0|$ from r along γ there exists a point q conjugate to p, assuming, of course, that γ extends that far.

In fact, the existence of a pair of conjugate points on a complete timelike geodesic γ can be proven under far weaker hypotheses than those of proposition 9.3.1. If $R_{ab}\xi^a\xi^b \geqq 0$ everywhere along the geodesic and $R_{ab}\xi^a\xi^b > 0$ at point $r \in \gamma$, then one can show that for p sufficiently far from r, the expansion of the timelike geodesic congruence emanating from p must be negative at r. Hence p will have a conjugate point q on γ. However, even if $R_{ab}\xi^a\xi^b = 0$ everywhere on γ, if the curvature terms on the right-hand side of equation (9.2.12) are nonzero at $r \in \gamma$, then σ_{ab} cannot vanish in a neighborhood of r. Since $-\sigma_{ab}\sigma^{ab}$ also appears on the right-hand side of

6. This formula previously was used in equation (3.4.8) above. It is closely related to the identity $\exp(\mathrm{tr}\, C) = \det[\exp(C)]$ for any matrix C, which can be proven by putting C in upper triangular form.

Raychaudhuri's equation, a similar argument establishes existence of conjugate points. Thus, all that is required for existence of conjugate points on γ is that $R_{ab}\xi^a\xi^b \geqq 0$ everywhere on γ and $R_{abcd}\xi^b\xi^d \neq 0$ at at least one point of γ. The full proof of this result can be found in proposition 4.4.2 of Hawking and Ellis (1973).

A spacetime (M, g_{ab}) is said to satisfy the *timelike generic condition* if each timelike geodesic possesses at least one point at which $R_{abcd}\xi^a\xi^d \neq 0$. Although the timelike generic condition may fail to hold in special, idealized spacetime models, it seems reasonable that it will hold in all physically realistic "generic" spacetimes. In view of the remarks of the previous paragraph, we have the following result which plays an important role in the proof of theorem 9.5.4 quoted in section 9.5.

PROPOSITION 9.3.2. Let (M, g_{ab}) satisfy the timelike generic condition and suppose $R_{ab}\xi^a\xi^b \geqq 0$ for all timelike ξ^a. Then every complete timelike geodesic possesses a pair of conjugate points.

Next, we turn our attention to the relation between conjugate points and the extremal length properties of timelike geodesics. Let $p, q \in M$ and consider a smooth one-parameter family of smooth timelike curves $\lambda_\alpha(t)$ from p to q, where the curve parameter t is chosen so that for all α we have $\lambda_\alpha(a) = p$, $\lambda_\alpha(b) = q$. We denote the tangent vectors, $(\partial/\partial t)^a$, by T^a and the deviation vectors, $(\partial/\partial\alpha)^a$, by X^a. Then X^a vanishes at both p and q and, as usual, we have $\pounds_T X^a = T^b\nabla_b X^a - X^b\nabla_b T^a = 0$ everywhere. The length of each curve is given by

$$\tau(\alpha) = \int_a^b f(\alpha, t)dt \quad , \tag{9.3.13}$$

where $f = (-T^aT_a)^{1/2}$. In section 3.3 we showed that the necessary and sufficient condition for the curve γ to extremize τ for all possible smooth families λ_α with $\lambda_0 = \gamma$ was that γ be a geodesic. We repeat, now, this calculation in a coordinate invariant form and then compute the second variation of arc length. We have

$$\begin{aligned}
\frac{d\tau}{d\alpha} &= \int_a^b \frac{\partial f}{\partial \alpha} dt \\
&= \int_a^b X^a\nabla_a(-T^bT_b)^{1/2}dt \\
&= -\int_a^b \frac{1}{f} T_b X^a \nabla_a T^b \, dt \\
&= -\int_a^b \frac{1}{f} T_b T^a \nabla_a X^b dt \\
&= -\int_a^b T^a\nabla_a \left[\frac{1}{f} T_b X^b\right] dt + \int_a^b X^b T^a \nabla_a (T_b/f) dt \\
&= \int_a^b X^b T^a \nabla_a (T_b/f) dt \quad ,
\end{aligned} \tag{9.3.14}$$

since $T^a\nabla_a(f^{-1}T_bX^b) = \partial(f^{-1}T_bX^b)/\partial t$ and X^b vanishes at the endpoints. Thus, setting $\alpha = 0$ we see that $d\tau/d\alpha = 0$ for arbitrary X^b if and only if $T^a\nabla_a(T_b/f) = 0$ at $\alpha = 0$, which is just the geodesic equation expressed in an arbitrary parameterization.

The second variation of arc length is

$$\frac{d^2\tau}{d\alpha^2} = \int_a^b X^c\nabla_c[X^bT^a\nabla_a(T_b/f)]dt \quad . \tag{9.3.15}$$

Evaluating this expression at $\alpha = 0$, assuming that λ_0 is a geodesic, we find

$$\begin{aligned}\left.\frac{d^2\tau}{d\alpha^2}\right|_{\alpha=0} &= \int_a^b X^b(X^c\nabla_cT^a)\nabla_a(T_b/f)dt \\ &\quad + \int_a^b X^bT^aX^c\nabla_c\nabla_a(T_b/f)dt \\ &= \int_a^b X^b(T^c\nabla_cX^a)\nabla_a(T_b/f)dt \\ &\quad + \int_a^b X^bT^aX^c\nabla_a\nabla_c(T_b/f)dt \\ &\quad + \int_a^b X^bT^aX^cR_{cab}{}^dT_d/f\,dt \\ &= \int_a^b X^bT^c\nabla_c[X^a\nabla_a(T_b/f)]dt \\ &\quad + \int_a^b X^bT^aX^cR_{cab}{}^dT_d/f\,dt \quad .\end{aligned} \tag{9.3.16}$$

The term in square brackets can be reexpressed as

$$\begin{aligned}X^a\nabla_a(T_b/f) &= \frac{1}{f}X^a\nabla_aT_b - \frac{1}{f^2}T_bX^a\nabla_af \\ &= \frac{1}{f}T^a\nabla_aX_b + \frac{1}{f^2}T_bT^a\nabla_a\left[\frac{1}{f}T_cX^c\right] \quad ,\end{aligned} \tag{9.3.17}$$

where the manipulations of equation (9.3.14) and the geodesic equation at $\alpha = 0$ were used. Equation (9.3.16) can be put in more easily recognizable form by choosing the curve parameterization so that $f = 1$ along the geodesic λ_0 and also choosing the deviation vector X^a to be orthogonal to T^a along λ_0. With these choices, we find

$$\begin{aligned}\left.\frac{d^2\tau}{d\alpha^2}\right|_{\alpha=0} &= \int_a^b X^b\{T^c\nabla_c(T^a\nabla_aX_b) + R_{cab}{}^dT^aT_dX^c\}dt \\ &= \int_a^b X^b(\mathcal{C}X)_bdt \quad ,\end{aligned} \tag{9.3.18}$$

where $\mathbb{O}$ is the operator appearing in the geodesic deviation equation (9.3.1). Using this formula, we may establish the following theorem.

THEOREM 9.3.3. *Let γ be a smooth timelike curve connecting two points $p, q \in M$. Then the necessary and sufficient condition that γ locally maximize the proper time between p and q over smooth one parameter variations is that γ be a geodesic with no point conjugate to p between p and q.*

Sketch of proof. If γ is not a geodesic, then using equation (9.3.14) we can construct a one-parameter family of curves λ_α with $d\tau/d\alpha > 0$ at $\alpha = 0$. If γ is a geodesic but has a conjugate point, r, between p and q, then we can find a deviation vector X_0^a such that $(\mathbb{O}X_0)^a = 0$ and X_0^a vanishes at p and r. By equation (9.3.18), the variation $X^a = X_0^a$ between p and r followed by $X^a = 0$ between r and q produces no change in τ to second order (see Fig. 9.3). By "rounding off the corner" at r for this variation,[7] one can produce a smooth deviation for which $d^2\tau/d\alpha^2 > 0$ at $\alpha = 0$.

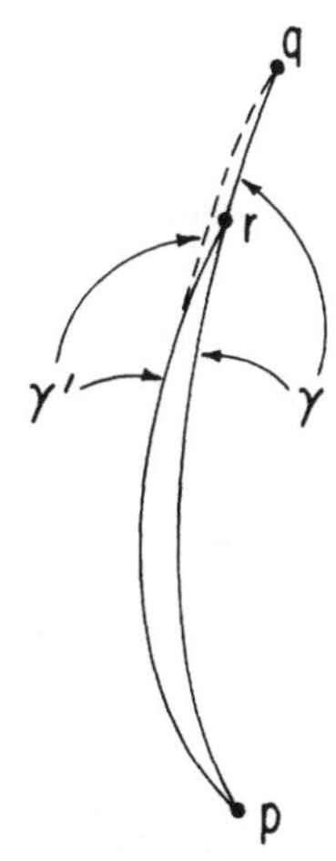

Fig. 9.3. A spacetime diagram illustrating the fact that if a timelike geodesic γ connecting points p and q has a point r conjugate to p lying between p and q, then a nearby timelike curve γ' connecting p and q can be constructed which has greater elapsed proper time than γ (see theorem 9.3.3).

Conversely, if γ is a geodesic with no point conjugate to p between p and q, then the matrix A defined above will be nonsingular between p and q, so we can define $Y^\mu = \sum_\nu (A^{-1})^\mu{}_\nu X^\nu$. Substituting $\sum_\nu A^\mu{}_\nu Y^\nu$ for X^μ in equation (9.3.18), we can show that $d^2\tau/d\alpha^2$ is manifestly negative definite at $\alpha = 0$. These calculations can be found in proposition 4.5.8 of Hawking and Ellis (1973). □

A notion similar to conjugacy of two points along a timelike geodesic can be defined for a point and a smooth (or, at least, C^2) spacelike hypersurface Σ. (By

7. Note that this "rounding off the corner" argument is made for the (infinitesimal) deviation vector, not for an actual (finite) geodesic. See Penrose (1972) for further discussion.

hypersurface, we mean an embedded, three-dimensional submanifold [see appendix B].) First, we define the *extrinsic curvature*, K_{ab}, of Σ. Let ξ^a be the unit tangent field of the congruence of timelike geodesics orthogonal to Σ. We define K_{ab} by

$$K_{ab} = \nabla_a \xi_b = B_{ba} \quad , \tag{9.3.19}$$

where evaluation of these tensors on Σ is understood. Note that K_{ab} is purely spatial, i.e., $K_{ab}\xi^a = K_{ab}\xi^b = 0$. Since this congruence is manifestly hypersurface orthogonal, we have $\omega_{ab} = 0$ so K_{ab} is symmetric, $K_{ab} = K_{ba}$. Hence, using equation (C.2.16), we can reexpress K_{ab} as

$$\begin{aligned} K_{ab} &= \frac{1}{2}\, \pounds_\xi g_{ab} \\ &= \frac{1}{2}\, \pounds_\xi (h_{ab} - \xi_a \xi_b) \\ &= \frac{1}{2}\, \pounds_\xi h_{ab} \quad , \end{aligned} \tag{9.3.20}$$

where h_{ab} was defined by equation (9.2.5) and the geodesic equation was used in the last step. Now, h_{ab} is the spatial metric induced on the hypersurfaces of constant proper time from Σ along the geodesic congruence orthogonal to Σ. Thus, K_{ab} measures the rate of change of this spatial metric as one moves along the congruence; i.e., it measures the "bending" of Σ in the spacetime (M, g_{ab}). Indeed, since Gaussian normal coordinates (see section 3.3) are adapted to ξ^a (see appendix C), in these coordinates we have

$$K_{\mu\nu} = \frac{1}{2} \frac{\partial h_{\mu\nu}}{\partial t} \quad , \tag{9.3.21}$$

which reinforces our interpretation of K_{ab} as representing the "time derivative" of h_{ab}. The trace of the extrinsic curvature is often denoted by K,

$$K \equiv K^a{}_a = h^{ab} K_{ab} \quad . \tag{9.3.22}$$

Thus, we have

$$K = \theta \quad , \tag{9.3.23}$$

where θ is the expansion of the geodesic congruence orthogonal to Σ.

A point p on a geodesic γ of the geodesic congruence orthogonal to Σ is said to be *conjugate* to Σ along γ if there exists an orthogonal deviation vector η^a of the congruence which is nonzero on Σ but vanishes at p. Thus, intuitively, p is conjugate to Σ if two "infinitesimally nearby" geodesics orthogonal to Σ cross at p as illustrated in Figure 9.4. By the same argument as given above for conjugate pairs of points, p will be conjugate to Σ if and only if the expansion, θ, of the congruence of geodesics orthogonal to Σ approaches $-\infty$ at p. Since this congruence is manifestly hypersurface orthogonal, lemma 9.2.1 immediately yields the following result.

PROPOSITION 9.3.4. Let (M, g_{ab}) be a spacetime satisfying $R_{ab}\xi^a\xi^b \geqq 0$ for all timelike ξ^a. Let Σ be a spacelike hypersurface with $K = \theta < 0$ at a point $q \in \Sigma$.

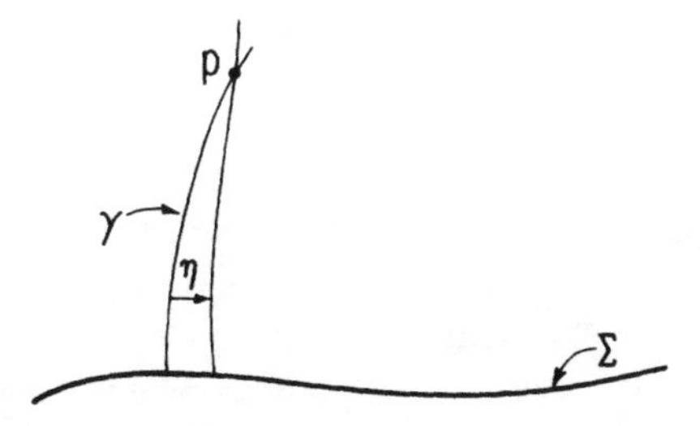

Fig. 9.4. A spacetime diagram depicting a point p conjugate to the hypersurface Σ along the geodesic γ. (The two geodesics shown are supposed to be "infinitesimally nearby.")

Then within proper time $\tau \leqq 3/|K|$ there exists a point p conjugate to Σ along the geodesic γ orthogonal to Σ and passing through q, assuming that γ can be extended that far.

Furthermore, by arguments similar to those used to prove theorem 9.3.3, the following theorem can be proven.

THEOREM 9.3.5. *Let γ be a smooth timelike curve connecting a point $p \in M$ to a point q on a smooth spacelike hypersurface Σ. Then the necessary and sufficient condition that γ locally maximize the proper time between p and Σ over smooth one-parameter variations is that γ be a geodesic orthogonal to Σ with no conjugate point to Σ between Σ and p.*

We turn our attention, now, to conjugate points of null geodesics. It follows directly from the geodesic deviation equation that for any Jacobi field η^a on a null geodesic μ with tangent k^a, we have

$$k^c \nabla_c [k^b \nabla_b (k^a \eta_a)] = 0 \quad , \tag{9.3.24}$$

which immediately implies that η^a cannot vanish at two points $p, q \in \mu$ unless $k^a \eta_a = 0$ everywhere along μ. Furthermore, if η^a is a Jacobi field, then so is $\eta^a + (a + b\lambda)k^a$, where a and b are constants, so p and q will be conjugate if and only if there exists a Jacobi field η^a which differs from zero by only a multiple of k^a at both p and q. Thus, along a null geodesic μ, the points $p, q \in \mu$ will be conjugate if and only if a vector $\hat{\eta}^a$ in $\hat{V}$ (see section 9.2) satisfies the geodesic deviation equation and vanishes at p and q. It is clear that all such $\hat{\eta}^a$ which vanish at p arise as deviation vectors of any null geodesic congruence containing the two-dimensional family of null geodesics emerging from p. By the same arguments as given above in the timelike case, q will be conjugate to p if and only if the expansion, θ, of such a null geodesic congruence approaches $-\infty$ at q. Thus, lemma 9.2.2 implies the following result.

PROPOSITION 9.3.6. Let (M, g_{ab}) be a spacetime satisfying $R_{ab}k^a k^b \geqq 0$ for all null k^a. Let μ be a null geodesic and let $p \in \mu$. Suppose the convergence, θ, of the null geodesics emanating from p attains the negative value θ_0 at $r \in \mu$. Then within affine length $\lambda \leqq 2/|\theta_0|$ from r, there exists a point q conjugate to p along μ, assuming that μ extends that far.

Again, this result can be strengthened to conclude that if $R_{ab}k^a k^b \geqq 0$ everywhere on a complete null geodesic μ and there exists at least one point $r \in \mu$ at which either $R_{ab}k^a k^b > 0$ or $k_{[e}C_{a]bc[d}k_{f]}k^b k^c \neq 0$, then μ possesses a pair of conjugate points. A spacetime (M, g_{ab}) is said to satisfy the *null generic condition* if every null geodesic possesses at least one point where either $R_{ab}k^a k^b \neq 0$ or $k_{[e}C_{a]bc[d}k_{f]}k^b k^c \neq 0$, i.e., at least one point where $k_{[e}R_{a]bc[d}k_{f]}k^b k^c \neq 0$. Thus, in analogy to proposition 9.3.2, we have:

PROPOSITION 9.3.7. Suppose (M, g_{ab}) satisfies the null generic condition and $R_{ab}k^a k^b \geqq 0$ for all null k^a. Then every complete null geodesic possesses a pair of conjugate points.

As discussed above, for timelike geodesics conjugate points signal when a timelike geodesic, γ, can be varied to yield a curve of greater length between two points $p, q \in M$. If a conjugate point, r, to p exists between p and q along γ, we can find a longer curve by perturbing toward an appropriate "infinitesimally close" geodesic between p and r, following γ from r to q, and then "rounding off the corner" at r. In an exactly similar manner, for null geodesics, conjugate points signal when a null geodesic μ connecting p and q can be varied to yield a timelike curve between p and q. Again, if we have a conjugate point, r, between p and q, we can perturb toward an "infinitesimally close" null geodesic from p to r, follow μ from r to q, and then "round off the corner" to produce a timelike curve from p to q. Thus, in analogy to theorem 9.3.3, we obtain the following result, the full proof of which can be found in Hawking and Ellis (1973).

THEOREM 9.3.8. *Let μ be a smooth causal curve and let $p, q \in \mu$. Then there does not exist a smooth one-parameter family of causal curves λ_α connecting p and q with $\lambda_0 = \mu$ and λ_α timelike for all $\alpha > 0$ (i.e., μ cannot be smoothly deformed to a timelike curve) if and only if μ is a null geodesic with no point conjugate to p along μ between p and q.*

For null geodesics, a notion of conjugacy also can be defined for a point and a *two*-dimensional spacelike surface (i.e., embedded submanifold) S. At each $q \in S$ there will exist precisely two future directed null vectors k_1^a, k_2^a which are orthogonal to S. If S is orientable, we can make a continuous choice of k_1^a and k_2^a over S and thereby define two families of null geodesics, which we may refer to as "outgoing" and "ingoing" families. (Even if S is not orientable, we still can define these two families locally in a neighborhood of any point.) We will refer to each of these families as congruences even though each spans only a null hypersurface rather than an open region of spacetime. The expansion θ, shear $\hat{\sigma}_{ab}$, and twist $\hat{\omega}_{ab} = 0$ of these congruences are well defined, since all deviation vectors orthogonal to the geodesic tangents k^a are included in the congruences. Let μ be a null geodesic in one of these congruences. A point $p \in \mu$ is said to be conjugate to S along μ if along μ there exists a deviation vector $\hat{\eta}^a$ of the congruence which is nonzero on S but vanishes at p. In analogy to proposition 9.3.4, we have

PROPOSITION 9.3.9. Let (M, g_{ab}) be a spacetime satisfying $R_{ab}k^a k^b \geqq 0$ for all null k^a. Let S be a smooth two-dimensional spacelike submanifold such that the expansion, θ, of, say, the "outgoing" null geodesics has the negative value θ_0 at $q \in S$. Then within affine parameter $\lambda \leqq 2/|\theta_0|$, there exists a point p conjugate to S along the outgoing null geodesic μ passing through q.

In analogy to theorem 9.3.5, we also have:

THEOREM 9.3.10. *Let S be a smooth two-dimensional spacelike submanifold and let μ be a smooth causal curve from S to p. Then the necessary and sufficient condition that μ cannot be smoothly deformed to a timelike curve connecting S and p is that μ be a null geodesic orthogonal to S with no point conjugate to S between S and p.*

As a consequence of this theorem, and the results of chapter 8 we obtain the following theorem which will be used in the proof of singularity theorem 9.5.3.

THEOREM 9.3.11. *Let (M, g_{ab}) be a globally hyperbolic spacetime and let K be a compact, orientable, two-dimensional spacelike submanifold of M. Then every $p \in \dot{I}^+(K)$ lies on a future directed null geodesic starting from K which is orthogonal to K and has no point conjugate to K between K and p.*

Proof. That p lies on a null geodesic from K follows from the remarks below theorem 8.3.11. If this null geodesic were not orthogonal to K or had a conjugate point, then by theorem 9.3.10 we would have $p \in I^+(K)$ and therefore $p \notin \dot{I}^+(K)$. □

9.4 Existence of Maximum Length Curves

In the previous two sections we established by means of "local" calculations necessary criteria for a timelike curve to be a curve of maximum length between two points (theorem 9.3.3) or between a point and a hypersurface Σ (theorem 9.3.5) as well as conditions under which these criteria could not be met (propositions 9.3.1 and 9.3.4). In this section, we shall use global arguments involving compactness of the spaces of causal curves $C(p, q)$ and $C(\Sigma, q)$ defined in section 8.3 to prove existence of maximum length curves in globally hyperbolic spacetimes. In the next section we shall see that, under further hypotheses, these two sets of results lead to contradictions if all geodesics are complete, thereby producing the singularity theorems.

The length, τ, of a smooth (or even C^1) causal curve λ between points $p, q \in M$ with tangent $T^a = (\partial/\partial t)^a$ is given by the formula

$$\tau[\lambda] = \int (-T^a T_a)^{1/2}\, dt \quad . \tag{9.4.1}$$

However, for the global arguments of this section, it is necessary to generalize this notion of length to continuous causal curves between p and q in order that τ be defined for all curves in $C(p,q)$. Let $\tilde{C}(p,q)$ denote the subset of $C(p,q)$ consisting of smooth timelike curves, with the topology induced by $C(p,q)$. Then—with the possible exception of cases where null geodesics connect p and q—$\tilde{C}(p,q)$ is dense in $C(p,q)$, i.e., except for certain null geodesics every continuous casual curve can be expressed as a limit [in the topology of $C(p,q)$] of a sequence of smooth timelike curves. If τ were continuous on $\tilde{C}(p,q)$, we could extend it to a continuous function on all of $C(p,q)$ by setting

$$\tau[\mu] = \lim_{n\to\infty} \tau[\lambda_n] \quad ,$$

where $\{\lambda_n\}$ is a sequence in $\tilde{C}(p,q)$ which approaches the continuous causal curve $\mu \in C(p,q)$. However, as Figure 9.5 illustrates, τ is not continuous on $\tilde{C}(p,q)$. Arbitrarily close in the topology of $C(p,q)$ to any smooth timelike curve we can find a "zigzag" smooth timelike curve of length arbitrarily close to zero. Nevertheless, we shall show below that τ is *upper semicontinuous* on $\tilde{C}(p,q)$, i.e., for each $\lambda \in \tilde{C}(p,q)$ given $\epsilon > 0$ there exists an open neighborhood $O \subset \tilde{C}(p,q)$ of λ such that for all $\lambda' \in O$ we have $\tau[\lambda'] \leqq \tau[\lambda] + \epsilon$. Given that τ is upper semicontinuous on $\tilde{C}(p,q)$, we can extend it to an upper semicontinuous function on $C(p,q)$ as follows. For $\mu \in C(p,q)$ and $O \subset C(p,q)$ an open neighborhood of μ, we define

$$T[O] = \text{l.u.b.}\{\tau[\lambda] \mid \lambda \in O,\ \lambda \in \tilde{C}(p,q)\} \quad , \tag{9.4.2}$$

where "l.u.b." denotes the least upper bound. Then we define $\tau[\mu]$ by,

$$\tau[\mu] = \text{g.l.b.}\{T[O] \mid O \text{ an open neighborhood of } \mu\} \quad , \tag{9.4.3}$$

where "g.l.b." denotes the greatest lower bound. Thus, the key fact that allows us to extend the definition of τ to $C(p,q)$ is expressed by the following proposition.

Fig. 9.5. A smooth timelike curve μ from p to q. Arbitrarily close to μ in the topology of $C(p,q)$ is a smooth timelike curve μ' with total elapsed proper time arbitrarily close to zero.

PROPOSITION 9.4.1. Let (M, g_{ab}) be a strongly causal spacetime. Let $p, q \in M$ with $q \in I^+(p)$. Then τ is upper semicontinuous on $\tilde{C}(p,q)$.

Proof. Let $\lambda \in \tilde{C}(p,q)$. We parameterize λ by proper time and denote its tangent by u^a. Within a normal neighborhood of each point $r \in \lambda$, the spacelike geodesics orthogonal to u^a will form a three-dimensional spacelike hypersurface. Within a sufficiently small open neighborhood $U \subset M$ of λ, these hypersurfaces will foliate U; i.e., a unique hypersurface will pass through each point of U. On U, we define the function F by setting $F(p)$ equal to the proper time value of λ at the intersection of λ with the hypersurface on which p lies. Then $\nabla^a F$ is timelike everywhere in U, and on λ we have $u^a = -\nabla^a F$, so $\nabla^a F \nabla_a F = -1$ on λ. Now, let $\rho \in \tilde{C}(p,q)$ with $\rho \subset U$. We parameterize ρ by F and denote its tangent by v^a. Thus, by our parameterization choice, we

$$v^a \nabla_a F = 1 \quad . \tag{9.4.4}$$

We decompose v^a as

$$v^a = \alpha \nabla^a F + m^a \quad , \tag{9.4.5}$$

where $m^a \nabla_a F = 0$, and thus m^a is spacelike. Contracting equation (9.4.5) with $\nabla_a F$ and using equation (9.4.4), we evaluate α and find

$$v^a = \nabla^a F / (\nabla_b F \nabla^b F) + m^a \quad . \tag{9.4.6}$$

Therefore, we find

$$v^a v_a = (\nabla^a F \nabla_a F)^{-1} + m^a m_a \quad , \tag{9.4.7}$$

and thus

$$(-v_a v^a)^{1/2} \leqq (-\nabla_a F \nabla^a F)^{-1/2} \tag{9.4.8}$$

since m^a is spacelike. Since $\nabla^a F$ is continuous, given $\epsilon > 0$, we can choose a neighborhood $U' \subset U$ of λ so that $(-\nabla^a F \nabla_a F)^{-1/2} \leqq 1 + \epsilon/\tau[\lambda]$ in U'. Then for all $\rho \in \tilde{C}(p,q)$ contained in the open neighborhood O' in $\tilde{C}(p,q)$ defined by U', we have

$$\tau[\rho] = \int (-v^a v_a)^{1/2} \, dF \leqq \tau[\lambda] + \epsilon \tag{9.4.9}$$

which proves upper semicontinuity. □

In section 8.3 we defined $C(\Sigma, p)$ for a Cauchy surface Σ in a globally hyperbolic spacetime, but, more generally, we may define $C(\Sigma, p)$ analogously for any achronal set Σ in a strongly causal spacetime. Essentially the same arguments as given above establish that τ is an upper semicontinuous function on the space $\tilde{C}(\Sigma, p)$ of smooth timelike curves from Σ to p. Therefore, by the same argument as given above, one can extend τ to an upper semicontinuous function defined on all of $C(\Sigma, p)$.

In the previous section, we showed that the necessary and sufficient condition for a *smooth* curve to locally maximize the length between two points or a point and a hypersurface was that it be a geodesic without conjugate points. Now that we have extended the definition of τ to continuous curves, there is a possibility that a continuous, nonsmooth curve between two points or a point and a hypersurface could have length greater than or equal to that of any geodesic. However, this possibility can be ruled out as follows. By direct calculation, one can prove that in any convex normal neighborhood U, the unique geodesic γ connecting two causally related points $r, s \in U$ has length strictly greater than that of any other piecewise smooth causal curve connecting those points (see proposition 4.5.3 of Hawking and Ellis 1973 for a proof). Therefore, by upper semicontinuity any continuous causal curve, μ, connecting r and s in U must satisfy $\tau[\mu] \leqq \tau[\gamma]$. However, if equality held with $\mu \neq \gamma$, let point q be such that $q \in \mu$ but $q \notin \gamma$. Let γ_1, γ_2 be the geodesic segments connecting r and q and q and s, respectively. Since by the above result γ_1 maximizes the length between r and q, while γ_2 maximizes the length between q and s, we have $\tau[\gamma_1] + \tau[\gamma_2] \geqq \tau[\mu] = \tau[\gamma]$, which contradicts the fact that γ has strictly greater length than any other piecewise smooth curve between r and s. Thus, within any convex normal neighborhood, the unique geodesic connecting any pair of causally related points has length strictly greater than that of any other continuous causal curve connecting the points. Thus, an arbitrary continuous causal curve connecting any two points cannot be a curve of maximum length between those points unless it is a geodesic, since if it failed to be a geodesic at any point, we could deform it in a convex normal neighborhood of that point to obtain a longer curve. Thus, in view of theorem 9.3.3 we have the following result.

THEOREM 9.4.2. *Let (M, g_{ab}) be a strongly causal spacetime. Let $p, q \in M$ with $q \in J^+(p)$, and consider the length function τ defined on $C(p, q)$. A necessary condition for τ to attain its maximum value at $\gamma \in C(p, q)$ is that γ be a geodesic with no point conjugate to p between p and q.*

Similarly, we have:

THEOREM 9.4.3. *Let (M, g_{ab}) be a strongly causal spacetime. Let $p \in M$, let Σ be an achronal, smooth spacelike hypersurface, and consider the length function τ defined on $C(\Sigma, p)$. A necessary condition for τ to attain its maximum value at $\gamma \in C(\Sigma, p)$ is that γ be a geodesic orthogonal to Σ with no point conjugate to Σ between Σ and p.*

The above results, of course, do not assert that τ must attain a maximum value. However, we conclude this section by proving two key results which show that a maximum is always attained in globally hyperbolic spacetimes.

THEOREM 9.4.4. *Let (M, g_{ab}) be a globally hyperbolic spacetime. Let $p, q \in M$ with $q \in J^+(p)$. Then there exists a curve $\gamma \in C(p, q)$ for which τ attains its maximum value on $C(p, q)$.*

Proof. By theorem 8.3.9, $C(p, q)$ is compact, and by proposition 9.4.1, τ is upper semicontinuous. Hence, by a slight generalization of theorem A.6 of appendix A, τ is bounded and attains its maximum. □

THEOREM 9.4.5. *Let (M, g_{ab}) be a globally hyperbolic spacetime. Let $p \in M$ and let Σ be a Cauchy surface. Then there exists a curve $\gamma \in C(\Sigma, p)$ for which τ attains its maximum value on $C(\Sigma, p)$.*

Proof. Again, $C(\Sigma, p)$ is compact and τ is upper semicontinuous. □

9.5 Singularity Theorems

We now have developed the machinery required to prove some of the singularity theorems. We shall give a complete proof of three theorems which establish the existence of singularities in the sense of timelike or null geodesic incompleteness under conditions relevant to cosmology and gravitational collapse. Then we shall simply quote a fourth theorem that establishes the existence of singularities under significantly weaker hypotheses.

The first theorem we shall prove can be interpreted as showing that if the universe is globally hyperbolic and at one instant of time is expanding everywhere at a rate bounded away from zero, then the universe must have begun in a singular state a finite time ago.

THEOREM 9.5.1. *Let (M, g_{ab}) be a globally hyperbolic spacetime with $R_{ab}\xi^a\xi^b \geqq 0$ for all timelike ξ^a, which will be the case if Einstein's equation is satisfied with the strong energy condition holding for matter. Suppose there exists a smooth (or, at least C^2) spacelike Cauchy surface Σ for which the trace of the extrinsic curvature (for the past directed normal geodesic congruence) satisfies $K \leqq C < 0$ everywhere, where C is a constant. Then no past directed timelike curve from Σ can have length greater than $3/|C|$. In particular, all past directed timelike geodesics are incomplete.*

Proof. Suppose there were a past directed timelike curve, λ, from Σ with length greater than $3/|C|$. Let p be a point on λ lying beyond length $3/|C|$ from Σ. By theorem 9.4.5, there exists a maximum length curve γ from p to Σ, which, clearly, also must have length greater than $3/|C|$. By theorem 9.4.3, γ must be a geodesic with no conjugate point between Σ and p. However, this contradicts proposition 9.3.4 which states that γ must have a conjugate point between Σ and p. Therefore, the original curve λ cannot exist. □

The strongest unwanted hypothesis in theorem 9.5.1 is that the universe be globally hyperbolic. Indeed, given only theorem 9.5.1, it might seem more reasonable to conclude that an everywhere expanding universe must fail to be globally hyperbolic rather than that it must be singular. Since theorem 9.4.5 plays a critical role in the proof, it might appear that the assumption of global hyperbolicity cannot be eliminated. However, we now shall prove a theorem due to Hawking (1967) which does remove this assumption. The main price paid for this removal is the additional hypothesis that Σ be compact (i.e., the universe is "closed") and the significantly weakened conclusion that only at least one past direct timelike geodesic (rather than all past directed timelike curves) must be incomplete.

THEOREM 9.5.2. *Let* (M, g_{ab}) *be a strongly causal*[8] *spacetime with* $R_{ab}\xi^a\xi^b \geqq 0$ *for all timelike* ξ^a, *as will be the case if Einstein's equation is satisfied with the strong energy condition holding for matter. Suppose there exists a compact, edgeless, achronal, smooth spacelike hypersurface S such that for the past directed normal geodesic congruence from S we have* $K < 0$ *everywhere on S. Let C denote the maximum value of K, so* $K \leqq C < 0$ *everywhere on S. Then at least one inextendible past directed timelike geodesic from S has length no greater than* $3/|C|$.

Proof. Suppose all past directed inextendible timelike geodesics from S had length greater than $3/|C|$. Since the spacetime $(\text{int}[D(S)], g_{ab})$ satisfies the hypotheses of theorem 9.5.1, all inextendible past directed timelike geodesics from S must leave $\text{int}[D(S)]$. Since $H(S)$ is the boundary of $D(S)$ (see proposition 8.3.6), all such geodesics must intersect $H^-(S)$ before their length becomes greater than $3/|C|$. In particular, this implies that $H^-(S) \neq \emptyset$. We shall prove that $H^-(S)$ must be compact and then show that this leads to a contradiction.

The key step in proving compactness of $H^-(S)$ is the demonstration that for each $p \in H^-(S)$ there exists a maximum length orthogonal geodesic from S to p. First, the length of any causal curve from S to $p \in H^-(S)$ is bounded from above by $3/|C|$, so the least upper bound, τ_0, of the length of all causal curves from S to p exists. We wish to find an orthogonal geodesic from S to p with length τ_0. Let $\{\lambda_n\}$ be a sequence of timelike curves from S to p such that

$$\lim_{n\to\infty} \tau[\lambda_n] = \tau_0 \quad .$$

Choose $q_n \in \lambda_n$ with $q_n \neq p$ such that the sequence $\{q_n\}$ converges to p (see Fig. 9.6). Since $q_n \in I^+(p)$, we have $q_n \in \text{int}[D^-(S)]$. Hence, by theorem 9.4.5, there exists a normal geodesic γ_n from S to q_n which maximizes the length of all causal curves from S to q_n. Clearly, we have

$$\lim_{n\to\infty} \tau[\gamma_n] = \tau_0 \quad .$$

Let r_n be the intersection point of γ_n with S. Since S is compact, there exists an accumulation point r of the sequence $\{r_n\}$. Let γ be the geodesic normal to S originating from r. Then, because of the continuous dependence of geodesics on their initial point and tangent vector, γ must intersect $H^-(S)$ at p and

$$\tau[\gamma] = \lim_{n\to\infty} \tau[\gamma_n] = \tau_0 \quad .$$

Thus, we have found the desired timelike geodesic orthogonal to S which maximizes the length from S to p.

To prove the compactness of $H^-(S)$, we show that every sequence $\{p_n\}$ in $H^-(S)$ has an accumulation point $p \in H^-(S)$. Let $\{\tilde{\gamma}_n\}$ be a sequence of maximum length

8. It has been shown by Hawking (1967) that the assumption of strong causality can be eliminated (see problem 3).

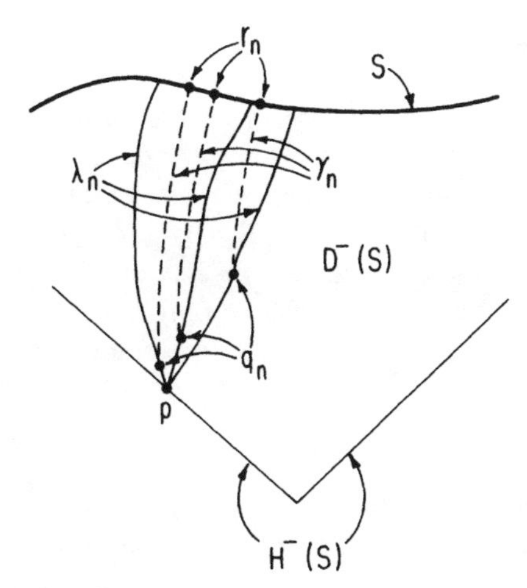

Fig. 9.6. A spacetime diagram showing a construction used in the proof of theorem 9.5.2.

orthogonal geodesics from S to p_n. We, now, in essence, repeat the argument given at the end of the previous paragraph. Let $\tilde{r}_n$ be the intersection point of $\tilde{\gamma}_n$ with S, and let $\tilde{r}$ be an accumulation point of $\{\tilde{r}_n\}$. Let $\tilde{\gamma}$ be the geodesic starting from $\tilde{r}$ orthogonal to S, and let p be the intersection point of $\tilde{\gamma}$ with $H^-(S)$. Then p is an accumulation point of $\{p_n\}$. Thus, $H^-(S)$ is compact.

However, since edge$(S) = \emptyset$, by theorem 8.3.5, $H^-(S)$ contains a future inextendible null geodesic. Since (M, g_{ab}) is strongly causal, by lemma 8.2.1 this is impossible if $H^-(S)$ is compact. Thus, our assumption that all past directed inextendible timelike geodesics from S have length greater than $3/|C|$ has led to a contradiction. □

The previous two theorems established timelike geodesic incompleteness in cosmological contexts. The next theorem proves null geodesic incompleteness in a context relevant to gravitational collapse. A compact, two-dimensional, smooth spacelike submanifold, T, having the property that the expansion, θ, of *both* sets (i.e., "ingoing" *and* "outgoing") of future directed null geodesics orthogonal to T is everywhere negative is called a *trapped surface*. In the extended Schwarzschild solution, all spheres inside the black hole (region II of Fig. 6.9) are trapped surfaces. As will be discussed further in chapter 12, it follows from this fact that trapped surfaces must form in any gravitational collapse whose initial conditions are sufficiently close to initial conditions for spherical collapse. The next theorem—which, historically, was the first general singularity theorem to be proven (Penrose 1965*a*)—shows that, under some further hypotheses, a singularity must occur after a trapped surface has formed.

THEOREM 9.5.3. *Let (M, g_{ab}) be a connected, globally hyperbolic spacetime with a noncompact Cauchy surface Σ. Suppose $R_{ab}k^a k^b \geqq 0$ for all null k^a, as will be the case if (M, g_{ab}) is a solution of Einstein's equation with matter satisfying the* *weak or strong energy condition. Suppose, further, that M contains a trapped surface T. Let* $\theta_0 < 0$ *denote the maximum value of* θ *for both sets of orthogonal geodesics on T. Then at least one inextendible future directed orthogonal null geodesic from T has affine length no greater than* $2/|\theta_0|$.

Proof. Suppose all future directed null geodesics from T have affine length $\geq 2/|\theta_0|$. Then we may define the map $f_+ : T \times [0, 2/|\theta_0|] \to M$ by setting $f(q, a)$ equal to the point of M lying at affine parameter a on the "outgoing" null geodesic normal to T starting at q. Similarly, we define f_- for the "ingoing" geodesics. Since $T \times [0, 2/|\theta_0|]$ is a compact set and f_+ and f_- are continuous, the images of f_+ and f_- and hence their union

$$A = f_+\{T \times [0, 2/|\theta_0|]\} \cup f_-\{T \times [0, 2/|\theta_0|]\}$$

must also be compact (see appendix A). However, by proposition 9.3.9 and theorem 9.3.11, $\dot{I}^+(T)$ is a subset of A, and since $\dot{I}^+(T)$ is closed, we conclude that $\dot{I}^+(T)$ is compact.

We show, now, that compactness of $\dot{I}^+(T)$ contradicts the existence of a noncompact Cauchy surface Σ. Using lemma 8.1.1, we choose a smooth timelike vector field t^a on M. Since $\dot{I}^+(T)$ is achronal, each integral curve of t^a can intersect $\dot{I}^+(T)$ at most once, while every integral curve of t^a intersects Σ precisely once. Thus, we may define a map $\psi : \dot{I}^+(T) \to \Sigma$ by following the integral curves of t^a from $\dot{I}^+(T)$ to Σ. Let $S \subset \Sigma$ denote the image, $\psi[\dot{I}^+(T)]$, of $\dot{I}^+(T)$ under ψ, and let S be given the topology induced by Σ. Then $\psi : \dot{I}^+(T) \to S$ is a homeomorphism. Since $\dot{I}^+(T)$ is compact, so is S, and, hence, viewed as a subset of Σ, S must be closed. On the other hand, since $\dot{I}^+(T)$ is a C^0-manifold (see theorem 8.1.3), each point of $\dot{I}^+(T)$ has a neighborhood homeomorphic to an open ball in $\mathbb{R}^3$. Since ψ is a homeomorphism, the same property holds for S, and, hence, viewed as a subset of Σ, S must be open. However, since M is connected, Σ also must be connected (see theorem 8.3.14). Therefore, since $\dot{I}^+(T) \neq \emptyset$, we must have $S = \Sigma$. However, this is impossible since S is compact but Σ is noncompact. □

Again, theorem 9.5.3 contains the unwanted hypothesis that (M, g_{ab}) is globally hyperbolic. However, with some additional assumptions, this hypothesis again can be eliminated by arguments similar to those given in the proof of theorem 9.5.2. We shall not attempt to give these arguments here, but will merely quote a singularity theorem of Hawking and Penrose (1970) which eliminates most of the unwanted assumptions, referring the reader to that reference or Hawking and Ellis (1973) for a proof.

THEOREM 9.5.4. *Suppose a spacetime* (M, g_{ab}) *satisfies the following four conditions.* (1) $R_{ab}v^a v^b \geq 0$ *for all timelike and null* v^a, *as will be the case if Einstein's equation is satisfied with the strong energy condition holding for matter.* (2) *The timelike and null generic conditions are satisfied (see section 9.3).* (3) *No closed timelike curve exists.* (4) *At least one of the following three properties holds: (a)* (M, g_{ab}) *possesses a compact achronal set without edge [i.e.,* (M, g_{ab}) *is a closed universe], (b)* (M, g_{ab}) *possesses a trapped surface, or (c) there exists a* *point $p \in M$ such that the expansion of the future (or past) directed null geodesics emanating from p becomes negative along each geodesic in this congruence. Then (M, g_{ab}) must contain at least one incomplete timelike or null geodesic.*

As compared with theorem 9.5.2, the above theorem adds only the hypothesis that the generic conditions are satisfied. However, it entirely eliminates the assumption that the universe is expanding everywhere. Similarly, as compared with theorem 9.5.3, the above theorem adds only the generic condition plus the condition that $R_{ab}\xi^a\xi^b \geqq 0$ for all timelike vectors. However, it eliminates entirely the assumption that (M, g_{ab}) be globally hyperbolic. Thus, theorem 9.5.4 has much wider applicability than the three previous theorems proven above. On the other hand, the conclusions of theorem 9.5.4 are slightly weaker in that no information is provided concerning which timelike or null geodesic is incomplete.

Gannon (1975) has strengthened theorem 9.5.4 by adding a fourth alternative to condition 4, namely that (M, g_{ab}) possesses a closed, achronal, edgeless set S which is non–simply connected[9] and is "regular near infinity." Here "regular near infinity" means that S can be expressed as a union of a nested family of sets, W_i (i.e., each W_i satisfies $W_i \subset W_{i+1}$), such that (i) each W_i is compact, (ii) its boundary, $\dot{W}_i$, in S is homeomorphic to a 2-sphere, (iii) $S - \text{int}(W_i)$ is homeomorphic to $S^2 \times \mathbb{R}^+$, where $\mathbb{R}^+ = [0, \infty)$, and (iv) the expansion, θ, of the inward directed null geodesics orthogonal to each W_i is negative everywhere on $\dot{W}_i$. Except for the compactness[10] of the W_i, the four conditions defining asymptotic regularity are satisfied by all asymptotically flat hypersurfaces in asymptotically flat spacetimes (see chapter 11). Thus, in essence, Gannon's theorem shows that an asymptotically flat spacetime satisfying conditions (1)–(3) of theorem 9.5.4. and which "initially" has appropriately nontrivial topology must develop a singularity. We refer the reader to Gannon (1975, 1976) for further details and results.

Theorem 9.5.4 gives us strong reason to believe that our universe is singular. As discussed in chapter 5, the observational evidence strongly suggests that our universe—or, at least, the portion of our universe within our causal past—is well described by a Robertson-Walker model (differing not too greatly from a $k = 0$ model), at least back as far as the decoupling time of matter and radiation. However, in these models, the expansion of the past directed null geodesics emanating from the event representing us at the present time becomes negative at a much more recent time than the decoupling time. Thus, there is strong reason to believe that condition 4(c) of theorem 9.5.4 is satisfied in our universe.[11] Since we expect that conditions (1)–(3) also are satisfied, it appears that our universe must be singular. Thus, it appears that we must confront the breakdown of classical general relativity expected to occur near singularities if we are to understand the origin of our universe.

9. See chapter 13 for the definition of simply connected.

10. A singularity theorem applicable to the case where the W_i are noncompact is given by Gannon (1976).

11. This conclusion also can be drawn from much weaker assumptions; see chapter 10 of Hawking and Ellis (1973).

Problems

1. Let (M, g_{ab}) be a spacetime with an everywhere timelike Killing field ξ^a. Let μ be an arbitrary timelike curve parameterized by proper time, τ, and let u^a denote the unit tangent to μ. Define $E = -u_a \xi^a$.

a) Show that $|dE/d\tau| \leqq aE$, where a denotes the magnitude of the acceleration of μ.

b) Use the result of part (*a*) to show that for a timelike curve with bounded integrated acceleration, i.e., $\int a \, d\tau < \infty$, E can change only by a finite amount. Use this result to prove that $|\xi_a \xi^a|$ is bounded along every timelike curve of bounded integrated acceleration.

The condition $\int a \, d\tau < \infty$ is satisfied by all world lines whose acceleration is produced by a physically realizable rocket ship. Thus, part (*b*) shows that even if observers are equipped with rocket ships, they cannot reach spacetime singularities where the norm of the timelike Killing field diverges to $-\infty$, as occurs, for example, in the negative mass Schwarzschild solution; see Chakrabarti, Geroch, and Liang (1983).

2. Let M be the torus $(S^1 \times S^1)$ and define the Lorentz metric g_{ab} by (Misner 1963) $ds^2 = \cos x(dy^2 - dx^2) + 2 \sin x \, dxdy$, where the angular coordinates x, y have ranges $0 \leqq x \leqq 2\pi$, $0 \leqq y \leqq 2\pi$. Show that the closed curves defined by $x = \pi/2$ and $x = 3\pi/2$ are null geodesics with affine parameter, λ, related to the y-coordinate by $\frac{1}{2} \lambda \, dy/d\lambda = 1$. Hence, show that if one traverses either of these curves infinitely many times in the negative y-direction, only a finite amount of affine parameter is used up. Note that when one parallelly transports the tangent to the geodesic around one cycle in the negative y-direction, the tangent comes back larger by the factor of e^{π}. Thus, this compact manifold with smooth metric is null geodesically incomplete. Timelike and spacelike geodesics similarly wind around the torus infinitely many times in the y-direction using up only a finite amount of affine parameter as they approach $x = \pi/2$ and $x = 3\pi/2$, so (M, g_{ab}) also is timelike and spacelike geodesically incomplete.

3. Eliminate the hypothesis of strong causality in theorem 9.5.2 as follows. Under the assumption that all past directed timelike geodesics have length greater than $3/|C|$, it was shown that (*a*) for $p \in H^-(S)$ there exists a maximum length geodesic, γ_p, from p to S and (*b*) $H^-(S)$ is compact. Define $T: H^-(S) \to \mathbb{R}$ by $T(p) = \tau[\gamma_p]$. (i) Show that T is continuous and, hence, achieves a minimum value on $H^-(S)$. (ii) Show that T strictly decreases along each future directed null geodesic generator of $H^-(S)$. Thus, obtain a contradiction with (i).

TEN

THE INITIAL VALUE FORMULATION

As discussed in chapter 4, general relativity asserts that spacetime structure and gravitation are described by a spacetime (M, g_{ab}) where M is a four-dimensional manifold and g_{ab} is a Lorentz metric satisfying Einstein's equation. In chapters 5 and 6 we obtained exact solutions of Einstein's equation which made highly successful physical predictions concerning cosmology and the structure and gravitation fields of spherical bodies. However, much more is required in order that general relativity be a physically viable theory. We see a wide variety of physical phenomena for which general relativity must account. Thus, it is essential that there exist a correspondingly wide class of solutions of Einstein's equation. For example, we know on physical grounds that there is a large class of possible gravitational fields of isolated bodies. If a correspondingly large class of solutions of Einstein's equation failed to exist, we would be forced to reject general relativity as a correct theory of nature.

A closely related issue concerns the fact that quite generally in classical physics we have a great deal of physical control over initial conditions of systems. If the system then is allowed to evolve freely, its behavior is completely determined by the initial conditions. For example, in ordinary particle mechanics we are able to control the initial positions and velocities of the particles. Given these initial conditions, if the system is permitted to evolve without outside interference, the dynamical evolution of the particles is determined. Similarly, as discussed in section 10.2, in electromagnetism we are able to arrange the initial values of the electric field, $\vec{E}$, and the magnetic field, $\vec{B}$, subject only to the constraints $\vec{\nabla} \cdot \vec{E} = 4\pi\rho$ and $\vec{\nabla} \cdot \vec{B} = 0$. Again, given these initial conditions, the subsequent evolution of the system is determined. Although our practical ability to control initial conditions in gravitational problems is far more limited, it seems natural to believe that (at least over regions much smaller than cosmological scales) we should, in principle, be able to control the initial conditions of the gravitational field and matter distribution, perhaps subject to some constraints as in electromagnetism. Thus, unless general relativity differs drastically from other theories of classical physics, it should permit a physically reasonable specification of initial data. Furthermore, given these initial data, Einstein's equation (possibly supplemented by additional equations for the matter) should determine the subsequent evolution.

If a theory can be formulated so that "appropriate initial data" may be specified (possibly subject to contraints) such that the subsequent dynamical evolution of the system is uniquely determined, we say that the theory possesses an initial value formulation. However, even if such a formulation exists, there remain further properties that a physically viable theory should satisfy. First, in an appropriate sense, "small changes" in initial data should produce only correspondingly "small changes" in the solution over any fixed compact region of spacetime. If this property were not satisfied, the theory would lose essentially all predictive power, since initial conditions can be measured only to finite accuracy. It is generally assumed that the pathological behavior which would result from the failure of this property does not occur in physics. Second, changes in the initial data in a region, S, of the initial data surface should not produce any changes in the solution outside the causal future, $J^+(S)$, of this region. If such changes occurred, we should be able to use them to propagate signals "faster than the speed of light." This would undermine the entire framework of relativity theory. If a theory possesses an initial value formulation which satisfies both of the above properties, we say that this initial value formulation is *well posed*. Note, however, that we have not attempted to give a mathematically precise definition of "well posed initial value formulation" here since the precise criteria depend on the type of theory considered.

The purpose of this chapter is to establish that general relativity admits a well posed initial value formulation. Thus, general relativity survives this rather stringent test of the physical viability of the theory. We begin in section 10.1 by discussing the initial value formulation of particle mechanics and the theory of the Klein-Gordon field in Minkowski spacetime. In section 10.2 we analyze the initial value formulation of Maxwell's equations in Minkowski spacetime, which shares many features analogous to general relativity with regard to initial constraints and gauge freedom. The initial value formulation of general relativity then is presented.

Our analysis of Einstein's equation and other field equations will be based on expressing them as second order hyperbolic systems. Analysis of Einstein's equation formulated as a first order hyperbolic system (Fischer and Marsden 1972) will not be discussed here. In addition, we shall consider only initial value formulations on spacelike Cauchy surfaces. An initial value formulation for general relativity on an initial surface formed by two intersecting null surfaces also can be given (Sachs 1962*a*; Muller zum Hagen and Seifert 1977).

10.1 Initial Value Formulation of Particles and Fields

A fundamental feature of Newton's second law of motion

$$\vec{F} = m\vec{a} \tag{10.1.1}$$

in ordinary, nonrelativistic, particle mechanics is that it relates the second time derivative, $\vec{a}$, of the spatial position of a particle to the force, $\vec{F}$, which—in usual cases—is a known function of the position and velocity of the particles in the system. Thus, for a system of particles interacting with themselves and/or external potentials with forces possibly dependent on positions and velocities but not on higher time derivatives of the particle positions, the laws of mechanics take the form

$$\frac{d^2 q_i}{dt^2} = F_i\left(q_1, \ldots, q_n; \frac{dq_1}{dt}, \ldots, \frac{dq_n}{dt}\right), \tag{10.1.2}$$

where $i = 1, \ldots, n$ and the number, n, of unknown positions is called the number of *degrees of freedom* of the system. Equation (10.1.2) is a system of n second order ordinary differential equations for the n quantities $q_1(t), \ldots, q_n(t)$. From the theory of such ordinary differential equations, it is well known (see, e.g., Coddington and Levinson 1955) that, given arbitrary initial values for the particle positions $q_{10}, \ldots, q_{n0}$, and velocities $(dq_1/dt)_0, \ldots, (dq_n/dt)_0$ at $t = t_0$ there always exists a unique solution of equation (10.1.2) over a finite time interval about t_0 with these initial values. Thus, ordinary particle mechanics possesses an initial value formulation. Furthermore, at a fixed time t the positions $q_1(t), \ldots, q_n(t)$ are continuous functions of the initial positions and velocities of the particles. Since the causal propagation of changes in the initial data is not an issue in nonrelativistic mechanics, we conclude that the initial value formulation of nonrelativistic particle mechanics is well posed.

The above simple example from particle mechanics contains the essential features of initial value formulations found in all physically reasonable field theories. Consider, for example, the massive Klein-Gordon field, ϕ, propagating in Minkowski spacetime,

$$\partial_a \partial^a \phi - m^2 \phi = 0 \quad . \tag{10.1.3}$$

Choosing global inertial coordinates t, x, y, z, we may write this equation in the form

$$\frac{\partial^2 \phi}{\partial t^2} = \frac{\partial^2 \phi}{\partial x^2} + \frac{\partial^2 \phi}{\partial y^2} + \frac{\partial^2 \phi}{\partial z^2} - m^2 \phi \quad . \tag{10.1.4}$$

The mathematical structure of equation (10.1.4) is markedly different from that of equation (10.1.2): Equation (10.1.2) is a system of ordinary differential equations while equation (10.1.4) is a single partial differential equation. Nevertheless, the essential content of these equations is quite similar. They both tell us how to compute the second time derivative of the unknown quantity (or quantities) at an instant of time, given the value and first time derivative of the quantity (or quantities) at that time. (Actually, for eq. [10.1.4] one does not even need to know $\partial\phi/\partial t$ to compute $\partial^2\phi/\partial t^2$, but in more general equations or with non-inertial time slicings, $\partial\phi/\partial t$ and its first spatial derivatives would appear on the right side of the equation corresponding to eq. [10.1.4].) Indeed, we may heuristically view equation (10.1.4) as arising from the limit as $N \to \infty$ of a system of N particles coupled by nearest neighbor harmonic oscillator interactions (see, e.g., Goldstein 1980). In this limit, the discrete index i goes over to the continuous label $\vec{x}$ and the finite set of variables $q_i(t)$ satisfying equation (10.1.2) goes over to the field variable $\phi(\vec{x}, t)$ satisfying equation (10.1.4).

The physical and mathematical analogy between equations (10.1.2) and (10.1.4) suggests that Klein-Gordon theory should have the following initial value formulation: We arbitrarily specify the values of ϕ and $\partial\phi/\partial t$ on a spatial hypersurface Σ_0 of constant inertial time, $t = t_0$. Then there should exist a unique solution of equation (10.1.4) having this initial data.

Indeed, it is not difficult to show that Klein-Gordon theory admits this formulation when one considers only analytic initial data, i.e., when ϕ and $\partial\phi/\partial t$ are analytic functions on Σ_0. To see this, we observe that from such initial data, we can compute all spatial derivatives of ϕ and $\partial\phi/\partial t$ at $t = t_0$. Equation (10.1.4) then gives $\partial^2\phi/\partial t^2$ at $t = t_0$ and by taking spatial derivatives of this equation, we may compute all the spatial derivatives of $\partial^2\phi/\partial t^2$ at $t = t_0$ in terms of previously computed quantities. We then may differentiate equation (10.1.4) with respect to t, and thereby compute $\partial^3\phi/\partial t^3$ and all its spatial derivatives at $t = t_0$. Continuing in this manner, we obtain *all* derivatives of ϕ at $t = t_0$. This enables us to write down a formal power series solution for ϕ. As proven by Cauchy and generalized by Kowalewski, for a wide class of partial differential equations—or even systems of partial differential equations—this power series has a finite radius of convergence. We state their theorem for second order partial differential equations (i.e., equations containing second partial derivatives of the unknown variables but none of higher order), but it is easily generalized to equations of arbitrary order. A proof of this theorem can be found in most books treating partial differential equations (e.g., Courant and Hilbert 1962).

THEOREM 10.1.1 (Cauchy-Kowalewski theorem): *Let $t, x^1, \ldots, x^{m-1}$ be coordinates of $\mathbb{R}^m$. Consider a system of n partial differential equations for n unknown functions $\phi_1, \ldots, \phi_n$ in $\mathbb{R}^m$, having the form*

$$\frac{\partial^2\phi_i}{\partial t^2} = F_i(t, x^\alpha;\ \phi_j;\ \partial\phi_j/\partial t;\ \partial\phi_j/\partial x^\alpha;\ \partial^2\phi_j/\partial t\partial x^\alpha;\ \partial^2\phi_j/\partial x^\alpha\partial x^\beta) \quad , \quad (10.1.5)$$

where each F_i is an analytic function of its variables. Let $f_i(x^\alpha)$ and $g_i(x^\alpha)$ be analytic functions. Then there is an open neighborhood O of the hypersurface $t = t_0$ such that within O there exists a unique analytic solution of equation (10.1.5) *such that* $\phi_i(t_0, x^\alpha) = f_i(x^\alpha)$ *and* $\dfrac{\partial\phi_i}{\partial t}(t_0, x^\alpha) = g_i(x^\alpha)$.

The Cauchy-Kowalewski theorem shows that Klein-Gordon theory has an initial value formulation, at least for analytic initial data. It shows that, in analogy to particle mechanics, the initial value of ϕ and its time derivative may be specified arbitrarily, and that these initial values determine the subsequent evolution of ϕ. It thereby also shows that there exists a large class of solutions to the Klein-Gordon equation, since there exist as many analytic solutions of equation (10.1.4) as there are pairs of arbitrary analytic functions of the three spatial variables x^α.

However, the Cauchy-Kowalewski analysis is not adequate for showing that the initial value formulation of Klein-Gordon theory is well posed. First, the analysis does not establish continuous dependence of solutions on initial data in a suitable sense. More precisely, we may define a topology on the space of initial data which makes two functions "close" if they and a finite number (or all) of their derivatives are close. For example, we may define a "distance" between two functions f_1 and f_2 on the $t = t_0$ initial data surface Σ_0, by summing the least upper bounds (l.u.b.) of the magnitude of $(f_1 - f_2)$ and all its derivatives up to order k,

$$\|f_1 - f_2\| = \underset{x\in\Sigma_0}{\text{l.u.b.}}\, |f_1(x) - f_2(x)| + \sum_{k_1,k_2,k_3} \underset{x\in\Sigma_0}{\text{l.u.b.}} \left| \frac{\partial^{k_1+k_2+k_3}(f_1 - f_2)}{\partial x^{k_1}\partial y^{k_2}\partial z^{k_3}} \right| \quad , \quad (10.1.6)$$

where $k_1 + k_2 + k_3 \leqq k$. We then may take the open balls in this norm as a basis of a topology (see appendix A). (Other reasonable choices of topology also can be given.) In a compact region of spacetime, we may define a similar topology on the solutions. The Cauchy-Kowalewski theorem gives no guarantee that—for any reasonable choices of topology on the space of initial data and the space of solutions—the map taking analytic initial data to the analytic solutions to which they give rise is continuous.

Furthermore, the Cauchy-Kowalewski analysis cannot even deal with the issue of causal propagation of the field. An analytic function is uniquely determined by the values of it and all its derivatives at one point and, thus, in particular, is uniquely determined by its value in an arbitrarily small open neighborhood of a point. This implies that in the analytic case if we alter the initial data in an arbitrary open region U of the initial surface Σ_0, we must, in fact, alter the initial data over the entire hypersurface Σ_0. Thus, to analyze the issue of causal propagation, we must consider nonanalytic initial data. However, the Cauchy-Kowalewski analysis does not even prove existence of a solution for C^∞, nonanalytic, initial data.

Thus, to show that the initial value formulation of Klein-Gordon theory is well posed, methods other than the Cauchy-Kowalewski analysis are required. We now shall outline a proof that the initial value formulation of the massive Klein-Gordon field in Minkowski spacetime is well posed. Then we shall state some much more general results obtained by this approach.

Let ϕ be a smooth solution of equation (10.1.3). Then the stress-energy-momentum tensor of ϕ,

$$T_{ab} = \partial_a\phi\partial_b\phi - \frac{1}{2}\eta_{ab}(\partial_c\phi\partial^c\phi + m^2\phi^2) \tag{10.1.7}$$

(see eq. [4.2.20]), is conserved,

$$\partial^a T_{ab} = 0 \quad . \tag{10.1.8}$$

Consequently, letting $\xi^a = (\partial/\partial t)^a$ denote the time translation Killing field orthogonal to the $t = t_0$ hypersurface Σ_0, we have

$$\partial^a(T_{ab}\xi^b) = 0 \quad . \tag{10.1.9}$$

[In fact, for any Killing field ξ^a and any conserved, symmetric T_{ab} in curved spacetime, we have $\nabla^a(T_{ab}\xi^b) = T_{ab}\nabla^a\xi^b = 0$ by eq. (C.3.1)] Let S_0 be a (three-dimensional) closed ball on the initial hypersurface Σ_0. Let Σ_1 denote the hypersurface $t = t_1$ (with $t_1 > t_0$), let $K = D^+(S_0) \cap J^-(\Sigma_1)$, and let $S_1 = D^+(S_0) \cap \Sigma_1$. Finally, let S_2 denote the "null portion" of the boundary of K (see Fig. 10.1). We integrate equation (10.1.9) over K and apply Gauss's law, thereby obtaining

$$\int_{S_1} T_{ab}\xi^a\xi^b + \int_{S_2} T_{ab}l^a\xi^b = \int_{S_0} T_{ab}\xi^a\xi^b \quad , \tag{10.1.10}$$

where l^a is the future directed normal to S_2. (The natural volume elements on S_0 and S_1 are understood in eq. [10.1.10]; the volume element on the null surface S_2 is explained in appendix B.) However, it is not difficult to verify from equation

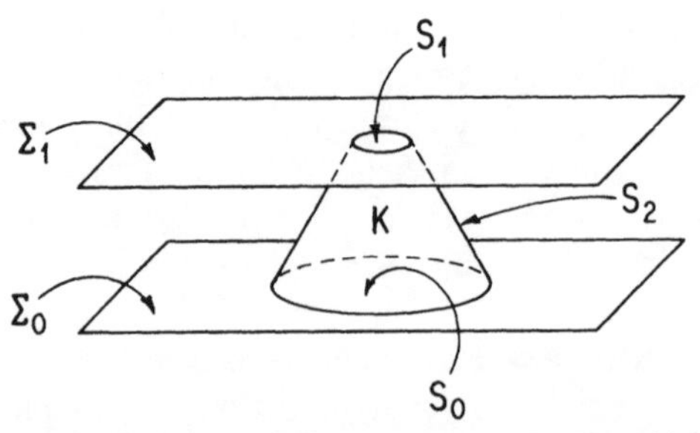

Fig. 10.1. A spacetime diagram showing the region K over which eq. (10.1.9) is integrated to obtain equation (10.1.10).

(10.1.7) that T_{ab} satisfies the dominant energy condition, i.e., if v^a is a future directed timelike vector, then $-T^a{}_b v^b$ is a future directed timelike or null vector. Consequently, we have $T_{ab} l^a \xi^b \geqq 0$. Hence, the second term on the left-hand side of equation (10.1.10) is nonnegative. Writing out the other terms explicitly, we obtain

$$\int_{S_1} \left[\left(\frac{\partial \phi}{\partial t} \right)^2 + |\vec{\nabla}\phi|^2 + m^2\phi^2 \right] \leqq \int_{S_0} \left[\left(\frac{\partial \phi}{\partial t} \right)^2 + |\vec{\nabla}\phi|^2 + m^2\phi^2 \right] \quad . \tag{10.1.11}$$

Equation (10.1.11) is the key equation needed to demonstrate the existence of a well-posed initial value formulation. First, it shows—without appealing to analyticity—that there can be at most one solution in $D^+(S_0)$ with given initial data $(\phi, \partial\phi/\partial t)$ on S_0. Namely, if ϕ_1 and ϕ_2 both are C^2 functions satisfying equation (10.1.3) with the same smooth (but not necessarily analytic) initial data, then their difference, $\psi = \phi_2 - \phi_1$, would be a C^2 solution of equation (10.1.3) with vanishing initial data. Hence, for ψ, the right-hand side of equation (10.1.11) vanishes, which implies (assuming $m \neq 0$) that $\psi = 0$ on S_1 and hence (since Σ_1 is arbitrary) $\psi = 0$ throughout $D^+(S_0)$. [If $m = 0$, we have $\vec{\nabla}\psi = 0$ and $\partial\psi/\partial t = 0$ in $D^+(S_0)$ and since $\psi = 0$ on S_0, this again implies that $\psi = 0$ throughout $D^+(S_0)$.] Similarly, ψ also vanishes throughout $D^-(S_0)$. This result also shows that the second requirement for a well posed initial value formulation is satisfied: A variation of the initial data outside of S_0 cannot affect the solution within $D(S_0)$.

Equation (10.1.11) also shows that solutions depend continuously on the initial data. To prove a useful form of this continuous dependence that also allows us to prove existence of solutions for arbitrary smooth initial data, we proceed as follows. For simplicity, we shall restrict attention to the massive case, $m \neq 0$. First, by differentiating equation (10.1.4) with respect to the coordinates x^μ we see that all partial derivatives of ϕ also satisfy the Klein-Gordon equation. Hence, we obtain inequalities of the form (10.1.11) bounding square integrals of higher space and time derivatives of ϕ on S_1 by the corresponding square integrals on S_0. Using equation (10.1.4), we can express all terms on S_0 containing more than one time derivative in terms of the initial data ϕ, $\partial\phi/\partial t$ and their spatial derivatives on S_0. Thus, we obtain inequalities of the form

$$\|\phi\|_{S_1,k} \leq C_{1,k} |||\phi|||_{S_0,k} + C_{2,k} |||\partial\phi/\partial t|||_{S_0,k-1} \quad , \tag{10.1.12}$$

where the norms $\|\phi\|_{S_1,k}$ and $|||\phi|||_{S_0,k}$ are defined by

$$\|\phi\|^2_{S_1,k} = \int_{S_1} \{|\phi|^2 + \cdots + \sum_i |\partial^{k_i}\phi|^2\} \quad , \tag{10.1.13}$$

$$|||\phi|||^2_{S_0,k} = \int_{S_0} \{|\phi|^2 + \cdots + \sum_i |D^{k_i}\phi|^2\} \quad , \tag{10.1.14}$$

where ∂^{k_i} denotes a kth order partial derivative with respect to space *and* time coordinates, while D^{k_i} denotes a kth order partial derivative with respect to space coordinates only. (In eqs. [10.1.13] and [10.1.14] we take the sums of all such partial derivatives of order less than or equal to k. The norms defined by eqs. [10.1.13] and [10.1.14] are called *Sobolev norms.*) By integrating equation (10.1.12) over t_1 from t_0 to the maximum value of t for which $D^+(S_0) \cap \Sigma_t \neq 0$, we obtain

$$\|\phi\|_{D^+(S),k} \leq C'_{1,k}|||\phi|||_{S_0,k} + C'_{2,k}|||\partial\phi/\partial t|||_{S_0,k-1} \quad . \tag{10.1.15}$$

Now, we apply the following key result. Let A be any subset of $\mathbb{R}^n$ (with the natural Euclidean metric) which satisfies the *uniform interior cone condition,* defined as follows: There exists a cone of fixed height h and fixed vertex angle θ such that for each $p \in A$ this cone can be mapped isometrically into A with vertex at p. Then, for $k > n/2$ the $\| \; \|_{A,k}$ norm of a smooth function f (defined by eq. [10.1.13] with the integration region taken to be A) bounds the numerical values of f in A, i.e., there exists a constant C such that

$$\underset{x\in A}{\text{l.u.b.}}\, |f(x)| \leq C\|f\|_{A,k} \quad . \tag{10.1.16}$$

A proof of this result is outlined in problem 1. Thus, taking $A = D^+(S_0)$ and $k = 3$, we find using equations (10.1.15) and (10.1.16),

$$\underset{x\in D^+(S_0)}{\text{l.u.b.}}\, |\phi| \leq C''_1|||\phi|||_{S_0,3} + C''_2|||\partial\phi/\partial t|||_{S_0,2} \quad . \tag{10.1.17}$$

Similarly, the numerical values of any mth order partial derivative of ϕ are bounded in terms of the initial data by

$$\underset{x\in D^+(S_0)}{\text{l.u.b.}}\, |\partial^m\phi| \leq C''_{1,m}|||\phi|||_{S_0,3+m} + C''_{2,m}|||\partial\phi/\partial t|||_{S_0,2+m} \quad . \tag{10.1.18}$$

The same type of bounds hold, of course, for $x \in D^-(S)$.

Equations (10.1.17) and (10.1.18) demonstrate the continuous dependence of ϕ and its derivatives on the initial data in a strong sense. More precisely, if we define a topology on solutions in $D(S_0)$ via a norm of the form equation (10.1.6) with $k = m$ and define a topology on the initial data on S_0 (or on all of Σ_0) via the $||| \; |||_{S_0,3+m}$ norm or the norm (10.1.6) with $k = m + 3$, then equation (10.1.18) demonstrates that the linear map from initial data to solutions is bounded and, hence, is continuous. More generally, the map from initial data on Σ_0 to solutions in any fixed compact region of spacetime is continuous in these topologies.

Finally, we use this continuity to prove existence of a smooth solution ϕ for arbitrary smooth initial data $(\phi, \partial\phi/\partial t)$ on Σ_0. We proceed by choosing a sequence $\{(\phi^m_i, \partial\phi^m_i/\partial t)\}$, $i = 1, 2, \ldots$, of analytic initial data on Σ_0 such that the functions in this sequence and their spatial derivatives up to order $(3 + m)$ converge uniformly to $(\phi, \partial\phi/\partial t)$ on S_0. By theorem 10.1.1 there exists a solution ϕ^m_i with initial data $(\phi^m_i, \partial\phi^m_i/\partial t)$ on Σ_0. (Actually, theorem 10.1.1 as stated guarantees existence of a solution only in an open neighborhood of Σ_0, but further analysis in the linear case establishes that a solution exists throughout $\mathbb{R}^n$.) However, according to equation (10.1.18) [and the corresponding inequality for $D^-(S_0)$], $\{\phi_i^m\}$ and its first m derivatives must converge uniformly in $D(S_0)$ to a function ϕ^m and its first m derivatives. Choosing $m \geq 2$, we easily verify that this limit function ϕ^m must satisfy equation (10.1.4). Thus, for all $m \geq 2$ we obtain a C^m solution in $D(S_0)$. However, we proved above that there is at most one C^2 solution in $D(S_0)$. Thus, $\phi^m = \phi^{m'} \equiv \phi$ for all $m, m' \geq 2$. Since ϕ is C^m for all $m \geq 2$, we have proven existence of a C^∞ solution throughout $D(S_0)$. Since S_0 is arbitrary, this solution exists on all of $\mathbb{R}^4$.

Thus, we have established that the massive Klein-Gordon field in Minkowski spacetime has a well posed initial value formulation. Note that, unlike the Cauchy-Kowalewski analysis, the proof of a well posed initial value formulation made use of detailed structure of the Klein-Gordon equation. In particular, the linearity of the equation was used in a number of places, and its "wave equation" character was essential to the construction of a conserved T_{ab} with positive energy. If we changed the sign of $\partial^2\phi/\partial t^2$ in equation (10.1.4), thereby turning equation (10.1.4) into a four-dimensional "Laplace type" equation, we would not be able to construct such a T_{ab}, and the method of proof would break down even though the applicability of theorem 10.1.1 would not be affected. Indeed, it is well known that equations of "Laplace type" do *not* have a well posed initial value formulation.

The above results for the Klein-Gordon field can be significantly generalized. In particular, we may replace the Klein-Gordon equation (10.1.3) in $\mathbb{R}^4$ by any equation on a manifold M of the form

$$g^{ab}\nabla_a\nabla_b\phi + A^a\nabla_a\phi + B\phi + C = 0 \quad , \qquad (10.1.19)$$

where ∇_a is any derivative operator, A^a is an arbitrary smooth vector field, B and C are arbitrary smooth functions, and g_{ab} is an arbitrary smooth Lorentz metric such that the spacetime (M, g_{ab}) is globally hyperbolic. (A second order linear partial differential equation is said to be *hyperbolic* if and only if it can be expressed in the form [10.1.19].) This equation will have a well posed initial value formulation for initial data $(\phi, n^a\nabla_a\phi)$ on any smooth, spacelike Cauchy surface Σ, where n^a is the unit normal to Σ. There are only a few significant additional complications to the proof of this much more general result. Specifically, for equation (10.1.19) in general we cannot construct a conserved T_{ab} satisfying the dominant energy condition, but we may construct a T_{ab} satisfying the dominant energy condition whose failure to be conserved can be bounded, so that inequalities of the form (10.1.12) (with $C_{1,k}$ and $C_{2,k}$ now functions of time) still hold. Also, in the proof of existence, if the coefficients g^{ab}, A^a, B, and C are nonanalytic, we must also approximate them by analytic coefficients before we can appeal to theorem 10.1.1.

These results can be further generalized to systems of equations, resulting in the following theorem, a complete proof of which can be found in Hawking and Ellis (1973).

THEOREM 10.1.2. *Let (M, g_{ab}) be a globally hyperbolic spacetime (or a globally hyperbolic region of an arbitrary spacetime) and let ∇_a be any derivative* *operator. Let Σ be a smooth, spacelike Cauchy surface. Consider the system of n linear equations for n unknown functions $\phi_1, \ldots, \phi_n$ of the form*

$$g^{ab}\nabla_a\nabla_b\phi_i + \sum_j (A_{ij})^a\nabla_a\phi_j + \sum_j B_{ij}\phi_j + C_i = 0 \quad . \tag{10.1.20}$$

(Eq. [10.1.20] is referred to as a linear, diagonal second order hyperbolic system.) Then equation (10.1.20) has a well posed initial value formulation on Σ. More precisely, given arbitrary smooth initial data, $(\phi_i, n^a\nabla_a\phi_i)$ for $i = 1, \ldots, n$ on Σ there exists a unique solution of equation (10.1.20) throughout M. Furthermore, the solutions depend continuously on the initial data in the sense described above for the Klein-Gordon equation in flat spacetime. Finally, a variation of the initial data outside of a closed subset, S, of Σ does not affect the solution in D(S).

We remark also that the differentiability assumptions on the coefficients g^{ab}, $(A_{ij})^a$, and B_{ij} as well as on Σ and the initial data $\phi_i, n^a\nabla_a\phi_i$ on Σ can be weakened significantly (see Hawking and Ellis 1973). For results on the initial value formulation of linear partial differential equations not of second order diagonal form, see, e.g., Bers, John, and Schechter (1964).

Finally we discuss generalizations of theorem 10.1.2 to some nonlinear systems of equations. Very few results other than theorem 10.1.1 are known concerning the initial value formulation of general, nonlinear systems of equations. However, for second order differential equations which are *quasilinear,* i.e., linear in the highest derivative terms, many of the results on linear systems apply locally. More precisely, we call a system of n second-order partial differential equations for n unknown functions $\phi_1, \ldots, \phi_n$ on a manifold M a *quasilinear, diagonal, second order hyperbolic system* if it can be put in the form

$$g^{ab}(x; \phi_j; \nabla_c\phi_j)\nabla_a\nabla_b\phi_i = F_i(x; \phi_j; \nabla_c\phi_j) \quad , \tag{10.1.21}$$

where ∇_a is any derivative operator, g^{ab} is a smooth Lorentz metric, and each F_i is a smooth function of its variables. (Eq. [10.1.21] differs from eq. [10.1.20] in that g^{ab} is now permitted to depend on the unknown variables and their first derivatives, and F_i now may have nonlinear dependence on these variables.) For equations of this type, we have the following theorem due to Leray (1952):

THEOREM 10.1.3. *Let $(\phi_0)_1, \ldots, (\phi_0)_n$ be any solution of the quasilinear hyperbolic system (10.1.21) on a manifold M and let $(g_0)^{ab} = g^{ab}(x; (\phi_0)_j; \nabla_c(\phi_0)_j)$. Suppose $(M, (g_0)_{ab})$ is globally hyperbolic (or, alternatively, consider a globally hyperbolic region of this spacetime). Let Σ be a smooth spacelike Cauchy surface for $(M, (g_0)_{ab})$. Then, the initial value formulation of equation (10.1.21) is well posed on Σ in the following sense: For initial data on Σ sufficiently close to the initial data for $(\phi_0)_1, \ldots, (\phi_0)_n$, there exists an open neighborhood O of Σ such that equation (10.1.21) has a solution, $\phi_1, \ldots, \phi_n$, in O and $(O, g_{ab}(x; \phi_j; \nabla_c\phi_j))$ is globally hyperbolic. The solution is unique in O and propagates causally in the sense that if the initial data for $\phi'_1, \ldots, \phi'_n$ agree with* *that of $\phi_1, \ldots, \phi_n$ on a subset, S, of Σ, then the solutions agree on $O \cap D^+(S)$. Finally, the solutions depend continuously on the initial data in the sense described above for the Klein-Gordon field.*

The basic idea of the proof of theorem 10.1.3 is to start by solving the *linear* system of the form (10.1.20) obtained by replacing ϕ_j and $\nabla_a \phi_j$ by $(\phi_0)_j$ and $\nabla_a(\phi_0)_j$ in g^{ab} and F_i in equation (10.1.21). Then we obtain a solution $(\phi_1)_j$ by theorem 10.1.2. We substitute this solution into the coefficients g^{ab} and F_i and repeat the procedure. In this way, we obtain a sequence $(\phi_n)_j$ of solutions to linear equations. For initial data sufficiently close to that for $(\phi_0)_j$ it can be shown that this sequence converges in a neighborhood of Σ and that the limit satisfies equation (10.1.21) and has the above desired properties. The proof of this convergence and the other properties quoted in theorem 10.1.3 is outlined in Hawking and Ellis (1973).

10.2 Initial Value Formulation of General Relativity

In this section, we shall show that general relativity has a well posed initial value formulation. We prove this by casting Einstein's equation into the form (10.1.21) for which theorem 10.1.3 applies. As we shall see, the analysis of Einstein's equation differs from that of the Klein-Gordon field in that there are initial value constraints and in that we will need to make a "gauge choice," i.e., a choice of coordinates, so that Einstein's equation takes the desired form. In order to gain more insight into the nature of these differences, we begin by analyzing the simpler problem of the initial value formulation of Maxwell's equations for the vector potential A_a in Minkowski spacetime. As we shall see, the initial value formulation of Einstein's equation is very closely analogous to the initial value formulation of these equations.

As discussed in section 4.2, the vacuum Maxwell's equations for the vector potential A_a in Minkowski spacetime take the form

$$\partial^a(\partial_a A_b - \partial_b A_a) = 0 \quad . \tag{10.2.1}$$

At first glance, equation (10.2.1) may appear satisfactory for determining A_a: We have four equations for the four unknown components of A_a. However, a closer examination of equation (10.2.1) indicates the possibility of serious trouble, since equation (10.2.1) is *not* of the form (10.1.20) for which a well posed initial value formulation is known to exist. Indeed, if we choose a surface, Σ_0, of constant inertial time, $t = t_0$, as our initial hypersurface, we see that the time component of equation (10.2.1) contains no second time derivatives at all. In ordinary vector notation,this equation is

$$\nabla^2 A_0 - \vec{\nabla} \cdot (\partial \vec{A} / \partial t) = 0 \quad , \tag{10.2.2}$$

i.e.,

$$\vec{\nabla} \cdot \vec{E} = 0 \quad , \tag{10.2.3}$$

where the electric field, $\vec{E}$, is defined by

$$\vec{E} = \vec{\nabla} A_0 - \partial \vec{A} / \partial t \quad , \tag{10.2.4}$$

i.e., in index notation,

$$E_a = (\partial_a A_b - \partial_b A_a)n^b = F_{ab}n^b \quad , \tag{10.2.5}$$

where n^a is the unit normal to Σ_0. Thus, equation (10.2.2) (or, equivalently, eq. [10.2.3]) gives an *initial value constraint* on the initial data $(A_\mu, \partial A_\mu/\partial t)$. Initial data which fail to satisfy (10.2.2) cannot possibly yield a solution of Maxwell's equations.

The remaining three components of Maxwell's equations do contain second time derivatives of the spatial components of A_a, so we can solve for $\partial^2 A_\mu/\partial t^2$ for $\mu = 1, 2, 3$ in the manner required by the Cauchy-Kowalewski theorem 10.1.1. (Note, however, that even these three equations are not of the form [10.1.20] for which a well posed initial value formulation is known to exist.) One might expect that by differentiating the initial value constraint (10.2.2), one could end up with an equation for $\partial^2 A_0/\partial t^2$ which then would permit an initial value formulation, at least in the sense of theorem 10.1.1. However, this is not the case. We have the identity

$$\partial^b \partial^a(\partial_a A_b - \partial_b A_a) = 0 \quad , \tag{10.2.6}$$

and this shows that the time derivative of equation (10.2.2) vanishes identically if the spatial components of Maxwell's equations are satisfied. Hence, the complete Maxwell's equations are equivalent to the spatial components of Maxwell's equations together with the initial value constraint (10.2.2). Thus, equation (10.2.1) is an *underdetermined* system for A_a; we really have only three equations (plus an initial value constraint) for four unknown functions. Indeed, by theorem 10.1.1 it is not difficult to see that in the analytic case we may specify A_0 arbitrarily throughout the spacetime and still obtain a solution. Thus Maxwell's equations do not admit an initial value formulation for A_a in the straightforward, mathematical sense discussed in the previous section.

However, this difficulty is not of a physical character. As already mentioned in section 4.2, two vector potentials which differ by the gradient, $\partial_a \chi$, of a function, χ, represent the same physical electromagnetic field. Thus, on account of this gauge arbitrariness, Maxwell's equations cannot possibly be expected to determine A_a from initial conditions. On the other hand, we now shall show that the initial values of A_μ and $\partial A_\mu/\partial t$ do uniquely determine a solution up to gauge, and that, physically, Maxwell's equations do admit a well posed initial value formulation.

The most direct way of showing this is to choose an appropriate gauge for A_a and show that Maxwell's equations for A_a in this gauge are of the form (10.1.20), thereby yielding a well posed initial value formulation. We choose the Lorentz gauge

$$\partial^a A_a = 0 \tag{10.2.7}$$

(see eq. [4.2.31] above). Maxwell's equations in this gauge are simply

$$\partial^a \partial_a A_b = 0 \quad . \tag{10.2.8}$$

Equation (10.2.8) together with equation (10.2.7) is physically equivalent to the original system (10.2.1) in the sense that solutions of (10.2.1) can differ from solutions of (10.2.7) and (10.2.8) only by a gauge transformation.

Given our initial data $(A_\mu, \partial A_\mu/\partial t)$, we make a gauge transformation so that $\partial^a A_a = 0$ on Σ_0. Now, equation (10.2.8) implies that

$$\partial^a \partial_a (\partial^b A_b) = \partial^b (\partial^a \partial_a A_b) = 0 \quad , \tag{10.2.9}$$

and thus, by theorem 10.1.2, if equation (10.2.8) is satisfied everywhere, then the gauge condition (10.2.7) also will be satisfied everywhere if and only if $\partial^b A_b = \partial(\partial^b A_b)/\partial t = 0$ on Σ_0. We already have ensured that $\partial^a A_a = 0$ on Σ_0, and using equation (10.2.8), we see that the initial condition $\partial(\partial^b A_b)/\partial t = 0$ is equivalent to the initial value constraint, equation (10.2.2). Thus, if $\vec{\nabla} \cdot \vec{E} = 0$ on Σ_0, then for our gauge transformed initial data, equation (10.2.7) will be satisfied throughout the spacetime if equation (10.2.8) holds. Thus, we need only solve equation (10.2.8) with the given values of the initial data. However, we can do this because equation (10.2.8) *does* have the form (10.1.20) for which a well posed initial value formulation has been established. Thus, by theorem 10.1.2, there exists a unique solution of (10.2.8) for the given (new) values of the initial data. Furthermore, this solution depends continuously on the initial data and has the desired domain of dependence property.

To obtain a solution with the original values $(A_\mu, \partial A_\mu/\partial t)$ of the initial data, we simply "undo" the gauge transformation which made $\partial^a A_a = 0$ on Σ_0. To show that this solution is physically unique, we note that two solutions of the original Maxwell's equations (10.2.1) with identical initial data, can be brought by gauge transformations into solutions of (10.2.8) with identical initial data. Since solutions of (10.2.8) with given initial data are unique, this shows that our two solutions can differ, at most, by a gauge transformation. Thus, Maxwell's equations in Minkowski spacetime possess a well posed initial value formulation in this sense.

Indeed, we may reformulate this result in a physically more satisfactory manner as follows: Let $\vec{E}$ and $\vec{B}$ be specified as arbitrary smooth vector fields on Σ_0, subject only to the constraints $\vec{\nabla} \cdot \vec{E} = \vec{\nabla} \cdot \vec{B} = 0$ on Σ_0. Then there exists a unique solution, F_{ab}, of Maxwell's equations with these initial data. Furthermore, F_{ab} depends continuously on the initial data $(\vec{E}, \vec{B})$, and F_{ab} at $p \in J^+(\Sigma_0)$ depends only on the initial data on $J^-(p) \cap \Sigma_0$. This result can be proven by introducing a vector potential, A_a, whose initial data on Σ_0 satisfies $\partial^a A_a = 0$ and reproduces the given values of $\vec{E}$ and $\vec{B}$ on Σ_0. (A particularly simple choice is to take $A_0 = 0$, $\partial \vec{A}/\partial t = -\vec{E}$, choose $\vec{A}$ to be any solution of $\vec{\nabla} \times \vec{A} = \vec{B}$, and set $\partial A_0/\partial t = \vec{\nabla} \cdot \vec{A}$.) Then we simply apply the results of the previous paragraph to construct a solution with the desired properties. Finally, we prove uniqueness by verifying that two different solutions, A_a and A'_a, of Maxwell's equations which reproduce the given $\vec{E}$ and $\vec{B}$ on Σ_0 can be brought by a gauge transformation into solutions of equation (10.2.8) with the same initial data. Hence, by uniqueness of solutions of equation (10.2.8), A_a and A'_a differ at most by a gauge transformation. Thus, they produce the same F_{ab}.

Thus, Maxwell's equations in Minkowski spacetime physically possess a well posed initial value formulation. This result may be generalized to curved spacetime (see problem 2).

We turn our attention, now, to Einstein's equation in a vacuum, $G_{ab} = 0$. The first issue to consider is the nature of the initial value formulation in this theory. In other theories of classical physics we are given the spacetime background and our task is to determine the time evolution of quantities in the background from their initial values and time derivatives. However, in general relativity we are solving for the spacetime itself. What should be the quantity or quantities to prescribe initially in general relativity in order that spacetime structure be determined?

In order to answer this question, we must view general relativity as describing the time evolution of some quantity. Let (M, g_{ab}) be a globally hyperbolic spacetime. (We consider only globally hyperbolic spacetimes since the initial value formulation should be relevant only in this case.) As proven in theorem 8.3.14, we can foliate (M, g_{ab}) by Cauchy surfaces, Σ_t, parameterized by a global time function, t. Let n^a be the unit normal vector field to the hypersurfaces Σ_t. The spacetime metric, g_{ab}, induces a spatial metric (i.e., a three-dimensional Riemannian metric) h_{ab} on each Σ_t by the formula

$$h_{ab} = g_{ab} + n_a n_b \quad . \tag{10.2.10}$$

Let t^a be a vector field on M satisfying $t^a \nabla_a t = 1$. We decompose t^a into its parts normal and tangential to Σ_t by defining the *lapse function*, N, and the *shift vector*, N^a, with respect to t^a by

$$N = -t^a n_a = (n^a \nabla_a t)^{-1} \quad , \tag{10.2.11}$$

$$N_a = h_{ab} t^b \tag{10.2.12}$$

(see Fig. 10.2).

We may interpret the vector field t^a as representing the "flow of time" throughout spacetime. As we "move forward in time" by parameter time t starting from the $t = 0$ surface Σ_0, we go to the surface Σ_t. If we identify the hypersurfaces Σ_0, Σ_t by the diffeomorphism resulting from following integral curves of t^a, we may view the effect of "moving forward in time" as that of changing the spatial metric on an abstract three-dimensional manifold Σ from $h_{ab}(0)$ to $h_{ab}(t)$. Thus, we may view a globally hyperbolic spacetime (M, g_{ab}) as representing the time development of a Riemannian metric on a fixed three-dimensional manifold. This suggests that we view the spatial metric on a three-dimensional hypersurface as the dynamical variable in general relativity. (The lapse function, N, and shift vector, N^a, are not

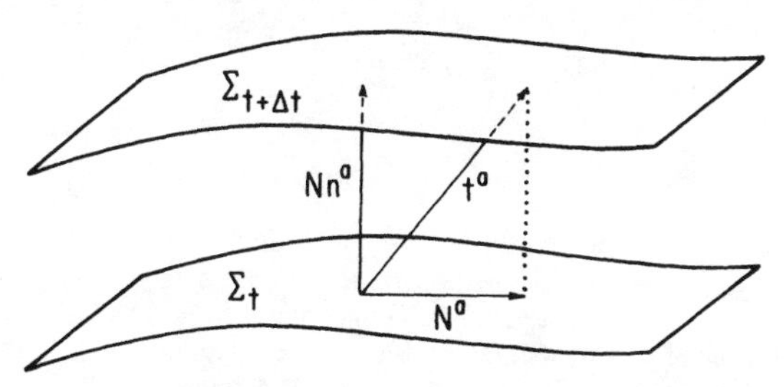

Fig. 10.2. A spacetime diagram illustrating the definition of the lapse function, N, and shift vector N^a.

considered dynamical, since they merely prescribe how to "move forward in time." Further motivation for viewing the spatial metric as the dynamical variable in general relativity arises from the Hamiltonian formulation given in appendix E.) Hence, we would expect appropriate initial data to consist of a Riemannian metric, h_{ab}, and its "time derivative" on a three-dimensional manifold Σ.

In section 9.3 above, we introduced the notion of extrinsic curvature as representing a well-defined notion of the "time derivative" of the spatial metric on a hypersurface Σ embedded in spacetime. In equation (9.3.19), ξ^a was the unit tangent to the congruence of timelike geodesics orthogonal to Σ. However, if n^a is any other unit timelike vector field which is normal to Σ, then its derivative along a direction tangential to Σ must agree on Σ with that of ξ^a, so we have

$$\begin{aligned} K_{ab} &= \nabla_a \xi_b = h_a{}^c \nabla_c \xi_b \\ &= h_a{}^c \nabla_c n_b \\ &= \frac{1}{2} \pounds_n h_{ab} \quad , \end{aligned} \tag{10.2.13}$$

where the reader is asked to verify the last equality in problem 3. This generalizes the formula for extrinsic curvature to nongeodesic normal slicings of spacetime. The relation between K_{ab} and the coordinate time derivative $\pounds_t h_{ab}$ is given by equation (E.2.30) of appendix E. Figure 10.3 illustrates the interpretation of K_{ab} in terms of the "bending" of Σ in spacetime.

The above considerations suggest that in general relativity, appropriate initial data should consist of a triple (Σ, h_{ab}, K_{ab}), where Σ is a three-dimensional manifold, h_{ab} is a Riemannian metric on Σ, and K_{ab} is a symmetric tensor field on Σ. We shall show below that given such initial data—subject to certain initial value constraints—there exists a globally hyperbolic spacetime (M, g_{ab}) satisfying Einstein's equation which possesses a Cauchy surface diffeomorphic to Σ on which the induced metric, equation (10.2.10), is h_{ab} and the induced extrinsic curvature, equation (10.2.13), is K_{ab}. Furthermore, this solution depends continuously on the initial data, satisfies the desired domain of dependence property, and is unique in a sense made precise below.

First, however, we establish some useful relations between the spacetime metric, derivative operator, and curvature, and the corresponding quantities they induce on a spacelike hypersurface Σ embedded in M. We already noted that the spacetime

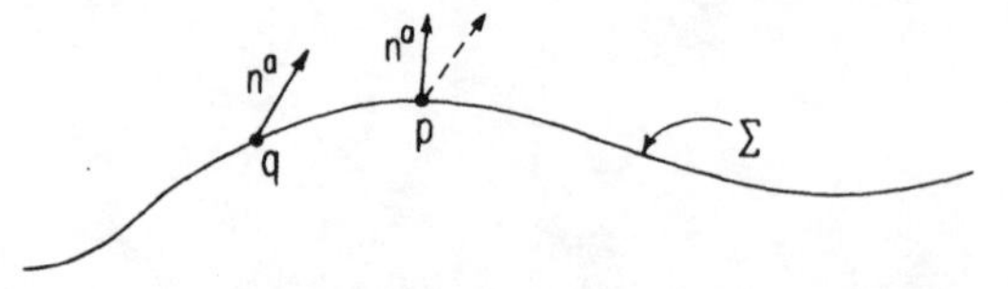

Fig. 10.3. A spacetime diagram illustrating the notion of the extrinsic curvature of a hypersurface Σ. The dashed arrow at p represents the parallel transport of the normal vector, n^a, at q along a geodesic connecting q to p. The failure of this vector to coincide with n^a at p corresponds intuitively to the bending of Σ in the spacetime in which it is embedded. The formula $K_{ab} = h_a{}^c \nabla_c n_b$ shows that K_{ab} directly measures this failure of the two vectors at p to coincide for q near p.

metric g_{ab} induces a Riemannian metric h_{ab} on Σ by equation (10.2.10). However, by theorem 3.1.1, h_{ab} uniquely determines a natural derivative operator on Σ, which we denote as D_a. Furthermore, the derivative operator, D_a, on Σ gives rise to a curvature tensor ${}^{(3)}R_{abc}{}^d$ on Σ. We obtain, now, formulas relating D_a and ${}^{(3)}R_{abc}{}^d$ to four-dimensional quantities.

Let v^a be a (spacetime) vector at a point $p \in \Sigma$. We may uniquely decompose v^a into components tangent to and perpendicular to Σ via

$$v^a = v_\perp n^a + v_\parallel^a \tag{10.2.14}$$

where n^a is the unit normal to Σ and $v_\parallel^a n_a = 0$. If $v_\perp = 0$ so that $v^a = v_\parallel^a$, we may view v^a as a vector lying in the tangent space to Σ at p. The condition that $v_\perp = 0$ is equivalent to

$$v^a = h^a{}_b v^b \tag{10.2.15}$$

with h_{ab} given by equation (10.2.10) and the first index of h_{ab} is raised by g^{ab}. More generally, we may view a spacetime tensor $T^{a_1 \cdots a_k}{}_{b_1 \cdots b_l}$ at $p \in \Sigma$ as a tensor over the tangent space to Σ at p if

$$T^{a_1 \cdots a_k}{}_{b_1 \cdots b_l} = h^{a_1}{}_{c_1} \cdots h^{a_k}{}_{c_k} h_{b_1}{}^{d_1} \cdots h_{b_l}{}^{d_l} T^{c_1 \cdots c_k}{}_{d_1 \cdots d_l} \tag{10.2.16}$$

Conversely, any tensor defined at point p on the manifold Σ uniquely gives rise to a spacetime tensor at p (i.e., a tensor over the tangent space to M at p) which satisfies equation (10.2.16). Note that $h^a{}_b$ plays the role of a projection operator from the tangent space to M at p to the tangent space to Σ at p.

Let $T^{a_1 \cdots a_k}{}_{b_1 \cdots b_l}$ be a tensor field on the manifold Σ. If we view $T^{a_1 \cdots a_k}{}_{b_1 \cdots b_l}$ as a spacetime tensor satisfying equation (10.2.16), we still cannot define $\nabla_c T^{a_1 \cdots a_k}{}_{b_1 \cdots b_l}$ since in order to calculate this quantity we would need to know how $T^{a_1 \cdots a_k}{}_{b_1 \cdots b_l}$ varies as we move off of Σ. However, $h_d{}^c \nabla_c T^{a_1 \cdots a_k}{}_{b_1 \cdots b_l}$ is well defined since for this quantity, no derivatives in directions pointing out of Σ are taken. This tensor need not satisfy equation (10.2.16), but we can project its indices using $h^a{}_b$ to obtain a tensor field on Σ. Then, we have the following result.

LEMMA 10.2.1. Let (M, g_{ab}) be a spacetime and let Σ be a smooth spacelike hypersurface in M. Let h_{ab} denote the induced metric on Σ, equation (10.2.10), and let D_a denote the derivative operator associated with h_{ab} (see theorem 3.1.1). Then D_a is given by the formula

$$D_c T^{a_1 \cdots a_k}{}_{b_1 \cdots b_l} = h^{a_1}{}_{d_1} \cdots h_{b_l}{}^{e_l} h_c{}^f \nabla_f T^{d_1 \cdots d_k}{}_{e_1 \cdots e_l} \quad , \tag{10.2.17}$$

where ∇_a is the derivative operator associated with g_{ab}.

Proof. It is straightforward to verify that D_a, defined by equation (10.2.17), satisfies the properties (1)–(5) of the definition of derivative operator given in section 3.1. Furthermore, we have

$$D_a h_{bc} = h_a{}^d h_b{}^e h_c{}^f \nabla_d (g_{ef} + n_e n_f) = 0 \tag{10.2.18}$$

since $\nabla_d g_{ef} = 0$ and $h_{ab} n^b = 0$. Thus, D_a is the unique derivative operator associated with h_{ab}. □

From equation (10.2.17) one can derive relations between the curvature ${}^{(3)}R_{abc}{}^{d}$ of Σ and the spacetime curvature $R_{abc}{}^{d}$. If ω_a is a dual vector field on Σ, we have

$$ {}^{(3)}R_{abc}{}^{d}\omega_d = D_a D_b \omega_c - D_b D_a \omega_c \quad . \tag{10.2.19} $$

However, we have

$$ \begin{aligned} D_a D_b \omega_c &= D_a(h_b{}^d h_c{}^e \nabla_d \omega_e) \\ &= h_a{}^f h_b{}^g h_c{}^k \nabla_f (h_g{}^d h_k{}^e \nabla_d \omega_e) \\ &= h_a{}^f h_b{}^d h_c{}^e \nabla_f \nabla_d \omega_e + h_c{}^e K_{ab} n^d \nabla_d \omega_e \\ &\quad + h_b{}^d K_{ac} n^e \nabla_d \omega_e \quad , \end{aligned} \tag{10.2.20} $$

where we have used the fact that

$$ h_a{}^b h_c{}^d \nabla_b h_d{}^e = h_a{}^b h_c{}^d \nabla_b (g_d{}^e + n_d n^e) = K_{ac} n^e \quad . \tag{10.2.21} $$

Now, the middle term on the right-hand side of equation (10.2.20) vanishes when antisymmetrized over a and b. Furthermore, we have

$$ h_b{}^d n^e \nabla_d \omega_e = h_b{}^d \nabla_d (n^e \omega_e) - h_b{}^d \omega_e \nabla_d n^e = -K_b{}^e \omega_e \quad . \tag{10.2.22} $$

Putting together all these results, we obtain

$$ {}^{(3)}R_{abc}{}^{d} = h_a{}^f h_b{}^g h_c{}^k h^d{}_j R_{fgk}{}^j - K_{ac} K_b{}^d + K_{bc} K_a{}^d \quad . \tag{10.2.23} $$

A similar calculation (problem 4) shows that

$$ D_a K^a{}_b - D_b K^a{}_a = R_{cd} n^d h^c{}_b \quad . \tag{10.2.24} $$

Equations (10.2.23) and (10.2.24) are known as the *Gauss-Codacci relations*.

We turn, now, to the analysis of the vacuum Einstein's equation. We will give initial data (h_{ab}, K_{ab}) on a three-dimensional manifold Σ and attempt to construct a globally hyperbolic spacetime (M, g_{ab}) for which Σ is a Cauchy surface on which the initial data are induced. Our strategy is to write down Einstein's equation for the metric components, $g_{\mu\nu}$, in a local coordinate system $\{y^\mu\}$ with the time coordinate, t, chosen so that the $t = 0$ surface corresponds to Σ (or, at least, that portion of Σ covered by the coordinate system). By casting the equations in the form (10.1.21), we will use theorem 10.1.3 to prove local existence of a solution with the desired properties. Then we shall outline how to "globalize" our local results to obtain the final conclusion, stated as theorem 10.2.2 below.

The components of the Einstein tensor, $G_{\mu\nu}$, can be expressed in terms of coordinate derivatives of the metric tensor components, $g_{\mu\nu}$, by the methods of section 3.4*a*. Einstein's equation in vacuum, $G_{ab} = 0$, yields a system of 10 second-order partial differential equations for the ten unknown metric components. Furthermore, these equations have a quasilinear form; i.e., they are linear in the second derivatives of the metric. Explicitly, from equations (3.4.5) and (3.1.30) we have

$$ R_{\mu\nu} = -\frac{1}{2} \sum_{\alpha,\beta} g^{\alpha\beta} \{ -2\partial_\beta \partial_{(\nu} g_{\mu)\alpha} + \partial_\alpha \partial_\beta g_{\mu\nu} + \partial_\mu \partial_\nu g_{\alpha\beta} \} + F_{\mu\nu}(g, \partial g) \tag{10.2.25} $$

and thus

$$G_{\mu\nu} = R_{\mu\nu} - \frac{1}{2} g_{\mu\nu} R$$
$$= -\frac{1}{2} \sum_{\alpha,\beta} g^{\alpha\beta}\{-2\partial_\beta \partial_{(\nu} g_{\mu)\alpha} + \partial_\alpha \partial_\beta g_{\mu\nu} + \partial_\mu \partial_\nu g_{\alpha\beta}\}$$
$$+\frac{1}{2} \sum_{\alpha,\beta,\rho,\sigma} g_{\mu\nu} g^{\alpha\beta} g^{\rho\sigma}\{-\partial_\beta \partial_\rho g_{\sigma\alpha} + \partial_\alpha \partial_\beta g_{\rho\sigma}\} + \tilde{F}_{\mu\nu}(g, \partial g) \qquad (10.2.26)$$

where F and $\tilde{F}$ are nonlinear functions of the metric components $g_{\alpha\beta}$ and their first derivatives. However, the right-hand side of equation (10.2.26) is not of the form (10.1.21). Indeed, from equation (10.2.26) one may show (problem 5) that the equations

$$\sum_\nu G_{\mu\nu} n^\nu = 0 \qquad (10.2.27)$$

(where n^a is the unit normal to the t = constant surfaces) contain no second time derivatives of any of the metric components; i.e., these components of $G_{ab} = 0$ at $t = 0$ depend only on the initial data. Thus, these equations provide initial value constraints, in close analogy with equation (10.2.2) in the electromagnetic case. We can express these equations in coordinate invariant form by using the Gauss-Codacci equations (10.2.23) and (10.2.24). From equation (10.2.24), we obtain the initial value constraint,

$$0 = h^b{}_a G_{bc} n^c = h^b{}_a R_{bc} n^c = D_b K^b{}_a - D_a K^b{}_b \quad . \qquad (10.2.28)$$

In addition, we have

$$R_{abcd} h^{ac} h^{bd} = R_{abcd}(g^{ac} + n^a n^c)(g^{bd} + n^b n^d)$$
$$= R + 2R_{ac} n^a n^c$$
$$= 2G_{ac} n^a n^c \quad . \qquad (10.2.29)$$

Thus, from equation (10.2.23) we obtain the additional constraint,

$$0 = G_{ab} n^a n^b$$
$$= \frac{1}{2} \{{}^{(3)}R + (K^a{}_a)^2 - K_{ab} K^{ab}\} \quad . \qquad (10.2.30)$$

Thus, equations (10.2.28) and (10.2.30) are the initial value constraint equations of general relativity expressed in a form analogous to equation (10.2.3). We shall return briefly to discuss some properties of these equations at the end of this section.

In the electromagnetic case, the identity (10.2.6) implied that if the constraint (10.2.3) is satisfied initially and the spatial components of Maxwell's equations are satisfied everywhere, then the constraint also is satisfied everywhere. A completely analogous result applies in general relativity. As a consequence of the Bianchi identity,

$$\nabla^a G_{ab} = 0 \quad , \qquad (10.2.31)$$

if the constraints (10.2.28) and (10.2.30) are satisfied initially and the spatial components of Einstein's equation are satisfied everywhere, then the constraints also are satisfied. To show this, we note that equation (10.2.31) relates the time derivative of the components $\Sigma\, G_{\mu\nu}n^{\nu}$, to non–time differentiated components of $G_{\mu\nu}$ and their spatial derivatives. Having solved the purely spatial components of Einstein's equation and obtained a solution, $g_{\mu\nu}$, we may set the spatial components of $G_{\mu\nu}$ equal to zero in equation (10.2.31) and view the metric components $g_{\mu\nu}$ as known functions, thus making equation (10.2.31) a linear, homogeneous system of four first order equations for the four unknown components $\Sigma\, G_{\mu\nu}n^{\nu}$. It then follows from the theory of first order partial differential equations (see, e.g., Courant and Hilbert 1962) that if these components vanish initially, they must vanish everywhere.

Thus, Einstein's equation $G_{ab} = 0$, is an underdetermined system of equations for the metric components $g_{\mu\nu}$. We have only six evolution equations (namely, the purely spatial components of $G_{ab} = 0$) for 10 unknown metric components. However, as in the Maxwell case this underdetermination is not physical. It results from the redundancy in the description of spacetime geometry by metric components, $g_{\mu\nu}$. As discussed in appendix C, if $\phi: M \to M$ is a diffeomorphism, then (M, g_{ab}) and $(M, \phi^* g_{ab})$ represent the same physical spacetime. Since the coordinate basis components of g_{ab} and $\phi^* g_{ab}$ are related by the coordinate transformation associated with ϕ, any two solutions of Einstein's equation whose coordinate basis metric components are related by the tensor transformation law (2.3.8) represent the same physical solution. Since four arbitrary functions appear in the transformation law, roughly speaking there should be only six "nongauge" functions in the 10 metric components $g_{\mu\nu}$. Thus, it is plausible that Einstein's equation contains the correct number of evolution equations, and that a well posed initial value formulation exists. We shall prove that this is the case by introducing—in close analogy with our treatment of Maxwell's equations—a convenient choice of "gauge" (i.e., coordinates) for which Einstein's equation has the form (10.1.21).

We shall employ *harmonic coordinates*, x^{μ}, i.e., coordinates which satisfy

$$H^{\mu} \equiv \nabla_a \nabla^a x^{\mu} = 0 \quad . \tag{10.2.32}$$

In a given spacetime (M, g_{ab}), we can construct harmonic coordinates in a neighborhood of that portion of Σ covered by our original set of coordinates $\{y^{\mu}\}$ as follows. For $\mu = 0, 1, 2, 3$ we give $\{y^{\mu}\}$ and its normal derivative on Σ as initial data for equation (10.2.32) (which is of the form [10.1.20] and thus possesses a well posed initial value formulation). Since the dual vectors $\{\nabla_a y^{\mu}\}$ are linearly independent on Σ, the solutions $\{x^{\mu}\}$ will have $\{\nabla_a x^{\mu}\}$ linearly independent in a neighborhood of Σ in M, so $\{x^{\mu}\}$ will yield a local coordinate system. Thus, there is no loss of generality in assuming local existence of harmonic coordinates.

Writing out the coordinate basis expression for equation (10.2.32) using the formulas of section 3.4*a*, we find that the harmonic coordinate condition takes the form

$$\begin{aligned} 0 = H^{\mu} &= \sum_{\alpha,\beta} \frac{1}{\sqrt{-g}}\, \partial_{\alpha}[\sqrt{-g}\, g^{\alpha\beta}\partial_{\beta}x^{\mu}] \\ &= \sum_{\alpha} \frac{1}{\sqrt{-g}}\, \partial_{\alpha}[\sqrt{-g}\, g^{\alpha\mu}] \\ &= \sum_{\alpha} [\partial_{\alpha}g^{\alpha\mu} + \frac{1}{2} g^{\alpha\mu} \sum_{\rho,\sigma} g^{\rho\sigma}\partial_{\alpha}g_{\rho\sigma}] \quad . \end{aligned} \tag{10.2.33}$$

Using equation (10.2.33), we see that most of the second derivative terms in equation (10.2.25) can be reexpressed in terms of H^μ and lower derivative terms, and thus in harmonic coordinates the vacuum Einstein equation becomes

$$0 = R^H_{\mu\nu} \equiv R_{\mu\nu} + \sum_\alpha g_{\alpha(\mu}\partial_{\nu)}H^\alpha$$
$$= -\frac{1}{2}\sum_{\alpha,\beta} g^{\alpha\beta}\partial_\alpha\partial_\beta g_{\mu\nu} + \hat{F}_{\mu\nu}(g, \partial g) \quad , \qquad (10.2.34)$$

where the superscript H on $R^H_{\mu\nu}$ denotes that this expression for the Ricci tensor is valid only for harmonic coordinates. Thus, Einstein's equation is equivalent to the system (10.2.34), together with the harmonic coordinate condition (10.2.32) or (10.2.33) (Choquet-Bruhat 1962). Equation (10.2.34) is known as the "reduced Einstein equation." The key point is that it *is* of the form (10.1.21) for which theorem 10.1.3 applies.

We now are in a position to prove local existence of a solution to Einstein's equation for initial data sufficiently near that of flat spacetime. Let the Riemannian metric h_{ab} and the symmetric tensor field K_{ab} be given on Σ satisfying the constraint equations (10.2.28) and (10.2.30). Choose a coordinate system on (a portion of) Σ and let $h_{\mu\nu}$ and $K_{\mu\nu}$ denote the coordinate basis components of h_{ab} and K_{ab}. We prescribe on Σ initial data $(g_{\mu\nu}, \partial g_{\mu\nu}/\partial t)$ such that $g_{\mu\nu} = h_{\mu\nu}$ for $\mu, \nu = 1, 2, 3$, and such that the extrinsic curvature computed from these initial data using equation (10.2.13) is K_{ab}. A particularly simple choice is to take $g_{00} = -1$, $g_{0\mu} = 0$ for $\mu = 1, 2, 3$, and $\partial g_{\mu\nu}/\partial t = K_{\mu\nu}$ for $\mu, \nu = 1, 2, 3$. Since $\partial g_{0\mu}/\partial t$ for $\mu = 0, 1, 2, 3$ is undetermined by these requirements, we may specify $\partial g_{0\mu}/\partial t$ such that $H^\mu = 0$ on Σ (see eq. [10.2.33]). If this initial data set is sufficiently near that of flat spacetime, then according to theorem 10.1.3 we can solve equation (10.2.34) in a neighborhood of the portion of Σ covered by our original coordinates, thereby producing a globally hyperbolic spacetime with this portion of Σ serving as a Cauchy surface. This solution of equation (10.2.34) will be a solution of Einstein's equation if $H^\mu = 0$ in this neighborhood. To prove this is the case, we note that the Einstein tensor $G_{\mu\nu}$ in arbitrary coordinates can be expressed in terms of $R^H_{\mu\nu}$ (defined by eq. [10.2.34]) and H^μ (given by eq. [10.2.33]) via

$$G_{\mu\nu} = R^H_{\mu\nu} - \frac{1}{2}R^H g_{\mu\nu} - \sum_\alpha \{g_{\alpha(\mu}\partial_{\nu)}H^\alpha - \frac{1}{2}g_{\mu\nu}\partial_\alpha H^\alpha\} \quad . \qquad (10.2.35)$$

Since $\sum_\nu G_{\mu\nu}n^\nu = 0$ on Σ and $H^\mu = 0$ on Σ, equation (10.2.35) implies that $\partial H^\mu/\partial t = 0$ on Σ if equation (10.2.34) is satisfied. Furthermore, when equation (10.2.34) is satisfied, the Bianchi identity yields

$$0 = \sum_\mu \nabla^\mu G_{\mu\nu} = -\sum_{\rho,\mu,\alpha} g^{\rho\mu}\nabla_\rho[g_{\alpha(\mu}\partial_{\nu)}H^\alpha - \frac{1}{2}g_{\mu\nu}\partial_\alpha H^\alpha]$$
$$= -\sum_{\rho,\mu,\alpha}\frac{1}{2}g_{\alpha\nu}g^{\rho\mu}\partial_\rho\partial_\mu H^\alpha + \{\text{lower order terms linear in } H^\alpha\} \quad . \qquad (10.2.36)$$

Thus, equation (10.2.36) (after multiplication by $g^{\lambda\nu}$ and summation over ν) takes the form (10.1.20) for which a unique solution exists according to theorem 10.1.2. Since $H^\mu = \partial H^\mu/\partial t = 0$ initially, this proves that $H^\mu = 0$ throughout the region where a solution to equation (10.2.34) exists. Thus, we have established local existence of a solution of Einstein's equation for initial data sufficiently near that of flat spacetime. Furthermore, theorem 10.1.3 shows that the solution depends continuously on the initial data and has the desired domain of dependence property.

The requirement that the initial data be "sufficiently near" that of flat spacetime can be removed by the following trick which uses the idea that any curved geometry appears "nearly flat" when examined on a sufficiently small scale. Suppose initial data $(g_{\mu\nu}, \partial g_{\mu\nu}/\partial t)$ which are not "sufficiently small" are given. By a coordinate transformation on Σ, we may assume that $g_{\mu\nu} = \text{diag}(-1, 1, 1, 1)$ at a point $p \in \Sigma$ and that p lies at the origin of coordinates, $x^\mu = 0$. Let $\lambda \in \mathbb{R}$. Suppose we scale the initial data by $(g_{\mu\nu}, \partial g_{\mu\nu}/\partial t) \to (\lambda^{-2} g_{\mu\nu}, \lambda^{-2} \partial g_{\mu\nu}/\partial t)$, and then make the coordinate transformation $x^\mu \to x'^\mu = \lambda^{-1} x^\mu$ for $\mu = 0, 1, 2, 3$. From the tensor transformation law (2.3.8), we see that under this combined transformation, the initial data become

$$g'_{\mu\nu}(x') = g_{\mu\nu}(\lambda x'), \qquad \frac{\partial g'_{\mu\nu}}{\partial t'}(x') = \lambda \frac{\partial g_{\mu\nu}}{\partial t}(\lambda x'),$$

where x' denotes the new spatial coordinates on Σ. As $\lambda \to 0$, we see that the new initial data and their derivatives become arbitrarily close to those of data for flat spacetime. Thus, there exists a (sufficiently small) λ_0 such that we can obtain a solution, $g^0_{\mu\nu}(x')$ to Einstein's equation in a neighborhood of p for the new initial data with $\lambda = \lambda_0$. The metric $g^0_{\mu\nu}(\lambda_0^{-1} x)$ then solves Einstein's equation in a neighborhood of p with the original initial data.

Local uniqueness of solutions with given initial data (h_{ab}, K_{ab}) can be proven as follows. Let (O^H, g^H_{ab}) be the spacetime solution of Einstein's equation constructed using harmonic coordinates from the initial data $(g^H_{\mu\nu}, \partial g^H_{\mu\nu}/\partial t)$ in the manner described above. Let (O, g_{ab}) be another solution of Einstein's equation (not necessarily in harmonic coordinates) which covers the same portion of Σ as (O^H, g^H_{ab}) and induces the same initial data (h_{ab}, K_{ab}) on Σ. We wish to find a diffeomorphism ψ from a neighborhood of Σ in O to a neighborhood of Σ in O^H such that ψ takes g_{ab} into g^H_{ab}, i.e., $\psi^* g_{ab} = g^H_{ab}$. We construct ψ as follows. Since g_{ab} and g^H_{ab} induce the same initial data on Σ, it can be verified that there exists a diffeomorphism ϕ such that the coordinate components of $\phi^* g_{ab}$ and its time derivative on Σ agree with the initial data $(g^H_{\mu\nu}, \partial g^H_{\mu\nu}/\partial t)$ of the solution g^H_{ab}. (In fact, there are many such diffeomorphisms, since ϕ can be arbitrary away from Σ.) We then may use ϕ^{-1} to bring the coordinates of O^H into O. We use the initial value and time derivative of these coordinates on Σ as initial data for a solution of equation (10.2.32) in O, thereby constructing harmonic coordinates in O. Using these harmonic coordinate labels, we define ψ to be the map which takes a point in O to the point of O^H with the same values of the harmonic coordinates. (In general, of course, ψ will be defined only on a neighborhood of Σ in O and will map only onto a neighborhood of Σ in O^H.) Then, in this neighborhood in O^H the harmonic coordinate components of $\psi^* g_{ab}$ will satisfy equation (10.2.34) and will have the same initial data on Σ as g^H_{ab}. Hence, by theorem

10.1.3 we must have $\psi^* g_{ab} = g^H_{ab}$ in a neighborhood of Σ, which is the desired local uniqueness result.

We outline, now, how to "globalize" these local existence and uniqueness results. We proved above that for $p \in \Sigma$ there exists a solution (O, g_{ab}) of Einstein's equation into which a neighborhood of Σ containing p can be embedded so that the given initial data set is induced on this portion of Σ. To show that there is a solution of Einstein's equation containing all of Σ, we cover Σ by neighborhoods for which local solutions exist. Using the paracompactness of Σ (see appendix A), we can ensure that these neighborhoods have only finite overlap at each point. Using the local uniqueness result proven above, it follows that for each $p \in \Sigma$ we can find a globally hyperbolic spacetime $(\tilde{O}, \tilde{g}_{ab})$ containing p (and with a portion of Σ serving as a Cauchy surface) which can be isometrically mapped into every local solution in the above family which contains p. By using the embeddings of the spacetimes $(\tilde{O}, \tilde{g}_{ab})$ into the local solutions to make identifications, we can consistently "patch together" the $(\tilde{O}, \tilde{g}_{ab})$ to construct a spacetime (M, g_{ab}) which solves Einstein's equation, which contains all of Σ, and which induces on Σ the given initial data. Furthermore, this spacetime will be globally hyperbolic with Cauchy surface Σ.

The spacetime (M, g_{ab}) constructed above clearly is not unique since any proper open subset of M containing Σ is another solution of Einstein's equation inducing the same initial data on Σ. However, we can consider the set $\mathcal{S}$ of all globally hyperbolic spacetimes modulo diffeomorphisms which are solutions of Einstein's equation and into which Σ with the given initial data can be embedded as a Cauchy surface. For two spacetimes (M^1, g^1_{ab}), (M^2, g^2_{ab}) in $\mathcal{S}$, we say $(M^1, g^1_{ab}) \geq (M^2, g^2_{ab})$ if (M^2, g^2_{ab}) can be isometrically mapped into (M^1, g^1_{ab}) keeping the Cauchy surface fixed. This relation yields a "partial order" on $\mathcal{S}$. [Here, a *partial order* on an arbitrary set S is a relation between elements satisfying, for all $a, b, c \in S$: (i) $a \geq a$, (ii) $a \geq b$ and $b \geq c$ implies $a \geq c$, and (iii) $a \geq b$ and $b \geq a$ implies $a = b$. The word "partial" refers to the fact that " $\geq$ " need not be defined for all pairs of elements.] Now, for any partially ordered set S, a subset $T \subset S$ for which the relation $\geq$ is defined between all pairs of elements is said to be *totally ordered*. An *upper bound* for T is an element $b \in S$ such that $b \geq a$ for all $a \in T$. Zorn's lemma (which is equivalent to the axiom of choice) asserts that if every totally ordered subset of S has an upper bound, then S has a *maximal element,* i.e., an element $m \in S$ such that, for all $c \in S$, the relation $c \geq m$ implies $c = m$. In our case, given any totally ordered subset $\mathcal{T}$ of $\mathcal{S}$, we obtain an upper bound by taking the union of all spacetimes occurring in $\mathcal{T}$ and then identifying points via the isometric embedding maps. Hence, by Zorn's lemma there exists a maximal element of $\mathcal{S}$, i.e., a spacetime $(\tilde{M}\tilde{g}_{ab})$ which cannot be isometrically mapped into any other spacetime in $\mathcal{S}$. In general, Zorn's lemma implies existence of a maximal element, but not its uniqueness. However, in our case, if we had a spacetime in $\mathcal{S}$ which could not be isometrically mapped into $(\tilde{M}, \tilde{g}_{ab})$, then it can be shown (Choquet-Bruhat and Geroch 1969) that we could "patch" the two spacetimes together to produce a "larger" solution, in violation of the maximality of $(\tilde{M}, \tilde{g}_{ab})$. This implies that $(\tilde{M}, \tilde{g}_{ab})$ is the unique spacetime having the property that every globally hyperbolic spacetime with the given initial data on the Cauchy surface Σ can be isometrically mapped into $(\tilde{M}, \tilde{g}_{ab})$.

Thus, putting together all the results proven or outlined above, we arrive at the following theorem.

THEOREM 10.2.2. *Let Σ be a three-dimensional C^∞ manifold, let h_{ab} be a smooth Riemannian metric on Σ, and let K_{ab} be a smooth symmetric tensor field on Σ. Suppose h_{ab} and K_{ab} satisfy the constraint equations (10.2.28) and (10.2.30). Then there exists a unique C^∞ spacetime, (M, g_{ab}), called the maximal Cauchy development of (Σ, h_{ab}, K_{ab}), satisfying the following four properties:* (i) *(M, g_{ab}) is a solution of Einstein's equation.* (ii) *(M, g_{ab}) is globally hyperbolic with Cauchy surface Σ.* (iii) *The induced metric and extrinsic curvature of Σ are, respectively, h_{ab} and K_{ab}.* (iv) *Every other spacetime satisfying* (i)–(iii) *can be mapped isometrically into a subset of (M, g_{ab}). Furthermore, (M, g_{ab}) satisfies the desired domain of dependence property in the following sense. Suppose (Σ, h_{ab}, K_{ab}) and $(\Sigma', h'_{ab}, K'_{ab})$ are initial data sets with maximal developments (M, g_{ab}) and (M', g'_{ab}). Suppose there is a diffeomorphism between $S \subset \Sigma$ and $S' \subset \Sigma'$ which carries (h_{ab}, K_{ab}) on S into (h'_{ab}, K'_{ab}) on S'. Then $D(S)$ in the spacetime (M, g_{ab}) is isometric to $D(S')$ in the spacetime (M', g'_{ab}). Finally, the solution g_{ab} on M depends continuously on the initial data (h_{ab}, K_{ab}) on Σ.* (A precise definition of the topologies on initial data and solutions which makes this map continuous is given in Hawking and Ellis 1973.)

Note that it may be possible to extend the "maximal development" (M, g_{ab}), i.e., isometrically map it into a proper subset of another spacetime. Theorem 10.2.2 asserts only that any such extension cannot have Σ as a Cauchy surface. It also should be noted that theorem 10.2.2 gives no information as to the "size" of (M, g_{ab}), other than the fact that it is maximal in the sense of property (iv). Indeed, the singularity theorems of chapter 9 show that in many cases (M, g_{ab}) cannot be geodesically complete. In particular, theorem 9.5.1 gives a stringent limit on "how large" (M, g_{ab}) can be for initial data with $K^a{}_a \geq C > 0$. On the other hand, recently it has been shown (Christodoulou and O'Murchadha 1981) that for asymptotically flat initial data (see chapter 11) the maximal development is "large enough" to include all "boosted" hypersurfaces in the asymptotic region.

Aside from showing that general relativity has the physically desirable property of possessing a well posed initial value formulation, theorem 10.2.2 also is very useful in that it puts globally hyperbolic spacetimes (M, g_{ab}) satisfying Einstein's equation into correspondence with initial data sets (Σ, h_{ab}, K_{ab}) satisfying the constraint equations. [The association of spacetimes with initial data sets, of course, is not one-to-one; many distinct initial data sets give rise to the same spacetime (M, g_{ab}), corresponding to the freedom of choosing a spacelike Cauchy surface in M.] It usually is far easier to solve the constraint equations on Σ than to solve Einstein's equation on M. Thus, for example, in arguments involving existence of certain types of solutions of Einstein's equation, great simplifications usually can be achieved if the question can be posed in terms of initial data sets. Furthermore, a number of issues in general relativity, such as the positivity of total energy of isolated systems, are formulated most naturally in terms of initial data sets (see section 11.2).

A relatively simple method exists for generating solutions of the constraint equations with $K^a{}_a = 0$ (Lichnerowicz 1944; York 1971). Given Σ, one prescribes an arbitrary Riemannian metric, h_{ab}, on Σ and solves the relatively simple constraint (10.2.28),

$$D^a K_{ab} = 0 \tag{10.2.37}$$

for a trace-free ($K^a{}_a = 0$) tensor field K_{ab}. Of course, (h_{ab}, K_{ab}) will not, in general, satisfy the additional constraint (10.2.30). However, we let $\tilde{h}_{ab} = \phi^4 h_{ab}$ and let $\tilde{D}_a$ be the derivative operator associated with $\tilde{h}_{ab}$. As shown in appendix D, if we define $\tilde{K}_{ab} = \phi^{-2} K_{ab}$, then equation (10.2.37) implies that

$$\tilde{D}^a \tilde{K}_{ab} = 0 \quad . \tag{10.2.38}$$

Furthermore, using equation (D.9) of appendix D, the constraint equation (10.2.30) for $(\tilde{h}_{ab}, \tilde{K}_{ab})$ can be expressed in terms of ϕ, h_{ab}, and K_{ab} as

$$D^a D_a \phi - \frac{1}{8} R\phi + \frac{1}{8} \phi^{-7} K^{ab} K_{ab} = 0 \quad , \tag{10.2.39}$$

where the indices here are raised by h^{ab} and R is the scalar curvature of h_{ab}. Equation (10.2.39) is a nonlinear elliptic equation for ϕ. Locally (i.e., in sufficiently small regions), solutions of equation (10.2.39) always exist, although global solutions (i.e., solutions defined over all of Σ) may not exist. (Results on the global existence of solutions of eq. [10.2.39], as well as of the similar equation which results when $K^a{}_a$ is a nonzero constant, are reviewed by Choquet-Bruhat and York 1980.) Thus, in the cases where equation (10.2.39) can be solved, an initial data set $(\Sigma, \tilde{h}_{ab}, \tilde{K}_{ab})$ satisfying the constraint equations (as well as $\tilde{K}^a{}_a = 0$) is generated from the set (Σ, h_{ab}, K_{ab}) satisfying merely equation (10.2.37) together with $K^a{}_a = 0$.

A particularly simple choice of K_{ab} which satisfies equation (10.2.37) and $K^a{}_a = 0$ is $K_{ab} = 0$. [If $K_{ab} = 0$ on Σ, then Σ is referred to as a *moment of time symmetry*. It is not difficult to see that the maximal development (M, g_{ab}) generated by initial data with $K_{ab} = 0$ will possess a reflection isometry about Σ.] Equation (10.2.39) then becomes a linear equation on ϕ. If, in addition, we choose h_{ab} to be flat, then equation (10.2.39) reduces to Laplace's equation in ordinary three-dimensional space. The monopole solution $\phi = 1 + M/2r$ yields initial data for the Schwarzschild solution. The solution of Laplace's equation obtained by superimposing two monopoles at different positions can be interpreted as initial data for two Schwarzschild black holes (Hahn and Lindquist 1964). Thus, the maximal development (M, g_{ab}) arising from these initial data is a spacetime where two black holes are initially at rest, and then, presumably, fall together and "collide." A portion of this spacetime has been obtained by numerical solution of Einstein's equation using computers by Smarr (1979).

An interesting issue that can be investigated (at least in a crude way) for a theory possessing an initial value formulation is how many "degrees of freedom" the theory has, i.e., "how many" distinct solutions of the equations exist. In particle mechanics, we defined above the number of degrees of freedom to be the dimension, n, of the configuration space. As discussed at the beginning of section 10.1, a proper initial data set for such a system consists of the $2n$ initial positions and velocities. Thus, an equivalent characterization of the number of degrees of freedom in ordinary particle mechanics is that it is the number of quantities that must be specified as initial data divided by 2. For the Klein-Gordon field, a proper initial data set consists of the value of the field and the value of its normal derivative on a Cauchy surface Σ, i.e., two arbitrary functions on Σ. Thus, by analogy with particle mechanics, we may say that the Klein-Gordon field has "one degree of freedom for each point of space."

How many degrees of freedom does the gravitational field have in general relativity? A proper initial data set for Einstein's equation consists of specifying 12 functions on Σ: the six independent components of h_{ab}, plus the six independent components of K_{ab}. However, the constraint equations (10.2.28) and (10.2.30) impose four relations on these 12 quantities, thus effectively reducing the number of "freely specifiable" functions on Σ to eight. Furthermore, many of the spacetimes generated by the initial data given by these eight "freely specifiable" functions are physically equivalent. In particular, if $\phi : \Sigma \to \Sigma$ is a diffeomorphism, then the data $\phi^* h_{ab}$ and $\phi^* K_{ab}$ on Σ generates the same physical spacetime as h_{ab} and K_{ab}. Thus, three of the eight freely specifiable functions correspond to diffeomorphisms on Σ and are not physically relevant. Furthermore, as mentioned above, initial data sets which cannot be taken into each other by a diffeomorphism still can correspond to different choices of Cauchy surface in the same spacetime and thus be physically equivalent. Since, roughly speaking, one arbitrary function is needed to specify a choice of Cauchy surface in a spacetime, the number of nongauge freely specifiable functions on Σ is reduced to four. Dividing by 2, we conclude that the gravitational field has two degrees of freedom per point of space. This is the same number of degrees of freedom as a linear spin-2 field propagating in flat spacetime, to which general relativity reduces in the weak field limit (see section 4.4*b*). Note that the above very crude "function counting" argument in no way singles out precisely which functions (i.e., which of the 12 metric or extrinsic curvature components or functions of them) can be freely specified, which functions are determined by the constraints, and which functions correspond to gauge transformations. Indeed, one of the major obstacles to developing a quantum theory of gravity (see chapter 14) is the inability to single out the physical degrees of freedom of the theory.

Finally, we comment briefly on the initial value formulation of Einstein's equation with matter sources, T_{ab}. First, the initial value constraints for the gravitational field now take the form

$$G_{ab} n^b = 8\pi T_{ab} n^b \quad . \tag{10.2.40}$$

From equations (10.2.28) and (10.2.30) we obtain

$$D^a (K_{ab} - K^c{}_c h_{ab}) = -8\pi J_b \quad , \tag{10.2.41}$$

$${}^{(3)}R + (K^a{}_a)^2 - K_{ab} K^{ab} = 16\pi\rho \quad , \tag{10.2.42}$$

where $\rho = T_{ab} n^a n^b$ and $J_b = -h_b{}^c T_{ca} n^a$.

The existence of a well posed initial value formulation for Einstein's equation with matter depends critically on the dynamical equations satisfied by the matter as well as on the formula for the stress-energy tensor in terms of the matter and spacetime metric. If the matter consists of fields $\phi_1, \ldots, \phi_n$ satisfying an equation of the form (10.1.21) (with g_{ab} the spacetime metric) and if T_{ab} depends only on the fields, the metric, and the first derivatives of the fields and metric, then the combined Einstein-matter field system in harmonic coordinates will be of the form (10.1.21), so a well posed initial value formulation will exist. Thus, the Einstein-Klein-Gordon equations and the Einstein-Maxwell equations have well posed initial value formulations. The existence of a well posed initial value formulation for a few systems not satisfying equations of the form (10.1.21) also has been established. In particular, the Einstein–perfect fluid system for appropriate choices of equation of state, $P = P(\rho)$, is known to possess a well posed initial value formulation (see Hawking and Ellis 1973). The existence of a well posed initial value formulation does not single out Einstein's equation from equations which occur in some alternative theories of gravity. In particular, the Brans-Dicke equations are equivalent to the Einstein-Klein-Gordon equations (see Dicke 1962) and hence possess a well posed initial value formulation. Some "higher derivative" theories of gravity also have been shown to have a well posed initial value formulation (Noakes 1983). However, it should be emphasized that the existence of a well posed initial value formulation is far from an automatic feature of most theories. In particular, the natural generalization to curved spacetime of the equations for linear fields of spin greater than 1 (see chapter 13) fail to have a well posed initial value formulation.

Problems

1. Show that the inequality (10.1.16) holds for any subset, A, of $\mathbb{R}^n$ satisfying the uniform interior cone condition by the following argument (Cantor 1973; Adams 1975): Let Q denote the solid closed cone in $\mathbb{R}^n$ of height H and solid angle, Ω, with vertex at the origin. Let $\psi : \mathbb{R} \to \mathbb{R}$ be a C^∞ function with $\psi(r) = 1$ for $r < H/3$ and $\psi(r) = 0$ for $r > 2H/3$.

a) for any C^∞ function $f: Q \to \mathbb{R}$, show that for all integers $k \geq 1$, we have,

$$f(0) = \frac{(-1)^k}{(k-1)!} \int_0^{R(\theta_0)} r^{k-1} \frac{d^k}{dr^k} [\psi(r) f(r, \theta_0)] \, dr \quad ,$$

where r is the usual spherical radial coordinate of $\mathbb{R}^n$, θ_0 denotes any fixed angle inside the cone, and $R(\theta_0)$ is the largest value of r in the cone at angle θ_0, i.e., the integral is taken over the portion of any ray through the origin lying within the cone.

b) By integrating the result of (*a*) over all angles inside the cone, show that

$$f(0) = C_1 \int_Q r^{k-n} \frac{d^k}{dr^k} (\psi f) \quad ,$$

where C_1 is a constant and the proper volume element of Q is understood in the integral.

c) Using the Schwartz inequality, show that for $k > n/2$ we have $|f(0)| \leq C \|f\|_{Q,k}$ and consequently, that equation (10.1.16) holds in the region A.

2. Let (M, g_{ab}) be a globally hyperbolic spacetime with spacelike Cauchy surface Σ.

Consider Maxwell's equations (4.3.12) and (4.3.13) in (M, g_{ab}) and define E_a and B_a on Σ by equations (4.2.21) and (4.2.22), with v^a taken to be the unit normal, n^a, to Σ. Note that $E_a n^a = B_a n^a = 0$.

a) Show that Maxwell's equations imply that $D_a E^a = 4\pi\rho$ and $D_a B^a = 0$ on Σ, where $\rho = -j_a n^a$ and D_a is the derivative operator on Σ. Note that the first relation implies that Gauss's law holds on Σ.

b) Show that the source-free ($j^a = 0$) Maxwell's equations have a well posed initial value formulation in the sense that given E^a and B^a on Σ subject to the constraints $D_a E^a = D_a B^a = 0$, there exists a unique solution, F_{ab}, of Maxwell's equations throughout M with these initial data and furthermore that this solution has the appropriate continuity and domain of dependence properties. To avoid "patching" arguments, you may assume global existence of a vector potential A_a.

3. Let n^a be a unit (i.e., $n^a n_a = -1$) hypersurface orthogonal vector field. Define h_{ab} by equation (10.2.10). Show that $h_a{}^c \nabla_c n_b = \frac{1}{2} \pounds_n h_{ab}$.

4. Derive the Gauss-Codacci relation (10.2.24). (Hint: Evaluate the left-hand side of eq. [10.2.24] by using formulas for K_{ab} and D_a.)

5. Show explicitly from equation (10.2.26) that the components $G_{\mu\nu} n^\nu$ contain no second time derivatives of the metric components.

6. Use "function counting" arguments like those given at the end of this chapter to conclude that the electromagnetic field has "two degrees of freedom for each point of space."

ELEVEN

ASYMPTOTIC FLATNESS

In general relativity one often is interested in studying the properties of isolated systems. Although no physical system truly can be isolated from the rest of the universe, it seems reasonable that if we wish to study, say, the structure of a condensed star, we should be able to ignore the influence of distant matter and cosmological curvature on the star and study the problem as though the star were situated in a spacetime which becomes flat (i.e., has a vanishing gravitational field) at large distances from it. Thus, asymptotically flat spacetimes represent ideally isolated systems in general relativity. The purpose of this chapter is to give an introduction to the analysis of these spacetimes.

In electromagnetism in special relativity one, similarly, is interested in the study of isolated charge distributions. In this case, one easily can give a precise definition of "isolated system" by specifying precise asymptotic falloff rates of the inertial coordinate components of the charge-current density, j^a, and electromagnetic field tensor, F_{ab}. For example, one may require that j^a vanishes outside a "world tube" of compact spatial support, that $F_{\mu\nu} = O(1/r^2)$ as $r \rightarrow \infty$ at fixed t, and that $F_{\mu\nu} = O(1/r)$ as $r \rightarrow \infty$ along any null geodesic. Maxwell's equations then say a great deal about the detailed structure and properties of the electromagnetic field at large distances. In particular, one has a multipole expansion of the electromagnetic field, which, in the stationary case, determines the precise asymptotic form of the electromagnetic field in terms of an infinite set of multipole coefficients which are related in a simple way to the charge-current distribution. In the dynamic case, one also has a multipole expansion which yields simple formulas for the energy radiated to infinity in terms of the multipole coefficients. Again, if no incoming radiation is present, one has a simple relation between the multipole coefficients and the charge-current distribution.

One would like to obtain similar results for isolated systems in general relativity. However, one immediately encounters a serious obstacle to carrying out even the first step of such an analysis. It no longer is straightforward to formulate a precise definition of "isolated system." The problem is that we no longer have a background flat metric, η_{ab}, in terms of which the falloff rates of the curvature of the spacetime metric, g_{ab}, can be specified. Thus, in particular, we have no natural global inertial coordinate system to define a preferred radial coordinate, r, for use in specifying falloff rates. One way around this problem is to define a spacetime to be asymptotically flat if there exists *any* system of coordinates x^0, x^1, x^2, x^3 such that the metric components in these coordinates behave in an appropriate way at large coordinate values, e.g., $g_{\mu\nu} = \eta_{\mu\nu} + O(1/r)$ as $r \rightarrow \infty$, along either spatial or null directions, where $r = [(x^1)^2 + (x^2)^2 + (x^3)^2]^{1/2}$. However, although this definition is adequate in many respects, it is very difficult to work with it since the coordinate invariance of all statements must be carefully checked. Furthermore, in many situations (such as calculations of the energy flux from the system) one is interested in going to the limit of large distances, "$r \rightarrow \infty$," but with the above notion of asymptotic flatness it is very difficult to specify precisely how such limits are to be taken in a meaningful, coordinate independent manner. In particular, many troublesome issues arise concerning the interchange of limits and derivatives.

The above difficulties have been solved by a formulation of the notion of asymptotic flatness which defines a spacetime to be asymptoticaly flat if an appropriate boundary representing "points at infinity" can be "added in" to the spacetime in a suitable way. This type of definition is manifestly coordinate independent, and, by providing definite boundary points representing infinity, it eliminates most of the difficulties associated with taking limits as one goes to infinity. We shall formulate this notion of asymptotic flatness in section 11.1.

Given the framework for analyzing isolated systems provided by this definition of asymptotic flatness, one would like to obtain results similar to those given by the multipole expansions of electromagnetism. In the stationary case, a satisfactory definition of multipole moments has been given (Hansen 1974) and it is known that the gravitational field outside a source is uniquely determined by these multipole moments (Beig and Simon 1980; Kundu 1981). However, in the nonstationary case, no useful general definition of multipole moments has been given. Furthermore, in either case one would not expect to obtain a simple relation between the distant gravitational field and the matter distribution since the nonlinearity of Einstein's equation effectively allows gravitation to act as its own source.

However, fully satisfactory definitions of the total energy of an isolated system and the energy carried away from the system by gravitational radiation have been given. As already mentioned in chapter 4, no notion of the local energy density of the gravitational field exists in general relativity. However, for isolated systems, the behavior of the gravitational field at large distances from the system provides a notion of "total gravitational mass," and this can be used to define total energy and radiated energy. This issue is discussed in section 11.2.

The first careful analysis of the energy flux of gravitational radiation was carried out by Bondi, van der Burg, and Metzner (1962) and Sachs (1962*b*), who specified the asymptotic falloff requirements by means of conditions on the coordinate components of the metric. Penrose (1963, 1965*b*) then introduced the notion of asymptotic flatness at "null infinity" (i.e., as one goes to large distances along null geodesics) by means of the type of boundary construction described below. A separate, coordinate independent, definition of asymptotic flatness at "spatial infinity" in terms of the "large distance" behavior of initial data on a Cauchy surface was introduced later (Geroch 1972*b*), based on earlier work of Arnowitt, Deser, and Misner (1962). These two notions of asymptotic flatness were combined into a single notion by

Ashtekar and Hansen (1978) and Ashtekar (1980). We shall follow the Ashtekar-Hansen approach in this chapter.

11.1 Conformal Infinity

As already indicated by the above discussion, we must overcome two problems in order to have a useful formalism for analyzing gravitational radiation and other aspects of the distant gravitational field of an isolated system. (i) We need a precise definition of the notion of asymptotic flatness. (ii) We need a meaningful notion of how to take "limits as one goes to infinity" and a precise framework for describing the mathematical entities these limits represent. We shall proceed by proposing a solution to problem (ii) for nongravitational fields in Minkowski spacetime. This solution then will be used to motivate solutions to problems (i) and (ii) for curved spacetimes.

In spherical coordinates, the metric of Minkowski spacetime takes the form

$$ds^2 = -dt^2 + dr^2 + r^2(d\theta^2 + \sin^2\theta\, d\phi^2) \quad . \tag{11.1.1}$$

Suppose we are interested in describing properties of radiation carried to infinity by a massless field such as a Klein-Gordon scalar field ϕ. Since this entails taking limits as one goes to infinity along null directions, it is convenient to introduce advanced and retarded null coordinates defined by

$$v = t + r \quad , \tag{11.1.2}$$

$$u = t - r \quad , \tag{11.1.3}$$

In the coordinates u, v, θ, ϕ the Minkowski metric components are

$$ds^2 = -du\, dv + \frac{1}{4}(v - u)^2(d\theta^2 + \sin^2\theta\, d\phi^2) \quad . \tag{11.1.4}$$

Suppose we are concerned with analyzing, say, outgoing radiation. With u fixed, we wish to take limits as $v \to \infty$ of our physical field ϕ and extract information about the radiation from the way this field approaches zero. In particular, the energy carried to infinity by the field is determined by the "$1/v$ piece" of ϕ in this limit. However, the taking of these limits is a rather awkward procedure which does not generalize easily to curved spacetime. It would make the analysis much easier if infinity were a "definite place" and one simply had to evaluate the fields and/or their derivatives at this "place."

A naive approach toward achieving this goal would be to introduce a new coordinate $V = 1/v$, so that "infinity" along outgoing null geodesics would correspond to the finite value, $V = 0$, of the new spacetime coordinate. However, the spacetime metric components in the new coordinates u, V, θ, ϕ are

$$ds^2 = \frac{1}{V^2} du\, dV + \frac{1}{4}\left(\frac{1}{V} - u\right)^2 (d\theta^2 + \sin^2\theta\, d\phi^2) \quad . \tag{11.1.5}$$

These components are singular at $V = 0$, so we cannot extend the spacetime metric there. Thus, we cannot do tensor analysis as $V = 0$ as though it were an ordinary "place." Of course, all we have done is introduce a bad coordinate whose behavior is much like the first example considered above in section 6.4.

However, suppose we consider a new, unphysical metric $\bar{g}_{ab}$ obtained by multiplying the Minkowski metric, η_{ab}, by $V^2 = 1/v^2$, i.e., $\bar{g}_{ab}$ is related to η_{ab} by a conformal transformation with conformal factor $\Omega = V$ (see appendix D). Then, in the coordinates u, V, θ, ϕ the components of $\bar{g}_{ab}$ are

$$d\bar{s}^2 = du\,dV + \frac{1}{4}(1 - uV)^2(d\theta^2 + \sin^2\theta\, d\phi^2) \quad , \tag{11.1.6}$$

and these components are well behaved at $V = 0$. Thus, let us extend the Minkowski manifold by "adding in" the points represented by $V = 0$. As seen above, the original flat metric, η_{ab}, cannot be smoothly extended to $V = 0$—Minkowski spacetime $(\mathbb{R}^4, \eta_{ab})$, of course, is inextendible as a spacetime—but the new, unphysical metric $\bar{g}_{ab}$ *can* be smoothly extended to $V = 0$. Hence, we may do ordinary tensor analysis at "infinity" (or, more precisely, at that portion of "infinity" represented by $v \to \infty$ at constant u) as a "place." In effect, we have brought in infinity to a finite distance by a conformal transformation. We now simply may evaluate fields and their covariant derivatives with respect to $\bar{g}_{ab}$ at infinity and thus avoid dealing with limits in the original physical spacetime.

In fact, our particular choice of $\bar{g}_{ab}$ is not ideal since our conformal factor $V = 1/v$ needlessly blows up at the events $v = 0$ of the original spacetime. Furthermore, although we have extended $\bar{g}_{ab}$ to "future null infinity" (i.e., the limit $v \to \infty$ at fixed u), we cannot similarly extend $\bar{g}_{ab}$ to "past null infinity" ($u \to -\infty$ at fixed v) or "spatial infinity" ($r \to \infty$ at fixed t). However, all of these drawbacks can be remedied by a more judicious choice of conformal factor. Let

$$\tilde{g}_{ab} = \Omega^2 \eta_{ab} \tag{11.1.7}$$

with

$$\Omega^2 = 4(1 + v^2)^{-1}(1 + u^2)^{-1} \quad . \tag{11.1.8}$$

Then $\tilde{g}_{ab}$ is a smooth metric on the original Minkowski manifold and $(\mathbb{R}^4, \tilde{g}_{ab})$ can be smoothly extended to a "larger" spacetime such that the boundary of the Minkowski region in this larger spacetime gives us a precise representation of "infinity." To see this, we define new coordinates T, R for Minkowski spacetime by

$$T = \tan^{-1}v + \tan^{-1}u \quad , \tag{11.1.9}$$

$$R = \tan^{-1}v - \tan^{-1}u \quad . \tag{11.1.10}$$

Then T and R have ranges restricted by the inequalities

$$-\pi < T + R < \pi \quad , \tag{11.1.11}$$

$$-\pi < T - R < \pi \quad , \tag{11.1.12}$$

$$0 \leqq R \quad . \tag{11.1.13}$$

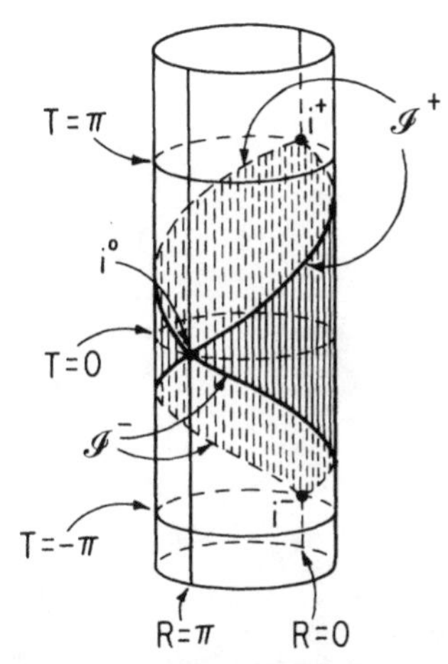

Fig. 11.1. A spacetime diagram of the Einstein static universe. As described in the text, Minkowski spacetime is conformally isometric to the region $O = I^+(i^-) \cap I^-(i^+)$ of this spacetime. The boundary of O—consisting of the points i^-, i^+, and i^0 and the null hypersurfaces $\mathscr{I}^-$ and $\mathscr{I}^+$—defines a precise notion of "infinity" for Minkowski spacetime.

The components of $\tilde{g}_{ab}$ in the coordinates T, R, θ, ϕ are given by

$$d\tilde{s}^2 = -dT^2 + dR^2 + \sin^2 R(d\theta^2 + \sin^2\theta \, d\phi^2) \quad . \tag{11.1.14}$$

Remarkably, this is precisely the natural Lorentz metric on $S^3 \times \mathbb{R}$, known as the Einstein static universe (see eq. [5.1.11] and problem 3 of chapter 5) except that the coordinate ranges are restricted by equations (11.1.11) and (11.1.12). Thus, we have obtained the following result: *There exists a conformal isometry*[1] *of Minkowski spacetime* $(\mathbb{R}^4, \eta_{ab})$ *into the open region* O *of the Einstein static universe* $(S^3 \times \mathbb{R}, \tilde{g}_{ab})$ *given by the coordinate restrictions (11.1.11) and (11.1.12).*

This result allows us to give a precise definition of infinity for Minkowski spacetime: We define the *conformal infinity of Minkowski spacetime* to be the boundary, $\dot{O}$, of O in the Einstein static universe. As illustrated in Figure 11.1, this boundary can be naturally divided into five parts: (1) The "bottom vertex point" i^-, called *past timelike infinity,* given by coordinates $R = 0, T = -\pi$. (2) The three-dimensional null surface $\mathscr{I}^-$, called *past null infinity,* given by $T = -\pi + R$ for $0 < R < \pi$. (3) The point i^0 at $R = \pi, T = 0$, called *spatial infinity*. (4) The three-dimensional null surface $\mathscr{I}^+$, called *future null infinity,* given by $T = \pi - R$ for $0 < R < \pi$. (5) The "top vertex point" i^+ at $R = 0, T = \pi$, called *future timelike infinity*. Note that all timelike geodesics of Minkowski spacetime begin at i^- and end at i^+, all spacelike geodesics begin and end at i^0, while all null geodesics begin at $\mathscr{I}^-$ and end at $\mathscr{I}^+$. Since it is difficult to draw complicated spacetime diagrams on $\overline{O}$ in Figure 11.1 and

1. As defined in appendix C, a conformal isometry of (M, g_{ab}) into (M', g'_{ab}) is a diffeomorphism $\psi: M \to M'$ such that $(\psi^* g)_{ab} = \Omega^2 g'_{ab}$.

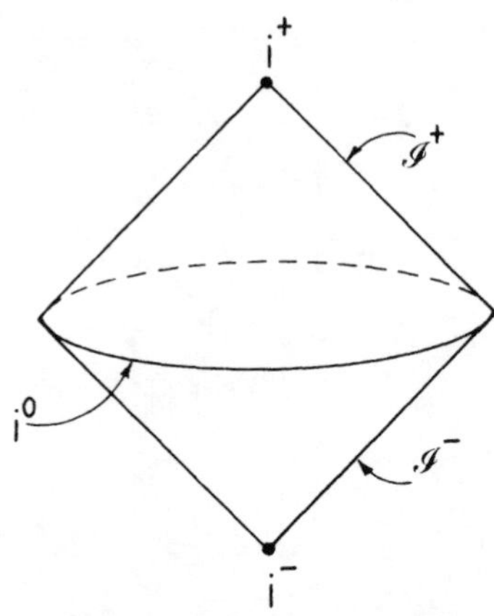

Fig. 11.2. The region $\bar{O}$ of the Einstein static universe represented as two null cones joined at their base. This representation of $\bar{O}$ is misleading since it shows i^0 as a 2-sphere rather than a point, but it often is much more convenient to use this representation for drawing spacetime diagrams than that of Figure 11.1.

since two spatial dimensions are suppressed in this diagram (i.e., S^3 is represented as S^1), one often represents $\bar{O}$ as two null cones joined at their base, as illustrated in Figure 11.2. However, it should be noted that this diagram inaccurately represents spatial infinity, i^0, as a sphere rather than as a point.

This definition of conformal infinity of Minkowski spacetime allows us to formulate precise asymptotic conditions on physical fields representing the exterior field of and/or the radiation resulting from an isolated source as follows: Depending on the physical field and the stringency of the asymptotic conditions we wish to impose, we require that a suitable power of Ω^{-1} times the field can be extended to conformal infinity, $\dot{O}$, in a sufficiently well behaved manner. With these conditions imposed, quantities which used to be represented in Minkowski spacetime as limits as $r \to \infty$ or $v \to \infty$ now can be represented as ordinary tensor fields on $\dot{O}$. This provides a highly satisfactory solution to problem (ii) for Minkowski spacetime.

We shall illustrate the power and utility of the conformal infinity construction for Minkowski spacetime by using it to give a remarkably simple proof of the following result.

PROPOSITION 11.1.1 Let ϕ be a solution of the massless Klein-Gordon equation (4.2.19) in Minkowski spacetime with smooth data ϕ_0, $\dot{\phi}_0$ of compact support at $t = 0$. Then $\phi = O(1/\lambda)$ as $\lambda \to \infty$ along every null geodesic and $\phi = O(1/\tau^2)$ as $\tau \to \infty$ along every timelike geodesic, where λ and τ denote, respectively, affine parameter along the null geodesic and proper time along the timelike geodesic.

Proof. According to appendix D, if ϕ satisfies the Klein-Gordon equation in Minkowski spacetime, then $\tilde{\phi} = \Omega^{-1}\phi$ satisfies the equation

$$\tilde{g}^{ab}\tilde{\nabla}_a\tilde{\nabla}_b\tilde{\phi} - \frac{1}{6}\tilde{R}\tilde{\phi} = 0 \tag{11.1.15}$$

in the region O of the Einstein static universe where $\tilde{\nabla}_a$ and $\tilde{R}$ are the derivative operator and scalar curvature of $\tilde{g}_{ab}$. Consider now, the initial data for equation (11.1.15) on the Cauchy surface $T = 0$ of the Einstein static universe obtained by extending $\tilde{\phi}_0$, $\dot{\tilde{\phi}}_0$ to "spatial infinity" by setting these quantities equal to zero at the point i^0. Since equation (11.1.15) is of the form for which theorem 10.1.2 applies, it follows that there exists a smooth solution $\tilde{\psi}$ throughout the *entire* Einstein static universe of equation (11.1.15) with the given initial data. By the uniqueness and domain of dependence properties of theorem 10.1.2, this solution $\tilde{\psi}$ must agree with $\tilde{\phi}$ in O. Thus, we have proven that $\tilde{\phi}$ can be smoothly extended to $\dot{O}$. The conclusions of the proposition follow immediately by translating this result back to statements about asymptotic limits in Minkowski spacetime. Indeed, the hypothesis of the theorem can be weakened significantly—all that is needed is that the data induced by $\tilde{\phi}$ at $T = 0$ can be smoothly extended to i^0—and stronger conclusions than those stated in the theorem can be drawn from the fact that $\tilde{\phi}$ can be smoothly extended to $\dot{O}$. □

We turn, now, to the issue of defining the notion of asymptotically flat curved spacetimes. The key idea is that—as will be seen below—our ability to perform the above construction of mapping Minkowski spacetime via a conformal isometry into a bounded region of the Einstein static universe crucially depended upon the structure of Minkowski spacetime "at infinity." This suggests that we *define* a spacetime to be asymptotically flat if a similar construction can be performed, i.e., if the physical spacetime can be mapped into a new, "unphysical" spacetime via a conformal isometry with properties similar to that of the Minkowski case. In this way we solve both problems (i) and (ii) mentioned at the beginning of this section, since we have a manifestly coordinate independent formulation of the notion of asymptotic flatness, and, as in the Minkowski case, the boundary in the unphysical spacetime of the image of the physical spacetime under the conformal isometry gives a precise framework for describing infinity.

However, there are two important properties of the construction of conformal infinity for Minkowski spacetime which do not carry over to curved spacetimes. First, we wish to consider spacetimes which become flat as one goes to "large distances in spacelike or null directions." However, we do not wish to require the spacetimes to become flat "at a fixed position at early or late times," since we may wish to describe spacetimes representing isolated bodies which may remain present at early and late times. Therefore, we cannot expect the conformal infinity of curved spacetime to be similar to that of Minkowski spacetime at past and future timelike infinity, i^- and i^+. For this reason, for curved spacetimes we do not impose any requirements on the structure of conformal infinity corresponding to the presence of these two points. Second, although the metric is required to become flat at spatial infinity, for the reasons discussed below, smoothness or even differentiability of the conformally rescaled metric $\tilde{g}_{ab}$ at spatial infinity is too strong a requirement. Thus, although conformal infinity of curved spacetime is required to contain a point i^0 representing spatial infinity, the smoothness properties which hold for Minkowski spacetime must be weakened significantly.

Our task, now, is to define asymptotic flatness by extracting features of the construction of conformal infinity of Minkowski spacetime—modulo the above modifications—which are strong enough that they can be implemented only for spacetimes which physically represent isolated systems but are not overly stringent to the extent that physically reasonable examples are excluded. The motivation for the particular choice of features given below comes from a study of some examples, from the stability of the properties under linear perturbations (see Geroch and Xanthopoulos 1978), and from the fact that it provides an appropriate framework for defining total energy and radiated energy (see section 11.2). Further discussion of the motivation is given by Geroch (1977). We state, now, the rather technical conditions of the definition of asymptotic flatness given by Ashtekar (1980) and then discuss the meaning of these conditions.

A vacuum spacetime (M, g_{ab}) is called to be *asymptotically flat at null and spatial infinity* if there exists a spacetime $(\tilde{M}, \tilde{g}_{ab})$—with $\tilde{g}_{ab}$ C^∞ everywhere except possibly at a point i^0 where it is $C^{>0}$ (defined below)—and a conformal isometry $\psi: M \to \psi[M] \subset \tilde{M}$ with conformal factor Ω (so that $\tilde{g}_{ab} = \Omega^2 \psi^* g_{ab}$ in $\psi[M]$) satisfying the following conditions:

(1) $\overline{J^+(i^0)} \cup \overline{J^-(i^0)} = \tilde{M} - M$. [Here $\overline{J^+(i^0)}$ is the closure of the causal future, $J^+(i^0)$, of i^0 (see chapter 8), and for notational simplicity we write M rather than $\psi[M]$ here and in the following.] Thus, i^0 is spacelike related to all points in M and the boundary, $\dot{M}$, of M consists of the union of i^0, $\mathscr{I}^+ \equiv \dot{J}^+(i^0) - i^0$ and $\mathscr{I}^- \equiv \dot{J}^-(i^0) - i^0$.

(2) There exists an open neighborhood, V, of $\dot{M} = i^0 \cup \mathscr{I}^+ \cup \mathscr{I}^-$ such that the spacetime $(V, \tilde{g}_{ab})$ is strongly causal.

(3) Ω can be extended to a function on all of $\tilde{M}$ which is C^2 at i^0 and C^∞ elsewhere.

(4) (*a*) On $\mathscr{I}^+$ and $\mathscr{I}^-$ we have $\Omega = 0$ and $\tilde{\nabla}_a \Omega \neq 0$. (Here $\tilde{\nabla}_a$ is the derivative operator associated with $\tilde{g}_{ab}$, although, clearly, this condition is independent of choice of derivative operator.) (*b*) We have $\Omega(i^0) = 0$, $\lim_{i^0} \tilde{\nabla}_a \Omega = 0$, and $\lim_{i^0} \tilde{\nabla}_a \tilde{\nabla}_b \Omega = 2\tilde{g}_{ab}(i^0)$. (We take limits at i^0 since $\tilde{g}_{ab}$ need not be C^1 there, and thus $\tilde{\nabla}_a$ need not be defined at i^0.)

(5) (*a*) The map of null directions at i^0 into the space of integral curves of $n^a \equiv \tilde{g}^{ab} \tilde{\nabla}_a \Omega$ on $\mathscr{I}^+$ and $\mathscr{I}^-$ is a diffeomorphism. (*b*) For a smooth function, ω, on $\tilde{M} - i^0$ with $\omega > 0$ on $M \cup \mathscr{I}^+ \cup \mathscr{I}^-$ which satisfies $\tilde{\nabla}_a(\omega^4 n^a) = 0$ on $\mathscr{I}^+ \cup \mathscr{I}^-$, the vector field $\omega^{-1} n^a$ is complete on $\mathscr{I}^+ \cup \mathscr{I}^-$.

We begin our discussion of this definition by explaining the meaning of the condition $C^{>0}$ imposed on $\tilde{g}_{ab}$ at i^0. A basic problem with the construction of conformal infinity is that all of spatial infinity is represented by only a single point. As one goes along curves in the physical spacetime which go to spatial infinity along different angular directions (or along directions differing by a boost), then physically the curves become farther and farther apart. However, in the unphysical spacetime, the different curves all terminate at the same point i^0. Therefore, while it is entirely reasonable to require that certain physical tensor fields (times suitable powers of Ω) have well defined limits along appropriate curves that go to spatial infinity, it is unreasonable to demand that these limits along curves going to infinity in different directions define the *same* limiting tensor at the point i^0. Because of the homogeneity and isotropy of Minkowski spacetime, it was possible in that case to define a conformal completion so that the unphysical metric and conformal factor were smooth at i^0. However, when one considers physical fields on Minkowski spacetime, one soon encounters this direction dependent phenomenon at i^0. For example, one easily may verify that for the Coulomb solution of Maxwell's equation, the tensor ΩF_{ab} approaches a well defined, finite limit as one approaches i^0 along any spacelike geodesic, but the tensor it defines at i^0 depends upon the choice of spacelike geodesic. (Roughly speaking, the "electric lines of force" of F_{ab} point radially inward at i^0.) The behavior of the spacetime metric at "large distances" in curved spacetimes such as the Schwarzschild solution is analogous to this behavior of the Coulomb field solution in flat spacetime. Therefore, it is unreasonable to demand that the unphysical metric, $\tilde{g}_{ab}$, be smooth at i^0.

It appears that an appropriate condition on $\tilde{g}_{ab}$ is that it be $C^{>0}$ at i^0. By $C^{>0}$ at i^0, we mean that $\tilde{g}_{ab}$ is continuous (C^0) at i^0 and, in addition, that its first derivatives have direction-dependent limits at i^0 which are smooth in their angular dependence. This latter notion can be made precise as follows. Let x^μ be a smooth coordinate system with origin at i^0. Define the "radial function" ρ by

$$\rho^2 = \sum_{\mu=1}^{4} (x^\mu)^2$$

and define the angular functions θ^α ($\alpha = 1, 2, 3$) in terms of the x^μ by the same formulas as used to define the 3-sphere coordinates in four-dimensional Euclidean space. We may use the coordinates θ^α to characterize each tangent direction at i^0. A function f is said to have a *regular direction-dependent limit* at i^0 if the following three properties are satisfied: (i) For each C^1 curve γ ending at i^0, the limit of f along γ exists at i^0. Furthermore, the value of this limit depends only on the tangent direction to γ at i^0. We define $F(\theta^\alpha) = \lim_{i^0} f$, where the limit is taken along a curve whose tangent direction at i^0 is characterized by θ^α. (ii) F is a smooth function on the 3-sphere. (iii) Along every C^1 curve ending at i^0, we have for all $n \geq 1$,

$$\lim_{i^0} \frac{\partial^n f}{\partial \theta^n} = \frac{\partial^n F}{\partial \theta^n} \quad \text{and} \quad \lim_{i^0} \rho^n \frac{\partial^n f}{\partial \rho^n} = 0 \quad .$$

(Here $\partial^n / \partial \theta^n$ denotes any nth order partial derivative with respect to the θ^α, with it understood that the same partial derivative occurs on both sides of the equation.) Thus, a precise statement of the condition that $\tilde{g}_{ab}$ is $C^{>0}$ at i^0 is that—in addition to $\tilde{g}_{ab}$ being continuous—all the first partial derivatives of the components of $\tilde{g}_{ab}$ in a smooth chart covering i^0 have regular direction dependent limits at i^0. It follows immediately that in a chart with $\{\nabla_a x^\mu\}$ orthonormal at i^0, the components of $\tilde{g}_{ab}$ take the form $\tilde{g}_{\mu\nu} = \eta_{\mu\nu} + \rho l_{\mu\nu}(\theta^\alpha) + o(\rho)$ as $\rho \rightarrow 0$. Further discussion of the $C^{>0}$ condition is given by Ashtekar and Hansen (1978) and Ashtekar (1980). Since the metric is only $C^{>0}$ at i^0, it is natural also to weaken the assumptions concerning differential structure of the manifold $\tilde{M}$ at i^0 to what Ashtekar and Hansen (1978) call $C^{>1}$ differential structure. (Otherwise, one would be able to define many inequivalent differential structures on $\tilde{M}$ at i^0 compatible with all the above conditions.) We refer the reader to that reference for the definition of this term.

Note that we have defined asymptotic flatness only for the case of vacuum spacetimes, $R_{ab} = 0$. However, since only properties of the spacetime "near infinity" will play any role in our analysis, we need only require $R_{ab} = 0$ "near infinity," e.g., in the intersection of M and the neighborhood V of condition (2). Indeed, one may weaken this requirement further to allow nonvanishing T_{ab} in V which goes to zero at infinity suitably rapidly—more precisely, such that $\Omega^{-2}T_{ab}$ is smooth on $\mathscr{I}^+$ and $\mathscr{I}^-$ and has appropriate limiting behavior at i^0.

According to equation (D.8) of appendix D (with the roles of $\tilde{g}_{ab}$ and g_{ab} reversed), the physical Ricci tensor, R_{ab}, is related to the unphysical Ricci tensor, $\tilde{R}_{ab}$, by

$$R_{ab} = \tilde{R}_{ab} + 2\Omega^{-1}\tilde{\nabla}_a\tilde{\nabla}_b\Omega + \tilde{g}_{ab}\tilde{g}^{cd}(\Omega^{-1}\tilde{\nabla}_c\tilde{\nabla}_d\Omega - 3\Omega^{-2}\tilde{\nabla}_c\Omega\tilde{\nabla}_d\Omega) \quad . \qquad (11.1.16)$$

Thus, the vanishing of the right-hand side of equation (11.1.16) is the vacuum Einstein field equation expressed in terms of the unphysical variables. Note that we have written, for example, $\tilde{g}^{cd}\tilde{\nabla}_c\tilde{\nabla}_d\Omega$ rather than $\tilde{\nabla}^c\tilde{\nabla}_c\Omega$ in order to avoid confusion over whether the physical or unphysical metric is used to raise and lower indices.

It should be emphasized that there is considerable arbitrariness in the association of an unphysical spacetime $(\tilde{M}, \tilde{g}_{ab})$ with an asymptotically flat physical spacetime (M, g_{ab}). If $(\tilde{M}, \tilde{g}_{ab})$ is an unphysical spacetime satisfying the properties of the definition with conformal factor Ω, then so is $(\tilde{M}, \omega^2\tilde{g}_{ab})$ with conformal factor $\omega\Omega$, provided only that the function ω is strictly positive, is smooth everywhere except possibly at i^0, is $C^{>0}$ at i^0, and satisfies $\omega(i^0) = 1$. Thus, there is considerable gauge freedom in the choice of the unphysical metric.

Let us now discuss the meaning of the five conditions appearing in the definition of asymptotic flatness. The first three conditions require in a fairly straightforward manner that $(\tilde{M}, \tilde{g}_{ab})$ possess some of the basic features of the conformal completion of Minkowski spacetime. The first, in effect, states that i^0 represents spatial infinity; the second states that no causal pathologies occur near infinity; and the third states that Ω is well behaved near infinity. The key condition in the definition is condition (4). The requirement that Ω vanish at $\mathscr{I}^+$, $\mathscr{I}^-$, and i^0 implies that at these places an "infinite amount of stretching" is involved in going from the unphysical metric $\tilde{g}_{ab}$ to the physical metric g_{ab}. This shows that $\mathscr{I}^+$, $\mathscr{I}^-$, and i^0 truly represent "infinity" of the physical spacetime. Furthermore, the requirements on the derivatives of Ω at $\mathscr{I}^+$, $\mathscr{I}^-$, and i^0 specified in conditions 4(a) and 4(b) imply that the physical metric, g_{ab}, becomes flat (and approaches flatness at the appropriate rate) as one goes to infinity.

To see this in more detail, we derive a coordinate form of the condition on the asymptotic behavior of the physical metric as one approaches future null infinity, $\mathscr{I}^+$. (A similar analysis applies, of course, to $\mathscr{I}^-$.) By multiplying the vacuum Einstein field equation (eq. [11.1.16] with $R_{ab} = 0$) by Ω and taking the limit at $\mathscr{I}^+$, we find (since $\tilde{g}_{ab}$, $\tilde{R}_{ab}$, and Ω are smooth there) that $\Omega^{-1}\tilde{g}^{cd}\tilde{\nabla}_c\Omega\tilde{\nabla}_d\Omega$ must be smooth at $\mathscr{I}^+$, i.e., more precisely (since $\Omega = 0$ at $\mathscr{I}^+$ and thus Ω^{-1} is not defined there) that this quantity can be smoothly extended to $\mathscr{I}^+$. In particular, this implies that $n^a = \tilde{g}^{ab}\tilde{\nabla}_b\Omega$ is null at $\mathscr{I}^+$, a result which also follows from the fact that n^a must be normal to $\mathscr{I}^+$ (since Ω is constant on $\mathscr{I}^+$) and $\mathscr{I}^+ = [\dot{J}^+(i^0) - i^0]$ is a null surface.

We show now that by using the "gauge freedom" in the conformal factor mentioned above, we can make $\Omega^{-1}\tilde{g}^{cd}\tilde{\nabla}_c\Omega\tilde{\nabla}_d\Omega$ vanish at $\mathscr{I}^+$. Namely, under the transformation $\Omega \to \Omega' = \omega\Omega$, $\tilde{g}_{ab} \to \tilde{g}'_{ab} = \omega^2\tilde{g}_{ab}$, we have

$$(\Omega')^{-1}(\tilde{g}')^{cd}\tilde{\nabla}'_c\Omega'\tilde{\nabla}'_d\Omega' =$$
$$\omega^{-3}\tilde{g}^{cd}[\Omega\tilde{\nabla}_c\omega\tilde{\nabla}_d\omega + 2\omega\tilde{\nabla}_c\Omega\tilde{\nabla}_d\omega + \omega^2\Omega^{-1}\tilde{\nabla}_c\Omega\tilde{\nabla}_d\Omega] \quad . \tag{11.1.17}$$

Therefore, by choosing ω to satisfy

$$n^a\tilde{\nabla}_a \ln \omega = -\frac{1}{2}\Omega^{-1}\tilde{g}^{cd}\tilde{\nabla}_c\Omega\tilde{\nabla}_d\Omega \tag{11.1.18}$$

on $\mathscr{I}^+$, we find a new unphysical metric and conformal factor which satisfies the desired relation. (Since eq. [11.1.18] is merely an ordinary differential equation along each integral curve of n^a, solutions always may be obtained.[2]) Thus, without loss of generality, we may assume that we have chosen the conformal factor and unphysical metric so that $\Omega^{-1}\tilde{g}^{cd}\tilde{\nabla}_c\Omega\tilde{\nabla}_d\Omega = 0$ on $\mathscr{I}^+$. The vacuum Einstein field equation then yields

$$2\tilde{\nabla}_a\tilde{\nabla}_b\Omega + \tilde{g}_{ab}\tilde{g}^{cd}\tilde{\nabla}_c\tilde{\nabla}_d\Omega = 0 \text{ on } \mathscr{I}^+ \quad , \tag{11.1.19}$$

which, in turn, implies

$$\tilde{\nabla}_a\tilde{\nabla}_b\Omega = 0 \text{ on } \mathscr{I}^+ \quad . \tag{11.1.20}$$

It follows trivially from equation (11.1.20) that the null tangent $n^a = \tilde{g}^{ab}\tilde{\nabla}_b\Omega$ to $\mathscr{I}^+$ satisfies the affinely parameterized geodesic equation on $\mathscr{I}^+$ with respect to the unphysical metric,

$$n^a\tilde{\nabla}_a n^b = 0 \quad . \tag{11.1.21}$$

Furthermore, since $B^b{}_a = \tilde{\nabla}_a n^b = 0$ on $\mathscr{I}^+$, we see that in the gauge (11.1.20), the expansion, shear, and twist of the null geodesic generators of $\mathscr{I}^+$ all vanish (see section 9.2).

The gauge choice (11.1.18) on ω which led to the gauge condition (11.1.20) still permits the additional freedom of choosing ω arbitrarily on any given cross section of $\mathscr{I}^+$, i.e., on a two-dimensional surface, $\mathscr{S}$, in $\mathscr{I}^+$ which intersects each null geodesic generator of $\mathscr{I}^+$ at precisely one point. It follows from condition 5(a) that $\mathscr{I}^+$ has topology $S^2 \times \mathbb{R}$, and thus, topologically, $\mathscr{S}$ must be a 2-sphere. The unphysical metric $\tilde{g}_{ab}$ induces a Riemannian metric $\tilde{h}_{ab}$ on $\mathscr{S}$. However, using arguments along the lines of problem 2 of chapter 3, it can be seen that every Riemannian metric on a two-dimensional sphere is equal to a conformal factor times the natural metric on the sphere, i.e., $\tilde{h}_{ab}$ is of the form $\tilde{h}_{ab} = f^2 h_{ab}$, with $(\mathscr{S}, h_{ab})$ isometric to the unit sphere in $\mathbb{R}^3$. Hence, we shall use the remaining freedom in the choice of

2. Note that solutions, ω, of equation (11.1.18) on $\mathscr{I}^+$ cannot have the correct limiting behavior at i^0. Thus, equation (11.1.18) really can be imposed only outside an (arbitrarily small) neighborhood of i^0. However, this need not concern us here since we presently deal only with the asymptotic form of the physical metric on $\mathscr{I}^+$.

ω to make $\mathcal{S}$ a metric sphere of unit radius. Since in the gauge (11.1.20) the expansion and shear of the null geodesic generators of $\mathcal{I}^+$ vanish, it follows that in this gauge, the induced metric on *every* cross section of $\mathcal{I}^+$ is that of a unit sphere.

Having made the above gauge choices, we introduce coordinates in a neighborhood of $\mathcal{I}^+$ in the unphysical spacetime as follows. Since $\tilde{\nabla}_a\Omega \neq 0$ on $\mathcal{I}^+$, we may use Ω itself as one of the coordinates. We introduce the natural spherical coordinates (θ, ϕ) on the spherical cross section, $\mathcal{S}$, and "carry" these coordinates to other points of $\mathcal{I}^+$ along the null geodesic generators of $\mathcal{I}^+$. We define the coordinate u on $\mathcal{I}^+$ to be the affine parameter (measured from $\mathcal{S}$) along the null geodesic generators of $\mathcal{I}^+$ with u scaled so that $n^a\nabla_a u = 1$. Finally, we extend (u, θ, ϕ) off of $\mathcal{I}^+$ by holding their values fixed along each null geodesic of the family (other than that which generates $\mathcal{I}^+$) orthogonal to the 2-spheres of constant u on $\mathcal{I}^+$.

It follows from the above that in these coordinates, the unphysical metric $\tilde{g}_{ab}$ takes on $\mathcal{I}^+$ the form

$$d\tilde{s}^2 \,|_{\mathcal{I}^+} = 2d\Omega\, du + d\theta^2 + \sin^2\theta\, d\phi^2 \quad . \tag{11.1.22}$$

Furthermore, the gauge condition (11.1.20) on Ω implies that $\tilde{g}_{uu}$, $\tilde{g}_{u\theta}$, and $\tilde{g}_{u\phi}$ all are $O(\Omega^2)$ as $\Omega \to 0$ (see problem 1). Thus, in a neighborhood of $\mathcal{I}^+$, the components of the physical metric, $g_{ab} = \Omega^{-2}\tilde{g}_{ab}$, take the form

$$\begin{aligned} ds^2 = {} & 2\Omega^{-2}d\Omega\, du + \Omega^{-2}(d\theta^2 + \sin^2\theta\, d\phi^2) \\ & + \text{terms } O(1) \text{ in } du^2,\, du\, d\theta,\, du\, d\phi \\ & + \text{terms } O(\Omega^{-1}) \text{ in } d\theta^2,\, d\theta\, d\phi,\, d\phi^2,\, d\Omega\, du,\, d\Omega^2,\, d\Omega\, d\theta,\, d\Omega\, d\phi \quad . \end{aligned} \tag{11.1.23}$$

We make, now, the coordinate transformation $v = 2/\Omega$. In the coordinates u, v, θ, ϕ, the physical metric takes the form as $v \to \infty$,

$$\begin{aligned} ds^2 = {} & -dv\, du + \frac{1}{4}v^2(d\theta^2 + \sin^2\theta\, d\phi^2) \\ & + \text{terms } O(1) \text{ in } du^2,\, du\, d\theta,\, du\, d\phi \\ & + \text{terms } O(v) \text{ in } d\theta^2,\, d\theta\, d\phi,\, d\phi^2 \\ & + \text{terms } O(1/v) \text{ in } dv\, du,\, dv\, d\theta,\, dv\, d\phi \\ & + O(1/v^3)\, dv^2 \quad . \end{aligned} \tag{11.1.24}$$

By making a further coordinate transformation $v \to v + f(u, \theta, \phi)$ the terms of $O(1)$ in du^2 can be eliminated at only the expense of introducing terms of $O(1)$ in $dv\, d\theta$ and $dv\, d\phi$. Having done this, we transform to "asymptotically Cartesian coordinates" defined by $t = \frac{1}{2}(u + v)$, $x = \frac{1}{2}(v - u)\sin\theta\cos\phi$, $y = \frac{1}{2}(v - u)\sin\theta\sin\phi$, $z = \frac{1}{2}(v - u)\cos\theta$. Then it is easily verified that the components of the physical metric in these coordinates differ from diag $(-1, 1, 1, 1)$ only by terms at most of order $1/v$ as $v \to \infty$. Thus, condition 4(*a*) together with Einstein's equation

(11.1.16) indeed requires the physical spacetime to become asymptotically Minkowskian as one goes toward null infinity. By judiciously imposing further coordinate conditions (Tamburino and Winicour 1966) one can further simplify the asymptotic formula for the metric to the type of expression originally postulated in the axisymmetric case by Bondi, van der Burg, and Metzner (1962).

Similarly, condition 4(*b*) together with the differentiability requirements on Ω and $\tilde{g}_{ab}$ at i^0 imply that the metric is asymptotically Minkowskian as one approaches spatial infinity. A precise coordinate form of this statement is given by Ashtekar and Hansen (1978) (see eq. [C21] of that reference). Ashtekar and Hansen also have proven that any three-dimensional spacelike submanifold of the unphysical spacetime which passes through i^0 and is sufficiently well behaved (namely, $C^{>1}$, i.e., its intrinsic metric $\tilde{h}_{ab}$ and unit normal $\tilde{n}^a$ are $C^{>0}$) at i^0 and smooth elsewhere yields an asymptotically flat initial data surface in the following sense (Geroch 1972*b*): An initial data set (Σ, h_{ab}, K_{ab}) consisting of a three-dimensional manifold, Σ, a Riemannian metric, h_{ab}, and a symmetric tensor K_{ab} is said to be *asymptotically flat* if there exists a manifold $\tilde{\Sigma}$ such that (i) there is a point $\Lambda \in \tilde{\Sigma}$ (called "the point at infinity") such that there exists a conformal isometry $\psi : \Sigma \to [\tilde{\Sigma} - \Lambda]$ with $\tilde{h}_{ab} = \Omega^2 \psi^* h_{ab}$ on $\tilde{\Sigma} - \Lambda$; (ii) Ω is C^2 at Λ and C^∞ elsewhere on $\tilde{\Sigma}$, while $\tilde{h}_{ab}$ is $C^{>0}$ at Λ and C^∞ elsewhere on $\tilde{\Sigma}$; (iii) Ω satisfies conditions at Λ analogous to condition 4(*b*) above, namely, $\Omega(\Lambda) = 0$ and $\lim_\Lambda \tilde{D}_a\Omega = 0$ while $\Omega^{-1}(\tilde{D}_a\tilde{D}_b\Omega - 2\tilde{h}_{ab})$ approaches a finite, direction-dependent limit at Λ, where $\tilde{D}_a$ is the derivative operator associated with h_{ab}; and (iv) the unphysical Ricci tensor ${}^{(3)}\tilde{R}_{ab}$ and $\psi^* K_{ab}$ have appropriate limiting behavior at Λ, namely $\Omega^{1/2}\,{}^{(3)}\tilde{R}_{ab}$ and $\Omega\psi^* K_{ab}$ approach finite, direction-dependent limits at Λ. For an asymptotically flat initial data set, one can construct an asymptotically Euclidean coordinate system on the physical initial data surface Σ such that the components of the physical metric differ from diag $(1, 1, 1)$ only by terms $O(1/r)$ as $r \to \infty$ and the first derivatives of these components are $O(1/r^2)$ (see problem 2). Furthermore, the components of the extrinsic curvature, K_{ab}, and the physical Ricci curvature tensor, ${}^{(3)}R_{ab}$, are, respectively, $O(1/r^2)$ and $O(1/r^3)$. Conditions for asymptotic flatness on an initial data surface formulated in this coordinate component manner were previously given by Arnowitt, Deser, and Misner (1962).

Finally, condition (5) of our above definition of asymptotic flatness is a technical condition which, roughly speaking, states that $\mathscr{I}^+$ and $\mathscr{I}^-$ have the right size. First, condition 5(*a*) states that all the null geodesic generators of $\mathscr{I}^+$ and $\mathscr{I}^-$ emanate from i^0 in the appropriate way. In particular, this implies that $\mathscr{I}^+$ and $\mathscr{I}^-$ have topology $S^2 \times \mathbb{R}$. If $\tilde{g}_{ab}$ were smooth at i^0, the null tangents at i^0 automatically would be diffeomorphic to the null geodesic generators of $\mathscr{I}^+$ and $\mathscr{I}^-$ if all these generators had endpoints at i^0, but since $\tilde{g}_{ab}$ is only $C^{>0}$ at i^0, this property must be imposed separately. Condition 5(*b*) requires "all of $\mathscr{I}^+$ and $\mathscr{I}^-$ to be present" in the conformally completed spacetime. It is equivalent to the statement that in the gauge (11.1.20) the null geodesic generators of $\mathscr{I}^+$ and $\mathscr{I}^-$ are complete. [The relation $\tilde{\nabla}_a(\omega^4 n^a) = 0$ is equivalent to equation (11.1.18) above.] The completeness of the null geodesic generators of $\mathscr{I}^+$ is equivalent in turn to the statement that in our coordinate system (u, v, θ, ϕ) constructed above, the retarded time, u, ranges from $-\infty$ to $+\infty$. Without a condition of the form 5(b), Geroch and Horowitz (1978) have shown that asymptotic flatness could cease to hold at a finite retarded time.

Note that our conditions (1)–(5) above define asymptotic flatness at *both* spatial and null infinity. However, for many purposes one requires only the notion of asymptotic flatness at spatial infinity alone or at null infinity alone, so it is useful to define these notions separately. These notions can be formulated by eliminating the irrelevant parts of our definition above. More precisely, a spacetime (M, g_{ab}) (not necessarily satisfying the vacuum field equation) is said to be *asymptotically flat at spatial infinity* if there is a spacetime $(\tilde{M}, \tilde{g}_{ab})$ with $\tilde{g}_{ab}$ C^{∞} everywhere except possibly at a point i^0 where it is $C^{>0}$ and a conformal isometry of M into $\tilde{M}$ with conformal factor Ω such that conditions 1, 3, and 4(b) are satisfied. A *vacuum* spacetime (M, g_{ab}) is said to be *asymptotically flat at null infinity* if there exists a manifold with boundary (see appendix B), $\overline{M}$, with smooth metric $\overline{g}_{ab}$ and a conformal isometry of M onto the interior of $\overline{M}$ with conformal factor Ω such that appropriate versions of conditions 1, 2, 3, 4(a), and 5 hold. More precisely, we define $\mathscr{I}$ to be the boundary of $\overline{M}$, and in place of 1 we require that $\mathscr{I}$ can be written as the disjoint union of two pieces, $\mathscr{I}^+$ and $\mathscr{I}^-$, such that $\mathscr{I}^+ \cap J^-[\text{int}(M)] = \emptyset$ and $\mathscr{I}^- \cap J^+[\text{int}(M)] = \emptyset$. We again require the unphysical spacetime to be strongly causal in a neighborhood of $\mathscr{I}$ (condition 2), that Ω have a C^{∞} extension to all of $\overline{M}$ (condition 3), and that conditions 4(a) and 5(b) hold on $\mathscr{I}$. Finally we replace 5(a) by the requirement that $\mathscr{I}^+$ and $\mathscr{I}^-$ each have the topology $S^2 \times \mathbb{R}$.

The above definition of asymptotic flatness at null infinity differs from the original formulation given by Penrose (1963, 1965b). Penrose defined a spacetime to be *asymptotically simple* if in place of the above versions of conditions 1, 2, 3, 4(a), and 5 we require that Ω have a C^{∞} extension to all of $\overline{M}$, that condition 4(a) hold, and that every maximally extended (in both directions) null geodesic passing through any point of int($\overline{M}$) has past and future endpoints on $\mathscr{I}$. This last condition is a very strong global condition on the physical spacetime, involving much more than asymptotic behavior at infinity. However, it can be modified to become an asymptotic condition—thereby defining the notion of a *weakly asymptotically simple spacetime*—by requiring that a neighborhood of $\mathscr{I}$ in $(\overline{M}, \overline{g}_{ab})$ for the given spacetime (M, g_{ab}) be isometric to a neighborhood of $\mathscr{I}$ for some asymptotically simple spacetime. Weak asymptotic simplicity then was taken as the criterion for asymptotic flatness at null infinity. Although this formulation is indirect and differs significantly in form from the one given above, it follows that a weakly asymptotically simple spacetime satisfies all the conditions of our definition except for 5(b); see Geroch and Horowitz (1978) for a demonstration that condition 5(b) need not be satisfied and see, e.g., Hawking and Ellis (1973) for the theorems needed to show that the other conditions are satisfied.

We conclude this section by mentioning two important properties of asymptotically flat spacetimes. The first has to do with the notion of asymptotic symmetries. Minkowski spacetime $(\mathbb{R}^4, \eta_{ab})$ has a 10-parameter group of isometries, the Poincaré group. This isometry group plays an important role in the analysis of the behavior of physical fields on Minkowski spacetime, in particular in the proof of conservation laws. In a general, curved spacetime one would not expect any exact isometries to be present. However, in an asymptotically flat spacetime, since the metric becomes flat as one approaches infinity, one might expect that it would be possible to define the notion of an asymptotic symmetry. It turns out that a natural notion of an asymptotic symmetry does indeed exist for asymptotically flat spacetimes, but the group of asymptotic symmetries is *not* the Poincaré group. Rather, it is a much larger group, containing an infinite-dimensional subgroup of "angle dependent translations" called supertranslations.

We shall give a brief discussion, here, of the nature of asymptotic symmetries at null infinity, referring the reader to Sachs (1962*c*) and Geroch (1977) for further discussion and details. Basically, we want an infinitesimal asymptotic symmetry at, say, future null infinity to be represented by a vector field ξ^a—or, more precisely, an equivalence class of vector fields—in the physical spacetime such that Killing's equation $\pounds_\xi g_{ab} = 0$ is satisfied to "as good an approximation as possible" as one goes to $\mathscr{I}^+$. The appropriate requirement for the rate of approach of $\pounds_\xi g_{ab}$ to zero can be determined by examining the general form of a metric of an asymptotically flat spacetime and imposing the strongest condition which generally admits solutions. We may formulate the resulting requirement as follows. First, we require that ξ^a, viewed as a vector field in the unphysical spacetime (i.e., $\psi^* \xi^a$), have a smooth extension to $\mathscr{I}^+$. Then we require in addition that the tensor field $\Omega^2 \pounds_\xi g_{ab}$ also have a smooth extension to $\mathscr{I}^+$ which vanishes on $\mathscr{I}^+$. Furthermore, two vector fields ξ^a and ξ'^a on the physical spacetime satisfying these conditions are considered to generate the same infinitesimal asymptotic symmetry if their extensions to $\mathscr{I}^+$ are equal there.

The asymptotic symmetry group thus defined is universal in the sense that one gets the same abstract group for all asymptotically flat spacetimes. It is initially surprising, however, that this group is *not* the 10-parameter Poincaré group but rather an infinite dimensional group known as the Bondi-Metzner-Sachs (BMS) group. Insight into the origin of these "extra" asymptotic symmetries can be obtained by verifying that in Minkowski spacetime in the coordinates of equation (11.1.4), for an arbitrary function $f = f(\theta, \phi)$ of the angular variables, the vector field

$$\xi^a = f(\partial/\partial u)^a + \frac{v}{2r^2}\frac{\partial f}{\partial \theta}(\partial/\partial\theta)^a + \frac{v}{2r^2 \sin^2\theta}\frac{\partial f}{\partial \phi}(\partial/\partial\phi)^a \quad (11.1.25)$$

is nonvanishing at $\mathscr{I}^+$ but $\Omega^2 \pounds_\xi \eta_{ab}$ vanishes at $\mathscr{I}^+$. Now, in a nonflat but asymptotically flat spacetime, one cannot, in general, find any vector field which satisfies Killing's equation near infinity to a better approximation than this. Hence, in a general curved spacetime an infinite dimensional family of "angle dependent translations" of the above type come just as close to satisfying Killing's equation as any other transformations.

We have introduced the BMS group here in terms of approximate symmetries in the physical spacetime as one approaches $\mathscr{I}^+$. However, the asymptotic symmetry transformations of the physical spacetime extend to transformations of $\mathscr{I}^+$ in the unphysical spacetime, and it turns out that one can give an equivalent characterization of the BMS group in terms of mappings of $\mathscr{I}^+$ into itself. In the unphysical spacetime, the unphysical metric $\tilde{g}_{ab}$ induces a degenerate metric $\tilde{h}_{ab}$ on the null hypersurface $\mathscr{I}^+$. Furthermore, since the vector field $n^a = \tilde{g}^{ab}\tilde{\nabla}_b\Omega$ is tangent to $\mathscr{I}^+$, n^a may be viewed as a vector field on $\mathscr{I}^+$. Under a change of conformal gauge $\Omega \to \omega\Omega$, $\tilde{h}_{ab}$ and n^a transform as $\tilde{h}_{ab} \to \tilde{h}'_{ab} = \omega^2\tilde{h}_{ab}$, $n^a \to (n')^a = \omega^{-1}n^a$. The manifold $\mathscr{I}^+$ together with $\tilde{h}_{ab}$ and n^a modulo conformal gauge are *universal,* i.e., they are "the same" for all asymptotically flat spacetimes in the following sense: Let $(\mathscr{I}_1^+, (\tilde{h}_1)_{ab}, (n_1)^a)$ be the future null infinity, induced degenerate metric, and induced vector field associated with the asymptotically flat (physical) spacetime $(M_1, (g_1)_{ab})$ and let $(\mathscr{I}_2^+, (\tilde{h}_2)_{ab}, (n_2)^a)$ be the corresponding structure for the asymptotically flat spacetime $(M_2, (g_2)_{ab})$. Then there exist choices of conformal gauge in the two unphysical spacetimes such that there is a diffeomorphism $\psi: \mathscr{I}_1^+ \to \mathscr{I}_2^+$ satisfying $\psi^*(\tilde{h}'_1)_{ab} = (\tilde{h}'_2)_{ab}$, $\psi^*(n'_1)^a = (n'_2)^a$. [*Proof.* We showed above that for any $\mathscr{I}^+$ there exists a conformal gauge and a choice of coordinates (u, θ, ϕ) such that $h_{ab} = (d\theta)_a(d\theta)_b + \sin^2\theta(d\phi)_a(d\phi)_b$ and $n^a = (\partial/\partial u)^a$. Let ψ be the map which associates with each point on $\mathscr{I}_1^+$ the point on $\mathscr{I}_2^+$ with the same value of these coordinates.] The BMS group may be characterized as the group of diffeomorphisms of $\mathscr{I}^+$ which preserves this universal structure; i.e., the BMS group consists of the diffeomorphisms $\psi: \mathscr{I}^+ \to \mathscr{I}^+$ such that $\psi^*\tilde{h}_{ab}$ and ψ^*n^a differ from $\tilde{h}_{ab}$ and n^a at most by a rescaling associated with a change of conformal gauge.

In terms of this characterization of the BMS group, the (infinitesimal) *supertranslations* are defined as the vector fields on $\mathscr{I}^+$ of the form $\xi^a = \alpha n^a$ where α is constant on each generator of $\mathscr{I}^+$ (i.e., $n^a\tilde{\nabla}_a\alpha = 0$) but otherwise is an arbitrary function. Thus, the "angle dependent translations" of Minkowski spacetime considered above give rise to BMS supertranslations. The supertranslations comprise an infinite-dimensional abelian normal subgroup of the BMS group, and the factor group obtained by quotienting the BMS group by the supertranslations is isomorphic to the Lorentz group.

Although for the reasons discussed above, the BMS group properly comprises the asymptotic symmetry group of an asymptotically flat spacetime, one still may ask if there is some natural procedure for "recovering" the Poincaré group by imposing extra conditions on the asymptotic symmetries. It turns out that one can partially succeed in doing this as follows. There exists a unique four-dimensional subgroup of the supertranslations which is a normal subgroup of the BMS group. In the case of Minkowski spacetime, this four-dimensional subgroup of the BMS group consists precisely of the asymptotic symmetries associated with the exact translational symmetries of Minkowski spacetime. Therefore, we may define the *asymptotic translations* of a general asymptotically flat spacetime as the BMS elements belonging to this subgroup. (On $\mathscr{I}^+$ in the gauge [11.1.20] with metric spherical cross sections, the infinitesimal translations are precisely the supertranslations of the form $\xi^a = \alpha n^a$ with α a linear combination of $l = 0$ and $l = 1$ spherical harmonics.) However, a similar procedure for the rotations and boosts fails. There exists no normal subgroup of the BMS group which is isomorphic to the Poincaré group. Thus, unless one makes use of more structure than the behavior of asymptotic symmetries on $\mathscr{I}^+$ (such as, e.g., the behavior of the asymptotic symmetries at i^0) or imposes stronger conditions on spacetimes than those given in our definition of asymptotic flatness, there exists no natural way of picking out a preferred Poincaré subgroup of asymptotic symmetries. The notion of a "pure translation" makes sense for a general asymptotically flat spacetime, but the notion of a "pure rotation" or "pure boost" (as opposed to a "rotation plus supertranslation" or "boost plus supertranslation") does not.

The group of asymptotic symmetries at spatial infinity—known as the *Spi group*—can be defined in a manner closely analogous to the BMS group. The structure of this group is very similar to that of the BMS group. We refer the reader to Ashtekar and Hansen (1978) and Ashtekar (1980) for a discussion of its properties. Interestingly, by restricting consideration to spacetimes which satisfy a stronger asymptotic falloff condition on the Weyl tensor at spatial infinity than that implied by our definition of asymptotic flatness, it is possible to pick out a preferred Poincaré subgroup of the Spi group (see Ashtekar 1980). A mathematically analogous condition on the Weyl tensor at null infinity also would allow one to obtain a preferred Poincaré subgroup of the BMS group. However, although this additional condition at spatial infinity appears to be physically reasonable, the analogous condition at null infinity excludes the possibility of gravitational radiation, and thus is much too stringent a condition to impose upon general isolated systems.

Finally we mention an important consequence of our definition of asymptotic flatness concerning the asymptotic behavior of the curvature as one goes to null infinity. The physical Ricci tensor vanishes near infinity so the physical Weyl tensor, $C_{abc}{}^d$, contains all the information about the physical curvature. Since the Weyl tensor is conformally invariant (see appendix D), we have $C_{abc}{}^d = \tilde{C}_{abc}{}^d$ on M, where $\tilde{C}_{abc}{}^d$ is the unphysical Weyl tensor. However, it is possible to show (see theorem 11 of Geroch 1977) that $\tilde{C}_{abc}{}^d$ must vanish at $\mathscr{I}^+$. Now, let γ be any null geodesic in $\tilde{M}$ from a point p in the physical spacetime to a point q on $\mathscr{I}^+$. Let λ be a (physical) affine parameter on γ—so that q is represented as the limit as λ goes to infinity—and let k^a denote the tangent to γ in this parameterization. The unphysical Weyl curvature on γ need not satisfy any special property except that it must vanish at q. However, in going from the unphysical to the physical spacetime, an infinite amount of "stretching" in the direction of k^a takes place. It turns out that this causes the physical Weyl tensor to display the following *peeling property:* As $\lambda \to \infty$, we have

$$C_{abcd} = \frac{C^{(1)}_{abcd}}{\lambda} + \frac{C^{(2)}_{abcd}}{\lambda^2} + \frac{C^{(3)}_{abcd}}{\lambda^3} + \frac{C^{(4)}_{abcd}}{\lambda^4} + O\left(\frac{1}{\lambda^5}\right) \quad , \tag{11.1.26}$$

where each $C^{(\mu)}_{abcd}$ has bounded components in an asymptotically Minkowskian basis, and, in the algebraic classification of section 7.3, $C^{(1)}_{abcd}$ is type IV, $C^{(2)}_{abcd}$ is type III, $C^{(3)}_{abcd}$ is type II (or type II–II), all with repeated principal null vector k^a, while $C^{(4)}_{abcd}$ is type I with k^a also being one of its principal null directions. For further discussion and a proof of this peeling property, we refer the reader to Penrose (1965*b*) and Geroch (1977).

11.2 Energy

The notion of energy and the law of conservation of energy play a key role in all physical theories. Already in the Newtonian theory of particle motion one has a notion of particle energy. In physical theories of broader scope and greater sophistication, the framework of the physical laws is markedly changed from Newtonian particle mechanics, but a notion of energy always has remained present. Thus, for example, as discussed in chapter 4, for the theory of a classical field on spacetime in special relativity, a stress-energy tensor, T_{ab}, is associated with the field. The total energy of the field associated with a time-translation Killing field t^a on a spacelike Cauchy hypersurface Σ is defined as $E = \int_\Sigma T_{ab} n^a t^b$, where n^a denotes the unit normal to Σ. The condition $\partial^a T_{ab} = 0$ then guarantees that total energy is conserved, i.e., is independent of the choice of Σ.

In general relativity, the energy properties of matter are, again, represented by a stress-energy tensor T_{ab}. Thus, the local energy density of matter as measured by a given observer remains well defined. However, although the condition $\nabla^a T_{ab} = 0$ may be interpreted as expressing local conservation of the energy-momentum of matter (see chapter 4), this condition does not, in general, lead to a global conservation law,[3] i.e., a law which states that the total energy (expressed as an integral involving T_{ab} over a spacelike hypersurface) is conserved. On physical grounds, this is not surprising since T_{ab} represents only the energy content of matter whereas "gravitational field energy" should make a contribution to total energy and thus should appear in any conservation law. However, as already mentioned in chapter 4, there is no known meaningful notion of the energy density of the gravitational field in general relativity. In Newtonian gravity, the energy density of the gravitational field[4] is $-(8\pi)^{-1}|\vec{\nabla}\phi|^2$. Since in the Newtonian limit, ϕ corresponds to a metric component (see section 4.4*a*), the most likely candidate for the energy density of the gravitational field in general relativity would be an expression quadratic in the first derivatives of the metric. However, since no tensor other than g_{ab} itself can be constructed locally from only the coordinate basis components of g_{ab} and their first derivatives, a meaningful expression quadratic in first derivatives of the metric can be obtained only if one has additional structure on spacetime, such as a preferred coordinate system or a decomposition of the spacetime metric into a "background part" and a "dynamical part" (so that, say, one could take derivatives of the "dynamical part" of the metric with respect to the derivative operator associated with the background part). Such additional structure would be completely counter to the spirit of general relativity, which views the spacetime metric as fully describing all aspects of spacetime structure and the gravitational field. While it remains conceivable that one could construct the needed extra structure nonlocally out of the spacetime metric, it seems highly unlikely that a generally applicable prescription exists for obtaining a physically meaningful local expression for gravitational energy density analogous to the Newtonian formula. No other reasonable candidates for a local expression for gravitational field energy density have been found. In particular, a local tensor expression involving the spacetime curvature in higher than linear

3. An exception occurs if a Killing vector field ξ^a is present in the spacetime. In that case $\nabla^a(T_{ab}\xi^b) = (\nabla^a T_{ab})\xi^b + T_{ab}\nabla^a\xi^b = 0$, so $\int_\Sigma T_{ab}\xi^b n^a$ is conserved, i.e., independent of choice of Cauchy surface Σ.

4. Newtonian gravity can be formulated in a manner where no preferred Galilean frames exist (Trautman 1965). When this is done, $\vec{\nabla}\phi$ becomes gauge dependent and the Newtonian formula for gravitational energy density suffers from many of the same difficulties as occur in general relativity.

order—such as the Bel-Robinson tensor (problem 6 of chapter 4)—does not even have the dimensions of energy density [i.e., (length)$^{-2}$ in units with $G = c = 1$], so one would need to introduce constants of nature other than G and c into such an expression if it were to represent gravitational field energy.

However, despite the absence of a notion of energy density of the gravitational field, there does exist a useful and meaningful notion of the *total* energy of an *isolated* system, i.e., more precisely, the total energy-momentum 4-vector present in an asymptotically flat spacetime. The first step toward obtaining such a notion is to view an isolated system in general relativity as being analogous to a particle in special relativity. In special relativity, a particle is assigned an energy-momentum 4-vector P^a. The energy of the particle is taken to be the time component of this vector, i.e., with respect to a time translation Killing field ξ^a, we have $E = -P_a \xi^a$. Furthermore, the mass of the particle is given by $M = (-P_a P^a)^{1/2}$, so if the particle is "at rest" with respect to ξ^a, we have $E = M$. Thus, a knowledge of the "rest frame" Killing field ξ^a together with the mass, M, of the particle determines its 4-momentum P^a. Our strategy for determining the 4-momentum of an isolated system in general relativity will be to begin with static spacetimes where a natural "rest frame" Killing field is given to us. Then we shall define the total mass of a static asymptotically flat spacetime by examining the influence of the gravitational field on distant test bodies. Indeed, we have done this already for the Schwarzschild spacetime in chapter 6. From the results of section 6.3, it follows that test bodies in the Schwarzschild spacetime in orbits with $r >> M$ behave just like test bodies in the gravitational field of a body of mass M in Newtonian gravity. Thus, in chapter 6 we identified the parameter M of the Schwarzschild metric as representing the total mass of the spacetime. We now shall generalize this result.

We begin by recalling that in the Newtonian theory of gravity, the Newtonian potential ϕ satisfies Laplace's equation in the exterior, vacuum region,

$$\nabla^2 \phi = 0 \quad . \tag{11.2.1}$$

Therefore, one has a multipole expansion for ϕ, and the mass, M, of an isolated system may be defined as minus the coefficient of the leading order term (i.e., the monopole, $1/r$, term) of this multipole expansion. An equivalent, but more physical, characterization of total mass—which also refers only to the asymptotic properties of the gravitational field—is given by the "Gauss's law" formula

$$M = \frac{1}{4\pi} \int_S \vec{\nabla} \phi \cdot \hat{N} \, dA \quad . \tag{11.2.2}$$

Here the integral is taken over any topological 2-sphere, S, which encloses all the sources and $\hat{N}$ is the unit outward normal to S. The integral is independent of choice of S because of equation (11.2.1), and it is easily seen to agree with the multipole expansion definition of M. Since $\vec{\nabla}\phi$ is the force that must be exerted on a unit test mass to "hold it in place," we see from equation (11.2.2) that $4\pi M$ is just the total outward force that must be applied to hold in place test matter with unit surface mass density distributed over S.

Consider, now, a static, asymptotically flat spacetime which is vacuum in a neighborhood of infinity and whose timelike Killing vector field ξ^a is normalized so that the "redshift factor," $V \equiv (-\xi_a \xi^a)^{1/2}$, approaches 1 at infinity. In a static spacetime the notion of "staying in place" is well defined; it means following an orbit of the Killing field ξ^a. The acceleration of such an orbit is

$$a^b = (\xi^a/V)\nabla_a(\xi^b/V) = \frac{1}{V^2}\xi^a\nabla_a\xi^b \quad . \tag{11.2.3}$$

Thus, the local force which must be exerted on a unit test mass to hold it in place is given by equation (11.2.3). However, if we choose to calculate the force which must be applied by a distant observer at infinity (e.g., by means of a long string), we find that this force differs from the local force by a factor of V (see problem 4 of chapter 6). Consider a topological 2-sphere S lying in the hypersurface orthogonal to ξ^a. The quantity

$$F = \int_S N^b(\xi^a/V)\nabla_a\xi_b \, dA \tag{11.2.4}$$

may be interpreted as the total outward force that must be exerted by a distant observer to keep in place a unit surface mass density distributed over S. Here we use the natural volume element on S induced by the spacetime metric, and N^a is the unit "outward pointing" normal to S which is orthogonal to ξ^a. Using Killing's equation $\nabla_a\xi_b = \nabla_{[a}\xi_{b]}$, we may rewrite equation (11.2.4) as

$$F = \frac{1}{2}\int_S N^{ab}\nabla_a\xi_b \, dA = -\frac{1}{2}\int_S \epsilon_{abcd}\nabla^c\xi^d \quad , \tag{11.2.5}$$

where, in the first line, $N^{ab} = 2V^{-1}\xi^{[a}N^{b]}$ is the normal "bi-vector" to S, and, in the second line ϵ_{abcd} is the volume element on spacetime associated with the spacetime metric and the integrand is viewed as two-form, $\boldsymbol{\alpha}$, to be integrated over the two-dimensional submanifold S (see appendix B). (The orientation of ϵ_{abcd} is chosen so that $\epsilon_{abcd} = -6N_{[ab}\epsilon_{cd]}$, where ϵ_{cd} is the volume element on S.) However, this integrand satisfies

$$\begin{aligned} \epsilon^{efab}\nabla_f[\epsilon_{abcd}\nabla^c\xi^d] &= \epsilon^{efab}\epsilon_{abcd}\nabla_f\nabla^c\xi^d \\ &= -4\nabla_f\nabla^{[e}\xi^{f]} \\ &= 4\nabla_f\nabla^f\xi^e \\ &= -4R^e{}_f\xi^f \quad , \end{aligned} \tag{11.2.6}$$

where equation (B.2.13) was used in the second line and equation (C.3.9) was used in the last line. Hence, by multiplying equation (11.2.6) by ϵ_{elmn} and contracting over e, we find

$$\nabla_{[l}\{\epsilon_{mn]cd}\nabla^c\xi^d\} = \frac{2}{3}R^e{}_f\xi^f\epsilon_{elmn} \quad , \tag{11.2.7}$$

which vanishes in the vacuum region, $R_{ab} = 0$. Thus, in the vacuum region, the differential form $\alpha_{ab} = \epsilon_{abcd}\nabla^c \xi^d$ satisfies

$$d\boldsymbol{\alpha} = 0 \quad . \tag{11.2.8}$$

Consequently, applying Stokes's theorem (theorem B.2.1) to the volume bounded by any two spheres S and S' in the exterior vacuum region, we see that the integral on the right-hand side of equation (11.2.5) is independent of choice of S, just as the integral in equation (11.2.2) was independent of S in the Newtonian case. Since both integrals represent the same physical quantity, it is natural to identify the left-hand sides of these equations as well. Thus, we are led to the following definition of the total mass of a static, asymptotically flat spacetime which is vacuum in the exterior region,

$$M = -\frac{1}{8\pi}\int_S \epsilon_{abcd}\nabla^c \xi^d \quad . \tag{11.2.9}$$

Note that the independence of the right-hand side of equation (11.2.9) on the choice of S depends only on the fact that ξ^a is a Killing field. Thus, we also may adopt equation (11.2.9) as the definition of total mass in all *stationary* asymptotically flat spacetimes which are vacuum near infinity. (If $T_{ab} \neq 0$ near infinity but approaches zero sufficiently rapidly for the spacetime to be asymptotically flat, eq. [11.2.9] may still be used to define M provided that the limit as the 2-sphere S "goes to infinity" is taken.) The definition (11.2.9)—first given by Komar (1959)—provides a fully satisfactory notion of the total mass in stationary, asymptotically flat spacetimes. Note, in addition, that it has the desirable feature of expressing M in stationary spacetimes as a conserved quantity associated with the time translation symmetry.

If the 2-sphere S is the boundary of a spacelike hypersurface Σ such that $\Sigma \cup S$ is a compact manifold with boundary (see appendix B), then we may apply Stokes's theorem to convert equation (11.2.9) to a volume integral over Σ,

$$\begin{aligned} M &= -\frac{1}{8\pi}\int_S \boldsymbol{\alpha} = -\frac{1}{8\pi}\int_\Sigma d\boldsymbol{\alpha} \\ &= -\frac{3}{8\pi}\int_\Sigma \nabla_{[e}\{\epsilon_{ab]cd}\nabla^c \xi^d\} \\ &= -\frac{1}{4\pi}\int_\Sigma R^d{}_f \xi^f \epsilon_{deab} \\ &= \frac{1}{4\pi}\int_\Sigma R_{ab} n^a \xi^b \, dV \\ &= 2\int_\Sigma \left(T_{ab} - \frac{1}{2}T g_{ab}\right) n^a \xi^b \, dV \quad . \end{aligned} \tag{11.2.10}$$

Here in the third line equation (11.2.7) was used, in the fourth line n^a is the unit future pointing normal to Σ (so that $\epsilon_{abc} = n^d \epsilon_{dabc}$ is the natural volume element on Σ, represented by dV), and Einstein's equation was used to get the final expression.

For a comparison of this formula with formulas holding in linearized gravity and in the static, spherically symmetric case, see problems 4 and 5.

We turn, now, to the definition of energy-momentum in the general, nonstationary, asymptotically flat case. Since there now is no notion of "holding a test mass in place" and thus no notion of "gravitational force," it is not obvious that a well-defined notion of mass or energy-momentum exists. Furthermore, it is far from clear to what vector space an "energy-momentum vector" of the spacetime should belong. However, we shall see that the properties of asymptotically flat spacetimes give just enough structure to define the energy-momentum vector, P_a, in terms of an asymptotic limit as one "goes to infinity" and that the properties of infinity provide an appropriate vector space structure for P_a. From general considerations, we expect that two types of limits as one goes to infinity should be of interest for the definition of energy-momentum: (1) One could try to obtain a notion of the total energy-momentum at a given "time" at spatial infinity by considering an asymptotically flat spacelike hypersurface Σ—i.e., a spacelike hypersurface which is $C^{>1}$ at i^0—and defining the total energy, E, and total momentum p_a on Σ in a manner analogous to the definition of M in a stationary spacetime. For this definition to lead to a meaningful notion of energy-momentum at spatial infinity, E and p_a should depend only on the asymptotic behavior of Σ, should be conserved (i.e., be unchanged if Σ undergoes a "time translation" near infinity), and should transform as a 4-vector if Σ undergoes a "Lorentz boost" near infinity. If these properties are satisfied, we may define the total energy-momentum, P_a, at spatial infinity as a vector in the cotangent space at i^0 such that total energy associated with any spacelike hypersurface Σ which is $C^{>1}$ at i^0 is given by $E = -P_a n^a$, where n^a is the unit normal to Σ at i^0, while the projection of P_a into Σ would yield the total momentum p_a at the "time" represented by Σ. (2) One could try to obtain a notion of the energy-momentum at a fixed "retarded time" by considering the behavior of the spacetime geometry as one goes to null infinity on an asymptotically null surface, Σ, such as the surfaces Σ_1 or Σ_2 shown in Figure 4.3 of chapter 4. This would enable one to study the energy and momentum carried away by gravitational radiation. Again, the total energy and momentum associated with Σ should depend only on the asymptotic properties of Σ. However, unlike the situation at i^0, a null or asymptotically null hypersurface is not naturally associated with any preferred "time direction," so we should have to specify an "asymptotic time translation" as additional input in order to obtain a value for the total energy E. Fortunately, as mentioned at the end of the previous section, the asymptotic symmetry group of null infinity has a preferred four-parameter subgroup of translations, so the notion of an "asymptotic time translation" is well defined. Again, E should transform as the time component of a vector under a change of the choice of asymptotic time translation. Thus, we seek a definition of energy-momentum such that associated with every cross section, $\mathcal{S}$, of $\mathcal{I}^+$ (or of $\mathcal{I}^-$), there should be a linear map from the four-dimensional vector space of BMS translations into $\mathbb{R}$. In other words, the energy-momentum of each cross section $\mathcal{S}$ at null infinity should be a vector, P_a, in the dual vector space of the BMS translations. The value of P_a applied to a given time translation then would be interpreted as the total energy associated with this time direction at the "retarded time" defined by $\mathcal{S}$, while its value for a spatial translation would be interpreted as the component of momentum in that spatial direction.

As mentioned above, both of these notions of energy-momentum exist for asymptotically flat spacetimes. We shall present the definition of energy-momentum at null infinity and discuss some of its properties, and then merely quote formulas for the energy-momentum at spatial infinity. The basic idea at null infinity is the following. As discussed above, in the case of a spacetime with an *exact* time translation symmetry, equation (11.2.9) provides a satisfactory definition of mass, and the integral in that formula is independent of the choice of 2-sphere S. Therefore, in the general, nonstationary case one might attempt to employ a formula like equation (11.2.9) to define energy, where ξ^a now is the generator of an *asymptotic* time translation symmetry, or more precisely, a vector field on the physical spacetime which is a member of the equivalence class associated with a BMS time translation. The integral now, of course, will depend on the choice of topologically spherical surface S. However, since Killing's equation is satisfied by ξ^a to a better and better approximation as one goes to infinity, one might expect that the dependence on S would become sufficiently weak that the limit as S "goes to infinity" would exist. More precisely, let $\{S_\alpha\}$ be a one-parameter family of spheres which, in the unphysical spacetime, approaches the cross section $\mathcal{S}$ of $\mathcal{I}^+$. We propose defining the energy, E, associated with the asymptotic time translation ξ^a by

$$E = -\lim_{S_\alpha \to \mathcal{S}} \frac{1}{8\pi} \int_{S_\alpha} \epsilon_{abcd} \nabla^c \xi^d \tag{11.2.11}$$

It turns out that the limit on the right-hand side of equation (11.2.11) always exists and is independent of the details of how S_α approaches $\mathcal{S}$. Indeed, this limit exists even when ξ^a is an arbitrary asymptotic symmetry; see Geroch and Winicour (1981) for a sketch of a proof of this result. Furthermore, E has the desired linear dependence on ξ^a. However, it turns out that E is *not* invariant under a change of the choice of representative ξ^a in the equivalence class associated with the given BMS time translation. Thus, E is not "gauge invariant" in this sense, and equation (11.2.11), as it stands, fails to define a satisfactory notion of energy at null infinity.

However, this deficiency can be remedied by imposing an additional condition on the representative vector field, ξ^a. As shown by Geroch and Winicour (1981), if one chooses ξ^a to satisfy

$$\nabla_a \xi^a = 0 \quad . \tag{11.2.12}$$

in a neighborhood of $\mathcal{I}^+$, then the formula for E becomes well defined, i.e., independent of the choice of ξ^a within the equivalence class satisfying condition (11.2.12). (Note that this gauge condition on ξ^a holds automatically when ξ^a is a Killing field.) We adopt this as our definition of energy at null infinity. The 4-momentum, P_a, is defined by allowing the linear map defined by the right-hand side of equation (11.2.11) (with gauge condition [11.2.12]) to act on arbitrary BMS translations.

The definition (11.2.11) of energy at null infinity agrees with the definition given in coordinate form, prior to the geometric formulation of the notion of asymptotic flatness, by Bondi, van der Burg, and Metzner (1962), and is referred to as the *Bondi energy*. It was reformulated in terms of the behavior at $\mathscr{I}^+$ of the Weyl curvature and the shear of the "outgoing" congruence of null geodesics orthogonal to $\mathscr{S}$ in the gauge (11.1.20) with metric spherical cross sections by Penrose (1963). The type of "linkage" formulation given above was introduced by Tamburino and Winicour (1966) and Winicour (1968), and a demonstration of equivalence with the Penrose formulation is given by Winicour (1968).

It is of great interest to examine how E changes with time, i.e., how the value of E associated with a given asymptotic time translation on a cross section $\mathscr{S}_1$ differs from the value of E associated with the same asymptotic time translation on a "later" cross section $\mathscr{S}_2$. It turns out that one can express the difference between $E[\mathscr{S}_2]$ and $E[\mathscr{S}_1]$ as an integral of a function, f, over the region, V, of $\mathscr{I}^+$ between $\mathscr{S}_1$ and $\mathscr{S}_2$,

$$E[\mathscr{S}_2] - E[\mathscr{S}_1] = -\int_V f \quad . \tag{11.2.13}$$

Hence, we may interpret f as the flux of energy carried away to infinity by gravitational radiation. Remarkably, as originally found by Bondi, van der Burg, and Metzner (1962) and reformulated in the conformal infinity framework by many authors (e.g., Penrose 1965*b*; Winicour 1968; Geroch 1977; Geroch and Winicour 1981), f is manifestly nonnegative, $f \geqq 0$. Thus, we find that E decreases with time, i.e., *gravitational radiation always carries positive energy away from a radiating system*.

In our discussion of linearized gravity in section 4.4*b*, we computed the energy carried away by gravitational radiation to the lowest nonvanishing approximation by using the second order Einstein tensor as an effective stress-energy tensor, t_{ab}, for the linearized gravitational field. How does this notion of energy flux in the approximation of section 4.4*b* compare with the lowest nonvanishing contribution to the energy flux calculated from the exact theory described above? First, the energy flux vector $-t^a{}_0$ we used in linearized gravity is not gauge invariant nor does it give a manifestly positive energy flux at $\mathscr{I}^+$, so it cannot agree with f. However, the *total* energy flux between stationary periods computed from $-t^a{}_0$ is gauge invariant and can be demonstrated to agree with the lowest nonvanishing approximation to the total radiated energy computed from f as follows: It has been shown (Persides and Papadopoulos 1979) that in the exact theory, the energy flux f agrees with the energy flux computed from the pseudotensor of Landau and Lifshitz (1962) with an appropriately chosen flat background metric. (Here a *pseudotensor field* is a tensor field which requires for its definition additional structure on spacetime, such as a preferred coordinate system or a preferred background metric.) However, we already remarked in section 4.4*b* that our t_{ab} differs from the lowest nonvanishing (i.e., quadratic in γ_{ab}) approximation to the Landau-Lifshitz pseudotensor by only a term that makes no contribution to the total energy flux between stationary periods. Thus, our calculation of the total energy radiated to infinity in the linearized approximation by systems which radiate only for a finite period of time is in accord with the general notion of the energy flux of gravitational radiation described here.

With regard to the definition of energy-momentum at spatial infinity, a notion of total energy and momentum on a hypersurface Σ was given in coordinate form by Arnowitt, Deser, and Misner (1962), motivated by the Hamiltonian formulation of general relativity. It was reformulated more geometrically in terms of spatial infinity by Geroch (1972*b*, 1977), Ashtekar and Hansen (1978), and Ashtekar (1980). We refer the reader to these references for these reformulations, and merely present here the formulas for energy and momentum in the original form given by Arnowitt, Deser, and Misner (ADM). Further discussion of the motivation for these formulas is given at the end of appendix E. Let (M, g_{ab}) be an asymptotically flat spacetime, and let Σ be a spacelike hypersurface which is $C^{>1}$ at i^0 so that (Σ, h_{ab}, K_{ab}) is an asymptotically flat initial data set as defined in section 11.1. Let x^1, x^2, x^3 be asymptotically Euclidean coordinates for Σ in the sense of problem 2. We define the total energy, E, and the coordinate components of total spatial momentum p_ν associated with Σ by

$$E = \frac{1}{16\pi} \lim_{r\to\infty} \sum_{\mu,\nu=1}^{3} \int \left(\frac{\partial h_{\mu\nu}}{\partial x^\mu} - \frac{\partial h_{\mu\mu}}{\partial x^\nu} \right) N^\nu \, dA \quad , \tag{11.2.14}$$

$$p_\nu = \frac{1}{8\pi} \lim_{r\to\infty} \sum_{\mu=1}^{3} \int (K_{\mu\nu} N^\mu - K^\mu{}_\mu N_\nu) dA \quad . \tag{11.2.15}$$

Here $r^2 = [(x^1)^2 + (x^2)^2 + (x^3)^2)]^{1/2}$, the integrals are taken over a sphere of constant r, and N^a is the unit outward normal to this sphere. Note that E does not depend explicitly on K_{ab}, although it should be remembered that h_{ab} and K_{ab} are related by the constraint equations (10.2.41) and (10.2.42). Equations (11.2.14) and (11.2.15) can be interpreted as yielding a number, E, and a vector, p^a, tangent to Σ at i^0, both of which can be shown to be independent of the choice of asymptotically Euclidean coordinates on Σ. Although it is not obvious from formula (11.2.14), in the stationary case if we choose Σ to be asymptotically orthogonal to the timelike Killing field, then this definition agrees with equation (11.2.9) (see Ashtekar and Magnon-Ashtekar 1979*a*). Furthermore, the Einstein evolution equations imply that (E, p_a) "transform" properly under changes of Σ, i.e., that the 4-vector P_a at i^0 defined by

$$P_a = -En_a + p_a \tag{11.2.16}$$

is independent of Σ, where n^a is the future-directed unit normal to Σ at i^0. Thus, we obtain a notion of energy-momentum at spatial infinity—known as the *ADM energy-momentum*—which satisfies all the desired properties mentioned above.

Given the above two notions of energy momentum, a number of interesting issues arise. One concerns the relation between the Bondi and ADM energies. From the above definitions, it is natural to interpret the ADM energy as the total energy available in the spacetime, and the Bondi energy as the energy remaining in the spacetime at the "retarded time" given by the cross section $\mathcal{S}$, after emission of gravitational radiation. If this is so, the Bondi energy should differ from the ADM energy by precisely the integral of the energy flux, f, over the portion of $\mathscr{I}^+$ to the causal past of $\mathscr{S}$. However, it is far from obvious from the definitions that this relation holds. Indeed the Bondi and ADM energy-momentum vectors are defined in different vector spaces, so it is not even obvious that a comparison can be made. However, it turns out that each BMS translation at $\mathscr{I}^+$ (or $\mathscr{I}^-$) can be naturally associated with a tangent vector at i^0, so a meaningful comparison is possible. Ashtekar and Magnon-Ashtekar (1979*b*) have shown that when the integrated energy flux to the past of $\mathscr{S}$ is finite, then the Bondi and ADM energies differ by precisely this amount. Thus, the expected relation between the Bondi and ADM energies does indeed hold.

An even more fundamental issue concerns the positivity of the Bondi and ADM energies. As mentioned above, the Bondi energy flux always is positive; i.e., the Bondi energy always decreases with time. However, we made no claim above that the value of the Bondi energy itself must be positive, i.e., that, even if it started out positive, it could not pass through zero and become negative. Nor did we assert that the ADM total energy could never be negative (so that, in view of the above result, the Bondi energy would start out negative). This raises the following two important physical issues: (i) Can an isolated system ever have negative total energy content? (ii) Even if the total energy content always is positive, is the total energy radiated away by the system bounded by its total energy content? In other words: (i) Is the ADM energy always positive? and (ii) Is the Bondi energy always positive?

In formulating these questions, one should keep in mind that the positivity of energy can be expected to hold in general only if the spacetime is nonsingular and conditions are imposed on the matter distribution. For example, the Schwarzschild solution with a negative value of M is an asymptotically flat solution of Einstein's equation with negative ADM and Bondi energies. However, the negative mass Schwarzschild spacetime has a singularity at "$r = 0$" and, unlike the Schwarzschild solution with positive M, it fails to be globally hyperbolic. Furthermore there is no obvious way of "covering up" this singularity with physically reasonable matter to produce a nonsingular interior solution which matches on to the negative mass Schwarzschild exterior. Indeed, equation (6.2.10) establishes that any static, spherically symmetric fluid interior solution for the negative mass Schwarzschild solution must have a negative energy density of matter. Thus, the example of the negative mass Schwarzschild solution appears to require either singularities or negative energy matter to produce a negative ADM or Bondi energy. Since the physical relevance of either of these possibilities is highly questionable, the issue remains: In a nonsingular, asymptotically flat spacetime with locally nonnegative matter energy density—i.e., more precisely, with T_{ab} satisfying the dominant energy condition (see chapter 9)—can the ADM or Bondi energy be negative? This issue is of great physical significance because if the energy of an isolated system need not be positive (or, at least, bounded from below[5]) it is unlikely that any isolated system could be

5. Under a scale transformation $g_{ab} \to \lambda^2 g_{ab}$ with λ constant, the energy scales as $E \to \lambda E$. Thus, if E can be negative, no lower bound for E exists.

absolutely stable, i.e., every isolated system eventually should decay to configurations of lower and lower energy.

It has proven remarkably difficult to establish the positivity of the ADM and Bondi energies in general relativity. Numerous authors have obtained proofs of the positivity of ADM energy in special cases or have given general arguments in favor of the conjecture which, however, fall short of providing a satisfactory, complete proof. Finally, Schoen and Yau (1981) succeeded in giving a complete proof of the positivity of the ADM energy. Shortly thereafter, a greatly simplified proof was given by Witten (1981) (see also Ashtekar and Horowitz 1982; Taubes and Parker 1982; and Reula 1982), who used a spinor field to express E as an integral over Σ of a manifestly nonnegative quantity. This proof of the positivity of the ADM energy has been extended to establish the positivity of Bondi energy (Horowitz and Perry 1982; Schoen and Yau 1982; Ludvigsen and Vickers 1982; Reula and Tod 1984). Thus, in general relativity the total energy of an isolated system is nonnegative, and this energy bounds the amount of energy that can be radiated away in the form of gravitational radiation.

Problems

1. Write out the coordinate components of the gauge condition $\tilde{\nabla}_a \tilde{\nabla}_b \Omega = 0$ on $\mathscr{I}^+$ in the coordinates $(\Omega, u, \theta, \phi)$ introduced in the text. Show that these relations imply that $\tilde{g}_{uu}$, $\tilde{g}_{u\theta}$, and $\tilde{g}_{u\phi}$ are all $O(\Omega^2)$ as $\Omega \to 0$.

2. Let (Σ, h_{ab}, K_{ab}) be an asymptotically flat initial data set. Introduce coordinates X, Y, Z in a neighborhood of Λ such that the gradients of these functions are orthonormal at Λ. Show that the condition that $\tilde{h}_{ab}$ is $C^{>0}$ at Λ implies that in these coordinates, the unphysical metric components are diag$(1, 1, 1) + O(R)$ as $R \to 0$, where $R = (X^2 + Y^2 + Z^2)^{1/2}$, and that the first derivatives of the metric components are $O(1)$ as $R \to 0$. Show further that condition (iii) of the definition of asymptotically flat initial data set implies $\Omega = R^2[1 + O(R)]$. Introduce the coordinates $x = X/R^2, y = Y/R^2, z = Z/R^2$ in the physical spacetime and show that the physical metric components take the form diag$(1, 1, 1,) + O(1/r)$ as $r \to \infty$ where $r = (x^2 + y^2 + z^2)^{1/2}$. Show further that the first coordinate derivatives of these components are $O(1/r^2)$ as $R \to \infty$. Finally, show that condition (iv) implies that the components of K_{ab} and the physical Ricci tensor ${}^{(3)}R_{ab}$ in this asymptotically Euclidean coordinate system are, respectively, $O(1/r^2)$ and $O(1/r^3)$.

3. *a*) Show that the Schwarzschild solution is asymptotically flat at future null infinity.

b) Show that every t = constant hypersurface in the Schwarzschild solution is an asymptotically flat initial data set.

(In fact, the Schwarzschild spacetime satisfies all the conditions of our definition of asymptotic flatness at spatial and null infinity, not just the above separate properties. For a proof, see Ashtekar and Hansen 1978.)

4. Consider a stationary solution with stress-energy T_{ab} in the context of linearized gravity (see section 4.4). Choose a global inertial coordinate system for the flat metric η_{ab} so that the "time direction" $(\partial/\partial t)^a$ of this coordinate system agrees with the stationary Killing field to zeroth order.

a) Show that the conservation equation, $\partial^a T_{ab} = 0$, implies $\int T_{\mu\nu} d^3x = 0$, where the integral is taken over a $t =$ constant slice and μ and ν take any values except $\mu = \nu = 0$ (where $x^0 = t$).

b) Show, therefore, that to first order in deviation from flatness, the general formula (11.2.10) for M in a stationary spacetime reduces to $M = \int T_{00}\, d^3x$, i.e., we obtain the usual formula of special relativity. Note, however, that the Komar integral (11.2.9) taken over a finite sphere lying within the matter distribution does not, in general, equal the mass contained within that sphere.

5. *a*) Calculate the right side of equation (11.2.9) for the Schwarzschild solution and show that we get the Schwarzschild mass parameter M. Thus, for a static, spherically symmetric fluid star we have two formulas for M, namely equation (6.2.10) and equation (11.2.10).

b) Show that the equality of the two formulas (6.2.10) and (11.2.10) for M yields the relation

$$\int_\Sigma (\rho + 3p) e^\phi \, dV = \int_\Sigma \rho \left(1 - \frac{2m(r)}{r}\right)^{1/2} dV \quad ,$$

where the integral is taken over a hypersurface Σ orthogonal to the static Killing field ξ^a, dV denotes the natural volume element on Σ, and $\phi \equiv \frac{1}{2} \ln(-\xi_a \xi^a)$ is the general relativistic analog of the Newtonian gravitational potential (see section 6.2).

c) Put in the G's and c's in the formula of part (*b*) and examine the Newtonian limit, $c \to \infty$. In this limit both sides of the above equation reduce to the rest mass of the fluid. Show that to the next nonvanishing approximation in $1/c$, the equation of part (*b*) yields

$$E_g = -2K \quad ,$$

where E_g is given by the usual Newtonian formula for gravitational potential energy, and $K \equiv \frac{3}{2} \int_\Sigma P \, dV$, so that in the case of a monatomic ideal gas, K is just the total thermal energy. Thus, the equation of part (*b*) may be viewed as the general relativistic version of the Newtonian virial theorem.

6. In an axisymmetric, asymptotically flat spacetime with axial Killing field ψ^a, we may define the total angular momentum, J, by

$$J = \frac{1}{16\pi} \int_S \epsilon_{abcd} \nabla^c \psi^d \quad ,$$

where the integral is taken over a sphere, S, in the asymptotic region where T_{ab} is assumed to vanish. (Note the factor of -2 difference between this formula and the formula [11.2.9] for M.)

a) Parallel the arguments given in section 11.2 for M to show that J is independent of S and that J is given by

$$J = -\int_{\Sigma} T_{ab} n^a \psi^b \, dV \quad ,$$

where the hypersurface Σ is chosen so that ψ^a is tangent to Σ. Note that the fact that J is independent of S can be interpreted as saying that in an axisymmetric spacetime angular momentum cannot be radiated away by gravitational radiation.

b) Show that $J = 0$ in any static, axisymmetric spacetime, i.e., a spacetime which possesses a hypersurface orthogonal timelike Killing field ξ^a with $\xi^a \psi_a = 0$.

c) Calculate J for the charged Kerr metric (eq. [12.3.1] of chapter 12) and show that $J = Ma$.

(For discussions of the definition of angular momentum in nonaxisymmetric asymptotically flat spacetimes, see Ashtekar 1980, Ashtekar and Streubel 1981, Geroch and Winicour 1981, and Ashtekar and Winicour 1982.)

TWELVE

BLACK HOLES

In chapter 6, we found that a sufficiently massive, cold, spherical fluid body cannot exist in hydrostatic equilibrium and, hence, must undergo complete gravitational collapse. The resulting spacetime structure was depicted in Figure 6.11. The most striking feature of this spacetime is that a black hole—i.e., a "region of no escape"—is produced. Any observer or light ray that enters region II of Figure 6.11 never will be able to escape from this region and, indeed, will end his existence in the spacetime singularity labeled by the coordinate value $r = 0$. Furthermore, this spacetime singularity at $r = 0$ produced by spherical collapse is contained within the black hole and, thus, cannot be "seen" by any observer who remains outside the black hole.

The purpose of this chapter is to generalize these ideas to the nonspherical case. Our analysis of spherical gravitational collapse was greatly aided by the fact that the only spherically symmetric, vacuum solution of Einstein's equation is the Schwarzschild solution, and thus the Schwarzschild metric must describe the spacetime geometry exterior to any spherical collapsing body. However, no such simplication occurs in the nonspherical case, and, indeed, the details of nonspherical collapse have been studied only for linear perturbations of spherical collapse (Price 1972*a, b*) and in numerical investigations. Nevertheless, there are reasons for believing that the basic picture of collapse will remain the same as in the spherical case, namely, that a black hole will form and the spacetime singularity resulting from the gravitational collapse will be hidden within the black hole. This issue is discussed in section 12.1. A precise definition of a black hole is given there. Some basic properties of black holes, including the law of area increase, then are proven in section 12.2.

Stationary black holes are of considerable physical interest since one would expect any black hole formed by gravitational collapse to "settle down" to a stationary final state. Remarkably, it has been possible to show that the two-parameter family of solutions—characterized by total mass, M, and total angular momentum, J—found by Kerr (1963) are the only vacuum solutions of Einstein's equation describing stationary black holes. The Kerr metric and its charged generalization are discussed in section 12.3, and the stationary black hole uniqueness results are summarized at the end of that section.

A surprising development in the theory of black holes was the discovery that energy can be extracted from a "rotating" black hole, i.e., a Kerr black hole with $J \neq 0$. Although nothing can escape from a black hole, it is possible to make a black hole "swallow" a particle or wave with negative total energy. We describe how this can be done in section 12.4 and also discuss the limitations on energy extraction imposed by the law of area increase.

Finally, one of the most intriguing aspects of the theory of black holes is the analogy between the laws of black hole physics and the ordinary laws of thermodynamics. At first sight, it would appear that the nature of these laws hardly could be more different. The laws of black hole physics are rigorous theorems in differential geometry, while the laws of thermodynamics are only macroscopic approximations to complicated, exact microscopic laws of physics. Nevertheless, a remarkably close mathematical analogy exists between these laws, highlighted by the analogy between the law of area increase for black holes and the law of entropy increase for thermodynamic systems. This analogy is developed in section 12.5. Further developments resulting from investigations of quantum effects near black holes are discussed in chapter 14.

12.1 Black Holes and the Cosmic Censor Conjecture

Our first task in the investigation of nonspherical collapse is to formulate a precise notion of a black hole. Basically, we wish to define a black hole as a "region of no escape" like region II of Figure 6.11, i.e., in physical terms a region of spacetime where gravity is so strong that any particle or light ray entering that region never can escape from it. However, this notion is *not* properly captured by defining a black hole in a spacetime (M, g_{ab}) to be simply a subset $A \subset M$ such that for any point $p \in A$ we have $J^+(p) \subset A$. With that definition the causal future of any set in any spacetime would be called a black hole. Thus, we must take much greater care to specify what portion of spacetime the impossibility of "escaping to" should be considered grounds for calling a region a black hole.

For asymptotically flat spacetimes, the impossibility of escaping to future null infinity, $\mathscr{I}^+$, provides an appropriate characterization of a black hole. The crucial property of region II of Figure 6.11 which distinguishes it from, say, the causal future of a point in Minkowski spacetime is that all of region II is confined to "small r," i.e., region II does not extend "out to infinity." This is illustrated by examination of a conformal diagram of the spherical collapse spacetime (M, g_{ab}) of Figure 6.11, that is, a spacetime diagram of the unphysical spacetime $(\tilde{M}, \tilde{g}_{ab})$ associated with (M, g_{ab}) (see chapter 11). This is shown in Figure 12.1. Another representation of $\tilde{M}$ is given in Figure 12.2. As illustrated by these diagrams, the causal past of future null infinity, $J^-(\mathscr{I}^+)$, is nonsingular, but it does not include the entire physical spacetime; region II is not contained in $J^-(\mathscr{I}^+)$. In contrast, for Minkowski spacetime $J^-(\mathscr{I}^+)$ includes the entire physical spacetime.

The idea that $J^-(\mathscr{I}^+)$ be "well behaved" but not include the entire spacetime leads to the following definition of a black hole. Let (M, g_{ab}) be an asymptotically flat spacetime with associated unphysical spacetime $(\tilde{M}, \tilde{g}_{ab})$. We say that (M, g_{ab}) is *strongly asymptotically predictable* if in the unphysical spacetime there is an open region $\tilde{V} \subset \tilde{M}$ with $\overline{M \cap J^-(\mathscr{I}^+)} \subset \tilde{V}$ such that $(\tilde{V}, \tilde{g}_{ab})$ is globally hyperbolic.

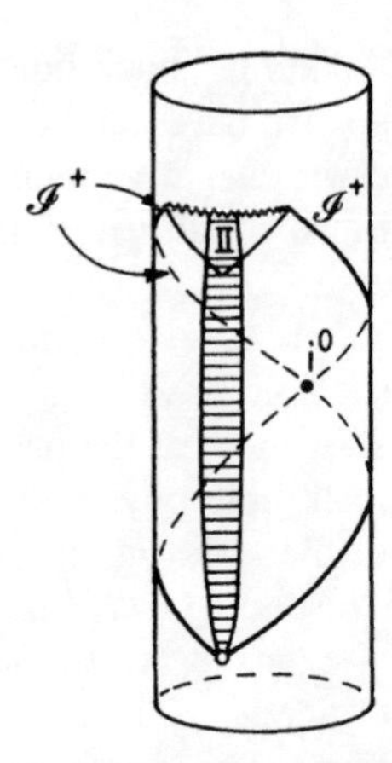

Fig. 12.1. A conformal diagram of the same spacetime as shown in Figures 6.11 and 6.12. From this conformal diagram, it is apparent that region II of the physical spacetime lies outside of $J^-(\mathscr{I}^+)$. In contrast, in Figure 11.1, $J^-(\mathscr{I}^+)$ includes the entire physical spacetime.

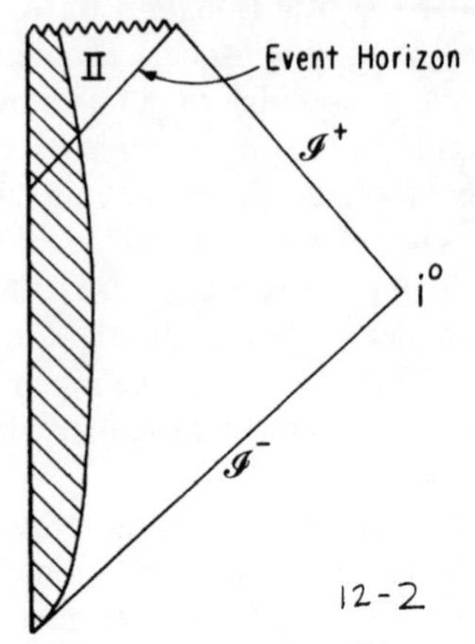

Fig. 12.2. Another representation of the closure, $\bar{M}$, of the physical spacetime depicted in Figure 12.1. As in Figure 12.1, the angular dimensions are suppressed so each point in this diagram (except those at $r = 0$ and the point i^0) represents a 2-sphere.

[Here, the closure of $M \cap J^-(\mathscr{I}^+)$ is taken in the unphysical spacetime $\tilde{M}$, so, in particular, $i^0 \in \tilde{V}$. Our definition of strong asymptotic predictability differs slightly from that given by Hawking and Ellis 1973 since, in particular, we use a different formulation of the notion of asymptotic flatness.] A strongly asymptotically predictable spacetime is said to contain a *black hole* if M is not contained in $J^-(\mathscr{I}^+)$. The *black hole region*[1], B, of such a spacetime is defined to be $B = [M - J^-(\mathscr{I}^+)]$, and the boundary of B in M, $H = \dot{J}^-(\mathscr{I}^+) \cap M$, is called the *event horizon*.

It should be noted that the requirement that $(\tilde{V}, \tilde{g}_{ab})$ be a globally hyperbolic region

1. The *white hole region* of a strongly asymptotically "retrodictable" spacetime is defined similarly by replacing $J^-(\mathscr{I}^+)$ with $J^+(\mathscr{I}^-)$.

of the unphysical spacetime implies that $(M \cap \tilde{V}, g_{ab})$ is a globally hyperbolic region of the physical spacetime. Namely, according to property (1) of the definition of asymptotic flatness given in chapter 11, we have $M = \tilde{M} - [J^+(i^0) \cup J^-(i^0)]$. Hence, a Cauchy surface for $(\tilde{V}, \tilde{g}_{ab})$ which passes through i^0 will be a Cauchy surface for $(M \cap \tilde{V}, \tilde{g}_{ab})$. However, since g_{ab} and $\tilde{g}_{ab} = \Omega^2 g_{ab}$ have the same causal structure, this implies that $(M \cap \tilde{V}, g_{ab})$ is globally hyperbolic.

By theorem 8.3.14, we can foliate $M \cap \tilde{V}$ with Cauchy surfaces Σ_t. For all $q \in M \cap \tilde{V}$ and all Σ_t with $q \in J^+(\Sigma_t)$, every past directed inextendible causal curve from q intersects Σ_t. This can be interpreted as saying that—apart from a possible "initial singularity" such as the white hole singularity of Figure 6.9—no singularities are "visible" to any observer in $[M \cap \tilde{V}] \supset [M \cap \overline{J^-(\mathscr{I}^+)}]$. In other words, in a strongly asymptotically predictable spacetime, all observers outside the black hole or on the event horizon cannot "see" any singularities develop at a finite "time." In contrast, an asymptotically flat spacetime which fails to be strongly asymptotically predictable is said to possess a *naked singularity*. Note that in a strongly asymptotically predictable spacetime it still is possible for the event horizon to be singular in the sense that there may exist incomplete causal geodesics (with respect to the physical metric, g_{ab}) in $M \cap \overline{J^-(\mathscr{I}^+)}$ which are inextendible in M. In particular, strong asymptotic predictability does not exclude the possibility that the null geodesic generators of the event horizon may be future incomplete. It is usually assumed in discussions of black holes that the event horizon is nonsingular, but since we shall not need any conditions beyond strong asymptotic predictability for the theorems proven below, we shall not impose any such further conditions here.

Note also that we have defined the notion of a black hole only for strongly asymptotically predictable spacetimes. The above definition could be given without modification for asymptotically flat spacetimes which are not strongly asymptotically predictable, but we choose not to do so since this case is not believed to be physically relevant (see below) and virtually all properties of black holes derived below require strong asymptotic predictability. The notion of a black hole also could be defined for some non-asymptotically flat spacetimes where a suitable notion of "infinity" can be introduced, such as spacetimes which asymptotically approach an "open" $(k = 0, -1)$ Robertson-Walker universe. On the other hand, there appears to be no natural notion of a black hole in a "closed" $(k = +1)$ Robertson-Walker universe which recollapses to a final singularity, since there is no natural region to which "escape" can be attempted. Of course, an approximate notion of a black hole still exists for any region of a closed Robertson-Walker universe that can be treated as an isolated system.

Now that we have given a precise definition of a black hole in a strongly asymptotically predictable spacetime, we may inquire as to the physical relevance of this definition. In the spherical case discussed in section 6.4 above, we have seen that gravitational collapse results in a strongly asymptotically predictable spacetime possessing a singularity contained within a black hole. What type of spacetime results from nonspherical collapse? First, we may invoke the singularity theorems to conclude that—at least for sufficiently small deviations from spherical symmetry—a spacetime singularity must occur in gravitational collapse; i.e., the occurrence of a singularity is not merely an artifact of the assumption of exact spherical symmetry.

To see this, we argue as follows: Starting from initial data (Σ, h_{ab}, K_{ab}) for spherical collapse, we find that trapped surfaces form in $D^+(\Sigma)$ since all the spheres with $r < 2M$ in the exterior Schwarzschild region are trapped surfaces contained within $D^+(\Sigma)$. It then follows from theorem 10.2.2 that for all initial data sufficiently near (Σ, h_{ab}, K_{ab}), trapped surfaces also must occur in the maximal Cauchy development of these data. (Even if departures from spherical symmetry are large, Schoen and Yau 1983 have proven that trapped surfaces always must occur when enough matter is condensed in a small region.) We then may appeal to theorem 9.5.3 or theorem 9.5.4 to establish the occurrence of a spacetime singularity.

However, the existence of a spacetime singularity does not establish the existence of a black hole, since the singularity could be naked, i.e., the spacetime could fail to be strongly asymptotically predictable. The strongest direct evidence that naked singularities do not occur comes from a study of linear perturbations of the Schwarzschild spacetime. Since the metric perturbation equations (see section 7.5) are of the form to which theorem 10.1.2 applies, we know that an initially well behaved perturbation on a Cauchy surface for the globally hyperbolic region $(\tilde{V} \cap M, g_{ab})$ will evolve to a nonsingular solution in $\tilde{V} \cap M$. However, if there exists a solution of the linear perturbation equations which "blows up"—i.e., is unbounded in $\tilde{V} \cap M$—this would indicate that the full, nonlinear equations might yield a naked singularity in $\tilde{V} \cap M$ even for data arbitrarily close to that for Schwarzschild. On the other hand, if all solutions of the linear perturbation equations are bounded in $\tilde{V} \cap M$ by an appropriate norm of the initial data, this would suggest that the solutions of the full, nonlinear equations would be nonsingular in $\tilde{V} \cap M$ for initial data sufficiently close to that for Schwarzschild. Studies of the behavior of solutions of the perturbation equations have shown that the solutions are bounded in $\tilde{V} \cap M$ (see Wald 1979*a*). This suggests that gravitational collapse with sufficiently small departures from spherical symmetry will produce a black hole rather than a naked singularity. In addition, numerical studies of linear perturbations of the Kerr black hole (see section 12.3) have indicated that it also is stable (Press and Teukolsky 1973).

Further support for the conclusion that gravitational collapse always produces a black hole rather than a naked singularity comes from the fact that a number of theoretical attempts to construct counterexamples all have failed in such a way as to suggest a conspiracy of nature against producing naked singularities (see, e.g., Wald 1974*a*; Jang and Wald 1977; Gibbons *et al*. 1983). In addition, although arguments based on aesthetic appeal always should be viewed with suspicion, as we shall attempt to illustrate in this chapter and in chapter 14, the study of the properties of black holes has led to so many remarkable theoretical developments that it is very difficult to believe that black holes are not relevant physically.

Thus, the above considerations have led to the conjecture that nature "censors" naked singularities. We may state this conjecture in physical terms as follows:

COSMIC CENSOR CONJECTURE *(version 1; physical formulation)*. The complete gravitational collapse of a body always results in a black hole rather than a naked singularity; i.e., all singularities of gravitational collapse are "hidden" within black holes, where they cannot be "seen" by distant observers.

The above formulation of the cosmic censor conjecture is imprecise because, among other reasons, we have not specified what conditions the matter fields must satisfy. Clearly, the conjecture is false without any conditions imposed on the matter fields, since one could write down any spacetime with a naked singularity and call it a solution of Einstein's equation by defining the stress-energy, T_{ab}, to be $(1/8\pi)G_{ab}$. Two natural conditions to impose on the matter sources are that (1) T_{ab} satisfy an energy condition (e.g., the dominant energy condition) and that (2) the coupled Einstein-matter field equations admit a well posed initial value formulation. However, perfect fluids satisfy both of these conditions, but violations of the cosmic censor conjecture with perfect fluid matter can be achieved. This can be understood from the fact that even in Minkowski spacetime the dynamical evolution of a perfect fluid can result in singularities such as those caused by "shell crossings" or the formation of shocks. In the coupled Einstein–perfect fluid system these types of singularities still may occur, and, indeed, because of Einstein's equation, the singularities in T_{ab} imply singularities of the spacetime curvature as well. Thus, naked singularities can occur in the gravitational collapse of a perfect fluid (Yodzis, Seifert, and Müller zum Hagen 1973, 1974). However, a perfect fluid is really a macroscopic approximation to the structure of matter rather than a fundamental description of it. Thus, the appearance of singularities in the dynamical evolution of a fluid—particularly the relatively "mild" types of singularities produced by shell crossings or shocks—may be viewed as representing simply the breakdown of our macroscopic approximation rather than the occurrence of a true, physical singularity of gravitational collapse like the $r = 0$ singularity of the Schwarzschild solution. This suggests that for a precise formulation of the cosmic censor conjecture we should require the matter fields to be "fundamental." We shall take this to mean that the coupled Einstein-matter field equations can be put in the form of a second order, quasilinear, diagonal, hyperbolic system (see section 10.1), since the fundamental classical fields which are known to accurately describe nature—namely, gravity and electromagnetism—are of this form. Thus, we are led to the following as one version of a mathematically precise formulation of the cosmic censor conjecture.

COSMIC CENSOR CONJECTURE *(version 1; precise formulation)*. Let (Σ, h_{ab}, K_{ab}) be an asymptotically flat initial data set (see chapter 11) for Einstein's equation with (Σ, h_{ab}) a complete Riemannian manifold. Let the matter sources be such that T_{ab} satisfies the dominant energy condition and the coupled Einstein-matter field equations are of the form (10.1.21). In addition let the initial data for the matter fields on Σ satisfy appropriate asymptotic falloff conditions at spatial infinity. Then the maximal Cauchy evolution of these initial data (see section 10.2) is an asymptotically flat, strongly asymptotically predictable spacetime.

The issue of whether the cosmic censor conjecture is correct remains the key unresolved issue in the theory of gravitational collapse. The physical relevance of black holes depends in large measure on the validity of this conjecture.

If the above version of the cosmic censor conjecture is correct, it is interesting to inquire as to whether its validity, in essence, stems from a more general, fundamental property of Einstein's equation. Penrose (1979) has suggested that this may be the case. He has conjectured what is, in essence, a stronger version of the cosmic censor conjecture which may be stated in physical terms as follows:

COSMIC CENSOR CONJECTURE *(version 2; physical formulation)*. All physically reasonable spacetimes are globally hyperbolic, i.e., apart from a possible initial singularity (such as the "big bang" singularity) no singularity is ever "visible" to any observer.

This version of the cosmic censor conjecture is stronger than version 1 in that it applies to any observer in any spacetime, not just to distant observers in asymptotically flat spacetimes. However, version 2 does not imply version 1, since, if, starting from asymptotically flat initial data, a singularity is formed which propagates out to null infinity and destroys asymptotic flatness while preserving global hyperbolicity, this would violate version 1 but not version 2.

The meaning of the second version of the cosmic censor conjecture may be reformulated more precisely as follows. First, it certainly is not true that every inextendible solution of Einstein's equation is globally hyperbolic. By cutting holes in Minkowski spacetime and then making topological identifications so that not all of the missing points can be restored, we easily may produce an inextendible, non–globally hyperbolic spacetime. To avoid the possibility of making "unnecessary" and "artificial" singularities of this sort, we may give a precise formulation of the notion that all reasonable spacetimes are globally hyperbolic as meaning that the maximal Cauchy development of nonsingular initial data (satisfying appropriate asymptotic conditions) always yields an inextendible spacetime. However, this property of solutions of Einstein's equation also is known to be false. In particular, the maximal Cauchy developments of initial data for the Taub universe (Misner 1967; Hawking and Ellis 1973) and the Kerr solution (see section 12.3) are known to be extendible. However, in the Taub case, the extension violates strong causality on the Cauchy horizon, and there is good reason to believe that if one slightly perturbs the initial data for the Taub universe in a suitable way, one would convert the Cauchy horizon to a singularity, thereby making the maximal Cauchy development inextendible. In the Kerr case, as discussed below, any observer on the Cauchy horizon of the extended spacetime can "see" the entire, noncompact initial data surface. One would expect that a suitably chosen small perturbation of the initial data which extends out to infinity (but does not violate asymptotic flatness) should produce an "infinite blueshift" singularity on the Cauchy horizon. Indeed, such behavior has been explicitly demonstrated to occur in the Reissner-Nordstrom spacetime (Chandrasekhar and Hartle 1982), whose properties are closely analogous to those of Kerr. Thus, although some violations of global hyperbolicity are known to occur, it appears that in these cases small perturbations may destroy the extendibility of the maximal Cauchy development, so that perhaps no "generic" violations are possible. Since it is difficult to give a precise definition of the term "generic," we formulate a precise statement of the second version of the cosmic censor conjecture as follows (Geroch and Horowitz 1979; Penrose 1979):

COSMIC CENSOR CONJECTURE *(version 2; precise formulation)*. Let (Σ, h_{ab}, K_{ab}) be an initial data set for Einstein's equation, with (Σ, h_{ab}) a complete Riemannian manifold and with the Einstein-matter equations of the form (10.1.21) with T_{ab} satisfying the dominant energy condition. Then, if the maximal Cauchy development of this initial data is extendible, for each $p \in H^+(\Sigma)$ in any extension, either strong causality is violated at p or $\overline{I^-(p) \cap \Sigma}$ is noncompact.

Apart from the absence of known counterexamples, there is virtually no evidence for or against the validity of this second version of the cosmic censor conjecture. Indeed, except for the singularity theorems (see chapter 9), very little is known about the general, global properties of solutions of Einstein's equation.

Let us return, now, to the subject of gravitational collapse and black holes. We shall assume the validity of the first version of the cosmic censor conjecture and shall briefly discuss three processes by which gravitational collapse to a black hole plausibly may occur in nature. The first process is the gravitational collapse of a star. The evolutionary history of a star was outlined briefly in section 6.2, where it was concluded that if a spherical star has mass greater than about 2 solar masses, it ultimately must undergo gravitational collapse, unless, of course, it can shed enough mass during the course of its evolution to drop below this upper mass limit. Rotation of the star may raise the upper mass limit, but since only one pulsar (i.e., neutron star) has been observed to be rotating rapidly enough to possibly affect its equilibrium structure significantly (Backer *et al.* 1982), it appears reasonable to take the spherical upper mass limit as generally applicable. Since many stars in our galaxy are observed to have mass greater than 2 solar masses and since the evolutionary lifetime of massive stars is much less than the age of our Galaxy, it would be very difficult to avoid the conclusion that many black holes have been produced in our Galaxy by this process. However, since there are great uncertainties as to the mass loss of stars—occurring either gradually during their evolution (particularly in the red giant phase) or violently in a supernova explosion at the endpoint of their evolution—there are no reliable estimates as to precisely how many such black holes should have been produced. Perhaps the best guess can be obtained from the observational estimate that supernovae occur in our Galaxy at the rate of several per century. Thus at least about 10^8 supernovae should have occurred during the lifetime of our Galaxy, so perhaps $\sim 10^8$ black holes may have formed. Of course, this estimate may be far too high since many supernovae may result in neutron stars rather than black holes, or it may be far too low since gravitational collapse to a black hole without the violent blowing off of the outer layers of a star may occur more frequently than supernovae. It should be noted that black holes formed by stellar collapse must lie in the relatively narrow mass range $2\, M_\odot \lesssim M \lesssim 100\, M_\odot$ since stars with $M \lesssim 2\, M_\odot$ should not collapse, while ordinary stars with $M \gtrsim 100\, M_\odot$ do not exist on account of pulsational instabilities.

A second process by which black holes may be formed in nature involves the gravitational collapse of the entire central core of a dense cluster of stars. During the dynamical evolution of a star cluster, gravitational encounters occasionally will result in a large transfer of energy to a single star, which then will "evaporate" from the cluster. The remaining stars thereby lose energy and become more tightly gravitationally bound. Thus, the central portion of a star cluster tends to become more and more dense as the cluster evolves. Eventually a stage will be reached where the core becomes so dense that tidal disruption of the stars will take place. It is difficult to predict precisely what will happen after this point, but many scenarios lead to the formation of a massive black hole (Rees 1978). (Massive black holes at the center of star clusters also could be produced by the direct collapse of part of the gas cloud out of which the star cluster originally formed, or by the coalescence and growth of black holes produced by stellar collapse.) The nuclei of galaxies provide the most likely sites for the formation of massive black holes by this process. However, again there are no reliable estimates as to how many galactic nuclei or cores of star clusters in other regions of a galaxy should contain a massive black hole. The mass of black holes formed by this process could range up to a sizable fraction of the mass of the star cluster, i.e., up to $\sim 10^{10} M_\odot$ for a black hole in the nucleus of a large galaxy.

A third, much more speculative, process by which black holes may have been produced is by the gravitational collapse of regions of enhanced density in the early universe. As discussed in chapter 5, the universe appears to be homogeneous and isotropic to an excellent approximation on large distance scales. However, the matter distribution we observe in the present universe certainly is not exactly homogeneous, so some initial inhomogeneities must have been present in the early universe. Some information about the initial spectrum of density inhomogeneities on galactic mass scales can be obtained by statistical studies of the present clustering behavior of galaxies (Peebles 1980), but very little is known about initial inhomogeneities on smaller scales. However, if sufficiently large inhomogeneities in the density were present in the early universe, the regions of enhanced density could collapse directly to a black hole rather than expand with the rest of the universe. Thus, large numbers of "primordial black holes" could have been produced in this way. Again, however, there are no reliable predictions of how many—if any—primordial black holes exist in our universe. Indeed, the best limits on the density inhomogeneities on small mass scales come from the requirement that they not overproduce small black holes.

An important feature of primordial black holes is that they could have been produced at any mass scale, including masses much smaller than a solar mass. No process in the present universe could produce such small black holes. For example, a Schwarzschild black hole of mass equal to that of the Earth, $M_E = 6 \times 10^{27}$ g, has $r_S = 2GM_E/c^2 \sim 1$ cm. Thus, to convert the Earth into a black hole, we would, presumably, have to compress it by nongravitational forces down to a radius of order r_S (and thus a density $\sim 10^{27}$ g/cm^{-3}) before self-gravitation would result in its collapse to a black hole. On the other hand, at $\tau \sim 10^{-11}$ s after the big bang, the density, ρ, of the universe was $\sim 10^{27}$ g/cm^{-3}. A density fluctuation with $\delta\rho \sim \rho$ over a scale of 1 cm at this time could have resulted in the formation of a primordial black hole with mass M_E. Thus, very soon after the big bang, black holes of small mass may have been produced.

How might black holes produced by the above three processes be detected? An important point to recognize is that black holes are extremely small objects. A black hole of one solar mass has a Schwarzschild radius of only 3 km; even a black hole of $10^{10} M_\odot$ in the nucleus of a large galaxy would have $r_S \sim 3 \times 10^{10}$ km $\sim 3 \times 10^{-3}$ light-years. When added to the fact that black holes are "black" (see problem 1), it is clear that direct detection of a black hole would be extremely difficult. The most promising possibility for indirect detection of a black hole comes from the fact that matter which accretes onto it will heat up and emit electromagnetic radiation before entering the black hole. For black holes formed by stellar collapse, the best opportunity for such accretion occurs if the black hole is in a close binary orbit with a star, so that matter can flow from the star toward the black hole. In this case, the matter would be expected to slowly spiral into the black hole, thereby forming an *accretion disk* around the black hole. Viscous heating in the accretion disk could result in the production of X-rays. A number of X-ray sources with an ordinary (i.e., uncollapsed) star in close binary orbit around an unseen (at optical frequencies) companion have been discovered, but accretion onto a neutron star (and possibly even a white dwarf) also could produce such X-rays, so one cannot conclude that these systems must contain a black hole. However, in the case of Cygnus X-1, the properties of the binary orbit yield a rather firm lower mass limit for the unseen companion of $\sim 9\ M_\odot$ (Paczyński 1974). This is far above the upper mass limit for neutron stars and white dwarfs, so there is very strong reason to believe that the unseen companion of the Cygnus X-1 system is a black hole. Very recently, the X-ray source LMC X-3 also has been determined to consist of a binary system, with the mass of the unseen companion well above the white dwarf and neutron star limits (Cowley *et al.* 1983; Paczyński 1983).

A massive black hole at the center of a star cluster or galactic nucleus could produce an observable effect by altering the equilibrium distribution of stars in the central portion of the cluster, causing more stars to be "drawn in" toward the center. Thus, if a massive black hole were present in a star cluster, one would expect to see a small brightness *enhancement* very near the center of the cluster. In addition, one would expect to see an increase in the average velocity of stars very near the center. Exactly such a brightness enhancement and increased "velocity dispersion" have been observed at the center of the galaxy M87 (Young *et al.* 1978; Sargent *et al.* 1978), thus providing strong evidence for the presence there of a black hole of mass $\sim 5 \times 10^9\ M_\odot$. Furthermore, the galaxy M87 is well known for the existence of a jet of highly energetic particles emanating from the center of the galaxy. Thus, there is a strong suggestion that a massive black hole may be involved in the mechanism which produces this jet, although at present there is no fully satisfactory theory of exactly how a black hole (or any other object) could produce such a jet. Since similar jets occur in other active galaxies and in quasars, massive black holes may be present in these bodies as well. Indeed, as already mentioned at the beginning of this book, the discovery of quasars in the early 1960s and the inability to explain their energy source without the presence of strong gravitational fields provided a great stimulus to the development of the theory of gravitational collapse and black holes.

Finally, we mention a possible means of detecting primordial black holes of very low mass. Classically such black holes would produce virtually no observable effects unless they were sufficiently numerous to provide a cosmologically significant contribution to the mass density of the universe or unless one of them happened to strike the Earth. However, as we shall see in chapter 14, particle creation occurs near black holes and is appreciable for black holes of very small mass. The effects of numerous primordial black holes with $M \lesssim 10^{15}$ g could significantly contribute to the γ-ray background (Page and Hawking 1976). However, no such contribution is seen, so one only may place an upper limit on the number of such small black holes by this means.

12.2 General Properties of Black Holes

In this section we shall use the global methods developed in chapters 8 and 9 to establish some general results in the theory of black holes. Most of these results are due to Hawking. Our discussion will differ slightly from that of Hawking and Ellis (1973) in that we shall not impose topological restrictions on the Cauchy surfaces for $M \cap \tilde{V}$, and some of our definitions (such as that of an outer trapped surface) are not identical to theirs.

First, we note that for any asymptotically flat spacetime (M, g_{ab}) with associated unphysical spacetime $(\tilde{M}, \tilde{g}_{ab})$, if $q \in \mathscr{I}^+$ and $p \in M \cap J^-(q)$, then any point $r \in \mathscr{I}^+$ lying beyond q on the future directed null geodesic generator of $\mathscr{I}^+$ passing through q, satisfies $p \in I^-(r)$. (This follows from the corollary to theorem 8.1.2, since p and r can be joined by a causal curve which is not an unbroken null geodesic.) Hence, we have $M \cap J^-(\mathscr{I}^+) = M \cap I^-(\mathscr{I}^+)$, so $J^-(\mathscr{I}^+)$ is open in M. Thus, the black hole region, $B = M - J^-(\mathscr{I}^+)$ is closed in M. In particular this means that the event horizon H is contained in B.

We define, now, the notion of a black hole at a given instant of time. Let (M, g_{ab}) be a strongly asymptotically predictable spacetime, with globally hyperbolic region $\tilde{V} \supset \overline{M \cap J^-(\mathscr{I}^+)}$ in the unphysical spacetime. Let $B = M - J^-(\mathscr{I}^+)$ be the black hole region of the spacetime. If Σ is a Cauchy surface for $\tilde{V}$, we shall refer to $\Sigma \cap B$ as the total black hole region at time Σ. Each connected component, $\mathscr{B}$, of $\Sigma \cap B$ will be called a *black hole at time* Σ. The number of black holes present in (M, g_{ab}) may vary with "time" (i.e., choice of Cauchy surface Σ), since new black holes may form and black holes present at one time may later coalesce. However, our first theorem may be interpreted as stating that a black hole may never disappear nor may it "bifurcate," i.e., split into more than one black hole at a later time.

THEOREM 12.2.1. *Let (M, g_{ab}) be a strongly asymptotically predictable spacetime and let Σ_1 and Σ_2 be Cauchy surfaces for $\tilde{V}$ with $\Sigma_2 \subset I^+(\Sigma_1)$. Let $\mathscr{B}_1$ be a nonempty connected component of $B \cap \Sigma_1$, i.e., $\mathscr{B}_1$ is a black hole at time Σ_1. Then $J^+(\mathscr{B}_1) \cap \Sigma_2 \neq \emptyset$ and is contained within a single connected component of $B \cap \Sigma_2$.*

Proof. That $J^+(\mathscr{B}_1) \cap \Sigma_2 \neq \emptyset$ follows immediately from the fact that Σ_2 is a Cauchy surface lying to the future of Σ_1. Clearly $J^+(\mathscr{B}_1) \subset B$, so $J^+(\mathscr{B}_1) \cap \Sigma_2$ is contained within $B \cap \Sigma_2$. Therefore, to complete the proof of the theorem, it suffices to show that $J^+(\mathscr{B}_1) \cap \Sigma_2$ is connected. Suppose this were not the case. Then we could find disjoint open sets $O, O' \subset \Sigma_2$ with $O \cap J^+(\mathscr{B}_1) \neq \emptyset$, $O' \cap J^+(\mathscr{B}_1) \neq \emptyset$, and $O \cup O' \supset J^+(\mathscr{B}_1) \cap \Sigma_2$. Then we would have $\mathscr{B}_1 \cap I^-(O) \neq \emptyset$, $\mathscr{B}_1 \cap I^-(O') \neq \emptyset$, and $\mathscr{B}_1 \subset I^-(O) \cup I^-(O')$. However, no point, p, of $\mathscr{B}_1$ can lie in both $I^-(O)$ and $I^-(O')$, for then we could divide the future directed timelike geodesics from p according to whether they intersected Σ_2 in O or O' and thereby divide the timelike vectors at p into nonempty, disjoint open sets. (This would contradict the fact that the interior of the future light cone of V_p is connected.) Thus, $I^-(O) \cap \Sigma_1$ and $I^-(O') \cap \Sigma_1$ must be disjoint open subsets of Σ_1, each of which intersects $\mathcal{B}_1$ and whose union contains $\mathcal{B}_1$. However, this contradicts the hypothesis that $\mathcal{B}_1$ is connected. □

One of the difficulties commonly encountered in working with the definition of a black hole is that one needs to know the entire future development of a spacetime in order to determine if a black hole is present. However, it may happen that one is given only initial data (Σ, h_{ab}, K_{ab}) for a spacetime and cannot explicitly solve the evolution equations. In that case, the general definition of a black hole is of little use for directly determining whether $B \neq \emptyset$ in the spacetime determined by these data and, if $B \neq \emptyset$, where on Σ the black hole region $\mathcal{B} = B \cap \Sigma$ may lie. Therefore, it is useful to develop criteria for the existence and location of a black hole without requiring knowledge of the global time development of the spacetime. The next three results give us such criteria.

In chapter 9, we proved that if a trapped surface is present in a spacetime, appropriate energy conditions are satisfied by matter, and a number of further hypotheses hold, then a singularity must occur (see theorems 9.5.3 and 9.5.4). Thus, trapped surfaces are associated with spacetime singularities. Our first result giving criteria for existence of a black hole states that in a strongly asymptotically predictable spacetime it is not only true that no singularities can be "seen" from null infinity, but it also is true that no trapped surfaces are visible from null infinity. In other words, all trapped surfaces must be entirely contained within black holes.

PROPOSITION 12.2.2. Let (M, g_{ab}) be a strongly asymptotically predictable spacetime for which $R_{ab}k^a k^b \geq 0$ for all null k^a, as will be the case if Einstein's equation holds with the weak or strong energy condition satisfied by matter. Suppose M contains a trapped surface, T. Then $T \subset B$, where B denotes the black hole region of the spacetime.

Proof. Suppose T were not entirely contained within B. Then in the unphysical spacetime, we would have $J^+(T) \cap \mathcal{I}^+ \neq \emptyset$. However, spatial infinity, i^0, is not to the causal future of any point in M, so clearly $i^0 \notin J^+(T)$. Furthermore, since the region $\tilde{V}$ of the unphysical spacetime is globally hyperbolic and T is compact, it follows from theorem 8.3.11 that $J^+(T)$ is closed in $\tilde{V}$. Therefore there is an open neighborhood of i^0 which fails to intersect $J^+(T)$, and, thus, an open region of $\mathcal{I}^+$ does not intersect $J^+(T)$. Since $\mathcal{I}^+$ is connected, this implies that there exists a point $q \in \mathcal{I}^+$ such that $q \in \dot{J}^+(T)$. Hence, according to theorem 9.3.11, in the unphysical spacetime there is a null geodesic γ from $p \in T$ to q which is orthogonal to T and has no conjugate point between T and q. With respect to the physical metric, g_{ab}, γ also is a null geodesic (see appendix D) orthogonal to T with no conjugate point, but now γ is future complete. However, this is impossible, because according to theorem 9.3.6, in the physical spacetime γ must have a conjugate point within affine parameter $2/|\theta_0|$ from p, where $\theta_0 < 0$ is the expansion at p of the orthogonal null geodesic congruence from T to which γ belongs. □

In fact, by a somewhat different argument, we can slightly generalize proposition 12.2.2 to apply to a *marginally trapped surface,* i.e., a compact, spacelike two-dimensional submanifold for which the expansion of both families of orthogonal geodesics is required only to be nonpositive, $\theta \le 0$, rather than strictly negative, $\theta < 0$. We have

PROPOSITION 12.2.3. Let T be a marginally trapped surface in a strongly asymptotically predictable spacetime for which $R_{ab}k^a k^b \ge 0$ for all null k^a. Then $T \subset B$.

Proof. As in the proof of the previous proposition, we know from theorem 9.3.11 that $\dot{J}^+(T)$ is generated by the null geodesics orthogonal to T. The expansion, θ, of these null geodesics in the physical spacetime is nonpositive initially. Therefore, we have $\theta \le 0$ everywhere on $\dot{J}^+(T)$, since $d\theta/d\lambda \le 0$ according to equation (9.2.32) and θ cannot become positive on $\dot{J}^+(T)$ by passing through $-\infty$ because that would imply existence of a conjugate point. Suppose, now, that $T \cap J^-(\mathscr{I}^+) \ne \emptyset$. Then, as in the proof of the previous proposition, in the unphysical spacetime there would exist a point $q \in \mathscr{I}^+$ with $q \in \dot{J}^+(T)$. Furthermore, all the points lying to the causal future of q along the null geodesic generator of $\mathscr{I}^+$ passing through q must lie in $I^+(T)$, since they can be joined to T by a causal curve which is not an unbroken null geodesic. Since $I^+(T)$ is open, all generators of $\mathscr{I}^+$ sufficiently near the one passing through q must enter $I^+(T) = \text{int}(J^+(T))$. On the other hand, all generators of $\mathscr{I}^+$ have past endpoints at i^0 and thus leave $J^+(T) = \overline{J^+(T)}$. Therefore, not only q but also an entire local cross section, $\mathscr{S}$, of $\mathscr{I}^+$ must lie on $\dot{J}^+(T)$. Furthermore, the null geodesic generators of $\dot{J}^+(T)$ must strike $\mathscr{S}$ orthogonally in order that there not exist any timelike curves from T to $\mathscr{S}$. However, in the physical spacetime, the expansion of the null geodesic congruence orthogonal to any cross section of $\mathscr{I}^+$ is positive near $\mathscr{I}^+$. This contradicts the previous result that the generators of $\dot{J}^+(T)$ have nonpositive expansion everywhere. □

Note that the only properties of T needed in the proofs of propositions 12.2.2 and 12.2.3 are that $J^+(T)$ is closed, that $i^0 \notin J^+(T)$, and that the expansion of the null geodesic generators of $\dot{J}^+(T)$ is initially nonpositive. Another important example of a set having these properties is the following. Let Σ be any asymptotically flat Cauchy surface for $\tilde{V}$, so that Σ passes through i^0 and is spacelike there. Let $C \subset \Sigma \cap M$ be a closed subset of Σ which forms a three-dimensional manifold with boundary (see appendix B) and suppose the two-dimensional boundary $S = \dot{C}$ of C has the property that the expansion, θ, of the outgoing family of null geodesics orthogonal to S is everywhere nonpositive, $\theta \le 0$. (Here, the *outgoing family* is defined to be the family of null geodesics orthogonal to S satisfying $k^a N_a \ge 0$, where k^a denotes a geodesic tangent and N^a is the normal to S in Σ which points outward from C.) A surface S satisfying these properties is called an *outer marginally trapped surface,* and C is called a *trapped region.* (Note that C need not be connected or compact. Note also that a trapped surface need not be outer trapped since it need not be the boundary of a three-dimensional volume.) Then C can be seen to satisfy the above desired properties as follows: By problem 8 of chapter 8, $J^+(C)$ is closed. Clearly, we have $i^0 \notin J^+(C)$. Finally, each of the null geodesic generators of $\dot{J}^+(C) \subset J^+(C)$ must have a past endpoint on C but cannot meet C in int(C), cannot meet $S = \dot{C}$ nonorthogonally, and cannot be an ingoing geodesic at S, since in any of these cases it would enter $I^+(C)$. Thus, $\theta \leq 0$ initially for the null geodesic generators of $\dot{J}^+(C)$. Hence, a repetition of the proof of proposition 12.2.3 yields the following result.

PROPOSITION 12.2.4. Let (M, g_{ab}) be a strongly asymptotically predictable spacetime for which $R_{ab}k^a k^b \geq 0$ for all null k^a. Let Σ be an asymptotically flat Cauchy surface for $\tilde{V}$, and let $C \subset \Sigma$ be a trapped region. Then $C \subset B \cap \Sigma$.

We define the *total trapped region*, $\mathcal{T}$, of a Cauchy surface, Σ, to be the closure of the union of all trapped regions, C, on Σ. We call the boundary, $\mathcal{A} = \dot{\mathcal{T}}$, of $\mathcal{T}$ the *apparent horizon* on Σ. It follows immediately from proposition 12.2.4 that in a strongly asymptotically predictable spacetime with $R_{ab}k^a k^b \geq 0$ for all null k^a, we have $\mathcal{T} \subset B \cap \Sigma$, so the apparent horizon always lies inside of (or coincides with) the true event horizon, $H \cap \Sigma$, on Σ. The apparent horizon, $\mathcal{A}$, satisfies the following property.

THEOREM 12.2.5. *If the total trapped region, $\mathcal{T}$, on a Cauchy surface Σ has the structure of a manifold with boundary, then the apparent horizon, $\mathcal{A}$, is an outer marginally trapped surface with vanishing expansion, $\theta = 0$.*

Proof. Clearly $\mathcal{T}$ is closed and we also have[2] $i^0 \notin \mathcal{T}$, so $\mathcal{T}$ satisfies the first two requirements for being a trapped region. To show that $\theta = 0$ on $\mathcal{A}$, we note that if $\theta > 0$ at $p \in \mathcal{A}$, then we could find a neighborhood, U, of p such that every surface, S, passing through U with $\theta \leq 0$ everywhere on S must leave $\mathcal{T}$. Hence, we could not have outer trapped surfaces passing arbitrarily close to p but staying within $\mathcal{T}$ as required by the fact that p is on the boundary of $\mathcal{T}$. On the other hand, given that $\theta \leqq 0$ everywhere on $\mathcal{A}$, if we had $\theta < 0$ at $q \in \mathcal{A}$, we could deform $\mathcal{A}$ outward in a neighborhood of q, preserving $\theta \leq 0$ everywhere. In this way, we would produce a trapped region C larger than $\mathcal{T}$, which is impossible. □

The final result we shall prove in this section concerns the evolution of the event horizon. Since the horizon, H, is the boundary of the past of $\mathcal{I}^+$, by theorem 8.1.3 it is an achronal, three-dimensional, embedded C^0 (in fact, C^{1-}) submanifold. Furthermore, by theorem 8.1.6, H is generated by future inextendible null geodesics,

2. To see this we note that in an asymptotically Euclidean coordinate system on Σ (see problem 2 of chapter 11) there exists a radius R such that all coordinate spheres with $r > R$ have everywhere positive expansion of the outgoing null geodesics. Hence no outer trapped surface, S, can enter the region $r > R$, since the expansion of the outgoing null geodesics from S would have to be at least as great as that from the coordinate sphere at the point where r takes its maximum value on S. Thus, $\mathcal{T}$ cannot enter the region $r > R$.

since no null geodesic generator of H can have a future endpoint on $\mathscr{I}^+$. Thus, if the intersection $\mathscr{H} = H \cap \Sigma$ of the horizon with a spacelike Cauchy surface, Σ, for $\tilde{V}$ is nonempty, it comprises a two-dimensional submanifold of Σ. The next theorem states that the area of $\mathscr{H}$ never decreases with time. As we shall see in sections 12.5 and 14.4, this result underlies what appears to be a profound relationship between black holes, thermodynamics, and quantum physics.

THEOREM 12.2.6 (black hole area theorem; Hawking 1971). *Let (M, g_{ab}) be a strongly asymptotically predictable spacetime satisfying $R_{ab}k^a k^b \geqq 0$ for all null k^a. Let Σ_1 and Σ_2 be spacelike Cauchy surfaces for the globally hyperbolic region $\tilde{V}$ with $\Sigma_2 \subset I^+(\Sigma_1)$ and let $\mathscr{H}_1 = H \cap \Sigma_1$, $\mathscr{H}_2 = H \cap \Sigma_2$, where H denotes the event horizon,* i.e., *the boundary of the black hole region of (M, g_{ab}). Then the area of $\mathscr{H}_2$ is greater than or equal to the area of $\mathscr{H}_1$.*

Proof. We establish, first, that the expansion, θ of the null geodesic generators of H is everywhere nonnegative, $\theta \geq 0$. Suppose $\theta < 0$ at $p \in H$. Let Σ be a spacelike Cauchy surface for $\tilde{V}$ passing through p and consider the two-surface $\mathscr{H} = H \cap \Sigma$. Since $\theta < 0$ at p, we can deform $\mathscr{H}$ outward in a neighborhood of p to obtain a surface $\mathscr{H}'$ on Σ which enters $J^-(\mathscr{I}^+)$ and has $\theta < 0$ everywhere in $J^-(\mathscr{I}^+)$. However, by the same type of argument as given in proposition 12.2.2, this leads to a contradiction as follows. Let $K \subset \Sigma$ be the closed region lying between $\mathscr{H}$ and $\mathscr{H}'$, and let $q \in \mathscr{I}^+$ with $q \in \dot{J}^+(K)$. Then the null geodesic generator of $\dot{J}^+(K)$ on which q lies must meet $\mathscr{H}'$ orthogonally. However, this is impossible, since $\theta < 0$ on $\mathscr{H}'$, and thus this generator will have a conjugate point before reaching q. Thus, $\theta \geq 0$ everywhere on H.

Now, as mentioned above, each $p \in \mathscr{H}_1$ lies on a future inextendible null geodesic, γ, contained in H. Since Σ_2 is a Cauchy surface, γ must intersect Σ_2 at a point $q \in \mathscr{H}_2$. Thus, we obtain a natural map from $\mathscr{H}_1$ into (a portion of) $\mathscr{H}_2$. Since $\theta \geqq 0$, the area of the portion of $\mathscr{H}_2$ given by the image of $\mathscr{H}_1$ under this map must be at least as large as the area of $\mathscr{H}_1$. In addition, since the map need not be onto—e.g., new black holes may form between Σ_1 and Σ_2—the area of $\mathscr{H}_2$ may be even larger. Thus, the area of $\mathscr{H}_2$ cannot be smaller than that of $\mathscr{H}_1$. □

12.3 The Charged Kerr Black Holes

Consider a body which undergoes complete gravitational collapse and—in accord with the first cosmic censor conjecture of section 12.1—forms a black hole. In the spherically symmetric case, the spacetime outside the body always is described by the Schwarzschild solution, and the final state of gravitational collapse will be a Schwarzschild black hole. However, in the nonspherical case, the spacetime geometry outside the collapsing body should vary with time and depend greatly on the details of the collapse. In particular, no gravitational radiation can be produced in a spherically symmetric spacetime, whereas large amounts of energy may be radiated away in nonspherical collapse. Nevertheless, one would expect on physical grounds that at sufficiently "late times," the spacetime geometry should "settle down" to a stationary final state. Furthermore, one would expect all the matter present to be rapidly "swallowed up" by the black hole, so the final state should be vacuum except, possibly, for the presence of electromagnetic fields associated with the black hole. (Even in cases—such as in the binary X-ray sources—where one has a steady flow of matter into the black hole, this matter normally would produce only a small perturbation on the structure of the black hole.) Thus, one expects that the final state of gravitational collapse will be a stationary, electrovac (i.e., vacuum except for electromagnetic fields) black hole. Therefore, it is of great interest to find all solutions of the Einstein-Maxwell equations which describe stationary black holes.

We have already discussed in detail the spherically symmetric, static black hole solution discovered by Schwarzschild (1916*a*). Shortly thereafter, a charged generalization of the Schwarzschild solution was discovered independently by Reissner (1916) and Nordstrom (1918) (see problem 3 of chapter 6). However, it was not until 1963 that another family of stationary, vacuum black hole solutions was discovered by Kerr. A charged generalization of the Kerr family was obtained shortly thereafter by Newman *et al.* (1965). These charged Kerr solutions form a three-parameter family, whose spacetime metric and electromagnetic vector potential are given by

$$ds^2 = -\left(\frac{\Delta - a^2 \sin^2\theta}{\Sigma}\right) dt^2 - \frac{2a \sin^2\theta(r^2 + a^2 - \Delta)}{\Sigma} dt d\phi$$

$$+ \left[\frac{(r^2 + a^2)^2 - \Delta a^2 \sin^2\theta}{\Sigma}\right] \sin^2\theta \, d\phi^2 + \frac{\Sigma}{\Delta} dr^2 + \Sigma \, d\theta^2 \quad , \qquad (12.3.1)$$

$$A_a = -\frac{er}{\Sigma}[(dt)_a - a \sin^2\theta(d\phi)_a] \quad , \qquad (12.3.2)$$

where

$$\Sigma = r^2 + a^2 \cos^2\theta \quad , \qquad (12.3.3)$$

$$\Delta = r^2 + a^2 + e^2 - 2Mr \quad , \qquad (12.3.4)$$

and e, a, and M are the three parameters of the family. When $e = 0$, we have $A_a = 0$ and the spacetime metric reduces to the vacuum Kerr family of solutions. When $a = 0$, we recover the Reissner-Nordstrom solutions, and when $e = a = 0$, equation (12.3.1) reduces to the Schwarzschild solution. Thus, all known stationary black hole solutions are encompassed by the three-parameter family (12.3.1), (12.3.2). As we shall see at the end of this section, no other stationary black hole solutions exist.

The charged Kerr metrics all are stationary and axisymmetric (see section 7.1), with Killing fields $\xi^a = (\partial/\partial t)^a$ and $\psi^a = (\partial/\partial\phi)^a$. They are asymptotically flat, as can be seen crudely from the fact that the metric components (12.3.1) approach those of the Minkowski metric in spherical polar coordinates as $r \to \infty$, and has been demonstrated in detail by Ashtekar and Hansen (1978). In the algebraic classification of section 7.3, they are type II–II solutions, with repeated principal null vectors,

$$l^a = \left(\frac{r^2 + a^2}{\Delta}\right)(\partial/\partial t)^a + \frac{a}{\Delta}(\partial/\partial\phi)^a + (\partial/\partial r)^a \quad , \qquad (12.3.5)$$

$$n^a = \frac{r^2 + a^2}{2\Sigma}(\partial/\partial t)^a + \frac{a}{2\Sigma}(\partial/\partial\phi)^a - \frac{\Delta}{2\Sigma}(\partial/\partial r)^a \quad . \qquad (12.3.6)$$

(Here l^a and n^a are normalized by a convenient choice due to Kinnersley 1969, with $l^a n_a = -1$.)

The three parameters e, a, and M appearing in the solutions all have a direct physical interpretation. For any 2-sphere, S, in the asymptotic region, we have

$$\frac{1}{2}\int_S \epsilon_{abcd} F^{cd} = 4\pi e \quad , \tag{12.3.7}$$

so, by problem 2 of chapter 4, we may interpret e as the total electric charge of the spacetime. Furthermore, we have

$$-\frac{1}{8\pi}\int_S \epsilon_{abcd} \nabla^c \xi^d = M \quad , \tag{12.3.8}$$

so, according to equation (11.2.9), M is the total mass. Finally, we have

$$\frac{1}{16\pi}\int_S \epsilon_{abcd} \nabla^c \psi^d = Ma \quad , \tag{12.3.9}$$

so, according to problem 6 of chapter 11, we have $a = J/M$, where J is the total angular momentum of the spacetime.

It should be noted that, in geometrized units, the charge to mass ratio of a proton is $q/m \sim 10^{18}$, and for an electron we have $q/m \sim 10^{21}$. Since the ratio of electromagnetic to gravitational force produced on a test body of charge q and mass m by a body of charge e and mass M is $\sim qe/mM$, it would be very difficult for any astrophysical body to achieve and/or maintain a charge to mass ratio of greater than $\sim 10^{-18}$, since a body with larger charge to mass ratio would selectively attract particles of opposite charge.[3] Hence, in astrophysically reasonable situations it appears that $e << M$, so we may neglect the effects of the electromagnetic field on the spacetime geometry and consider only the Kerr family of black holes.

The coordinate basis components of the charged Kerr metrics, equation (12.3.1) are nonsingular and define a nondegenerate metric everywhere except where $\Sigma = 0$ and where $\Delta = 0$. Evaluation of curvature invariants such as $R_{abcd}R^{abcd}$ shows that the singularity at

$$\Sigma = r^2 + a^2 \cos^2\theta = 0 \tag{12.3.10}$$

is a true, curvature singularity when $M \neq 0$. If one were to interpret r, θ, and ϕ as representing spherical polar coordinates, the fact that there is a singularity at the origin, $r = 0$, only for the angular value $\theta = \pi/2$ would appear rather puzzling. In particular, if we interpret the singularity in this way—i.e., if we define the charged Kerr metrics on the manifold $\mathbb{R}^4$ with the origin $r = 0$ removed—we then would have incomplete geodesics (such as those on the axis, $\sin\theta = 0$) which terminate at $r = 0$ but along which the curvature remains finite. In fact, this spacetime would be extendible. This provides a good illustration of the impropriety of making a choice of the manifold structure on the basis of a naive interpretation of the coordinate

3. One mechanism to obtain a net charge on an astrophysical body which, in principle, could overcome this limit is to have a rotating body placed in a magnetic field. However, for a Kerr black hole, this mechanism also would lead to a negligible charge buildup in astrophysically reasonable situations (Wald 1974*b*).

system in which the metric is given. Some insight into the true nature of the singularity of the charged Kerr metrics at $\Sigma = 0$ can be obtained by consideration of the case $e = M = 0$, $a \neq 0$. In that case, the Kerr metric (12.3.1) actually is nothing more than the metric of Minkowski spacetime expressed in spheroidal coordinates. Here the singularity at $\Sigma = 0$ is, of course, merely a coordinate singularity, and it is located on the ring of radius a in the plane $z = 0$. This suggests that when $M \neq 0$ and $a \neq 0$ we interpret the true singularity at $\Sigma = 0$ as a ring singularity, i.e., that we define the charged Kerr metrics on a manifold whose structure in a neighborhood of this singularity has the topology of $\mathbb{R}^4$ with the set $S^1 \times \mathbb{R}$—that is, a ring, S^1, cross "time," $\mathbb{R}$—removed. This can be implemented explicitly by transforming to the quasi-Cartesian coordinates given by Kerr and Schild (1965), where $\Sigma = 0$ takes the coordinate form of a ring. However, a problem still remains in that when $M \neq 0$ the metric components fail to be smooth across the coordinate disk enclosed by the ring singularity in the $z = 0$ plane. This problem can be remedied by defining the charged Kerr metrics on a manifold with the following relatively complicated topology in a neighborhood of the singularity at $\Sigma = 0$. We take two copies, M_1 and M_2, of $\mathbb{R}^4$ with the "ring" $z = 0$, $x^2 + y^2 = a^2$, removed. We then attach M_1 to M_2 by identifying the "top side" of the disk $z = 0$, $x^2 + y^2 < a^2$, of M_1 with the "bottom side" of the corresponding disk of M_2, and, similarly, identify the "bottom side" of the disk of M_1 with the "top side" of the disk of M_2. The charged Kerr metrics with $M \neq 0$, $a \neq 0$ then may be smoothly defined on this manifold in such a way that the curvature scalar $R_{abcd}R^{abcd}$ blows up along every incomplete geodesic, thereby guaranteeing that the spacetime is inextendible. Details of this construction can be found in Hawking and Ellis (1973).

It should be noted that when one passes "through the ring" in going from M_1 to M_2 in this spacetime, this corresponds in the original coordinates to passing through $r = 0$ into negative values of r. However, for negative values of r of sufficiently small magnitude and for θ sufficiently close to $\pi/2$ we have $\psi_a \psi^a = g_{\phi\phi} < 0$ (see eq. [12.3.1]). Thus, $\psi^a = (\partial/\partial\phi)^a$ becomes timelike near the ring singularity. However, the orbits of ψ^a must be closed (i.e., the coordinate ϕ must be periodically identified with period 2π) in order that the charged Kerr spacetime be asymptotically flat as $r \to \infty$. Thus, closed timelike curves exist in a neighborhood of the ring singularity.

When $e^2 + a^2 > M^2$, there are no solutions of the equation $\Delta = 0$ so the true singularity at $\Sigma = 0$ is the only singularity of the coordinate components (12.3.1). In this case, the ring singularity is "naked," i.e., the charged Kerr metrics fail to be strongly asymptotically predictable, and thus they do *not* describe black holes. Furthermore, one may make use of the causality violation occurring near the ring singularity to go "backwards in time" by an arbitrarily large amount as measured by the t coordinate of (12.3.1) and thereby produce closed timelike curves passing through any point in the spacetime.

In the case

$$e^2 + a^2 \leqq M^2 \tag{12.3.11}$$

Δ vanishes at the r-coordinate values

$$r_\pm = M \pm (M^2 - a^2 - e^2)^{1/2} \quad . \tag{12.3.12}$$

As shown by Boyer and Lindquist (1967) and Carter (1968*a*), the singularities in the metric components at $r = r_+$ and (for $a \neq 0$ or $e \neq 0$) at $r = r_-$ are coordinate singularities of the same nature as the singularity at $r = 2M$ in the Schwarzschild spacetime. Thus, we may extend the spacetime through these coordinate singularities much as in the Schwarzschild case. When these extensions are patched together, a remarkable global structure of the extended charged Kerr spacetimes is obtained. A conformal diagram of the extended Schwarzschild spacetime is shown in Figure 12.3, and a conformal diagram of the extended charged Kerr spacetime with $a \neq 0$ is shown in Figure 12.4 for the "non-extreme" case $a^2 + e^2 < M^2$. Region I of Figure 12.4 is the asymptotically flat region covered in a nonsingular fashion by the original coordinates (12.3.1) with $r > r_+$. By extending through the coordinate singularity at $r = r_+$, we obtain region II representing a black hole, region III representing a white hole, and region IV representing another asymptotically flat region, just as in the Schwarzschild case, Figures 6.9 and 12.3. However, unlike the Schwarzschild case, instead of encountering a true singularity at the "top boundary" of region II and "bottom boundary" of region III, we encounter merely another coordinate singularity at $r = r_-$. Thus, we can extend region II through $r = r_-$ to obtain regions V and VI. These regions contain the ring singularity at $\Sigma = 0$ and, as described above, one can pass through the ring singularity to obtain another asymptotically flat region with $r \to -\infty$. (In this asymptotically flat region the ring singularity is a naked singularity of negative mass. With respect to the original asymptotically flat region I, the ring singularity, of course, lies within a black hole.) One may then continue to extend the charged Kerr spacetime "upward" ad infinitum to obtain a region VII, identical in structure to region III, and obtain regions VIII and IX, identical in structure to regions IV and I, etc. Similarly, one may extend the charged Kerr solutions "downward" ad infinitum. The structure of the extended Reissner-Nordstrom spacetime ($a = 0$, $e \neq 0$) is very similar except that the true singularity at $\Sigma = 0$ no longer has a ring structure, and one cannot extend to negative values of r. The global structure of the "extreme" charged Kerr case $e^2 + a^2 = M^2$ (where $r_+ = r_- = M$) differs from Figure 12.4 but has a similar structure consisting of "blocks" with $r > M$ and $r < M$ patched together in an infinite chain.

Thus, an observer starting in region I of the extended charged Kerr spacetime of Figure 12.4 may cross the event horizon at $r = r_+$ and enter the black hole region II. However, instead of inevitably falling into a singularity within a finite proper time as occurs in the Schwarzschild spacetime, the observer may pass through the "inner

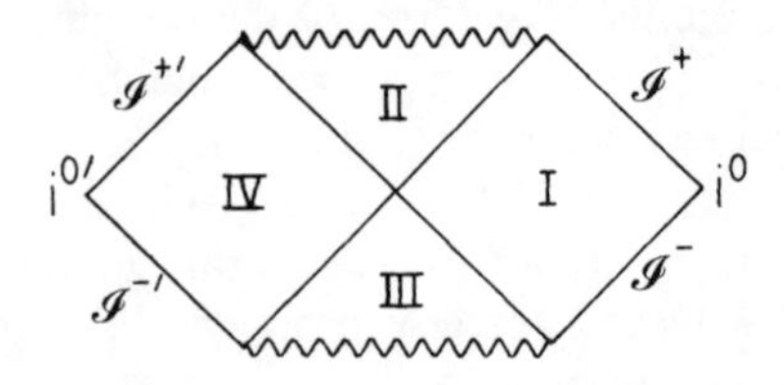

Fig. 12.3. A conformal diagram of the extended Schwarzschild spacetime (see Fig. 6.9), represented in the same manner as used in Figure 12.2. Note that since the extended Schwarzschild spacetime has two distinct asymptotically flat regions, two distinct conformal boundaries are shown.

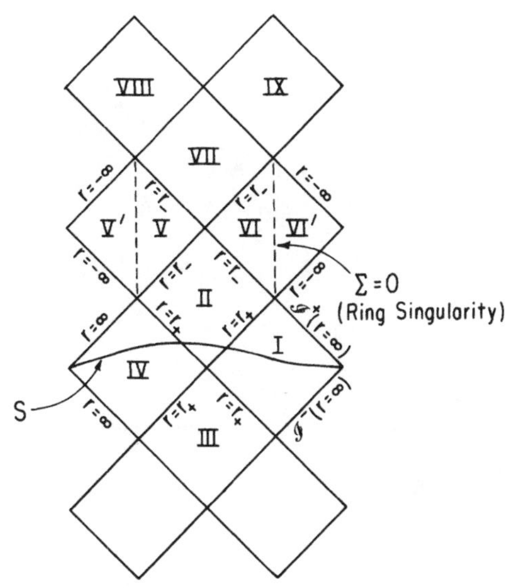

Fig. 12.4. A conformal diagram of the extended charged Kerr spacetime in the case $a \neq 0$, $a^2 + e^2 < M^2$.

horizon" $r = r_-$ (which is a Cauchy horizon for the hypersurface S shown in Fig. 12.4), thereby entering region V or VI. From there, he may end his existence in the ring singularity but he also may pass through the ring singularity to a new asymptotically flat region, or he may enter the "white hole region" VII and from there enter the new asymptotically flat region VIII or IX. From there he may enter the new black hole associated with these asymptotic regions, and continue his journey.

How much of this extended charged Kerr spacetime should be taken seriously? What portion of this spacetime would be produced by a physically realistic gravitational collapse, starting from "non-exotic" initial data, i.e., from an asymptotically flat initial data surface S with topology $\mathbb{R}^3$? Unlike the Schwarzschild case, we have no reason to believe that the exterior gravitational field of any physically reasonable collapsing body will be described by the Kerr metric, since, as mentioned above, in the nonspherical case we would, in general, expect a complicated dynamical evolution which only "settles down" to a stationary geometry at late times in $J^-(\mathscr{I}^+)$. Thus, we are not in a position to follow the dynamical evolution of the gravitational collapse of a body which forms a Kerr black hole and thereby determine the detailed spacetime geometry inside the black hole. However, in the case of spherical collapse of a charged body ($e \neq 0$), the spacetime geometry exterior to the matter *is* described by the Reissner-Nordstrom solution since Birkhoff's theorem can be generalized to show that the Reissner-Nordstrom spacetime is the unique spherical electrovac solution. The dynamical evolution of a simple system like a charged, spherical shell of dust can be obtained explicitly. In this case, spacetime is flat inside the shell, and the flat interior of the shell entirely "covers up" regions III and IV of Figure 12.4. Part or all of regions II and V (including, in all cases, the singularity at $r = 0$ in region V) also are covered up. The behavior of the shell for the various choices of total mass M, total charge e, and total rest mass $\mathcal{M}$ are chronicled in detail

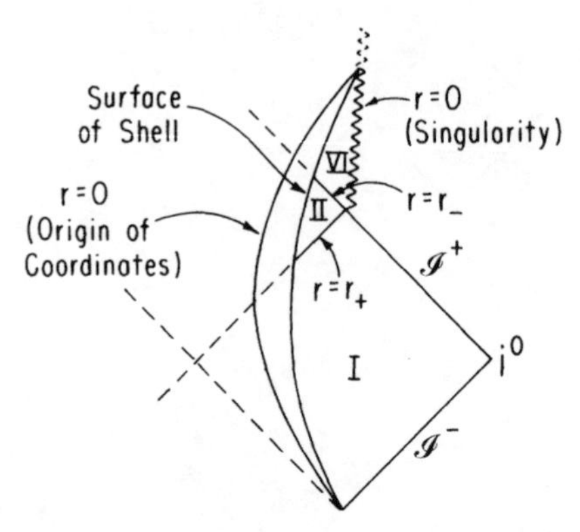

Fig. 12.5. A conformal diagram of a spacetime in which a charged spherical shell with $M > \mathcal{M} > |e|$ undergoes gravitational collapse. The dashed lines refer to the extended Reissner-Nordstrom spacetime.

by Boulware (1973). We show the resulting spacetime for the case $M > \mathcal{M} > |e|$ in Figure 12.5. As shown there, the shell crosses the surface $r = r_-$, which is a Cauchy horizon for region I. The spacetime is extendible across r_-, but the extension is not determined by Einstein's equation since it is outside the domain of dependence of the initial data surface. However, if one assumes that the extension is spherically symmetric, the extension is uniquely given by the Reissner-Nordstrom geometry, and we find that the shell crashes into the singularity at $r = 0$ (which formed *outside* the shell) in region VI, as shown in Figure 12.5. For other choices of the parameters M, $\mathcal{M}$, and e, the shell may reexpand into region VII of the Reissner-Nordstrom geometry (Boulware 1973). Thus, one may obtain spacetimes from the collapse of a charged body which have features differing greatly from that of Figure 6.11. However, as already mentioned in our discussion of cosmic censorship in section 12.1, it has been shown that the Cauchy horizon of Figure 12.5 is unstable (Chandrasekhar and Hartle 1982). Linear perturbations of the initial data for the Einstein-Maxwell equations on a Cauchy surface for region I of Figure 12.5 become singular at $r = r_-$. The basic reason for this is that an observer crossing $r = r_-$ in Figure 12.5 "sees" all of region I. Oscillations in the gravitational and electromagnetic field which occur at finite frequency in region I are "seen" to occur at infinite frequency by an observer crossing r_-; i.e., there is an "infinite blueshift" effect, which makes the perturbation singular there. Thus, there is good reason to believe that in a physically realistic case where the shell is not exactly spherical, the Cauchy horizon $r = r_-$ in Figure 12.5 will become a true, physical singularity, thereby producing an "all encompassing" singularity inside the black hole formed by the collapse of the shell. It is believed that similar phenomena will occur for any physically realistic collapse to a charged Kerr black hole. Thus, the spacetime produced by physically realistic collapse is expected to be qualitatively similar to the spherical case, Figure 6.11, rather than that suggested by the extended charged Kerr solutions, Figure 12.4.

Another feature of the charged Kerr spacetime worthy of note is that the norm of the timelike Killing field,

$$\xi^a \xi_a = g_{tt} = \frac{a^2 \sin^2\theta - \Delta}{\Sigma} \quad , \tag{12.3.13}$$

becomes positive in the region where

$$r^2 + a^2 \cos^2\theta + e^2 - 2Mr < 0 \quad , \tag{12.3.14}$$

part of which lies *outside* the black hole if $a \neq 0$. Thus, in the region

$$r_+ < r < M + (M^2 - e^2 - a^2 \cos^2\theta)^{1/2} \quad , \tag{12.3.15}$$

called the *ergosphere* and depicted in Figure 12.6, the asymptotic time translation Killing field $\xi^a = (\partial/\partial t)^a$ becomes spacelike. Hence, an observer in the ergosphere would have to "go faster than light" to follow an orbit of ξ^a; i.e., he cannot remain stationary, even though he is outside the black hole. The nature of this nonstationarity can be seen from the equation

$$\sum_{\mu,\nu} g_{\mu\nu} u^\mu u^\nu < 0 \tag{12.3.16}$$

satisfied by the components of the tangent vector, u^a, to any timelike curve. Inside the ergosphere, all terms on the left-hand side of equation (12.3.16) are manifestly positive except the term $2g_{t\phi}u^t u^\phi = 2g_{t\phi}(dt/d\tau)(d\phi/d\tau)$, where τ denotes proper time along the curve. Since $\nabla^a t$ is past directed timelike in the ergosphere, we have $dt/d\tau = u^a \nabla_a t > 0$. Thus, since $g_{t\phi} < 0$ in the ergosphere, we must have

$$d\phi/d\tau > 0 \tag{12.3.17}$$

for all timelike curves in the ergosphere. In other words, all observers in the ergosphere are forced to rotate in the direction of rotation of the black hole. This may be viewed as an extreme case of the "dragging of inertial frames" effect, thereby providing a dramatic example of how some aspects of Mach's principle are incorporated into general relativity.

The closest analog to a family of static observers outside the black hole in the charged Kerr geometry are the "locally nonrotating observers" (see problem 3 of chapter 7) whose 4-velocity is given by $u^a = -\nabla^a t/[-\nabla_b t \nabla^b t]^{1/2}$. These observers rotate with coordinate angular velocity

$$\Omega = \frac{d\phi}{dt} = -\frac{g_{t\phi}}{g_{\phi\phi}} = \frac{a(r^2 + a^2 - \Delta)}{(r^2 + a^2)^2 - \Delta a^2 \sin^2\theta} \quad . \tag{12.3.18}$$

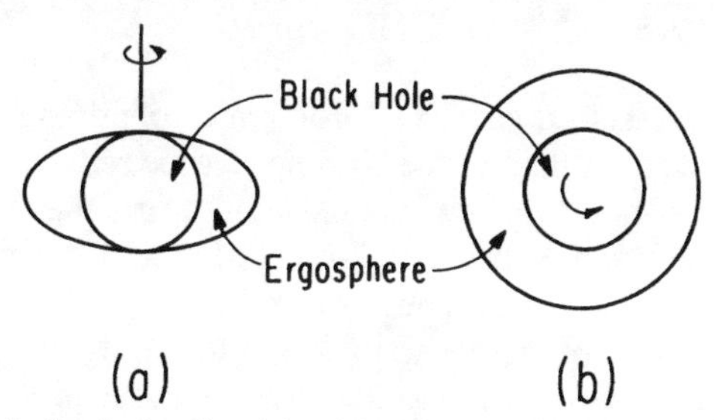

Fig. 12.6. A sketch showing (*a*) a "side view" and (*b*) a "top view" of the ergosphere of a Kerr black hole.

In the limit as one approaches the black hole event horizon, $r \to r_+$, this coordinate angular velocity becomes

$$\Omega_H = \frac{a}{r_+^2 + a^2} \quad . \tag{12.3.19}$$

This is closely related to the fact that it is the Killing field

$$\chi^a = (\partial/\partial t)^a + \Omega_H(\partial/\partial\phi)^a \tag{12.3.20}$$

[rather than $(\partial/\partial t)^a$] which is tangent to the null geodesic generators of the horizon of the charged Kerr black hole. Equation (12.3.20) can be interpreted as saying that the event horizon of the charged Kerr black hole rotates with angular velocity Ω_H.

We turn, now, to a brief discussion of geodesic motion. For simplicity, we shall consider only the Kerr geometry, $e = 0$. (See problem 2 for the case $e \neq 0$.) As in the Schwarzschild case, the timelike Killing field ξ^a and the axial Killing field ψ^a yield via proposition C.3.1 a conserved energy, E, and angular momentum, L, per unit rest mass for geodesics,

$$E = -u^a\xi_a = \left(1 - \frac{2Mr}{\Sigma}\right)\dot{t} + \frac{2Mar\sin^2\theta}{\Sigma}\dot{\phi} \quad , \tag{12.3.21}$$

$$L = u^a\psi_a = -\frac{2Mar\sin^2\theta}{\Sigma}\dot{t} + \frac{(r^2+a^2)^2 - \Delta a^2\sin^2\theta}{\Sigma}\sin^2\theta\,\dot{\phi} \quad , \tag{12.3.22}$$

where $\dot{x}^\mu = dx^\mu/d\tau$. In addition, we have

$$g_{ab}u^a u^b = -\kappa \quad , \tag{12.3.23}$$

where $\kappa = 1$ for timelike geodesics and $\kappa = 0$ for null geodesics. One may use equations (12.3.21) and (12.3.22) to eliminate $\dot{t}$ and $\dot{\phi}$ in terms of E and L, and the result may be substituted into equation (12.3.23). In the case of equatorial geodesics, $\theta = \pi/2$, one obtains

$$\frac{1}{2}\dot{r}^2 + V(E,L,r) = 0 \quad , \tag{12.3.24}$$

where

$$V = -\kappa\frac{M}{r} + \frac{L^2}{2r^2} + \frac{1}{2}(\kappa - E^2)\left(1 + \frac{a^2}{r^2}\right) - \frac{M}{r^3}(L - aE)^2 \quad . \tag{12.3.25}$$

Thus, as in the Schwarzschild case, the problem of obtaining the timelike and null geodesics in the equatorial plane of the Kerr spacetime reduces to solving a problem of ordinary nonrelativistic, one-dimensional motion in an effective potential, with the only significant additional complication here being the fact that V now depends nontrivially on E as well as on L. Thus, we may find the behavior of freely falling test bodies and light rays by methods similar to those of section 6.3. In particular, the circular orbits are given by the simultaneous solutions of $V = 0$ and $dV/dr = 0$. Their properties are discussed by Bardeen, Press, and Teukolsky (1972). It is noteworthy that binding energies significantly higher than in the Schwarzschild case can be achieved for circular orbits around a Kerr black hole. For a Kerr black hole with $a = M$, the last stable circular orbit (with positive L) has $E = 1/\sqrt{3}$. Thus, if a test body with positive L spirals in to the last stable circular orbit as a result of energy loss via gravitational radiation, it will have radiated away $1 - 1/\sqrt{3} \approx 42\%$ of its original rest energy, as opposed to only $\sim 6\%$ in the Schwarzschild case, equation (6.3.23).

The constants of motion E and L, equations (12.3.21) and (12.3.22), do *not* provide a sufficient number of first integrals to determine nonequatorial motion. Therefore, in that case we might expect to have to return to the geodesic equation, $u^a \nabla_a u^b = 0$, and solve a coupled set of second order nonlinear ordinary differential equations to obtain $r(\tau)$ and $\theta(\tau)$. Remarkably, however, it turns out that this is not necessary. The Kerr metric possesses the Killing tensor,

$$K_{ab} = 2\Sigma l_{(a} n_{b)} + r^2 g_{ab} \tag{12.3.26}$$

(Walker and Penrose 1970), where l^a and n^a are given by equations (12.3.5) and (12.3.6). Thus, an additional constant of the motion

$$C = K_{ab} u^a u^b \tag{12.3.27}$$

is obtained, as discussed at the end of appendix C. This allows one to integrate the geodesic equation explicitly, as was first done by Carter (1968*a*), who used the separability of the Hamilton-Jacobi equation for geodesics rather than the existence of K_{ab} to obtain this additional constant of motion. In addition, the Kerr metric—as well as all other type II–II vacuum spacetimes—possesses a "conformal Killing spinor" (see Walker and Penrose 1970) which enables one to determine the parallel propagation of "polarization vectors" along null geodesics in a simple manner.

A remarkable simplification also occurs when one studies the propagation of a Klein-Gordon scalar test field, equation (4.3.9), in the Kerr spacetimes. One can take advantage of the stationary and axial Killing fields of Kerr to expand the scalar field in a Fourier series in the angular coordinate ϕ and a Fourier integral over the time coordinate t. This effectively reduces the Klein-Gordon equation to a partial differential equation in the two remaining variables, r and θ. However, it turns out that this equation can be solved by separation of variables (Carter 1968*b*), so the problem of determining the behavior of Klein-Gordon test fields in the Kerr spacetimes is reduced to solving *ordinary* differential equations.

Even more remarkable simplifications occur when one studies Maxwell's equations and the linearized Einstein equation in the Kerr background. Here, after making use of the symmetries in ϕ and t, one would expect to be left with a complicated, coupled system of partial differential equations in r and θ for, respectively, the components of the electromagnetic vector potential, A_μ, and the components of the metric perturbation, $\gamma_{\mu\nu}$. Indeed, this is what happens when one writes down the equations for A_μ or $\gamma_{\mu\nu}$, even when simplifying gauge choices such as the Lorentz gauge for A_μ or the transverse traceless gauge for $\gamma_{\mu\nu}$ (see section 7.5) are made. However, one may write down the equations in the Newman-Penrose (1962) formalism (see section 3.4*b*), using the repeated principal null vectors of the Kerr metric

l^a and n^a, equations (12.3.5) and (12.3.6), as the real null vectors and

$$m^a = \frac{1}{2^{1/2}(r + ia\cos\theta)}[ia\sin\theta(\partial/\partial t)^a + (\partial/\partial\theta)^a + \frac{i}{\sin\theta}(\partial/\partial\phi)^a] \quad (12.3.28)$$

as the complex null vector of the Newman-Penrose basis. Then, as discovered by Teukolsky (1972), a decoupled equation can be derived for

$$\Phi_0 = F_{ab} l^a m^b \quad (12.3.29)$$

in the Maxwell case, and for

$$\Psi_0 = -C_{abcd} l^a m^b l^c m^d \quad (12.3.30)$$

in the linearized Einstein case. (Decoupled equations similarly may be obtained for $\Phi_2 = F_{ab}\overline{m}^a n^b$ and $\Psi_4 = -C_{abcd} n^a \overline{m}^b n^c \overline{m}^d$.) Furthermore, Teukolsky showed that these equations also may be solved by separation of variables. In addition, a knowledge of Φ_0 (or Φ_2) determines a Maxwell perturbation modulo the trivial "change in charge" solution obtained by linearizing equation (12.3.2) off the Kerr background, while a knowledge of Ψ_0 (or Ψ_4) determines a gravitational perturbation modulo the trivial perturbations obtained by variation of the Kerr parameters M, a (Wald 1973). Finally, the complete vector potential, A_a, and metric perturbation, γ_{ab}, associated, respectively, with Φ_0 and Ψ_0, can be obtained explicitly (Cohen and Kegeles 1974; Wald 1978*a*; Chandrasekhar 1983). Thus, the problem of determining the behavior of electromagnetic and gravitational perturbations of a Kerr black hole also reduces to solving ordinary differential equations, thereby making tractable many problems of interest. Similar simplifications of the coupled linearized Einstein-Maxwell system occur in the Reissner-Nordstrom case, $a = 0$, $e \neq 0$ (Moncrief 1975; Chandrasekhar 1979). However, it appears that no such simplifications occur in the general charged Kerr case, $a \neq 0$, $e \neq 0$. For a complete discussion of the electromagnetic and gravitational perturbations of a Kerr black hole, we refer the reader to Chandrasekhar (1983).

Thus, as summarized above, a great deal is known about the properties of the Kerr black holes. These are the only known stationary vacuum black hole solutions of Einstein's equation. However, since they comprise only a two-parameter family, one might expect that many more stationary vacuum black hole solutions should exist. After all, as mentioned in the introduction to chapter 11, the exterior gravitational field of a stationary body is characterized by an infinite set of multipole coefficients. Why should all the higher multipole moments of a stationary black hole be related in a unique way to its mass and angular momentum? Remarkably, as a result of theorems of Israel, Carter, Hawking, and Robinson obtained between 1967 and 1975, a virtually complete proof has been given that the Kerr black holes are the only possible stationary vacuum black holes. Thus, if the first cosmic censor conjecture is correct and if the spacetime resulting from gravitational collapse always "settles down" to a stationary, vacuum final state, the end product of collapse must always be a Kerr black hole. The complete gravitational collapse of two bodies differing greatly from each other in composition, shape, and structure will produce indistinguishable final states provided only that their end products have the same total mass and total angular momentum.

Since we are far from having the general stationary vacuum solution of Einstein's equation in explicit form,[4] the proof of the uniqueness of the Kerr black holes has proceeded by a relatively long chain of arguments. Logically, the first step—although, historically, one of the last steps—in this chain is the proof by Hawking that the two-dimensional surface formed by the intersection of the horizon of a stationary black hole with a Cauchy surface must have topology S^2. This is established by showing that if it had any other topology, it would be possible to deform it outward into $J^-(\mathscr{I}^+)$ such that the expansion, θ, of the outgoing null geodesics satisfies $\theta \leq 0$ everywhere. This would contradict proposition 12.2.4. Details of the proof can be found in Hawking and Ellis (1973). (See also Gannon 1976 for some results in the nonstationary case.)

The next step in the uniqueness proof, also due to Hawking, is the demonstration that a stationary vacuum black hole must be static or axisymmetric. First, we note that in a stationary spacetime containing a black hole, the time translation isometry must leave the horizon invariant. Hence, the Killing field ξ^a must lie tangent to the horizon and, hence, always must be spacelike or null on the horizon. Now, one of the following three possibilities must hold: (i) No ergosphere is present in the spacetime, i.e., the stationary Killing field ξ^a is everywhere timelike or null outside the black hole. In this case, ξ^a must be null on the horizon. (ii) An ergosphere is present but is disjoint from the horizon, and ξ^a is null on the horizon. (iii) An ergosphere is present and intersects the horizon, as happens for a Kerr black hole. In this case, ξ^a is spacelike on (a portion of) the horizon. In case (i), a generalization of a theorem of Lichnerowicz (1955) establishes that the spacetime must be static (Hawking and Ellis 1973). Under some additional assumptions, results of Hajicek (1973) show that the outer boundary of the ergosphere in a stationary vacuum spacetime always must intersect the horizon. Thus, it appears that case (ii) cannot occur. Plausibility arguments against case (ii) also are given in Hawking and Ellis (1973). Finally, in case (iii), using the properties of the horizon in a stationary spacetime and using the analyticity of stationary vacuum solutions (Müller zum Hagen 1970), Hawking proved existence of a one-parameter group of isometries which commute with the stationary isometries and whose orbits on the horizon coincide with the null geodesic generators of the horizon. Thus, one obtains a Killing field χ^a distinct from ξ^a, and by taking a linear combination of χ^a and ξ^a, one obtains a Killing field ψ^a whose orbits are closed, i.e., an axial Killing field.[5] Again, details of this proof are given in Hawking and Ellis (1973).

4. Indeed, until the early 1970s the Kerr solutions were virtually the only known stationary, nonstatic, asymptotically flat vacuum solutions. As discussed in sections 7.1 and 7.4, great progress has been made in obtaining the general stationary, axisymmetric vacuum solution, but even so we are far from having the solutions in sufficiently explicit form to determine if they represent black holes.

5. The proof that a stationary, nonstatic black hole must be axisymmetric continues to hold in the case where a distribution of matter is placed outside a rotating black hole. This leads to an apparent paradox since one would expect it to be possible to "hold in place" a nonaxisymmetric distribution of matter far from the black hole, thereby producing a stationary nonaxisymmetric spacetime. The resolution of this paradox is that such a matter distribution will produce an effective "tidal friction" causing the black hole to "spin down" and thus be nonstationary until it reaches a final static state. A discussion of this process is given by Hawking and Hartle (1972).

The case of a static, vacuum, topologically spherical black hole was analyzed by Israel (1967), who proved that the only such black holes are the Schwarzschild solutions. Some additional assumptions were made in the proof, but the most notable of them—that all the surfaces of constant $\xi^a \xi_a$ are topologically spheres—has been eliminated by Müller zum Hagen, Robinson, and Seifert (1973) and Robinson (1977).

Finally the case of a stationary, axisymmetric, vacuum topologically spherical black hole was analyzed by Carter (1971) and Robinson (1975) using the methods described in section 7.1 to cast Einstein's equation and the black hole boundary conditions into a relatively simple form. They succeeded in proving that all stationary axisymmetric black holes are uniquely characterized by two parameters which appear in the boundary conditions. Since the Kerr solutions exhaust all possible values of these parameters, it follows that they are the only possible stationary axisymmetric black holes.

The above results have been generalized to the electrovac case. Hawking's theorem on the spherical topology of a stationary black hole still applies since it requires only that the dominant energy condition be satisfied by matter. The proof that a stationary black hole with no ergosphere must be static generalizes to the electrovac case (Carter 1973), and the proof of existence of an axial Killing field if ξ^a is spacelike on the horizon depends only on the general form, (10.1.21), of the equations and thus also applies to the electrovac case. Israel's theorem has been generalized to show that the only possible static, electrovac black holes are the Reissner-Nordstrom solutions (Israel 1968). Finally, Mazur (1982) and Bunting (unpublished) have generalized the Carter-Robinson proof of uniqueness of Kerr to show that the charged Kerr solutions (together with their generalizations possessing magnetic charge, which are obtained by applying a duality rotation to the charged Kerr electromagnetic field) are the only stationary, axisymmetric electrovac solutions. Further generalizations to exclude the possible presence of other types of classical fields around a black hole also have been given (Bekenstein 1972; Hartle 1972; Teitelboim 1972). In addition, numerous examples have been given (see, e.g., Wald 1972*a*) to illustrate how the final black hole state resulting from gravitational collapse can lose all information about the collapsing body except its mass, angular momentum, and charge.

12.4 Energy Extraction from Black Holes

By definition, a black hole is a "region of no escape." No material body or light ray ever can be extracted from a black hole. Therefore, it came as a great surprise when Penrose (1969) noted that energy can be extracted from a black hole with an ergosphere. The mechanism proposed by Penrose can be understood as follows. The Killing field ξ^a which becomes a time translation asymptotically at infinity is spacelike in the ergosphere. Thus, for a test particle of 4-momentum $p^a = mu^a$, the energy

$$E = -p^a \xi_a \tag{12.4.1}$$

need not be positive in the ergosphere. Therefore, by making a black hole absorb a particle with negative total energy, we can extract energy from a black hole! To see this in more detail, suppose we start in our laboratory far from the black hole by throwing a particle toward the black hole. If we denote the 4-momentum of this particle by p_0^a, its total energy measured in the laboratory will be

$$E_0 = -p_0^a \xi_a \quad . \tag{12.4.2}$$

As it falls freely toward the black hole, E_0 will remain constant. Suppose, when the particle enters the ergosphere, we arrange—e.g., by means of explosives and a timing device—to have it break up into two fragments as illustrated in Figure 12.7. By local conservation of energy-momentum, we have

$$p_0^a = p_1^a + p_2^a \quad , \tag{12.4.3}$$

where p_1^a and p_2^a are the 4-momenta of the two fragments. Contracting equation (12.4.3) with ξ_a, we obtain

$$E_0 = E_1 + E_2 \quad . \tag{12.4.4}$$

However, inside the ergosphere, we can arrange the breakup so that one of the fragments has negative total energy,

$$E_1 < 0 \quad . \tag{12.4.5}$$

Therefore, if the other fragment returns to our laboratory in free (i.e., geodesic) motion, it will have an energy E_2 which is *greater* than the initial energy E_0.

In the case of a Kerr black hole of mass M with $a \neq 0$, one may explicitly verify that the breakup process can be done so that the second fragment does, indeed, escape to infinity. One also may verify that the negative energy fragment always falls into the black hole. Thus, at the end of the process, one has energy $E_0 + |E_1|$ in the laboratory, and the mass of the black hole must be reduced to $M - |E_1|$. Thus, the energy $|E_1|$ has been extracted from the black hole!

How much energy can be extracted from a Kerr black hole in this manner? As we shall see below shortly, the energy extraction process is self-limiting because the negative energy particles which enter the black hole also carry negative angular momentum, i.e., angular momentum opposite that of the black hole. As a result, the angular momentum $J = Ma$ of the black hole will be reduced to zero while M is still finite. However, when $J = 0$ the ergosphere no longer is present, and no further energy extraction can occur.

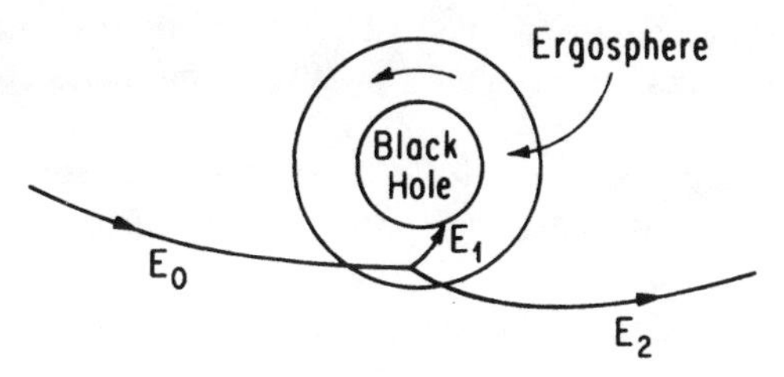

Fig. 12.7. A diagram illustrating the Penrose process for extracting energy from a Kerr black hole.

To see this limit on energy extraction in detail, we use the fact that the Killing field χ^a, defined by equation (12.3.20), is future directed null on the horizon. Hence for any particle which enters the black hole (which includes all negative energy particles), we have

$$0 > p^a \chi_a = p^a(\xi_a + \Omega_H \psi_a) = -E + \Omega_H L \quad , \tag{12.4.6}$$

where $L = p^a \psi_a$ and Ω_H was defined by equation (12.3.19). Thus, we find that

$$L < E/\Omega_H \quad , \tag{12.4.7}$$

which verifies the above statement that a negative-energy particle entering the black hole carries negative angular momentum. After the black hole "swallows" a particle, it should settle back down to a Kerr solution with parameters modified by $\delta M = E$, $\delta J = L$. Thus, from equation (12.4.7), the change in black hole parameters is restricted by

$$\delta J < \delta M/\Omega_H \quad , \tag{12.4.8}$$

which can be rewritten as (Christodoulou 1970)

$$\delta M_{\text{irr}} > 0 \quad , \tag{12.4.9}$$

where the *irreducible mass*, M_{irr}, is defined by

$$M_{\text{irr}}^2 = \frac{1}{2}[M^2 + (M^4 - J^2)^{1/2}] \quad . \tag{12.4.10}$$

Inverting equation (12.4.10), we find

$$\begin{aligned} M^2 &= M_{\text{irr}}^2 + \frac{1}{4}\frac{J^2}{M_{\text{irr}}^2} \\ &\geq M_{\text{irr}}^2 \quad . \end{aligned} \tag{12.4.11}$$

Thus, the mass of a black hole cannot be reduced below the initial value of M_{irr} via the Penrose process. If we start with a Kerr black hole of mass M_0 and angular momentum J_0, then even assuming that equality is achieved in equation (12.4.7), by the time the energy $M_0 - M_{\text{irr}}(M_0, J_0)$ has been extracted, the angular momentum of the black hole will have been reduced to zero. Since one can come arbitrarily close to achieving equality in equation (12.4.7), we can come arbitrarily close to extracting energy $M_0 - M_{\text{irr}}$ from the black hole via the Penrose process. We may interpret $M_0 - M_{\text{irr}}$ as the rotational energy of the black hole. For a maximally rotating black hole, $J_0 = M_0^2$, it represents $(1 - 1/\sqrt{2}) \approx 29\%$ of the mass-energy of the black hole.

The universal nature of the limit on energy extraction implied by (12.4.11)—obtained above only for the specific process proposed by Penrose—can be seen from the area theorem 12.2.6. The area of the event horizon of a Kerr black hole is given by

$$A = \int_{r=r_+} \sqrt{g_{\theta\theta} g_{\phi\phi}}\, d\theta\, d\phi$$

$$= \int (r_+^2 + a^2)\sin\theta \, d\theta \, d\phi$$

$$= 4\pi(r_+^2 + a^2)$$

$$= 16\pi M_{\text{irr}}^2 \quad . \tag{12.4.12}$$

Thus, from the area theorem we obtain the completely general result that M_{irr} can never decrease, from which follows the above energy extraction limit for all possible processes. The consistency of the area theorem, proven by general, abstract arguments, with the result (12.4.11) obtained by the calculation of a specific process for a Kerr black hole, lends further aesthetic support for the validity of the first cosmic censor conjecture, which is used in the above argument when we assume that the infalling particle merely changes the black hole parameters and does not convert the black hole to a naked singularity.

It should be noted that the area theorem also can be used to obtain an interesting upper limit on the energy radiated away in the form of gravitational waves when two black holes collide (Hawking 1971). Consider, for example, initial data representing two widely separated Schwarzschild black holes initially "at rest," with masses M_1 and M_2. (See the paragraph below eq. [10.2.39] of section 10.2 for a brief discussion of how to construct such data explicitly.) Presumably, the dynamical evolution of these data will yield a spacetime where the two black holes fall toward each other, coalesce, and "settle down" to a single Schwarzschild black hole of mass M. The total energy radiated away in this process is

$$E_{\text{rad}} = M_1 + M_2 - M \quad . \tag{12.4.13}$$

An upper limit on E_{rad} can be obtained by noting that the initial black hole area is

$$A_i = A_1 + A_2 = 16\pi(M_1^2 + M_2^2) \quad . \tag{12.4.14}$$

The final area is

$$A_f = 16\pi M^2 \quad ; \tag{12.4.15}$$

and, by the area theorem, we have

$$A_f \geqq A_i \quad . \tag{12.4.16}$$

Putting together equations (12.4.13)–(12.4.16), we obtain

$$E_{\text{rad}} \leqq M_1 + M_2 - (M_1^2 + M_2^2)^{1/2} \quad . \tag{12.4.17}$$

For the case $M_1 = M_2$, this implies that at most $(1 - 1/\sqrt{2}) \approx 29\%$ of the original mass can be radiated away. Numerical calculations (Smarr 1979) indicate that far less energy than this upper limit actually will be radiated in this process.

Although the Penrose process is of great importance for demonstrating that, in principle, the maximum amount of energy permitted by the area theorem can be extracted from a rotating black hole, the process requires a precisely timed breakup of the incident particle at relativistic velocities and it is not a practical energy extraction method (Bardeen, Press, and Teukolsky 1972; Wald 1974*c*). Interestingly there is a wave analog of the Penrose process, known as *superradiant scattering*

(Misner, unpublished; Zel'dovich 1972; Starobinskii 1973), which allows energy to be extracted from a black hole in a relatively simple manner. If a scalar, electromagnetic, or gravitational wave is incident upon a black hole, part of the wave (the "transmitted wave") will be absorbed by the black hole and part of the wave (the "reflected wave") will escape back to infinity. Normally the transmitted wave will carry positive energy into the black hole, and the reflected wave will have less energy than the incident wave. However, for a wave of the form $\phi = \mathrm{Re}[\phi_0(r, \theta)e^{-i\omega t}e^{im\phi}]$ with

$$0 < \omega < m\Omega_H \quad , \tag{12.4.18}$$

the transmitted wave will carry negative energy into the black hole (analogous to the negative energy fragment in the Penrose process for particles) and the reflected wave will return to infinity with greater amplitude and energy than the incident wave. This is demonstrated most easily for the case of scalar waves. By contracting the stress tensor, equation (4.3.10), of a Klein-Gordon scalar field ϕ with the timelike Killing field ξ^a of the Kerr background, we obtain an "energy current,"

$$J_a = -T_{ab}\xi^b \quad , \tag{12.4.19}$$

which is conserved since $\nabla^a J_a = -(\nabla^a T_{ab})\xi^b - T_{ab}\nabla^a\xi^b = 0$. Hence, if we integrate $\nabla_a J^a$ over the region K of spacetime shown in Figure 12.8, we find by Gauss's law that the difference between the incoming and outgoing energies (i.e., the integrated flux of J^a over the "large sphere") equals the integrated flux of J^a on the horizon. However, on the horizon the time averaged flux is given by

$$\begin{aligned}\langle J_a n^a\rangle = -\langle J_a \chi^a\rangle = \langle T_{ab}\chi^a\xi^b\rangle &= \langle(\chi^a\nabla_a\phi)(\xi^b\nabla_b\phi)\rangle \\ &= \frac{1}{2}\omega(\omega - m\Omega_H)\,|\phi_0|^2 \quad ,\end{aligned} \tag{12.4.20}$$

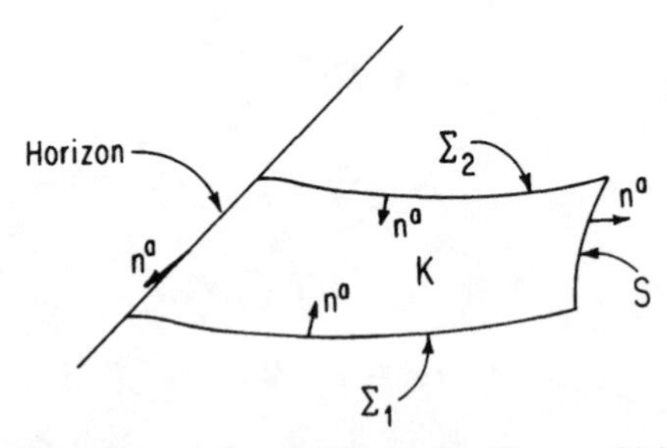

Fig. 12.8. A spacetime diagram showing the region K over which $\nabla_a J^a = 0$ is integrated to derive the conclusion in the text concerning superradiant scattering. Here the spacelike hypersurface Σ_2 is a "time translate" of Σ_1 by Δt, and the timelike hypersurface S represents a "large sphere" at infinity. The integral of $J_a n^a$ over S represents the net energy flow out of K to infinity (i.e., the outgoing minus incoming energy) during the time Δt, whereas the integral of $J_a n^a$ over the horizon represents the net energy flow into the black hole. By Gauss's law (see appendix B) the integral of $J_a n^a$ over the entire boundary vanishes. (The appropriate directions of n^a on each part of the boundary are shown.) However, for a wave with time dependence $e^{-i\omega t}$ the integrals over Σ_1 and Σ_2 cancel by time translation symmetry. Hence, the integral of $J_a n^a$ over S equals minus the integral of $J_a n^a$ over the horizon.

where $n^a = -\chi^a$ (with χ^a given by eq. [12.3.20]) is the appropriately directed normal to the horizon. (The term in T_{ab} proportional to g_{ab} does not contribute since $\chi^a \xi_a = 0$ on the horizon.) Thus, in the frequency range of equation (12.4.18) the energy flux through the horizon is negative, and hence superradiance is obtained. This conclusion also can be derived by considering the "particle number current,"

$$j^a = -i(\overline{\phi}\nabla^a \phi - \phi \nabla^a \overline{\phi}) \tag{12.4.21}$$

for the complex field $\phi_0 e^{-i\omega t} e^{im\phi}$. This current is conserved by virtue of the Klein-Gordon equation. Again, $-j^a \chi_a$ is negative in the range (12.4.18), thus establishing superradiance.

The superradiance of electromagnetic waves in the range (12.4.18) can be proven similarly by consideration of the energy current, $-T_{ab}\xi^b$, or a (non–gauge invariant) "particle number current" analogous to (12.4.21), although the demonstration of the negative integrated flux of the energy current through the horizon is less straightforward (see problem 5). In the gravitational case, we also can form a conserved "effective energy current", $(1/8\pi)G^{(2)}_{ab}\xi^b$, where $G^{(2)}_{ab}$ is the second order Einstein tensor (see section 4.4*b* for the expression for $G^{(2)}_{ab}$ for perturbations off flat spacetime), and a conserved "particle number current." The formula for the energy current is rather complicated, and neither current is gauge invariant. However, one can avoid dealing with them by working, instead, with the decoupled Teukolsky equation for the variables Ψ_0 or Ψ_4 defined in section 12.3. By doing so, superradiance in the range (12.4.18) for gravitational waves can be established in a relatively straightforward manner (Teukolsky and Press 1974). Superradiance of electromagnetic waves also can be proven from the Teukolsky equation for Φ_0 or Φ_2. Numerical computations (Teukolsky and Press 1974) have shown that the largest superradiance effects occur for gravitational waves, where an amplification factor of up to 1.38 can be achieved for a Kerr black hole with $a = M$.

The necessity of superradiance in the regime (12.4.18) also can be seen directly from the area theorem (Bekenstein 1973*a*). For a wave with frequency ω and azimuthal number m in a stationary axisymmetric background, the ratio of angular momentum flux to energy flux at infinity is

$$\mathcal{L}/\mathcal{E} = m/\omega \quad . \tag{12.4.22}$$

Thus, by conservation of energy and angular momentum, when such a wave is incident upon a black hole, the change in energy, δM, and angular momentum, δJ, of the black hole must be related by

$$\delta J/\delta M = m/\omega \quad . \tag{12.4.23}$$

However, in the range (12.4.18), we have

$$\delta J/\delta M > 1/\Omega_H \quad . \tag{12.4.24}$$

If $\delta M > 0$, this would violate equation (12.4.8), which, by equations (12.4.9) and (12.4.12), would violate the area theorem. Hence, we must have $\delta M < 0$—i.e., superradiance must occur—when $0 < \omega < m\Omega_H$.

Interestingly, fermion fields do *not* display superradiance (Unruh 1973; Güven 1977). The "particle number current," j^a, associated with neutrino or Dirac fields (see chapter 13) is manifestly a timelike or null vector, and hence $j_a n^a$ is manifestly nonnegative on the horizon for all waves including those in the range (12.4.18). Thus, the reflected wave never has larger amplitude than the incident wave. In this case the argument given in the previous paragraph is inapplicable because the stress tensor of these fields fails to satisfy the weak energy condition, so the area theorem does not hold.

The behavior of both boson and fermion fields incident upon a Kerr black hole with $a \neq 0$ is in very close mathematical analogy to a well studied effect in non-gravitational physics known as the Klein "paradox." If a Klein-Gordon field of charge Q and mass M in one spatial dimension is incident upon an electrostatic potential, V, such that $V \to 0$ as $x \to -\infty$ but $V \to V_0 > 2M/Q$ for $x \to +\infty$, then when $\omega + M < QV_0$ the reflected wave has greater amplitude and energy than the incident wave. For a Dirac field, the reflected wave will be smaller than the incident wave—again the Dirac "particle number current" is always timelike—but the transmitted wave still has a negative "kinetic energy" (Klein 1929). In quantum field theory, the interpretation of the Klein paradox is that in both the boson and fermion cases particle-antiparticle pairs are spontaneously created in the strong electrostatic field associated with the potential, V. When incoming particles also are present, stimulated emission occurs in the boson case, and in the classical limit this gives rise to the larger "reflected wave" obtained in the classical analysis. The close analogy between the Klein paradox and the scattering of waves by a Kerr black hole suggests that spontaneous particle creation should occur near a Kerr black hole. Indeed, this is the case, but we shall postpone our discussion of this phenomenon until chapter 14.

12.5 Black Holes and Thermodynamics

The area theorem 12.2.6 states that in any physically allowed process, the total area of all black holes in the universe cannot decrease, $\delta A \geqq 0$. This law bears a strong resemblance to the second law of thermodynamics, which states that in any physically allowed process the total entropy of all matter in the universe cannot decrease, $\delta S \geqq 0$. It might appear that this similarity is of a very superficial nature. After all, the area theorem is a mathematically rigorous consequence of general relativity, whereas the second law of thermodynamics is believed not to be a rigorous consequence of the laws of nature but rather a law which holds with overwhelming likelihood for systems with a large number of degrees of freedom. Nevertheless, we shall show in this section that this formal analog for black holes of the second law of thermodynamics extends to the other laws of thermodynamics as well. We will return to this issue in chapter 14, where further evidence will be presented that the relationship between the laws of black hole physics and the laws of thermodynamics is of a fundamental nature.

Our first task is to introduce a quantity, κ, defined on the horizon of an arbitrary stationary black hole (not necessarily vacuum or electrovac in its exterior region) and derive a number of properties of κ. As mentioned at the end of section 12.3, for a stationary black hole, there exists a Killing field, χ^a, which is normal to the horizon of the black hole. If χ^a does not coincide with the stationary Killing field ξ^a, we obtain an axial Killing field ψ^a in the spacetime by taking a linear combination of χ^a and ξ^a. Thus, in general we may write χ^a as

$$\chi^a = \xi^a + \Omega_H \psi^a \quad , \tag{12.5.1}$$

where (as in the case of a Kerr black hole, eq. [12.3.20]) the constant Ω_H is called the angular velocity of the horizon. Since the horizon is a null surface and χ^a is normal to the horizon, we have $\chi^a \chi_a = 0$ on the horizon, so, in particular, $\chi^a \chi_a$ is constant on the horizon. Hence $\nabla^a(\chi^b \chi_b)$ also is normal to the horizon, so on the horizon there exists a function κ such that

$$\nabla^a(\chi^b \chi_b) = -2\kappa \chi^a \quad . \tag{12.5.2}$$

Taking the Lie derivative of equation (12.5.2) with respect to the Killing field χ^a we find

$$\pounds_\chi \kappa = 0 \quad , \tag{12.5.3}$$

i.e., κ is a constant on the orbits of χ^a. In fact, we shall prove below that κ is constant over the horizon, i.e., its value does not change from orbit to orbit. For a charged Kerr black hole, the value of κ is

$$\kappa = \frac{(M^2 - a^2 - e^2)^{1/2}}{2M[M + (M^2 - a^2 - e^2)^{1/2}] - e^2} \quad . \tag{12.5.4}$$

We may rewrite equation (12.5.2) in the form

$$\chi^b \nabla_a \chi_b = -\chi^b \nabla_b \chi_a = -\kappa \chi_a \quad , \tag{12.5.5}$$

which is just the geodesic equation in a non-affine parameterization. Thus, κ measures the failure of the Killing parameter, v, defined by

$$\chi^a \nabla_a v = 1 \quad , \tag{12.5.6}$$

to agree with the affine parameter, λ, along the null geodesic generators of the horizon. If we define k^a on the horizon by

$$k^a = e^{-\kappa v} \chi^a \quad , \tag{12.5.7}$$

we find

$$k^b \nabla_b k^a = e^{-2\kappa v}[\chi^b \nabla_b \chi^a - \chi^a \chi^b \nabla_b(\kappa v)] = 0 \quad ; \tag{12.5.8}$$

i.e., k^a is the affinely parameterized tangent to the null geodesic generators of the horizon. This shows that on the horizon the relation between affine parameter λ and Killing parameter v, is given by

$$\frac{d\lambda}{dv} \propto e^{\kappa v} \quad , \tag{12.5.9}$$

so that, if $\kappa \neq 0$, we have

$$\lambda \propto e^{\kappa v} \quad . \tag{12.5.10}$$

Since χ^a is hypersurface orthogonal at the horizon, by Frobenius's theorem (see appendix B) we have on the horizon

$$\chi_{[a}\nabla_b\chi_{c]} = 0 \quad . \tag{12.5.11}$$

Using Killing's equation $\nabla_b\chi_c = -\nabla_c\chi_b$, this implies

$$\chi_c\nabla_a\chi_b = -2\chi_{[a}\nabla_{b]}\chi_c \tag{12.5.12}$$

on the horizon. Contracting with $\nabla^a\chi^b$, we find

$$\begin{aligned} \chi_c(\nabla^a\chi^b)(\nabla_a\chi_b) &= -2(\chi_a\nabla^a\chi^b)(\nabla_b\chi_c) \\ &= -2\kappa\chi^b\nabla_b\chi_c \\ &= -2\kappa^2\chi_c \quad . \end{aligned} \tag{12.5.13}$$

Thus, we obtain a simple explicit formula for κ,

$$\kappa^2 = -\frac{1}{2}(\nabla^a\chi^b)(\nabla_a\chi_b) \quad , \tag{12.5.14}$$

where evaluation on the horizon is understood.

Equation (12.5.14) provides us with a physical interpretation of κ as follows. We have everywhere (i.e., not just on the horizon)

$$3(\chi^{[a}\nabla^b\chi^{c]})(\chi_{[a}\nabla_b\chi_{c]}) = \chi^a\chi_a(\nabla^b\chi^c)(\nabla_b\chi_c) - 2(\chi^a\nabla^b\chi^c)(\chi_b\nabla_a\chi_c) \quad . \tag{12.5.15}$$

Since $\chi_{[a}\nabla_b\chi_{c]} = 0$ on the horizon, the gradient of the left-hand side vanishes on the horizon. On the other hand, by equation (12.5.2), $\nabla_b(\chi^a\chi_a) \neq 0$ on the horizon, provided that $\kappa \neq 0$. Hence, by l'Hospital's rule, the left-hand side of equation (12.5.15) divided by $\chi^a\chi_a$ must approach zero on the horizon. Thus, using equation (12.5.14), we find

$$\kappa^2 = \lim\{-(\chi^b\nabla_b\chi^c)(\chi^a\nabla_a\chi_c)/\chi^d\chi_d\} \quad , \tag{12.5.16}$$

where "lim" stands for the limit as one approaches the horizon. Now,

$$a^c = (\chi^b\nabla_b\chi^c)/(-\chi^a\chi_a) \tag{12.5.17}$$

is just the acceleration of an orbit of χ^a. Thus, we have

$$\kappa = \lim(Va) \quad , \tag{12.5.18}$$

where $a = (a^c a_c)^{1/2}$ and $V = (-\chi^a\chi_a)^{1/2}$. In the case of a static black hole, we have $\chi^a = \xi^a$. Then V is just the redshift factor, and, by problem 4 of chapter 6, Va is the force that must be exerted at infinity to hold a unit test mass in place. Thus, κ is the limiting value of this force at the horizon; i.e., it is the *surface gravity* of the black hole. (Of course, the locally exerted force, a, becomes infinite on the horizon.) For a rotating black hole, $\Omega_H \neq 0$, a test mass cannot be held stationary with respect to infinity near the black hole, but we shall continue to refer to κ as the surface gravity.

Using equations (12.5.7) and (12.5.11), we find on the horizon

$$k_{[a}\nabla_{b]}k_c = -e^{-2\kappa v}\chi_c\left[\frac{1}{2}\nabla_a\chi_b + \chi_{[a}\nabla_{b]}(\kappa v)\right] \quad . \tag{12.5.19}$$

Contracting equation (12.5.19) with any two vectors m^b, n^c tangent to the horizon (i.e., $\chi^a m_a = \chi^a n_a = 0$), we obtain

$$m^b n^c \nabla_b k_c = 0 \quad , \tag{12.5.20}$$

i.e., in the notation of section 9.2, we have

$$\widehat{\nabla_a k_b} = 0 \quad . \tag{12.5.21}$$

Thus, the expansion θ, twist $\hat{\omega}_{ab}$, and shear $\hat{\sigma}_{ab}$ of the null geodesic generators of the horizon vanish. From equations (9.2.32) and (9.2.33), we see also that on the horizon

$$R_{ab} k^a k^b = 0 \tag{12.5.22}$$

and

$$\widehat{C_{abcd} k^a k^d} = 0 \quad . \tag{12.5.23}$$

The latter equation states that k^a is a principal null vector of the Weyl tensor (see section 7.3). In fact, we shall see below that k^a must be a repeated principal null vector.

It should be emphasized that most of the above equations—in particular, the relation (12.5.2) defining κ—hold only on the horizon. Thus, we may not simply apply ∇_a to, say, equation (12.5.2) since we may differentiate equation (12.5.2) only in directions tangent to the horizon. If the horizon were a spacelike surface, we could apply $h^{ab}\nabla_a$ to all equations, where the projection operator h_{ab} was defined by equation (10.2.10). However, the horizon is a null surface and has no natural projection operator associated with it. Nevertheless, the tensor $\epsilon^{abcd}\chi_d$, where ϵ_{abcd} is the spacetime volume element (see appendix B) is tangent to the horizon, since $(\epsilon^{abcd}\chi_d)\chi_c = 0$. Thus, we may apply $\epsilon^{abcd}\chi_d \nabla_c$ to any equation holding on the horizon. Equivalently, we may apply $\chi_{[d}\nabla_{c]}$ to any such equation.

We now shall prove that the surface gravity, κ, is constant over the event horizon. Applying $\chi_{[d}\nabla_{c]}$ to equation (12.5.5), we obtain

$$\begin{aligned} \chi_a \chi_{[d}\nabla_{c]}\kappa + \kappa\, \chi_{[d}\nabla_{c]}\chi_a &= \chi_{[d}\nabla_{c]}(\chi^b \nabla_b \chi_a) \\ &= (\chi_{[d}\nabla_{c]}\chi^b)(\nabla_b \chi_a) + \chi^b \chi_{[d}\nabla_{c]}\nabla_b \chi_a \\ &= (\chi_{[d}\nabla_{c]}\chi^b)(\nabla_b \chi_a) - \chi^b R_{ba[c}{}^e \chi_{d]}\chi_e \quad , \end{aligned} \tag{12.5.24}$$

where equation (C.3.6) in appendix C was used in the last step. However, using equations (12.5.12) and (12.5.5), the first term on the right-hand side of the last line of equation (12.5.25) is found to be

$$\begin{aligned} (\chi_{[d}\nabla_{c]}\chi^b)(\nabla_b \chi_a) &= -\frac{1}{2}(\chi^b \nabla_d \chi_c)\nabla_b \chi_a \\ &= -\frac{1}{2}\kappa \chi_a \nabla_d \chi_c \\ &= \kappa\, \chi_{[d}\nabla_{c]}\chi_a \quad , \end{aligned} \tag{12.5.25}$$

which cancels the second term on the left-hand side of equation (12.5.24). Thus, we obtain

$$\chi_a \chi_{[d} \nabla_{c]} \kappa = \chi^b R_{ab[c}{}^e \chi_{d]} \chi_e \quad . \tag{12.5.26}$$

On the other hand, if we apply $\chi_{[d} \nabla_{e]}$ to equation (12.5.12), we obtain

$$\begin{aligned}(\chi_{[d} \nabla_{e]} \chi_c) \nabla_a \chi_b &+ \chi_c \chi_{[d} \nabla_{e]} \nabla_a \chi_b \\ &= -2(\chi_{[d} \nabla_{e]} \chi_{[a}) \nabla_{b]} \chi_c - 2(\chi_{[d} \nabla_{e]} \nabla_{[b} \chi_{|c|}) \chi_{a]} \quad . \end{aligned} \tag{12.5.27}$$

Using equation (12.5.12) repeatedly, we find that the first term on the left-hand side of equation (12.5.27) cancels the first term on the right-hand side. Thus, using equation (C.3.6), we obtain

$$-\chi_c R_{ab[e}{}^f \chi_{d]} \chi_f = 2 \chi_{[a} R_{b]c[e}{}^f \chi_{d]} \chi_f \quad . \tag{12.5.28}$$

If we multiply by g^{ce} and contract over c and e, the left-hand side vanishes, and we find

$$-\chi_{[a} R_{b]}{}^f \chi_f \chi_d = \chi_{[a} R_{b]cd}{}^f \chi^c \chi_f \quad . \tag{12.5.29}$$

However, the term on the right-hand side of this equation is of the same form as the right-hand side of equation (12.5.26). Thus, comparison of these two equations yields

$$\chi_{[d} \nabla_{c]} \kappa = -\chi_{[d} R_{c]}{}^f \chi_f \quad . \tag{12.5.30}$$

Up to this point we have not used Einstein's equation anywhere in the above analysis. However, we show now that Einstein's equation together with the dominant energy condition (defined in section 9.2) implies that the right-hand side of equation (12.5.30) vanishes. Namely, the dominant energy condition states that $-T^a{}_b \chi^b$ must be a future directed timelike or null vector. However, Einstein's equation together with equation (12.5.22) implies $T^a{}_b \chi^b \chi_a = 0$. This implies that $-T^a{}_b \chi^b$ must point in the direction of χ^a, i.e., $\chi_{[c} T_{a]b} \chi^b = 0$. Hence using Einstein's equation again, we find that the right-hand side of equation (12.5.30) also must vanish, and we conclude that

$$\chi_{[d} \nabla_{c]} \kappa = 0 \quad , \tag{12.5.31}$$

which states that κ is constant on the horizon. Note, incidentially, that the vanishing of both sides of equation (12.5.29) implies by equation (3.2.28) that $\chi_{[a} C_{b]cdf} \chi^c \chi^f = 0$, i.e., that χ^a is a twice repeated principal null vector of the Weyl tensor on the horizon.

A simple formula for the mass of a stationary, axisymmetric spacetime containing a black hole now may be obtained (Bardeen, Carter, and Hawking 1973). Let Σ be an asymptotically flat spacelike hypersurface which intersects the horizon, H, on a 2-sphere, $\mathcal{H}$, which forms the boundary of Σ. The calculation which led to equation (11.2.10) of the previous chapter is modified only by the presence of an additional boundary term due to $\mathcal{H}$. We obtain

$$M = 2 \int_\Sigma \left(T_{ab} - \frac{1}{2} T g_{ab} \right) n^a \xi^b dV - \frac{1}{8\pi} \int_{\mathcal{H}} \epsilon_{abcd} \nabla^c \xi^d \quad . \tag{12.5.32}$$

We may evaluate this boundary term by using equation (12.5.1) to write

$$\int_{\mathcal{H}} \epsilon_{abcd} \nabla^c \xi^d = \int_{\mathcal{H}} \epsilon_{abcd} \nabla^c \chi^d - \Omega_H \int_{\mathcal{H}} \epsilon_{abcd} \nabla^c \psi^d$$

$$= \int_{\mathcal{H}} \epsilon_{abcd} \nabla^c \chi^d - 16\pi \Omega_H J_H \quad , \tag{12.5.33}$$

where we may interpret $J_H \equiv (1/16\pi) \int_{\mathcal{H}} \epsilon_{abcd} \nabla^c \psi^d$ as the angular momentum of the black hole (see problem 6 of chapter 11). On the other hand, we may express the volume element, ϵ_{ab}, on $\mathcal{H}$ as

$$\epsilon_{ab} = \epsilon_{abcd} N^c \chi^d \quad , \tag{12.5.34}$$

where N^a is the "ingoing" future directed null normal to $\mathcal{H}$, normalized so that $N^a \chi_a = -1$. Thus, we have

$$\epsilon^{ab} \epsilon_{abcd} \nabla^c \chi^d = N_e \chi_f \epsilon^{abef} \epsilon_{abcd} \nabla^c \chi^d = -4 N_c \chi_d \nabla^c \chi^d = -4\kappa \quad , \tag{12.5.35}$$

and hence

$$\int_{\mathcal{H}} \epsilon_{abcd} \nabla^c \chi^d = \frac{1}{2} \int_{\mathcal{H}} (\epsilon^{ef} \epsilon_{efcd} \nabla^c \chi^d) \epsilon_{ab} = -2\kappa A \quad , \tag{12.5.36}$$

where $A = \int_{\mathcal{H}} \epsilon_{ab}$ is the area of the event horizon. Thus, we obtain the following formula for M:

$$M = 2 \int_{\Sigma} \left(T_{ab} - \frac{1}{2} T g_{ab} \right) n^a \xi^b dV + \frac{1}{4\pi} \kappa A + 2\Omega_H J_H \quad . \tag{12.5.37}$$

Of greatest interest for the development of the analogy between the laws of black hole physics and the laws of thermodynamics is the derivation of a differential formula for M, i.e., a formula for how M changes when a small stationary, axisymmetric change is made in the solution. We may use the freedom of applying diffeomorphisms to solutions to ensure that ξ^a, ψ^a, and the location of the horizon on the manifold remain unchanged when we vary the spacetime metric. For simplicity, we shall treat only the vacuum case[6], $T_{ab} = 0$. A generalization of the differential mass formula to the case of fluid matter outside the black hole was given in the original paper of Bardeen, Carter, and Hawking (1973), and further generalizations have been given by Carter (1973).

A formula for δM in the vacuum case can be obtained by varying equation (12.5.37),

$$\delta M = \frac{1}{4\pi} (A \delta\kappa + \kappa \delta A) + 2(J_H \delta\Omega_H + \Omega_H \delta J_H) \quad . \tag{12.5.38}$$

6. Actually, as discussed at the end of section 12.3, in the vacuum case, the only possible black hole solutions are the Kerr metrics, so our formula (12.5.44) below could be derived simply by verifying that it holds for the Kerr metrics. However, as mentioned in the text, the derivation we give generalizes to the case where matter is present outside the black hole.

However, this is not the desired formula. Another formula for δM can be derived as follows. First, we note that in any spacetime if v^a and w^a commute and satisfy $\nabla_a v^a = \nabla_a w^a = 0$, then

$$\nabla_a(v^{[a}w^{b]}) = 0 \tag{12.5.39}$$

and hence

$$\nabla_{[a}(\epsilon_{bc]de}v^d w^e) = 0 \tag{12.5.40}$$

Consequently, if we apply Stokes's theorem to a three-dimensional volume bounded by two spheres S and S', we find that $\int_S \epsilon_{abcd}v^c w^d = \int_{S'} \epsilon_{abcd}v^c w^d$. We apply this general result here, choosing S to be a "sphere at infinity," S' to be a sphere, $\mathcal{H}$, on the horizon, $w^a = \xi^a$, and

$$v^a = \nabla_b(\gamma^{ab} - g^{ab}\gamma) \quad , \tag{12.5.41}$$

where γ_{ab} is the perturbed metric and $\gamma = g^{ab}\gamma_{ab}$. The above required properties are satisfied by ξ^a and v^a since for any stationary perturbation we have $\pounds_\xi v^a = 0$ (i.e., v^a and ξ^a commute), we have $\nabla_a \xi^a = 0$ by Killing's equation, and $\nabla_a v^a = 0$ is just the trace of the perturbed vacuum Einstein equation, $\dot{R} = 0$ (see eq. [7.5.15]). Thus, we find

$$\int_S \epsilon_{abcd}\xi^d\nabla_e(\gamma^{ce} - g^{ce}\gamma) = \int_{\mathcal{H}} \epsilon_{abcd}\xi^d\nabla_e(\gamma^{ce} - g^{ce}\gamma) \quad . \tag{12.5.42}$$

If both the ADM and Komar expressions for total mass, equations (11.2.14) and (11.2.9), are used, the left-hand side is evaluated to be $8\pi\delta M$. A significantly longer computation reveals that the right-hand side is equal to $-2A\delta\kappa - 16\pi J_H\delta\Omega_H$ (see Bardeen, Carter, and Hawking 1973). Thus, we obtain

$$\delta M = -\frac{1}{4\pi}A\delta\kappa - 2J_H\delta\Omega_H \quad . \tag{12.5.43}$$

Adding equations (12.5.38) and (12.5.43), we obtain the desired formula,

$$\delta M = \frac{1}{8\pi}\kappa\delta A + \Omega_H\delta J_H \quad . \tag{12.5.44}$$

The close mathematical analogy between laws of black hole physics derived above and the ordinary laws of thermodynamics is displayed in Table 12.1. We noted at the beginning of this section that the black hole area theorem is analogous to the second law of thermodynamics. Now we have obtained the formula (12.5.44) for δM, which is closely analogous to the first law of thermodynamics. In particular, the term $\Omega_H\delta J_H$ is analogous to the "work term" $P\delta V$ of the first law; indeed, for an ordinary rotating body a term of the form $\Omega\delta J$ would be present in the thermodynamic formula. The term δA appears in equation (12.5.44) in the same manner as δS appears in the first law of thermodynamics, except that it is multiplied by $(1/8\pi)\kappa$ rather than T, so κ plays the role of temperature in the black hole laws. But we proved above that κ satisfies an important property analogous to the property of temperature

Table 12.1
BLACK HOLES AND THERMODYNAMICS

LAW	CONTEXT	
	Thermodynamics	Black Holes
Zeroth	T constant throughout body in thermal equilibrium	κ constant over horizon of stationary black hole
First	$dE = TdS +$ work terms	$dM = \frac{1}{8\pi}\kappa dA + \Omega_H dJ$
Second	$\delta S \geqq 0$ in any process	$\delta A \geqq 0$ in any process
Third	Impossible to achieve $T = 0$ by a physical process	Impossible to achieve $\kappa = 0$ by a physical process

in the zeroth law of thermodynamics: it is uniform over an "equilibrium" (i.e., stationary) black hole. Finally, we see from equation (12.5.4) that for the charged Kerr black holes, κ vanishes only for the "extreme" case $M^2 = a^2 + e^2$. Explicit calculations (see, e.g., Wald 1974*a*) show that the closer one gets to an "extreme" black hole, the harder it is to get a further step closer, in a manner similar to the third law of thermodynamics. (However, the analog of the alternate version of the third law of thermodynamics, which states that $S \to 0$ as $T \to 0$, is not satisfied in black hole physics, since A may remain finite as $\kappa \to 0$.)

Note that the analogous quantities in Table 12.1 are $E \leftrightarrow M$, $T \leftrightarrow \alpha\kappa$, and $S \leftrightarrow (1/8\pi\alpha)A$, where α is a constant. A hint that the relation between black hole laws and thermodynamic laws might be more than just an analogy comes from the fact that E and M are not merely analogs in the formulas but represent the same physical quantity: total energy. However, the thermodynamic temperature of a black hole in classical general relativity is absolute zero since a black hole is a perfect absorber but does not emit anything. Thus, it would appear that κ could not physically represent a temperature. Nevertheless, in 1974 Hawking discovered that quantum particle creation effects result in an effective "emission" of particles from a black hole with a blackbody spectrum at temperature $T = \hbar\kappa/2\pi$. Thus, κ does physically represent the thermodynamic temperature of a black hole! This suggests that the relationship between laws of black hole physics and thermodynamics may be much more than an analogy: The black hole laws of Table 12.1 may be precisely the ordinary laws of thermodynamics applied to a black hole. We shall discuss this issue further in the last section of chapter 14.

Problems

1. Since an observer outside a black hole does not lie within the causal future of the black hole, such an observal literally cannot "see" the black hole. As is apparent from Figure 6.11, an observer looking at a region where gravitational collapse has oc-

curred would, in principle, see the collapsing matter at a stage where it is just outside the black hole. Consider a particle—such as a particle on the surface of the collapsing body—which falls into a Schwarzschild black hole. Show that for any smooth, timelike curve, $dU/d\tau$ must have a finite, nonzero value on the horizon, where U denotes the Kruskal coordinate (6.4.26) and τ is the proper time along the curve. Show, therefore, that if the particle emits photons radially outward at a constant rate (with respect to its proper time), the rate at which photons will be received by a distant static observer will vary as $e^{-t/4M}$ at late times, where t is the Schwarzschild coordinate time ($\approx$ proper time of the observer). The frequency of each photon also will be redshifted by this factor. Since $4M \approx 2 \times 10^{-5}(M/M_{\odot})$ s, this means that the region where collapse has occurred will appear black on a very rapid time scale (Ames and Thorne 1968).

2. *a*) Let (M, g_{ab}) be a spacetime with a Killing field, w^a, and suppose A^a is a vector potential which respects this symmetry, i.e., $\pounds_w A^a = 0$. Show that for a particle of charge q moving under the Lorentz force law (4.3.2), $w^a(mu_a + qA_a)$ is constant along the world line of the particle.

b) Obtain the constants of motion, E and L, for charged particle motion in the charged Kerr spacetime. Use this result to derive an effective potential for radial motion in the equatorial plane, thereby generalizing equation (12.3.25) to the charged case.

3. Show that the energy (defined as in problem 2 above) of a particle of mass m and charge q held fixed at radius r outside a Reissner-Nordstrom black hole of mass M and charge e is $E = m(1 - 2M/r + e^2/r^2)^{1/2} + qe/r$. Hence, if q has the opposite sign of e, we will have $E < 0$ for r sufficiently close to r_+. Thus, we may extract energy from a Reissner-Nordstrom black hole by lowering a charged particle to near the horizon and then dropping it into the black hole. By paralleling the derivation of equation (12.4.9) in the Kerr case, obtain an upper limit for the amount of energy that can be extracted by this process (Christodoulou and Ruffini 1971). Show that this upper limit agrees with that obtained from the area theorem.

4. Suppose two widely separated Kerr black holes with parameters (M_1, J_1) and (M_2, J_2) initially are at rest in an axisymmetric configuration, i.e., their rotation axes are aligned along the direction of their separation. Assume that these black holes fall together and coalesce into a single black hole. Since angular momentum cannot be radiated away in an axisymmetric spacetime (see problem 6 of chapter 11), the final black hole will have angular momentum $J = J_1 + J_2$. Derive an upper limit for the energy radiated away in this process. Note that this upper limit is larger when J_1 and J_2 are antiparallel rather than parallel. This suggests the existence of a gravitational "spin-spin" force which is attractive for antiparallel spins. (The existence of a force of the correct magnitude and sign to account for this effect can be demonstrated directly from the equation of motion for a spinning test body [Wald 1972*b*].)

5. *a*) Let F_{ab} be a closed two-form (i.e., satisfy equation (4.3.13)) and let w^a be an arbitrary vector field. Show that

$$\pounds_w F_{ab} = -2\nabla_{[a}(F_{b]c}w^c) \quad .$$

b) Show that the time averaged flux of the Maxwell energy current, $J^a = -T^a{}_b\xi^b$, across the horizon of a Kerr black hole is negative when ω and m satisfy equation (12.4.18), and thus that superradiance of electromagnetic waves occurs in that regime. (Hint: Use part (*a*) to relate $F_{ab}\xi^b$ to $F_{ab}\chi^b$.)

THIRTEEN

SPINORS

In chapter 4 we briefly considered the issue of what types of quantities appear in physical laws. We noted that tensor fields—i.e., multilinear maps associated with each spacetime point, taking vectors and dual vectors into numbers—encompass a very general class of mathematical entities, and this helps to account for why essentially all physical quantities in spacetime are represented by tensor fields. In the first section of this chapter we shall reinvestigate this question from a more systematic point of view, using the "special covariance" of the laws of physics in special relativity. This will motivate us to define and investigate the properties of more general entities called spinor fields.

In essence, a spinor at a point x of spacetime is an ordered pair of complex numbers associated with an orthonormal basis of the tangent space V_x which transforms in a specified way under a continuous change of basis. The most unusual aspect of this transformation law—which contrasts sharply with the analogous transformation laws for ordinary tensors—is that a spinor changes sign when the basis completes a rotation of 2π radians about a fixed axis and thereby returns to its original configuration. Thus, the numerical values of a spinor in a given orthonormal basis cannot be directly physically measurable since it has two possible distinct values in that basis. However, real bilinear products of spinors and complex conjugate spinors may be identified with ordinary vectors and thus have a direct physical interpretation. Indeed every null vector can be expressed as the tensor product of a spinor and its complex conjugate. In this sense, a spinor may be viewed as a "square root" of a null vector.

Spinors arise most naturally in the context of quantum theory. In quantum mechanics, the numerical value of a wave function ψ is not physically measurable since ψ and $e^{i\alpha}\psi$ represent the same physical state. Consequently, no contradiction results from having a wave function be represented by a spinor field. Indeed, we shall see that spinors arise naturally when one considers from a general viewpoint the types of fields which can occur in quantum theory.

However, we should emphasize that the notion of spinors has proven to be an extremely powerful tool for analyzing purely classical problems. Perhaps the most dramatic example of this is Witten's (1981) spinorial proof of the positive mass conjecture. In section 13.2 we shall give further examples of this by deriving a useful spinorial decomposition of the curvature tensor and obtaining the existence and properties of the principal null directions of the Weyl tensor in a manner far simpler than can be achieved by tensor methods.

We begin our discussion in section 13.1 by arguing from a general viewpoint that the isometry group of a spacetime acts in a natural way on the states of a physical theory defined on that spacetime. For a quantum theory defined on Minkowski spacetime, this leads us to examine the unitary representations of the Poincaré group on a Hilbert space. However, because state vectors which differ by a phase factor represent the same physical state, representations "up to phase" of the Poincaré group also are allowed. These representations are in one-to-one correspondence with true representations of the covering group of the Poincaré group, namely, the group[1] ISL(2, $\mathbb{C}$) composed of all translations and linear maps of unit determinant acting on a two-dimensional, complex vector space. Ordinary tensor fields on Minkowski spacetime arise as realizations of the true representations of the Poincaré group. Spinors and spinorial tensor fields arise as realizations of representations of ISL(2, $\mathbb{C}$). The relation between spinors and vectors also is obtained in section 13.1, and other basic properties of spinors and spinorial tensors are established. We conclude section 13.1 by defining the notion of the derivative of spinor fields in Minkowski spacetime and giving the linear equations for fields in Minkowski spacetime associated with the irreducible representations of ISL(2, $\mathbb{C}$) of mass m and spin s.

In section 13.2 we consider the generalization of the notion of spinors to curved spacetime. Since the presentation of spinors in section 13.1 is based heavily on the Poincaré group, we must significantly reformulate the notion of spinors in order to define them in curved spacetime. We do so by means of a construction involving fiber bundles. As explained in section 13.2, it turns out that the spacetime manifold must satisfy certain topological properties in order to admit a notion of spinor fields, and that more than one inequivalent spinor structure may exist in a spacetime which is not simply connected.

The derivative operator acting on ordinary tensor fields associated with g_{ab} can be generalized to act on spinorial tensor fields. This allows us to obtain a spinorial decomposition on the Riemann curvature tensor. As applications of spinor methods, we conclude section 13.2 by deriving the algebraic classification of the Weyl tensor and demonstrating the inconsistency of the natural generalization to curved spacetime of the Minkowski spacetime equations for a massless field of spin greater than 1.

We take this opportunity to bring two points concerning terminology and conventions to the attention of the reader. First, the term "spinor" in this chapter refers to an SL(2,$\mathbb{C}$) (2-component) spinor. As mentioned at the end of section 13.1, a Dirac (4-component) spinor is simply an SL(2, $\mathbb{C}$) spinor together with a complex conjugate SL(2, $\mathbb{C}$) spinor. In this sense, SL(2, $\mathbb{C}$) spinors may be viewed as being more fundamental objects than Dirac spinors, and it is more natural for us to work

1. The group SL (2, $\mathbb{C}$) consists of the "*s*pecial" (i.e., unit determinant) *l*inear maps on $\mathbb{C}^2$. The group ISL (2, $\mathbb{C}$) contains, in addition, the (*i*nhomogeneous) translation maps.

with them. However, we emphasize that in most references on quantum theory, the term "spinor" means "Dirac spinor."

Second, for the reason explained below equation (13.1.18), *in this chapter we use the metric signature convention* $+ - - -$. Thus, in the formulas of this chapter which involve the spacetime metric, a change of sign for each appearance of the metric must be made in order to obtain agreement with the formulas appearing in the other chapters of this book. Further remarks on these sign changes are given in the section on notation and conventions at the beginning of this book.

13.1 Spinors in Minkowski Spacetime

The main purpose of this section is to motivate the definition of spinor fields on Minkowski spacetime and establish some of their basic properties. We shall do so by investigating the general issue of what mathematical entities may represent physical fields in Minkowski spacetime. Our approach will be group-theoretic in nature. We first shall argue that if "special covariance" of the physical laws holds, then the isometry group of a spacetime acts in a natural way on the states of a physical system. In the case of a quantum theory in Minkowski spacetime, we thereby obtain a unitary representation up to phase of the Poincaré group. The study of these representations leads to consideration of the group, $SL(2, \mathbb{C})$, consisting of linear maps of unit determinant acting on a complex two-dimensional vector space, W. The notion of spinor fields in Minkowski spacetime then is obtained by assigning vectors in W to points of spacetime in an appropriate way.

To begin, we give a general argument that there should exist an action of the isometry group of a spacetime (M, g_{ab}) on the collection, $\mathcal{S}$, of states of a physical theory defined on that spacetime. We shall assume that the physical properties of each state in $\mathcal{S}$ can be characterized by local measurements made at each spacetime event by a family of observers. A good example of an $\mathcal{S}$ satisfying this property is the collection of states of a physical system that can be described by tensor fields of a specified type on M (with the components of these fields corresponding to the local physically measurable quantities); but since our purpose is to investigate the possibilities for what $\mathcal{S}$ may consist of, we leave $\mathcal{S}$ unspecified. Consider a family of observers, O, equipped with measuring apparatus on M. We shall assume that these observers can be characterized by specifying an orthonormal basis $(e_\alpha)^a$ with $\alpha = 0, 1, 2, 3$ for the tangent space at each point of M. Here the first vector $(e_0)^a$ in the basis at each event is chosen tangent to the world line of the observer at that event, and the remaining basis vectors $(e_\alpha)^a$ for $\alpha = 1, 2, 3$ serve as references for how the apparatus he carries is aligned. Since all experiments in physics measure numbers, associated with each $s \in \mathcal{S}$ there should be a collection of numbers corresponding to the outcomes of a complete set of measurements on the state s made by these observers. We shall assume for simplicity that each $x \in M$ only a finite number, k, of measurements need be made. Then for the given spacetime (M, g_{ab}) and the given family, O, of observers, we obtain a map $f_O: M \times \mathcal{S} \rightarrow \mathbb{R}^k$ which uniquely characterizes each $s \in \mathcal{S}$ in terms of the measurements made by these observers. A different family, $\tilde{O}$, of observers would, in general, obtain a different map, $f_{\tilde{O}}$, i.e., the numerical results of the measurements on s may depend on how the observers move and how they orient their measuring apparatus. Consider, now, a diffeomorphism $\phi: M \to M$ and allow ϕ to map the basis fields $(e_\alpha)^a$ into $\phi^*(e_\alpha)^a$ in the manner described in appendix C. In general, the basis $\phi^*(e_\alpha)^a$ will not be orthonormal at each point and thus will not correspond to a physically realizable family of observers. However, when (and only when) ϕ is an isometry, $\phi^*(e_\alpha)^a$ will be orthonormal, and we can use ϕ to map our original family, O, of physical observers associated with the basis field $(e_\alpha)^a$ into a new family $\tilde{O}$ of physical observers associated with $\phi^*(e_\alpha)^a$.

If the laws of physics are "specially covariant" under the isometries of (M, g_{ab}) (see chapter 4), then any physically possible result of a set of measurements made by O must be a physically possible result for a set of measurements made by $\tilde{O}$. In other words, given any $s \in \mathcal{S}$, there must exist an $\tilde{s} \in \mathcal{S}$ such that the results of measurements by O on s are identical to the results of measurements by $\tilde{O}$ on $\tilde{s}$, i.e., for each $x \in M$ we have $f_O(x, s) = f_{\tilde{O}}(\phi(x), \tilde{s})$. Thus, associated with each isometry ϕ, we obtain a map $\tilde{\phi}: \mathcal{S} \to \mathcal{S}$, defined by the requirement that $\tilde{\phi}(s)$ "look the same" to the observers $\tilde{O}$ as s "looks" to O. In the case where $\mathcal{S}$ consists of tensor fields, this map is simply the map ϕ^* defined in appendix C.

The isometries on (M, g_{ab}) form a Lie group (see section 7.2). We shall denote the abstract group isomorphic to the group of isometries by G and denote the isometry associated with $g \in G$ as ϕ_g. By the above remarks, for each $g \in G$ we obtain a map $\tilde{\phi}_g: \mathcal{S} \to \mathcal{S}$. Furthermore, from the physical criteria which defined the map $\tilde{\phi}_g$, it is clear that for all $g_1, g_2 \in G$, we have

$$\tilde{\phi}_{g_1} \circ \tilde{\phi}_{g_2} = \tilde{\phi}_{g_1 g_2} \quad . \tag{13.1.1}$$

We specialize, now, to the case of Minkowski spacetime $(\mathbb{R}^4, \eta_{ab})$. The isometry group of Minkowski spacetime is the extended Poincaré group, but since, as mentioned in section 4.2, the laws of physics in Minkowski spacetime are believed to be "specially covariant" only under proper Poincaré transformations, we shall take G to be the group of proper Poincaré transformations. In order to proceed further, we must specify the nature of the physical theory in more detail. We shall take the framework of our physical theory to be that of quantum theory. As we shall see below, spinors will emerge as candidates for physical fields in quantum theory.

In quantum theory, states of a system are represented as vectors with unit norm in a Hilbert space[2] $\mathcal{H}$. However, two vectors which differ by an overall phase factor, i.e., a complex number c with $|c| = 1$, represent the same physical state. Thus, the physical states, $\mathcal{S}$, are the unit *rays* in the Hilbert space, i.e., the equivalence classes of unit norm vectors differing only by a phase factor. Let $\tilde{\phi}_g: \mathcal{S} \to \mathcal{S}$ denote the map of $\mathcal{S}$ into itself associated with the isometry ϕ_g. We may associated with $\tilde{\phi}_g$ a map $U_g: \mathcal{H} \to \mathcal{H}$, where the phase of $U_g(\psi)$ for all $\psi \in \mathcal{H}$ can be chosen arbitrarily. The requirement that $\tilde{\phi}_g$ take the states in $\mathcal{S}$ characterized by observer O to states which "look the same" to observer $\tilde{O}$ implies that all transition probabilities must be preserved by $\tilde{\phi}_g$. This implies that each U_g must satisfy $|(U_g \psi_1, U_g \psi_2)| = |(\psi_1, \psi_2)|$

2. See the beginning of section 14.2 for the definition of a Hilbert space. An introduction to some of the basic properties of a Hilbert space also is given there, but this discussion is not essential for the present chapter.

for all $\psi_1, \psi_2 \in \mathcal{H}$. As shown by Wigner (1959), this implies that U_g can be re-phased so that it is either unitary or antiunitary. Since all proper Poincaré transformations can be continuously deformed to the identity element, the continuous dependence of $\tilde{\phi}_g$ with g requires that U_g be unitary. The requirement (13.1.1) on $\tilde{\phi}_g$ implies that U_g must satisfy

$$U_{g_1} U_{g_2} = \omega(g_1, g_2) U_{g_1 g_2} \quad , \tag{13.1.2}$$

where ω is a phase factor, $|\omega(g_1, g_2)| = 1$.

We take this opportunity to introduce some terminology. Let G and G' be groups. A map $h: G \to G'$ is said to be a *homomorphism* if for all $g_1, g_2 \in G$ we have $h(g_1 g_2) = h(g_1) h(g_2)$. Now, the collection, $GL(V)$, of one-to-one, onto linear maps of a (not necessarily finite-dimensional) vector space, V, into itself has a natural group structure. A homomorphism $h: G \to GL(V)$ is called a *representation* of the group G, and V is said to be its *representation space*. A map of G into $GL(V)$ satisfying a relation of the form (13.1.2) is called a *projective representation* or a "representation up to phase."

Our strategy for obtaining physical fields on spacetime now may be explained. We have seen above that the Hilbert space of quantum states is the representation space for a unitary representation up to phase of the proper Poincaré group. Therefore, we may pose the mathematical problem of finding all the unitary representations up to phase of the Poincaré group on a Hilbert space which depend continuously on the Poincaré group elements in the sense described by Wigner (1939). We then may seek to define fields on spacetime corresponding to all the representations we have found.

The problem of finding the continuous unitary representations up to phase of the Poincaré group was systematically analyzed by Wigner (1939). The first key result of Wigner's analysis is that the unitary maps U_g can be redefined by multiplication by phase factors in such a way as to make $\omega(g_1, g_2) = \pm 1$. Thus, the U_g may be chosen so that they yield a "representation up to sign." (The proof of this result is nontrivial and comprises a substantial portion of Wigner's analysis.) The next key result (see Bargmann 1954) is that the representations up to sign of the Poincaré group correspond precisely to the true representations of its universal covering group. We digress, now, to define the term "universal covering group."

Let M be a connected manifold and let $p, q \in M$. Let $\gamma: [0,1] \to M$ and $\gamma': [0,1] \to M$ be continuous curves with $\gamma(0) = \gamma'(0) = p$ and $\gamma(1) = \gamma'(1) = q$. We say that γ and γ' are *homotopic* if they can be continuously deformed into each other keeping their endpoints fixed, i.e., if there exists a continuous function $F: [0,1] \times [0,1] \to M$ such that $F(0,t) = \gamma(t)$ and $F(1,t) = \gamma'(t)$ for all $t \in [0,1]$ and $F(s,0) = p$, $F(s,1) = q$ for all $s \in [0,1]$. It is easy to check that homotopy defines an equivalence relation between curves from p to q. M is said to be *simply connected* if every closed curve in M [i.e., every curve with $\gamma(0) = \gamma(1)$] is homotopic to the trivial curve $\gamma(t) = \gamma(0)$ for all $t \in [0,1]$. Equivalently, M is simply connected if for each $p, q \in M$ all the curves connecting p and q are homotopic. Note that the number of homotopy equivalence classes of curves between $p, q \in M$ is independent of the choice of p and q. Note also that the set of homotopy equivalence classes of closed curves through p can be given a natural group structure as follows. If Γ_1 denotes the equivalence class of curve γ_1, and Γ_2 denotes the equivalence class of curve γ_2, we define the product $\Gamma_1\Gamma_2$ to be the equivalence class, Γ, of the curve γ defined by

$$\gamma(t) = \begin{cases} \gamma_1(2t) & (0 \leqq t \leqq \frac{1}{2}) \\ \gamma_2(1 - 2t) & (\frac{1}{2} \leqq t \leqq 1) \end{cases} , \tag{13.1.3}$$

i.e., γ goes first around γ_1 and then around γ_2. It is easy to check that Γ is independent of the choice of representative curves γ_1 and γ_2 in Γ_1 and Γ_2 and that the above product law defines a group structure, with the identity element consisting of the equivalence class of the trivial closed curve. This group is called the *fundamental group* of M and is denoted $\pi_1(M)$. Thus, M is simply connected if and only if $\pi_1(M)$ is the trivial group consisting only of the identity element.

Now, fix $p \in M$ and define $\hat{M}$ to be the set of equivalence classes of curves between p and q as q ranges over M. There is a natural map $f:\hat{M} \to M$ which takes each $\hat{q} \in \hat{M}$ into the endpoint $q \in M$ of the equivalence class of curves $\hat{q}$. This map will be one-to-one if and only if M is simply connected. However, even if M is not simply connected, we always can find a simply connected neighborhood, U, of $q \in M$ and look at the equivalence classes of curves obtained by following a curve in $\hat{q}$ from p to q and adjoining to it a curve lying entirely within U. In this way, we obtain a one-to-one, onto map from a region, $\hat{U}$, of $\hat{M}$ into U. By requiring that all such maps be diffeomorphisms we define a manifold structure on $\hat{M}$. The manifold $\hat{M}$ thus obtained is called *universal covering manifold* of M. (Note that $\hat{M}$ is independent of the choice of $p \in M$ used in the construction, i.e., different choices of p result in diffeomorphic universal covering manifolds.) It follows directly from the construction of $\hat{M}$ that we have a "covering map" $f:\hat{M} \to M$ such that for every simply connected open set $U \subset M$, f is a diffeomorphism between each connected component of $f^{-1}[U]$ and U. Furthermore, $\hat{M}$ is simply connected, as can be seen from the fact that f maps every closed curve in $\hat{M}$ into a closed curve in M which is homotopically trivial. In fact, the preceding two properties characterize $\hat{M}$.

Since a Lie group, G, also is a manifold (see section 7.2), the above construction can be performed on G to obtain a universal covering manifold, $\hat{G}$. Furthermore, a group structure can be defined on $\hat{G}$ as follows. We choose the fixed point p in the construction of $\hat{G}$ to be the identity element, e, of G. Let $\hat{g}_1, \hat{g}_2 \in \hat{G}$ and let γ_1, γ_2 be curves in G belonging, respectively, to the equivalence classes $\hat{g}_1, \hat{g}_2$ of curves from e to points of G. We define $\hat{g}_1\hat{g}_2$ to be the equivalence class of the curve $\gamma(t) = \gamma_1(t)\gamma_2(t)$, where $\gamma_1(t)\gamma_2(t)$ is defined by the group product in G. The equivalence class of γ is independent of the choice of representative curves γ_1, γ_2, so $\hat{g}_1\hat{g}_2$ is well defined. This product makes $\hat{G}$ a Lie group, and it also makes the covering map $f:\hat{G} \to G$ a group homomorphism. $\hat{G}$ is called the *universal covering group* of G, and it is characterized by the conditions that $\hat{G}$ is the universal covering manifold of G and that its covering map, f, is a homomorphism.

The Poincaré Lie group fails to be simply connected. A rotation by 2π cannot be continuously deformed to the identity transformation. More precisely, the curve γ in the group manifold defined by $\gamma(t)$ = rotation about a fixed axis by the angle $(2\pi t)$ begins and ends at the identity element, e, but is not homotopic to the trivial curve

$\gamma'(t) = e$ for all t. What is perhaps more surprising is that the homotopy class of the 2π-rotation is the only nontrivial homotopy class of closed curves through e in the Poincaré group, i.e., every closed curve passing through e either is homotopic to the 2π-rotation curve γ or the trivial curve γ'. In particular, a rotation by 4π about a fixed axis can be continuously deformed to γ' as illustrated in Figure 13.1. Thus, the universal covering group, $\hat{G}$, of the Poincaré group, G, will yield a twofold covering of G. As we shall see below, $\hat{G}$ is isomorphic to the group ISL(2, $\mathbb{C}$) of translations and linear maps of unit determinant on a two-dimensional complex vector space. Our next major task therefore, will be to establish some properties of the groups SL(2, $\mathbb{C}$) and ISL(2, $\mathbb{C}$) and the vector space upon which they naturally act. These properties are basic to the notion of spinors, and we now shall make a lengthy digression to study these properties.

Let W be a two-dimensional vector space over the complex numbers. Following the index notation conventions discussed in chapter 2, we shall use latin superscripts to denote vectors in W and greek superscripts to denote components of vectors in W with respect to a basis. However, in order to distinguish vectors in W from tangent vectors in spacetime, we shall use *capital* letters in the superscripts. Thus, for example, ξ^A denotes an element of W, and ξ^Σ denotes a component of ξ^A. As in the

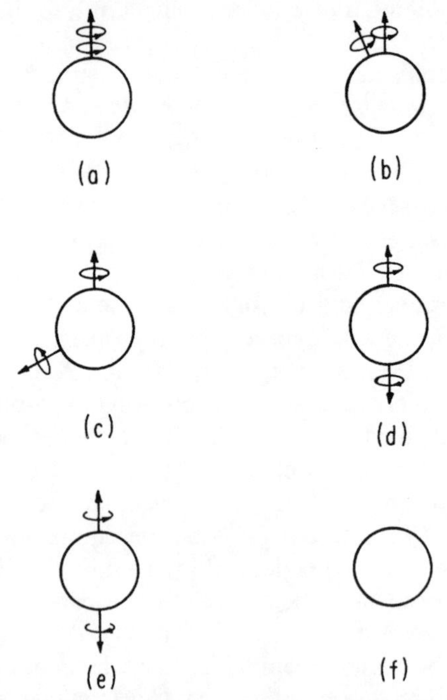

Fig. 13.1. A diagram illustrating how a rotation of a sphere by 4π about a given axis can be continuously deformed to no rotation. We break up the 4π-rotation into two 2π-rotations and continuously change the axis of the second 2π-rotation until it is opposite the axis of the first rotation, as shown in the sequence (a)–(d). Then we continuously decrease the angle of *both* rotations to zero, as shown in the sequence (d)–(f).

case of a real vector space, starting with W we can construct the dual space, W^*, composed of linear maps from W into $\mathbb{C}$. Then W^* is a two-dimensional vector space over $\mathbb{C}$, and we shall denote its elements with subscripts. Thus, for example, λ_A denotes an element of W^*. However, for a vector space, W, over $\mathbb{C}$, we also can form the *complex conjugate dual space* $\overline{W}^*$, composed of the antilinear maps from W into $\mathbb{C}$. (Here a map $f: W \to \mathbb{C}$ is said to be *antilinear* if $f(\xi_1^A + \xi_2^A) = f(\xi_1^A) + f(\xi_2^A)$ and $f(c\xi^A) = \bar{c}f(\xi^A)$ for all ξ_1^A, ξ_2^A, $\xi^A \in W$ and for all $c \in \mathbb{C}$.) Then $\overline{W}^*$ also is a two-dimensional vector space over $\mathbb{C}$, and we shall denote its elements with a *primed* lower index, e.g., $\psi_{A'} \in \overline{W}^*$. Finally, we define the *complex conjugate space,* $\overline{W}$, to be the dual space of $\overline{W}^*$, and we denote elements of $\overline{W}$ with a primed upper index, e.g., $\phi^{A'} \in \overline{W}$. Note that there is a natural *antilinear* one-to-one correspondence between elements $\xi \in W$ and $\phi \in \overline{W}$ defined by the requirement that $\psi(\xi) = \phi(\psi)$ for all $\psi \in \overline{W}^*$. (We omitted the indices here since their presence could cause confusion.) We call this map of W onto $\overline{W}$ (as well as the inverse map of $\overline{W}$ onto W) *complex conjugation.* We denote the image of $\xi^A \in W$ under complex conjugation as $\bar{\xi}^{A'} \in \overline{W}$, and similarly denote the image of $\phi^{A'} \in \overline{W}$ as $\bar{\phi}^A \in W$.

A tensor, T, of type $(k, l; k', l')$ over W is defined as a multilinear map,

$$T: \underbrace{W^* \times \cdots \times W^*}_{k} \times \underbrace{W \times \cdots \times W}_{l} \times \underbrace{\overline{W}^* \times \cdots \overline{W}^*}_{k'} \times \underbrace{\overline{W} \times \cdots \overline{W}}_{l'} \to \mathbb{C} \quad .$$

We shall use a natural generalization of the index notation for tensors over a real vector space to denote tensors over W. Thus, for example, $T^{AB}{}_C{}^{D'}$ denotes a tensor of type $(2, 1; 1, 0)$. The relative ordering of the primed and unprimed indices is irrelevant, e.g., $T^{AD'B}{}_C$ denotes the same tensor as $T^{AB}{}_C{}^{D'}$. However, the ordering within the unprimed indices and within the primed indices is as relevant as in the case of tensors over a real vector space. The complex conjugation map of vectors extends to tensors and maps a tensor T of type $(k, l; k', l')$ into a tensor, denoted $\overline{T}$, of type $(k', l'; k, l)$. Note that $\overline{\overline{T}} = T$. Finally, we now have two distinct notions of contraction: contraction over unprimed indices, taking tensors of type $(k, l; k', l')$ into tensors of type $(k - 1, l - 1; k', l')$, and contraction over primed indices, taking tensors of type $(k, l; k', l')$ into tensors of type $(k, l; k' - 1, l' - 1)$. However, contraction over one primed index and one unprimed index is not defined. Again, we shall adopt the notation of using the same letter twice in the indices to denote contraction.

For a two-dimensional vector space, W, over $\mathbb{C}$, the vector space of antisymmetric tensors of type $(0, 2; 0,0)$ is one-dimensional. If a particular such tensor $\epsilon_{AB} = -\epsilon_{BA}$ is chosen, the pair (W, ϵ_{AB}) is called a *spinor space*. The elements of W are called *spinors,* and tensors over W are called *spinorial tensors*. We can use ϵ_{AB} to map spinors into dual spinors via $\xi^A \to \epsilon_{AB}\xi^A$. Since ϵ_{AB} is nondegenerate (i.e., $\epsilon_{AB}\xi^A \neq 0$ unless $\xi^A = 0$) we obtain from ϵ_{AB} an isomorphism of W and W^* much like the isomorphism that would be obtained from a metric on W. We shall take advantage of this similarity by employing notational conventions for ϵ_{AB} similar to those employed for metrics in chapter 2. First, we shall denote the dual vector $\epsilon_{AB}\xi^A$ as simply ξ_B, and more generally, use ϵ_{AB} to lower unprimed indices on all spinorial tensors.

Note, however, that since ϵ_{AB} is antisymmetric in A and B, it makes a difference which index of ϵ_{AB} is contracted with ξ^C in the index-lowering operation. We follow the standard convention of using contraction over the *first* index of ϵ_{AB} to lower indices. Thus, we have

$$\xi_B = \epsilon_{AB}\xi^A = -\epsilon_{BA}\xi^A \quad . \tag{13.1.4}$$

Following standard conventions, we define ϵ^{AB} to be *minus* the inverse of ϵ_{AB}, i.e., ϵ^{AB} is the antisymmetric tensor of type $(2,0;0,0)$ which satisfies,

$$\epsilon^{AB}\epsilon_{BC} = -\delta^A{}_C \quad , \tag{13.1.5}$$

where $\delta^A{}_C$ denotes the identity map on W. In order to compensate for the minus sign in equation (13.1.5) we use contraction over the *second* index of ϵ^{AB} to raise an index; e.g., for $\mu_A \in W^*$ we have

$$\mu^A = \epsilon^{AB}\mu_B = -\epsilon^{BA}\mu_B \quad . \tag{13.1.6}$$

Thus, it is essential to pay careful attention to index positions in order to prevent sign errors. Note that we have

$$\begin{aligned} \xi_A\phi^A &= (\epsilon_{BA}\xi^B)\phi^A \\ &= -\epsilon_{AB}\xi^B\phi^A \\ &= -\xi^B\phi_B \quad . \end{aligned} \tag{13.1.7}$$

In particular, for any spinor ξ^A, we have $\xi_A\xi^A = 0$. It also should be noted that confusion can arise with the symbol $\delta^A{}_C$, which could be interpreted either as (*a*) the identity operator on W or as (*b*) the identity operator, $\delta_B{}^D$, on W^* with its first index raised and its second index lowered. [Tensors (*a*) and (*b*) differ by a minus sign.] Thus, it is preferable to use $\epsilon_C{}^A$ to denote the identity operators on both W and W^*, since no confusion in interpretation arises for this symbol. Finally, we denote the tensors obtained from ϵ_{AB} and ϵ^{AB} via complex conjugation as $\bar{\epsilon}_{A'B'}$ and $\bar{\epsilon}^{A'B'}$ and use them to lower and raise *primed* indices, with the same convention of contraction over the first index of $\bar{\epsilon}_{A'B'}$ and the second index of $\bar{\epsilon}^{A'B'}$.

A linear map $L: W \to W$ is represented by a tensor $L^A{}_B$. The determinant of L is defined by

$$\det(L) = \frac{1}{2}\epsilon_{AB}\epsilon^{CD}L^A{}_C L^B{}_D \quad . \tag{13.1.8}$$

It is well known that $L^A{}_B$ is one-to-one, onto, and hence invertible if and only if its determinant is nonvanishing. We define $\mathrm{SL}(2,\mathbb{C})$ to be the group of linear maps of W into itself which have unit determinant. Here the group product is defined by composition, i.e., $(LM)^A{}_B = L^A{}_C M^C{}_B$, and the group inverse is given by the inverse of the linear map. Since the determinant condition is one complex equation on the four complex components of $L^A{}_B$, it takes six *real* parameters to specify an element of $\mathrm{SL}(2,\mathbb{C})$. Indeed, using the polar decomposition theorem—which states that with respect to any inner product introduced on W, every element of $\mathrm{SL}(2,\mathbb{C})$ can be written uniquely as the composition of a unitary map of determinant one [i.e., an element of SU(2)] and a positive, self-adjoint map—it can be shown that SL(2, $\mathbb{C}$) has the natural manifold structure $S^3 \times \mathbb{R}^3$. Thus, since the group operations are smooth with respect to this manifold structure, SL(2, $\mathbb{C}$) is a six-dimensional Lie group. Since S^3 and $\mathbb{R}^3$ are simply connected, it also follows that SL(2, $\mathbb{C}$) is simply connected. Note that the condition that $L^A{}_B$ have unit determinant is equivalent to

$$L^A{}_C L^B{}_D \epsilon_{AB} = \epsilon_{CD} \quad , \tag{13.1.9}$$

which states that ϵ_{AB} is preserved under the action of $L^A{}_B$.

The relation between SL(2, $\mathbb{C}$) and the Lorentz group now may be established. The tensors of type (1, 0; 1,0) comprise a four (complex) dimensional vector space, Y. A convenient basis of Y can be defined as follows. Let o^A, ι^A be a basis of W satisfying

$$o_A \iota^A = \epsilon_{AB} o^A \iota^B = 1 \quad . \tag{13.1.10}$$

Then the tensors

$$t^{AA'} = \frac{1}{\sqrt{2}}(o^A \bar{o}^{A'} + \iota^A \bar{\iota}^{A'}) \quad , \tag{13.1.11}$$

$$x^{AA'} = \frac{1}{\sqrt{2}}(o^A \bar{\iota}^{A'} + \iota^A \bar{o}^{A'}) \quad , \tag{13.1.12}$$

$$y^{AA'} = \frac{i}{\sqrt{2}}(o^A \bar{\iota}^{A'} - \iota^A \bar{o}^{A'}) \quad , \tag{13.1.13}$$

$$z^{AA'} = \frac{1}{\sqrt{2}}(o^A \bar{o}^{A'} - \iota^A \bar{\iota}^{A'}) \tag{13.1.14}$$

comprise a basis of Y. Now, complex conjugation maps Y into itself, and the elements $\phi^{AA'}$ of Y which are taken into themselves under complex conjugation, $\bar{\phi}^{A'A} = \phi^{AA'}$, are called *real*. It is straightforward to check that the above basis elements of Y are real and, further, that the real elements of Y are precisely those which can be written as sums of these basis elements with real coefficients. Thus, the real elements of Y comprise a four-dimensional vector space over $\mathbb{R}$, which we shall denote as V.

The tensor

$$g_{AA'BB'} = \epsilon_{AB} \bar{\epsilon}_{A'B'} \tag{13.1.15}$$

yields a multilinear map $V \times V \to \mathbb{R}$, since it is easily verified that $g_{AA'BB'} \phi^{AA'} \psi^{BB'}$ is real for $\phi^{AA'}$, $\psi^{BB'} \in V$. Furthermore, $g_{AA'BB'}$ is nondegenerate with signature $+ - - -$ and thus defines a Lorentz metric on V. (This can be verified explicitly by checking that the basis vectors [13.1.11]–[13.1.14] are orthogonal with respect to $g_{AA'BB'}$, that $g_{AA'BB'} t^{AA'} t^{BB'} = 1$, and that the similar norms of $x^{AA'}$, $y^{AA'}$, and $z^{AA'}$ are -1.)

Now, associated with each map $L^A{}_B \in$ SL(2, $\mathbb{C}$) is the map $\lambda: V \to V$ defined by

$$\lambda^{AA'}{}_{BB'} = L^A{}_B \bar{L}^{A'}{}_{B'} \quad . \tag{13.1.16}$$

Since by equation (13.1.9) $L^A{}_B$ preserves ϵ_{AB}, it follows that $\lambda^{AA'}{}_{BB'}$ preserves $g_{AA'BB'}$, i.e., we have

$$\lambda^{AA'}{}_{CC'}\lambda^{BB'}{}_{DD'}g_{AA'BB'} = g_{CC'DD'} \quad . \tag{13.1.17}$$

But, by definition, the extended Lorentz group—denoted $O(3, 1)$—consists precisely of the metric preserving linear maps on a four-dimensional real vector space with a Lorentz signature metric. Thus, $\lambda^{AA'}{}_{BB'}$ is a Lorentz transformation on V! Indeed, the Lorentz transformations which arise in this way comprise the proper Lorentz group—denoted Λ—as can be verified explicitly from the component form of this correspondence which we shall obtain below (see eqs. [13.1.31] and [13.1.32]). Hence, we have found that associated with every element $L^A{}_B$ of SL(2, $\mathbb{C}$) is a proper Lorentz transformation $\lambda^{AA'}{}_{BB'}$, defined by equation (13.1.16). Furthermore, $L^A{}_B$ and $M^A{}_B$ give rise to the same Lorentz group element if and only if $M^A{}_B = \pm L^A{}_B$. Indeed, the map $f: \mathrm{SL}(2, \mathbb{C}) \to \Lambda$ obtained from equation (13.1.16) satisfies all the properties required of the universal covering map; namely, it is a homomorphism from the simply connected Lie group SL(2, $\mathbb{C}$) onto Λ, and for any simply connected open set $U \subset \Lambda$ it is a diffeomorphism between each connected component of $f^{-1}[U]$ and U. Thus, we conclude that SL(2, $\mathbb{C}$) is the universal covering group of the Lorentz group. Similarly, the group ISL(2, $\mathbb{C}$)—defined by composing in the natural way the elements of SL(2, $\mathbb{C}$) with the elements of the two-complex-dimensional translation group of W—can be seen to be the universal covering group of the proper Poincaré group.

Since $g_{AA'BB'}$ is a metric on V, we can define its inverse metric $g^{AA'BB'}$. From equations (13.1.5) and (13.1.15), it follows that

$$g^{AA'BB'} = \epsilon^{AB}\bar{\epsilon}^{A'B'} \quad . \tag{13.1.18}$$

Let T be a tensor of type (k, l) over V. Then, viewed as a tensor over W, T is of type $(k, l; k, l)$, i.e., T has k upper primed-unprimed index pairs and l lower primed-unprimed index pairs. If we view T as a tensor over V, it would be natural to use $g_{AA'BB'}$ and $g^{AA'BB'}$ to raise and lower "V-indices," i.e., primed-unprimed index pairs. On the other hand, if we view T as a tensor over W, we already have defined the raising and lowering of indices via ϵ_{AB}, ϵ^{AB}, $\bar{\epsilon}_{A'B'}$, and $\bar{\epsilon}^{A'B'}$. Fortunately, it is easily verified that these two distinct notions of raising and lowering indices for tensors over V always yield the same result. This is the reason why we choose to use metric signature $+ - - -$ in this chapter. We could have conformed to our previous signature conventions by defining $g_{AA'BB'}$ to be $-\epsilon_{AB}\bar{\epsilon}_{A'B'}$, but then the two notions of raising and lowering V-indices would differ by a sign and in each calculation we would need to specify which raising and lowering convention we are using.

It is instructive to express some of the above relations in basis component form. With respect to a basis o^A, ι^A satisfying equation (13.1.10), it is easy to check that

$$\epsilon_{AB} = o_A\iota_B - \iota_A o_B \tag{13.1.19}$$

since both sides give the same result when applied to o^A or ι^A. Thus, we may represent the components $\epsilon_{\Sigma\Omega}$ of ϵ_{AB} by the matrix,

$$\epsilon_{\Sigma\Omega} = \begin{pmatrix} 0 & 1 \\ -1 & 0 \end{pmatrix} \quad . \tag{13.1.20}$$

Similarly, we have

$$\epsilon^{AB} = o^A \iota^B - \iota^A o^B \tag{13.1.21}$$

and may represent its components as

$$\epsilon^{\Sigma\Omega} = \begin{pmatrix} 0 & 1 \\ -1 & 0 \end{pmatrix} \quad . \tag{13.1.22}$$

The components of an SL(2, $\mathbb{C}$) transformation $L^A{}_B$ are

$$L^\Sigma{}_\Omega = \begin{pmatrix} a & b \\ c & d \end{pmatrix} \quad , \tag{13.1.23}$$

where a, b, c, d are complex numbers satisfying

$$ad - bc = 1 \quad . \tag{13.1.24}$$

The components of the basis elements (13.1.11)–(13.1.14) of V are

$$t^{\Sigma\Sigma'} = \frac{1}{\sqrt{2}}\begin{pmatrix} 1 & 0 \\ 0 & 1 \end{pmatrix} \quad , \tag{13.1.25}$$

$$x^{\Sigma\Sigma'} = \frac{1}{\sqrt{2}}\begin{pmatrix} 0 & 1 \\ 1 & 0 \end{pmatrix} \quad , \tag{13.1.26}$$

$$y^{\Sigma\Sigma'} = -\frac{1}{\sqrt{2}}\begin{pmatrix} 0 & -i \\ i & 0 \end{pmatrix} \quad , \tag{13.1.27}$$

$$z^{\Sigma\Sigma'} = \frac{1}{\sqrt{2}}\begin{pmatrix} 1 & 0 \\ 0 & -1 \end{pmatrix} \quad . \tag{13.1.28}$$

Note that the right-hand sides of equations (13.1.25)–(13.1.28) are —apart from the factor of $1/\sqrt{2}$ and the minus sign in equation (13.1.27)—just the Pauli spin matrices. An arbitrary vector $v^{AA'} \in V$ can be written as

$$v^{AA'} = tt^{AA'} + xx^{AA'} + yy^{AA'} + zz^{AA'} \quad , \tag{13.1.29}$$

and thus its components with respect to the basis of V associated with o^A, ι^A may be represented by the matrix

$$v^{\Sigma\Sigma'} = \frac{1}{\sqrt{2}}\begin{pmatrix} t + z & x + iy \\ x - iy & t - z \end{pmatrix} \quad . \tag{13.1.30}$$

The transformation on the components of $v^{AA'}$ induced via equation (13.1.16) by the SL(2, $\mathbb{C}$) transformation $L^A{}_B$ is given by

$$\begin{pmatrix} t' + z' & x' + iy' \\ x' - iy' & t' - z' \end{pmatrix} = \begin{pmatrix} a & b \\ c & d \end{pmatrix}\begin{pmatrix} t + z & x + iy \\ x - iy & t - z \end{pmatrix}\begin{pmatrix} \bar{a} & \bar{c} \\ \bar{b} & \bar{d} \end{pmatrix} \quad , \tag{13.1.31}$$

where ordinary matrix multiplication is understood on the right-hand side. By rewriting equation (13.1.31) in the form

$$x'^{\mu} = \sum_{\nu} \lambda^{\mu}{}_{\nu} x^{\nu} \quad , \tag{13.1.32}$$

where $x^{\mu} = (t, x, y, z)$, one obtains in explicit, component form the map f taking elements of SL(2, $\mathbb{C}$) (represented as 2×2 complex matrices, $L^{\Sigma}{}_{\Omega}$) into elements of Λ (represented as 4×4 real matrices, $\lambda^{\mu}{}_{\nu}$).

Given a spinor $\psi^A \in W$, we can construct a vector $k^{AA'} \in V$ by

$$k^{AA'} = \psi^A \bar{\psi}^{A'} \quad . \tag{13.1.33}$$

We have

$$g_{AA'BB'} k^{AA'} k^{BB'} = (\epsilon_{AB} \psi^A \psi^B)(\bar{\epsilon}_{A'B'} \bar{\psi}^{A'} \bar{\psi}^{B'}) = 0 \quad , \tag{13.1.34}$$

so $k^{AA'}$ is a null vector with respect to the Lorentz metric on V. We thus may view ψ^A as a "square root" of the null vector $k^{AA'}$. Note that for any two spinors ψ^A, ϕ^A we have

$$g_{AA'BB'} \psi^A \bar{\psi}^{A'} \phi^B \bar{\phi}^{B'} = (\psi_B \phi^B)(\bar{\psi}_{B'} \bar{\phi}^{B'}) = |\psi_B \phi^B|^2 \quad . \tag{13.1.35}$$

Thus, the null vector associated with the arbitrary spinors ψ^A and ϕ^A have manifestly nonnegative inner product which means (with our new metric signature $+ - - -$) that these null vectors lie on the same half of the light cone. By convention we call this the future light cone. Thus, the vector space V has a natural time orientation. Furthermore, the real spinorial tensor,

$$e_{AA'BB'CC'DD'} = i(\epsilon_{AB}\epsilon_{CD}\bar{\epsilon}_{A'C'}\bar{\epsilon}_{B'D'} - \bar{\epsilon}_{A'B'}\bar{\epsilon}_{C'D'}\epsilon_{AC}\epsilon_{BD}) \quad , \tag{13.1.36}$$

defines a totally antisymmetric tensor of type $(0, 4)$ over V and thus yields an orientation of V.

If ϕ^A differs by a phase factor from ψ^A, i.e., $\phi^A = c\psi^A$ with $|c| = 1$, then ϕ^A defines the same null vector $k^{AA'}$ as ψ^A, so the same $k^{AA'}$ is associated with a one-parameter family of spinors. However, we can define the real tensor $F^{AA'BB'}$ by

$$F^{AA'BB'} = \psi^A \psi^B \bar{\epsilon}^{A'B'} + \bar{\psi}^{A'} \bar{\psi}^{B'} \epsilon^{AB} \quad . \tag{13.1.37}$$

Viewed as a tensor of type $(2, 0)$ over V, $F^{AA'BB'}$ is antisymmetric, i.e., $F^{AA'BB'} = -F^{BB'AA'}$. Furthermore, $F^{AA'BB'}$ satisfies

$$F^{AA'BB'} F_{AA'BB'} = 0 \tag{13.1.38}$$

and

$$F^{AA'BB'} k_{BB'} = 0 \quad , \tag{13.1.39}$$

where $k^{AA'} = \psi^A \bar{\psi}^{A'}$ as before. Thus, viewed as a tensor over V, $F^{AA'BB'}$ is a null bi-vector with null vector $k^{AA'}$, i.e., $F^{AA'BB'}$ is of the form

$$F^{AA'BB'} = k^{AA'} m^{BB'} - k^{BB'} m^{AA'} \quad , \tag{13.1.40}$$

where $k^{AA'}$ is null and $m^{AA'}$ is orthogonal to $k^{AA'}$. We call $F^{AA'BB'}$ the *null flag* associated with spinor ψ^A. Two spinors ψ^A and ϕ^A give rise to the same null flag if and only if they differ at most by sign, $\phi^A = \pm\psi^A$. No tensor over V can be constructed from a spinor ψ^A which distinguishes between ψ^A and $-\psi^A$.

We return, now, to the general issue of obtaining the possible fields on spacetime which may arise in quantum theory. We were led by the considerations discussed above to seek unitary representations of the group ISL(2, $\mathbb{C}$) on a Hilbert space. We now shall define spinor fields and spinorial tensor fields on spacetime which have well-defined "transformation laws" under ISL(2, $\mathbb{C}$). By constructing Hilbert spaces out of these fields (as we shall do later), these transformation laws yield the desired representations of ISL(2, $\mathbb{C}$).

We define a spinor field on Minkowski spacetime ($\mathbb{R}^4$, η_{ab}) to be simply a map of spacetime into spinor space, W. Similarly, a spinorial tensor field of a specified type is defined as a map of spacetime into the tensor space over W of that type. We define an action of ISL(2, $\mathbb{C}$) on spinor fields as follows. Associated with any $g \in$ ISL(2, $\mathbb{C}$) is the transformation $\psi^A(x) \to L^A{}_B\psi^B[P^{-1}(x)]$ where $L^A{}_B \in$ SL(2, $\mathbb{C}$) is the "homogeneous part" of g and P is the Poincaré group element associated with g. In this way, we obtain by brute force a representation of ISL(2, $\mathbb{C}$) on the vector space of spinor fields. However, this representation does not correspond to a true representation of the Poincaré group. Since there are two ISL(2, $\mathbb{C}$) elements for every Poincaré element, P, when we attempt to define a transformation, P^*, on spinor fields associated with a Poincaré element, P, we encounter a sign ambiguity,

$$(P^*\psi)^A(x) = \pm L^A{}_B\psi^B[P^{-1}(x)] \quad . \tag{13.1.41}$$

This ambiguity is resolved if we are given not only P but a continuous curve in the Poincaré group from e to P, since such a curve is uniquely associated with an element of ISL(2, $\mathbb{C}$). Thus, in this well-defined sense, a spinor at point x changes sign under a rotation by 2π about a fixed axis at x. However, given only P (and not a curve from e to P), we cannot resolve the sign ambiguity in equation (13.1.41) in a natural way. No choice of sign can be made so that we obtain a true representation of the Poincaré group. Note that although we have defined an action up to sign of the Poincaré group on spinor fields, we have *not* defined the action of an arbitrary diffeomorphism on spinor fields. Thus, in particular, the Lie derivative of a spinor field with respect to a non–Killing vector field is undefined.

The relation between spinor fields and ordinary vector fields now may be established. As discussed above, the real tensors of type $(1,0;1,0)$ over W form a four-dimensional real vector space V, on which the Lorentz metric (13.1.15) is defined. Let t^a, x^a, y^a, z^a be an orthonormal basis field in Minkowski spacetime associated with a global family of inertial observers O. Let o^A, ι^A be a basis for W satisfying (13.1.10). For each point x in Minkowski spacetime, we define the linear map σ which takes vectors in V to vectors in the tangent space, V_x, at x by identifying t^a, x^a, y^a, z^a with the basis (13.1.11)–(13.1.14) of V. In other words, we define the hybrid vector/spinorial tensor field $\sigma^a{}_{AA'}$ by

$$\sigma^a{}_{AA'} = t^a t_{AA'} - x^a x_{AA'} - y^a y_{AA'} - z^a z_{AA'} \quad . \tag{13.1.42}$$

Then at each x, $\sigma^a{}_{AA'}$ is a vector space isomorphism between V and V_x which preserves the Lorentz metric defined on these spaces; i.e., we have

$$\eta_{ab} = \sigma_a{}^{AA'}\sigma_b{}^{BB'}g_{AA'BB'} \quad . \tag{13.1.43}$$

(Here all lower case indices are raised and lowered with η^{ab} and η_{ab}, while capital letter indices are raised and lowered with ϵ^{AB}, ϵ_{AB}, and their complex conjugates.) Using $\sigma^a{}_{AA'}$, we may map real spinorial tensor fields of type $(1,0;1,0)$ on Minkowski spacetime into vector fields. We find, then, that the action of the Poincaré group preserves this association; namely, the action (13.1.41) of Poincaré group element P on spinor fields induces on a spinorial tensor $v^{AA'}$ of type $(1,0;1,0)$ the transformation

$$(P^*v)^{AA'}(x) = L^A{}_B\bar{L}^{A'}{}_{B'}v^{BB'}[P^{-1}(x)] = \lambda^{AA'}{}_{BB'}v^{BB'}[P^{-1}(x)] \quad , \tag{13.1.44}$$

where $\lambda^{AA'}{}_{BB'}$ is the Lorentz transformation corresponding to the "homogeneous part" of P. Hence, we have

$$\begin{aligned}\sigma^a{}_{AA'}(P^*v)^{AA'}(x) &= \sigma^a{}_{AA'}\lambda^{AA'}{}_{BB'}v^{BB'}[P^{-1}(x)] \\ &= (\sigma^a{}_{AA'}\lambda^{AA'}{}_{BB'}\sigma_b{}^{BB'})\sigma^b{}_{CC'}v^{CC'}[P^{-1}(x)] \\ &= \lambda^a{}_b v^b[P^{-1}(x)] \\ &= (P^*v)^a(x) \quad ,\end{aligned} \tag{13.1.45}$$

where in the last line P^* is the natural action of P on the vector field $v^a = \sigma^a{}_{AA'}v^{AA'}$ as defined in appendix C for a general diffeomorphism. Consequently, by selecting—once and for all—a map $\sigma^a{}_{AA'}$ of the form (13.1.42), *we may consistently identify real spinorial tensor fields of type (1, 0;1, 0) with vector fields*. It is customary to literally make this identification[3] in our notation by omitting the map $\sigma^a{}_{AA'}$ from our expressions and denoting, for example, the vector field associated with $v^{AA'}$ by simply $v^{AA'}$ rather than by $\sigma^a{}_{AA'}v^{AA'}$. We shall follow this practice here. Because of our new metric signature convention no confusion will arise as to whether indices are raised and lowered with the metric or with ϵ^{AB} and ϵ_{AB}.

Our physical interpretation of a spinor field ψ^A now may be given. As described above, associated with every spinor is a null flag $F^{AA'BB'}$ defined by equation (13.1.37). Using the map $\sigma^a{}_{AA'}$, we may view $F^{AA'BB'}$ as a tensor field of type $(2,0)$ on spacetime. However, tensor fields are objects whose interpretation and measurability are well understood. We take the physically measurable properties of ψ^A to be the quantities determinable from $F^{AA'BB'}$. Note that this implies that ψ^A and $-\psi^A$ are physically indistinguishable. It may seem strange that we have gone to a great deal of trouble to define spinor fields only to interpret them by associating them with tensor fields. However, the dynamical evolution of physical fields represented by spinors (see below) is given by differential equations involving the spinor fields, not

3. It also is customary and convenient to make the identifications $a = AA'$, $b = BB'$, etc., in the labeling indices (see Penrose and Rindler 1984), so that, for example, one may write $\eta_{ab} = \epsilon_{AB}\bar{\epsilon}_{A'B'}$. However, we shall not follow this practice here because we wish to maintain the spinor indices so that equations in this chapter (where the metric signature is $+---$) will easily be distinguished from those in other chapters (where the signature is $-+++$).

just their null flags. If the region where a spinor field, ψ^A, initially is nonvanishing is disconnected, then a knowledge of the initial value of its null flag $F^{AA'BB'}$ (as well as time derivatives of $F^{AA'BB'}$) will not suffice to determine the subsequent evolution of ψ^A or $F^{AA'BB'}$ since there will be a sign ambiguity in ψ^A in each of the initial regions where it is nonvanishing, and relative sign differences in these regions will affect its subsequent evolution. Thus, in this sense, a spinor field contains more physically relevant information than is present in its null flag. Furthermore, even in the case where the set where $\psi^A \neq 0$ is connected so that it can be recovered (up to sign) from its null flag, the formulation of the dynamical laws in terms of null flags would be extremely cumbersome. On the other hand, these laws take a simple and natural form when formulated in terms of spinor fields.

We take this opportunity to point out several identities which are very useful in calculations involving spinors. Let T_{ab} be a (real) tensor of type (0, 2), and let $T_{AA'BB'}$ be the corresponding spinorial tensor. It is straightforward to verify that

$$\frac{1}{2}(T_{AA'BB'} - T_{BB'AA'}) = T_{(AB)[A'B']} + T_{[AB](A'B')} \quad , \tag{13.1.46}$$

where, as in the case of ordinary tensors, round and square brackets denote, respectively, symmetrization and antisymmetrization and we remind the reader that the relative order of primed and umprimed indices is irrelevant. However, since $\bar{\epsilon}_{A'B'}$ spans the one-dimensional vector space of antisymmetric spinorial tensors of type (0, 0; 0, 2), we must have

$$T_{(AB)[A'B']} = \phi_{AB}\bar{\epsilon}_{A'B'} \quad , \tag{13.1.47}$$

where ϕ_{AB} is symmetric. Contracting equation (13.1.47) with $\bar{\epsilon}^{A'B'}$, we find

$$\phi_{AB} = \frac{1}{2}T_{(AB)A'}{}^{A'} \quad . \tag{13.1.48}$$

Similarly, we have

$$T_{[AB](A'B')} = \epsilon_{AB}\psi_{A'B'} \quad . \tag{13.1.49}$$

However, reality of $T_{AA'BB'}$ implies that $\bar{\psi}_{AB} = \phi_{AB}$. Thus, we find

$$\frac{1}{2}(T_{AA'BB'} - T_{BB'AA'}) = \phi_{AB}\bar{\epsilon}_{A'B'} + \bar{\phi}_{A'B'}\epsilon_{AB} \quad , \tag{13.1.50}$$

where ϕ_{AB} is given by equation (13.1.48). In particular, every antisymmetric tensor $T_{ab} = T_{[ab]}$ can be written in the form given by the right-hand side of equation (13.1.50). Similarly, we have

$$\begin{aligned}\frac{1}{2}(T_{AA'BB'} + T_{BB'AA'}) &= T_{(AB)(A'B')} + T_{[AB][A'B']} \\ &= T_{(AB)(A'B')} + \frac{1}{4}\epsilon_{AB}\bar{\epsilon}_{A'B'}T \quad ,\end{aligned} \tag{13.1.51}$$

where $T = T_A{}^A{}_{A'}{}^{A'} = T_a{}^a$. In particular, every symmetric tensor $T_{ab} = T_{(ab)}$ can be written in the form given by the right-hand side of equation (13.1.51). Contracting equation (13.1.51) with $\bar{\epsilon}^{A'B'}$, we obtain

$$T_{[AB]A'}{}^{A'} = \frac{1}{2}\epsilon_{AB}T \quad . \tag{13.1.52}$$

If T_{ab} is symmetric, the square brackets can be omitted from the left-hand side of equation (13.1.52).

Derivatives of spinor fields on Minkowski spacetime may be defined as follows. Since a spinor field ψ^A is a map of spacetime points into W, we may take the ordinary partial derivatives of ψ^A with respect to global inertial coordinates of Minkowski spacetime. We define $\partial_{BB'}\psi^A$ to be the spinorial tensor field of type $(1, 1; 0, 1)$ whose components with respect to the basis o^A, ι^A are

$$\partial_{\Lambda\Lambda'}\psi^\Gamma = \sum_\mu \sigma^\mu{}_{\Lambda\Lambda'}\frac{\partial\psi^\Gamma}{\partial x^\mu} \quad . \tag{13.1.53}$$

For a given fixed choice of $\sigma^a{}_{AA'}$, the spinorial tensor field $\partial_{BB'}\psi^A$ determined in this manner is independent of the choice of global inertial coordinates, x^μ, and of spin basis o^A, ι^A, so $\partial_{BB'}\psi^A$ is well defined. The derivative of a spinorial tensor field of arbitrary rank is defined similarly by taking partial derivatives of its components with respect to global inertial coordinates and applying $\sigma^a{}_{AA'}$. It is easily verified that $\partial_{AA'}$ is linear, satisfies the Leibnitz rule, commutes with contraction, and also satisfies

$$\partial_{AA'}\epsilon_{BC} = 0 \quad . \tag{13.1.54}$$

Furthermore, for a vector field $v^{BB'}$ we have

$$\partial_{AA'}v^{BB'} = \sigma^a{}_{AA'}\sigma_b{}^{BB'}\partial_a v^b \quad . \tag{13.1.55}$$

More generally, for an ordinary tensor field of arbitrary rank, the action of $\partial_{AA'}$ agrees with that of the usual derivative operator, ∂_a. Thus, $\partial_{AA'}$ may be viewed as a generalization to spinorial tensor fields of the usual derivative operator ∂_a on Minkowski spacetime.

Note that the derivative operators commute when applied to an arbitrary spinorial tensor field on Minkowski spacetime,

$$\partial_{AA'}\partial_{BB'} = \partial_{BB'}\partial_{AA'} \quad . \tag{13.1.56}$$

Consequently, the same derivation as led to equation (13.1.52) above now yields

$$\partial_{AA'}\partial_B{}^{A'} = \frac{1}{2}\epsilon_{AB}\,\Box \tag{13.1.57}$$

where

$$\Box = \partial_{AA'}\partial^{AA'} \tag{13.1.58}$$

Our motivation for introducing spinor fields arose from seeking unitary representations up to phase of the Poincaré group on a Hilbert space. We finally are ready to return to this issue now and show how Hilbert spaces can be built out of spinorial tensor fields satisfying certain equations such that the transformation law (13.1.41) leads to the desired unitary representations. Recall, first, that a representation is said to be *reducible* if all the linear maps occurring in the representation take a fixed proper subspace of the vector space into itself. It is easy to show that on a finite-dimensional vector space, V, every unitary representation can be decomposed into a direct sum of irreducible representations, i.e., V can be written as a direct sum of subspaces each of which is invariant but has no proper invariant sub-subspaces. For an infinite-dimensional Hilbert space this result does not always hold, but Wigner (1939) has shown that the unitary representations of ISL$(2, \mathbb{C})$ can be decomposed into irreducible representations. Thus, it suffices to consider only the irreducible representations, since all representations can be constructed out of these. This simplifies the analysis considerably.

The irreducible representations are conveniently labeled by the values of the Casimir operators,[4] m^2 and S^2, of the Lie algebra of ISL$(2, \mathbb{C})$ (which is isomorphic to the Poincaré Lie algebra), which can be interpreted as representing, respectively, squared 4-momentum (i.e., mass squared), and squared angular momentum about the center of mass. We may classify the representations into the following four cases according to the value of m^2: (*a*) $m^2 > 0$, (*b*) $m^2 = 0$ but the translations are not all represented by the identity operator, I, (*c*) $m^2 = 0$ and the translations are all represented by I, and (*d*) $m^2 < 0$.

In the representations (*c*), all states are translationally invariant. Thus, these representations appear to be of no physical significance. The "tachyonic" representations, (*d*), also do not appear to be of physical significance, although some tachyon field theories have been investigated (Feinberg 1967). The representations in these classes (*c*) and (*d*) have been obtained by Bargmann (1947), but a systematic construction of fields on Minkowski spacetime which realize these representations does not appear to have been given. The representations of classes (*a*) and (*b*) were obtained by Wigner (1939), and a realization of all these representations as fields on spacetime was first given in a systematic way by Bargmann and Wigner (1948). The representations of class (*a*) are characterized by the values of m^2 and S^2 with $S^2 = s(s+1)$, where the *spin* s takes the values $s = 0, \frac{1}{2}, 1, \ldots$. The representations of class (*b*) can be divided into two subclasses: (*b*1) representations characterized by a helicity parameter s (whose magnitude also is called the *spin*) with values $s = 0, \pm\frac{1}{2}, \pm 1, \ldots$, and (*b*2) the so-called "continuous spin" representations.

4. *The universal enveloping algebra*, $\mathcal{U}$, of a Lie algebra, L, is obtained by taking the direct sum, $\bigoplus_{k=0}^{\infty} \mathcal{T}(k, 0)$, of all "upper index" tensors over L and defining two elements to be equivalent if they can be reduced to each other by any formal calculation in which for any $v^a, w^a \in L$ we replace $v^a w^b - v^b w^a$ by $[v, w]^a$. An element $X \in \mathcal{U}$ which commutes with all $v \in L$ is called a *Casimir element*. By Stone's theorem (see, e.g., Reed and Simon 1972) and results of Gårding (1947) every unitary representation of a Lie group G gives rise to a self-adjoint representation of its Lie algebra, L. The representatives of the Casimir elements, called *Casimir operators*, commute with all representatives of L and hence with all representatives of G. Therefore, by Schur's lemma, in an irreducible representation every Casimir operator must be a multiple of the identity operator. These numbers provide convenient labels of the irreducible representation. The universal enveloping algebra of the Poincaré group possesses two independent Casimir elements. Their interpretation in terms of mass and spin arises from identifying the Poincaré Lie algebra with the Killing fields of Minkowski spacetime.

The fields on spacetime associated with the representations ($b2$) do not appear to have any physical significance or mathematical utility. However, the representations of classes (a) and ($b1$) describe all physical fields known to occur in quantum theory.

The equations which select the subspaces of spinorial tensor fields which realize the representations of classes (a) and ($b1$) can be given in many equivalent forms. A convenient choice for the representations of class (a) of mass m and spin s is

$$(\Box + m^2)\phi^{A_1 \cdots A_n} = 0 \quad , \tag{13.1.59}$$

where $\phi^{A_1 \cdots A_n}$ is totally symmetric, $\phi^{A_1 \cdots A_n} = \phi^{(A_1 \cdots A_n)}$, and the number of indices is $n = 2s$. For $s > 0$, equation (13.1.59) also can be expressed in the following form. We define the auxiliary variable $\sigma_{A_1'}{}^{A_2 \cdots A_n}$ by

$$\partial_{A_1' A_1} \phi^{A_1 \cdots A_n} = \frac{m}{\sqrt{2}} \sigma_{A_1'}{}^{A_2 \cdots A_n} \quad . \tag{13.1.60}$$

From equations (13.1.59) and (13.1.57) we obtain

$$\partial^{A_1' A_1} \sigma_{A_1'}{}^{A_2 \cdots A_n} = -\frac{m}{\sqrt{2}} \phi^{A_1 \cdots A_n} \quad . \tag{13.1.61}$$

Furthermore, using equation (13.1.57) again, one may verify that the pair of equations (13.1.60) and (13.1.61) imply equation (13.1.59), so the coupled first order system (13.1.60), (13.1.61) is equivalent to equation (13.1.59). Indeed, by repeated differentiation of $\phi^{A_1 \cdots A_n}$ and contraction over the unprimed indices, a whole hierarchy of auxiliary variables may be defined, each of which is coupled to the preceding variable by equations of the form (13.1.60), (13.1.61).

Many equivalent forms also may be given of an inner product which gives the solutions of equation (13.1.59) the structure of a Hilbert space. A convenient expression is obtained as follows.[5] For two solutions $\phi^{A_1 \cdots A_n}$ and $\psi^{A_1 \cdots A_n}$ of equation (13.1.59) with auxiliary variables $\sigma_{A_1'}{}^{A_2 \cdots A_n}$, $\rho_{A_1'}{}^{A_2 \cdots A_n}$, respectively, we define the particle current vector $j^{AA'}(\phi, \psi)$ by

$$\begin{aligned} j^{AA'} = (-i)^{n-1}\{&\overline{\phi}^{A'A_2' \cdots A_n'} \partial_{A_2' A_2} \cdots \partial_{A_n' A_n} \psi^{AA_2 \cdots A_n} \\ &+ \overline{\sigma}^{AA_2' \cdots A_n'} \partial_{A_2' A_2} \cdots \partial_{A_n' A_n} \rho^{A'A_2 \cdots A_n}\} \quad . \end{aligned} \tag{13.1.62}$$

It follows from equations (13.1.60) and (13.1.61) that $j^{AA'}$ is conserved, $\partial_{AA'} j^{AA'} = 0$. We define the inner product of ϕ and ψ by integrating the normal component of $j^{AA'}$ over a Cauchy surface,

$$(\phi, \psi) = \int_\Sigma j^{AA'} n_{AA'} \, dV \quad . \tag{13.1.63}$$

Although it is not obvious from this expression, the inner product (13.1.63) is positive definite when n is odd (i.e., half-integral s) and is positive definite for positive frequency solutions when n is even (i.e., integral s). (This result can be proven by reexpressing our inner product as an integral in Fourier-transform space, where it can be written in a manifestly positive definite form; see Bargmann and

5. I am indebted to P. Yip for providing me with this form of the inner product.

Wigner 1948.) Thus in all cases, the positive frequency solutions of equation (13.1.59) with finite norm in the inner product (13.1.63) form a Hilbert space. The natural action of ISL(2, $\mathbb{C}$) on $\phi^{A_1 \cdots A_n}$ gives rise to the irreducible representations of class (*a*) characterized by m^2 and $s = n/2$. For integral s, these representations are true representations of the Poincaré group, and the above construction could be reformulated using only ordinary tensor fields. However, for half-integral s, these representations of ISL(2, $\mathbb{C}$) are only representations up to sign of the Poincaré group, and the use of spinorial tensor fields is essential.

When $m = 0$, the representation selected by equation (13.1.59) becomes reducible except in the case $s = 0$. The irreducible representations in class (*b*1) with $s > 0$ are obtained from the equation (Penrose 1965*b*)

$$\partial_{A_1'A_1} \phi^{A_1 \cdots A_n} = 0 \quad , \tag{13.1.64}$$

where again $\phi^{A_1 \cdots A_n}$ is totally symmetric and $n = 2s$. (The negative s representations are obtained from the complex conjugate of eq. [13.1.64].) For $s = \frac{1}{2}$, the current

$$j^{AA'}(\phi, \psi) = \bar{\phi}^{A'} \psi^A \tag{13.1.65}$$

can be used to define an inner product as before. For $s > \frac{1}{2}$ we must introduce potentials (Penrose 1965*b*), and a gauge independent expression for a current vector cannot be given. However, a simple expression for the inner product in momentum space can be obtained (Bargmann and Wigner 1948). Again, the unitary representations thus obtained are true representations of the Poincaré group if and only if s is integral.

We comment that in the case $s = \frac{1}{2}$ equation (13.1.59) written in the form (13.1.60) and (13.1.61) is known as the *Dirac equation* and the pair of spinors $(\phi^A, \sigma_{A'})$ is called a *Dirac spinor*. By choosing a basis of spinor space and denoting the four components of $(\phi^A, \sigma_{A'})$ as ψ_0, ψ_1, ψ_2, ψ_3, respectively, the component form of equations (13.1.60) and (13.1.61) yields the usual form of the Dirac equation (problem 2). Similarly, equation (13.1.64) in the case $s = \frac{1}{2}$ is known as the *(Weyl) neutrino equation*. Note that as claimed in section 12.4, the Dirac current vector (13.1.62) with $\phi^A = \psi^A$ is the sum of two future directed null vectors and hence is a future directed timelike vector, while the neutrino current vector (13.1.65) is future directed and null. Finally, we point out that equation (13.1.64) in the case $s = 1$ is equivalent to Maxwell's equations (problem 3) while in the case $s = 2$ it is equivalent to the linearized Einstein equation (see problem 6).

In summary, we have found that spinorial tensor fields give rise to all the unitary representations up to phase of the Poincaré group which are believed to have physical relevance. This suggests that spinorial tensor fields may be the most general type of fields in Minkowski spacetime which can arise in a "specially covariant" quantum theory. In any case, we have completed the primary task of motivating the introduction of spinor fields from a general and systematic viewpoint.

13.2 Spinors in Curved Spacetime

In the previous section, we defined the notion of spinor fields on Minkowski spacetime. Our definition was motivated by the fact that the natural action of the Poincaré group on spinor fields and spinorial tensor fields gave rise to the desired representations up to phase of the Poincaré group. Thus, the "transformation property" (13.1.41) under Poincaré isometries was an essential ingredient of our definition of spinor fields on Minkowski spacetime and was used to identify real spinorial tensor fields of type $(1, 0; 1, 0)$ with vector fields. However, the Poincaré group does not act in a natural way on a curved spacetime, so clearly this characteristic property of spinor fields cannot be carried over in a direct manner to curved spacetime. Thus, we seek to reformulate the notion of spinor fields so that it applies in curved spacetimes and, of course, such that it reduces in Minkowski spacetime to the notion of spinor fields given in the previous section.

Since a general, curved spacetime possesses no isometries or any other preferred classes of diffeomorphisms and since even in Minkowski spacetime there is no natural action of the full group of diffeomorphisms on spinor fields, we cannot expect to define a "transformation law" of the type (2.2.10) under diffeomorphisms for spinor fields in curved spacetime. However, as in Minkowski spacetime, we may represent an observer together with his measuring apparatus at an event, x, in a curved spacetime (M, g_{ab}) by an orthonormal tetrad at x. Hence, associated with two different observers O_1 and O_2 at x is a Lorentz transformation which rotates the tetrad of O_1 into that of O_2. Note that this Lorentz transformation acts on the tangent space, V_x, rather than the spacetime manifold, M. We shall seek to define spinors at x so that associated with each Lorentz transformation is the spinor transformation

$$\psi^A(x) \to \pm L^A{}_B \psi^B(x) \tag{13.2.1}$$

such that the results of all measurements by O_2 on the spinor $\pm L^A{}_B \psi^B$ at x are identical to the results of all measurements by O_1 on ψ^A. (Again, the sign ambiguity in [13.2.1] is resolved if a continuous curve connecting the Lorentz transformation to the identity element is specified.) Thus, in formulating a notion of spinors in curved spacetime we shall replace the action (13.1.41) of the Poincaré group of isometries on Minkowski spacetime by the action (13.2.1) of the Lorentz group on the tangent space at each point.

Fiber bundles provide a precise mathematical framework for defining spinor fields in curved spacetime. We shall proceed, therefore, by defining the general notion of a principal fiber bundle and its associated fiber bundles. The construction of the spinor bundle then will be described.

Let G be a Lie group, let B be a manifold, and consider a C^∞ map $\phi: G \times B \to B$. We shall write $\phi(g, p)$ as $\phi_g(p)$. The map ϕ is said to be a *left action* of G on B if (i) for each fixed $g \in G$, the map $\phi_g: B \to B$ is a diffeomorphism and (ii) for all $g_1, g_2 \in G$, we have $\phi_{g_1} \circ \phi_{g_2} = \phi_{g_1 g_2}$. An example of a left action of G on the manifold $B = G$ is provided by the left translation map considered in section 7.2. For a general left action ϕ it follows from (ii) that $\phi_e \circ \phi_e = \phi_e$, where e is the identity element, which (composing with ϕ_e^{-1}) implies that ϕ_e is just the identity map on B. This implies further that $\phi_{g^{-1}} = \phi_g{}^{-1}$ for all $g \in G$. A left action ϕ is said to be *free* if for each $g \neq e$, ϕ_g leaves no point of B fixed, i.e., if for all $p \in B$ and $g \neq e$ we have $\phi_g(p) \neq p$. [On the other hand, since ϕ_e is the identity map, we have $\phi_e(p) = p$ for all $p \in B$.] Thus, for example, left translation is a free action of G on the manifold G. For each $p \in B$, the set $O = \{\phi_g(p) \mid g \in G\}$ is called the *orbit* of p under G. It is easily seen that the condition that two points of B lie on the same orbit of g defines an equivalence relation between points of B, so B can be expressed as a disjoint union of orbits.

In essence, a principal fiber bundle is a manifold which locally (but not necessarily globally) "looks like" the product, $G \times M$, of a Lie group G and a manifold M. More precisely, a *principal fiber bundle* (B, G, M, ϕ) consists of a manifold B (called the *bundle manifold*), a Lie group G (called the *fiber group*), a manifold M (called the *base manifold*), and a free left action $\phi: G \times B \to B$ satisfying the following two properties: (i) The orbits of G are in one-to-one, onto correspondence with the points of M, and the projection map $\pi: B \to M$ which assigns to each $p \in B$ the point of M associated with the orbit of p is C^∞. (ii) For each $x \in M$ there exists an open neighborhood, U, of x such that there is a diffeomorphism, ψ, taking $\pi^{-1}[U] \subset B$ into $G \times U$ such that the action of G on $\pi^{-1}[U]$ corresponds to left multiplication on $G \times U$; i.e., if $\psi(p) = (g, x)$, then $\psi[\phi_{g'}(p)] = (g'g, x)$. Figure 13.2 illustrates the nature of a principal fiber bundle. Note that we always have $\dim(B) = \dim(G) + \dim(M)$.

Thus, a particularly simple example of a principal fiber bundle is obtained by taking the product manifold, $B = G \times M$, of a Lie group G and a manifold M, with the left action of G on B defined by left multiplication, i.e., $\phi_{g'}(g, x) = (g'g, x)$. A bundle of this form is said to be trivial. One of the simplest examples of a nontrivial principal fiber bundle is obtained by taking B to be the circle, S^1, and G to be the group, Z_2, consisting of the two elements e, a with $a^2 = e$. (Thus, G is a zero-dimensional, disconnected Lie group.) We define a left action of Z_2 on the circle, $\phi: Z_2 \times B \to B$, by $\phi_e(\theta) = \theta$ and $\phi_a(\theta) = \theta + \pi$. Thus, the orbits of Z_2 are the opposite points on the circle B, and the collection of orbits, M, can be given the manifold structure S^1 so that the projection $\pi: B \to M$ is smooth. It then may be verified that (S^1, Z_2, S^1, ϕ) is a principal fiber bundle. This bundle is nontrivial since B is not diffeomorphic to the product manifold $G \times M$, which consists of two disconnected circles. The bundles B and $G \times M$ are illustrated in Figure 13.3.

A particularly important example of a principal fiber bundle is the *bundle of bases*, defined as follows. Let M be an n-dimensional manifold and consider the collection, B, of pairs $(x, (v_\mu)^a)$ where $x \in M$ and $(v_\mu)^a$ (where $\mu = 1, \ldots, n$) is a basis of the

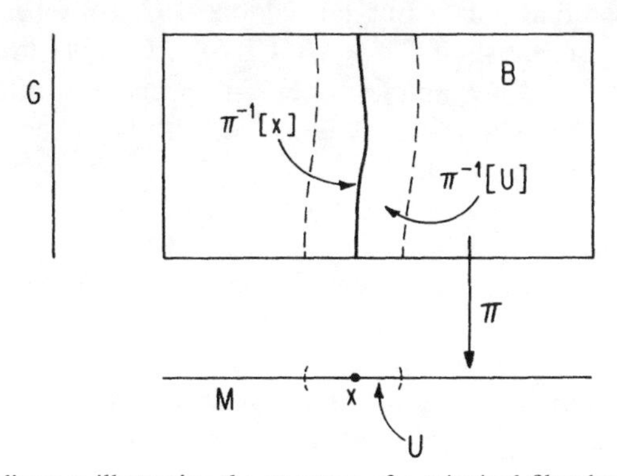

Fig. 13.2. A diagram illustrating the structure of a principal fiber bundle (see text).

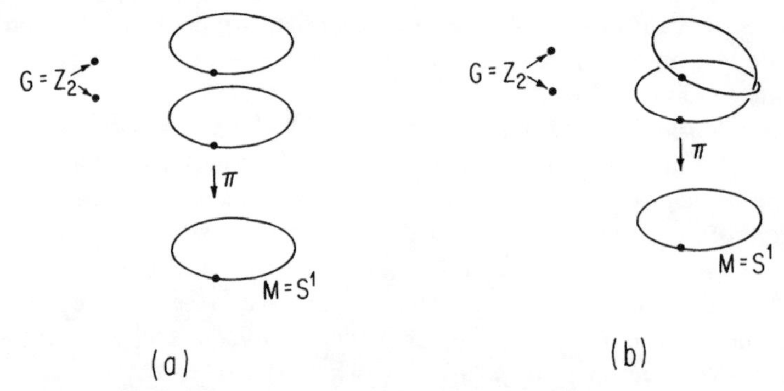

Fig. 13.3. (*a*) The trivial principal fiber bundle $Z_2 \times S^1$. (*b*) The principal bundle (S^1, Z_2, S^1, ϕ) constructed in the text.

tangent space, V_x. We can put a manifold structure on B as follows. If $\psi : U \to \mathbb{R}^n$ is a chart on a region, U, of M, we obtain a coordinate basis $(\partial/\partial x^\mu)^a$ associated with ψ. At each $x \in U$ we can expand each $(v_\mu)^a$ in terms of this basis, $(v_\mu)^a = \sum_\nu \alpha_\mu{}^\nu (\partial/\partial x^\nu)^a$. Let $\tilde{U}$ consist of all points $(x, (v_\mu)^a)$ of B with $x \in U$. We define $\tilde{\psi} : \tilde{U} \to \mathbb{R}^{n+n^2}$ by

$$\tilde{\psi}(x, (v_\mu)^a) = (x^\sigma, \alpha_\mu{}^\nu) \quad . \tag{13.2.2}$$

We use the collection of all maps of this form to define a family of charts on B. It is easily checked that these charts are compatible, thus making B a manifold of dimension $n + n^2$. The general linear group in n dimensions, GL(n)—which consists of the invertible linear maps taking $\mathbb{R}^n$ into itself with group multiplication defined by composition—acts on the left on B as follows. We view each $A \in \mathrm{GL}(n)$ as a matrix, $A^\mu{}_\nu$, of components of the linear map A in the natural basis of $\mathbb{R}^n$. We define $\phi_A : B \to B$ by

$$\phi_A(x, (v_\mu)^a) = (x, \sum_\nu (A^{-1})^\nu{}_\mu (v_\nu)^a) \quad ; \tag{13.2.3}$$

i.e., we use the inverse map, A^{-1}, to transform the basis of the tangent space at x. Then it is easily seen that ϕ is a free left action of GL(n) on B and that the above properties (i) and (ii) are satisfied. Thus, $(B, \mathrm{GL}(n), M, \phi)$ is a principal fiber bundle. For some manifolds, M—for example, for $\mathbb{R}^n$—the bundle manifold has the structure $\mathrm{GL}(n) \times M$, i.e., the bundle of bases is trivial. On the other hand, the bundle of bases of other manifolds, such as the 2-sphere, S^2, is nontrivial.

Similarly, for a manifold M on which a metric, g_{ab}, of signature (p, q) is defined, we may construct the *bundle of orthonormal bases*. The only changes from the above construction are that we take B to consist of pairs $(x, (e_\mu)^a)$, where $x \in M$ and $(e_\mu)^a$ now is an orthonormal basis for V_x, and the group which acts on B is now the orthogonal group, $O(p, q)$, rather than GL(n). In particular, for a spacetime (M, g_{ab}) the bundle of orthonormal bases has fiber group equal to the (improper) Lorentz group $O(3, 1)$. Similarly for an orientable, time orientable spacetime, we may construct the bundle of oriented, time oriented bases with the proper Lorentz group, Λ, as the fiber group.

Given a principal fiber bundle (B, G, M, ϕ), another manifold F, and a (not necessarily free) left action $\chi: G \times F \to F$ of G on F, there is a general procedure for constructing from B a new manifold E, in which, in effect, the fiber group G is replaced by the manifold F. To do so, we take the product manifold $B \times F$ and define a left action, ψ, of G on $B \times F$ by

$$\psi_g[b,f] = [\phi_g(b), \chi_g(f)] \quad . \tag{13.2.4}$$

We define E to be the set of orbits of G on $B \times F$. There is a natural projection map $p: E \to M$ given by $p(y) = \pi(b)$, where $b \in B$ is such that $(b,f) \in B \times F$ lies on the orbit of $y \in E$ and $\pi: B \to M$ is the projection map on the principal fiber bundle. For every neighborhood $U \subset M$ satisfying property (ii) of the definition of principal fiber bundle, it follows that $p^{-1}[U] \subset E$ is homeomorphic to $F \times U$. We define a manifold structure on E by requiring that these homeomorphisms be diffeomorphisms. Thus, E locally "looks like" $F \times M$ in the same sense as B locally "looks like" $G \times M$. The manifold E together with the large amount of structure on E determined by B, G, M, ϕ, F, and χ is called a *fiber bundle* or, more precisely, the fiber bundle associated to (B, G, M, ϕ) with fiber manifold F and group action χ. For each $x \in M$, the subset $p^{-1}(x) \subset E$— called the *fiber over x*—is diffeomorphic to F.

It may appear from the above rather complicated constructions and definitions that fiber bundles comprise an extremely specialized class of manifolds. In fact, however, a large variety of manifolds can be expressed in a natural and very useful way as fiber bundles. The utility of the fiber bundle viewpoint for proving theorems on the topology of manifolds can be seen, for example, in Steenrod (1951).

Given a principal fiber bundle $(B, G, M\, \phi)$ we may take $F = G$ and χ_g to be left translation. The resulting associated fiber bundle E is diffeomorphic to B, so every principal fiber bundle also can be viewed as an associated fiber bundle. Another particularly simple example of a fiber bundle associated to a principal bundle (B, G, M, ϕ) is obtained by letting F be any manifold and taking χ to be the trivial action, $\chi_g(f) = f$ for all $g \in G$ and $f \in F$. The resulting fiber bundle E is diffeomorphic to the product manifold, $F \times M$, of the fiber manifold F with the base manifold M. A fiber bundle which is diffeomorphic to $F \times M$ is called *trivial*. A simple example of a nontrivial fiber bundle, associated to the principal bundle (S^1, Z_2, S^1, ϕ) discussed above, with fiber manifold $F = \mathbb{R}$ is obtained from the following action of $Z_2 = \{e, a\}$ on $\mathbb{R}$. We define $\chi_e(x) = x, \chi_a(x) = -x$ for all $x \in \mathbb{R}$. The resulting fiber bundle E "locally looks like" the cylinder $\mathbb{R} \times S^1$, but has the property that the $\mathbb{R}$-fibers "flip upside down" when one goes once around a curve which projects down to a circle in the base manifold $M = S^1$. The manifold E is called the *Möbius strip* and also could be constructed by identifying the points in the plane, $\mathbb{R}^2$, via $(x, y) = (x + 1, -y)$.

An important example of a fiber bundle associated to the bundle of bases of an n-dimensional manifold M is obtained as follows. We take $F = \mathbb{R}^n$ and let $\mathrm{GL}(n)$ act on $\mathbb{R}^n$ in the natural way, i.e., for $a = (a^1, \ldots, a^n) \in \mathbb{R}^n$ and $A \in GL(n)$ we define

$$(\chi_A(a))^\mu = \sum_{\nu=1}^{n} A^\mu{}_\nu a^\nu \quad . \tag{13.2.5}$$

Now, corresponding to each point $(x,(v_\mu)^a; a^\nu)$ in the manifold $B \times \mathbb{R}^n$ we can associate the vector

$$v^a = \sum_{\mu=1}^{n} a^\mu (v_\mu)^a$$

at $x \in M$. Given the actions (13.2.3) and (13.2.5) of G on B and on $\mathbb{R}^n$, it is easily seen that two points of $B \times \mathbb{R}^n$ correspond to the same vector v^a at x if and only if they lie on the same orbit (13.2.4). Thus, the points of the associated fiber bundle E are in one-to-one correspondence with the pairs (x, v^a), where $x \in M$ and $v^a \in V_x$. We call E the *tangent bundle* of M and denote it by $T(M)$. If $T(M)$ is trivial, we say that M is *parallelizable*. The bundle $T_{k,l}(M)$ of tensors of type (k, l) over M can be constructed similarly by taking F to be the tensor product, $F_{k,l}$, of k copies of $\mathbb{R}^n$ and l copies of its dual space $(\mathbb{R}^n)^*$ and taking χ to be the natural action of GL(n) on $F_{k,l}$ given by

$$(\chi_A(b))^{\mu_1 \cdots \mu_k}{}_{\nu_1 \cdots \nu_l} =$$

$$\sum_{\alpha_i, \beta_j=1}^{n} A^{\mu_1}{}_{\alpha_1} \cdots A^{\mu_k}{}_{\alpha_k} (A^{-1})^{\beta_1}{}_{\nu_1} \cdots (A^{-1})^{\beta_l}{}_{\nu_l} b^{\alpha_1 \cdots \alpha_k}{}_{\beta_1 \cdots \beta_l} \quad . \tag{13.2.6}$$

If a metric, g_{ab}, of signature (p, q) is given on M we also may construct $T_{k,l}(M)$ by restricting attention to orthonormal bases, i.e., by starting with the principal bundle of orthonormal bases and defining an action of $O(p, q)$ on $F_{k,l}$, via equation (13.2.6).

Thus, the above construction gives us a new viewpoint on tangent vectors: A tangent vector at a point x on a manifold M is a point of the tangent bundle $T(M)$ lying in the fiber over x, i.e., a point $y \in T(M)$ such that $p(y) = x$. More generally, a tensor of type (k, l) at $x \in M$ is a point in the fiber over x of the bundle $T_{k,l}(M)$ of tensors of type (k, l). A smooth tensor field on M is a *cross section* of $T_{k,l}(M)$, i.e., a C^∞ map $\Sigma : M \to T_{k,l}(M)$ such that $p \circ \Sigma$ is the identity map on M. Note that this viewpoint on tensors is rather close in spirit to the characterization of tensors mentioned in chapter 2 in terms of the tensor transformation law (2.3.8). There it was remarked that the change in the coordinate basis components of a tensor under a coordinate transformation could be used to characterize it. Here, a point in $B \times F_{k,l}$ is a collection of n^{k+l} numbers associated with a point $x \in M$ and a (not necessarily coordinate) basis of the tangent space of x. The group action (13.2.6) can be viewed as telling us how this collection of numbers changes when we make a change of basis. The equivalence class of the collection of numbers under changes of basis at x is a point of $T_{k,l}(M)$, i.e., a tensor at x. Both the tensor transformation law and the fiber bundle characterization of tensors have little advantage over the direct definition of tensors as multilinear maps given in chapter 2. However, since there is no analogous direct definition of spinors, we must proceed to define spinors by the relatively indirect route of specifying their behavior under SL(2, $\mathbb{C}$) transformations associated with changes of orthonormal basis. We turn, now, to this task.

The basic idea for constructing spinor fields in a curved spacetime (M, g_{ab}) is to start with the principal fiber bundle of oriented, time oriented orthonormal bases, so that each fiber is diffeomorphic to the proper Lorentz group. Then we "unwrap" each fiber to produce a principal SL(2, $\mathbb{C}$) bundle over M. The spinor bundle then is constructed as the fiber bundle associated to this principal bundle with fiber $F = \mathbb{C}^2$, where χ is taken to be the natural action of SL(2, $\mathbb{C}$) on $\mathbb{C}^2$. A spinor at $x \in M$ may then be defined as a point in the fiber over x of the spinor bundle.

However, complications may arise in this construction if the topology of the spacetime M is nontrivial. Consider, first, the case where M is simply connected. Then M is orientable and time orientable, so there is no difficulty in constructing the principal bundle, (B, Λ, M, ϕ) of oriented, time oriented bases. Consider, now, a closed curve, γ, in B. Since M is simply connected, the curve $\pi \circ \gamma$ obtained by projecting γ down onto M can be continuously deformed to the trivial curve through x, i.e., the curve $e(t) = x$ for all t. This implies that in B the curve γ can be continuously deformed to a curve which lies entirely in the fiber $\pi^{-1}[x]$ over x. However, this fiber is diffeomorphic to the proper Lorentz group, Λ, and $\pi_1(\Lambda) = Z_2$. Thus, in general, we would expect there to exist precisely two distinct homotopy equivalence classes of closed curves in B. However, it is possible for there to exist only one homotopy equivalence class of curves, i.e., for B to be simply connected. Although the curve in the fiber over x corresponding to a 2π rotation of the basis of V_x cannot be deformed to the trivial curve while remaining within the fiber over x, for certain manifolds this curve *can* be continuously deformed to the trivial curve in B by moving it out of the fiber over x. In other words, it may happen that a 2π-rotation of a tetrad can be "undone" by transporting the tetrad along a one-parameter sequence of closed curves through x in M. *If B is simply connected, the notion of spinors on M cannot be defined.*[6] This is because we cannot consistently assign a change of sign to a spinor under a 2π-rotation of the basis if this 2π-rotation can be continuously deformed to the identity. Examples of manifolds for which B is simply connected are given by Geroch (1968*a*). It is known (Milnor 1963; Clarke 1971) that B will fail to be simply connected—i.e., $\pi_1(B) = Z_2$ and spinors can be defined—if and only if the second Stiefel-Whitney class (defined, e.g., in Steenrod 1951) of M vanishes. Geroch (1968*a*) has proven[7] that if M is noncompact, then $\pi_1(B) = Z_2$ if and only if M is parallelizable. Further equivalent criteria are given by Geroch (1968*a*, 1970*a*) and Clarke (1971).

If $\pi_1(B) = Z_2$, we define spinors on M as follows. The universal covering manifold, $\hat{B}$, of B will have the natural structure of a principal fiber bundle with fiber group SL(2, $\mathbb{C}$). We define the *spinor bundle*, $S(M)$, to be the fiber bundle associ-

6. However, one still may define "generalized spin structures"; see Avis and Isham (1980) and the references cited therein.

7. Since $O(n)$ and $O(n, 1)$ have fundamental group Z_2 for all $n > 2$, one can define analogous notions of spinors for Riemannian and Lorentzian spaces of dimension greater, respectively, than 2 and 3. (See problem 1 for the Riemannian case with $n = 3$; see Yip (1983) for a discussion of spinors in two-dimensional spacetimes, where $O(1, 1)$ is simply connected.) Geroch's parallelizability criterion applies only to four-dimensional spacetimes.

ated to $(\hat{B}, \text{SL}(2,\mathbb{C}), M, \hat{\phi})$ with fiber manifold $\mathbb{C}^2$, with the action χ_L of $L \in \text{SL}(2,\mathbb{C})$ on $\in \mathbb{C}^2$ taken to be the natural action given by

$$[\chi_L(c)]^\Sigma = \sum_{\Gamma=1}^{2} L^\Sigma{}_\Gamma c^\Gamma \quad . \tag{13.2.7}$$

A *spinor* at $x \in M$ is defined to be a point of $S(M)$ lying in the fiber over x. Similarly, the bundle, $S_{k,l;k',l'}(M)$ of spinorial tensors of type $(k,l;k',l')$ is defined by taking F to be the tensor product of k copies of $\mathbb{C}^2$, l copies of its dual space $\mathbb{C}^{2*}$, k' copies of its complex conjugate space $\overline{\mathbb{C}}^2$, and l' copies of its complex conjugate dual space $\overline{\mathbb{C}}^{2*}$. The action of χ on

$$c^{\Sigma_1 \cdots \Sigma_k}{}_{\Lambda_1 \cdots \Lambda_l}{}^{\Sigma'_1 \cdots \Sigma'_{k'}}{}_{\Lambda'_1 \cdots \Lambda'_{l'}} \in F$$

is given by

$$[\chi_L(c)]^{\Sigma_1 \cdots}{}_{\cdots \Lambda'_{l'}} = \sum_{\Gamma_1 \cdots \Delta'_{l'}} L^{\Sigma_1}{}_{\Gamma_1} \cdots (\bar{L}^{-1})^{\Delta'_{l'}}{}_{\Lambda'_{l'}} c^{\Gamma_1 \cdots}{}_{\cdots \Delta'_{l'}} \quad . \tag{13.2.8}$$

Equivalently, once the notion of spinors has been given, spinorial tensors may be defined in terms of linear (and antilinear) maps on spinors in a manner analogous to the way ordinary tensors were constructed from vectors in chapter 2. As in the previous section we shall use capital latin indices to denote spinorial tensors.

The above definition of spinors is sufficiently abstruse that some words of explanation might be helpful. Recall, first, that a point of B consists of a point $x \in M$ together with an oriented, time oriented orthornormal basis at x. If we fix a particular basis $\{(e_\mu)^a\}$ at x, a point of $\hat{B}$ may be viewed as consisting of a point $x \in M$ together with a continuous, one-parameter family of oriented, time oriented bases at x starting from $\{(e_\mu)^a\}$, where two such families are considered equivalent if they have the same endpoint and are homotopic. In other words, the fiber of $\hat{B}$ over x can be viewed as consisting of all oriented, time oriented orthonormal bases at x together with their 2π-rotations. A point of $\hat{B} \times \mathbb{C}^2$ is a pair of complex numbers associated with x and a basis or 2π-rotated basis at x. The group action χ_g of equation (13.2.7) tells us how these numbers "transform" under a change of basis. Each equivalence class of transformed numbers and bases at x—i.e., each orbit of $\hat{\psi}_g$, equation (13.2.4)—defines a spinor at x. Thus, our fiber bundle construction is just a precise way of implementing the notion that "a spinor at x is an ordered pair of complex numbers which transforms by the natural representation of SL$(2,\mathbb{C})$ under a change of basis."

Note that $\mathbb{C}^2$ has a great deal of structure beyond that of a complex two-dimensional vector space; for example, it has a natural inner product and a notion of the real and imaginary parts of vectors. However, only those properties of $\mathbb{C}^2$ which are preserved under the action of SL$(2,\mathbb{C})$ survive to yield structure on the fibers of the spinor bundle $S(M)$. Thus, for example, since there are elements of SL$(2,\mathbb{C})$ which take real vectors in $\mathbb{C}^2$ to complex vectors, there is no natural notion of the real and imaginary parts of a spinor at x. However, since the SL$(2,\mathbb{C})$ maps are linear, they preserve addition and scalar multiplication, so the spinors at x do have the natural structure of a two-dimensional complex vector space. Beyond that, only the element $\epsilon_{\Sigma\Gamma} \in \mathbb{C}^{2*} \otimes \mathbb{C}^{2*}$ (as well as objects constructed from it) given by the matrix array (13.1.20) is preserved under the action of SL(2, $\mathbb{C}$), so the only additional natural structure of spinors is that of a tensor ϵ_{AB} of type (0, 2; 0, 0). Thus, each fiber of $S(M)$ has precisely the natural structure of spinor space (W, ϵ_{AB}). Note that there is no natural way of identifying the different fibers of $S(M)$, i.e., there is no way of saying that a spinor at $x \in M$ is "the same" as one at $y \in M$, just as there is no way of saying that tangent vectors at different points are "the same." However, we do know what it means for a spinor to vary smoothly from point to point: A smooth spinor field is simply a (smooth) cross section of $S(M)$.

The relation between spinorial tensor fields of type (1, 0; 1, 0) and vector fields may be seen as follows. Fix an isomorphism, σ, between the real elements of $\mathbb{C}^2 \otimes \overline{\mathbb{C}}^2$ and $\mathbb{R}^4$—such as that given by equation (13.1.30)—for which the Lorentz metric $\epsilon_{\Sigma\Lambda}\bar{\epsilon}_{\Sigma'\Lambda'}$ on $\mathrm{Re}(\mathbb{C}^2 \otimes \overline{\mathbb{C}}^2)$ is taken into the Lorentz metric $\mathrm{diag}(1, -1, -1, -1)$ on $\mathbb{R}^4$. Then the natural 2 to 1 map of $\hat{B}$ onto B gives rise to a 2 to 1 map σ' of $\hat{B} \times \mathrm{Re}(\mathbb{C}^2 \otimes \overline{\mathbb{C}}^2)$ onto $B \times \mathbb{R}^4$. Furthermore, σ' commutes with the natural group actions $\hat{\psi}$ and ψ on these spaces, i.e., we have $\sigma' \circ \hat{\psi}_L = \psi_\Lambda \circ \sigma$, where Λ is the Lorentz group element associated with $L \in \mathrm{SL}(2, \mathbb{C})$. Hence, we obtain a (one-to-one) correspondence, $\tilde{\sigma}$, between the SL(2, $\mathbb{C}$) orbits of $\hat{B} \times \mathrm{Re}(\mathbb{C}^2 \otimes \overline{\mathbb{C}}^2)$ and the Lorentz group orbits of $B \times \mathbb{R}^4$. Thus, we obtain in turn an identification of the real subset of $S_{1,0;1,0}(M)$ with $T(M)$; i.e., real spinorial tensors of type (1, 0; 1, 0) are identified with vectors. As in the previous section, we shall incorporate this identification into our notation by allowing $v^{AA'}$ to denote a vector rather than explicitly writing $\tilde{\sigma}^a{}_{AA'}v^{AA'}$. Again the spacetime metric is related to ϵ_{AB} by

$$g_{AA'BB'} = \epsilon_{AB}\bar{\epsilon}_{A'B'} \quad . \tag{13.2.9}$$

In the case where the spacetime manifold, M, is not simply connected, the analysis of whether the notion of spinors can be defined is somewhat more complicated. First, M may fail to be orientable or time oriented. *In such a case, the bundle, B, of oriented, time oriented orthonormal bases does not exist, and the notion of spinors cannot be defined.*[8] If M is orientable and time orientable, we may construct B, but we cannot obtain $\hat{B}$ by taking the universal covering manifold of B since this would "unwrap" M as well as the fibers of B. In order to make sense of the idea of unwrapping only the fibers of B, it is necessary that the fundamental group of B be of the direct product form

$$\pi_1(B) = G_1 \times G_2 \quad , \tag{13.2.10}$$

8. We should reemphasize that the term "spinors" here and elsewhere in this chapter means SL(2, $\mathbb{C}$) spinors. Other types of spinors can be defined by a similar construction starting from other bundles of bases. In particular, in a time orientable but not necessarily orientable spacetime, let $\tilde{B}$ denote the bundle of time oriented orthonormal bases. The fiber group is then the (disconnected) group obtained by composing proper Lorentz transformations with a "parity" transformation. Its covering group can be made to act on a four-dimensional complex vector space via the usual transformation formulas for Dirac spinors. Thus, a notion of (4-component) Dirac spinors can be defined on nonorientable spacetimes. In the case of an orientable spacetime, Dirac spinors are naturally isomorphic to a pair $(\phi^A, \sigma_{A'})$ consisting of a two-component spinor ϕ^A and a complex conjugate dual spinor $\sigma_{A'}$, as mentioned at the end of section 13.1.

where G_1 is the fundamental group, $\pi_1(\Lambda) = Z_2$, of the Lorentz group and G_2 is the fundamental group, $\pi_1(M)$, of M. [Here the direct product $G = G_1 \times G_2$ of groups G_1 and G_2 is defined as the group consisting of ordered pairs (g_1, g_2) with $g_1 \in G_1$, $g_2 \in G_2$ with composition law $(g_1, g_2)(g'_1, g'_2) = (g_1 g'_1, g_2 g'_2)$.] Furthermore, it is understood in equation (13.2.10) that each element of the subgroup of $\pi_1(B)$ of the form (g_1, e_2) with e_2 the identity element of G_2 is homotopic to the closed curve g_1 lying within a single Lorentz group fiber, whereas each element of the subgroup (e_1, g_2) projects down to the closed curve $g_2 \in \pi_1(M)$ in M. Note that in the case where M is simply connected, equation (13.2.10) reduces to our previous criterion $\pi_1(B) = \pi_1(\Lambda) = Z_2$. *If* $\pi_1(B)$ *cannot be expressed in the form* (13.2.10), *the notion of spinors cannot be defined on* M. On the other hand, if $\pi_1(B)$ is of the form (13.2.10), we may construct a (nonuniversal) covering space, $\hat{B}$, of B by defining two closed curves through a point $b \in B$ to be equivalent if their composition is homotopic to a closed curve of the form (e_1, g_2). Then $\hat{B}$ has the natural structure of a principal SL(2, $\mathbb{C}$) bundle over M, and one can proceed to define spinors on M in the same manner as in the case where M is simply connected. Again—assuming that M is orientable and time orientable—spinors can be defined [i.e., $\pi_1(B)$ is of the form (13.2.10)] if and only if the second Stiefel-Whitney class of M vanishes. Furthermore, if in addition M is noncompact, spinors can be defined if and only if M is parallelizable (Geroch 1968*a*; see also Clarke 1971).

One further point is worthy of note in the case where M is not simply connected. The decomposition of $\pi_1(B)$ as a direct product of subgroups of the form (13.2.10) need not be unique, and each distinct decomposition gives rise to a distinct notion of spinors on M. The reason is as follows. Each element of $\pi_1(B)$ corresponds to an equivalence class of transports of tetrads around closed curves in M. When we express $\pi_1(B)$ in the form (13.2.10), we, in effect, state that tetrad transports of the form (e_1, g_2) correspond to no net rotation of the tetrad, while those of the form (a_1, g_2)—where a_1 is the non-identity element of $\pi_1(\Lambda) = Z_2$—correspond to a 2π rotation of the tetrad. Suppose, now, that $\pi_1(M)$ has a normal subgroup[9] H with factor group[10] isomorphic to Z_2. Then we can define G'_2 to be the subset of $\pi_1(B)$ consisting of all elements either of the form (e, h) or of the form (a, bh), where $h \in H$ and $b \notin H$. Then it is easy to check that G'_2 is a subgroup of $\pi_1(B)$ isomorphic to $\pi_1(M)$ and that $\pi_1(B)$ is of the form $G_1 \times G'_2$. We may use this decomposition to define the covering space $\hat{B}'$ and use $\hat{B}'$ to define spinors. With this new decomposition of $\pi_1(B)$, closed curves in B of the form (a, bh) now correspond to no net rotation of the tetrad in M, whereas curves of the form (e, bh) now correspond to a 2π-rotation. Spinors on M defined with respect to $\hat{B}$ will not change sign if transported around a closed curve $bh \in \pi_1(M)$ in M in such a way that the com-

9. A subgroup H of a group G is said to be *normal* if $ghg^{-1} \in H$ for all $g \in G$ and $h \in H$.

10. For a normal subgroup H, the orbits obtained by the natural left action of H on G—called *right cosets* of H—can be given a group structure via the composition law $c_1 c_2 = c$, where c is the coset of H containing $g_1 g_2$, where $g_1 \in c_1$ and $g_2 \in c_2$. (It is necessary that H be a normal subgroup in order that this composition law be independent of the choice of representative elements g_1, g_2.) This group of cosets is called the *factor group* of H and is denoted G/H.

ponents of its null flag (13.1.37) with respect to the basis associated with $(e, bh) \in \pi_1(B)$ remain constant during the transport. Spinors defined with respect to $\hat{B}'$ *will* change sign under the same transport. Thus, depending on the group structure of $\pi_1(M)$, there may be several[11] distinct ways of defining spinors in the sense that there may exist freedom to decide whether or not a spinor changes sign when transported around certain closed curves in M. For example, if $\pi_1(M) = Z_2$, then if spinors can be defined at all, there exist two inequivalent definitions. On the other hand, if $\pi_1(M) = Z_3$ (i.e., the cyclic group consisting of three elements), then the definition of spinors is unique.

We turn, now, to the definition of derivatives of spinors in curved spacetime. As mentioned above, in curved spacetime there is no natural identification of the spinor spaces at different points, just as the tangent spaces at different points cannot be identified in a natural way. However, in chapter 3 we introduced the notion of a derivative operator on ordinary tensor fields and found that there is a unique derivative operator, ∇_a, associated with a metric g_{ab} via the requirement $\nabla_a g_{bc} = 0$. Furthermore, given a derivative operator, we obtain a notion of parallel transport, i.e., a curve-dependent identification of the tangent spaces at different points. This notion of parallel transport of tensor fields obtained from ∇_a gives rise to a unique notion of parallel transport of spinors,[12] namely, a spinor will be said to be parallelly transported along a curve if its null flag (13.1.37) is parallelly transported and, in addition, the spinor itself does not change sign discontinuously. (Recall that the continuity of a spinor field is a well-defined notion.) This notion of parallel transport then can be used to take derivatives of spinors along a curve γ—since a spinor at $\gamma(t)$ now can be compared with a spinor at $\gamma(t + \delta t)$ via parallel transport—and this, in turn, uniquely gives rise to a notion of a spinor derivative operator $\nabla_{AA'}$ taking spinorial tensor fields of type $(k, l; k', l')$ into spinorial tensor fields of type $(k, l + 1; k', l' + 1)$. This derivative operator $\nabla_{AA'}$ will satisfy the analogs of properties (1)–(3) listed in section 3.1 for a derivative operator on ordinary tensor fields where, in property (3), contraction now is with respect to spinor indices. Furthermore, when applied to ordinary tensor fields, $\nabla_{AA'}$ will agree with the derivative operator ∇_a associated with g_{ab}, which implies, in particular, that properties (4) and (5) are satisfied by $\nabla_{AA'}$. In addition, $\nabla_{AA'}$ satisfies two further conditions: (i) for all spinorial tensor fields ψ we have $\overline{(\nabla\psi)} = \nabla\bar{\psi}$, i.e., $\nabla_{AA'}$ is "real," and (ii) $\nabla_{AA'}\epsilon_{BC} = 0 = \nabla_{AA'}\epsilon^{BC}$. Note that although the derivative operator ∇_a satisfying $\nabla_a g_{bc} = 0$ is but one of a wide class of derivative operators on tensor fields, most of these other derivative operators cannot be generalized to apply to spinor fields, since for these notions of derivative the parallel transport of a null flag will not, in general, retain the form of a null flag.

11. The number of distinct ways of defining spinors equals the number of normal subgroups of $\pi_1(M)$ with factor group Z_2. This, in turn, is equal to the number of generators of the cohomology group $H^1(M; Z_2)$; see Isham (1978) for further discussion.

12. This can be demonstrated more systematically by reformulating the notion of parallel transport in terms of the theory of connections on a principal fiber bundle and its associated bundles (see, e.g., Bishop and Crittenden 1964). The connection on B gives rise to a connection on $\hat{B}$, which in turn yields a connection on $S(M)$.

If we take two spinor derivatives of a dual spinor field α_C and antisymmetrize over the derivative indices, the same argument used to define the Riemann tensor in chapter 3 shows that the resulting spinorial tensor field at a point $x \in M$ depends only on the value of α_C at x. Thus, there exists a spinorial tensor field $\chi_{AA'BB'C}{}^D$ such that for all dual spinor fields α_C, we have

$$(\nabla_{AA'}\nabla_{BB'} - \nabla_{BB'}\nabla_{AA'})\alpha_C = \chi_{AA'BB'C}{}^D\alpha_D \quad . \tag{13.2.11}$$

We apply this commutator to the dual vector $\alpha_C\bar{\alpha}_{C'}$, using the Leibnitz rule and the reality property (i) of $\nabla_{AA'}$. Comparison with formula (3.2.3) for all $\alpha_C\bar{\alpha}_C$ establishes that

$$R_{AA'BB'CC'}{}^{DD'} = \chi_{AA'BB'C}{}^D\bar{\epsilon}_{C'}{}^{D'} + \bar{\chi}_{AA'BB'C'}{}^{D'}\epsilon_C{}^D \quad , \tag{13.2.12}$$

where $R_{AA'BB'CC'}{}^{DD'}$ is the spinorial equivalent of the Riemann tensor. Arguments similar to those which led to equation (3.2.12) show that, for an arbitrary "down index" spinorial tensor field, $\alpha_{C_1\cdots C_l C'_1\cdots C'_{l'}}$, we have

$$(\nabla_{AA'}\nabla_{BB'} - \nabla_{BB'}\nabla_{AA'})\alpha_{C_1\cdots C'_{l'}} = \sum_{i=1}^{l} \chi_{AA'BB'C_i}{}^D\alpha_{C_1\cdots D\cdots C'_{l'}}$$

$$+ \sum_{i=1}^{l'} \bar{\chi}_{AA'BB'C'_i}{}^{D'}\alpha_{C_1\cdots D'\cdots C'_{l'}} \quad . \tag{13.2.13}$$

The generalization of equation (13.2.13) to spinorial tensor fields with "up indices" is easily obtained by raising indices on both sides of equation (13.2.13) with ϵ^{EC_j} and $\bar{\epsilon}^{E'C'_j}$.

The antisymmetry of R_{abcd} in its last two indices implies via equations (13.2.12) and (13.1.46) that $\chi_{AA'BB'CD}$ is symmetric in C and D. Since R_{abcd} is antisymmetric in its first two indices, the argument which led to equation (13.1.50) implies further that $\chi_{AA'BB'CD}$ can be expressed in the form

$$\chi_{AA'BB'CD} = \Lambda_{ABCD}\bar{\epsilon}_{A'B'} + \Phi_{A'B'CD}\epsilon_{AB} \quad , \tag{13.2.14}$$

where

$$\Lambda_{ABCD} = \Lambda_{(AB)(CD)} \tag{13.2.15}$$

and

$$\Phi_{A'B'CD} = \Phi_{(A'B')(CD)} \quad . \tag{13.2.16}$$

Substituting these results in equation (13.2.12), we find that the Riemann tensor symmetry $R_{abcd} = R_{cdab}$ implies that $\Phi_{A'B'CD}$ is real:

$$\bar{\Phi}_{ABC'D'} = \Phi_{C'D'AB} \quad , \tag{13.2.17}$$

and Λ_{ABCD} satisfies

$$\Lambda_{ABCD} = \Lambda_{CDAB} \quad . \tag{13.2.18}$$

Because of equation (13.2.18) $\epsilon^{AC}\Lambda_{ABCD}$ is antisymmetric in B and D and hence must be a multiple of ϵ_{BD}. We define Ψ_{ABCD} by

$$\Psi_{ABCD} = \Lambda_{ABCD} - \Lambda(\epsilon_{AC}\epsilon_{BD} + \epsilon_{BC}\epsilon_{AD}) \quad , \tag{13.2.19}$$

where

$$\Lambda = \frac{1}{6} \epsilon^{AC} \epsilon^{BD} \Lambda_{ABCD} \quad . \tag{13.2.20}$$

Then by construction Ψ_{ABCD} satisfies

$$\epsilon^{AC} \Psi_{ABCD} = 0 \quad . \tag{13.2.21}$$

Together with the other symmetries of Λ_{ABCD}, this implies that Ψ_{ABCD} is totally symmetric,

$$\Psi_{ABCD} = \Psi_{(ABCD)} \quad . \tag{13.2.22}$$

Finally, a calculation using equation (13.1.36) establishes that the Riemann tensor symmetry $R_{a[bcd]} = 0$ implies the reality of Λ,

$$\overline{\Lambda} = \Lambda \quad . \tag{13.2.23}$$

Thus, we have found that

$$\chi_{AA'BB'CD} = \Psi_{ABCD} \bar{\epsilon}_{A'B'} + \Phi_{A'B'CD} \epsilon_{AB} + \Lambda(\epsilon_{AC} \epsilon_{BD} + \epsilon_{BC} \epsilon_{AD}) \bar{\epsilon}_{A'B'} \quad , \tag{13.2.24}$$

and hence we have obtained the following spinorial decomposition of the Riemann tensor:

$$\begin{aligned} R_{AA'BB'CC'}{}^{DD'} = {} & \Psi_{ABC}{}^{D} \bar{\epsilon}_{A'B'} \bar{\epsilon}_{C'}{}^{D'} + \Phi_{A'B'C}{}^{D} \epsilon_{AB} \bar{\epsilon}_{C'}{}^{D'} \\ & + \Lambda(\epsilon_{AC} \epsilon_{B}{}^{D} + \epsilon_{BC} \epsilon_{A}{}^{D}) \bar{\epsilon}_{A'B'} \bar{\epsilon}_{C'}{}^{D'} + \text{C.C.} \quad , \end{aligned} \tag{13.2.25}$$

where C.C. denotes the complex conjugate of the preceding terms. To obtain the Ricci tensor, we contract over B and D and over B' and D'. We obtain

$$R_{AA'CC'} = -2\Phi_{A'C'AC} + 6\Lambda \, \bar{\epsilon}_{A'C'} \epsilon_{AC} \quad . \tag{13.2.26}$$

This shows that $-2\Phi_{A'C'AC}$ is just the spinor equivalent of the trace-free Ricci tensor $R_{ab} + \frac{1}{4} R g_{ab}$, where now we define $R = -R_{ab} g^{ab}$ in order to compensate for our change in metric signature and thereby agree with our previous definition given in chapter 3. (The definitions of $R_{abc}{}^{d}$ and R_{ab} are unaffected by the change of metric signature made in this chapter.) Furthermore, equation (13.2.26) shows that $\Lambda = -R/24$. Using these results and comparing the decomposition (13.2.25) with that of (3.2.28), we find that the spinor equivalent, $C_{AA'BB'CC'DD'}$, of the Weyl tensor C_{abcd} is simply

$$C_{AA'BB'CC'DD'} = \Psi_{ABCD} \bar{\epsilon}_{A'B'} \bar{\epsilon}_{C'D'} + \overline{\Psi}_{A'B'C'D'} \epsilon_{AB} \epsilon_{CD} \quad . \tag{13.2.27}$$

We refer to Ψ_{ABCD} as the *Weyl spinor*.

The decomposition (13.2.24) of $\chi_{AA'BB'C}{}^{D}$ allows us to reexpress equation (13.2.11) in such a way as to separate the effect of Ricci and Weyl curvature on spinors. If we contract equation (13.2.11) with ϵ^{AB}, we find

$$\nabla_{A(A'} \nabla_{B')}{}^{A} \alpha_C = \Phi_{A'B'C}{}^{D} \alpha_D \quad , \tag{13.2.28}$$

where we used the identity (13.1.46) to obtain the left-hand side of this equation and formula (13.2.24) to obtain the right-hand side. On the other hand, if we contract equation (13.2.11) with $\bar{\epsilon}^{A'B'}$, we get

$$\nabla_{A'(A}\nabla_{B)}{}^{A'}\alpha_C = \Lambda_{ABC}{}^D\alpha_D = \Psi_{ABC}{}^D\alpha_D - 2\Lambda\epsilon_{C(A}\alpha_{B)} \quad . \qquad (13.2.29)$$

The two equations (13.2.28) and (13.2.29) are equivalent to equation (13.2.11). For future reference, we note that the generalization of equation (13.2.29) to an n index spinor $\phi_{A_1\cdots A_n}$ is

$$\nabla_{A'(A}\nabla_{B)}{}^{A'}\phi_{C_1\cdots C_n} = \sum_{i=1}^{n}\{\Psi_{ABC_i}{}^D\phi_{C_1\cdots D\cdots C_n} - 2\Lambda\epsilon_{C_i(A}\phi_{|C_1\cdots C_{i-1}|B)C_{i+1}\cdots C_n}\} \quad . \qquad (13.2.30)$$

Note also that the general identity (13.1.52) implies that

$$\nabla_{A'[A}\nabla_{B]}{}^{A'} = \frac{1}{2}\epsilon_{AB}\Box \quad , \qquad (13.2.31)$$

where $\Box = \nabla_{AA'}\nabla^{AA'}$.

We now are in a position to explain the spinor motivation of the Newman-Penrose (1962) formalism mentioned in section 3.4. Instead of choosing an orthonormal basis of the tangent space at each point, we choose a basis $\xi_0^A = o^A$, $\xi_1^A = \iota^A$ of spinor space at each point, normalized so that

$$o_A\iota^A = 1 \quad . \qquad (13.2.32)$$

Instead of defining the 24 real Ricci rotation coefficients by equation (3.4.14), we define the 12 complex *spin coefficients* by

$$\gamma_{\Gamma\Delta'\Sigma\Lambda} = (\xi_\Gamma)^A(\bar{\xi}_{\Delta'})^{A'}(\xi_\Sigma)_B\nabla_{AA'}(\xi_\Lambda)^B \quad . \qquad (13.2.33)$$

Here Γ, Δ', Σ, Λ take the values 0, 1 and using the Leibnitz rule and equation (13.2.32) one verifies that $\gamma_{\Gamma\Delta'\Sigma\Lambda}$ is symmetric in Σ and Λ, so equation (13.2.33) indeed defines 12 independent complex quantities. For the specific notation customarily used for each spin coefficient, see Newman and Penrose (1962). Finally, instead of using equation (3.4.17) to express the tetrad basis components of the Riemann tensor in terms of the Ricci rotation coefficients, we use equations (13.2.11) and (13.2.24) to express the spinor basis components of Ψ_{ABCD}, $\Phi_{A'B'CD}$, and Λ in terms of the spin coefficients. In this way, we obtain equations equivalent to equation (3.4.21) of chapter 3, but the explicit form of these equations written out term by term (see eq. [4.2] of Newman and Penrose 1962) is far simpler in appearance than equation (3.4.21) would be if a separate letter were used for each Ricci rotation coefficient and each curvature component.

How can one evaluate the spin coefficients defined by equation (13.2.33)? More generally, how does one compute derivatives, $\nabla_{AA'}$, of arbitrary spinorial tensors? Recall that the spinor derivative operator $\nabla_{AA'}$ corresponding to the tensor derivative operator ∇_a satisfying $\nabla_a g_{bc} = 0$ is the only sensible derivative operator for spinors, so no analog of an "ordinary derivative" operator $\partial_{AA'}$ exists. Thus, there is no analog for spinors of the coordinate basis methods used for tensors. However, we can evaluate the spin coefficients (13.2.33) by using the formula (Friedman, unpublished)

$$\gamma_{AA'\Sigma\Lambda} \equiv (\xi_\Sigma)_B \nabla_{AA'}(\xi_\Lambda)^B = \frac{1}{2} \sum_{\Gamma',\Delta'=0}^{1} \bar{\epsilon}^{\Gamma'\Delta'}(\bar{\xi}_{\Gamma'})_{B'}(\xi_\Sigma)_B \nabla_{AA'}[(\xi_\Lambda)^B(\bar{\xi}_{\Delta'})^{B'}] \quad , \quad (10.2.34)$$

which may be verified by expanding the derivative of the term in square brackets via the Leibnitz rule and using equation (13.2.32). However, the term in square brackets is a vector quantity, i.e.,

$$l^{AA'} = \iota^A \bar{\iota}^{A'} \quad (13.2.35)$$

and

$$n^{AA'} = o^A \bar{o}^{A'} \quad (13.2.36)$$

are null vectors, while

$$m^{AA'} = \iota^A \bar{o}^{A'} \quad (13.2.37)$$

and

$$\bar{m}^{AA'} = o^A \bar{\iota}^{A'} \quad (13.2.38)$$

are (complex) linear combinations of the spacelike unit vectors

$$x^{AA'} = \frac{1}{\sqrt{2}}(\bar{\iota}^{A'} o^A + \iota^A \bar{o}^{A'}) \quad , \quad (13.2.39)$$

$$y^{AA'} = \frac{i}{\sqrt{2}}(\bar{\iota}^{A'} o^A - \iota^A \bar{o}^{A'}) \quad . \quad (13.2.40)$$

Thus, the spin coefficients can be evaluated in terms of $\nabla_a l^b$, $\nabla_a n^b$, $\nabla_a x^b$, and $\nabla_a y^b$, which can be calculated by standard methods. Indeed, the entire spinor calculus can be reexpressed as a tetrad calculus using the complex null tetrad l^a, n^a, m^a, *and* $\bar{m}^a$. Once the spin coefficients have been found, the derivative of an arbitrary spinorial tensor field ψ can be evaluated by first expanding ψ in a basis formed out of ι^A, o^A and their complex conjugates. Then, using the Leibnitz rule, we can express $\nabla\psi$ in terms of derivatives of the (scalar) basis components and the spin coefficients.

We illustrate, now, the utility of spinor methods by deriving the algebraic classification of the Weyl tensor (Penrose 1960). In section 7.3 we asserted that in algebraically general spacetimes the Weyl tensor admits four distinct principal null directions satisfying equation (7.3.1), whereas in algebraically special spacetimes some of these null directions coincide and satisfy the stronger conditions listed in Table 7.1. We did not attempt to prove these assertions in chapter 7 because a direct proof by tensor methods is relatively difficult. However, the algebraic classification of the Weyl tensor by spinor methods is remarkably simple. Consider the Weyl spinor Ψ_{ABCD} and fix a basis ι^A, o^A of spinor space with $o_A \iota^A = 1$ and with ι^A chosen so that

$$\Psi_{ABCD}\iota^A \iota^B \iota^C \iota^D = 1 \quad . \quad (13.2.41)$$

Let $z \in \mathbb{C}$, set

$$\alpha^A = z\iota^A + o^A \quad , \tag{13.2.42}$$

and consider the quantity

$$f(z) = \Psi_{ABCD}\alpha^A\alpha^B\alpha^C\alpha^D \quad . \tag{13.2.43}$$

Since f is a fourth degree polynomial in z and by equation (13.2.41) the coefficient of z^4 is equal to 1, we can factor it into the form

$$f(z) = (z - c_1)(z - c_2)(z - c_3)(z - c_4) \quad . \tag{13.2.44}$$

Now, for $i = 1, 2, 3, 4$, let

$$(\kappa_i)^A = o^A + c_i\iota^A \quad . \tag{13.2.45}$$

Then we have

$$z - c_i = (\kappa_i)_A\alpha^A \quad , \tag{13.2.46}$$

and hence we have established that there exist four spinors $(\kappa_i)^A$, called *principal spinors,* such that the equation

$$\Psi_{ABCD}\alpha^A\alpha^B\alpha^C\alpha^D = (\kappa_1)_A(\kappa_2)_B(\kappa_3)_C(\kappa_4)_D\alpha^A\alpha^B\alpha^C\alpha^D \tag{13.2.47}$$

holds for all α^A of the form (13.2.42), and, consequently, for all spinors α^A. However, since the Weyl spinor is totally symmetric (see eq. [13.2.22]), this can be true if and only if

$$\Psi_{ABCD} = (\kappa_1)_{(A}(\kappa_2)_B(\kappa_3)_C(\kappa_4)_{D)} \quad . \tag{13.2.48}$$

Thus, we obtain a general decomposition of the Weyl spinor as the symmetrized product of four principal spinors. Note that κ^A is a principal spinor if and only if

$$\Psi_{ABCD}\kappa^A\kappa^B\kappa^C\kappa^D = 0 \quad . \tag{13.2.49}$$

Furthermore, κ^A is 2, 3, or 4 times repeated—i.e., it appears 2, 3, or 4 times in the decomposition (13.2.48)—if and only if, respectively,

$$\Psi_{ABCD}\kappa^A\kappa^B\kappa^C = 0 \quad \text{(two times)} \quad , \tag{13.2.50}$$

$$\Psi_{ABCD}\kappa^A\kappa^B = 0 \quad \text{(three times)} \quad , \tag{13.2.51}$$

$$\Psi_{ABCD}\kappa^A = 0 \quad \text{(four times)} \quad . \tag{13.2.52}$$

If κ^A is a principal spinor, the null vector

$$k^{AA'} = \kappa^A\bar{\kappa}^{A'} \tag{13.2.53}$$

is called a principal null vector. The conditions (13.2.49)–(13.2.52) on κ^A now can be translated to the conditions on k^a listed in Table 7.1 (problem 5). Thus, we obtain the algebraic classification of section 7.3.

We conclude this section by examining the generalization to curved spacetime of the equations in Minkowski spacetime for fields of mass m and spin s. The most natural generalization of these equations is obtained by the "minimal coupling" prescription of replacing $\partial_{AA'}$ everywhere by $\nabla_{AA'}$. However, for $m > 0$, this prescription gives inequivalent results when aplied to equation (13.1.59) as opposed to equations (13.1.60) and (13.1.61). The curved spacetime version of equation (13.1.59) has a well posed initial value formulation for all s, but the current (13.1.62) is no longer conserved. On the other hand, for $s = \frac{1}{2}$ the equations

$$\nabla_{AA'}\phi^A = \frac{m}{\sqrt{2}}\sigma_{A'} \quad , \tag{13.2.54}$$

$$\nabla^{AA'}\sigma_{A'} = -\frac{m}{\sqrt{2}}\phi^A \tag{13.2.55}$$

have a well posed initial value formulation (see problem 8), and the current (13.1.62) generalized to curved spacetime is conserved. Hence, we adopt equations (13.2.54) and (13.2.55) as the generalization to curved spacetime for the Dirac equation. Similar results hold for $s = 1$. However, when $s > 1$, the curved spacetime versions of equations (13.1.60) and (13.1.61) do not have a well posed initial value formulation (Buchdahl 1962). Similarly, the curved spacetime spin s equation for $m = 0$,

$$\nabla_{A_1A_1'}\phi^{A_1\cdots A_n} = 0 \tag{13.2.56}$$

has a well posed initial value formulation for $s = \frac{1}{2}$ and for $s = 1$ (where it is equivalent to Maxwell's equations in curved spacetime), but, as we shall see below, it fails to have a well posed initial value formulation for $s > 1$. Thus, the natural curved spacetime generalization of the Minkowski equations for $s > 1$ do not yield physically viable models for fields in curved spacetime.

To prove that equation (13.2.56) does not have a well posed initial value formulation when $s > 1$, we apply $\nabla^{A_1'}{}_B$ to equation (13.2.56), we write $\nabla^{A_1'}{}_B\nabla_{A_1A_1'} = \nabla^{A_1'}{}_{[B}\nabla_{A_1]A_1'} + \nabla^{A_1'}{}_{(B}\nabla_{A_1)A_1'}$, and then use equations (13.2.30) and (13.2.31). We thereby find that $\phi^{A_1\cdots A_n}$ satisfies a wave equation of the form

$$0 = \Box\phi^{A_1\cdots A_n} + 2(n-1)\Psi_{BC}{}^{A_1(A_2}\phi^{A_3\cdots A_n)BC} - 2(n+2)\Lambda\phi^{A_1\cdots A_n} \quad . \tag{13.2.57}$$

However, unlike the situation in flat spacetime, if equation (13.2.57) is satisfied everywhere in spacetime and equation (13.2.56) holds on an initial surface, this does not imply that equation (13.2.56) holds everywhere. Indeed, contracting equation (13.2.57) with $\epsilon_{A_1A_2}$ and using the fact that $\phi^{A_1\cdots A_n} = \phi^{(A\cdots A_n)}$ we find that for $s > 1$, $\phi^{A_1\cdots A_n}$ must satisfy

$$0 = \Psi_{BCD}{}^{(A_3}\phi^{A_4\cdots A_n)BCD} \quad . \tag{13.2.58}$$

Equation (13.2.58) is a purely algebraic condition which must be satisfied by $\phi^{A_1\cdots A_n}$ throughout spacetime. Note that it is *not* preserved under evolution by equation (13.2.57); i.e., even if equation (13.2.58) and its time derivative hold for data for $\phi^{A_1\cdots A_n}$ on an initial surface, there is no reason why equation (13.2.58) will hold for the solution of (13.2.57) evolved from these data. Thus, few, if any, solutions of equation (13.2.56) will exist in a general curved spacetime. Thus, for $s > 1$ there is no natural generalization to curved spacetime of the notion of a "pure" massless spin s field.

Problems

1. SU(2) Spinors: As mentioned in a footnote to the text, analogous notions of spinors exist for all Riemannian and Lorentzian spaces of sufficiently high dimension. In particular, we can define spinors for ordinary three-dimensional Euclidean space as follows. Let W be a two-dimensional complex vector space on which a Hermitian metric is given, i.e., for which we have a real tensor $G_{A'A}$ which satisfies $G_{A'A}\bar{\psi}^{A'}\psi^A > 0$ for all $\psi^A \neq 0$. We define $G^{A'A}$ by

$$G^{A'A}G_{A'B} = \epsilon_B{}^A$$

and rescale ϵ_{AB} if necessary so that it satisfies

$$\epsilon^{AB}G_{A'A}G_{B'B} = \bar{\epsilon}_{A'B'} \quad .$$

We use ϵ^{AB} and ϵ_{AB} to raise and lower indices as before, and now we may use $G_{A'A}$ and $G^{A'A}$ to eliminate all primed indices in favor of unprimed indices. We define the $\dagger$ operation by

$$(\phi^\dagger)^{A\cdots}{}_{\cdots B} = G_{A'}{}^{A} \cdots G^{B'}{}_{B}\bar{\phi}^{A'\cdots}{}_{\cdots B'} \quad .$$

The group SU(2) is defined as the group of unitary maps on W with unit determinant, i.e., as the subgroup of SL(2,$\mathbb{C}$) consisting of maps $U^A{}_B$ satisfying the additional condition $(U^\dagger)^A{}_B = (U^{-1})^A{}_B$. Parallel the discussion of section 13.1 to show that (*a*) the two-index symmetric spinors $\phi^{AB} = \phi^{BA}$ which are self-adjoint, $(\phi^\dagger)^{AB} = \phi^{AB}$, form a three-dimensional, real vector space on which $\epsilon_{A_1A_2}\epsilon_{B_1B_2}$ is a negative definite metric and (*b*) SU(2) is the universal covering group of the rotation group SO(3).

We comment that if one is given a spacelike hypersurface in spacetime, one may use the normal vector $n_{A'A}$ at each point to define the Hermitian metric $G_{A'A} = \sqrt{2}\, n_{A'A}$. Hence, on the hypersurface one may associate the SL(2, $\mathbb{C}$) spinor space of spacetime with the SU(2) spinor space of the hypersurface. This enables one to obtain a "3 + 1 decomposition" of spinors (Sen 1982).

2. Choose a spinor basis o^A, ι^A satisying $o_A\iota^A = 1$ and write out the components of the Dirac equation (eqs. [13.1.60] and [13.1.61] for $n = 1$). Show, thereby, that our version of the Dirac equation is equivalent to the form found in most books.

3. According to equation (13.1.50), a real antisymmetric tensor F_{ab} can be written in the form

$$F_{AA'BB'} = \phi_{AB}\bar{\epsilon}_{A'B'} + \bar{\phi}_{A'B'}\epsilon_{AB} \quad ,$$

where $\phi_{AB} = \phi_{(AB)}$. Show that F_{ab} satisfies the source-free Maxwell equations (4.2.23) and (4.2.24) if and only if ϕ^{AB} satisfies equation (13.1.64).

4. Show that equation (13.1.64) has a well posed initial value formulation in Minkowski spacetime as follows:

a) Show that equation (13.1.64) implies $\Box\phi^{A_1\cdots A_n} = 0$.

b) Using theorem 10.1.2, show that if $\Box\phi^{A_1\cdots A_n} = 0$ throughout spacetime and equation (13.1.64) and its normal derivative $n^{BB'}\partial_{BB'}(\partial_{A_1A_1'}\phi^{A_1\cdots A_n}) = 0$ are satisfied on an initial hypersurface, then equation (13.1.64) is satisfied.

c) Show that if $\Box\phi^{A_1\cdots A_n} = 0$ and equation (13.1.64) holds initially, then $\partial_{BB'}\partial_{A_1A_1'}\phi^{A_1\cdots A_n} = 0$ automatically holds intitially. (Hint: Use the fact that $n_{[b}\partial_{c]}$ applied to eq. [13.1.64] must hold initially to conclude that initially $\partial_{BB'}\partial_{A_1A_1'}\phi^{A_1\cdots A_n}$ is of the form $n_{BB'}\alpha_{A_1'}{}^{A_2\cdots A_n}$. Show that an expression of this form can be nonvanishing for timelike $n^{BB'}$ if and only if it remains nonvanishing when contracted with $\bar{\epsilon}^{B'A_1'}$.)

The results (*a*), (*b*) and (*c*) show that equation (13.1.64) is equivalent to the equation $\Box\phi^{A_1\cdots A_n} = 0$ (which has a well posed initial value formulation) together with equation (13.1.64) holding only as an initial value constraint.

5. Show that the conditions (13.2.49)–(13.2.52) on principal spinors translate into the conditions of Table 7.1 on the corresponding principal null directions.

6. Show that in a vacuum spacetime, $R_{ab} = 0$, the Bianchi identity $\nabla_{[a}R_{bc]de} = 0$ takes the form $\nabla_{AA'}\Psi^{ABCD} = 0$. In particular, this shows that the linearized Einstein equation off Minkowski spacetime implies that the linearized Weyl spinor satisfies the equation for a massless, spin-2 field. [In fact, given a solution of eq. (13.1.64) in the case $s = 2$, one can find a metric perturbation γ_{ab} (unique up to gauge transformations) satisfying the source-free linearized Einstein equation, (4.4.11) and (4.4.12), whose linearized Weyl spinor equals the given solution. Thus, eq. (13.1.64) with $s = 2$ is equivalent to eqs. (4.4.11) and (4.4.12).]

7. Solve problem 6 of chapter 4 using the spinor decomposition of the Weyl tensor. Include a proof that $T_{abcd} = T_{(abcd)}$.

8. Show that equations (13.2.54) and (13.2.55) are equivalent to the equation $(\Box + m^2 - 6\Lambda)\phi^A = 0$. This equation is of the general form for which theorem 10.1.2 applies, so this establishes that the Dirac equation in curved spacetime has a well posed initial value formulation.

FOURTEEN

QUANTUM EFFECTS IN STRONG GRAVITATIONAL FIELDS

As described in chapter 4, the theory of general relativity put forth a revolutionary new viewpoint on spacetime structure and gravitation. However, in one important sense, this theory is not revolutionary enough. It is well established that all known physical fields must be described on a fundamental level by the principles of quantum theory. In quantum theory, states of a system are represented by vectors in a Hilbert space, $\mathcal{H}$, and observable quantities are represented by self-adjoint linear maps acting on $\mathcal{H}$. Unless the state of the system happens to be in an eigenstate of the observable, the observable will not have a definite value and one can predict only probabilities for the outcomes of measurements. However, general relativity is a purely classical theory, since in the framework of general relativity the observable quantities—in particular, the spacetime metric—always have definite values. Thus, if the principles of quantum theory are to apply to the gravitational field, general relativity must at best be only an approximation to a truly fundamental theory of gravity, perhaps in the same sort of way as Maxwell's theory of electromagnetism is only an approximation to quantum electrodynamics.

Classical descriptions of ordinary matter are, in general, excellent approximations for describing phenomena which occur on macroscopic scales, although even here quantum effects can be important for suitably prepared states. However, the classical description of matter becomes wholly inadequate on atomic and smaller scales. In this case, the scale at which the classical description breaks down is determined by the masses and charges of the fundamental particles as well as the two fundamental constants of nature which enter the theory, namely Planck's constant, $\hbar$, and the speed of light, c. Similarly, in a quantum theory of gravitation based on general relativity, one would expect that the fundamental scale at which the classical description becomes wholly inadequate should be set by $\hbar$, c, and the gravitational constant, G. There is a unique combination of these constants which has the dimensions of length, namely the quantity $l_P \equiv (G\hbar/c^3)^{1/2}$, called the *Planck length*. As might be expected, the Planck length arises naturally in attempts to formulate a quantum theory of gravity. Thus, dimensional arguments suggest that a classical description of spacetime structure should break down at scales of the order of the Planck length and smaller.

In cgs units the magnitude of the Planck length is only $\sim 10^{-33}$ cm. (The corresponding Planck scales of other quantities such as time and energy are given in appendix F.) The smallness of l_P compared with typical length scales occurring in atomic, nuclear, and elementary particle physics is directly related to the weakness of the gravitational force between two elementary particles as compared with the other fundamental forces, namely the strong, weak, and electromagnetic interactions. The Planck scales lie many orders of magnitude beyond what we presently are able to probe with high energy accelerators. Therefore, it might appear that the development of a quantum theory of gravity—while undoubtedly a laudable goal—would be unlikely to have much relevance to any presently observable phenomena.

However, there are at least two good reasons for believing that the predictions of a quantum theory of gravity could be very relevant to phenomena at presently observable scales. The first reason arises from the theory of elementary particles. It is widely believed that the true, fundamental theory of nature will achieve a unification of the forces of nature by describing them simply as different aspects of a single entity. A unification of the classical theories of electricity and magnetism was achieved over a century ago by Maxwell. More recently, the Weinberg-Salam theory has given a successful unified description of the weak and electromagnetic interactions. It is presently believed that the "grand unified models" may successfully unify the strong and electroweak interactions. The unification of gravitation with the strong-electroweak interaction would be the next logical step in this program. Interestingly, the natural length scale which arises in the grand unified theories is only a few orders of magnitude larger than l_P. Thus, it is quite possible that a quantum theory of gravitation may even play an important role in the unification of the strong and electroweak interactions. A unified theory of all forces undoubtedly would yield many new predictions of phenomena at presently observable scales.

The second reason arises directly from general relativity. As discussed in chapter 9, spacetime singularities occur in the solutions of classical general relativity relevant to gravitational collapse and cosmology. Thus, in these situations, the classical description of spacetime structure must break down. In particular, one cannot expect the homogeneous, isotropic models of chapter 5 to be an adequate description of our universe in the regime where they predict curvature of magnitude l_P^{-2} or greater, i.e., for $t < t_P \equiv (G\hbar/c^5)^{1/2} \sim 10^{-43}$ s. Thus, it appears that the development of a quantum theory of gravitation will be an essential requirement for our understanding of the initial state of our universe. It is not implausible that phenomena which occur in the very early universe and which can be understood only in the framework of quantum gravity will lead to observationally verifiable predictions about the structure of the present universe.

However, even if quantum gravity leads to no predictions of phenomena which can be observed with present day technology, the formulation of a quantum theory of gravity undoubtedly would be of major significance for theoretical physics. After all, classical general relativity has provided us with major new insights into the workings of nature even though the new observationally verifiable predictions it makes are relatively meager. The new fundamental insights provided bv a quantum theory of gravity certainly should be no less significant than those provided by general relativity.

As discussed in section 14.1, all of the known procedures for formulating a quantum field theory associated with a classical theory run into difficulties when applied to general relativity. Thus, the formulation of a viable quantum theory of gravity remains a goal for the future. However, a completely satisfactory theory exists for a free (i.e., linear) quantum matter field propagating in a fixed background curved spacetime. Although such a theory is, at best, only an approximation to a full quantum theory of gravity with quantum matter, the effects thereby predicted should at least give a good indication of the types of quantum effects which may occur in strong gravitational fields. In particular, as described in section 14.2, the creation of particles by a gravitational field is predicted by this theory. Remarkably, when applied to the case of a black hole in section 14.3, one finds that particle creation causes the effective "emission" by a black hole of a thermal spectrum of particles at temperature $kT = \hbar\kappa/2\pi$, where κ is the surface gravity of the black hole. The implications of this result for the relationship between black holes and thermodynamics are explored in section 14.4.

14.1 Quantum Gravity

It is generally believed that the correct, fundamental description of all physical fields is given by the general framework of quantum field theory. In quantum field theory, states of a system are described by vectors in a Hilbert space $\mathcal{H}$, and the physical field is described by an operator (i.e., a linear map) on $\mathcal{H}$ defined at each spacetime point. However, unlike ordinary, nonrelativistic Schrödinger quantum mechanics where well defined—albeit probabilistic—predictions always can be made once one is given the Hamiltonian of the system, serious difficulties arise when one attempts to formulate quantum field theories. Many of these difficulties can be traced to the fact that, even for a free field, the expression obtained for the field operator does not make mathematical sense as an operator defined at each spacetime point but must be interpreted as a distribution on spacetime (see section 14.2 below). This corresponds to the physical fact that the field cannot be measured at a single point; only averages of the field over spacetime regions are physically well defined. In the case of a free field this causes no serious problems and a well defined quantum field theory can be constructed. However, for the more interesting case of a theory with interactions (i.e., for a field or fields satisfying nonlinear equations) one is led unavoidably to consider the products of field operators at the same spacetime point. Such quantities have no natural mathematical meaning, and consequently, with the exception of some simple models in lower spacetime dimensions, there are at present no known examples of quantum field theories of physically reasonable interacting fields which are mathematically well defined.

However, although one does not know how to formulate an exact theory of an interacting quantum field, one can formally treat the interactions as perturbations of the well defined free field theory. One thereby may obtain a formal expansion for physical quantities as power series in the "coupling constant" (i.e., the coefficient of the nonlinear interaction term in the equation). However, in general the formal expression for the individual terms in the power series will yield divergent results when one attempts to evaluate them, as should be expected from the fact that the exact theory upon which the perturbation series is based is ill defined. Nevertheless, one can introduce cutoffs into the divergent expressions in order to obtain finite answers. This, of course, leaves the theory in an unsatisfactory state, since the predicted values of all physical quantities depend upon the chosen values of the cutoff parameters. However, it may happen that in the limit of large cutoff parameters, the dependence of all physical quantities on the cutoff parameters may be identical to their dependence on the so-called bare parameters (such as masses, charges, and the coupling constants) which originally were present in the theory. If this occurs, then one can take the limit as the cutoff parameters go to infinity while readjusting the bare parameters so as to produce the desired finite answers for certain physical quantities. In this way, one obtains finite expressions for each term in the perturbation series expansions for all physical quantities with only the same number of free parameters in the theory as originally were present classically. If this occurs, the theory is called *renormalizable,* and it is widely believed that physically viable quantum field theories must be renormalizable (or, at least, satisfy properties closely akin to renormalizability; see Weinberg 1979). Quantum electrodynamics (i.e., the quantum field theory of a Dirac field interacting with an electromagnetic field) is renormalizable, and the agreement to high accuracy of the predictions of the first few terms of its perturbation series with experiments provides the best quantitative evidence for the belief that quantum field theory provides a correct description of nature. The Weinberg-Salam theory of the electroweak interactions and "quantum chromodynamics" (i.e., the theory of quarks interacting with gluons) also are renormalizable quantum field theories which are believed to accurately describe nature. However, the state of affairs with regard to giving an exact formulation even of renormalizable field theories is unsatisfactory. Only the individual terms in the perturbation series are defined precisely, and there is good reason to believe that the full perturbation series does not converge.

Given a classical field theory formulated in terms of a Lagrangian or Hamiltonian, there exist a number of procedures for formulating a quantum field theory associated with the classical theory. However, general relativity is sufficiently different from other classical field theories that, as we shall see in more detail below, the approaches which have been tried for formulating quantum general relativity all have encountered fundamental difficulties. The essential difference between general relativity and other classical theories appears to be the dual role played by the field g_{ab} as both the quantity which describes the dynamical aspects of gravity and the quantity which describes the background spacetime structure. Thus, it would appear that in order to "quantize" the dynamical degrees of freedom of the gravitational field, one must also give a quantum mechanical description of spacetime structure. This latter problem has no analog for other quantum field theories which are formulated on a fixed, background spacetime, which is treated classically.

The nature of the difficulties caused by the dual role of g_{ab} are perhaps best illustrated by the following simple example. It is a fundamental property of quantum field theory in Minkowski spacetime that the field operator, $\hat{\psi}$, corresponding to an integer-spin classical field, ψ, evaluated at spacelike related points must commute with itself, i.e., for x and x' spacelike related[1] we have

$$[\hat{\psi}(x), \hat{\psi}(x')] \equiv \hat{\psi}(x)\hat{\psi}(x') - \hat{\psi}(x')\hat{\psi}(x) = 0 \quad . \tag{14.1.1}$$

(This equation expresses the fact that a measurement of ψ at x' cannot influence the value of ψ at x.) Now, as mentioned in section 4.4 and in problem 6 of chapter 13, linearized gravity is just the theory of a massless, spin-2 field in Minkowski spacetime. Thus, we may view general relativity as the theory of a self-interacting spin-2 field. By analogy with flat spacetime quantum field theories it would be natural to expect the metric field operator $\hat{g}_{ab}$ to satisfy the commutation relation

$$[\hat{g}_{ab}(x), \hat{g}_{cd}(x')] = 0 \tag{14.1.2}$$

for x and x' spacelike related. However, this equation makes no sense since we do not know if x and x' are spacelike related until we know the metric, and equation (14.1.2) is an operator equation which, if valid, must hold independently of the state of the gravitational field, i.e., independently of the value of (or probability distribution for) the metric. More generally, the entire notion of causality becomes ill defined when the notion of a classical spacetime metric is abandoned. Thus, some of the fundamental results which are believed to hold for all other quantum field theories appear to be inapplicable to general relativity.

The differences between general relativity and other field theories and the difficulties with causality that arise when the spacetime metric does not have a definite value suggest the possibility that perhaps the principles of quantum theory do *not* apply to gravity—that classical general relativity is correct at the fundamental level. However, this viewpoint appears to be untenable, because the spacetime metric is coupled to matter sources. Suppose spacetime structure is described by a classical spacetime (M, g_{ab}) and quantum theory applies to these matter sources. What is the curvature of spacetime associated with a given quantum state of the matter? If the classical Einstein equation is to hold in the limit where the matter distribution can be described classically, the most natural candidate for a quantum version of Einstein's equation (with gravity treated classically), is[2]

$$G_{ab} = 8\pi\langle\hat{T}_{ab}\rangle \quad , \tag{14.1.3}$$

where $\langle\hat{T}_{ab}\rangle$ denotes the expectation value of the stress-energy operator $\hat{T}_{ab}$ in the given quantum state. Now consider a state of matter where, with probability $1/2$, all the matter is located in a certain region, O_1, of spacetime and, with probability $1/2$, the matter is located in a region O_2 disjoint from O_1. According to equation (14.1.3), the gravitational field will behave like half of the matter is in O_1 and the other half is in O_2. Suppose, now, that we make a measurement of the location of the matter. We then will find the matter to be either entirely in O_1 or entirely in O_2. If equation (14.1.3) continues to hold after we have resolved the quantum state of the matter by

1. As mentioned above, we must in fact treat $\hat{\psi}$ as a distribution. Strictly speaking, equation (14.1.1) should be replaced by $[\hat{\psi}(f), \hat{\psi}(g)] = 0$ whenever the supports of the test functions f and g are spacelike separated.

2. Note that in postulating a semiclassical equation like (14.1.3), the superposition principle for matter states is lost, since different matter states are associated with different spacetimes.

this measurement, then the gravitational field must change in a discontinuous, acausal manner. Thus, the attempt to treat gravity classically leads to serious difficulties. These difficulties apparently can be avoided only by treating the spacetime metric in a probabilistic fashion—i.e., by quantizing the gravitational field—so that in the initial state it has probability 1/2 of corresponding to the gravitational field of matter in O_1, and has probability 1/2 of corresponding to the gravitational field of matter in O_2.

The issue of how to formulate a quantum theory of gravitation is presently under active investigation by many researchers. We shall confine our discussion here to a very brief mention of some of the main approaches that have been tried. A more detailed discussion of these approaches as well as a much more complete description of the range of topics related to quantum gravity which presently are under investigation can be found in Isham, Penrose, and Sciama (1975, 1981) and the final five chapters of Hawking and Israel (1979).

The *covariant perturbation method* is perhaps the most straightforward approach to formulating a quantum theory of gravity. Here one writes the spacetime metric g_{ab} as

$$g_{ab} = \eta_{ab} + \gamma_{ab} \quad , \tag{14.1.4}$$

where η_{ab} is a flat metric and we shall assume that $M = \mathbb{R}^4$ so that (M, η_{ab}) is Minkowski spacetime. [More generally, one could replace $(\mathbb{R}^4, \eta_{ab})$ by any solution, $(M, {}^0g_{ab})$, of Einstein's equation.] To first order in γ_{ab}, the classical Einstein equation is just the equation for a free, spin-2 field. Therefore, as already mentioned above, we may view the full Einstein equation (with γ_{ab} *not* assumed to be "small") as the sum of this "free" piece plus a nonlinear, self-interaction term; i.e., we may view Einstein's equation as an equation for a self-interacting spin-2 field γ_{ab} in Minkowski spacetime $(\mathbb{R}^4, \eta_{ab})$. The covariant method treats gravity by viewing it in this manner as an ordinary "Poincaré covariant" field theory. The dynamical variable, γ_{ab}, has considerable gauge arbitrariness, but it is known how to obtain a perturbation series expansion for the quantum field theory of "non-abelian" gauge fields of this type (Fadde'ev and Popov 1967; DeWitt 1967*a*,*b*). Thus, in this approach there is no obstacle, in principle, to obtaining a formal perturbation series for the type of physical quantities usually calculated for field theories in Minkowski spacetime. Note that the flat, background metric η_{ab} introduced in equation (14.1.4) is treated in an entirely classical manner in this approach.

The main difficulty which arises in this approach is that the perturbation theory one obtains for γ_{ab} is non-renormalizable. Indeed, the non-renormalizability of quantum gravity in the covariant perturbation approach can be seen purely from dimensional arguments. The expansion parameter in the perturbation series one obtains is the squared Planck length,[3] l_P^2. Thus, the terms in the perturbation series of successively higher order in l_P^2 must have the dimensions of correspondingly higher powers of inverse length and the "superficial degree of divergence" (see, e.g., Coleman 1973)

3. The classical vacuum Einstein equation, $G_{ab} = 0$, does not involve Newton's constant G. However, G enters the quantum theory through the normalization of γ_{ab}, which accounts for the presence of G (via l_P) in the quantum theory.

of these terms increases. Hence, there is no possibility of canceling the divergent terms with the "bare" terms. Thus, the only hope for obtaining a well defined perturbation series without the introduction of new parameters is that the perturbation series be *finite* in each order, i.e., that when all the contributions to order l_P^{2n} are added together, the divergences will cancel each other. In fact, this happens in the lowest ("one loop") order. However, finiteness does *not* occur in lowest order for gravity coupled to matter fields, nor is it expected to occur in higher orders in pure quantum gravity; see Deser, van Nieuwenhuizen, and Boulware (1975) and the references cited therein for further discussion.

Thus, in the covariant perturbation approach to formulating a quantum theory of gravity, it appears that meaningful physical predictions cannot be made. In addition, this approach has a number of other unappealing features. The breakup of the metric into a background metric which is treated classically and a dynamical field γ_{ab}, which is quantized, is unnatural from the viewpoint of classical general relativity. Furthermore, the perturbation theory one obtains from this approach will, in each order, satisfy causality conditions with respect to the background metric η_{ab} rather than the true metric g_{ab}. Although the summed series (if it were to converge) still could satisfy appropriate causality conditions, the covariant perturbation approach would provide a very awkward way of displaying the role of the spacetime metric in causal structure. Finally, in this approach it is very difficult even to formulate questions about such issues as the quantum effects occurring near the initial singularity of the universe, since the usual procedures for formulating quantum field theories in Minkowski spacetime effectively assume that the interactions are "turned off" in the distant past. On the other hand, the covariant approach has the advantage that one can obtain concrete expressions for physical quantities in perturbation theory without having to develop an entirely new conceptual framework.

An important approach which does not rely on a breakup of the spacetime metric such as (14.1.4) is the *canonical quantization method*. This approach for formulating a quantum theory is applicable if the classical theory has been put in Hamiltonian form. The basic idea here is (i) to take the states of the system to be described by wave functions $\Psi(q)$ of the configuration variables, (ii) to replace each momentum variable by differentiation with respect to the conjugate configuration variable, and (iii) to determine the time evolution of Ψ via the Schrödinger equation, $i\hbar\, \partial\Psi/\partial t = \hat{H}\Psi$, where $\hat{H}$ is an operator corresponding to the classical Hamiltonian $H(p,q)$. A precise formulation of these rules for simple quantum mechanical systems can be found in Ashtekar and Geroch (1974).

As discussed in appendix E, general relativity can be cast in Hamiltonian form, so one can attempt to apply the canonical quantization rules to general relativity. However, a serious difficulty arises because of the presence of the constraint (E.2.33). Attempts to solve this constraint (so that one can obtain configuration variables representing only the "true dynamical degrees of freedom") or to impose this constraint as an additional condition on the state vector have not been successful. Thus, the difficulties caused by the constraint (E.2.33) in the Hamiltonian formulation of general relativity have proven to be a serious obstacle to obtaining a quantum theory of gravity via the canonical approach. We refer the reader to Ashtekar and Geroch (1974) and Kuchař (1981) for further discussion.

Another important approach to formulating a quantum theory of gravity is the *path integral method*. The viewpoint here is to stress the amplitudes for physical processes (rather than states or operators) as the fundamental entities of the theory. For the quantum theory of a field ψ on Minkowski spacetime derived from an action $S[\psi]$ (see appendix E), one expresses the amplitude for the field to go from the function $\psi_1(\vec{x})$ on the hypersurface $t = t_1$ to the function $\psi_2(\vec{x})$ on the hypersurface $t = t_2$ as a path integral:

$$\langle \psi_1, t_1 \,|\, \psi_2, t_2 \rangle = \int e^{iS[\psi]} d\mu[\psi] \quad . \tag{14.1.5}$$

Here, the integral is to be taken over *all* field configurations ψ (not just those satisfying the classical field equation) in the spacetime region between t_1 and t_2 which "interpolate" between ψ_1 and ψ_2, and $d\mu[\psi]$ denotes a measure on the space of field configurations. The major difficulty which arises in the path integral approach is the definition of the measure $d\mu[\psi]$. One simply does not know how to make mathematical sense of $d\mu[\psi]$, except in the context of perturbation theory about a free field. (In that context, one can define the integral [14.1.5] so as to obtain the standard perturbation series, and, indeed, the path integral approach provides a simple formal derivation of this series which is particularly useful in the case of gauge theories; see Abers and Lee 1973.) However, despite the inability to define (14.1.5) rigorously, the formal manipulations suggested by the path integral viewpoint have provided valuable insights[4] and useful approximation schemes.

General relativity can be derived from an action principle (see appendix E), so one can attempt to formulate a quantum theory of gravity by the path integral approach. It would be natural to write down an integral of the form

$$Z = \int e^{iS[g_{ab}]} d\mu[g_{ab}] \quad , \tag{14.1.6}$$

with S given by equation (E.1.13) or equation (E.1.42), to represent the amplitude, $\langle (h_1)_{ab}, t_1 \,|\, (h_2)_{ab}, t_2 \rangle$, for going from spatial metric $(h_1)_{ab}$ at time t_1 to $(h_2)_{ab}$ at time t_2. However, in general relativity the "times" t_1 and t_2 are merely coordinate labels and do not have physical significance. Thus, the amplitude $\langle (h_1)_{ab}, t_1 \,|\, (h_2)_{ab}, t_2 \rangle$ is not a physically meaningful quantity. The underlying reason for this difficulty is closely related to the difficulty of the canonical quantization approach: The "true dynamical degrees of freedom" of the gravitational field have not been isolated. As mentioned near the end of appendix E, general relativity is much like a "parameterized theory" where time is treated as a dynamical variable. Thus, it is redundant to insert the

4. In particular, the path integral viewpoint provides a simple explanation of how the classical limit arises. Namely, in appropriate circumstances the dominant contributions to the integral (14.1.5) should arise from field configurations where the phase factor $S[\psi]$ is extremized, since then the nearby configurations will add coherently. But these field configurations are precisely the solutions of the classical equations (see appendix E).

"label times" t_1 and t_2 in the amplitudes, as, in effect, an "intrinsic time" already is present in the canonical variables h_{ab}, π^{ab} describing the gravitational field. However, since one does not have a well defined decomposition of these variables into "true dynamical degrees of freedom" and "intrinsic time variables," it is far from clear precisely what physical amplitude Z is supposed to represent. Furthermore, another fundamental difficulty which arises is the usual problem in the path integral approach of giving precise meaning to $d\mu[g_{ab}]$ so that the right-hand side of equation (14.1.6) is well defined. On the other hand, in the path integral approach one can envision analyzing issues such as the probability for a change of spatial topology by including in the integral (14.1.6) spacetime metrics for which such a spatial topology change occurs. It is difficult to see how this issue could even be formulated in the canonical approach (since the presence of a Hamiltonian requires, in essence, global hyperbolicity, and hence no change of spatial topology) or in the covariant approach.

An important variant of the above path integral approach is the *Euclidean path integral approach*. In field theories in Minkowski spacetime, a number of quantities which arise—in particular the vacuum expectation values of products of field operators—are holomorphic functions of the global inertial coordinates t, x, y, z in a domain that includes negative imaginary values of the time coordinate, i.e., $t = -i\tau$, where τ is real and positive (see Streater and Wightman 1964). It is useful to view this type of analytic continuation in the following manner. We define *complexified Minkowski spacetime* to be the four-complex-dimensional manifold $\mathbb{C}^4$ (which also may be viewed as an eight-real-dimensional manifold) with complex metric η_{ab} defined in terms of the complex Cartesian coordinates t, x, y, z of $\mathbb{C}^4$ by

$$ds^2 = -dt^2 + dx^2 + dy^2 + dz^2 \quad . \tag{14.1.7}$$

Thus by restricting to real values of t, x, y, z, we recover ordinary Minkowski spacetime as a four-real-dimensional submanifold of $(\mathbb{C}^4, \eta_{ab})$. However, by restricting x, y, z to be real but t to be pure imaginary, we obtain another four-real-dimensional submanifold, now with a real, Euclidean metric

$$ds^2 = +d\tau^2 + dx^2 + dy^2 + dz^2 \tag{14.1.8}$$

where $\tau = it$. This submanifold is referred to as a *Euclidean section* of $(\mathbb{C}^4, \eta_{ab})$. Thus, we may view the above analytic continuation of functions to negative imaginary values of t as the evaluation of these functions on the positive τ region of a Euclidean section of complexified Minkowski spacetime. We may perform our analysis of the field theory in this Euclidean section and then analytically continue the functions back to the Minkowskian section to obtain the physical predictions. In the path integral approach, there is considerable potential advantage to proceeding in this manner, since in many theories the Euclidean action, $S_E[\psi] \equiv -iS[\psi]|_{t=-i\tau}$ is positive definite and the analytically continued integral is exponentially damped at "large ψ" rather than oscillatory. Thus, the possibility of making mathematical sense of the integral in equation (14.1.5) appears to be greatly enhanced in the Euclidean section.

Many difficulties arise when one attempts to apply the Euclidean path integral approach to quantum gravity. In general relativity, one does not have a natural background flat spacetime $(\mathbb{R}^4, \eta_{ab})$ which can be "complexified" to allow analytic continuation to be performed. Nevertheless, one could try to analytically continue Lorentzian signature metrics to Riemannian metrics and work with a path integral over Riemannian metrics. However, except in special cases such as static spacetimes, it is generally impossible to represent an analytic spacetime (M, g_{ab}) as a "Lorentzian section" of a four-complex-dimensional manifold with complex metric which possesses a "Euclidean section," i.e., a four-real-dimensional submanifold with real, Riemannian metric. Thus, one does not have a general prescription for analytically continuing Lorentz signature metrics to Riemannian metrics. (Furthermore, even if one did, one does not have any theorems guaranteeing the analyticity of any quantities arising in quantum gravity.) Nevertheless, one can postpone the resolution of interpretational issues and study the properties of the path integral (14.1.6) over Riemannian metrics, with $iS[g_{ab}]$ replaced by $-S_E[g_{ab}]$, where S_E is given by the analog of equation (E.1.42) for Riemannian metrics.[5] A number of highly suggestive results have been obtained in this manner (Hawking 1979). Perhaps the most dramatic success of the Euclidean approach is that, as explained in section 14.3, the creation of particles by a Schwarzschild black hole can be related in a direct and simple manner to properties of the Euclidean Schwarzschild solution.

A considerably more unconventional approach toward the formulation of a quantum theory of gravity is provided by the *twistor approach*. A *twistor* in Minkowski spacetime may be defined as a pair,

$$Z = (\omega^A, \pi_{A'}) \quad , \tag{14.1.9}$$

consisting of a spinor field, ω^A, and a complex conjugate spinor field $\pi_{A'}$ satisfying the twistor equation,

$$\partial_{AA'} \omega^B = -i\, \epsilon_A{}^B \pi_{A'} \quad . \tag{14.1.10}$$

We refer the reader to such references as Penrose (1967), Penrose and MacCallum (1972), Penrose (1975), and Penrose and Ward (1980) for the motivation for introducing twistors and a discussion of their properties. The collection of twistors on Minkowski spacetime forms a four-complex-dimensional vector space, and the projective twistors—i.e., the equivalence classes of twistors which differ by a nonzero complex multiple—comprise the three-complex-dimensional manifold $\mathbb{C}P^3$. There exist a number of natural correspondences between Minkowski spacetime and twistor space. For example, a null geodesic in Minkowski spacetime corresponds to a point in null projective twistor space (i.e., the space of projective twistors satisfying $Z \cdot Z \equiv \omega^A \overline{\pi}_A + \overline{\omega}^{A'} \pi_{A'} = 0$), whereas a point in Minkowski spacetime corresponds to a 2-sphere of null projective twistors. Further correspondences between compactified, complexified Minkowski spacetime and twistor space also may be obtained. Furthermore, certain differential equations in Minkowski spacetime can be reformulated as analyticity conditions in twistor space (see Ward 1981).

5. The action S_E is not positive definite. However, it is positive definite for asymptotically Euclidean metrics with $R = 0$ (Schoen and Yau 1979), and Hawking (1979) has proposed a further analytic continuation of the "conformal degree of freedom" of the metric which, in effect, makes S_E positive definite.

The basic starting point of the twistor quantization approach is to use twistor space rather than spacetime as the underlying classical manifold structure upon which the quantum fields are defined. In view of the correspondences between spacetime and twistor space, this should result in a quantum theory where, roughly speaking, certain causal relationships in spacetime retain their classical properties, but the notion of spacetime points becomes "fuzzy." (In essentially all other approaches, the notion of spacetime points is well defined but, as discussed above, since the spacetime metric is described probabilistically, the notion of causality becomes "fuzzy.") Hence, in the twistor approach one can envision incorporating causality into the quantum theory in a natural way. Thus far, twistors have proven to be useful mathematical tools (see e.g., Atiyah and Ward 1977). However, at present, little progress has been made toward formulating a concrete quantum theory of gravity via the twistor approach.

Given the present lack of success of the above approaches to formulating a quantum version of general relativity, it is natural to consider modifications of classical general relativity which may lead to better behavior of the quantum theory. In particular, one may seek to modify the Einstein field equation so that the quantum theory becomes renormalizable in the covariant perturbation approach. Perhaps the simplest attempt along these lines is to modify the Einstein Lagrangian (E.1.12) by adding terms quadratic in the curvature. The coordinate component form of the new field equation then involves fourth derivatives of the metric, so this theory often is referred to as a "higher derivative" theory of gravity. Stelle (1977) has shown that the quantum version of this theory is formally renormalizable. However, other serious difficulties arise in the theory, so it does not appear that this theory is physically viable.

A much more ambitious attempt to modify general relativity is given by *supergravity theories*. Supergravity theories are a class of models involving interacting fields of spin 0, 1/2, 1, 3/2, and 2. (The prefix "super" refers to the fact that these models have a certain type of symmetry called "supersymmetry" between the boson [i.e., integer spin] and fermion [i.e., half-integer spin] degrees of freedom.) The main goals of supergravity theories are (i) to improve the renormalizability (or finiteness) properties of quantum gravity and (ii) to unify gravity with the other basic interactions of nature. With regard to the first goal, it has been shown that supergravity is finite at least to "two loop" order in perturbation theory. A present, it is not known how far in perturbation theory this finiteness extends, and it is believed to be possible that supergravity could be finite to all orders. The second goal is particularly ambitious since one asks supergravity to account for all interactions observed in high energy particle physics. At present, there appear to be serious difficulties in reaching this goal. We refer the reader to van Nieuwenhuizen (1981) and the articles in Hawking and Roček (1981) for an introduction to supergravity, a discussion of some recent research, and references to earlier work on supergravity and supersymmetry.

In summary, although many approaches have been tried, there presently does not exist a demonstrably viable quantum theory of gravity. It is possible that the difficulties are basically technical in nature and that, for example, a better procedure for dealing with the constraint (E.2.33) will lead to a satisfactory formulation of quantum gravity via the canonical approach, or that supergravity will be shown to be a fully satisfactory theory. Alternatively, it is possible that the difficulties may be of a very fundamental nature, and that the quantum theory of gravity simply does not fit into the framework established for other quantum theories.

However, the lack of a satisfactory quantum theory of gravity does not mean that we cannot perform any reliable calculations of quantum effects occurring in strong gravitational fields. In atomic physics, in appropriate circumstances one can reliably calculate electromagnetically induced transition rates of electrons in atoms using a classical treatment of the electromagnetic field. Similarly, in quantum field theory, using a classical treatment of the electromagnetic field one can, in appropriate circumstances, reliably calculate the spontaneous creation of electron-positron pairs in a strong electric field. Thus, it is expected that by treating gravity in the classical framework of general relativity, it still should be possible to reliably calculate some of the quantum effects that gravity produces on other fields, such as the creation of particle-antiparticle pairs. It is to the study of these effects that we now turn.

14.2 Quantum Fields in Curved Spacetime

As indicated at the end of the previous section, in analogy with quantum field theory in an external potential (see, e.g., Wightman 1971), we seek to formulate a theory of a quantum field which propagates in a classically describable spacetime (M, g_{ab}). For simplicity, we will restrict attention to the case of a real Klein-Gordon scalar field ϕ. The analysis of other linear fields of spin $s \leq 1$ is very similar (see, e.g., Wald 1979*b*), although there are important, well known differences which occur in the fermion ($s = 1/2$) case. As mentioned at the end of chapter 13, fields of spin $s > 1$ do not have a natural generalization to curved spacetime. We shall not consider nonlinear (i.e., self-interacting) fields.[6] Much of the discussion below is based on Wald (1975, 1979*b*). Further discussion of many of the effects predicted by the theory of quantum fields in curved spacetime and a detailed bibliography of the original papers is given by Birrell and Davies (1982).

As has been indicated several times above, Hilbert spaces play a fundamental role in quantum theory. For the benefit of the reader who is not familiar with Hilbert spaces but has studied chapter 2 and the discussion of spinor space given in chapter 13, we give a very brief introduction to Hilbert spaces here, emphasizing some of the differences between the finite-dimensional case which was considered previously and the infinite-dimensional case which is relevant here. For a much more complete and systematic discussion, we refer the reader to Riesz and Sz.-Nagy (1955) and Reed and Simon (1972).

Let V be a (not necessarily finite-dimensional) vector space over the complex numbers, $\mathbb{C}$. An *inner product* on V is a map $i: V \times V \to \mathbb{C}$—where we denote the complex number $i(v_1, v_2)$ as simply (v_1, v_2)—satisying the following three proper-

6. Even in Minkowski spacetime the quantum theory of nonlinear fields is well defined only in the context of perturbation theory. An interesting issue which arises is whether a renormalizable theory in Minkowski spacetime remains renormalizable in curved spacetime. See Birrell (1981) for a review of work on this issue.

ties: (*a*) i is linear in the second variable, (*b*) $\overline{(v_1, v_2)} = (v_2, v_1)$, where the bar denotes complex conjugation, and (*c*) $(v, v) \geq 0$ with equality holding if and only if $v = 0$. Note that properties (*a*) and (*b*) imply that i is antilinear in the first variable. A vector space equipped with an inner product is called an *inner product space*. If V is an inner product space, we define the *norm* of each $v \in V$ by $\|v\| = \sqrt{(v, v)}$. We obtain a natural topology on V (called the *strong* topology) by defining a subset of V to be open if and only if it can be expressed as a union of "open balls," i.e., sets of the form $B_{R,v_0} = \{v \in V \mid \|v - v_0\| < R\}$. This choice of topology will be understood in discussions of convergence and continuity below.

A sequence $\{v_n\}$ of vectors in V is said to be a *Cauchy sequence* if given $\epsilon > 0$ there exists an integer N such that for all m, $n > N$ we have $\|v_n - v_m\| < \epsilon$. It follows directly that all convergent sequences are Cauchy sequences. Conversely, for finite-dimensional inner product spaces, all Cauchy sequences converge. However, in infinite dimensions, it is easy to construct examples of inner product spaces which possess Cauchy sequences which do not converge (see problem 1). A space where all Cauchy sequences converge is said to be *complete,* and a complete inner product space is called a *Hilbert space*. Thus, in particular, all finite-dimensional inner product spaces are Hilbert spaces. If an infinite-dimensional inner product space, V, fails to be complete there exists a standard procedure for constructing a unique Hilbert space $\mathcal{H}$—called the Hilbert space completion of V—such that V is isomorphic to a subspace $W \subset \mathcal{H}$ with $\overline{W} = \mathcal{H}$ (see, e.g., Reed and Simon 1972) where $\overline{W}$ denotes the closure of W.

An important difference between finite and infinite-dimensional Hilbert spaces is that in the infinite-dimensional case linear maps need not be continuous. Indeed, it is straightforward to show that a linear map $A: \mathcal{H}_1 \to \mathcal{H}_2$ between two Hilbert spaces is continuous if and only if it is bounded, i.e., there exists a $C \in \mathbb{R}$ such that $\|Av\| \leq C\|v\|$ for all $v \in \mathcal{H}_1$. It is easy to construct examples of unbounded linear maps when $\mathcal{H}_1$ is infinite-dimensional.

In order to get a space with properties similar to the dual of a finite-dimensional vector space, we define the *dual,* $\mathcal{H}^*$, of a Hilbert space, $\mathcal{H}$, to be the vector space of *continuous* linear maps from $\mathcal{H}$ into $\mathbb{C}$. Similarly, the complex conjugate dual space $\overline{\mathcal{H}}^*$, and complex conjugate space $\overline{\mathcal{H}}$ are defined as in our discussion of spinor space in chapter 13, with the additional proviso that the antilinear and linear maps involved be continuous.

Now, for a general topological vector space V (i.e., a vector space with a topology defined on it such that the operations of addition and scalar multiplication are continuous) the argument which in finite dimensions proves that V is naturally isomorphic to all of V^{**} now shows only that V is naturally isomorphic to a subspace of V^{**}, i.e., $V \subset V^{**}$. Similarly, in general, complex conjugation maps V only into a subspace of $\overline{V}$. (Note, however, that the natural antilinear correspondence between V^* and $\overline{V}^*$—which associates $\alpha \in V^*$ with $\beta \in \overline{V}^*$ via $\alpha(v) = \overline{\beta(v)}$ for all $v \in V$—always is one-to-one and onto.) However, for a Hilbert space $\mathcal{H}$, although the simple proof used in finite dimensions breaks down, it turns out that the inner product—which is essentially a nondegenerate tensor of type $(0, 1; 0, 1)$ over $\mathcal{H}$—yields a one-to-one, onto, linear correspondence between $\overline{\mathcal{H}}$ and $\mathcal{H}^*$. This result is known as the *Riesz lemma*. Thus, for a Hilbert space $\mathcal{H}$, it follows that $\mathcal{H}^*$ is naturally isomorphic to $\overline{\mathcal{H}}$, that $\mathcal{H}$ and $\mathcal{H}^{**}$ are naturally isomorphic, and that the antilinear correspondence between $\mathcal{H}$ and $\overline{\mathcal{H}}$ is one-to-one and onto.

The *span* of a collection of vectors $\{v_\alpha\}$ in a Hilbert space $\mathcal{H}$ is defined to be the subspace of vectors that can be expressed as linear combinations of *finitely* many of the $\{v_\alpha\}$. The collection $\{v_\alpha\}$ is said to be a *basis* of $\mathcal{H}$ if the closure of the span of $\{v_\alpha\}$ equals $\mathcal{H}$, but the closure of the span of any proper subset of the $\{v_\alpha\}$ fails to equal $\mathcal{H}$. It can be shown that every Hilbert space admits an orthonormal basis $\{e_\alpha\}$ (see, e.g., Reed and Simon 1972). In general, $\{e_\alpha\}$ may consist of uncountably many elements, but in the case of a *separable* Hilbert space—i.e., a Hilbert space which possesses a countable subset of vectors whose closure is $\mathcal{H}$—then $\{e_\alpha\}$ must be countable. It is generally assumed that the Hilbert spaces arising in quantum theory are separable, and we will restrict attention to separable Hilbert spaces below.

A further difference in infinite dimensions occurs in the definition of tensor products. As in finite dimensions, $\mathcal{H} \otimes \mathcal{H}$ consists of bilinear maps $T: \mathcal{H}^* \times \mathcal{H}^* \to \mathbb{C}$. However, in order to obtain a natural Hilbert space structure on $\mathcal{H} \otimes \mathcal{H}$, we impose the additional requirement that T satisfy

$$\sum_{i,j=1}^{\infty} |T(e_i^*, e_j^*)|^2 < \infty \quad , \tag{14.2.1}$$

where $\{e_i^*\}$ is an orthonormal basis of $\mathcal{H}^*$. The other tensor product spaces are defined similarly. Note that not all continuous linear maps $A: \mathcal{H} \to \mathcal{H}$ satisfy the analog of equation (14.2.1) and hence not all continuous linear maps can be viewed as elements of $\mathcal{H} \otimes \mathcal{H}^*$. (Those which can are called *Hilbert-Schmidt maps*.) An index notion analogous to that used in finite dimensions could be employed for tensors over $\mathcal{H}$ (Geroch, unpublished), but to avoid confusion with spacetime tensors we shall not use this notation here.

A linear map $L: \mathcal{H} \to \mathcal{H}$ is called an *operator*. If L is bounded, we define the *adjoint* of L, denoted $L^\dagger$, to be the bounded operator which satisfies

$$(L^\dagger w, v) = (w, Lv) \tag{14.2.2}$$

for all v, $w \in \mathcal{H}$. (The existence of an operator $L^\dagger$ satisfying eq. [14.2.2] is guaranteed by the Riesz lemma.) We say L is *self-adjoint* if $L^\dagger = L$, and we say that L is *unitary* if $L^\dagger L = LL^\dagger = I$, where I is the identity map on $\mathcal{H}$. In the case where L is unbounded, the definition of $L^\dagger$ is not as straightforward. First, in general it may only be possible to define L on a dense *domain* $\mathcal{D}(L)$, i.e., a subspace of vectors whose closure equals $\mathcal{H}$. Consider the equation

$$(u, v) = (w, Lv) \quad . \tag{14.2.3}$$

For each fixed pair of vectors u, w such that equation (14.2.3) is satisfied for all $v \in \mathcal{D}(L)$, we say $w \in \mathcal{D}(L^\dagger)$ and we define $L^\dagger w = u$. [If $\mathcal{D}(L^\dagger)$ fails to be dense—i.e., if its closure is not all of $\mathcal{H}$—then $L^\dagger$ is not defined.] We call an unbounded operator, L, *self-adjoint*[7] if $\mathcal{D}(L^\dagger) = \mathcal{D}(L)$ and $L^\dagger v = Lv$ for all $v \in \mathcal{D}(L)$.

7. The precise equality of the domains of $L^\dagger$ and L is an important part of the definition, since it is an essential ingredient in the proof of the spectral theorem (see, e.g., Reed and Simon 1972). If $L^\dagger v = Lv$ for all $v \in \mathcal{D}(L)$ but $\mathcal{D}(L^\dagger) \supset \mathcal{D}(L)$, then L is said to be *hermitian*.

We present, now, some of the basic ingredients of the theory of a free (i.e., linear) real, Klein-Gordon scalar field ϕ in Minkowski spacetime. We shall present this theory in "Heisenberg representation" form, i.e., the operators representing observables will evolve with time but the states do not. Classically, ϕ satisfies

$$\nabla_a \nabla^a \phi - m^2 \phi = 0 \quad . \tag{14.2.4}$$

Our first major task is to construct the Hilbert space of states of the quantum theory. It is natural to construct the space of states of a single scalar particle out of the vector space of solutions of equation (14.2.4). The conserved current (12.4.21) yields a promising candidate for an inner product on this space. If α and β are solutions of equation (14.2.4), we define their Klein-Gordon "inner product" by

$$(\alpha, \beta)_{\text{KG}} = -\int_\Sigma j_a[\alpha, \beta] n^a dV = i \int_\Sigma (\bar{\alpha} \nabla_a \beta - \beta \nabla_a \bar{\alpha}) n^a dV \quad , \tag{14.2.5}$$

where the integral is taken over a Cauchy surface Σ, and we put quotes around "inner product" (to be dropped hereafter) because $(\ , \ )_{\text{KG}}$ is not positive definite. However, if we restrict attention to the subspace of *positive frequency solutions*—i.e., solutions whose time Fourier transform

$$\tilde{\phi}(\omega, \vec{x}) = (2\pi)^{-1/2} \int_{-\infty}^{\infty} e^{i\omega t} \phi(t, \vec{x}) dt \tag{14.2.6}$$

vanishes for $\omega < 0$—then $(\ , \ )_{\text{KG}}$ is positive definite. We define the *one-particle Hilbert space*, $\mathcal{H}$, to be the vector space composed of positive frequency solutions of equation (14.2.4) whose Klein-Gordon norm is finite, with inner product on $\mathcal{H}$ defined by (14.2.5). (More precisely, we define V to be the vector space of smooth, positive frequency solutions which vanish rapidly at spatial infinity, with inner product [14.2.5]. We define $\mathcal{H}$ to be the Hilbert space completion of V.) By taking Fourier transforms, $\mathcal{H}$ can be shown to be isomorphic to the Hilbert space, $L^2(M_+)$, of square integrable functions on the positive mass shell, M_+, of Fourier transform space. Note that the negative frequency solutions can be put in natural linear correspondence with vectors in $\overline{\mathcal{H}} = \mathcal{H}^*$.

In general, if $\mathcal{H}_1$ is the Hilbert space of states of one quantum system and $\mathcal{H}_2$ is that of a second system, then the tensor product $\mathcal{H}_1 \otimes \mathcal{H}_2$ represents states of the total (combined) system. In the case of a Klein-Gordon scalar field, the symmetric tensor product $\mathcal{H} \otimes_S \mathcal{H}$—consisting of symmetric linear maps from $\mathcal{H}^* \times \mathcal{H}^*$ into $\mathbb{C}$ which satisfy equation (14.2.1)—represents the possible states of two scalar particles. The use of only this subspace (rather than all of $\mathcal{H} \otimes \mathcal{H}$) to describe the possible two-particle states reflects the indistinguishability of elementary particles; an interchange of particles produces the same physical state. The choice of the symmetric tensor product (used for all bosons, i.e., integer spin fields) rather than the antisymmetric tensor product (used for all fermions, i.e., half-integer spin fields) is closely related to the properties of these fields required by the spin-statistics theorem. Similarly, the Hilbert space of n free scalar particles is taken to be the n-fold symmetrized tensor product $\otimes_S^n \mathcal{H}$. The space of states where *no* particles are present is assumed to be one-dimensional and hence may be taken to be $\mathbb{C}$.

The Hilbert space of all possible states of the Klein-Gordon scalar field is taken to be the symmetric Fock space, $\mathcal{F}_S(\mathcal{H})$, constructed from $\mathcal{H}$. Here $\mathcal{F}_S(\mathcal{H})$ is defined as the direct sum of the complex numbers, $\mathbb{C}$, with all the symmetrized tensor products of $\mathcal{H}$,

$$\mathcal{F}_S(\mathcal{H}) = \mathbb{C} \oplus [\bigoplus_{n=1}^{\infty} (\otimes_S^n \mathcal{H})] \tag{14.2.7}$$

Here, the direct sum

$$\bigoplus_{i=1}^{\infty} \mathcal{H}_i$$

of a collection $\{\mathcal{H}_i\}$ of Hilbert spaces is defined to be the Hilbert space obtained from the collection of sequences of the form $(v_1, v_2, \ldots)$ with each $v_i \in \mathcal{H}_i$ and

$$\sum_{i=1}^{\infty} \|v_i\|^2 < \infty \quad ,$$

with addition, scalar multiplication, and inner product defined in the obvious way. Thus, each $\Psi \in \mathcal{F}_S(\mathcal{H})$ can be written as

$$\Psi = (\alpha_0, \alpha_1, \alpha_2, \ldots) \quad , \tag{14.2.8}$$

where $\alpha_0 \in \mathbb{C}$, $\alpha_1 \in \mathcal{H}$, $\alpha_2 \in \mathcal{H} \otimes_S \mathcal{H}$, etc. The state

$$|0> \ \equiv (1, 0, 0, \ldots) \tag{14.2.9}$$

represents the *vacuum state* of the field, i.e., the state in which no particles are present. Hence for the general Fock space state (14.2.8), α_0 gives the amplitude for finding the field to be in the vacuum state, α_1 is the "one-particle amplitude" [i.e., the probability of finding only a single particle present in state $\beta \in \mathcal{H}$ is $|(\beta, \alpha_1)|^2$), α_2 is the two-particle amplitude, etc. Thus, every state in $\mathcal{F}_S(\mathcal{H})$ has a direct physical interpretation in terms of the probabilities for finding various numbers of particles in the various possible states.

The most important observable in the theory of a scalar field is the value of the scalar field itself. Since observables in quantum theory are represented by self-adjoint operators, we seek an operator $\hat{\phi}(x)$ defined at each spacetime point x which describes the scalar field. Classically, the field ϕ can be decomposed via Fourier transforms into modes of spatial wave vector $\vec{k}$, so that the amplitude of each mode satisfies the same equation as a classical harmonic oscillator. Analogy with the quantization of the ordinary harmonic oscillator then suggests the following definition of $\hat{\phi}$. First, for each one-particle state $\sigma \in \mathcal{H}$, we define the *annihilation operator* $a(\bar{\sigma}): \mathcal{F}_S(\mathcal{H}) \to \mathcal{F}_S(\mathcal{H})$ as follows. For $\Psi \in \mathcal{F}_S(\mathcal{H})$ given by equation (14.2.8), we set

$$a(\bar{\sigma})\Psi = (\bar{\sigma} \cdot \alpha_1, \sqrt{2}\, \bar{\sigma} \cdot \alpha_2, \sqrt{3}\, \bar{\sigma} \cdot \alpha_3, \ldots) \quad . \tag{14.2.10}$$

Here $\bar{\sigma}$ is the vector $\overline{\mathcal{H}}$ associated with σ under the complex conjugation map, and $\bar{\sigma} \cdot \alpha_n$ is the element of $\otimes_S^{n-1} \mathcal{H}$ obtained by inserting $\bar{\sigma}$ into one of the "slots" of the map α_n. Note that the vacuum state is uniquely characterized (up to phase) by the condition

$$a(\bar{\sigma})|0> \ = 0 \tag{14.2.11}$$

for all $\sigma \in \mathcal{H}$. The adjoint of $a(\overline{\sigma})$ is the *creation operator*, $a^\dagger(\sigma)$, given by

$$a^\dagger(\sigma)\Psi = (0, \alpha_0\sigma, \sqrt{2}\, \alpha_1 \otimes_S \sigma, \sqrt{3}\, \alpha_2 \otimes_S \sigma, \ldots) \quad . \qquad (14.2.12)$$

In terms of a and $a^\dagger$, the quantum field operator $\hat{\phi}(x)$ is defined by

$$\hat{\phi}(x) = \sum_{i=1}^{\infty} [\sigma_i(x)\, a(\overline{\sigma}_i) + \overline{\sigma}_i(x)\, a^\dagger(\sigma_i)] \quad , \qquad (14.2.13)$$

where the sum runs over an orthonormal basis $\{\sigma_i\}$ of $\mathcal{H}$. Thus, $\hat{\phi}$ satisfies the Klein-Gordon equation (14.2.4) in x, and the operator coefficients of the expansion of $\hat{\phi}$ in terms of a basis of positive frequency solutions and their complex conjugates are just the annihilation and creation operators. In fact, the sum in equation (14.2.13) does not converge pointwise and must be interpreted in a distributional sense, i.e., $\hat{\phi}$ can be defined only as an operator-valued distribution on spacetime. (See, e.g., Reed and Simon 1972 for the definition of a distribution, and see, e.g., eq. [2.7] of Wald 1979*b* for the distributional version of eq. [14.2.13].) For calculations involving only linear operations on $\hat{\phi}$, this presents only a minor technical nuisance, but for nonlinear operations it presents a serious obstacle to making mathematical sense out of the expressions which result, since the product of two distributions evaluated at the same spacetime point does not, in general, have any natural mathematical interpretation. This completes our brief introduction to the theory of the Klein-Gordon quantum field in Minkowski spacetime.

Consider, now, the quantum field theory of a Klein-Gordon field in a curved spacetime background (M, g_{ab}). The states of the field still are described as vectors in a Hilbert space $\mathfrak{L}$, but in general there may be no unambiguous physical interpretation of these states in terms of particles. The field ϕ again is described by an operator $\hat{\phi}$ defined on spacetime (or, more precisely, an operator-valued distribution) which satisfies the curved spacetime Klein-Gordon equation (14.2.4).

Perhaps the simplest case to consider is that of a globally hyperbolic spacetime which is nearly isometric to Minkowski spacetime except in a limited region of space over a limited duration of time. Such a spacetime could be produced by focusing matter (or gravitational radiation) onto a small region of space and then allowing the matter to disburse back to infinity. In order to avoid dealing with detailed asymptotic falloff conditions on the gravitational field, we shall consider the highly idealized case of a spacetime which is flat outside a compact spacetime region, as illustrated in Figure 14.1.

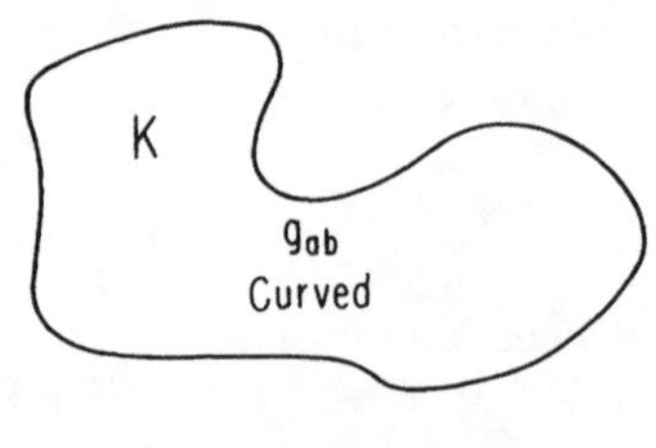

Fig. 14.1. A spacetime diagram of a spacetime $(\mathbb{R}^4, g_{ab})$ which is isometric to Minkowski spacetime $(\mathbb{R}^4, \eta_{ab})$ outside of a compact region K.

Outside the future of the region K of Figure 14.1 an observer (or family of observers) would be unaware that he was not in Minkowski spacetime. Therefore, he would associate with each state of the field $\Psi \in \mathfrak{X}$ a vector in the Fock space $\mathfrak{F}_S(\mathcal{H}_{\text{in}})$ constructed from the one-particle Hilbert space, $\mathcal{H}_{\text{in}}$, of positive frequency solutions of the free Klein-Gordon field in Minkowski spacetime. Let $U: \mathfrak{X} \rightarrow \mathfrak{F}_S(\mathcal{H}_{\text{in}})$ denote this isomorphism of $\mathfrak{X}$ with Fock space obtained by characterizing each state in $\mathfrak{X}$ by how it "looks" to such an observer in the past. To such an observer, the field operator $\hat{\phi}$ must be physically indistinguishable from the Minkowski field operator on Minkowski spacetime, so for x outside the future of K we have

$$U\hat{\phi}(x)U^{-1} = \sum_{i=1}^{\infty} [\sigma_i(x)a_{\text{in}}(\overline{\sigma}_i) + \overline{\sigma}_i(x)a_{\text{in}}^{\dagger}(\sigma_i)] \quad \text{if } x \not\in J^+(K) \quad , \tag{14.2.14}$$

where $\{\sigma_i\}$ is an orthonormal basis of $\mathcal{H}_{\text{in}}$. However, the fact that $\hat{\phi}$ satisfies equation (14.2.4) throughout the spacetime implies that for all $x \in M$ we have

$$U\hat{\phi}(x)U^{-1} = \sum_{i=1}^{\infty} [\sigma_i'(x)a_{\text{in}}(\overline{\sigma}_i) + \overline{\sigma}_i'(x)a_{\text{in}}^{\dagger}(\sigma_i)] \quad , \tag{14.2.15}$$

where σ_i' is the solution of equation (14.2.4) in the curved spacetime which coincides with σ_i outside the future of K.

Similarly, we have an isomorphism $W: \mathfrak{X} \rightarrow \mathfrak{F}_S(\mathcal{H}_{\text{out}})$ which associates with each state in $\mathfrak{X}$ the Minkowski spacetime state it "looks like" in the future.[8] Furthermore, we have

$$W\hat{\phi}(x)W^{-1} = \sum_{i=1}^{\infty} [\rho_i'(x)a_{\text{out}}(\overline{\rho}_i) + \overline{\rho}_i'(x)a_{\text{out}}^{\dagger}(\rho_i)] \quad , \tag{14.2.16}$$

where $\{\rho_i\}$ is an orthonormal basis of $\mathcal{H}_{\text{out}}$ and ρ_i' is the solution of equation (14.2.4) which coincides with ρ_i outside the past of K.

One of the most important issues to consider is how the characterization of the states of the field as "in" states compares with their characterization as "out" states. This is given by the *S-matrix*, $S = WU^{-1}$. Given any "in" state $\Psi \in \mathfrak{F}_S(\mathcal{H}_{\text{in}})$ describing how the state "looks" at early times, the "out" state $S\Psi \in \mathfrak{F}_S(\mathcal{H}_{\text{out}})$ describes how the state "looks" at late times. In particular, for $\Psi = |0_{\text{in}}\rangle$, $\Psi_0 = S|0_{\text{in}}\rangle$ tells us the spontaneous creation of particles by the gravitational field.

Equations (14.2.15) and (14.2.16) allow us to solve for Ψ_0. First, we compose equation (14.2.15) with S on the left and S^{-1} on the right and equate the right-hand side of the resulting equation with the right-hand side of equation (14.2.16). Let $\sigma \in \mathcal{H}_{\text{in}}$ and let σ' be the solution of the curved spacetime Klein-Gordon equation which coincides with σ outside the future of K. Taking the Klein-Gordon inner product with σ', we find that a_{in} and a_{out} are related by

$$Sa_{\text{in}}(\overline{\sigma})S^{-1} = a_{\text{out}}(\overline{C\sigma}) - a_{\text{out}}^{\dagger}(\overline{D\sigma}) \tag{14.2.17}$$

8. Here $\mathcal{H}_{\text{out}}$ is again the Minkowski single-particle Hilbert space and hence is isomorphic to $\mathcal{H}_{\text{in}}$, but it is useful to view $\mathcal{H}_{\text{in}}$ and $\mathcal{H}_{\text{out}}$ as distinct spaces. In more general spacetimes, there may be no natural way of identifying $\mathcal{H}_{\text{in}}$ and $\mathcal{H}_{\text{out}}$.

for all $\sigma \in \mathcal{H}_{\text{in}}$. Here the maps $C:\mathcal{H}_{\text{in}} \to \mathcal{H}_{\text{out}}$ and $D:\mathcal{H}_{\text{in}} \to \overline{\mathcal{H}}_{\text{out}}$ are defined as follows. Outside the past of K, $\sigma'(x)$ must again coincide with some solution, $f(x)$, of the Klein-Gordon equation in Minkowski spacetime. Let μ be the positive frequency part of f in Minkowski spacetime, and let λ be the negative frequency part f. We may view μ as an element of $\mathcal{H}_{\text{out}}$, and λ as an element $\overline{\mathcal{H}}_{\text{out}}$. We define $C\sigma = \mu$ and $D\sigma = \lambda$. Relations satisfied by C and D which are necessary for the consistency of equation (14.2.17) are derived in problem 3. A relation of the form (14.2.17) with C and D satisfying the conditions of problem 3 is called a *Bogoliubov transformation*.

We now can solve for $\Psi_0 \equiv S\,|\,0_{\text{in}}\rangle \in \mathfrak{F}_S(\mathcal{H}_{\text{out}})$ by applying both sides of equation (14.2.17) to Ψ_0. Since $a_{\text{in}}(\overline{\sigma})\,|\,0_{\text{in}}\rangle = 0$ for all $\sigma \in \mathcal{H}_{\text{in}}$, we find

$$\{a_{\text{out}}(\overline{C\sigma}) - a^{\dagger}_{\text{out}}(\overline{D\sigma})\}\Psi_0 = 0 \tag{14.2.18}$$

for all $\sigma \in \mathcal{H}_{\text{in}}$. Writing

$$\Psi_0 = (\alpha_0, \alpha_1, \alpha_2, \ldots) \tag{14.2.19}$$

and using the definitions (14.2.10) and (14.2.12) of a_{out} and $a^{\dagger}_{\text{out}}$, we may solve equation (14.2.18) inductively for α_n. The result is (see Wald 1979*b*)

$$\alpha_n = \begin{cases} 0 & (n \text{ odd}) \\ c\dfrac{(n!)^{1/2}}{2^{n/2}(n/2)!} \bigotimes_S^{n/2} \epsilon & (n \text{ even}) \end{cases} \tag{14.2.20}$$

Here c is a constant determined up to phase by the requirement that $\|\Psi_0\| = 1$, and ϵ is the following element of $\mathcal{H}_{\text{out}} \otimes_S \mathcal{H}_{\text{out}}$. The map DC^{-1} takes vectors in $\mathcal{H}_{\text{out}}$ to vectors in $\overline{\mathcal{H}}_{\text{out}}$ and hence may be viewed as a map from $\mathcal{H}_{\text{out}} \times \mathcal{H}_{\text{out}}$ into $\mathbb{C}$. Thus, the map $E = \overline{D}\,\overline{C}^{-1}$ obtained from DC^{-1} by complex conjugation can be viewed as a map from $\mathcal{H}^*_{\text{out}} \times \mathcal{H}^*_{\text{out}}$ into $\mathbb{C}$. It follows from problem 3 that this map is symmetric. Furthermore, it is proved in Wald (1979*b*) and Dimock (1979) that this map satisfies equation (14.2.1) in the case of a spacetime which is flat outside a compact region. Thus, it defines an element of $\mathcal{H}_{\text{out}} \otimes \mathcal{H}_{\text{out}}$, which we have denoted by ϵ. Given our solution for Ψ_0, the action of S on all other states of $\mathfrak{F}_S(\mathcal{H}_{\text{in}})$ can be determined by using the adjoint of equation (14.2.17) to express the action of S on an arbitrary product of "in" creation operators applied to $|\,0_{\text{in}}\rangle$ in terms of products of "out" creation and annihilation operators applied to Ψ_0. Since all elements of $\mathfrak{F}_S(\mathcal{H}_{\text{in}})$ can be expressed as limits of sums of vectors of this form, this suffices to determine S.

Two features of our solution for Ψ_0 should be emphasized. First, according to equation (14.2.20), the amplitude for producing an odd number of particles from the vacuum vanishes. This can be interpreted as saying that particles always are produced in pairs. In the theory of a real scalar field considered here, antiparticles are the same as particles. However, for a complex field, where antiparticles are distinct from particles, one finds that equal numbers of antiparticles and particles are created, i.e., spontaneous creation occurs via the production of particle-antiparticle pairs.[9]

9. An exception to this statement can occur in certain circumstances for fermion fields; see Christ (1980), Gibbons (1979), and Wald (1979*b*) for further discussion. Expressions for Ψ_0 which display more explicitly the "pairing" of the particles and antiparticles also can be found in Wald (1979*b*).

Second, it is clear from equation (14.2.20) and the definition of ϵ that the necessary and sufficient condition that no spontaneous particle creation occur is that the operator D vanish. In other words, no particle creation occurs in quantum field theory if and only if in the (classical) scattering of a positive frequency wave through the curvature, no negative frequency parts ever are picked up. Thus, with only a few important exceptions (such as given in problem 4), particle creation generally occurs in any time varying gravitational field. Of course, the amount of particle creation and/or its physical effects are generally negligible unless one goes to strong field regimes.

In the discussion of particle creation above, we restricted attention to spacetimes of the form illustrated in Figure 14.1. The Hilbert spaces $\mathcal{H}_{\text{in}}$ and $\mathcal{H}_{\text{out}}$ and the "classical scattering" given by the operators C and D are well defined under weaker conditions than the requirement that the curvature be exactly zero outside a compact region (see Kay 1982), but it has not yet been proven that ϵ satisfies equation (14.2.1) under these weaker conditions. More generally, one can ask if the above particle creation results can be extended to globally hyperbolic spacetimes which do not become flat in the past or future. However, in general, such spacetimes will not possess asymptotic "in" and "out" regimes where the notion of "particles" is physically meaningful. In this regard, it should be emphasized that in ordinary high energy particle physics, the notion of a particle is undefined while interactions are occurring. However, in ordinary particle physics, the interactions normally take place over such microscopic time scales that the particle description of events is adequate for essentially all purposes. In the gravitational case, however, the interactions of the quantum field with the gravitational field may take place over macroscopic spacetime regions. Thus, even in the simple case of a spacetime of the form shown in Figure 14.1, the notion of "particle" is not well defined in the region of nonvanishing curvature. In the case of a spacetime which does not become flat in the past or future, the interactions take place over all time, and, in general, it may not even make sense to talk of incoming or outgoing particle states. We will return to the issue of the physical meaning of particles in curved spacetime at the end of the next section.

In more mathematical detail, the difficulty in defining particle states in a general curved spacetime can be seen to arise in the following manner. In Minkowski spacetime, two key ingredients were used in our construction of the Hilbert space $\mathcal{H}$ of particle states from the vector space, V, of solutions of the Klein-Gordon equation. First, the current (12.4.21) gave us a natural nondegenerate (but not positive definite) map from $V \times V$ into numbers, namely, the Klein-Gordon inner product defined by equation (14.2.5). Second, the decomposition of solutions into positive and negative frequency parts gave us the additional structure needed to pick out a preferred subspace of V on which the Klein-Gordon inner product is positive definite so that it truly defines an inner product. In a general curved spacetime, the current (12.4.21) still is conserved, so the analog of the Klein-Gordon inner product (14.2.5) exists. However, there is no natural analog of the positive and negative frequency decomposition, so in general there is no natural choice of a suitable subspace on which the Klein-Gordon inner product is positive definite. The problem is not that there are no such subspaces but rather that there are many and, in general, none is preferred. Of course, a good approximate notion of the positive and negative frequency parts of a solution exists if the scales of spacetime variation of the solution are much smaller than the scales defined by the spacetime curvature. Thus, for example, in the present universe it is meaningful to talk of a particle provided only that the particle in question has wavelength much smaller than the radius of spatial curvature of the universe and has inverse frequency much smaller than the Hubble time.

An important case where a natural positive and negative frequency decomposition does exist is that of stationary spacetime, since here the Killing parameter provides a preferred time coordinate with respect to which Fourier transforms can be taken. Ashtekar and Magnon (1975) and Kay (1978) have shown that for the Klein-Gordon field (with $m > 0$) in any globally hyperbolic stationary spacetime (with the norm of the timelike Killing field bounded away from zero) one can define a natural Fock space of particle states and define the field operator $\hat{\phi}$ on this Hilbert space in analogy with (14.2.13). Thus, the notion of particles is mathematically (and physically) well defined in stationary spacetimes.[10] The above discussion of the S-matrix and the above results on particle creation can be carried over to spacetimes which merely become stationary (rather than flat) in the past and future, except that equation (14.2.1) need not be satisfied[11] by the "two-particle amplitude" ϵ.

As emphasized above, in spacetimes which fail to be asymptotically stationary in the past and/or future it may not be meaningful to try to characterize states of the field in terms of ingoing and/or outgoing particle states. However, we mention one important general approach toward enabling one to do so. Consider a spacetime (M, g_{ab}) which is asymptotically stationary in the past and future and for which the two-particle amplitude ϵ satisfies equation (14.2.1). Then the "in" Fock space $\mathfrak{F}_S(\mathcal{H}_{\text{in}})$ and the "out" Fock space $\mathfrak{F}_S(\mathcal{H}_{\text{out}})$ are well defined, as is the S-matrix relating the two spaces. We define the *Feynman propagator*, $\Delta(x, y)$, of the scalar field in the spacetime (M, g_{ab}) by

$$i\,\Delta(x, y) = \frac{\langle 0_{\text{out}} | T(\hat{\phi}(x)\hat{\phi}(y)) | 0_{\text{in}} \rangle}{\langle 0_{\text{out}} | 0_{\text{in}} \rangle} \quad . \tag{14.2.21}$$

Here we use Dirac notation for inner products, and the "time ordered product" of field operators is defined by

$$T(\hat{\phi}(x)\hat{\phi}(y)) = \begin{cases} \hat{\phi}(x)\hat{\phi}(y) & \text{if } y \notin J^+(x) \\ \hat{\phi}(y)\hat{\phi}(x) & \text{if } x \notin J^+(y) \end{cases} \quad . \tag{14.2.22}$$

[If x and y are spacelike separated, then $\hat{\phi}(x)$ and $\hat{\phi}(y)$ commute, so the two expressions on the right-hand side of eq. (14.2.22) agree.] One can show that $\Delta(x, y)$

10. However, see the discussion at the end of section 14.3 below.

11. In the case of a closed universe with initial and final static regimes, it is known that ϵ satisfies equation (14.2.1) (Fulling, Narcowich, and Wald 1982). If ϵ does not satisfy equation (14.2.1), then the S-matrix does not exist, i.e., the "in" and "out" Fock spaces cannot be viewed as representing the same Hilbert space of quantum states. However, in such cases as in the case of the "infrared catastrophe" of quantum electrodynamics, it may still be possible to make meaningful physical predictions. The algebraic approach to quantum field theory (Haag and Kastler 1964) provides a framework for dealing with this difficulty by generalizing the notion of states from that of vectors in a fixed Hilbert space to that of positive linear functionals on the algebra of observables.

is a Green's function for the Klein-Gordon equation, i.e., in each variable it is a distributional solution of the Klein-Gordon equation with a δ-function source at the value of the other variable. Furthermore, it follows directly from the general form (14.2.15) and (14.2.16) of the field operator as well as the property (14.2.11) of the vacuum state that $\Delta(x, y)$ "propagates" only positive frequencies into the future—i.e., for a function f which vanishes outside a compact set, the quantity $\int \Delta(x, y) f(y) \sqrt{-g}\, d^4y$ is a purely positive frequency solution in the future—and it propagates only negative frequencies into the past. Thus, given the notion of asymptotic states for the quantum field theory, one can define a Green's function for the Klein-Gordon equation by equation (14.2.21) from which the asymptotic notions of positive and negative frequency decompositions can be recovered directly. Conversely, given a Green's function $G(x, y)$ which satisfies appropriate properties—in particular, which "propagates" into the future only functions which have positive Klein-Gordon norm, and propagates into the past only functions with negative Klein-Gordon norm—then we may use G to *define* the notions of "positive frequency" in the past and future. These definitions then may be used to define the notion of asymptotic particle states of the quantum field so that (formally, at least) G becomes the Feynman propagator of the theory. Thus, the problem of defining asymptotic states is equivalent to the problem of selecting an appropriate Green's function for the Klein-Gordon equation to play the role of Feynman propagator. Hence, in cases where natural candidates exist for the Feynman propagator (see, e.g., Hartle and Hawking 1976; Rumpf 1976), the notion of asymptotic particle states can be defined.

One important regime where gravity should be sufficiently strong and time dependent to cause significant particle creation is in the early universe. Unfortunately, since the universe is not believed to be asymptotically stationary in the past and since there is no generally agreed upon prescription for defining the Feynman propagator, one does not have a natural, unambiguous notion of incoming particle states. Furthermore, since one would not expect the approximation of treating the spacetime metric classically to be adequate very near the "big bang" singularity, the applicability of any notion of incoming particle states defined on a classical spacetime model would be questionable. Despite these difficulties, quantum effects occurring in the early universe have been investigated by many authors, with full fledged efforts beginning in the 1960s with the work of Parker (1969) and others. Indeed, much of the theory of quantum fields in curved spacetime was developed in conjunction with these investigations. We shall not attempt to discuss this work here but refer the reader to Birrell and Davies (1982) for a summary and bibliography. Instead, we turn to the consideration of particle creation near black holes.

14.3 Particle Creation near Black Holes

An important regime where spontaneous particle creation might be expected to occur is in the vicinity of a black hole. Indeed, we already noted in chapter 12 that superradiant scattering is closely analogous to stimulated emission. This strongly suggests that spontaneous "emission" from a Kerr black hole should occur. It turns out that such spontaneous particle creation near a Kerr black hole does indeed take place, and superradiant scattering indeed is the classical limit of the stimulated emission associated with this particle creation (Starobinskii 1973; Unruh 1974; Wald 1976). However, by far the most dramatic result arising from the investigation of particle creation near black holes was Hawking's discovery that particle creation also occurs near a Schwarzschild black hole, resulting in the "emission" of a thermal spectrum of particles. We turn, now, to a derivation and discussion of this result, referring the reader to Hawking (1975) and Wald (1975) for further details.

Consider, first, the extended Schwarzschild spacetime of Figure 6.9, or Figure 12.3. Suppose we are interested in the *classical* wave propagation of a Klein-Gordon scalar field in region I of this spacetime. Intuitively, one might expect that any solution of the source-free Klein-Gordon equation in region I either must have "started from infinity" or must have entered region I from the white hole region III. Similarly, at "late times," one might expect that every solution will propagate into the black hole region II and/or propagate back to infinity. To investigate whether this is true, we expand ϕ in spherical harmonics and write the wave equation (14.2.4) for each mode of the form $r^{-1}f(r,t)Y_{lm}(\theta,\phi)$. We obtain

$$\frac{\partial^2 f}{\partial t^2} - \frac{\partial^2 f}{\partial {r_*}^2} + \left(1 - \frac{2M}{r}\right)\left[\frac{l(l+1)}{r^2} + \frac{2M}{r^3} + m^2\right] f = 0 \quad , \tag{14.3.1}$$

where the coordinate r_* was defined by equation (6.4.20), M is the mass of the black hole, and m is the mass of the Klein-Gordon field. But equation (14.3.1) has precisely the form of the wave equation for a massless scalar field in a two-dimensional flat spacetime (with Cartesian coordinates t, r_*) with a scalar potential

$$V(r_*) = \left(1 - \frac{2M}{r}\right)\left[\frac{l(l+1)}{r^2} + \frac{2M}{r^3} + m^2\right] \quad . \tag{14.3.2}$$

(Similar results hold for electromagnetic, gravitational, neutrino, and Dirac perturbations of Schwarzschild spacetime, and these results also generalize to the Kerr black hole; see Chandrasekhar 1983 for a complete discussion.) As $r_* \to -\infty$ (i.e., $r \to 2M$), the potential $V(r_*)$ behaves as $(1 - 2M/r) \sim \exp(r_*/2M)$, i.e., it falls off exponentially in r_*. As $r_* \to \infty$ (i.e., $r \to \infty$), in the massive case $V(r_*)$ behaves as $\sim(m^2 - 2Mm^2/r_*)$, whereas if $m = 0$ we have $V(r_*) \sim l(l+1)/r_*^2$. Hence, the theory of scattering by potentials in flat spacetime (see, e.g., Reed and Simon 1979) suggests that the following results should hold—in confirmation of the above intuitive expectations—although a complete proof has not yet been given: If $m \neq 0$, then every wave packet should, in the asymptotic past, behave like a free ($V = 0$) massless solution in (t, r_*)–space propagating in from $r_* \to -\infty$ (i.e., from the white hole horizon) together with a massive solution (distorted by a $1/r_*$ potential) propagating in from $r_* \to \infty$. Similarly, in the asymptotic future, every wave packet should behave as a free massless wave propagating to $r_* \to -\infty$ together with a (distorted) massive wave propagating to $r_* \to \infty$. If $m = 0$, then every wave packet should approach a free massless solution in both the asymptotic past and future and hence should be of the form $f_+(u) + g_+(v)$ as $t \to +\infty$ and $f_-(u) + g_-(v)$ as $t \to -\infty$, where $u = t - r_*$ and $v = t + r_*$. This would imply that a massless Klein-Gordon field in Schwarzschild spacetime is determined by its value on the white hole horizon and $\mathscr{I}^-$ [which fixes $f_-(u)$ and $g_-(v)$, respectively] or, equivalently, by its value on the black hole horizon and $\mathscr{I}^+$ [which fixes $g_+(v)$ and $f_+(u)$]. We shall assume the validity of these conclusions. In the massive case, the value of the field at the white hole horizon should provide an appropriate characterization of the part of the wave incoming from the white hole, but some characterization other than the value of the field at $\mathscr{I}^-$ must be used to describe the part of the wave incoming from infinity. Similarly, a characterization other than the value of the field at $\mathscr{I}^+$ must be used to describe the part of the wave which propagates out to infinity at late times. In order to take advantage of the precise description of ingoing and outgoing parts of a massless field at infinity of the Schwarzschild spacetime[12] in terms of "data" on $\mathscr{I}^-$ and $\mathscr{I}^+$, we shall explicitly treat the massless Klein-Gordon case below. However, all the results should apply to the massive case, as well as to other fields propagating in the Schwarzschild background. The modifications which arise in the case of a Kerr black hole will be mentioned below.

We have concluded that the solutions of the massless Klein-Gordon equation in the asymptotic past can be put into correspondence with functions on the white hole horizon together with functions on $\mathscr{I}^-$. Therefore, we expect the one-particle Hilbert space, $\mathscr{H}_{\text{in}}$, of incoming states to consist of the direct sum of a Hilbert space of particles incoming from infinity and a Hilbert space of particles coming from the white hole,

$$\mathscr{H}_{\text{in}} = \mathscr{H}_{\text{in},\infty} \oplus \mathscr{H}_{\text{in,wh}}$$

We have a well defined, natural notion of "positive frequency" solutions at $\mathscr{I}^-$, namely those solutions whose Fourier transform on $\mathscr{I}^-$ with respect to a BMS time translation parameter—such as the Schwarzschild advanced time coordinate v, equation (6.4.22)—contains only positive frequencies. This allows us to define a Hilbert space $\mathscr{H}_{\text{in},\infty}$ of incoming particle states from infinity. However, it is considerably less obvious how to define the notion of "positive frequency" for solutions emanating from the white hole. Indeed, there appear to be two distinct natural candidates for such a definition—namely, a "white hole incoming solution" could be defined to be "positive frequency" if its Fourier transform on the white hole horizon has only positive frequencies, where the Fourier transform is taken either with respect to (i) the Schwarzschild retarded time coordinate u, equation (6.4.21), (i.e., the Killing parameter along the generators of the horizon) or (ii) the Kruskal retarded time U, equation (6.4.26), (i.e., the affine parameter along the generators of the horizon). Since the Killing and affine parameters are distinct—quite generally, they are related by equation (12.5.10)—these two definitions lead to distinct notions of "positive frequency" and, hence, distinct constructions of the one particle Hilbert space of white hole states, $\mathscr{H}_{\text{in, wh}}$. Thus, we have an important ambiguity in the definition of $\mathscr{H}_{\text{in}}$ and, hence, in the Hilbert space $\mathfrak{F}_S(\mathscr{H}_{\text{in}})$ of all incoming states. A change in the definition of "positive frequency" on the white hole horizon corresponds to a change in the choice of incoming state representing the vacuum. Thus, the ambiguity in the calculation of particle creation that would result from the ambiguity in the definition

12. In other asymptotically flat spacetimes, $\mathscr{I}^-$ and $\mathscr{I}^+$ need not be good "initial data surfaces" for massless fields; see Geroch (1978*b*).

of $\mathcal{H}_{\text{in, wh}}$ may be viewed as resulting from the ambiguity in defining the condition that no "particles" emanate from the white hole.

Similarly, there is an ambiguity in the definition of the Hilbert space, $\mathcal{H}_{\text{out,bh}}$, of particles propagating into the black hole, and hence in

$$\mathcal{H}_{\text{out}} = \mathcal{H}_{\text{out},\infty} \oplus \mathcal{H}_{\text{out,bh}}$$

and in $\mathfrak{F}_S(\mathcal{H}_{\text{out}})$. In view of these ambiguities, it might appear that no physically meaningful predictions about particle creation near black holes can be made. However, the ambiguities in the definitions of both incoming and outgoing states can be treated in such a way as to extract unambiguous physical predictions as follows.

First, the ambiguities in the definition of $\mathcal{H}_{\text{in,wh}}$ can be eliminated (Hawking 1975) by replacing the extended Schwarzschild spacetime by the spacetime appropriate to a collapsing spherical body, Figure 6.11 or Figure 12.2. This eliminates the white hole horizon and thus makes $\mathcal{H}_{\text{in}}$ be simply $\mathcal{H}_{\text{in},\infty}$, which is unambiguously defined. Furthermore, the spacetimes representing gravitational collapse—as opposed to the extended Schwarzschild solution—are believed to describe black holes occurring in nature, so in any case the calculation of particle creation in the spacetime of Figure 6.11 or Figure 12.2 is of greater physical relevance than that of particle creation in the extended Schwarzschild solution.

The black hole, of course, is the essential feature of the problem we wish to study, so the ambiguity in the definition of $\mathcal{H}_{\text{out,bh}}$ remains. However, we still can make important physical predictions which avoid this ambiguity as follows. The "out" Hilbert space $\mathfrak{F}_S(\mathcal{H}_{\text{out},\infty} \oplus \mathcal{H}_{\text{out,bh}})$ is naturally isomorphic to the tensor product space $\mathfrak{F}_S(\mathcal{H}_{\text{out},\infty}) \otimes \mathfrak{F}_S(\mathcal{H}_{\text{out,bh}})$ and thus may be viewed as a "joint state" of two systems: (i) particles propagating out to infinity and (ii) particles propagating into the black hole. Now, whenever in quantum mechanics one has a state Ψ lying in a tensor product $\mathcal{K}_1 \otimes \mathcal{K}_2$ of two Hilbert spaces $\mathcal{K}_1, \mathcal{K}_2$, one can form from Ψ the *density matrix* $\rho \in \mathcal{K}_1 \otimes \overline{\mathcal{K}}_1$ by taking the trace of $\Psi \otimes \overline{\Psi} \in (\mathcal{K}_1 \otimes \mathcal{K}_2) \otimes (\overline{\mathcal{K}}_1 \otimes \overline{\mathcal{K}}_2) \cong (\mathcal{K}_1 \otimes \overline{\mathcal{K}}_1) \otimes (\mathcal{K}_2 \otimes \overline{\mathcal{K}}_2)$ with respect to a basis of $\mathcal{K}_2$. More precisely, we define ρ—viewed now as a linear map from $\mathcal{K}_1$ into $\mathcal{K}_1$—by the formula

$$(w, \rho v) = \sum_{i=1}^{\infty}(w \otimes e_i, \Psi)(\Psi, v \otimes e_i) \tag{14.3.3}$$

for all $v, w \in \mathcal{K}_1$, where $\{e_1\}$ is an orthonormal basis of $\mathcal{K}_2$ and the inner product on the left-hand side of equation (14.3.3) is taken in $\mathcal{K}_1$, while the inner products on the right side are taken in $\mathcal{K}_1 \otimes \mathcal{K}_2$. It follows immediately that for any observable O for the first system we have

$$(\Psi, O\Psi) = \text{tr}(\rho O) \quad , \tag{14.3.4}$$

where tr denotes the trace of the linear map ρO. Since the probabilities for the possible outcomes of any measurement can be expresssed in terms of the expectation values of projection operators, it follows that all information about the first system can be recovered from ρ. In our case, we can "trace out" the degrees of freedom corresponding to particles which enter the black hole and thereby obtain a density matrix describing the particles which propagate out to infinity. This density matrix does not depend on the choice of definition of "positive frequency" used in the construction of $\mathcal{H}_{\text{out,bh}}$ for the following reason. A change in the definition of positive frequency on the black hole horizon will induce a Bogoliubov transformation on the annihilation and creation operators associated with the states representing particles which enter the black hole, but will leave unchanged the annihilation and creation operators associated with the states representing particles which propagate to infinity. This will cause the expression for the "out" state vector $\Psi \in \mathfrak{F}_{\text{out}} = \mathfrak{F}_S(\mathcal{H}_{\text{out},\infty}) \otimes \mathfrak{F}_S(\mathcal{H}_{\text{out,bh}})$ to change to $\Psi' = \mathcal{S}\Psi$ with $\mathcal{S}$ of the form $\mathcal{S} = I_1 \times S_2$, where I_1 is the identity operator on $\mathfrak{F}_S(\mathcal{H}_{\text{out},\infty})$ and S_2 is a unitary operator on $\mathfrak{F}_S(\mathcal{H}_{\text{out,bh}})$. Consequently, by equation (14.3.3) the density matrix ρ' associated with Ψ' will agree with the density matrix ρ associated with Ψ. Hence one obtains unambiguous physical predictions for the particle creation seen by a distant observer at late times.

Thus, the density matrix, ρ, describing the spontaneously created particles which escape to infinity can be unambiguously determined as follows. First, we choose any convenient definition of "positive frequency" on the black hole horizon and construct $\mathcal{H}_{\text{out,bh}}$. Then we obtain the operators C and D by solving for the classical scattering of Klein-Gordon waves in the spacetime of Figure 12.2. Then we obtain $\Psi_0 = S|0_{\text{in}}\rangle$ from equation (14.2.20). Finally, we calculate ρ from Ψ_0 as described above.

It might be expected that particles would be created during the dynamic phase of the collapse (the details of which would depend upon the details of the collapse) but that after the Schwarzschild black hole has "settled down" to its final static state, the particle creation would cease. However, this is not what Hawking (1975) found. An observer at infinity always "sees" dynamical aspects of the collapse, and the classical scattering of positive frequency waves to negative frequencies continues to occur at arbitrarily late times.

We begin our analysis of this particle creation effect by noting the following behavior of solutions of the Klein-Gordon equation in the *extended* Schwarzschild spacetime of Figure 6.9 or Figure 12.3 which have time dependence $e^{-i\omega t}$ in region I. From equation (14.3.1) it follows that near the horizon ($r_* \to -\infty$) each such solution behaves as a "free wave" $a\exp(-i\omega u) + b\exp(-i\omega v)$ in (t, r_*)-space where $u = t - r_*$ and $v = t + r_*$. The solutions with $b = 0$ are referred to as purely "outgoing" at the horizon. By superposing solutions with $b = 0$ with different frequencies, we may produce nonsingular wave packets which vanish on the black hole horizon ($u \to \infty$). However, each individual outgoing mode of frequency ω possesses an "infinite oscillation" singularity on the black hole horizon. Indeed, in terms of the Kruskal coordinate $U = -e^{-\kappa u}$ [where $\kappa = 1/(4M)$ is the surface gravity of the black hole], the behavior of these solutions is $a\exp[i\omega\kappa^{-1}\ln(-U)]$. Let γ be any geodesic which enters the black hole from region I and let λ be its affine parameter, with $\lambda = 0$ chosen to correspond to the intersection of the geodesic with the horizon. Since λ depends smoothly on U and satisfies $dU/d\lambda \neq 0$ on the horizon ($U = 0$), it follows that near $\lambda = 0$ each outgoing mode oscillates as a function of λ as $\exp[i\omega\kappa^{-1}\ln(-\alpha\lambda)]$, where $\alpha = dU/d\lambda|_{\lambda=0}$. Thus, the frequency of each mode as locally determined by a freely falling observer entering the black hole diverges at the horizon in this characteristic manner. This divergence is related to the fact that even for static observers the gravitational redshift effect (calculated in section 6.3) becomes infinite on the horizon.

The nature of the classical scattering relevant to the quantum particle creation effect is best analyzed by considering the propagation of waves *backward* in time. In particular, consider the solutions $\phi_{\omega lm}$ in extended Schwarzschild spacetime which have time dependence $e^{-i\omega t}$, angular dependence $Y_{lm}(\theta, \phi)$ and are purely outgoing at the horizon. We may view this solution as "starting" from $\mathscr{I}^+$. As this wave propagates into the past, part of it will be scattered back to $\mathscr{I}^-$ and part of it will be scattered into the white hole as illustrated in Figure 14.2. Now consider the same "initial" wave at $\mathscr{I}^+$ propagating in the spacetime of Figure 12.2, where collapse to a Schwarzschild black hole has occurred, rather than in the extended Schwarzschild spacetime. Again, part of the wave will be scattered directly back to $\mathscr{I}^-$, but we are mainly interested in the portion of the wave corresponding to the part which goes into the white hole in the extended Schwarzschild spacetime. Now this part of the wave will propagate through the collapsing matter and end up at $\mathscr{I}^-$, as illustrated in Figure 14.3. We can obtain the approximate form of this wave at $\mathscr{I}^-$ as follows. Let μ be a null geodesic generator of the horizon and, for convenience, set equal to zero the advanced time v_0 at which its continuation into the past intersects $\mathscr{I}^-$, $v_0 = 0$. Since the locally measured frequency of the wave becomes infinite at μ in the "Schwarzschild portion" of the spacetime, the geometrical optics approximation (see chapter 4) will hold in the vicinity of μ for the propagation of the wave from the black hole horizon back to $\mathscr{I}^-$. Thus, to an approximation which becomes more and more exact as one approaches μ, the wave will have the form $\phi_0 e^{iS}$ where ϕ_0 is constant, and the surfaces of constant phase, S, are null. Hence, the pattern made by the wave near $v = 0$ at $\mathscr{I}^-$ can be obtained by continuing the null geodesic generators of the surfaces of constant S back to $\mathscr{I}^-$. However, the behavior of the geodesics

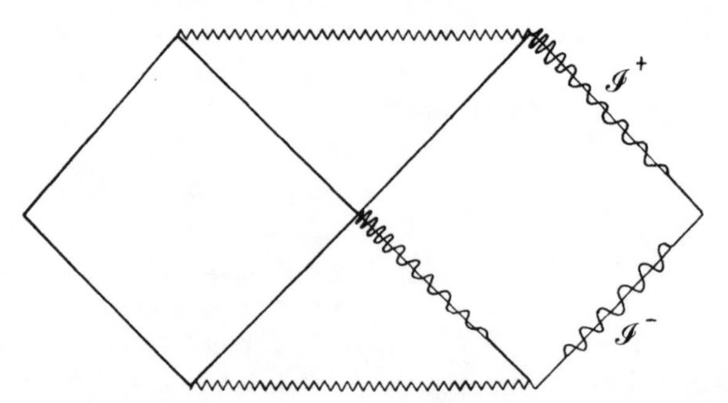

Fig. 14.2. A conformal diagram of the extended Schwarzschild spacetime (see Fig. 12.3) showing the oscillations of a wave of frequency ω, i.e., $\pounds_\xi \phi = -i\omega\phi$. The apparent increase in oscillation frequency shown in the figure on $\mathscr{I}^+$ at late retarded time (which also occurs on $\mathscr{I}^+$ at early retarded time and on $\mathscr{I}^-$ at late and early advanced times) is merely an artifact of the conformal diagram caused by the behavior of the conformal factor there. However, the increase in frequency shown on the white hole horizon near its crossing point with the black hole horizon represents a physical "infinite blueshift" singularity of the wave.

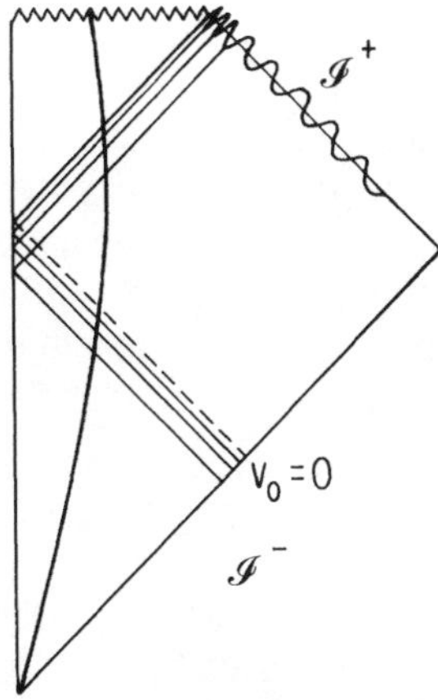

Fig. 14.3. A conformal diagram of a spherically symmetric spacetime in which gravitational collapse to a Schwarzschild black hole occurs, showing the oscillations of a wave which behaves as $e^{-i\omega u}$ on $\mathscr{I}^+$. If one propagates this wave backward into the past starting from $\mathscr{I}^+$, the part of the wave which propagates through the collapsing matter will produce an "infinite blueshift" singularity—of the same nature as shown on the white hole horizon in Figure 14.2—at advanced time v_0 on $\mathscr{I}^-$.

sufficiently near to μ will be accurately described by a geodesic deviation vector η^a, i.e., the deviation from μ of neighboring geodesics propagates linearly along μ. Consequently, choosing the direction of η^a at $\mathscr{I}^-$ to be along the null geodesic generator of $\mathscr{I}^-$, we see that near $v = 0$ the solution ϕ will behave as a function of advanced time v on $\mathscr{I}^-$ in the same way as ϕ behaves as a function of affine parameter λ along the geodesic tangent to η^a at any other point of μ. However, this latter behavior has already been determined above on the horizon of the "Schwarzschild portion" of the spacetime. Hence, we conclude that near $v = 0$ on $\mathscr{I}^-$, the time dependence of the solution is given by

$$\phi(v) = \begin{cases} 0 & (v > 0) \\ \phi_0 \exp\left[\dfrac{i\omega}{\kappa} \ln(-\alpha v)\right] & (v < 0) \end{cases} \quad . \tag{14.3.5}$$

The crucial point is that although we started with the purely positive frequency mode $e^{-i\omega u}$ at $\mathscr{I}^+$, the solution at $\mathscr{I}^-$ is *not* purely positive frequency. Indeed, it is not difficult to show that the Fourier transform, $\tilde{\phi}$, of ϕ with respect to v satisfies for $\sigma > 0$

$$\tilde{\phi}(-\sigma) = -e^{-\pi\omega/\kappa}\tilde{\phi}(\sigma) \tag{14.3.6}$$

(see, e.g., appendix A of Wald 1975), so the magnitude of the negative frequency part of ϕ at $\mathscr{I}^-$ is the factor $e^{-\pi\omega/\kappa}$ times the magnitude of the positive frequency part of ϕ. Note that the derivation of (14.3.5) is independent of the details of collapse. Note also that the decomposition of ϕ into positive and negative frequency parts at $\mathscr{I}^-$ obtained in the case of collapse to a black hole is equivalent to the decomposition one would obtain in the extended Schwarzschild spacetime using the affine parameter (rather than the Killing parameter, or any other choice) to define the notion of positive frequency on the white hole horizon.

To obtain the scattering operators on the one-particle Hilbert space, one should work with normalized wave packets. Consider a positive frequency wave packet at $\mathscr{I}^+$ composed mainly of frequencies near ω and centered on retarded time u. For large u (i.e., after the black hole appears to an observer at infinity to have "settled down" to its final static state), the analysis given above can be applied to this packet and equation (14.3.6) will continue to hold.[13] It follows directly from general properties derived in problems 2, 3, and 5 that the expected number of particles spontaneously created in the state represented by this packet is (Hawking 1975)

$$\langle N \rangle = |t|^2 \frac{e^{-2\pi\omega/\kappa}}{1 - e^{-2\pi\omega/\kappa}} \quad . \tag{14.3.7}$$

Here $|t|^2$ is the square of the Klein-Gordon norm of the part of the wave packet which would enter the white hole in the extended Schwarzschild solution. But this is equal to the absorption cross section of the black hole for that mode. Thus, equation (14.3.7) is precisely the formula which would hold for a perfect blackbody emitter at temperature given by

$$kT = \frac{\hbar\kappa}{2\pi c} = \frac{\hbar c^3}{8\pi G M} \quad , \tag{14.3.8}$$

i.e.,

$$T \approx 6 \times 10^{-8}(M_\odot/M)\ \mathrm{K} \quad , \tag{14.3.9}$$

where k is Boltzman's constant and we have restored the G's and c's. This similarity of black hole "emission" via particle creation with blackbody emission extends well beyond the agreement (14.3.7) in expected numbers of particles. A complete analysis of the density matrix describing the outgoing state at infinity shows that it is identical in all aspects to a thermal density matrix at temperature (14.3.8) (Wald 1975; Parker 1975; Hawking 1976). Furthermore, in the presence of incoming particles the black hole continues to behave exactly like a blackbody (Panangaden and Wald 1977). The significance of the fact that black holes behave like perfect blackbodies will be explored further in the next section.

Note that the temperture of the particles "emitted" by the black hole is inversely proportional to the mass of the black hole. Thus, if energy is added to a black hole, the temperature decreases, i.e., a black hole has negative specific heat. Negative specific heats are typical of self-gravitating systems. For example, in Newtonian gravity, a self-gravitating star composed of an ideal gas has a negative specific heat.

The main modification of the above analysis needed to treat the case of the Kerr black hole arises from the fact that the Killing field $\chi^a = (\partial/\partial t)^a + \Omega_H(\partial/\partial\phi)^a$, equation (12.3.20), rather than the stationary Killing field $(\partial/\partial t)^a$, is normal to the

13. One rather disturbing feature of the analysis should be pointed out. The mean frequency of the wave packet at $\mathscr{I}^-$ increases with advanced time v as $\kappa e^{\kappa v}$ (see Wald 1976), which rapidly becomes very large, in particular, much larger than the Planck frequency. These ultrahigh frequencies only enter the intermediate stages of calculation—they do not appear in any of the final physical predictions—but even so, one may feel uncomfortable that during the calculation one considers conditions so extreme that the classical wave propagation used would appear to be very difficult to justify.

horizon in this case. Consequently, for an "initial" wave at $\mathscr{I}^+$ with time dependence $e^{-i\omega u}$ and angular dependence $e^{im\phi}$, the dependence of the solution on affine parameter along a geodesic which enters the black hole is $\exp[i(\omega - m\Omega_H)\kappa^{-1} \ln(-\alpha\lambda)]$. For the non-superradiant modes, the only change in the final expression for the matrix is the replacement of ω by $\omega - m\Omega_H$. For the superradiant modes, the form of the density matrix changes (Wald 1975), but equation (14.3.7) continues to hold with $\omega \to \omega - m\Omega_H$ provided that $|t|^2$ is interpreted to be negative. Because of the frequency shift $\omega \to \omega - m\Omega_H$ in equation (14.3.7), the Kerr black hole preferentially loses angular momentum (see Page 1976*b* for quantitative details).

It is noteworthy that the thermal nature of black hole emission can be related in a direct manner to the properties of the Euclidean Schwarzschild solution. As discussed briefly in section 14.1, in the Euclidean approach to quantum field theory one seeks to define quantities on a "Euclidean section" and then obtain the physical, spacetime quantities by analytic continuation. In particular, the Feynman propagator for a field on spacetime is obtained by analytic continuation of the Green's function on the Euclidean section. (Since the equation for a free field is elliptic rather than hyperbolic on the Euclidean section, there often will exist a unique Green's function on the Euclidean section satisfying natural boundary conditions.) In our case, one is naturally led to examine the properties of the *Euclidean Schwarzschild solution* obtained by analytically continuing t to real values of $\tau = it$ in equation (6.1.44),

$$ds_E^2 = +\left(1 - \frac{2M}{r}\right)d\tau^2 + \left(1 - \frac{2M}{r}\right)^{-1} dr^2 + r^2 d\Omega^2 \quad . \qquad (14.3.10)$$

In Euclidean Schwarzschild space, we again encounter a singularity in the coordinate components of the metric at $r = 2M$. As we shall see below, this singularity is again merely a coordinate singularity but its nature is quite different from the Lorentzian case. To analyze it, we define a new coordinate R by

$$R = 4M(1 - 2M/r)^{1/2} \quad . \qquad (14.3.11)$$

Then, we have

$$ds_E^2 = R^2 d(\tau/4M)^2 + \left(\frac{r}{2M}\right)^4 dR^2 + r^2 d\Omega^2 \quad , \qquad (14.3.12)$$

where now r is understood to be the function of R determined by equation (14.3.11). From equation (14.3.12), it is manifest that the coordinate singularity at $R = 0$ (i.e., $r = 2M$) is of the same nature as the coordinate singularity that occurs at the origin of polar coordinates on the plane, where now R plays the role of the radial coordinate and $\tau/4M$ plays the role of the angular coordinate. Therefore, a natural choice of manifold structure for the Euclidean Schwarzschild solution is to periodically identify $\tau/4M$—with period 2π—in the region $r > 2M$ and then "add in" a single point[14] in the "R-τ plane" to extend the space to $R = 0$. In this way, one obtains a complete Riemannian manifold (M, g^E_{ab}) with topology $M = \mathbb{R}^2 \times S^2$. Note that the Euclidean

14. Since the Schwarzschild manifold is the "R-τ plane" crossed with the "θ-ϕ 2-sphere," one, of course, really is "adding in" a 2-sphere to the manifold when one adds a point to the R-τ plane.

Schwarzschild manifold has no region corresponding to the region $r < 2M$ in the Lorentzian spacetime.

With the above interpretation of the Euclidean Schwarzschild solution, clearly every continuous function on M is periodic in τ with period $8\pi M$. In particular, any Green's function will satisfy this property in each of its variables. Hence, if the Feynman propagator in the Lorentzian Schwarzschild solution is defined by analytic continuation of a Euclidean Green's function, it automatically will be periodic in "imaginary time" $\tau = it$. However, this property of periodicity in imaginary time is characteristic of a thermal state, as we now shall explain.

Suppose we have an ordinary quantum mechanical system with a time independent Hamiltonian operator H. The state of thermal equilibrium of the system at temperature $kT \equiv \beta^{-1}$ is defined to be that described by the density matrix

$$\rho = e^{-\beta H}/Z \quad , \tag{14.3.13}$$

where

$$Z = \mathrm{tr}(e^{-\beta H}) \quad . \tag{14.3.14}$$

Furthermore, in the Heisenberg representation, every observable O evolves with time by

$$O(t_0 + t) = e^{iHt/\hbar} O(t_0) e^{-iHt/\hbar} \quad . \tag{14.3.15}$$

In quantum field theory, one in general cannot rigorously define a Hamiltonian operator H and, in any case, $e^{-\beta H}$ would not define a normalizable density matrix. However, at least in a formal sense, equation (14.3.13) still should describe a state of thermal equilibrium and the field operator $\hat{\phi}$ should evolve with time via equation (14.3.15). We define the *thermal Feynman propagator* at temperature $kT = \beta^{-1}$ by

$$i\Delta_T(x_1, x_2) = \mathrm{tr}[\rho T(\hat{\phi}(x_1)\hat{\phi}(x_2))] = Z^{-1}\mathrm{tr}[e^{-\beta H} T(\hat{\phi}(x_1)\hat{\phi}(x_2))] \tag{14.3.16}$$

where, on the right side of this equation, T denotes the time ordered product (see eq. [14.2.22]). Now, suppose that we can analytically continue t to imaginary values, $t = -i\tau$, such that equations (14.3.15) and (14.3.16) continue to hold. Consider the case where $\tau_2 - \beta\hbar < \tau_1 < \tau_2$ and let x_1' denote the "imaginary time translate" of x_1 by $\beta\hbar$, i.e., x_1' has the same spatial coordinates as x_1 but has imaginary time coordinate $\tau_1 + \beta\hbar$. Then, we have

$$\begin{aligned} i\Delta_T(x_1', x_2) &= Z^{-1}\mathrm{tr}[e^{-\beta H} T(\hat{\phi}(x_1')\hat{\phi}(x_2))] \\ &= Z^{-1}\mathrm{tr}[e^{-\beta H} \hat{\phi}(x_1')\hat{\phi}(x_2)] \\ &= Z^{-1}\mathrm{tr}[e^{-\beta H}\{e^{\beta H}\hat{\phi}(x_1)e^{-\beta H}\}\hat{\phi}(x_2)] \\ &= Z^{-1}\mathrm{tr}[\hat{\phi}(x_1)e^{-\beta H}\hat{\phi}(x_2)] \\ &= Z^{-1}\mathrm{tr}[e^{-\beta H}\hat{\phi}(x_2)\hat{\phi}(x_1)] \\ &= Z^{-1}\mathrm{tr}[e^{-\beta H} T(\hat{\phi}(x_1)\hat{\phi}(x_2))] \\ &= i\Delta_T(x_1, x_2) \quad , \end{aligned} \tag{14.3.17}$$

where the cyclic property of the trace was used in the fifth line. More generally, when restricted to the strip $|\tau_1 - \tau_2| < \beta\hbar$, $\Delta_T(x_1, x_2)$ is periodic in each time variable with period $P = \beta\hbar$.

Thus, the Feynman propagator on Schwarzschild spacetime obtained from the Euclidean Schwarzschild solution is most naturally interpreted as a thermal Feynman propagator for the field at temperature

$$kT = \beta^{-1} = \hbar/P = \hbar/8\pi M \quad . \tag{14.3.18}$$

This suggests the existence of a state of thermal equilibrium of the quantum field at temperature (14.3.18) in Schwarzschild spacetime. This state—or, more precisely, density matrix—is known as the *Hartle-Hawking vacuum* (Hartle and Hawking 1976; Israel 1976) and it corresponds to what would result at late times if a thermal distribution at temperature (14.3.18) rather than $|0_{\text{in}}\rangle$ was sent in from $\mathscr{I}^-$. However, thermal equilibrium in Schwarzschild spacetime should not be possible unless the Schwarzschild black hole behaves like a perfect blackbody. Thus, the thermal nature of particle creation by a Schwarzschild black hole is strongly suggested by the properties of the Euclidean Schwarzschild solution. Note that this argument for the thermal properties of a Schwarzschild black hole is applicable to the case of a nonlinear (i.e., self-interacting) field (Gibbons and Perry 1976; Sewell 1982).

As a result of the thermal particle creation, it is clear that the quantum field carries energy away from a Schwarzschild black hole. The full stress-energy properties of the quantum field in Schwarzschild spacetime can be obtained from its stress-energy operator $\hat{T}_{ab}$. Since the problem of determining $\hat{T}_{ab}$ and using it to calculate "back reaction" effects of the quantum field on the spacetime is of interest in many applications (particularly, in cosmology), we first shall discuss these issues in the general context of quantum field theory in curved spacetime, and then return to the particular case of a black hole.

It is natural to postulate that the stress-energy operator of a quantum field in curved spacetime is given in terms of the field operator by the same formula as applies classically, i.e., for the Klein-Gordon quantum field,

$$\hat{T}_{ab} = \nabla_a \hat{\phi} \nabla_b \hat{\phi} - \frac{1}{2} g_{ab}[m^2 \hat{\phi}^2 + \nabla_c \hat{\phi} \nabla^c \hat{\phi}] \quad . \tag{14.3.19}$$

Unfortunately, this formula requires us to compose two field operators at the same spacetime point. Since, as mentioned above, $\hat{\phi}$ is well defined only as a distribution on spacetime, this product is ill defined. Indeed, if one formally substitutes the mode sum (14.2.15) for $\hat{\phi}$ into equation (14.3.19), the resulting expression one obtains for $\hat{T}_{ab}$ diverges.

This divergence in the expression for $\hat{T}_{ab}$ also occurs in Minkowski spacetime. In that case, it is interpreted as arising from the sum of the "zero-point energies" of the infinite number of modes of oscillation of the field. The divergence is cured by subtracting this zero-point energy from the formal expression for $\hat{T}_{ab}$ so that the expected stress-energy of the vacuum state vanishes, $\langle 0|\hat{T}_{ab}|0\rangle = 0$. This is equivalent to *normal ordering* the expression for $\hat{T}_{ab}$, i.e., placing all annihilation operators to the right of creation operators in the formal mode sum obtained from equation (14.3.19). In this manner, one obtains well defined, finite expressions for matrix elements of $\hat{T}_{ab}$ between physically reasonable states.

In curved spacetime, there is, in general, no meaningful notion of an "instantaneous vacuum state" with respect to which one could "normal order" the expression for $\hat{T}_{ab}$. Nevertheless, there exist a number of procedures for "regularizing" $\hat{T}_{ab}$, i.e., separating it in a natural way into the sum of a divergent part and a finite part. Many of these procedures—such as dimensional regularization (see, e.g., Brown 1977) and zeta-function regularization (Hawking 1977)—are rigorously defined only on a Riemannian manifold (see Wald 1979*c*), but at least one of them—namely, "point-splitting"—also is well defined on Lorentzian spacetimes (see Fulling 1983 for a summary of results). Unfortunately, the "divergent part" of $\hat{T}_{ab}$ cannot be "absorbed" into parameters already present in the theory; i.e., the determination of $\hat{T}_{ab}$ suffers from the same "non-renormalizability" difficulty as is present in full quantum gravity described in section 14.1 above. Consequently, one finds that one must introduce two new, nonclassical parameters into the expression for $\hat{T}_{ab}$ corresponding to the freedom of adding in the identity operator times multiples of the two conserved local curvature terms[15] of dimension $(\text{length})^{-4}$. Thus, there is a two-parameter ambiguity in the expression for $\hat{T}_{ab}$. However, $\hat{T}_{ab}$ satisfies a list of physically reasonable properties which uniquely determine it up to this two-parameter ambiguity (Wald 1977*b*, 1978*b*), so it appears that this is the only ambiguity present.

An important feature of the expectation value $\langle\Psi|\hat{T}_{ab}|\Psi\rangle$ of the quantum stress-energy operator $\hat{T}_{ab}$ in a state Ψ is that it need not satisfy any of the energy conditions that may be satisfied by the classical stress-energy tensor. Indeed, even in flat spacetime one can find states where the expectation value of the normal ordered Klein-Gordon stress-energy operator has negative energy density in a region of spacetime, even though the energy density of the classical stress-energy tensor of a Klein-Gordon field is manifestly positive definite everywhere for all field configurations (see problem 6). Thus, properties which hold in classical general relativity by virtue of energy conditions satisfied by matter will not necessarily hold for quantum fields. This has important consequences for the black hole area theorem, as will be discussed further in the next section.

In the context of quantum field theory in curved spacetime, it is natural to postulate that the back-reaction effects of the quantum field on the gravitational field will be governed by the semiclassical Einstein equation,

$$G_{ab} = 8\pi\langle\Psi|\hat{T}_{ab}|\Psi\rangle \quad ; \tag{14.3.20}$$

i.e., it is physically possible for the spacetime to be (M, g_{ab}) and for the quantum field to be in state Ψ on (M, g_{ab}) if and only if equation (14.3.20) is satisfied. Actually, equation (14.3.20) would *not* be expected to arise as the lowest approximation to a

15. These terms are the ones obtained by variation of the actions $\int R^2$ and $\int R_{ab}R^{ab}$ with respect to the metric (see appendix E). Most regularization prescriptions yield a precise value for one combination of these parameters, but since different prescriptions sometimes lead to different values, it probably is wisest to regard both of them as undetermined.

quantum field theory of gravity coupled to a matter field. This is because in the full theory one would expect to have $\langle \hat{G}_{ab} \rangle = 8\pi \langle \hat{T}_{ab} \rangle$ hold exactly, where $\hat{G}_{ab}$ is the Einstein operator and the state implicit in the expectation values now includes the degrees of freedom of the gravitational field. Furthermore, one would expect that $\hat{G}_{ab}$ would be given in terms of the metric operator by the same formula as holds classically, $\hat{G}_{ab} = G_{ab}[\hat{g}_{cd}]$. However, since G_{ab} is a nonlinear function of g_{cd} we expect $\langle \hat{G}_{ab} \rangle \neq G_{ab}[\langle \hat{g}_{cd} \rangle]$. Indeed, if we write $\hat{g}_{ab} = g^C_{ab}\hat{I} + \hat{\gamma}_{ab}$— where g^C_{ab} is a classical solution of Einstein's equation and $\hat{I}$ is the identity operator—and if we keep only terms quadratic in $\hat{\gamma}_{ab}$ in the formula for $\hat{G}_{ab}$, then $\langle \hat{G}_{ab} \rangle$ and $G_{ab}[\langle \hat{g}_{ab} \rangle]$ will differ by $-8\pi\langle \hat{t}_{ab} \rangle$, where $\hat{t}_{ab}$ is given in terms of $\hat{\gamma}_{ab}$ by a formula very similar to that for $\hat{T}_{ab}$ in terms of $\hat{\phi}$. (See eqs. [4.4.54] and [4.4.51] for an explicit formula for t_{ab} in the case $g^C_{ab} = \eta_{ab}$.) Consequently, in the lowest approximation to a full quantum field theory of gravity coupled to matter, one would expect to get the additional term $8\pi\langle \hat{t}_{ab} \rangle$ appearing on the right-hand side of equation (14.3.20), and the contribution from this term should be comparable to that of $\langle \hat{T}_{ab} \rangle$. One can interpret this fact as saying that the quantum back-reaction effects caused by *gravitons* (i.e., the quantized degrees of freedom of the linearized gravitational field) are as important as that of any other quantum field, and thus should not be neglected in equation (14.3.20). Nevertheless, one can justify equation (14.3.20) in terms of a systematic approximation to a full quantum field theory including gravitation as follows. If we have N matter fields present, then, roughly speaking, the effects of the matter fields will be N times as important as that of the gravitons. Hence, in the limit of large N, the neglect of the gravitons should be justified, and one will obtain equation (14.3.20) (with a coefficient of N on the right-hand side) as the lowest approximation in a "$1/N$ expansion" of the full theory of quantum gravity coupled to matter. In any case, equation (14.3.20) should at least provide a qualitative indication of the back reaction effects produced by quantum fields on the gravitational field.

Unfortunately, the dynamics predicted by equation (14.3.20) is drastically different from classical general relativity. There exist many unphysical solutions of equation (14.3.20) in which the spacetime curvature is initially small but exponentially grows with time with time scale of order of the Planck time. Indeed, Horowitz (1980) has shown that for any massless quantum field, equation (14.3.20) predicts this type of instability of Minkowski spacetime. Thus, the situation is very similar to that which occurs in the classical electrodynamics of a point charge when radiation reaction is taken into account (see, e.g., Jackson 1962). There, the "small" correction to the equations of motion caused by radiation reaction leads to new "runaway" solutions. In the case of electrodynamics, one still can make physically sensible predictions by simply disregarding these runaway solutions, although doing so requires putting constraints on the initial conditions of the point charge which depend upon what external forces are to be applied in the future. It appears that it will be much more difficult to develop procedures for extracting the physically sensible solutions of equation (14.3.20).

In the case of a Schwarzschild black hole (or, more generally, in any vacuum spacetime, $R_{ab} = 0$) the two local curvature terms which enter the formula for $\hat{T}_{ab}$ with undetermined coefficients vanish identically. Hence $\hat{T}_{ab}$ is completely well defined in the Schwarzschild spacetime. The value of $\langle \hat{T}_{ab} \rangle$ in the "Hartle-Hawking vacuum" has been calculated analytically on the horizon by Candelas (1980) and computed numerically near the black hole by Fawcett (1983). The magnitude of the Kruskal coordinate components of $\langle \hat{T}_{ab} \rangle$ near the black hole are found to be of order $1/M^4$ in Planck units $G = c = \hbar = 1$, as expected on dimensional grounds. Since the background curvature is of order $1/M^2$, for $M >> 1$ (i.e., in cgs units, $M >> 10^{-5}$ g), the quantum field should make only a small correction to the structure of the black hole. The energy flux into the black hole is found to be negative, as must be the case since the "Hartle-Hawking vacuum" is time independent and the energy flux at future null infinity is positive. Such a negative energy flux is possible since, as mentioned above, $\langle \hat{T}_{ab} \rangle$ need not satisfy any of the classical energy conditions.

As discussed above, serious difficulties arise when one tries to use equation (14.3.20) to calculate the back-reaction effects. However, on physical grounds, one expects that the main back-reaction effect of the quantum field will be to cause the black hole to lose mass at the same rate at which energy is radiated to infinity by particle creation. This can be calculated by multiplying $\langle N \rangle$, equation (14.3.7), by ω and summing over modes. Since the emission is of a blackbody nature, if the absorption cross section $|t|^2$ for each mode were the same as for a black sphere of radius R in Minkowski spacetime and if $kT >> \hbar c R^{-1}$, the energy flux for a massless field with two degrees of freedom would be given by Stefan's law, $dE/dt = \sigma T^4 A = \sigma T^4 4\pi R^2$, where $\sigma = \pi^2 k^4 / 60 \hbar^3 c^2$ is the Stefan-Boltzmann constant. In the geometric optics approximation, the black hole does absorb just like a black sphere of radius $R = 3^{3/2} M$ (see chapter 6). However, since the relevant modes in black hole particle creation have frequencies of order M^{-1} (i.e., one has $kT \sim \hbar c R^{-1}$) one must use physical optics[16] (i.e., exact wave propagation) to compute $|t|^2$ accurately. Nevertheless, Stefan's law provides a correct order of magnitude estimate for the energy flux from a black hole. Hence, ignoring numerical factors, we find that in Planck units, the mass loss rate of the black hole is approximately

$$\frac{dM}{du} \sim -AT^4 \sim -M^2(1/M)^4 = -\frac{1}{M^2} \quad . \tag{14.3.21}$$

Thus, as the black hole loses mass, the increase in temperature more than compensates for the decrease in area, and the energy loss due to particle creation occurs at a faster rate. Integrating equation (14.3.21), we obtain the striking conclusion that a black hole should radiate *all* of its mass in a finite time τ given in Planck units by

$$\tau \sim M^3 \quad . \tag{14.3.22}$$

In cgs units, M^3 is $\sim 10^{71} (M/M_\odot)^3$ s, so the lifetime for total "evaporation" of a black hole of a solar mass or larger is enormously greater than the age of the universe.

16. The failure of geometrical optics to accurately describe wave propagation well outside the horizon (where most of the scattering occurs) should be distinguished from the validity of geometrical optics to describe wave propagation very near the horizon, as used above in the calculation of particle creation.

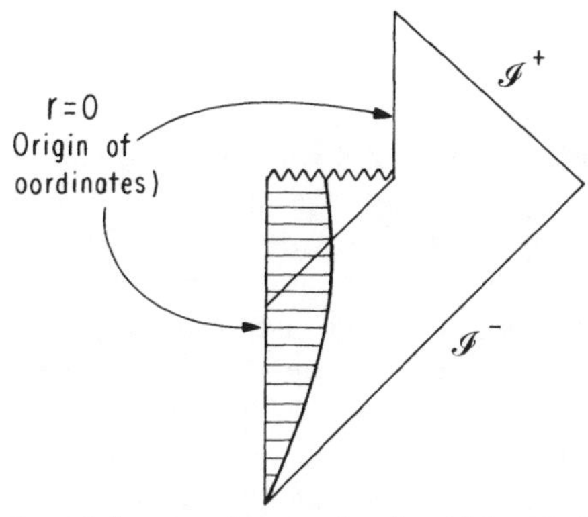

Fig. 14.4. A conformal diagram of a spacetime in which a black hole is produced by the collapse of matter and "evaporates" as a result of the particle creation process.

However, if any primordial black holes of mass $\sim 5 \times 10^{14}$ g were produced in the early universe, they would be undergoing the final stages of evaporation now (Page 1976*a*), and any primordial black holes of smaller mass would have already evaporated. Since the temperature of a black hole becomes large as evaporation occurs, a significant amount of γ-radiation would result if many black holes of mass $\lesssim 5 \times 10^{14}$ g were produced in the early universe. As already mentioned at the end of section 12.1, no such contribution to the γ-ray background has been observed (Page and Hawking 1976), so it appears that we do not have a sufficient number of low mass primordial black holes in our universe to be able to observe the effects of black hole evaporation.

By the time the black hole has evaporated down to the Planck mass, we cannot expect the approximation of treating gravity classically to be valid, and a description of what occurs at that stage will have to await to complete theory of quantum gravity. However, it seems natural to expect that the black hole will totally disappear at the end of the process, rather than, say, leaving behind a Planck-size black hole remnant. The causal structure of a spacetime in which a black hole is formed and then evaporates completely is illustrated in Figure 14.4.

Assuming that total evaporation occurs, we mention two important consequences. First, it appears that conservation of baryons (and/or leptons) can be grossly violated in the process of collapse to a black hole followed by evaporation even if baryon number (and/or lepton number) is conserved locally; namely, we can form the black hole out of matter with a large net baryon number (e.g., purely out of neutrons, with no antineutrons). The black hole uniqueness theorems discussed at the end of section 12.3 strictly apply only to the case where certain classically describable fields are present, but they strongly suggest that the black hole produced by the collapse of baryons will be indistinguishable from one produced by the collapse of antibaryons. Hence, in the particle creation process, there should be no preferential production of baryons over antibaryons and, indeed, in any case most of the mass of the black hole should be radiated away by massless particles carrying no baryon number. Thus, at the end of the evaporation process, the net baryon number should be very nearly zero and a large change in baryon number will have occurred.

Second, as discussed above, the particles reaching infinity during the evaporation process are described by a density matrix. The use of a density matrix here is purely for convenience, since the joint state of the particles entering the black hole together with those reaching infinity is pure, provided, of course, that the incoming state is pure. However, if the black hole evaporates completely as in Figure 14.4, then the *total* state of the system should be described by a density matrix. Roughly speaking, the correlations with the particles reaching infinity should propagate into the black hole singularity and be lost forever when the black hole disappears. Thus, it appears that in the process of black hole formation and evaporation, an initial pure state can evolve to a final density matrix (Hawking 1976; Wald 1980; see Page 1980 for a differing view). This shows that a form of time reversal asymmetry must be present, since an initial density matrix never can evolve to a final pure state, although a physically meaningful form of time reversal symmetry may still hold (Wald 1980). The existence of time reversal asymmetry in the laws of quantum gravity has been suggested on other grounds by Penrose (1979). Furthermore, the evolution of a pure state to a density matrix in the process of black hole formation and evaporation may be closely related to issues in quantum measurement theory (Penrose 1981).

Finally, we briefly describe a further result which shows a connection between particle creation near a black hole and properties of the ordinary vacuum state of a quantum field in Minkowski spacetime. This result also gives insight into the physical meaning of "particles" in curved spacetime and the reasons for the mathematical ambiguities in their definition.

The close mathematical analogy between the "Rindler wedge" of Minkowski spacetime and the region $r > 2M$ of Schwarzschild spacetime already has been displayed at the end of chapter 6. Indeed, there we used the extension of Rindler spacetime to Minkowski spacetime to motivate the Kruskal extension of Schwarzschild spacetime. One can define a quantum field theory in Rindler spacetime by using the Rindler time coordinate, t_R, rather than a Minkowski time translation coordinate, t_M, to define the notion of "positive frequency." In other words, we may "quantize" the field in (a wedge of) Minkowski spacetime by using a "boost" rather than an ordinary time translation as our timelike Killing vector field. These two notions of "positive frequency" differ, so one is led to distinct notions of "Rindler particles" and "Minkowski particles" (Fulling 1973). The relation between the characterization of states of the quantum field in terms of Rindler particles as compared with their characterization in terms of Minkowski particles can be found by obtaining the Bogoliubov transformation between Rindler and Minkowski annihilation and creation operators, using the same procedure as in the case of true particle creation described in the previous section. Since the relation between Rindler time and Minkowski time is essentially the same as that holding near the horizon between Schwarzschild time and Kruskal time, it should not be surprising that the Minkowski positive and negative frequency parts of a solution which oscillates in Rindler time like $e^{-i\omega t_R}$ are, again, related by equation (14.3.6). In close mathematical analogy with the calculation of true particle creation in the Schwarzschild spacetime, it is found here that the ordinary Minkowski vacuum $|0\rangle$ is formally represented by a thermal density matrix of Rindler particles.

A physical interpretation of this result was provided by Unruh (1976). He showed that a particle detector which uniformly accelerates—i.e., travels along a trajectory of a boost Killing field—is sensitive to the Rindler particles associated with that Killing field. This is most easily seen by considering a simple model of a particle detector consisting of an ordinary nonrelativistic quantum mechanical system (e.g., a particle in a box) linearly coupled to the quantum field. Detection of a particle is said to occur if a transition to a higher energy state occurs in the quantum mechanical system. From ordinary time-dependent perturbation theory, when the quantum field is in the ordinary Minkowski vacuum state, $|0\rangle$, the rate (i.e., probability per unit time) for a transition upward in energy by E for an accelerating point detector is found to be (see Unruh 1976 or DeWitt 1979)

$$R(E) \propto \lim_{T\to\infty} \frac{1}{T} \int_{-T}^{T} \int_{-T}^{T} e^{-iE(t_R - t_R')} \langle 0 | \hat{\phi}[x(t_R)] \hat{\phi}[x(t_R')] | 0 \rangle \, dt_R dt_R' \quad , \qquad (14.3.23)$$

where t_R has been scaled so that it agrees with proper time along the world line, $x(t_R)$, of the accelerating particle detector. Thus, the transition probability is directly related to a Fourier transform of the vacuum expectation value of the product of field operators, where this Fourier transform picks out a positive frequency component with respect to t_R' and a negative frequency component with respect to t_R. However, for an accelerating detector these "positive and negative frequency parts" are measured with respect to Rindler time rather than Minkowski time. Although by equation (14.2.13), $\hat{\phi}|0\rangle$ has no positive frequency part with respect to Minkowski time and $\langle 0|\hat{\phi}$ has no negative frequency part (which accounts for why an inertial particle detector does not detect particles), they *do* have such positive and negative frequency parts with respect to Rindler time. This corresponds precisely to the above representation of the Minkowski vacuum as a thermal state of Rindler particles. Thus, Unruh (1976) found that an accelerating particle detector in Minkowski spacetime would behave as though it were placed in a thermal bath of "real" particles, with temperature given by

$$kT = \frac{\hbar a}{2\pi c} \quad . \qquad (14.3.24)$$

Thus, the ordinary vacuum state in Minkowski spacetime is seen by an accelerating observer to possess thermal properties which are formally very similar to the thermal effects resulting from true particle creation by a black hole. Unfortunately, in cgs units, equation (14.3.24) reads

$$T \approx 4 \times 10^{-23} a \text{ K} \quad , \qquad (14.3.25)$$

so it is clear that this effect is much too small to be perceived by an ordinary laboratory detector. However, the effect of this thermal bath on the spin of accelerating electrons may be measurable (Bell and Leinaas 1983). Further discussion of this effect from the viewpoint of an inertial observer is given by Unruh and Wald (1984).

Thus, the "Rindler particles" defined by using a nonstandard notion of positive frequency in Minkowski spacetime are seen to be associated with true physical effects. Rindler particles are "real" to accelerating observers! This shows that different notions of "particle" are useful for different purposes. Therefore, in a general curved spacetime where no timelike Killing field is present, it should not be surprising that there is no natural, universally applicable procedure for defining particles and that, in a given spacetime, different constructions (such as constructions using affine parameter or Killing parameter on the horizon of a black hole) will define different notions of particles. Although the notion of particles is very convenient for many purposes, it is not an essential ingredient of quantum theory. Physical predictions always can be expressed in terms of matrix elements $\langle\Phi| O |\Psi\rangle$ of an operator O between states Φ and Ψ, such as in equation (14.3.23) or equation (14.3.20). These predictions will not depend upon how Φ and Ψ are labeled in terms of particle states.

A similar analysis of the behavior of a particle detector in Schwarzschild spacetime shows that when the field is in the Hartle-Hawking vacuum state, a stationary detector at any radius $r > 2M$ will behave as though immersed in a thermal bath at temperature

$$kT = \frac{\hbar\kappa}{2\pi c V} \quad , \tag{14.3.26}$$

where $V = (-\xi^a \xi_a)^{1/2}$ is the redshift factor appropriate to that radius. Far from the black hole ($r \to \infty$) we have $V \to 1$, and the effect may be interpreted as the response of the detector to the "real" particles produced by the black hole (as well as those incoming from $\mathscr{I}^-$). Near the black hole ($r \to 2M$) we have $\kappa/V \to a$, where a is the proper acceleration of the detector (see eq. [12.5.18]), and the effect (which becomes divergent at the horizon) may be interpreted as being due to the acceleration of the detector. A freely falling detector sees essentially no particles near the black hole (Unruh 1977), in accordance with the negligible stress-energy $\langle\hat{T}_{ab}\rangle$ of the quantum field near the horizon.

14.4 Black Hole Thermodynamics

In section 12.5 we obtained a remarkable mathematical analogy between the ordinary laws of thermodynamics and laws applying to black holes which were derived from classical general relativity. As can be seen from Table 12.1, if one makes the formal replacements $E \to M$, $T \to \alpha\kappa$, and $S \to A/8\pi\alpha$ in the laws of thermodynamics, one obtains valid laws applying to black holes. A hint that this relationship might be more than just a formal analogy already arose from the fact that the analogous quantities E and M actually represent the same physical quantity, namely, total energy. However, since the thermodynamic temperature of a black hole in classical general relativity is absolute zero, the physical analogy appeared to end there.

However, we now have seen that when quantum effects are taken into account, then in a very real sense the thermodynamic temperature of a black hole is not zero but is $\kappa/2\pi$ in units where $k = \hbar = G = c = 1$. A black hole absorbs and "emits" particles exactly like a perfect blackbody at that temperature. Thus, T and $\kappa/2\pi$ are not merely analogous quantities but again represent the same physical quantity. Setting $\alpha = (2\pi)^{-1}$, we see that the remaining analogous quantities are S and $\frac{1}{4}A$. Does $\frac{1}{4}A$ physically represent the entropy of a black hole?

We now shall argue that the answer to this question appears to be yes. The ordinary second law of thermodynamics states that the total entropy of matter in the universe never decreases. However, if a black hole is present, one would like to restrict attention to matter outside black holes, since matter that falls in is "swallowed up" by the singularity within the black hole and, in any case, it cannot be measured by an external observer. However, one easily can make the total entropy, S, of matter outside black holes decrease by dropping matter into a black hole. On the other hand, the area theorem of classical general relativity states that the surface area, A, of a black hole never decreases. However, on account of the quantum particle creation process, this law can be violated since $\langle \hat{T}_{ab} \rangle$ does not satisfy the energy condition assumed in the proof of the area theorem. Indeed, as discussed in the previous section, an isolated black hole eventually "evaporates" completely, thereby decreasing its area to zero. Thus, when black holes are present and quantum effects are taken into account, both the ordinary second law and the area theorem can be violated. However, in the processes where $\delta S < 0$ due to loss of matter into a black hole, we increase the black hole area, $\delta A > 0$. Similarly, in the evaporation process where $\delta A < 0$, we increase the entropy of matter outside black holes, $\delta S > 0$, by the emission of thermal radiation. Therefore, let us define the *generalized entropy*, S', by (Bekenstein 1973*b*, 1974)

$$S' = S + \frac{1}{4} k \frac{c^3 A}{G\hbar} \quad , \tag{14.4.1}$$

where we have restored the constants k, $\hbar$, G, and c in this formula. The fact that a decrease in S seems always to be compensated by an increase in A and, similarly, a decrease in A seems always to be compensated by an increase in S, suggests that in any process the *generalized second law*,

$$\delta S' \geqq 0 \quad , \tag{14.4.2}$$

may be valid.

In fact, the law (14.4.2) was first proposed by Bekenstein (with an undetermined numerical factor in the coefficient of A in eq. [14.4.1]) prior to the discovery of the quantum particle creation effects. However, in the context of purely classical general relativity, equation (14.4.2) could be violated by putting a black hole in a thermal bath at a temperature lower than that formally assigned to the black hole, thereby producing heat flow from a cold body (the bath) to a hotter body (the black hole). Alternatively, one could put matter in a box and carefully lower the box to the horizon of the black hole before dropping it in. By this latter procedure, the entropy of matter inside the box still would be lost, but by "redshifting away" the energy in the box, the area increase of the black hole could be kept arbitrarily small. However, when the quantum effects are taken into account, these methods for violating the generalized second law do not work. In the first example, the black hole will radiate more than it will absorb, so heat actually flows from the black hole to the bath. In the second example, the effects of the radiation—described at the end of the previous section—which is felt by any stationary body near the black hole alters the transfer of energy into the black hole in just such a way as to ensure that the area increase of the black hole always is large enough to keep equation (14.4.2) satisfied (Unruh and Wald 1982). Thus, the generalized second law appears to hold, at least insofar as it can be tested by gedankenexperiments.

However, the generalized second law has a natural and simple interpretation. It can be viewed as nothing more than the *ordinary* second law of thermodynamics applied to a system containing a black hole. In order to maintain this view, one must take the final step of assigning $\frac{1}{4}A$ as the *physical* entropy of a black hole. If this final step is taken, the analogy between the laws of thermodynamics and the laws of black hole physics no longer would be an analogy at all. The laws of black hole thermodynamics (including the generalized second law) would be seen as being nothing more than the *ordinary* laws of thermodynamics applied to a self-gravitating quantum system containing a black hole.

Thus, the apparent validity of the generalized second law strongly suggests that $\frac{1}{4}A$ is the thermodynamic entropy of a black hole. However, the underlying physical basis by which $\frac{1}{4}A$ arises as the black hole entropy remains unclear. Since the entropy of an ordinary physical system is essentially the logarithm of the number of microscopic states compatible with the observed macroscopic state, the assignment of $\frac{1}{4}A$ as the black hole entropy seems to indicate that in a full quantum theory of gravity the number of "internal states" of a black hole will be $N \sim e^{A/4}$. However, general arguments for the validity of the second law of thermodynamics for ordinary systems are based on notions of the "fraction of time" a system spends in a given macroscopic state. Since the nature of time in general relativity is drastically different from that in nongravitational physics, it is not clear precisely how the generalized second law will arise even if $\frac{1}{4}A$ is a measure of the number of internal states of a black hole.

Thus, we appear to be in a situation with regard to black hole thermodynamics which is very similar to the situation with regard to ordinary thermodynamics prior to the discovery of the underlying basis of these laws arising from statistical physics. We have discovered the laws of black hole thermodynamics—in this case by calculations and gedankenexperiments rather than by laboratory experiments—but the underlying basis of these laws is not known and presumably will not be fully understood until we have a quantum theory of gravitation. Nevertheless, the existence of the laws of black hole thermodynamics indicates the likelihood of a deep connection between gravitation, quantum theory, and statistical physics. It remains for future investigations to explore this connection further.

Problems

1. Let V be the collection of infinite sequences of complex numbers $\{a_i\} = (a_1, a_2, \ldots ,)$ such that only *finitely* many of the a_i are nonzero. Define addition and scalar multiplication on V by $\{a_i\} + \{b_i\} = \{a_i + b_i\}$ and $c\{a_i\} = \{ca_i\}$ to make V a vector space over $\mathbb{C}$. Define $(\{a_i\}, \{b_i\}) = \sum_{i=1}^{\infty} \overline{a}_i b_i$.

a) Show that (,) is an inner product, thus making V an inner product space.

b) Show that V is incomplete and thus is *not* a Hilbert space.

c) Let V' be the collection of sequences satisfying $\sum_{i=1}^{\infty}|a_i|^2 < \infty$. Define a vector space structure and inner product on V' in the same way as for V. Show that V' is complete and thus defines a Hilbert space, usually denoted l_2. (You may use the fact that $\mathbb{C}$ is complete in your proof.)

2. *a*) Show that the annihilation and creation operators defined by equations (14.2.10) and (14.2.12) satisfy the commutation relations

$$[a(\overline{\sigma}), a(\rho)] = 0 \quad ,$$

$$[a^\dagger(\sigma), a^\dagger(\rho)] = 0 \quad ,$$

$$[a(\overline{\sigma}), a^\dagger(\rho)] = (\sigma, \rho)I$$

for all $\sigma, \rho \in \mathcal{H}$, where the commutator of two operators is defined by $[A, B] = AB - BA$, and I is the identity operator on $\mathcal{F}_S(\mathcal{H})$.

b) Show that the operator

$$N(\sigma) = a^\dagger(\sigma)a(\overline{\sigma})$$

can be interpreted as the number operator for the (normalized) state $\sigma \in \mathcal{H}$, i.e., the eigenstates of $N(\sigma)$ are the states with a definite number of particles in state σ, and the eigenvalues of $N(\sigma)$ in these eigenstates are the number of such particles.

3. Define the operators $A: \mathcal{H}_{\text{out}} \to \mathcal{H}_{\text{in}}$, $B: \mathcal{H}_{\text{out}} \to \overline{\mathcal{H}_{\text{in}}}$ by the "time reverse" of the definitions of C and D given in the text, i.e., for $\tau \in \mathcal{H}_{\text{out}}$, $A\tau$ is the state in $\mathcal{H}_{\text{in}}$ obtained by propagating into the past the solution $\tilde{\tau}$ which agrees with τ in the future and taking its positive frequency part in the past, while $B\tau$ is the state in $\overline{\mathcal{H}_{\text{in}}}$ associated with its negative frequency part in the past. Use the independence of the Klein-Gordon inner product, equation (14.2.5), on the choice of Cauchy surface together with the fact that in the asymptotic future or past the Klein-Gordon inner product of any positive frequency solution with any negative frequency solution vanishes to prove that the following relations are satisfied by A, B, C, D:

$$A^\dagger A - B^\dagger B = I \quad ,$$

$$B^\dagger \overline{A} = A^\dagger \overline{B} \quad ,$$

$$C^\dagger C - D^\dagger D = I \quad ,$$

$$D^\dagger \overline{C} = C^\dagger \overline{D} \quad ,$$

$$A^\dagger = C, \; B^\dagger = -\overline{D} \quad .$$

Note that the relation $D^\dagger \overline{C} = C^\dagger \overline{D}$ implies that $E = \overline{D}\,\overline{C}^{-1}$ is symmetric, $E^\dagger = \overline{E}$.

4. Consider a conformally invariant field (see appendix D) in a spacetime of the form illustrated in Figure 14.1 which, in addition, is conformally flat, i.e., $g_{ab} = \Omega^2 \eta_{ab}$ everywhere. Show that no particle creation can occur.

5. With A and B defined as in problem 3, it is clear that by "time reversal symmetry" equation (14.2.17) must continue to hold if we interchange "in" and "out," replace C and D by A and B, and replace S by S^{-1}.

a) Using this modification of equation (14.2.17) and the results of problem 2(b), show that the expected number of particles in state $\sigma \in \mathcal{H}_{\text{out}}$ spontaneously created from the "in" vacuum state is given by

$$\langle N(\sigma)\rangle = \|B\sigma\|^2 \quad .$$

b) Using equation (14.3.6) as well as the result of part (a) and the relations proven in problem 3, derive equation (14.3.7) for particle creation by black holes.

6. The classical Klein-Gordon stress-energy tensor (4.3.10) has manifestly positive energy density everywhere; i.e., for any timelike vector t^a at any point we always have $T_{ab}\,t^a t^b \geq 0$. Show that this property does *not* carry over to quantum field theory by finding a state, Ψ, of the free Klein-Gordon field in Minkowski spacetime for which the normal ordered stress-energy operator satisfies $\langle\Psi|\,\hat{T}_{ab}\,|\Psi\rangle\, t^a t^b < 0$ over a region of spacetime. (Hint: Consider a state Ψ obtained by superposing the vacuum state with a small admixture of a two-particle state.) Note, however, that *total* energy $E = \int_\Sigma \langle \hat{T}_{ab}\rangle \xi^a n^b$, where ξ^a is a time translation Killing field, always is nonnegative for the free Klein-Gordon field in Minkowski spacetime.

7. Consider a box of volume V filled with blackbody radiation and possibly also containing a black hole. Ignore the influence of the black hole on the distribution of radiation in the box. Suppose that the total energy contained in the box (i.e., the mass of the black hole plus the energy of the radiation) is E.

a) Write down the formula for the total generalized entropy, S', of the box as a function of the mass-energy M apportioned to the black hole.

b) By extremizing S' with respect to M, determine the conditions which must be satisfied by E and V in order that equilibrium between the black hole and radiation be possible (Gibbons and Perry 1978). For $V = 1$ meter^3, evaluate the minimum value of E needed for equilibrium. In the case where equilibrium is possible, determine which of the extrema of S' are locally stable by computing the second derivative of S'.

c) Estimate the conditions on E and V such that the configuration of (globally) maximum generalized entropy will contain a black hole rather than be pure radiation.

APPENDIX A

TOPOLOGICAL SPACES

In mathematics, many different proofs and arguments are based on the same or a similar set of ideas. An important goal of mathematics is to isolate the key ideas in as general a form as possible, since if one can derive general results about these ideas in the abstract, one then may apply them to the cases of interest without duplication of effort. The notion of a topological space provides a beautiful illustration of this program. One starts with a very simple, abstract definition and ends up proving many powerful results that have a very wide range of application. Our main interest here in topological spaces arises from the fact that in general relativity spacetime has the structure of a topological space. Indeed, spacetime has a great deal more structure, and we circumvented the mention of topological spaces in chapter 2 by defining the notion of a manifold directly. However, the purely topological arguments and results discussed in this section play an important role in many of the constructions and proofs of chapters 8 and 9. Therefore, we collect in this appendix many of the key definitions and theorems concerning topological spaces.

A *topological space* $(X, \mathcal{T})$ consists of a set X together with a collection $\mathcal{T}$ of subsets of X satisfying the following three properties:
(1) The union of an arbitrary collection of subsets, each of which is in $\mathcal{T}$, is in $\mathcal{T}$: If $O_\alpha \in \mathcal{T}$ for all α, then $\bigcup_\alpha O_\alpha \in \mathcal{T}$.
(2) The intersection of a finite number of subsets in $\mathcal{T}$ is in $\mathcal{T}$: If $O_1, \ldots, O_n \in \mathcal{T}$, then

$$\bigcap_{i=1}^{n} O_i \in \mathcal{T}.$$

(3) The entire set X and the empty set $\emptyset$ are in $\mathcal{T}$.
$\mathcal{T}$ is referred to as a *topology* on X, and subsets of X which are listed in the collection $\mathcal{T}$ are called *open sets*.

Any set X can easily be made into a topological space by taking $\mathcal{T}$ = {all subsets of X} (the *discrete topology*) or by taking $\mathcal{T} = \{X, \emptyset\}$ (the *indiscrete topology*). A much more interesting example of a topological space is obtained by taking $X = \mathbb{R}$, the set of real numbers, and defining $\mathcal{T}$ to consist of all subsets of $\mathbb{R}$ which can be expressed as unions of open intervals (a, b). (Thus, with this choice of $\mathcal{T}$ on $\mathbb{R}$ an open interval is an open set; historically, this example is the reason why the terminology "open set" is used in the discussion of an abstract topological space.) More generally, for any metric space, the collection of all subsets which can be expressed as unions of open balls yields a topology.

If $(X, \mathcal{T})$ is a topological space and A is any subset of X, we may make A itself into a topological space by defining the topology, $\mathcal{S}$, on A to consist of all subsets of A which can be expressed as intersections of elements of $\mathcal{T}$ with A, $\mathcal{S} = \{U \mid U = A \cap O, O \in \mathcal{T}\}$. $\mathcal{S}$ is called the *induced* (or *relative*) topology.

If $(X_1, \mathcal{T}_1)$ and $(X_2, \mathcal{T}_2)$ are topological spaces, we can make the product space $X_1 \times X_2 \equiv \{(x_1, x_2) \mid x_1 \in X_1, \ x_2 \in X_2\}$ into a topological space $(X_1 \times X_2, \mathcal{T})$ by defining $\mathcal{T}$ to consist of all subsets of $X_1 \times X_2$ which can be expressed as unions of sets of the form $O_1 \times O_2$ with $O_1 \in \mathcal{T}_1$ and $O_2 \in \mathcal{T}_2$. $\mathcal{T}$ is called the *product* topology, and using the above definition of a topology on $\mathbb{R}$, we can use the construction of the product topology to define a topology on $\mathbb{R}^n$. The topology we get is the same one as would be obtained by directly defining $\mathcal{T}$ to consist of all subsets of $\mathbb{R}^n$ which can be expressed as unions of open balls.

If $(X, \mathcal{T})$ and $(Y, \mathcal{S})$ are topological spaces, a map $f: X \to Y$ is said to be *continuous* if the inverse image, $f^{-1}[O] \equiv \{x \in X \mid f(x) \in O\}$, of every open set O in Y is an open set in X. For functions from $\mathbb{R}$ into $\mathbb{R}$, it is easy to verify that with the above definition of a topology on $\mathbb{R}$, this definition of continuity is equivalent to the usual $\epsilon - \delta$ definition. If f is continuous, one-to-one, onto, and its inverse is continuous, f is called a *homeomorphism* and $(X, \mathcal{T})$ and $(Y, \mathcal{S})$ are said to be homeomorphic. Homeomorphic topological spaces have identical topological properties.

If $(X, \mathcal{T})$ is a topological space, a subset C of X is said to be *closed* if its complement $X - C \equiv \{x \in X \mid x \notin C\}$ is open. Thus, for example, a closed interval $[a, b]$ of $\mathbb{R}$ (with the standard topology on $\mathbb{R}$) is a closed set. From the topological space axioms it follows immediately that the intersection of an arbitrary collection of closed sets is closed and the union of a finite number of closed sets is closed. Note that a subset may be neither open nor closed (e.g., the half-open interval $[a, b)$ in $\mathbb{R}$) or may be both open and closed (as are all subsets in the discrete topology). Indeed, the possibility of having subsets which are both open and closed gives rise to a topological definition of connectedness: A topological space $(X, \mathcal{T})$ is said to be *connected* if the only subsets which are both open and closed are the entire space X and the empty set $\emptyset$. $\mathbb{R}^n$ with the standard topology defined above is connected.

If $(X, \mathcal{T})$ is a topological space and A is an arbitrary subset of X, the *closure*, $\bar{A}$, of A is defined as the intersection of all closed sets containing A. Clearly, $\bar{A}$ is closed, contains A, and equals A if and only if A is closed. The *interior* of A is defined as the union of all open sets contained within A. Clearly the interior of A is open, is contained in A, and equals A if and only if A is open. The *boundary* of A, denoted $\dot{A}$, consists of all points which lie in $\bar{A}$ but not in the interior of A.

A topological space $(X, \mathcal{T})$ is said to be *Hausdorff* if for each pair of distinct points $p, q \in X, p \neq q$, one can find open sets $O_p, O_q \in \mathcal{T}$ such that $p \in O_p, q \in O_q$, and $O_p \cap O_q = \emptyset$. It is easy to check that $\mathbb{R}^n$ with the standard topology is Hausdorff.

One of the most powerful notions in topology is that of compactness, which is defined as follows. If $(X, \mathcal{T})$ is a topological space and A is a subset of X, a collection $\{O_\alpha\}$ of open sets is said to be an *open cover* of A if the union of these sets contains A. A subcollection of the sets $\{O_\alpha\}$ which also covers A is referred to as a *subcover*. The set A is said to be *compact* if every open cover of A has a finite subcover (i.e., a subcover consisting of only a finite number of sets). Thus, for example, in any topological space a set consisting of a single point is compact. On the other hand, the open interval $(0, 1)$ in $\mathbb{R}$ (with the standard topology) is not compact since the sets $O_n = (1/n, 1)$ for $n = 2, 3, \ldots$, yield an open cover of $(0, 1)$ which admits no finite subcover.

The following theorems describe the implications of compactness and show the utility of this notion. Proofs of the theorems can be found in nearly any textbook on topology (e.g., Hocking and Young 1961; Kelley 1955).

Perhaps the most important theorem concerning compact subsets of $\mathbb{R}$ is the Heine-Borel theorem:

THEOREM A.1 (Heine-Borel). *A closed interval $[a, b]$ of real numbers is compact (with the standard topology on $\mathbb{R}$).*

The general relation between compact and closed sets is described by the following two theorems, the proofs of which are straightforward:

THEOREM A.2. *Let $(X, \mathcal{T})$ be Hausdorff and let $A \subset X$ be compact. Then A is closed.*

THEOREM A.3. *Let $(X, \mathcal{T})$ be compact and let $A \subset X$ be closed. Then A is compact.*

Combining the above three theorems, we arrive at the following strengthened statement on the compactness of subsets of $\mathbb{R}$:

THEOREM A.4. *A subset A of real numbers is compact if and only if it is closed and bounded.*

The property of compactness is easily proven to be preserved under continuous maps. We have:

THEOREM A.5. *Let $(X, \mathcal{T})$ and $(Y, \mathcal{S})$ be topological spaces. Suppose $(X, \mathcal{T})$ is compact and $f: X \to Y$ is continuous. Then $f[X] \equiv \{y \in Y \mid y = f(x)\}$ is compact.*

Because of the properties of compact subsets of $\mathbb{R}$ given by Theorem A.4, we have as a corollary

THEOREM A.6. *A continuous function from a compact topological space into $\mathbb{R}$ is bounded and attains its maximum and minimum values.*

The following theorem gives an immediate extension of results on compactness for $\mathbb{R}$ to results for $\mathbb{R}^n$.

THEOREM A.7 (Tychonoff theorem). *Let $(X_1, \mathcal{T}_1)$ and $(X_2, \mathcal{T}_2)$ be compact topological spaces. Then the product space $X_1 \times X_2$ is compact in the product topology.*

Theorem A.7 can be generalized to apply to the product of infinitely many topological spaces, but the axiom of choice is needed for this generalization.

A corollary of this and the above theorems is

THEOREM A.8. *A subset, A, of* $\mathbb{R}^n$ *is compact if and only if it is closed and bounded.*

Thus, for example, the n-dimensional sphere S^n (defined as the set of points in $\mathbb{R}^{n+1}$ satisfying $x_1^2 + \cdots + x_{n+1}^2 = 1$) in the induced topology is compact, since it is easily seen to be a closed and bounded subset of $\mathbb{R}^{n+1}$.

A further notion we shall need is that of convergence of sequences. A sequence $\{x_n\}$ of points in a topological space $(X, \mathcal{T})$ is said to *converge* to point x if given any open neighborhood O of x (i.e., an open set O containing x) there is an N such that $x_n \in O$ for all $n > N$. The point x is said to be the *limit* of this sequence. It is easy to check that for $\mathbb{R}$ (with the standard topology) this agrees with the usual definition of convergence. A point $y \in X$ is said to be an *accumulation point* (or *limit point*) of $\{x_n\}$ if every open neighborhood of y contains infinitely many points of the sequence. If $\{x_n\}$ converges to x, then x clearly also is an accumulation point of the sequence. However, in a general topological space, if y is an accumulation point of $\{x_n\}$, it may not even be possible to find a subsequence $\{y_n\}$ of points of the sequence $\{x_n\}$ such that $\{y_n\}$ converges to y. However, the extraction of a subsequence convergent to y will always be possible if $(X, \mathcal{T})$ is *first countable*, that is, if for each $p \in X$ there is a countable collection $\{O_n\}$ of open sets such that every open neighborhood, O, of p contains at least one member of this collection. $\mathbb{R}^n$ with the standard topology is first countable; indeed, it satisfies the stronger requirement of *second countability*: there is a countable collection of open sets such that every open set can be expressed as a union of sets in this collection. For $\mathbb{R}^n$, the open balls with rational radii centered on points with rational coordinates compose such a countable collection of open sets.

An important relation between compactness and convergence of sequences is expressed by the Bolzano-Weierstrass theorem:

THEOREM A.9 (Bolzano-Weierstrass theorem). *Let* $(X, \mathcal{T})$ *be a topological space and let* $A \subset X$. *If A is compact, then every sequence* $\{x_n\}$ *of points in A has an accumulation point lying in A. Conversely, if* $(X, \mathcal{T})$ *is second countable and every sequence in A has an accumulation point in A, then A is compact. Thus, in particular, if* $(X, \mathcal{T})$ *is second countable, A is compact if and only if every sequence in A has a convergent subsequence whose limit lies in A.*

Finally, we define the notion of paracompactness, a property which manifolds are required to satisfy in order to prevent them from being "too large." Let $(X, \mathcal{T})$ be a topological space and let $\{O_\alpha\}$ be an open cover of X. An open cover $\{V_\beta\}$ is said to be a *refinement* of $\{O_\alpha\}$ if for each V_β there exists an O_α such that $V_\beta \subset O_\alpha$. The cover $\{V_\beta\}$ is said to be *locally finite* if each $x \in X$ has an open neighborhood W such that only finitely many V_β satisfy $W \cap V_\beta \neq \emptyset$. The topological space $(X, \mathcal{T})$ is said to be *paracompact* if every open cover $\{O_\alpha\}$ of X has a locally finite refinement $\{V_\beta\}$.

It is not difficult to show (see, e.g., Hocking and Young 1961) that any Hausdorff topological space which is locally compact (i.e., such that every point has an open neighborhood with compact closure) and which can be expressed as a countable union of compact subsets is paracompact. Thus, $\mathbb{R}^n$, S^m and their products are easily verified to be paracompact. Indeed, it is not easy to construct examples of topological spaces which satisfy all the requirements for a manifold but are not paracompact; the "long line" (see Hocking and Young 1961) is perhaps the simplest example, although the axiom of choice is required to define it.

For a manifold, M, paracompactness has a number of important consequences. It can be shown (see Kobayashi and Nomizu 1963) that paracompactness implies that (1) M admits a Riemannian metric and (2) M is second countable. This latter result implies, incidentally, that we can cover M by a locally finite, countable family of charts (ψ_i, O_i) with each $\overline{O}_i$ compact. Conversely, if M satisfies all the requirements of a manifold (see chapter 2), then either of properties (1) or (2) implies that M is paracompact.

Probably the most important consequence of paracompactness for a manifold is the existence of a partition of unity. Given a locally finite open cover $\{O_\alpha\}$ of M, a *partition of unity* subordinate to $\{O_\alpha\}$ is a collection of smooth functions $\{f_\alpha\}$ such that (i) the support of f_α (i.e., the closure of the set where f_α is nonvanishing) is contained within O_α, (ii) $0 \leq f_\alpha \leq 1$, and (iii) $\sum_\alpha f_\alpha = 1$. [Since only finitely many f_α are nonvanishing at any point, the sum in property (iii) is only a finite sum.] It can be shown (Kobayashi and Nomizu 1963) that every locally finite open cover $\{O_\alpha\}$ of M such that each $\overline{O}_\alpha$ is compact admits a subordinate partition of unity. The existence of a partition of unity allows us to globalize many local results. For example, we can prove the above mentioned result that a paracompact manifold admits a Riemannian metric by covering it with a locally finite family of charts (ψ_α, O_α) with $\overline{O}_\alpha$ compact, defining a Riemannian metric $(g_\alpha)_{ab}$ on each local coordinate neighborhood, and setting $g_{ab} = \Sigma f_\alpha (g_\alpha)_{ab}$, where $\{f_\alpha\}$ is a partition of unity subordinate to $\{O_\alpha\}$. Similarly, as discussed in appendix B, the existence of a partition of unity allows us to define integration over a paracompact manifold.

APPENDIX B

DIFFERENTIAL FORMS, INTEGRATION, AND FROBENIUS'S THEOREM

In this appendix we shall collect a number of results related to differential forms and integration. Most of these results require only manifold structure; specifically, they do not require the presence of a metric or a preferred derivative operator. Thus, they are basic results of very general applicability in differential geometry.

B.1 Differential Forms

Let M be an n-dimensional manifold. A differential p-form is a totally antisymmetric tensor of type $(0, p)$, i.e., $\omega_{a_1 \cdots a_p}$ is a p-form if

$$\omega_{a_1 \cdots a_p} = \omega_{[a_1 \cdots a_p]} \tag{B.1.1}$$

We denote the vector space of p-forms at a point x by Λ_x^p and the collection of p-form fields by Λ^p. Note that $\Lambda_x^p = \{0\}$ if $p > n$ and dim $\Lambda_x^p = n!/p!(n-p)!$ for $0 \leqq p \leqq n$. If we take the outer product of a p-form $\omega_{a_1 \cdots a_p}$ and a q-form $\mu_{b_1 \cdots b_q}$, we will get a tensor of type $(0, p + q)$; but since this tensor will not, in general, be totally antisymmetric, it is not a $(p + q)$-form. However, we can totally antisymmetrize this tensor, thus producing a map $\wedge : \Lambda_x^p \times \Lambda_x^q \rightarrow \Lambda_x^{p+q}$ via

$$(\omega \wedge \mu)_{a_1 \cdots a_p b_1 \cdots b_q} = \frac{(p+q)!}{p!q!} \omega_{[a_1 \cdots a_p} \mu_{b_1 \cdots b_q]} \quad . \tag{B.1.2}$$

(If $p + q > n$, this tensor, of course, will be zero.) We define the vector space of all differential forms at x to be the direct sum of the Λ_x^p,

$$\Lambda_x = \bigoplus_{p=0}^{n} \Lambda_x^p \quad . \tag{B.1.3}$$

The map $\wedge : \Lambda_x \times \Lambda_x \rightarrow \Lambda_x$ gives Λ_x the structure of a Grassmann algebra[1] over the vector space of one-forms.

If we are given a derivative operator, ∇_a, we could define a map from smooth p-form fields to $(p + 1)$-form fields by

$$\omega_{a_1 \cdots a_p} \rightarrow (p+1) \nabla_{[b} \omega_{a_1 \cdots a_p]} \quad . \tag{B.1.4}$$

If instead we were given another derivative operator $\tilde{\nabla}_a$, we would obtain the map

$$\omega_{a_1 \cdots a_p} \rightarrow (p+1) \tilde{\nabla}_{[b} \omega_{a_1 \cdots a_p]} \quad . \tag{B.1.5}$$

1. See, e.g., Bishop and Crittenden (1964) for the definition of a Grassmann algebra.

However, according to equation (3.1.14), we have

$$\nabla_{[b}\omega_{a_1\cdots a_p]} - \tilde{\nabla}_{[b}\omega_{a_1\cdots a_p]} = \sum_{j=1}^{p} C^d{}_{[ba_j}\omega_{a_1\cdots|d|\cdots a_p]} = 0 \qquad \text{(B.1.6)}$$

since $C^c{}_{ab}$ is symmetric in a and b. Thus the map defined by equation (B.1.4) is independent of derivative operator, i.e., it is well defined without the presence of a preferred derivative operator on M. We denote this map by d. In particular, we may use the ordinary derivative, ∂_a, associated with any coordinate system to calculate d.

Since the index structure of differential forms is trivial, it is customary to drop the indices when writing them; e.g., we write $\boldsymbol{\omega}$ instead of $\omega_{a_1\cdots a_p}$ and write $\boldsymbol{\omega} \wedge \boldsymbol{\mu}$ instead of $(\omega \wedge \mu)_{a_1\cdots b_q}$. (The only disadvantage in doing so is that we must remember the dimensionality of the forms with which we are dealing.) We shall use boldface letters for forms to avoid confusion with functions. We denote the $(p+1)$-form resulting from the action of the map $d: \Lambda^p \to \Lambda^{p+1}$ on the p-form $\boldsymbol{\omega}$ by $d\boldsymbol{\omega}$.

An important property of the map d is that $d^2 = d \circ d = 0$. This result, known as the *Poincaré lemma,* follows from the fact that we can compute d using an ordinary derivative operator. Indeed, restoring the indices, we have for an arbitrary smooth p-form $\boldsymbol{\omega}$,

$$(d^2\omega)_{bca_1\cdots a_p} = (p+2)(p+1)\partial_{[b}\partial_c\omega_{a_1\cdots a_p]} = 0 \quad , \qquad \text{(B.1.7)}$$

because of the equality of mixed partial derivatives in $\mathbb{R}^n$.

Conversely, it can be shown (see, e.g., Flanders 1963) that if one has a *closed* p-form, i.e., a p-form $\boldsymbol{\alpha}$ satisfying $d\boldsymbol{\alpha} = 0$, then *locally* (i.e., in any open region diffeomorphic to $\mathbb{R}^n$) this form is *exact*, i.e., there exists a $(p-1)$-form $\boldsymbol{\beta}$ such that $\boldsymbol{\alpha} = d\boldsymbol{\beta}$. However, in general this result is not valid globally. Indeed, an important theorem in algebraic topology due to de Rham establishes that the dimension of the vector space of closed p-forms modulo the exact p-forms equals a topological quantity: the pth Betti number of the manifold.[2]

B.2 Integration

Let M be an n-dimensional manifold. At each point $x \in M$, the vector space of n-forms will be one-dimensional. If it is possible to find a continuous, nowhere vanishing n-form field $\boldsymbol{\epsilon} = \epsilon_{[a_1\cdots a_n]}$ on M, then M is said to be *orientable* and $\boldsymbol{\epsilon}$ is said to provide an orientation.[3] Two orientations $\boldsymbol{\epsilon}$ and $\boldsymbol{\epsilon}'$ are considered equivalent if $\boldsymbol{\epsilon} = f\boldsymbol{\epsilon}'$, where f is a (strictly) positive function, so any orientable manifold

2. Roughly speaking, the pth Betti number of M is the number of independent p-dimensional boundaryless surfaces in M which are not themselves boundaries of $(p+1)$-dimensional regions. For more details, including a complete statement and proof of de Rham's theorem, see, e.g., Warner (1971).

3. For the case where M is an n-dimensional surface in the Euclidean space $\mathbb{R}^{n+1}$, this definition of orientability is equivalent to the more intuitive notion that there exists a consistent (i.e., continuous) choice of normal vector u^a to M: If a continuous nonvanishing u^a exists, then $\boldsymbol{\epsilon} = \tilde{\epsilon}_{a_1\ldots a_{n+1}}u^{a_1}$ provides an orientation of M, where $\tilde{\boldsymbol{\epsilon}}$ is an orientation of $\mathbb{R}^{n+1}$. Conversely, if M is orientable, then $\tilde{\epsilon}^{a_1\ldots a_{n+1}}\epsilon_{a_1\ldots a_n}$ provides a continuous normal vector.

possesses two inequivalent orientations, usually referred to as "right handed" and "left handed." It is easy to check that the manifolds $\mathbb{R}^n$ and S^m are orientable. Indeed, it is not difficult to show that every simply connected manifold is orientable. (As discussed further in chapter 13, a topological space is said to be *simply connected* if every closed curve can be continuously deformed to a point. $\mathbb{R}^n$ and S^m for $m \geq 2$ are simply connected.) Furthermore, the product of any two orientable manifolds is orientable. Thus, we obtain a wide class of examples of orientable manifolds. On the other hand, the Möbius strip [defined as $\mathbb{R}^2$ with the identification $(x, y) = (x + 1, -y)$] provides a simple example of a nonorientable manifold.

We will define the integral of a continuous (or, more generally, a measurable[4]) n-form field $\boldsymbol{\alpha}$ over an n-dimensional orientable manifold (with respect to the orientation $\boldsymbol{\epsilon}$) as follows. We begin by considering an open region $U \subset M$ covered by a single coordinate system ψ. If we expand $\boldsymbol{\epsilon}$ in the coordinate basis of ψ, we will obtain

$$\boldsymbol{\epsilon} = h dx^1 \wedge \ldots \wedge dx^n \tag{B.2.1}$$

(i.e. $\epsilon_{a_1 \cdots a_n} = n!\, h(dx^1)_{[a_1} \cdots (dx^n)_{a_n]}$), where the function h is nonvanishing. If $h > 0$, the coordinate system, ψ, is said to be *right handed* with respect to $\boldsymbol{\epsilon}$; if $h < 0$, ψ is called *left handed*. We may also expand $\boldsymbol{\alpha}$ in the coordinate basis, thereby obtaining

$$\boldsymbol{\alpha} = a(x^1, \ldots, x^n) dx^1 \wedge \ldots \wedge dx^n \quad . \tag{B.2.2}$$

If ψ is right handed, we define the integral of $\boldsymbol{\alpha}$ over the region U by

$$\int_U \boldsymbol{\alpha} = \int_{\psi[U]} a dx^1 \ldots dx^n \quad , \tag{B.2.3}$$

where the right-hand side is the standard Riemann (or Lebesgue) integral in $\mathbb{R}^n$. If ψ is left-handed, we define $\int_U \boldsymbol{\alpha}$ to be minus the right-hand side of equation (B.2.3).

First, we note that $\int_U \boldsymbol{\alpha}$ is independent of the choice of coordinate system, ψ, covering U; namely, if we had used a different coordinate system ψ' to cover U, then the expansion of $\boldsymbol{\alpha}$ in the new coordinate basis would be

$$\boldsymbol{\alpha} = a' dx'^1 \wedge \ldots \wedge dx'^n \quad . \tag{B.2.4}$$

But it follows from the tensor transformation law, equation (2.3.8), that

$$a' = a \det\left(\frac{\partial x^\mu}{\partial x'^\nu}\right) \quad . \tag{B.2.5}$$

The standard law for transformation of integrals in $\mathbb{R}^n$ then shows that our definition, equation (B.2.3), is coordinate independent.

To define the integral of $\boldsymbol{\alpha}$ over all of M, we use the paracompactness property of M. As discussed at the end of appendix A, a paracompact manifold can be covered by a countable collection $\{O_i\}$ of locally finite coordinate neighborhoods such that

4. $\boldsymbol{\alpha}$ is said to be measurable if for all charts its coordinate basis components are Lebesgue measurable functions in $\mathbb{R}^n$.

each $\overline{O}_i$ is compact. Furthermore, a partition of unity $\{f_i\}$ subordinate to this covering will exist. If $\sum_i \int_{\psi_i[O_i]} f_i |a_i| dx^1 \ldots dx^n < \infty$, we say $\boldsymbol{\alpha}$ is integrable and we define

$$\int_M \boldsymbol{\alpha} = \sum_i \int_{O_i} f_i \boldsymbol{\alpha} \quad . \tag{B.2.6}$$

It can be shown that this definition is independent of the choice of covering $\{O_i\}$ and partition of unity $\{f_i\}$ and thus properly defines $\int_M \boldsymbol{\alpha}$.

We can use the above definition of integration on manifolds to define the integral of p-forms on M over well behaved, orientable p-dimensional surfaces in M. First, we must define more precisely the notion of a "well behaved surface." Let S be a manifold of dimension $p < n$. If $\phi : S \to M$ is C^∞, is locally one to one—i.e., each $q \in S$ has an open neighborhood O such that ϕ restricted to O is one-to-one—and $\phi^{-1} : \phi[O] \to S$ is C^∞, then $\phi[S]$ is said to be an *immersed submanifold* of M. If, in addition, ϕ is globally one-to-one (i.e., $\phi[S]$ does not "intersect itself"), then $\phi[S]$ is said to be an *embedded submanifold* of M. (In some references the additional condition is imposed on embedded submanifolds that $\phi : S \to \phi[S]$ is a homeomorphism with the topology on $\phi[S]$ induced from M. Roughly speaking, this additional condition ensures that $\phi[S]$ does not come arbitrarily close to intersecting itself.) We shall use the notion of an embedded submanifold as our precise notion of a "well behaved surface" in M. An embedded submanifold of dimension $(n - 1)$ is called a *hypersurface*.

For an embedded submanifold, there is a natural manifold structure on $\phi[S]$ obtained via ϕ from the manifold structure on S. Thus, at each $q \in \phi[S]$, the tangent space W_q for $\phi[S]$ is defined. This tangent space is naturally identified with a p-dimensional subspace of V_q, the tangent space of q in M. Thus, a p-form $\boldsymbol{\beta}$ in M at q naturally gives rise to a p-form $\tilde{\boldsymbol{\beta}}$ on $\phi[S]$ by restriction of the action of $\boldsymbol{\beta}$ to vectors lying in W_q. The integral of $\boldsymbol{\beta}$ over the surface $\phi[S]$ may then be defined as simply the integral of the p-form $\tilde{\boldsymbol{\beta}}$ over the p-dimensional manifold $\phi[S]$.

An important special case of an embedded submanifold arises when $\phi[S]$ is the $(n - 1)$-dimensional boundary, $\dot{N}$, of a closed region $N \subset M$ such that N is a "manifold with boundary." Here, the notion of an n-dimensional *manifold with boundary*, N, can be defined in the abstract in the same way as a manifold (see chapter 2) except that $\mathbb{R}^n$ is replaced by "half of $\mathbb{R}^n$," i.e., by the portion of $\mathbb{R}^n$ with $x^1 \leqq 0$. The boundary, $\dot{N}$, of N is composed of the set of points of N which are mapped into $x^1 = 0$ by the chart maps. Note that these chart maps of N with x^1 set to zero give $\dot{N}$ the structure of an $(n - 1)$-dimensional manifold without boundary. Note also that $\text{int}(N) \equiv N - \dot{N}$ is an n-dimensional manifold without boundary.

If N is an orientable manifold with boundary, then an orientation on N induces a natural orientation on the boundary as follows: We consider the coordinate systems on $\dot{N}$ which arise from deleting the first coordinate, x^1, of a right-handed coordinate system on N in the family of charts that makes N into a manifold with boundary. We wish to define an orientation on $\dot{N}$ which makes these coordinate systems be "right handed." In order to do so, we verify first that the Jacobian, $\det(\partial x'^\mu / \partial x^\nu)$, is positive in the overlap region of any two such coordinate systems. Then, we choose a partition of unity (F_i, U_i) of $\dot{N}$, where each U_i is a coordinate neighborhood of this type. Finally, we define $\dot{\epsilon}$ on $\dot{N}$ by $\dot{\epsilon} = \sum_i F_i \, dx_i^2 \cdots dx_i^n$. Then $\dot{\epsilon}$ is continuous and nonvanishing and thus defines the desired orientation of $\dot{N}$. Having defined the orientation of $\dot{N}$, we may now state one of the most important results concerning integration on manifolds, the proof of which can be found in many references (see, e.g., Flanders 1963).

THEOREM B.2.1 (Stokes's theorem). *Let N be an n-dimensional compact oriented manifold with boundary and let $\boldsymbol{\alpha}$ be an $(n - 1)$-form on M which is C^1. Then*

$$\int_{\text{int}(N)} d\boldsymbol{\alpha} = \int_{\dot{N}} \boldsymbol{\alpha} \quad . \tag{B.2.7}$$

Integration of functions on an orientable manifold M can be accomplished if one is given a *volume element*, that is, continuous nonvanishing n-form $\boldsymbol{\epsilon}$. (A volume element differs from an orientation in that orientations are considered equivalent if they differ by positive multiples whereas volume elements are not.) The integral of f over M is defined by

$$\int_M f = \int_M f\boldsymbol{\epsilon} \quad , \tag{B.2.8}$$

where the integral of the n-form $f\boldsymbol{\epsilon}$ was defined previously.[5]

If one is given only the structure of a manifold, M, there is no natural choice of volume element. However, if M has a metric, g_{ab}, defined on it, then a natural choice of $\boldsymbol{\epsilon}$ is specified up to sign (i.e., up to choice of orientation) by the condition

$$\epsilon^{a_1 \cdots a_n} \epsilon_{a_1 \cdots a_n} = (-1)^s n! \quad , \tag{B.2.9}$$

where s is the number of minuses appearing in the signature of g_{ab}. (Thus, $s = 0$ for a Riemannian metric, while $s = 1$ for a Lorentzian metric.) Note that differentiation of equation (B.2.9) using the derivative operator, ∇_a, associated with the metric implies that

$$2\epsilon^{a_1 \cdots a_n} \nabla_b \epsilon_{a_1 \cdots a_n} = 0 \quad , \tag{B.2.10}$$

which, in turn, implies that

$$\nabla_b \epsilon_{a_1 \cdots a_n} = 0 \tag{B.2.11}$$

since $\nabla_b \epsilon_{a_1 \cdots a_n}$ is totally antisymmetric in its last n indices and $\epsilon^{a_1 \cdots a_n}$ is nonvanishing. It is also worth noting that

$$\epsilon^{a_1 \cdots a_n} \epsilon_{b_1 \cdots b_n} = (-1)^s n! \; \delta^{[a_1}{}_{b_1} \delta^{a_2}{}_{b_2} \cdots \delta^{a_n]}{}_{b_n} \quad , \tag{B.2.12}$$

where $\delta^a{}_b$ is the identity map on the tangent space. Equation (B.2.12) follows from the fact that the tensors of type (n, n) on an n-dimensional manifold which are totally antisymmetric in all lower and all upper indices form a one-dimensional vector space and thus must be proportional to the antisymmetrized product of $\delta^a{}_b$ tensors; the

5. Integration of functions on a nonorientable manifold can be defined by choosing a continuous "n-form modulo sign" $\boldsymbol{\epsilon}'$ and performing integrals of $f\boldsymbol{\epsilon}'$ over each of the local coordinate neighborhoods by choosing the sign of $\boldsymbol{\epsilon}'$ which makes the coordinate system "right handed" with respect to it.

constant of proportionality is fixed by the normalization condition (B.2.9). Contraction of equation (B.2.12) over j of its indices yields

$$\epsilon^{a_1 \cdots a_j a_{j+1} \cdots a_n} \epsilon_{a_1 \cdots a_j b_{j+1} \cdots b_n} = (-1)^s (n-j)!\, j!\, \delta^{[a_{j+1}}{}_{b_{j+1}} \cdots \delta^{a_n]}{}_{b_n} \quad . \tag{B.2.13}$$

Equation (B.2.9) implies that the components of ϵ in a right handed orthonormal basis are

$$\epsilon_{\mu_1 \cdots \mu_n} = \begin{cases} (-1)^P & \text{if all } \mu_i \text{ are distinct} \\ 0 & \text{otherwise} \end{cases} \quad , \tag{B.2.14}$$

where P is the signature of the permutation $(1, \ldots, n) \to (\mu_1, \ldots, \mu_n)$. In a coordinate basis, the components of ϵ satisfy

$$\sum_{\mu_1, \ldots, \nu_n} g^{\mu_1 \nu_1} \ldots g^{\mu_n \nu_n} \epsilon_{\mu_1 \cdots \mu_n} \epsilon_{\nu_1 \cdots \nu_n} = (-1)^s n! \tag{B.2.15}$$

But the left-hand side of this expression is just $(n!)(\epsilon_{12 \cdots n})^2$ times the determinant of the matrix $(g^{\mu\nu})$, and $\det(g^{\mu\nu}) = 1/\det(g_{\mu\nu})$. Thus, choosing the plus sign appropriate for a right handed coordinate system, we find

$$\epsilon_{12 \cdots n} = [(-1)^s \det(g_{\mu\nu})]^{1/2} = \sqrt{|g|} \quad , \tag{B.2.16}$$

where $g = \det(g_{\mu\nu})$. Thus, in any (right handed) coordinate basis, the natural volume element defined by equation (B.2.9) takes the form

$$\epsilon = \sqrt{|g|}\, dx^1 \wedge \ldots \wedge dx^n \quad . \tag{B.2.17}$$

Using the natural volume element ϵ associated with a metric, we can convert Stokes's theorem, equation (B.2.7), into a "Gauss's law" form. Let N be an oriented, compact n-dimensional manifold with boundary. Let g_{ab} be a metric on N with associated volume element ϵ. Given any C^1 vector field v^a, we obtain an $(n-1)$-form α by

$$\alpha_{a_1 \cdots a_{n-1}} = \epsilon_{b a_1 \cdots a_{n-1}} v^b \quad . \tag{B.2.18}$$

We have

$$\begin{aligned} (d\alpha)_{c a_1 \cdots a_{n-1}} &= n \nabla_{[c} (\epsilon_{|b| a_1 \cdots a_{n-1}]} v^b) \\ &= n \epsilon_{b [a_1 \cdots a_{n-1}} \nabla_{c]} v^b \quad , \end{aligned} \tag{B.2.19}$$

where equation (B.2.11) was used. On the other hand any totally antisymmetric tensor of type $(0, n)$ must be proportional to ϵ, so

$$\epsilon_{b [a_1 \cdots a_{n-1}} \nabla_{c]} v^b = h \epsilon_{c a_1 \cdots a_{n-1}} \quad . \tag{B.2.20}$$

The function h may be evaluated by contracting with $\epsilon^{c a_1 \cdots a_{n-1}}$ and using equation (B.2.13). We obtain

$$\nabla_b v^b = nh \quad . \tag{B.2.21}$$

Thus, we find

$$d\alpha = (\nabla_a v^a) \epsilon \quad , \tag{B.2.22}$$

and thus Stokes's theorem states that

$$\int_{\text{int}(N)} \nabla_a v^a = \int_{\dot{N}} \epsilon_{ba_1 \cdots a_{n-1}} v^b \quad , \tag{B.2.23}$$

where the natural volume element $\boldsymbol{\epsilon}$ on N is understood in the integral on the left-hand side of equation (B.2.23).

The right-hand side of equation (B.2.23) can be reexpressed as follows. The metric g_{ab} on N induces a tensor field h_{ab} on $\dot{N}$ by restriction of g_{ab} to vectors tangent to $\dot{N}$. If h_{ab} is nondegenerate—which will be the case if $\dot{N}$ is not a null surface—we may use it to define a volume element $\tilde{\boldsymbol{\epsilon}}$ on $\dot{N}$. It is not difficult to show that

$$\frac{1}{n} \epsilon_{a_1 \cdots a_n} = n_{[a_1} \tilde{\epsilon}_{a_2 \cdots a_n]} \quad , \tag{B.2.24}$$

where n^b is the unit normal to $\dot{N}$ and is chosen to be "outward pointing" if spacelike and "inward pointing" if timelike in order that $\tilde{\boldsymbol{\epsilon}}$ be of the orientation class used in Stokes's theorem. Contracting v^a into both sides of equation (B.2.24) and restricting the resulting $(n-1)$-forms to vectors tangent to $\dot{N}$, we obtain

$$\epsilon_{ba_1 \cdots a_{n-1}} v^b = (n_b v^b) \tilde{\epsilon}_{a_1 \cdots a_{n-1}} \quad , \tag{B.2.25}$$

where we view both sides of this equation as forms on $\dot{N}$. Thus, if $\dot{N}$ is not null, we can express Stokes's theorem in the form.

$$\int_{\text{int}(N)} \nabla_a v^a = \int_{\dot{N}} n_a v^a \tag{B.2.26}$$

for all C^1 vector fields v^a where the natural volume elements $\boldsymbol{\epsilon}$ and $\tilde{\boldsymbol{\epsilon}}$ on int(N) and $\dot{N}$, respectively, are understood. Of course, if $\dot{N}$ is null, equation (B.2.23) still applies. Furthermore, in the null case, if we choose any $\tilde{\boldsymbol{\epsilon}}$ on $\dot{N}$ in the orientation class used in Stokes's theorem and define n^a to be the normal to $\dot{N}$ such that equation (B.2.24) holds, then Stokes's theorem again takes the form (B.2.26).

B.3 Frobenius's Theorem

Let M be an n-dimensional manifold. An issue which arises frequently is the following: At each point $x \in M$ we are given a subspace $W_x \subset V_x$ of the tangent space V_x with dim $W_x = m < n$. The subspace W_x is required to vary smoothly with x in the sense that for each $x \in M$ we can find an open neighborhood O of x such that in O, W is spanned by C^∞ vector fields. We denote the collection of subspaces W_x by W. We wish to know whether we can find integral submanifolds of W, i.e., whether through each point x we can find an embedded submanifold S such that the tangent space to this submanifold at each $y \in S$ coincides with W. An important special case of this general problem arises when one has a metric on M and wishes to know if a vector field ξ^a is orthogonal to a family of hypersurfaces (see, e.g., section 6.1), i.e., whether the $(n-1)$-dimensional subspaces, W, orthogonal to ξ^a are integrable.

If the subspaces W are one-dimensional, the above problem reduces to that of finding integral curves of a smooth vector field v^a. As discussed in section 2.2, such integral curves always can be found. However, if dim $W > 1$, it is possible for the W-planes to "twist around" so that integral submanifolds cannot be found. To see that this is the case, we note that if we could find integral submanifolds, we could span W in a neighborhood of any point by coordinate vector fields $X_1^a, \ldots, X_m^a$ in M such that $[X_\mu, X_\nu] = 0$. Since any two vector fields Y^a, Z^a which lie in W can be expressed as linear combinations of these coordinate vector fields, this implies that for all Y^a, $Z^a \in W$ we have

$$[Y, Z] = \sum_{\mu,\nu} [f_\mu X_\mu, g_\nu X_\nu] = \sum_{\mu,\nu} (f_\mu X_\mu(g_\nu) - g_\mu X_\mu(f_\nu)) X_\nu \in W \quad . \tag{B.3.1}$$

If W satisfies the property that $[Y, Z]^a \in W$ for all Y^a, $Z^a \in W$, then W is said to be *involute*. We have just shown that a necessary condition for W to possess integral submanifolds is that it be involute. Conversely, it can be shown (see, e.g., Bishop and Crittenden 1964) that this condition is also sufficient. This result is known as Frobenius's theorem.

THEOREM B.3.1 (Frobenius's theorem; vector form). *A necessary and sufficient condition for a smooth specification, W, of m-dimensional subspaces of the tangent space at each $x \in M$ to possess integral submanifolds is that W be involute, i.e., for all Y^a, $Z^a \in W$ we have $[Y, Z]^a \in W$.*

Frobenius's theorem also has a dual formulation in terms of differential forms. Given $W_x \subset V_x$ as above, we can consider the one-forms $\boldsymbol{\omega} \in V_x^*$ which satisfy

$$\omega_a X^a = 0 \tag{B.3.2}$$

for all $X^a \in W_x$. It is not difficult to see that such $\boldsymbol{\omega}$'s span an $(n - m)$-dimensional subspace, $T_x^* \subset V_x^*$, of the dual tangent space at x. Conversely, an $(n - m)$-dimensional subspace T_x^* of V_x^* defines an m-dimensional subspace W_x of V_x via equation (B.3.2). Thus, we may reformulate our above question in terms of T^*: Under what conditions does a smooth specification, T^*, of $(n - m)$-dimensional subspaces of one-forms at each point have the property that the associated tangent subspaces W (consisting at each x of all vectors X^a satisfying $\omega_a X^a = 0$ for all $\omega_a \in T_x^*$) admit integral submanifolds?

According to Frobenius's theorem, integral submanifolds will exist if and only if for all $\omega_a \in T^*$ and all Y^a, $Z^a \in W$ (so that $\omega_a Y^a = \omega_a Z^a = 0$), we have

$$\omega_a [Y, Z]^a = 0 \quad . \tag{B.3.3}$$

To see what this implies for ω_a, we substitute our expression (3.1.2) for the commutator in terms of an arbitrary derivative operator ∇_a to obtain

$$\begin{aligned} 0 &= \omega_a (Y^b \nabla_b Z^a - Z^b \nabla_b Y^a) \\ &= -Z^a Y^b \nabla_b \omega_a + Y^a Z^b \nabla_b \omega_a \\ &= 2 Y^a Z^b \nabla_{[b} \omega_{a]} \quad . \end{aligned} \tag{B.3.4}$$

However, equation (B.3.4) can hold for Y^a and Z^a in the subspace annihilated by T^* if and only if $\nabla_{[a}\omega_{b]}$ can be expressed as

$$\nabla_{[a}\omega_{b]} = \sum_{\alpha=1}^{n-m} \mu^{\alpha}{}_{[a}v^{\alpha}{}_{b]} \quad , \tag{B.3.5}$$

where each $\boldsymbol{v}^{\alpha}$ is an arbitrary one-form and each $\boldsymbol{\mu}^{\alpha} \in T^*$. Thus, we can reformulate Frobenius's theorem in terms of differential forms as follows:

THEOREM B.3.2 (Frobenius's theorem; dual formulation). *Let T^* be a smooth specification of an $(n - m)$-dimensional subspace of one-forms. Then the associated m-dimensional subspace W of the tangent space admits integral submanifolds if and only if for all $\boldsymbol{\omega} \in T^*$ we have $d\boldsymbol{\omega} = \sum_{\alpha} \boldsymbol{\mu}^{\alpha} \wedge \boldsymbol{v}^{\alpha}$, where each $\boldsymbol{\mu}^{\alpha} \in T^*$.*

The dual formulation of Frobenius's theorem gives a useful criterion for when a vector field ξ^a is hypersurface orthogonal. Letting T^* be the one-dimensional subspace spanned by $\xi_a = g_{ab}\xi^b$, we see that ξ^a will be hypersurface orthogonal if and only if $\nabla_{[a}\xi_{b]} = \xi_{[a}v_{b]}$ (where we have set $\mu_a = \xi_a$ since T^* is one-dimensional). This latter condition is equivalent to $\xi_{[a}\nabla_b\xi_{c]} = 0$, and thus we see that the necessary and sufficient condition that ξ^a be hypersurface orthogonal is

$$\xi_{[a}\nabla_b\xi_{c]} = 0 \quad . \tag{B.3.6}$$

APPENDIX C

MAPS OF MANIFOLDS, LIE DERIVATIVES, AND KILLING FIELDS

This appendix deals with topics related to the maps induced on tensor fields by maps between manifolds. As will be shown in section C.1, if we have a map, $\phi: M \rightarrow N$, between manifolds M and N, we can use ϕ to bring upper index tensor fields from M to N and lower index tensor fields from N to M. If ϕ is a diffeomorphism, all types of tensor fields can be carried from M to N or from N to M. An important special case of this result occurs when $\phi_t: M \rightarrow M$ is a one-parameter family of diffeomorphisms generated by a vector field v^a. We can compare a given tensor field with the new tensor field that arises from the action of ϕ_t for small t. As will be shown in section C.2, this gives rise to the notion of the Lie derivative with respect to the vector field v^a. Finally, a vector field which generates a one-parameter group of isometries is called a Killing vector field. Using the general formulas for Lie derivatives, an equation for Killing fields is easily obtained and some important properties of them are derived in section C.3.

C.1 Maps of Manifolds

Let M and N be manifolds (*not* necessarily of the same dimension) and let $\phi: M \rightarrow N$ be a C^∞ map. In a natural manner, ϕ "pulls back" a function $f: N \rightarrow \mathbb{R}$ on N to the function $f \circ \phi: M \rightarrow \mathbb{R}$ obtained by composing f with ϕ. Similarly, in a natural way, ϕ "carries along" tangent vectors at $p \in M$ to tangent vectors at $\phi(p) \in N$—i.e., it defines a map $\phi^*: V_p \rightarrow V_{\phi(p)}$—as follows: For $v \in V_p$ we define $\phi^* v \in V_{\phi(p)}$ by

$$(\phi^* v)(f) = v(f \circ \phi) \tag{C.1.1}$$

for all smooth $f: N \rightarrow \mathbb{R}$, where we have dropped the vector indices on v and $\phi^* v$ since that notation is inconvenient here. It is easy to check that $\phi^* v$ satisfies the properties required of a tangent vector at $\phi(p)$ and thus equation (C.1.1) properly defines the map ϕ^*. Note that ϕ^* is linear and may be viewed as the "derivative of ϕ" at p. [The matrix of components of ϕ^* in the coordinate bases of a coordinate system $\{x^\nu\}$ at p and a coordinate system $\{y^\mu\}$ at $\phi(p)$ equals the Jacobian matrix of the map ϕ between the coordinates, i.e., $(\phi^*)^\mu{}_\nu = \partial y^\mu / \partial x^\nu$.] By the implicit function theorem, $\phi: M \rightarrow N$ will be one-to-one in a neighborhood of p if $\phi^*: V_p \rightarrow V_{\phi(p)}$ is one-to-one.

Similarly, we can use ϕ to "pull back" dual vectors at $\phi(p)$ to dual vectors at p. We define the map $\phi_*: V^*_{\phi(p)} \to V^*_p$ by requiring that for all $v^a \in V_p$,

$$(\phi_* \mu)_a v^a = \mu_a (\phi^* v)^a \quad . \tag{C.1.2}$$

We can extend the action of ϕ_* to map tensors of type $(0, l)$ at $\phi(p)$ to tensors of type $(0, l)$ at p by

$$(\phi_* T)_{a_1 \cdots a_l} (v_1)^{a_1} \cdots (v_l)^{a_l} = T_{a_1 \cdots a_l} (\phi^* v_1)^{a_1} \cdots (\phi^* v_l)^{a_l} \quad . \tag{C.1.3}$$

Similarly, we can extend the action of ϕ^* to map tensors of type $(k, 0)$ at p to tensors of type $(k, 0)$ at $\phi(p)$ by

$$(\phi^* T)^{b_1 \cdots b_k} (\mu_1)_{b_1} \cdots (\mu_k)_{b_k} = T^{b_1 \cdots b_k} (\phi_* \mu_1)_{b_1} \cdots (\phi_* \mu_k)_{b_k} \quad . \tag{C.1.4}$$

(By eq. [C.1.2], this is consistent with our original definition of ϕ^* on vectors.) However, in general we cannot extend ϕ^* or ϕ_* to mixed tensors since ϕ^* does not know how to "carry along" lower index tensors, while ϕ_* does not know how to "pull back" upper index tensors.

As defined in chapter 2, a C^∞ map $\phi: M \to N$ is said to be a diffeomorphism if it is one-to-one, onto, and its inverse is C^∞. If ϕ is a diffeomorphism (which necessarily implies dim M = dim N), then we can use ϕ^{-1} to extend the definition of ϕ^* to tensors of all types by using the fact that $(\phi^{-1})^*$ goes from $V_{\phi(p)}$ to V_p. If $T^{b_1 \cdots b_k}{}_{a_1 \cdots a_l}$ is a tensor of type (k, l) at p, we define the tensor $(\phi^* T)^{b_1 \cdots b_k}{}_{a_1 \cdots a_l}$ at $\phi(p)$ by,

$$(\phi^* T)^{b_1 \cdots b_k}{}_{a_1 \cdots a_l} (\mu_1)_{b_1} \cdots (\mu_k)_{b_k} (t_1)^{a_1} \cdots (t_l)^{a_l}$$
$$= T^{b_1 \cdots b_k}{}_{a_1 \cdots a_l} (\phi_* \mu_1)_{b_1} \cdots ([\phi^{-1}]^* t_l)^{a_l} \quad . \tag{C.1.5}$$

Similarly, we could extend the map ϕ_* to all tensors. However, it is not difficult to show that $\phi_* = (\phi^{-1})^*$, so we need only consider ϕ^* and $(\phi^{-1})^*$.

If $\phi: M \to M$ is a diffeomorphism and T is a tensor field on M, we can compare T with $\phi^* T$. If $\phi^* T = T$, then even though we have "moved T" via ϕ, it has "stayed the same." In other words, ϕ is a *symmetry transformation* for the tensor field T. In the case of the metric g_{ab}, a symmetry transformation—i.e., a diffeomorphism ϕ such that $(\phi^* g)_{ab} = g_{ab}$—is called an *isometry*.

We have already remarked in chapter 2 that if $\phi: M \to N$ is a diffeomorphism, than M and N have identical manifold structure. If a theory describes nature in terms of a spacetime manifold, M, and tensor fields, $T^{(i)}$, defined on the manifold, then if $\phi: M \to N$ is a diffeomorphism, the solutions $(M, T^{(i)})$ and $(N, \phi^* T^{(i)})$ have physically identical properties. Any physically meaningful statement about $(M, T^{(i)})$ will hold with equal validity for $(N, \phi^* T^{(i)})$. On the other hand, if $(N, T'^{(i)})$ is not related to $(M, T^{(i)})$ by a diffeomorphism and if the tensor fields $T^{(i)}$ represent measurable quantities, then $(N, T'^{(i)})$ will be physically distinguishable from $(M, T^{(i)})$. Thus, the diffeomorphisms comprise the gauge freedom of any theory formulated in terms of tensor fields on a spacetime manifold. In particular, diffeomorphisms comprise the gauge freedom of general relativity.

It is worth noting that an alternative viewpoint on diffeomorphisms can be taken. Above, we have discussed diffeomorphisms without introducing or making any reference to coordinate systems. We have taken an "active" point of view by associating with ϕ a map from tensors at p to tensors at $\phi(p)$. However, if we are given a coordinate system $\{x^\mu\}$ covering a neighborhood, U, of p and a coordinate system $\{y^\mu\}$ covering a neighborhood, V, of $\phi(p)$, we may take the following "passive" point of view. We may use ϕ to define a new coordinate system x'^μ in the neighborhood $O = \phi^{-1}[V]$ of p by setting $x'^\mu(q) = y^\mu(\phi(q))$ for $q \in O$. We then may view the effect of ϕ as leaving p and all tensors at p unchanged, but inducing the coordinate transformation $x^\mu \to x'^\mu$. This "passive" point of view on diffeomorphisms is, philosophically, drastically different from the above "active" viewpoint, but, in practice, these viewpoints are really equivalent since the components of the tensor $\phi^* T$ at $\phi(p)$ in the coordinate system $\{y^\mu\}$ in the active viewpoint are precisely the components of T at p in the coordinate system $\{x'^\mu\}$ in the passive viewpoint.

C.2 Lie Derivatives

Let M be a manifold and let ϕ_t be a one-parameter group of diffeomorphisms. As discussed in section 2.2, ϕ_t will be generated by a vector field, v^a. By the results of the previous section, we can use ϕ_t^* to carry along a smooth tensor field $T^{a_1 \cdots a_k}{}_{b_1 \cdots b_l}$. Comparison of $T^{a_1 \cdots a_k}{}_{b_1 \cdots b_l}$ and $\phi^*_{-t} T^{a_1 \cdots a_k}{}_{b_1 \cdots b_l}$ for small t gives rise to the notion of the Lie derivative, $\pounds_v$, with respect to v^a. More precisely, we define $\pounds_v$ by

$$\pounds_v T^{a_1 \cdots a_k}{}_{b_1 \cdots b_l} = \lim_{t \to 0} \left\{ \frac{\phi^*_{-t} T^{a_1 \cdots a_k}{}_{b_1 \cdots b_l} - T^{a_1 \cdots a_k}{}_{b_1 \cdots b_l}}{t} \right\} \quad , \tag{C.2.1}$$

where all tensors appearing in equation (C.2.1) are evaluated at the same point p. Note that the vector index on v^a is dropped in the symbol $\pounds_v$ since its presence could lead to confusion.

It follows immediately from its definition, equation (C.2.1), that $\pounds_v$ is a linear map from smooth tensor fields of type (k, l) to smooth tensor fields of type (k, l). It also is not difficult to show (see eq. [C.2.4] below) that $\pounds_v$ satisfies the Leibnitz rule on outer products of tensors. Furthermore, since v^a is tangent to the integral curves of ϕ_t, for functions $f: M \to \mathbb{R}$ we have

$$\pounds_v(f) = v(f) \quad . \tag{C.2.2}$$

Note also that $\pounds_v T^{a_1 \cdots a_k}{}_{b_1 \cdots b_l} = 0$ everywhere if and only if for all t, ϕ_t is a symmetry transformation for $T^{a_1 \cdots a_k}{}_{b_1 \cdots b_l}$.

To analyze the action of $\pounds_v$ on an arbitrary tensor field, it is helpful to introduce a coordinate system on M where the parameter t along the integral curves of v^a is chosen as one of the coordinates x^1, so that $v^a = (\partial/\partial x^1)^a$. (This always can be done locally in any region where $v^a \neq 0$.) The action of ϕ_{-t} then corresponds to the coordinate transformation $x^1 \to x^1 + t$, with $x^2, \ldots, x^n$ held fixed. From the parenthetical remark below equation (C.1.1), we have $(\phi^*)^\mu{}_\nu = \delta^\mu{}_\nu$ and hence, the coordinate basis components of $\phi^*_{-t} T^{a_1 \cdots a_k}{}_{b_1 \cdots b_l}$ at the point p whose coordinates are $(x^1, \ldots, x^n)$ are

$$(\phi^*_{-t} T^{\mu_1 \cdots \mu_k}{}_{\nu_1 \cdots \nu_l})(x^1, \ldots, x^n) = T^{\mu_1 \cdots \mu_k}{}_{\nu_1 \cdots \nu_l}(x^1 + t, x^2, \ldots, x^n) \quad . \tag{C.2.3}$$

Consequently, the components of the Lie derivative of $T^{a_1 \cdots a_k}{}_{b_1 \cdots b_l}$ in a coordinate system adapted to v^a are simply

$$£_v T^{\mu_1 \cdots \mu_k}{}_{\nu_1 \cdots \nu_l} = \frac{\partial T^{\mu_1 \cdots \mu_k}{}_{\nu_1 \cdots \nu_l}}{\partial x^1} \quad . \tag{C.2.4}$$

Thus, in particular, ϕ_t will be a symmetry transformation of $T^{a_1 \cdots a_k}{}_{b_1 \cdots b_l}$ if and only if the components $T^{\mu_1 \cdots \mu_k}{}_{\nu_1 \cdots \nu_l}$ in a coordinate system adapted to v^a are independent of the integral curve coordinate x^1.

We can obtain a coordinate independent expression for the Lie derivative of a vector field w^a by noting that in an adapted coordinate system we have by equation (C.2.4),

$$£_v w^\mu = \frac{\partial w^\mu}{\partial x^1} \quad . \tag{C.2.5}$$

On the other hand, since $v^a = (\partial/\partial x^1)^a$ and $w^a = \sum_\mu w^\mu (\partial/\partial x^\mu)^a$, the commutator of v^a and w^a is given by

$$\begin{aligned} [v, w]^\mu &= \sum_\nu \left(v^\nu \frac{\partial w^\mu}{\partial x^\nu} - w^\nu \frac{\partial v^\mu}{\partial x^\nu} \right) \\ &= \frac{\partial w^\mu}{\partial x^1} \quad . \end{aligned} \tag{C.2.6}$$

Thus, we find that the components of $£_v w^a$ and $[v, w]^a$ are equal in an adapted coordinate system. However, since both of these quantities are defined in a coordinate-independent manner, we obtain

$$£_v w^a = [v, w]^a \quad , \tag{C.2.7}$$

which is the coordinate-independent formula we sought for the Lie derivative of a vector field.

The action of the Lie derivative on all other types of tensor fields is determined by equations (C.2.2), (C.2.7) and the Leibnitz rule. For example, for a dual vector field, μ_a, we have by equation (C.2.2)

$$£_v(\mu_a w^a) = v(\mu_a w^a) \quad , \tag{C.2.8}$$

where w^a is an arbitrary field. On the other hand, by the Leibnitz rule and equation (C.2.7), we have,

$$£_v(\mu_a w^a) = w^a £_v \mu_a + \mu_a [v, w]^a \quad . \tag{C.2.9}$$

From the equality of the right sides of equations (C.2.8) and (C.2.9) we obtain a formula which determines $£_v \mu_a$. This formula is most conveniently expressed in terms of a derivative operator. If ∇_a is an arbitrary derivative operator on M, we have by properties (4) and (2) of the definition of derivative operator (see section 3.1)

$$\begin{aligned} v(\mu_a w^a) &= v^b \nabla_b (\mu_a w^a) \\ &= v^b w^a \nabla_b \mu_a + v^b \mu_a \nabla_b w^a \quad . \end{aligned} \tag{C.2.10}$$

On the other hand, we showed previously (see eq. [3.1.2]) that

$$[v, w]^a = v^b \nabla_b w^a - w^b \nabla_b v^a \quad . \tag{C.2.11}$$

Thus, we find

$$v^b w^a \nabla_b \mu_a + v^b \mu_a \nabla_b w^a = w^a \pounds_v \mu_a + \mu_a v^b \nabla_b w^a - \mu_a w^b \nabla_b v^a \quad , \tag{C.2.12}$$

i.e.,

$$\pounds_v \mu_a = v^b \nabla_b \mu_a + \mu_b \nabla_a v^b \quad . \tag{C.2.13}$$

More generally for an arbitrary tensor field $T^{a_1 \cdots a_k}{}_{b_1 \cdots b_l}$ we find by induction that

$$\pounds_v T^{a_1 \cdots a_k}{}_{b_1 \cdots b_l} = v^c \nabla_c T^{a_1 \cdots a_k}{}_{b_1 \cdots b_l} - \sum_{i=1}^{k} T^{a_1 \cdots c \cdots a_k}{}_{b_1 \cdots b_l} \nabla_c v^{a_i}$$
$$+ \sum_{j=1}^{l} T^{a_1 \cdots a_k}{}_{b_1 \cdots c \cdots b_l} \nabla_{b_j} v^c \quad . \tag{C.2.14}$$

Again, we emphasize that equation (C.2.14) holds for any derivative operator ∇_a.

Finally, we already remarked in section C.1 above that if $\phi: M \to M$ is a diffeomorphism, then (M, g_{ab}) and $(M, \phi^* g_{ab})$ represent the same physical spacetime. If we consider a one-parameter family of spacetimes $(M, g_{ab}(\lambda))$, then $(M, \phi^*_\lambda g_{ab}(\lambda))$ represents the same physical one-parameter family, where ϕ_λ is an arbitrary one-parameter group of diffeomorphisms. If, as in sections 4.4 and 7.5, we consider the first order perturbation of $g_{ab}|_{\lambda=0}$ obtained by differentiating $g_{ab}(\lambda)$ with respect to λ at $\lambda = 0$, we find that $\gamma_{ab} = dg_{ab}/d\lambda|_{\lambda=0}$ and $\gamma'_{ab} = d(\phi^*_\lambda g_{ab})/d\lambda|_{\lambda=0}$ represent the same physical perturbation. But, it is not difficult to see that

$$\gamma'_{ab} = \gamma_{ab} - \pounds_v g_{ab} \quad , \tag{C.2.15}$$

where v^a is the vector field which generates ϕ_λ and $g_{ab} = g_{ab}(\lambda = 0)$. Thus, the gauge freedom in perturbations, γ_{ab}, is given by $\pounds_v g_{ab}$, where v^a is an arbitrary vector field. Furthermore, by equation (C.2.14) we have

$$\begin{aligned} \pounds_v g_{ab} &= v^c \nabla_c g_{ab} + g_{cb} \nabla_a v^c + g_{ac} \nabla_b v^c \\ &= \nabla_a v_b + \nabla_b v_a \quad , \end{aligned} \tag{C.2.16}$$

where the second line of equation (C.2.16) holds when ∇_a is the derivative operator associated with g_{ab}. Thus, the gauge transformations of linearized general relativity about a solution g_{ab} are

$$\gamma_{ab} \to \gamma'_{ab} = \gamma_{ab} - \nabla_a v_b - \nabla_b v_a \quad . \tag{C.2.17}$$

This is closely analogous to the gauge freedom $A_a \to A'_a = A_a - \nabla_a \chi$ of electromagnetism.

C.3 Killing Vector Fields

If $\phi_t: M \to M$ is one-parameter group of isometries, $\phi^*_t g_{ab} = g_{ab}$, the vector field ξ^a which generates ϕ_t is called a *Killing vector field*. As already remarked below equation (C.2.2), the necessary and sufficient condition for ϕ_t to be a group of isometries is $\pounds_\xi g_{ab} = 0$. Thus, according to equation (C.2.16), the necessary and sufficient condition that ξ^a be a Killing field is that it satisfy *Killing's equation*

$$\nabla_a \xi_b + \nabla_b \xi_a = 0 \quad , \tag{C.3.1}$$

where ∇_a is the derivative operator associated with g_{ab}.

One of the most useful properties of Killing vector fields is given in the following proposition.

PROPOSITION C.3.1. Let ξ^a be a Killing vector field and let γ be a geodesic with tangent u^a. Then $\xi_a u^a$ is constant along γ.

Proof. We have

$$\begin{aligned} u^b \nabla_b (\xi_a u^a) &= u^b u^a \nabla_b \xi_a + \xi_a u^b \nabla_b u^a \\ &= 0 \quad , \end{aligned} \tag{C.3.2}$$

since the first term vanishes by Killing's equation (C.3.1) and the second term vanishes by the geodesic equation. □

Since in general relativity timelike geodesics represent the spacetime motions of freely falling particles and null geodesics represent the paths of light rays, proposition C.3.1 can be interpreted as saying that every one-parameter family of symmetries gives rise to a conserved quantity for particles and light rays. This conserved quantity enables one to determine the gravitational redshift in stationary spacetimes and is extremely useful for integrating the geodesic equation when symmetries are present (see section 6.3).

Another useful formula relates the second derivative of a Killing field to the Riemann tensor. By definition of the Riemann tensor, we have

$$\nabla_a \nabla_b \xi_c - \nabla_b \nabla_a \xi_c = R_{abc}{}^d \xi_d \quad . \tag{C.3.3}$$

On the other hand, by Killing's equation, we can rewrite equation (C.3.3) as

$$\nabla_a \nabla_b \xi_c + \nabla_b \nabla_c \xi_a = R_{abc}{}^d \xi_d \quad . \tag{C.3.4}$$

If we write down the same equation with cyclic permutations of the indices (abc), and then add the (abc) equation to the (bca) equation and subtract the (cab) equation, we obtain

$$\begin{aligned} 2\nabla_b \nabla_c \xi_a &= (R_{abc}{}^d + R_{bca}{}^d - R_{cab}{}^d)\xi_d \\ &= -2R_{cab}{}^d \xi_d \quad , \end{aligned} \tag{C.3.5}$$

where the symmetry property (3.2.14) of the Riemann tensor was used in the last equality. Thus, for any Killing field ξ^a, we obtain the formula

$$\nabla_a \nabla_b \xi_c = -R_{bca}{}^d \xi_d \quad . \tag{C.3.6}$$

An important consequence of equation (C.3.6) is that a Killing field, ξ^a, is completely determined by the values of ξ^a and $L_{ab} \equiv \nabla_a \xi_b$ at any point $p \in M$; namely, if we are given (ξ^a, L_{ab}) at p, then (ξ^a, L_{ab}) at any other point q is determined by integration of the system of ordinary differential equations

$$v^a \nabla_a \xi_b = v^a L_{ab} \quad , \tag{C.3.7}$$

$$v^a \nabla_a L_{bc} = -R_{bca}{}^d \xi_d v^a \quad , \tag{C.3.8}$$

along any curve connecting p and q, where v^a denotes the tangent to the curve. Immediate corollaries of this result are (i) if a Killing field and its derivative vanish at a point, then the Killing field vanishes everywhere, and (ii) on a manifold of dimension n, there can be at most $n + n(n-1)/2 = n(n+1)/2$ linearly independent Killing fields [and, thus, at most an $n(n+1)/2$ parameter group of isometries], since this is the dimension of the space of initial data for (ξ^a, L_{ab}).

It is worth noting that if we contract equation (C.3.6) over a and b, we find

$$\nabla^a \nabla_a \xi_c = -R_c{}^d \xi_d \quad . \tag{C.3.9}$$

Thus, in a vacuum spacetime, $R_c{}^d = 0$, ξ^a satisfies the source-free Maxwell equation (4.3.15) for a vector potential in the Lorentz gauge. (There is a sign difference in the Ricci tensor term between eqs. [4.3.15] and [C.3.9], so Maxwell's equation is not satisfied when $R_{ab} \neq 0$.) The Lorentz gauge condition $\nabla_a \xi^a = 0$ is also satisfied because of Killing's equation, and thus all Killing fields in vacuum spacetimes give rise to solutions of Maxwell's equation. Some solutions of physical interest can be obtained in this way (Wald 1974*b*).

In the case of a hypersurface orthogonal Killing vector field, χ^a, we can obtain a simple formula for $\nabla_a \chi_b$. By Frobenius's theorem B.3.2, there exists a vector field v^a such that

$$\nabla_a \chi_b = \nabla_{[a} \chi_{b]} = \chi_{[a} v_{b]} \quad . \tag{C.3.10}$$

Assuming that χ^a is not null, we may choose v^a to be orthogonal to χ^a. Contracting equation (C.3.10) with χ^b, we obtain

$$\frac{1}{2} \nabla_a (\chi^b \chi_b) = -\frac{1}{2} v_a \chi^b \chi_b \quad . \tag{C.3.11}$$

Hence, by solving equation (C.3.11)for v^a and substituting the result in equation (C.3.10), we find that an arbitrary hypersurface orthogonal Killing field χ^a with $\chi^a \chi_a \neq 0$ satisfies

$$\nabla_a \chi_b = -\chi_{[a} \nabla_{b]} \ln |\chi^c \chi_c| \quad . \tag{C.3.12}$$

Finally, we mention two generalizations of the notion of Killing vector fields. First, a *conformal isometry*, ϕ, on a manifold, M, with metric, g_{ab}, is defined to be a diffeomorphism $\phi: M \to M$ for which there is a function Ω such that $\phi^* g_{ab} = \Omega^2 g_{ab}$. (The fact that ϕ is a diffeomorphism implies that Ω is nonvanishing. The case $\Omega = 1$, of course, corresponds to an ordinary isometry.) The infinitesimal generator, ψ^a, of a one-parameter group, ϕ_t, of conformal isometries is called a *conformal Killing vector field*. Clearly, the Lie derivative of g_{ab} with respect to ψ^a must be proportional to g_{ab}. Thus, ψ^a satisfies

$$\nabla_a \psi_b + \nabla_b \psi_a = \alpha g_{ab} \quad , \tag{C.3.13}$$

where ∇_a is the derivative operator associated with g_{ab}. Taking the trace of equation (C.3.13), we evaluate the function α, thus obtaining

$$\nabla_a \psi_b + \nabla_b \psi_a = \frac{2}{n} (\nabla^c \psi_c) g_{ab} \quad , \tag{C.3.14}$$

where $n = \dim M$. Equation (C.3.14) is known as the *conformal Killing equation*.

In Proposition C.3.1, we proved that for any geodesic with tangent u^a and for any Killing field ξ^a, the inner product, $\xi_a u^a$, is constant along the geodesic. The same calculation for a conformal Killing field yields

$$u^b \nabla_b (\psi_a u^a) = \frac{1}{n} (\nabla^c \psi_c) u^b u_b \quad . \tag{C.3.15}$$

Thus, in general, $\psi_a u^a$ is *not* constant along a geodesic. However, for a null geodesic we have $u^b u_b = 0$, so the right-hand side of equation (C.3.15) vanishes. Thus, conformal Killing fields give rise to constants of motion for null geodesics.

The second generalization we mention of a Killing vector is that of a Killing tensor. A *Killing tensor field* of order m on a manifold M with derivative operator ∇_a is defined to be a totally symmetric m-index tensor field, $K_{a_1 \cdots a_m} = K_{(a_1 \cdots a_m)}$, which satisfies the equation

$$\nabla_{(b} K_{a_1 \cdots a_m)} = 0 \quad . \tag{C.3.16}$$

Although equation (C.3.16) is a natural generalization of Killing's equation (C.3.1), it should be noted that (aside from Killing vectors or Killing tensors formed from products of Killing vectors) Killing tensor fields do not arise in any natural way from groups of diffeomorphisms of M. However, Killing tensors share with Killing vectors the property of giving rise to constants of the motion: A repetition of the proof of Proposition C.3.1 shows that for any geodesic γ with tangent u^a, the quantity $K_{a_1 \cdots a_m} u^{a_1} \ldots u^{a_m}$ is constant along γ. The Kerr metric (see chapter 12) possesses a nontrivial Killing tensor K_{ab}, and the constant of motion to which it gives rise (together with the constants obtained from the two Killing vectors) enables one to obtain all the geodesics explicitly.

CONFORMAL TRANSFORMATIONS

Let M be an n-dimensional manifold with metric g_{ab} of any signature. If Ω is a smooth, strictly positive function, then the metric $\tilde{g}_{ab} = \Omega^2 g_{ab}$ is said to arise from g_{ab} via a *conformal transformation*. It should be emphasized that a conformal transformation is not, in general, associated with a diffeomorphism of M. [As discussed at the end of Appendix C, a diffeomorphism $\psi: M \to M$ for which $(\psi^* g)_{ab} = \Omega^2 g_{ab}$ is called a conformal isometry.] Conformal transformations occur in many contexts in general relativity, in particular, in the definiton of asymptotic flatness (chapter 11). The derivative operator and curvature of $\tilde{g}_{ab}$ are related in a relatively simple way to those of g_{ab}. In this appendix we derive these relations and also discuss the behavior under conformal transformations of solutions to some equations.

First, we note that in the case where g_{ab} is a Lorentz metric, a vector v^a is timelike, null, or spacelike with respect to the metric g_{ab} if and only if it satisfies the same property with respect to $\tilde{g}_{ab}$. Thus, (M, g_{ab}) and $(M, \tilde{g}_{ab})$ have identical causal structure. Conversely, if the light cones of two Lorentz metrics g_{ab} and $\bar{g}_{ab}$ coincide at a point $p \in M$, then at p, $\bar{g}_{ab}$ must be a multiple of g_{ab}, $\bar{g}_{ab} = \Omega^2 g_{ab}$. [*Proof*—Let t^a, $x_1^a, \ldots, x_{n-1}^a$, be an orthonormal basis of g_{ab}. Then $t^a \pm x_i^a$ is null with respect to g_{ab} and hence with respect to $\bar{g}_{ab}$, which implies that, with respect to $\bar{g}_{ab}$, t^a and x_i^a are orthogonal and their norms have the same magnitude. The fact that $t^a + 2^{-1/2}(x_i^a + x_j^a)$ is null for $i \neq j$ then shows that x_i^a is orthogonal to x_j^a. Thus, apart from a constant multiple, $t^a, x_1^a, \ldots, x_{n-1}^a$ is an orthonormal basis of $\bar{g}_{ab}$.] Consequently, if the spacetimes (M, g_{ab}) and $(M, \bar{g}_{ab})$ have identical causal structure, then $\bar{g}_{ab}$ must be related to g_{ab} by a conformal transformation.

Since in the situation under consideration two metrics, g_{ab} and $\tilde{g}_{ab}$, are present, confusion can arise as to which metric is being used to raise and lower indices. We shall deal with this problem by explicitly writing the metric in all formulas in which indices are raised or lowered. We shall denote the inverse metric to g_{ab} as g^{ab} and the inverse metric to $\tilde{g}_{ab}$ as $\tilde{g}^{ab}$. Clearly, we have $\tilde{g}^{ab} = \Omega^{-2} g^{ab}$, since then $\tilde{g}^{ab}\tilde{g}_{bc} = g^{ab}g_{bc} = \delta^a{}_c$. Note that $\tilde{g}^{ab}$ is *not* equal to $\tilde{g}_{ab}$ with indices raised by g^{ab}.

Let ∇_a denote the derivative operator associated with g_{ab}, and let $\tilde{\nabla}_a$ denote the derivative operator associated with $\tilde{g}_{ab}$. The relation between $\tilde{\nabla}_a$ and ∇_a is given by equations (3.1.7) and (3.1.28). Reversing the roles of ∇_a and $\tilde{\nabla}_a$ in these equations (so that $C^c{}_{ab}$ now is defined by $\tilde{\nabla}_a \omega_b = \nabla_a \omega_b - C^c{}_{ab}\omega_c$), we find

$$C^c{}_{ab} = \frac{1}{2}\tilde{g}^{cd}\{\nabla_a \tilde{g}_{bd} + \nabla_b \tilde{g}_{ad} - \nabla_d \tilde{g}_{ab}\} \quad . \tag{D.1}$$

However, since $\nabla_a g_{bc} = 0$, we have

$$\nabla_a \tilde{g}_{bc} = \nabla_a(\Omega^2 g_{bc}) = 2\Omega g_{bc} \nabla_a \Omega \quad . \tag{D.2}$$

Hence, we obtain

$$\begin{aligned} C^c{}_{ab} &= \Omega^{-1} g^{cd}\{g_{bd}\nabla_a\Omega + g_{ad}\nabla_b\Omega - g_{ab}\nabla_d\Omega\} \\ &= 2\delta^c{}_{(a}\nabla_{b)}\ln\Omega - g_{ab}g^{cd}\nabla_d \ln\Omega \quad , \end{aligned} \tag{D.3}$$

which expresses $C^c{}_{ab}$ in terms of Ω and g_{ab}.

We can use equation (D.3) to compare the geodesics with respect to ∇_a with those with respect to $\tilde{\nabla}_a$. The tangent, v^a, to an affinely parameterized geodesic γ with respect to ∇_a satisfies

$$v^a \nabla_a v^b = 0 \quad . \tag{D.4}$$

Hence, we have

$$v^a \tilde{\nabla}_a v^b = v^a \nabla_a v^b + v^a C^b{}_{ac} v^c = 2v^b v^c \nabla_c \ln\Omega - (g_{ac}v^a v^c) g^{bd}\nabla_d \ln\Omega \quad . \tag{D.5}$$

Thus, in general γ fails to be a geodesic with respect to $\tilde{\nabla}_a$. However, in the case of a null geodesic, $g_{ac}v^a v^c = 0$, equation (D.5) is just the (non-affinely parameterized) geodesic equation (3.3.2) with $\alpha = 2v^c\nabla_c \ln\Omega$. Thus, null geodesics are conformally invariant, i.e., the null geodesics with respect to ∇_a coincide with those with respect to $\tilde{\nabla}_a$, with the affine parameter $\tilde{\lambda}$ for $\tilde{\nabla}_a$-geodesics related to the affine parameter λ for ∇_a-geodesics by

$$\frac{d\tilde{\lambda}}{d\lambda} = c\Omega^2 \quad , \tag{D.6}$$

where c is a constant.

The relation between the curvature, $\tilde{R}_{abc}{}^d$, associated with $\tilde{\nabla}_a$ and the curvature, $R_{abc}{}^d$, associated with ∇_a is given by equation (7.5.8). Hence, using our formula (D.3) for $C^c{}_{ab}$, we find

$$\begin{aligned} \tilde{R}_{abc}{}^d &= R_{abc}{}^d - 2\nabla_{[a}C^d{}_{b]c} + 2C^e{}_{c[a}C^d{}_{b]e} \\ &= R_{abc}{}^d + 2\delta^d{}_{[a}\nabla_{b]}\nabla_c \ln\Omega - 2g^{de}g_{c[a}\nabla_{b]}\nabla_e \ln\Omega \\ &\quad + 2(\nabla_{[a}\ln\Omega)\delta^d{}_{b]}\nabla_c \ln\Omega - 2(\nabla_{[a}\ln\Omega)g_{b]c}g^{df}\nabla_f \ln\Omega \\ &\quad - 2g_{c[a}\delta^d{}_{b]}g^{ef}(\nabla_e \ln\Omega)\nabla_f \ln\Omega \quad . \end{aligned} \tag{D.7}$$

Contracting over b and d, we obtain

$$\begin{aligned} \tilde{R}_{ac} = R_{ac} &- (n-2)\nabla_a\nabla_c \ln\Omega - g_{ac}g^{de}\nabla_d\nabla_e \ln\Omega \\ &+ (n-2)(\nabla_a \ln\Omega)\nabla_c \ln\Omega - (n-2)g_{ac}g^{de}(\nabla_d \ln\Omega)\nabla_e \ln\Omega \quad . \end{aligned} \tag{D.8}$$

Contracting equation (D.8) with $\tilde{g}^{ac} = \Omega^{-2}g^{ac}$, we obtain

$$\begin{aligned} \tilde{R} = \Omega^{-2}\{R &- 2(n-1)g^{ac}\nabla_a\nabla_c \ln\Omega \\ &- (n-2)(n-1)g^{ac}(\nabla_a \ln\Omega)\nabla_c \ln\Omega\} \quad , \end{aligned} \tag{D.9}$$

where $\tilde{R} \equiv \tilde{g}^{ab}\tilde{R}_{ab}$ and $R \equiv g^{ab}R_{ab}$. Finally, from the definition of the Weyl tensor, equation (3.2.28), we find that $C_{abc}{}^d$ is unchanged by a conformal transformation of the metric,

$$\tilde{C}_{abc}{}^{d} = C_{abc}{}^{d} \quad . \tag{D.10}$$

(Note, however, that because it would be natural to use different metrics to raise and lower indices on $\tilde{C}_{abc}{}^d$ and $C_{abc}{}^d$, the equality of $\tilde{C}_{abc}{}^d$ and $C_{abc}{}^d$ depends crucially on the index positions. For example, we have $\tilde{C}_{abcd} \equiv \tilde{g}_{de}\tilde{C}_{abc}{}^e = \Omega^2 g_{de}C_{abc}{}^e = \Omega^2 C_{abcd}$.) Equations (D.8)–(D.10) are the desired formulas expressing how curvature is changed by conformal transformations.

Next, we analyze the conformal invariance of certain equations involving the metric. An equation for a field Ψ is said to be *conformally invariant* if there exists a number $s \in \mathbb{R}$ (called the *conformal weight* of the field) such that Ψ is a solution with metric g_{ab} if and only if $\tilde{\Psi} = \Omega^s\Psi$ is a solution with metric $\tilde{g}_{ab} = \Omega^2 g_{ab}$. Many equations for physical fields are conformally invariant, and the study of the behavior of equations under conformal transformations also is useful for many mathematical purposes.

As a first example, we show that the equation

$$g^{ab}\nabla_a\nabla_b\phi = 0 \tag{D.11}$$

for a scalar field is *not* conformally invariant if $\dim M \neq 2$. (In the case of a Riemannian metric, eq. [D.11] is a natural generalization of Laplace's equation to curved space. For a Lorentz metric, eq. [D.11] is the massless Klein-Gordon equation considered in chapters 4 and 10.) We have

$$\begin{aligned}
\tilde{g}^{ab}\tilde{\nabla}_a\tilde{\nabla}_b(\tilde{\phi}) &= \Omega^{-2}g^{ab}\tilde{\nabla}_a[\nabla_b(\Omega^s\phi)] \\
&= \Omega^{-2}g^{ab}[\nabla_a\nabla_b(\Omega^s\phi) - C^c{}_{ab}\nabla_c(\Omega^s\phi)] \\
&= \Omega^{s-2}g^{ab}\nabla_a\nabla_b\phi + (2s + n - 2)\Omega^{s-3}g^{ab}\nabla_a\Omega\nabla_b\phi \\
&\quad + s\Omega^{s-3}\phi g^{ab}\nabla_a\nabla_b\Omega \\
&\quad + s(n + s - 3)\Omega^{s-4}\phi g^{ab}\nabla_a\Omega\nabla_b\Omega \quad .
\end{aligned} \tag{D.12}$$

Thus, if $n = 2$, we may choose $s = 0$ and equation (D.12) then implies $\tilde{g}^{ab}\tilde{\nabla}_a\tilde{\nabla}_b\tilde{\phi} = 0$ if and only if $g^{ab}\nabla_a\nabla_b\phi = 0$. However, if $n \neq 2$, no choice of s will make $\tilde{g}^{ab}\tilde{\nabla}_a\tilde{\nabla}_b\tilde{\phi}$ vanish whenever $g^{ab}\nabla_a\nabla_b\phi$ vanishes. Thus, equation (D.11) is not conformally invariant except in two dimensions.

However, for $n > 1$, it is possible to modify equation (D.11) in a simple manner so that it becomes conformally invariant. First, if we choose $s = 1 - n/2$ then $\nabla_a\Omega\nabla_b\phi$ term in equation (D.12) will be eliminated. Using this choice of s and the behavior of the scalar curvature, R, under conformal transformations given by equation (D.9), we find that for $n \neq 1$ the addition of the term $\alpha R\phi$ to equation (D.11) will cancel the $\phi g^{ab}\nabla_a\nabla_b\Omega$ term *and* the $\phi g^{ab}\nabla_a\Omega\nabla_b\Omega$ in equation (D.12) provided we choose $\alpha = -(n - 2)/4(n - 1)$. Thus, the equation

$$g^{ab}\nabla_a\nabla_b\phi - \frac{n-2}{4(n-1)}R\phi = 0 \tag{D.13}$$

is conformally invariant, with weight $s = 1 - n/2$, since we have

$$\left(\tilde{g}^{ab}\tilde{\nabla}_a\tilde{\nabla}_b - \frac{n-2}{4(n-1)}\tilde{R}\right)\left[\Omega^{1-n/2}\phi\right] =$$

$$\Omega^{-1-n/2}\left[g^{ab}\nabla_a\nabla_b - \frac{n-2}{4(n-1)}R\right]\phi \quad . \tag{D.14}$$

Thus, equation (D.13) provides a conformally invariant generalization to curved geometry of the Laplace and Klein-Gordon equations in flat spaces.

Next, we demonstrate that Maxwell's equations,

$$g^{ac}\nabla_c F_{ab} = 0 \quad , \tag{D.15}$$

$$\nabla_{[a}F_{bc]} = 0 \quad , \tag{D.16}$$

are conformally invariant in four dimensions. We have

$$\begin{aligned}\tilde{g}^{ac}\tilde{\nabla}_c(\Omega^s F_{ab}) &= \Omega^{-2}g^{ac}\{\nabla_c(\Omega^s F_{ab}) - C^d{}_{ca}\Omega^s F_{db} - C^d{}_{cb}\Omega^s F_{ad}\} \\ &= \Omega^{s-2}g^{ac}\nabla_c F_{ab} + (n - 4 + s)\Omega^{s-3}g^{ac}F_{ab}\nabla_c\Omega \quad . \end{aligned}\tag{D.17}$$

On the other hand, we have

$$\tilde{\nabla}_{[a}(\Omega^s F_{bc]}) = \Omega^s\nabla_{[a}F_{bc]} + s\Omega^{s-1}(\nabla_{[a}\Omega)F_{bc]} \quad . \tag{D.18}$$

Thus, inspection of equations (D.17) and (D.18) shows that for $n \neq 4$ Maxwell's equations fail to be conformally invariant, but in the physically relevant case of four dimensions, conformal invariance holds with conformal weight[1] $s = 0$.

Finally, we note the conformal invariance properties of the conservation equation,

$$\nabla_a T^{ab} = 0 \tag{D.19}$$

for a symmetric tensor field $T^{ab} = T^{ba}$. We have

$$\begin{aligned}\tilde{\nabla}_a(\Omega^s T^{ab}) &= \nabla_a(\Omega^s T^{ab}) + C^a{}_{ac}\Omega^s T^{cb} + C^b{}_{ac}\Omega^s T^{ac} \\ &= \Omega^s\nabla_a T^{ab} + (s + n + 2)\Omega^{s-1}T^{ab}\nabla_a\Omega - \Omega^{s-1}g^{ba}T\nabla_a\Omega \quad , \end{aligned}\tag{D.20}$$

where $T = g_{cd}T^{cd}$. Thus, we see that equation (D.19) is not conformally invariant. However, if we impose in addition to equation (D.19) and $T^{ab} = T^{ba}$ the requirement that $T = 0$, then equation (D.19) becomes conformally invariant with $s = -n - 2$. Conversely, suppose that the stress-energy tensor of a conformally invariant field is itself conformally invariant in the sense that $T^{ab} \to \tilde{T}^{ab} = \Omega^w T^{ab}$ under conformal transformations of the metric and field variables. (This will be the case if T^{ab} is obtained by functional differentiation of a conformally invariant action with respect to the metric [see appendix E]. We use the notation w rather than s here because the conformal weight of T^{ab} need not be equal to the conformal weight of the field.) Then

1. Note that for a tensor field, the assignment of conformal weight depends on index positions; i.e., the conformal weight of F^{ab} would be $s = -4$. An invariant notion of conformal weight is $s' = s - N_l + N_u$, where N_l is the number of "lower indices" and N_u is the number of "upper indices" of the tensor. Thus, the invariant conformal weight of the Maxwell field is -2.

the conservation equation must be satisfied in both the original and conformally transformed spaces. Hence, equation (D.20) shows that we must have $T = 0$ (as well as $w = -n - 2$). Thus, if the stress tensor of a conformally invariant field is conformally invariant, its trace must vanish identically.

APPENDIX E

LAGRANGIAN AND HAMILTONIAN FORMULATIONS OF GENERAL RELATIVITY

The dynamical content of general relativity is fully expressed by Einstein's field equation, $G_{ab} = 8\pi T_{ab}$. Nevertheless, even in a purely classical (i.e., non-quantum) context, it is convenient and useful for many purposes to have Lagrangian and Hamiltonian formulations of general relativity. For example, as we shall see below, Einstein's equation can be derived from a very simple and natural Lagrangian, thus contributing further to the aesthetic appeal of general relativity. The Hamiltonian formulation yields insights into the nature of dynamics of general relativity. Indeed, although we analyzed the initial value formulation of general relativity in chapter 10 using only the field equation, the viewpoint that Einstein's equation describes the evolution of the spatial metric, h_{ab}, with "time" is perhaps best motivated via the Hamiltonian formulation. In addition, the Hamiltonian formulation also motivates the definition given in chapter 11 of the total energy at spatial infinity of an asymptotically flat spacetime.

However, an even stronger reason for studying the Lagrangian and Hamiltonian formulations of general relativity arises from the desire to obtain a quantum theory of gravitation. Although the entire content of a classical field theory is expressed by the field equation, most prescriptions for formulating a quantum field theory associated with the classical theory require that the classical theory be expressed in a Lagrangian or Hamiltonian form. In particular, the path integral formulation requires that one have an action principle at the classical level—i.e., it requires a Lagrangian formulation of the classical theory—while the canonical quantization procedure requires that the classical field theory be cast in Hamiltonian form. Thus, the Lagrangian and Hamiltonian formulations of general relativity may play an important role in the development of a quantum theory of gravity (see chapter 14).

E.1 Lagrangian Formulation

We begin our discussion by explaining what we mean by a Lagrangian formulation of a field theory. Consider a theory involving a tensor field (or collection of tensor fields) defined on a manifold M. We shall suppress all indices and denote the field (or fields) by ψ. Let $S[\psi]$ be a functional of ψ, i.e., S is a map from field configurations on M into numbers. Let ψ_λ be a smooth one parameter family of field configurations starting from ψ_0 which satisfy appropriate boundary conditions. We denote $d\psi_\lambda/d\lambda\,|_{\lambda=0}$ by $\delta\psi$. Suppose $dS/d\lambda$ at $\lambda = 0$ exists for all such one-parameter families starting from ψ_0. Suppose, furthermore, that there exists a smooth tensor field χ [which is dual to ψ, i.e., if ψ is a tensor field of type (k, l), then χ will be of type (l, k)] such that for all such families we have

$$\frac{dS}{d\lambda} = \int_M \chi \, \delta\psi \quad , \tag{E.1.1}$$

where contraction of all indices in the integral is understood. Then we say that S is *functionally differentiable* at ψ_0. We call χ the *functional derivative*[1] of S and denote it as

$$\chi = \left.\frac{\delta S}{\delta\psi}\right|_{\psi_0} \quad . \tag{E.1.2}$$

Consider, now, a functional S of the form

$$S[\psi] = \int_M \mathcal{L}[\psi] \tag{E.1.3}$$

where $\mathcal{L}$ is a local function of ψ and a finite number of its derivatives, i.e.,

$$\mathcal{L}|_x = \mathcal{L}(\psi(x), \nabla\psi(x), \ldots, \nabla^k\psi(x)) \quad . \tag{E.1.4}$$

Suppose that S is functionally differentiable and that the field configurations ψ which extremize S,

$$\left.\frac{\delta S}{\delta\psi}\right|_{\psi} = 0 \quad , \tag{E.1.5}$$

are precisely the ones which are solutions of the field equation for ψ. Then S is called an *action,* $\mathcal{L}$ is called a *Lagrangian density,* and the specification of such an $\mathcal{L}$ is what we mean by a Lagrangian formulation of the field theory.

Thus the notion of a Lagrangian formulation of a field theory is closely analogous to that of a Lagrangian formulation in ordinary particle mechanics. In particle mechanics one specifies an action functional of the particle path as an integral of a Lagrangian function over the path. The variational problem analogous to (E.1.5) is made precise by focusing attention on paths of finite length and extremizing the action wih respect to path variations which keep the path endpoints fixed. By analogy, to make our variational problem in the field case precise, we shall focus attention on a compact region, U, of M and will consider one-parameter families ψ_λ which keep fixed the value of ψ on the boundary, $\dot{U}$.

As a simple example, we give a Lagrangian formulation of the theory of a Klein-Gordon scalar field ϕ in Minkowski spacetime. We define

$$\mathcal{L}_{\mathrm{KG}} = -\frac{1}{2}(\partial_a\phi\partial^a\phi + m^2\phi^2) \quad . \tag{E.1.6}$$

1. More generally, if there exists a tensor distribution χ such that $dS/d\lambda|_{\lambda=0} = \chi[\delta\psi]$, we also say that S is functionally differentiable and refer to χ as the functional derivative of S at ψ_0.

(Here, the normalization of $\mathcal{L}_{KG}$ is chosen for later convenience in defining conjugate momenta [see section E.2].) We obtain

$$\left.\frac{dS_{KG}}{d\lambda}\right|_{\lambda=0} = -\int [\partial_a\phi_0\partial^a(\delta\phi) + m^2\phi_0(\delta\phi)]$$

$$= \int [\partial_a\partial^a\phi_0 - m^2\phi_0]\delta\phi \quad , \tag{E.1.7}$$

where the natural volume element on Minkowski spacetime is understood in the integrals, and the boundary term in the integration by parts does not contribute on account of our boundary condition on ϕ_λ, which requires $\delta\phi = 0$ on the boundary. Thus, S_{KG} is functionally differentiable and we have

$$\frac{\delta S_{KG}}{\delta\phi} = \partial_a\partial^a\phi - m^2\phi \quad . \tag{E.1.8}$$

Consequently, equation (E.1.5) is just the Klein-Gordon equation (4.2.19), as desired. Similarly, the function

$$\mathcal{L}_{EM} = -\frac{1}{4}F_{ab}F^{ab} = -\partial_{[a}A_{b]}\partial^{[a}A^{b]} \tag{E.1.9}$$

of the field variable A_a is a Lagrangian density for Maxwell's equations in Minkowski spacetime. Indeed, equations (E.1.6) and (E.1.9) often are taken as the starting point for the study of Klein-Gordon and Maxwell fields: One writes down the simple and natural scalar functions $\mathcal{L}_{KG}$ or $\mathcal{L}_{EM}$ and obtains the field equations via equation (E.1.5).

For general relativity, the field variable is the spacetime metric, g_{ab}, defined on a four-dimensional manifold, M. In this case, a slight awkwardness results from the fact the natural volume element to use in the integrals (E.1.1) and (E.1.3) is the volume element ϵ_{abcd} determined from g_{ab} via equation (B.2.9). Consequently, the volume element itself depends on the field variable, and hence its variation must be taken into account when calculating functional derivatives. One way to handle this situation would be to define $\mathcal{L}$ to be a totally antisymmetric four-index tensor rather than a scalar, i.e., to incorporate the volume element into $\mathcal{L}$. This would require us to make a similar modification of our definition of functional derivatives. Instead we shall follow the considerably less cumbersome procedure of introducing a fixed volume element $e_{abcd} = e_{[abcd]}$ on M and defining all integrals over M to be with respect to e_{abcd} rather than ϵ_{abcd}. One way to do this (at least over a portion of M) would be to choose a coordinate system and take e_{abcd} to be the associated coordinate volume element, but we emphasize that the introduction of a coordinate system is not necessary. Since any two volume elements differ at each point by at most a scalar factor, we have

$$\epsilon_{abcd} = f e_{abcd} \quad . \tag{E.1.10}$$

In any basis where the nonvanishing components of e_{abcd} have the values ± 1, the calculation which led to equation (B.2.16) for the case of a coordinate basis and its associated volume element shows that $f = \sqrt{-g}$, where g denotes the determinant of the matrix of components, $g_{\mu\nu}$, of the metric in that basis. Hence, we shall follow the same notational convention as used for coordinate bases in section 3.4a and denote f as $\sqrt{-g}$. Given the volume element e_{abcd} on M, we define a *tensor density* $T^{a\cdots b}{}_{c\cdots d}$ to be a tensor which can be expressed in the form

$$T^{a\cdots b}{}_{c\cdots d} = \sqrt{-g}\,\tilde{T}^{a\cdots b}{}_{c\cdots d} \quad , \tag{E.1.11}$$

where $\tilde{T}^{a\cdots b}{}_{c\cdots d}$ is a tensor whose value does not depend on the choice of e_{abcd}. In order that the action, S, of general relativity be independent of e_{abcd}, it is necessary that the Lagrangian density $\mathcal{L}$ be a scalar density. Similarly, in order for $dS/d\lambda$ to be independent of e_{abcd}, the functional derivatives of S must be tensor densities.

We now shall demonstrate that—except for boundary terms to be dealt with later—the scalar density

$$\mathcal{L}_G = \sqrt{-g}R \tag{E.1.12}$$

is a Lagrangian density for the vacuum Einstein equation. The corresponding action

$$S[g^{ab}] = \int \mathcal{L}_G \boldsymbol{e} \tag{E.1.13}$$

is known as the *Hilbert action*. Here we have written $\boldsymbol{e}$ (using the differential forms notation of appendix B) in order to emphasize our use of this volume element. In addition, for convenience, we have taken the inverse metric g^{ab} as the field variable rather than g_{ab}. For a one-parameter variation we define δg^{ab} to be $dg^{ab}/d\lambda$. However, in order to use without modification the results of section 7.5 where g_{ab} was used as the independent variable, we shall define $\delta g_{ab} = dg_{ab}/d\lambda$. Thus, we warn the reader that, since $g^{ac}g_{cb} = \delta^a{}_b$, we have $\delta g_{ab} = -g_{ac}g_{bd}\delta g^{cd}$, i.e., in this section we will *not* use the unperturbed metric to raise and lower indices of the metric perturbations. Note also that since g^{ab} and hence δg^{ab} must be symmetric, one can add an antisymmetric tensor to any functional derivative with respect to g^{ab} without affecting equation (E.1.1). We eliminate this arbitrariness by requiring that all such functional derivatives be symmetric.

For a one-parameter family starting from g^{ab}, we have

$$\frac{d\mathcal{L}_G}{d\lambda} = \sqrt{-g}\,(\delta R_{ab})g^{ab} + \sqrt{-g}\,R_{ab}\delta g^{ab} + R\delta(\sqrt{-g}) \quad . \tag{E.1.14}$$

From equation (7.5.14) of chapter 7, we have

$$g^{ab}\delta R_{ab} = \nabla^a v_a \quad , \tag{E.1.15}$$

where

$$v_a = \nabla^b(\delta g_{ab}) - g^{cd}\nabla_a(\delta g_{cd}) \quad . \tag{E.1.16}$$

In addition, using equation (9.3.11), we have

$$\begin{aligned}\delta(\sqrt{-g}) &= \frac{1}{2}\sqrt{-g}\,g^{ab}\delta g_{ab} \\ &= -\frac{1}{2}\sqrt{-g}\,g_{ab}\delta g^{ab} \quad .\end{aligned} \tag{E.1.17}$$

Thus we obtain

$$\frac{dS_G}{d\lambda} = \int \frac{d\mathcal{L}_G}{d\lambda} \boldsymbol{e} = \int \nabla^a v_a \sqrt{-g}\, \boldsymbol{e} + \int \left(R_{ab} - \frac{1}{2} R g_{ab} \right) \delta g^{ab} \sqrt{-g}\, \boldsymbol{e} \quad . \tag{E.1.18}$$

The first term in equation (E.1.18) is the integral of a divergence, $\nabla^a v_a$, with respect to the natural volume element $\boldsymbol{\epsilon} = \sqrt{-g}\,\boldsymbol{e}$. Hence, by Stokes's theorem this integral contributes only a boundary term. In fact, this term does not vanish for general variations where g^{ab} is held fixed on the boundary, although it does vanish for variations where the first derivatives of g^{ab} also are held fixed. However, in order to simplify the discussion here, we shall ignore this contribution for the present. (At the end of this section we shall calculate this term and show how to modify S_G to cancel its contribution.) Thus discarding the boundary term, we find

$$\frac{\delta S_G}{\delta g^{ab}} = \sqrt{-g} \left(R_{ab} - \frac{1}{2} R g_{ab} \right) \quad , \tag{E.1.19}$$

and equation (E.1.5) is seen to be equivalent to Einstein's equation in vacuum, as desired. Thus, from the Lagrangian viewpoint, Einstein's equation arises in a very natural way, since the Lagrangian density of equation (E.1.12) is unquestionably one of the simplest scalar densities which can be constructed from the spacetime metric.

It is interesting to note that instead of viewing the metric alone as the field variable for general relativity, we could view the metric and the derivative operator ∇_a as independent variables. Remarkably, if we use the same Lagrangian density (E.1.12) but now view R_{ab} as a function of the derivative operator alone (i.e., independent of g^{ab}) and vary the *Palatini action*,

$$\mathcal{S}_G[g^{ab}, \nabla_a] = \int \sqrt{-g}\, R_{ab} g^{ab} \boldsymbol{e} \quad , \tag{E.1.20}$$

with respect to both g^{ab} and ∇_a, we recover Einstein's equation (E.1.19) together with the metric compatibility condition $\nabla_c g^{ab} = 0$ on the derivative operator. To prove this we begin by noting that since ∇_a can be expressed in terms of an arbitrary fixed derivative operator $\tilde{\nabla}_a$ and a tensor field $C^c{}_{ab}$ (see section 3.1), variation of ∇_a is equivalent to variation of $C^c{}_{ab}$. In considering one parameter variations of g^{ab} and ∇_a, it will be convenient to choose $\tilde{\nabla}_a$ to be the derivative operator compatible with g^{ab} at $\lambda = 0$. The key change from the previous calculation is that we must use equation (7.5.10) rather than equation (7.5.14) to evaluate δR_{ab}. We find at $\lambda = 0$

$$\begin{aligned}
\frac{d\mathcal{S}_G}{d\lambda} &= -2 \int g^{ab} \nabla_{[a} \delta C^c{}_{c]b} \sqrt{-g}\, \boldsymbol{e} + \int \left(R_{ab} - \frac{1}{2} R g_{ab} \right) \delta g^{ab} \sqrt{-g}\, \boldsymbol{e} \\
&= -2 \int g^{ab} \tilde{\nabla}_{[a} \delta C^c{}_{c]b} \sqrt{-g}\, \boldsymbol{e} \\
&\quad + \int g^{ab} [C^d{}_{ab} \delta C^c{}_{cd} + C^c{}_{cd} \delta C^d{}_{ab} - 2 C^d{}_{cb} \delta C^c{}_{ad}] \sqrt{-g}\, \boldsymbol{e} \\
&\quad + \int \left(R_{ab} - \frac{1}{2} R g_{ab} \right) \delta g^{ab} \sqrt{-g}\, \boldsymbol{e}
\end{aligned}$$

$$= \int [C^{bd}{}_d \delta^a{}_c + C^d{}_{dc} g^{ab} - 2C^b{}_c{}^a] \delta C^c{}_{ab} \sqrt{-g}\, \boldsymbol{e}$$

$$+ \int \left(R_{ab} - \frac{1}{2} R g_{ab} \right) \delta g^{ab} \sqrt{-g}\, \boldsymbol{e} \quad . \tag{E.1.21}$$

Here, the first term on the right side of the second line vanishes by Stokes's theorem, since $\tilde{\nabla}_a$ is the metric compatible derivative operator. (In this case there is no boundary term since we require $\delta C^c{}_{ab}$ to vanish on the boundary.) The vanishing of $\delta \mathcal{S}_G / \delta C^c{}_{ab}$ requires the term in brackets in the final equality of equation (E.1.21) to vanish when symmetrized over a and b, which, after some algebra, implies $C^c{}_{ab} = 0$, i.e., $\nabla_a = \tilde{\nabla}_a$. The vanishing of $\delta \mathcal{S}_G / \delta g^{ab}$ yields Einstein's equation as before.

The non-vacuum Einstein equation with matter fields such as the Klein-Gordon scalar field or the Maxwell field also can be obtained from a Lagrangian formulation in a very simple and natural manner. First, we must find a suitable Lagrangian density $\mathcal{L}_M$ for the matter fields in curved spacetime. In particular, for the Klein-Gordon field it is easily verified that functional differentiation with respect to ϕ of the action, S_{KG}, obtained from the Lagrangian density,

$$\mathcal{L}_{\text{KG}} = -\frac{1}{2} \sqrt{-g}\, (g^{ab} \nabla_a \phi \nabla_b \phi + m^2 \phi^2) \quad , \tag{E.1.22}$$

yields the Klein-Gordon equation (4.3.9) in curved spacetime. Similarly,

$$\mathcal{L}_{\text{EM}} = -\frac{1}{4} \sqrt{-g}\, g^{ac} g^{bd} F_{ab} F_{cd} = -\sqrt{-g}\, g^{ac} g^{bd} \nabla_{[a} A_{b]} \nabla_{[c} A_{d]} \tag{E.1.23}$$

yields Maxwell's equations in curved spacetime. To obtain the coupled Einstein-matter field equations, we construct a (total) Lagrangian density, $\mathcal{L}$, by adding together the Einstein Lagrangian density $\mathcal{L}_G$ with a multiple of the Lagrangian density, $\mathcal{L}_M$, for the matter field,

$$\mathcal{L} = \mathcal{L}_G + \alpha_M \mathcal{L}_M \quad , \tag{E.1.24}$$

where α_M is a constant. Since $\mathcal{L}_G$ does not depend on the matter field, variation of the total action, S, with respect to it will yield the same equation as variation of S_M alone. Variation of S with respect to g^{ab} yields the equation

$$G_{ab} = R_{ab} - \frac{1}{2} R g_{ab} = 8\pi T_{ab} \quad , \tag{E.1.25}$$

where the tensor T_{ab} is given by

$$T_{ab} = -\frac{\alpha_M}{8\pi} \frac{1}{\sqrt{-g}} \frac{\delta S_M}{\delta g^{ab}} \quad . \tag{E.1.26}$$

For the Lagrangian densities (E.1.22) and (E.1.23), it is easily verified that for appropriate choices of α_M, T_{ab} agrees, respectively, with equations (4.3.10) and (4.3.14). Thus, the Lagrangian density (E.1.24) with $\mathcal{L}_M = \mathcal{L}_{\text{KG}}$ and $\alpha_{\text{KG}} = 16\pi$ yields the coupled Einstein-Klein-Gordon equations, whereas (E.1.24) with

$\mathcal{L}_M = \mathcal{L}_{\text{EM}}$ and $\alpha_{\text{EM}} = 4$ yields the Einstein-Maxwell equations. More generally, if one takes a Lagrangian density $\mathcal{L}_M$ as the starting point in the definition of a matter field theory, then equation (E.1.26) may be used to *define* the stress-energy tensor, T_{ab}, of that field. If the Lagrangian density for the matter field does not depend on the choice of derivative operator ∇_a, then the Einstein-matter equations also may be derived from variation of the sum of the Palatini action and matter action.

The matter action S_M must be invariant under diffeomorphisms, i.e., if $f_\lambda : M \to M$ is a one-parameter family of diffeomorphisms, we have $S_M[g^{ab}, \psi] = S_M[f_\lambda{}^* g^{ab}, f_\lambda{}^* \psi]$, where $f_\lambda{}^*$ is defined in appendix C. Hence, for such variations, we have

$$0 = \frac{dS_M}{d\lambda} = \int \frac{\delta S_M}{\delta g^{ab}} \delta g^{ab} + \int \frac{\delta S_M}{\delta \psi} \delta \psi \quad . \tag{E.1.27}$$

Recall from appendix C that for such variations, δg^{ab} has the general form $\pounds_w g^{ab} = 2\nabla^{(a} w^{b)}$, where w^a is an arbitrary vector field. Suppose, now, that ψ satisfies the matter field equations. Then $\delta S_M / \delta \psi |_\psi = 0$ and the second term in equation (E.1.27) makes no contribution. Thus, using the definition (E.1.26), we find that if ψ satisfies the matter field equations, then for all smooth w^a of compact support, we have

$$\begin{aligned} 0 &= \int \sqrt{-g}\, T_{ab} \nabla^{(a} w^{b)} \boldsymbol{e} \\ &= \int T_{ab} \nabla^a w^b \boldsymbol{\epsilon} \\ &= -\int (\nabla^a T_{ab}) w^b \boldsymbol{\epsilon} \quad , \end{aligned} \tag{E.1.28}$$

which implies that

$$\nabla^a T_{ab} = 0 \quad . \tag{E.1.29}$$

Thus, for a diffeomorphism invariant action, T_{ab} always is conserved by virtue of the matter field equation. This reinforces the interpretation of T_{ab} as representing the stress-energy-momentum tensor of the matter field. Note also that by applying the above argument to S_G, it follows that (independent of any field equation) we have

$$\nabla^a G_{ab} = 0 \quad . \tag{E.1.30}$$

Thus, in the Lagrangian formulation of general relativity, the contracted Bianchi identity may be viewed as a consequence of the invariance of the Hilbert action under diffeomorphisms.

In Minkowski spacetime $(\mathbb{R}^4, \eta_{ab})$ there exists an alternative procedure for defining a stress-energy tensor associated with a field ψ starting from its Lagrangian $\mathcal{L}$. Consider, for simplicity, the case where $\mathcal{L}$ is a local function of η_{ab}, ψ, and $\partial_a \psi$ but no higher derivatives of ψ. Then, the equation of motion for ψ is

$$0 = \frac{\delta S}{\delta \psi} = \frac{\partial \mathcal{L}}{\partial \psi} - \partial_a \left[\frac{\partial \mathcal{L}}{\partial(\partial_a \psi)} \right] \quad . \tag{E.1.31}$$

Here, in the case where ψ is a scalar field $\partial\mathcal{L}/\partial(\partial_a\psi)$ means the vector field V^a which at each point x satisfies

$$\frac{d\mathcal{L}}{d\lambda} = V^a\delta(\partial_a\psi) \tag{E.1.32}$$

for all smooth one-parameter variations of ψ which keep fixed at x the value of all quantities upon which $\mathcal{L}$ depends except $\partial_a\psi$. (The components of V^a in a basis are just the partial derivatives of $\mathcal{L}$ with respect to the corresponding dual basis components of $\partial_a\psi$.) If ψ is a tensor of type (k, l), then $V^a = \partial\mathcal{L}/\partial(\partial_a\psi)$ will be a tensor of type $(l + 1, k)$, with contraction of all indices understood in equation (E.1.32). Partial derivatives of $\mathcal{L}$ with respect to other tensor variables are defined similarly. For an arbitrary smooth one-parameter family $(\psi_\lambda, (\eta_\lambda)_{ab})$ of fields ψ_λ and flat metrics $(\eta_\lambda)_{ab}$, we have

$$\delta\mathcal{L} \equiv \frac{d\mathcal{L}}{d\lambda} = \frac{\partial\mathcal{L}}{\partial\psi}\delta\psi + \frac{\partial\mathcal{L}}{\partial(\partial_a\psi)}\delta(\partial_a\psi) + \frac{\partial\mathcal{L}}{\partial\eta_{ab}}\delta\eta_{ab} \quad , \tag{E.1.33}$$

where no boundary conditions on the varied quantities need be assumed here. Consider, now, the variations of $\mathcal{L}$ produced by a one-parameter family of diffeomorphisms generated by a vector field ξ^a. Then equation (E.1.33) becomes

$$\delta\mathcal{L} = \pounds_\xi\mathcal{L} = \xi^a\partial_a\mathcal{L} = \frac{\partial\mathcal{L}}{\partial\psi}\pounds_\xi\psi + \frac{\partial\mathcal{L}}{\partial(\partial_a\psi)}\pounds_\xi\partial_a\psi + \frac{\partial\mathcal{L}}{\partial\eta_{ab}}\pounds_\xi\eta_{ab} \quad . \tag{E.1.34}$$

Now restrict attention to the case where ξ^a is a Killing field. Then the last term in equation (E.1.34) vanishes, and in the second term we have $\pounds_\xi\partial_a\psi = \partial_a(\pounds_\xi\psi)$. Using equation (E.1.31) to substitute for $\partial\mathcal{L}/\partial\psi$, we obtain

$$\partial_a\left[\frac{\partial\mathcal{L}}{\partial(\partial_a\psi)}\pounds_\xi\psi - \xi^a\mathcal{L}\right] = 0 \quad . \tag{E.1.35}$$

This result is known as *Noether's theorem* as applied to the Poincaré group of symmetries. In particular, the validity of equation (E.1.35) for all translational Killing fields implies that the tensor

$$S^{ab} = \frac{\partial\mathcal{L}}{\partial(\partial_a\psi)}\partial^b\psi - g^{ab}\mathcal{L} \quad , \tag{E.1.36}$$

known as the *canonical energy-momentum tensor*, is conserved, $\partial_a S^{ab} = 0$. For the Klein-Gordon field, we find that S_{ab} agrees with T_{ab} (up to a numerical factor), where T_{ab} is defined by equation (E.1.26) evaluated at Minkowski spacetime using the curved space Lagrangian density (E.1.22). However, this agreement does not occur for higher spin fields. Indeed, in the case of a Maxwell field, S^{ab} is not even gauge invariant. Furthermore, in general S_{ab} is not symmetric, nor does it naturally generalize to a conserved tensor in curved spacetime (Kuchař 1976). Thus we adopt T_{ab} as our definition of the stress-energy tensor. It is the quantity which naturally appears on the right-hand side of Einstein's equation (E.1.25) in a Lagrangian formulation of the Einstein-matter field equations.

We conclude this section by evaluating the boundary term occurring in the variation (E.1.18) of the Hilbert action when δg^{ab} is required to vanish on the boundary but no conditions are placed on the derivatives of δg^{ab}. We have

$$\int_U \nabla_a v^a \boldsymbol{\epsilon} = \int_{\dot{U}} v_a n^a \quad , \tag{E.1.37}$$

where n^a is the unit normal to the boundary $\dot{U}$ (which is assumed to be non-null) and the natural volume element on $\dot{U}$ is understood (see appendix B). Using equation (E.1.16), we have on $\dot{U}$

$$\begin{aligned} v_a n^a &= n^a g^{bc}[\nabla_c(\delta g_{ab}) - \nabla_a(\delta g_{bc})] \\ &= n^a h^{bc}[\nabla_c(\delta g_{ab}) - \nabla_a(\delta g_{bc})] \\ &= -n^a h^{bc}\nabla_a(\delta g_{bc}) \quad , \end{aligned} \tag{E.1.38}$$

where $h_{ab} = g_{ab} \pm n_a n_b$ is the induced metric on $\dot{U}$ (see chapter 10) and we have $h^{bc}\nabla_c(\delta g_{ab}) = 0$ because $\delta g_{ab} = 0$ on $\dot{U}$. However, the right-hand side of equation (E.1.38) is related to the variation of the trace of the extrinsic curvature of the boundary. We have

$$K \equiv K^a{}_a = h^a{}_b \nabla_a n^b \tag{E.1.39}$$

and hence

$$\begin{aligned} \delta K &= h^a{}_b (\delta C)^b{}_{ac} n^c \\ &= \frac{1}{2} n^c h^a{}_b g^{bd}[\nabla_a(\delta g_{cd}) + \nabla_c(\delta g_{ad}) - \nabla_d(\delta g_{ac})] \\ &= \frac{1}{2} n^c h^{ad} \nabla_c(\delta g_{ad}) \quad . \end{aligned} \tag{E.1.40}$$

Thus, under variations of the metric for which $\delta g_{ab} = 0$ on $\dot{U}$ we obtain from equations (E.1.18), (E.1.38), and (E.1.40)

$$\frac{dS_G}{d\lambda} = -2\int_{\dot{U}} \delta K + \int_U G_{ab}\delta g^{ab} \boldsymbol{\epsilon} \quad . \tag{E.1.41}$$

In fact, equation (E.1.41) continues to hold if we allow variations of g_{ab} for which only the induced metric on the boundary is held fixed, $\delta h_{ab} = 0$. This can be verified directly or deduced from the fact that if $\delta h_{ab} = 0$ on the boundary, we can find a gauge transformation $\nabla_{(a} l_{b)}$ with $l_b = 0$ on the boundary which makes $\delta g_{ab} = 0$. Since equation (E.1.41) holds for all variations with $\delta g_{ab} = 0$ on $\dot{U}$ and since all terms in equation (E.1.41) are invariant under such gauge transformations, this equation must continue to hold for variations which merely satisfy $\delta h_{ab} = 0$.

Thus, the extremization of S_G with respect to variations with $\delta g_{ab} = 0$ or $\delta h_{ab} = 0$ on the boundary contains an additional, unwanted term. However, this can be remedied by modifying S_G. We define

$$S'_G = S_G + 2\int_{\dot{U}} K \quad . \tag{E.1.42}$$

Then extremization of S'_G yields the desired result since variation of the boundary term in equation (E.1.42) cancels the boundary term in (E.1.41). Thus, when boundary terms are taken into account, S'_G is the appropriate action to use for general relativity.

E.2 Hamiltonian Formulation

A Lagrangian formulation of a field theory is "spacetime covariant." One specifies on the spacetime manifold an action functional of the field ψ whose extremization yields the field equations. On the other hand, a Hamiltonian formulation of a field theory requires a breakup of spacetime into space and time. Indeed, the first step in producing a Hamiltonian formulation of a field theory consists of choosing a time function t and a vector field t^a on a spacetime such that the surfaces, Σ_t, of constant t are spacelike Cauchy surfaces and such that $t^a \nabla_a t = 1$. The vector field t^a may be interpreted as describing the "flow of time" in the spacetime and can be used to identify each Σ_t with the initial surface Σ_0. In Minkowski spacetime the choice of t and t^a usually is made via a global inertial coordinate system, but in curved spacetime there may not be any preferred choice. In performing integrals of functions over M it would be natural for most purposes to use the volume element ϵ_{abcd} associated with the spacetime metric. Similarly, in performing integrals over Σ_t, it would be natural in most cases to use the volume element ${}^{(3)}\epsilon_{abc} = \epsilon_{dabc} n^d$, where n^d is the unit normal to Σ_t. However, these volume elements will, in general, be "time dependent" in the sense that $\pounds_t \epsilon_{abcd} \neq 0$ and $\pounds_t {}^{(3)}\epsilon_{abc} \neq 0$. The use of a time-dependent volume element on Σ_t is particularly inconvenient if we wish to identify Σ_t with Σ_0 in order to view dynamical evolution as the change of fields on the fixed manifold Σ_0. Therefore, we shall introduce a fixed volume element e_{abcd} on M satisfying $\pounds_t e_{abcd} = 0$. [One way to do this—at least locally—would be to introduce coordinates x^1, x^2, x^3 in addition to t such that $t^a = (\partial/\partial t)^a$ and to take $\boldsymbol{e}$ to be the coordinate volume element $dt \wedge dx^1 \wedge dx^2 \wedge dx^3$.] On each Σ_t, we define ${}^{(3)}e_{abc} = e_{dabc} t^d$. Unless otherwise stated, all integrals over M will be performed using the volume element e_{abcd} and all integrals over Σ_t will be with respect to the volume element ${}^{(3)}e_{abc}$. Thus, in particular, in order that our results be independent of our choice of e_{abcd}, the Lagrangian density must be a scalar density on M and the momentum π (defined below) must be a tensor density on Σ_t. As previously noted in our discussion of the Einstein Lagrangian, the introduction of e_{abcd} could be avoided by incorporating the volume element into the definition of $\mathcal{L}$, π, and other quantities, but we choose not to do so since this procedure is rather cumbersome.

The next step in giving a Hamiltonian formulation is to define a configuration space for the field by specifying what tensor field (or fields) q on Σ_t physically describes the instantaneous configuration of the field ψ. The space of possible momenta of the field at a given configuration q then is taken to be the "cotangent space," V_q^*, of the configuration space at q. Since the set of possible configurations of the field is infinite-dimensional, we shall not attempt here to give a precise definition of V_q^*. However, in the case where the allowed infinitesimal variations (i.e., "tangent vectors") δq at q are represented by tensor fields on Σ_t of type (k, l), we shall take the space of momenta to consist of tensor fields, π, of type (l, k) on Σ_t, so that π maps δq into $\mathbb{R}$ via $\delta q \rightarrow \int_{\Sigma_t} \pi \delta q$, where contraction of indices is understood. A prescription must then be given for associating a momentum π to the field ψ on Σ_t. The final and most nontrivial step required for a Hamiltonian formulation of a field theory is the specification of a functional $H[q, \pi]$ on Σ_t, called the *Hamiltonian,* which is of the form

$$H = \int_{\Sigma_t} \mathcal{H} \quad , \tag{E.2.1}$$

where the *Hamiltonian density* $\mathcal{H}$ is the local function of q, π and of their spatial derivatives up to a finite order, such that the pair of equations,

$$\dot{q} \equiv \pounds_t q = \frac{\delta H}{\delta \pi} \quad , \tag{E.2.2}$$

$$\dot{\pi} \equiv \pounds_t \pi = -\frac{\delta H}{\delta q} \quad , \tag{E.2.3}$$

is equivalent to the field equation satisfied by ψ.

Given a Lagrangian formulation of a field theory, there is a standard prescription for obtaining a Hamiltonian formulation which is closely analogous to the well known procedure in ordinary particle mechanics. First, one takes q to be simply the field ψ evaluated on Σ_t. Then one views the Lagrangian density as a function of q, its time derivatives, and its space derivatives. Assuming that $\mathcal{L}$ does not depend on time derivatives of q higher than first order, we take the momentum, π, associated with ψ on Σ_t to be

$$\pi = \frac{\partial \mathcal{L}}{\partial \dot{q}} \quad . \tag{E.2.4}$$

Next, we attempt to solve equation (E.2.4) for $\dot{q}$ as a function of q and π. If this can be done, we define

$$\mathcal{H}(q, \pi) = \pi \dot{q} - \mathcal{L} \quad , \tag{E.2.5}$$

where $\dot{q} = \dot{q}(q, \pi)$ is understood in this equation both in its explicit appearance and in its implicit appearance in $\mathcal{L}$. With this choice of $\mathcal{H}$, equations (E.2.2) and (E.2.3) are equivalent to equation (E.1.5). To see this, we define

$$J = \int_{t_1}^{t_2} H\, dt = \int_{t_1}^{t_2} dt \int_{\Sigma_t} \mathcal{H} = -S + \int_{t_1}^{t_2} dt \int_{\Sigma_t} \pi \dot{q} \quad . \tag{E.2.6}$$

Then, for a smooth one-parameter variation of ψ which satisfies $\delta\psi = 0$ at $t = t_1$ and $t = t_2$, we have

$$\begin{aligned} \frac{dJ}{d\lambda} &= \int_{t_1}^{t_2} dt \int_{\Sigma_t} \left[\frac{\delta H}{\delta q} \delta q + \frac{\delta H}{\delta \pi} \delta \pi \right] \\ &= \int_{t_1}^{t_2} dt \int_{\Sigma_t} [\pi \delta \dot{q} + \dot{q} \delta \pi] - \frac{dS}{d\lambda} \end{aligned}$$

$$= \int_{t_1}^{t_2} dt \int_{\Sigma_t} [-\dot{\pi}\delta q + \dot{q}\delta\pi] - \frac{dS}{d\lambda} \quad , \tag{E.2.7}$$

where an integration by parts was performed in the last line. Thus, comparing the first and last line of this equation, we see that $\delta S/\delta\psi = 0$ if and only if equations (E.2.2) and (E.2.3) are satisfied. Thus, $\mathcal{H}$ is a Hamiltonian density for ψ.

The above procedure yields in a straightforward manner a Hamiltonian formulation of the theory of a Klein-Gordon field in Minkowski spacetime. We choose a global inertial coordinate system to obtain t and t^a and choose e_{abcd} to be the natural volume element ϵ_{abcd} since $\pounds_t \epsilon_{abcd} = 0$. We choose q on Σ_t to be ϕ evaluated on Σ_t and write $\mathcal{L}_{\text{KG}}$, equation (E.1.6), as

$$\mathcal{L}_{\text{KG}} = \frac{1}{2}(\dot{\phi}^2 - \vec{\nabla}\phi \cdot \vec{\nabla}\phi - m^2\phi^2) \quad , \tag{E.2.8}$$

where we use ordinary three-dimensional vector notation on Σ_t. We find

$$\pi = \frac{\partial \mathcal{L}_{\text{KG}}}{\partial \dot{\phi}} = \dot{\phi} \quad . \tag{E.2.9}$$

Hence, we define the Hamiltonian density by

$$\mathcal{H}_{\text{KG}} = \pi\dot{\phi} - \mathcal{L}_{\text{KG}} = \frac{1}{2}(\pi^2 + \vec{\nabla}\phi \cdot \vec{\nabla}\phi + m^2\phi^2) \quad . \tag{E.2.10}$$

It then may be verified that for $H_{\text{KG}} = \int_{\Sigma_t} \mathcal{H}_{\text{KG}}$, equations (E.2.2) and (E.2.3) indeed are equivalent to the Klein-Gordon equation. Note also that the numerical value of H_{KG} is just the total energy of the Klein-Gordon field.

For the case of the electromagnetic field in Minkowski spacetime, it is not as straightforward to obtain a Hamiltonian formulation by this procedure. We provisionally take q to be the vector potential A_a evaluated on Σ_t and decompose it into its normal and tangential parts,

$$V = -A_a n^a \quad , \tag{E.2.11}$$

$${}^{(3)}A_a = h_a{}^b A_b \quad , \tag{E.2.12}$$

where n^a is the unit normal to Σ_t and $h_{ab} = \eta_{ab} + n_a n_b$ is the induced spatial metric on Σ_t. In ordinary three-dimensional vector notation, the Lagrangian density, equation (E.1.9) is

$$\mathcal{L}_{\text{EM}} = \frac{1}{2}(\dot{\vec{A}} + \vec{\nabla}V) \cdot (\dot{\vec{A}} + \vec{\nabla}V) - \frac{1}{2}(\vec{\nabla} \times \vec{A}) \cdot (\vec{\nabla} \times \vec{A}) \quad . \tag{E.2.13}$$

Hence the momentum conjugate to $\vec{A}$ is

$$\vec{\pi} = \dot{\vec{A}} + \vec{\nabla}V \equiv -\vec{E} \quad . \tag{E.2.14}$$

However, $\dot{V}$ does not appear in $\mathcal{L}_{\text{EM}}$, so the momentum π_V conjugate to V vanishes identically,

$$\pi_V = 0 \quad . \tag{E.2.15}$$

Thus, we do not obtain an invertible relation between π and $\dot{q}$. Consequently, if we define $\mathcal{H}$ by equation (E.2.5), we will not be able to eliminate $\dot{q}$ in favor of π and q, and our general prescription for obtaining a Hamiltonian formulation breaks down. This difficulty is directly related to the fact that there is gauge arbitrariness in A_a, and hence we cannot expect to get deterministic dynamics for A_a of the form (E.2.2), (E.2.3).

However, this difficulty can be resolved by the following considerations. The fact that π_V vanishes identically suggests that we should not view V as a dynamical variable. It suggests that we should take the configuration field q to be simply $\vec{A}$. Therefore, we define $\mathcal{H}_{\text{EM}}$ by

$$\begin{aligned} \mathcal{H}_{\text{EM}} &= \vec{\pi} \cdot \dot{\vec{A}} - \mathcal{L}_{\text{EM}} \\ &= \frac{1}{2}\vec{\pi} \cdot \vec{\pi} + \frac{1}{2}\vec{B} \cdot \vec{B} - \vec{\pi} \cdot \vec{\nabla} V \\ &= \frac{1}{2}\vec{\pi} \cdot \vec{\pi} + \frac{1}{2}\vec{B} \cdot \vec{B} + V\vec{\nabla} \cdot \vec{\pi} - \vec{\nabla} \cdot (V\vec{\pi}) \quad , \end{aligned} \tag{E.2.16}$$

where $\vec{B} \equiv \vec{\nabla} \times \vec{A}$. The last term on the right-hand side of equation (E.2.16) is a total divergence and thus contributes only a boundary term to $H_{\text{EM}} = \int_{\Sigma_t} \mathcal{H}_{\text{EM}}$ which vanishes in the limit as the boundary goes to infinity for the asymptotic conditions usually imposed on V and $\vec{\pi} = -\vec{E}$. Hence we shall discard this term.

We now view H_{EM} as a functional of $\vec{A}$ and $\vec{\pi}$, with V effectively playing the role of a Lagrange multiplier, i.e., we append the equation

$$\frac{\delta H_{\text{EM}}}{\delta V} = 0 \tag{E.2.17}$$

to the equations (E.2.2) and (E.2.3) for $\dot{\vec{A}}$ and $\dot{\vec{\pi}}$. Equation (E.2.17) yields

$$\vec{\nabla} \cdot \vec{E} = 0 \quad , \tag{E.2.18}$$

whereas equations (E.2.2) and (E.2.3) yield, respectively,

$$\dot{\vec{A}} = \frac{\delta H_{\text{EM}}}{\delta \vec{\pi}} = \vec{\pi} - \vec{\nabla} V = -\vec{E} - \vec{\nabla} V \tag{E.2.19}$$

$$\dot{\vec{\pi}} = -\dot{\vec{E}} = -\frac{\delta H_{\text{EM}}}{\delta \vec{A}} = -\vec{\nabla} \times (\vec{\nabla} \times \vec{A}) \quad . \tag{E.2.20}$$

Thus, we see that the system of equations (E.2.18)–(E.2.20) is equivalent to Maxwell's equations. Furthermore, we obtain from this formulation a natural breakup of Maxwell's equations into the constraint (E.2.18) and the evolution equations (E.2.19), (E.2.20). Note that, again, the numerical value of H_{EM} for a solution of Maxwell's equations is proportional to the total energy of the electromagnetic field.

Thus, we have obtained a Hamiltonian formulation of Maxwell's equations in Minkowski spacetime which has the feature that a non–dynamical variable appears in $\mathcal{H}$ and effectively plays the role of a Lagrange multiplier enforcing the constraint (E.2.17). This type of Hamiltonian formulation is called a *constrained Hamiltonian formulation*. As will be discussed further below, it can be expected to arise in any theory where the field variables have a gauge arbitrariness.

We turn, now, to the task of obtaining a Hamiltonian formulation of Einstein's equations. As in the previous cases, we choose a time function t and a "time flow" vector field t^a on M satisfying $t^a \nabla_a t = 1$. Note, however, that in this case one cannot interpret t and t^a in terms of physical measurements using clocks until one knows the spacetime metric, which, of course, is the unknown field variable in Einstein's equation. Given a metric g_{ab}, it is convenient to decompose t^a into its normal and tangential parts with respect to the surfaces, Σ_t, of constant t. As in chapter 10, we define the *lapse function*, N, by

$$N = -g_{ab} t^a n^b = (n^a \nabla_a t)^{-1} \tag{E.2.21}$$

and the *shift vector* N^a by

$$N^a = h^a{}_b t^b \quad , \tag{E.2.22}$$

where again n^a is the unit normal to Σ_t and $h_{ab} = g_{ab} + n_a n_b$ is the induced spatial metric on Σ_t. Thus, N measures the rate of flow of proper time, τ, with respect to coordinate time, t, as one moves normally to Σ_t, whereas N^a measures the amount of "shift" tangential to Σ_t contained in the time flow vector field t^a (see Fig. 10.2 of chapter 10). In terms of N, N^a, and t^a, we have

$$n^a = \frac{1}{N}(t^a - N^a) \quad , \tag{E.2.23}$$

and hence the inverse spacetime metric can be written as

$$g^{ab} = h^{ab} - n^a n^b = h^{ab} - N^{-2}(t^a - N^a)(t^b - N^b) \quad . \tag{E.2.24}$$

It is convenient to choose as our field variables the spatial metric, h_{ab}, the lapse function N, and the covariant form of the shift vector, $N_a = h_{ab} N^b$ rather than the inverse metric, g^{ab}, which was used as the field variable in the previous section. The requirements that $h^{ac} h_{cb}$ be the identity operator on the tangent space to Σ_t and that $h^{ab} \nabla_b t = 0$ allow us to compute h^{ab} from h_{ab} and thence obtain $N^a = h^{ab} N_b$. Thus, from equation (E.2.24) we see that the information contained in (h_{ab}, N, N_a) is equivalent to that contained in g^{ab}.

Again, we shall use a fixed volume element e_{abcd} on spacetime satisfying $\pounds_t e_{abcd} = 0$ and will use the volume element ${}^{(3)}e_{abc} = e_{dabc} t^d$ on Σ_t. We note in analogy with the remarks below equation (E.1.10), we have ${}^{(3)}\epsilon_{abc} = \sqrt{h}\, {}^{(3)}e_{abc}$, where h is the determinant of the matrix of components, $h_{\mu\nu}$, of h_{ab} in a basis where the nonvanishing components of ${}^{(3)}e_{abc}$ have the values ± 1. It then follows that

$$\sqrt{-g} = N \sqrt{h} \quad . \tag{E.2.25}$$

The first step in obtaining a Hamiltonian functional for general relativity is to express the gravitational action in terms of (h_{ab}, N, N_a) and their time and space derivatives. To simplify the discussion here, we will defer the analysis of boundary terms until the end of this section. Thus, we start with the Hilbert action (E.1.13) rather than (E.1.42) and for the present will discard the boundary terms which arise in subsequent calculations. We express the scalar curvature, R, as

$$R = 2(G_{ab}n^a n^b - R_{ab}n^a n^b) \quad . \tag{E.2.26}$$

From equation (10.2.30) we have

$$G_{ab}n^a n^b = \frac{1}{2}[{}^{(3)}R - K_{ab}K^{ab} + K^2] \quad , \tag{E.2.27}$$

where K_{ab} is the extrinsic curvature of Σ_t and $K = K^a{}_a$. On the other hand, from the definition of the Riemann tensor, we have

$$\begin{aligned} R_{ab}n^a n^b &= R_{acb}{}^c n^a n^b \\ &= -n^a(\nabla_a \nabla_c - \nabla_c \nabla_a)n^c \\ &= (\nabla_a n^a)(\nabla_c n^c) - (\nabla_c n^a)(\nabla_a n^c) \\ &\quad - \nabla_a(n^a \nabla_c n^c) + \nabla_c(n^a \nabla_a n^c) \\ &= K^2 - K_{ac}K^{ac} - \nabla_a(n^a \nabla_c n^c) + \nabla_c(n^a \nabla_a n^c) \quad . \end{aligned} \tag{E.2.28}$$

The last two terms on the right-hand side of equation (E.2.28) are divergences and thus will be discarded. Hence, from equations (E.1.12) and (E.2.25)–(E.2.28), we obtain

$$\mathcal{L}_G = \sqrt{h}N[{}^{(3)}R + K_{ab}K^{ab} - K^2] \quad . \tag{E.2.29}$$

The extrinsic curvature, K_{ab}, is related to the "time derivative," $\dot{h}_{ab} \equiv h_a{}^c h_b{}^d \pounds_t h_{cd}$ of h_{ab} by

$$\begin{aligned} K_{ab} &= \frac{1}{2}\pounds_n h_{ab} = \frac{1}{2}[n^c\nabla_c h_{ab} + h_{ac}\nabla_b n^c + h_{cb}\nabla_a n^c] \\ &= \frac{1}{2}N^{-1}[Nn^c\nabla_c h_{ab} + h_{ac}\nabla_b(Nn^c) + h_{cb}\nabla_a(Nn^c)] \\ &= \frac{1}{2}N^{-1}h_a{}^c h_b{}^d[\pounds_t h_{cd} - \pounds_N h_{cd}] \\ &= \frac{1}{2}N^{-1}[\dot{h}_{ab} - D_a N_b - D_b N_a] \quad , \end{aligned} \tag{E.2.30}$$

where D_a is the derivative operator on Σ_t associated with h_{ab} (see chapter 10) and equation (E.2.23) was used to go from the second line to the third line. Thus, substitution of equation (E.2.30) into equation (E.2.29) expresses the gravitational action in the desired form given by Arnowitt, Deser, and Misner (1962).

The momentum canonically conjugate to h_{ab} is

$$\pi^{ab} = \frac{\partial \mathcal{L}_G}{\partial \dot{h}_{ab}} = \sqrt{h}(K^{ab} - Kh^{ab}) \quad . \tag{E.2.31}$$

However, $\mathscr{L}_G$ does not contain any time derivatives of N or N_a, so their conjugate momenta vanish identically. In analogy with the electromagnetic case, we interpret this fact as telling us that N and N_a should not be viewed as dynamical variables. Hence, we redefine our configuration space to consist of Riemannian metrics, h_{ab} on Σ_t. We define our Hamiltonian density by

$$\begin{aligned}\mathscr{H}_G &= \pi^{ab}\dot{h}_{ab} - \mathscr{L}_G \\ &= -h^{1/2}N\,{}^{(3)}R + Nh^{-1/2}\left[\pi^{ab}\pi_{ab} - \frac{1}{2}\pi^2\right] + 2\pi^{ab}D_aN_b \\ &= h^{1/2}\left\{N\left[-{}^{(3)}R + h^{-1}\pi^{ab}\pi_{ab} - \frac{1}{2}h^{-1}\pi^2\right] - 2N_b[D_a(h^{-1/2}\pi^{ab})] \right. \\ &\quad \left. + 2D_a(h^{-1/2}N_b\pi^{ab})\right\} \quad , \end{aligned} \tag{E.2.32}$$

where $\pi = \pi^a{}_a$. Again, the last term in equation (E.2.32) contributes only a boundary term to $H_G = \int \mathscr{H}_G\,{}^{(3)}\boldsymbol{e}$ and will be dropped. Variation of H_G with respect to N and N_a yields the equations

$$-{}^{(3)}R + h^{-1}\pi^{ab}\pi_{ab} - \frac{1}{2}h^{-1}\pi^2 = 0 \quad , \tag{E.2.33}$$

$$D_a(h^{-1/2}\pi^{ab}) = 0 \quad , \tag{E.2.34}$$

which, with the substitution (E.2.31), can be recognized as the initial value constraint equations (10.2.28) and (10.2.30) found in chapter 10. The dynamical equations (E.2.2) and (E.2.3) obtained from H_G are (Arnowitt, Deser, and Misner 1962)

$$\dot{h}_{ab} = \frac{\delta H_G}{\delta \pi^{ab}} = 2h^{-1/2}N\left(\pi_{ab} - \frac{1}{2}h_{ab}\pi\right) + 2D_{(a}N_{b)} \quad , \tag{E.2.35}$$

$$\begin{aligned}\dot{\pi}^{ab} = -\frac{\delta H_G}{\delta h_{ab}} &= -Nh^{1/2}\left({}^{(3)}R^{ab} - \frac{1}{2}{}^{(3)}Rh^{ab}\right) \\ &\quad + \frac{1}{2}Nh^{-1/2}h^{ab}\left(\pi_{cd}\pi^{cd} - \frac{1}{2}\pi^2\right) \\ &\quad - 2Nh^{-1/2}\left(\pi^{ac}\pi_c{}^b - \frac{1}{2}\pi\pi^{ab}\right) \\ &\quad + h^{1/2}(D^aD^bN - h^{ab}D^cD_cN) \\ &\quad + h^{1/2}D_c(h^{-1/2}N^c\pi^{ab}) - 2\pi^{c(a}D_cN^{b)} \quad , \end{aligned} \tag{E.2.36}$$

where, again, boundary terms have been ignored and equation (E.2.34) was used. Equations (E.2.33)–(E.2.36) are equivalent to the vacuum Einstein equation, $R_{ab} = 0$. Thus, we have succeeded in giving a constrained Hamiltonian formulation of Einstein's equation.

The presence of constraints in our Hamiltonian formulations of Maxwell's equations and Einstein's equations indicates that we have not isolated the "true dynamical degrees of freedom" in our choice of configuration space. Even though we already have eliminated V and N and N_a as dynamical variables, the constraints tell us that our phase space still is "too large." This, in turn, is directly related to the gauge freedom present in our configuration variables $\vec{A}$ and h_{ab}, respectively. In the Maxwell case, $\vec{A}$ and $\vec{A} - \vec{\nabla}\chi$ represent the same physical configuration. This suggests that we should take our configuration space to be not simply the space of vector potentials, $\vec{A}$, but the space of equivalence classes, $\tilde{\vec{A}}$, of vector potentials, where two vector potentials are equivalent if they differ only by a gauge transformation. The "cotangent space" at $\tilde{\vec{A}}$ then would be the space of linear functions of variations of $\vec{A}$ which depend only on the equivalence class. Thus, the momenta would be represented by vector fields $\vec{\pi}$ having the property that

$$\int \vec{\pi} \cdot [\delta\vec{A} - \vec{\nabla}(\delta\chi)] = \int \vec{\pi} \cdot \delta\vec{A} \quad . \tag{E.2.37}$$

However, this property holds if and only if

$$\vec{\nabla} \cdot \vec{\pi} = 0 \quad . \tag{E.2.38}$$

Thus, with our new choice of configuration space, the momentum space consists precisely of the divergence-free vector fields on Σ_t. But this means that the constraint (E.2.18) is automatically satisfied! We may drop the term $V(\vec{\nabla} \cdot \vec{\pi})$ from equation (E.2.16) and take the Hamiltonian density to be simply

$$\mathcal{H}_{\mathrm{EM}} = \frac{1}{2}(\vec{\pi} \cdot \vec{\pi} + \vec{B} \cdot \vec{B}) \quad , \tag{E.2.39}$$

where, again, $\vec{B} \equiv \vec{\nabla} \times \vec{A}$. (Note that $\vec{B}$ depends only on the equivalence class of $\vec{A}$ and hence is a well defined function of $\tilde{\vec{A}}$.) Hamilton's equations of motion (E.2.2) and (E.2.3) yield

$$\dot{\tilde{\vec{A}}} = \frac{\delta \tilde{H}_{\mathrm{EM}}}{\delta \vec{\pi}} = \tilde{\vec{\pi}} \quad , \tag{E.2.40}$$

$$\dot{\vec{\pi}} = -\frac{\delta \tilde{H}_{\mathrm{EM}}}{\delta \tilde{\vec{A}}} = -\vec{\nabla} \times \vec{B} \quad . \tag{E.2.41}$$

The equivalence classes appearing on both sides of equation (E.2.40) can be eliminated by taking the curl of this equation. It then easily may be verified that equations (E.2.40) and (E.2.41) with $\vec{E} \equiv -\vec{\pi}$ are equivalent to Maxwell's equations, where we remind the reader that $\vec{\nabla} \cdot \vec{B} = 0$ follows automatically from the definition of $\vec{B}$, whereas $\vec{\nabla} \cdot \vec{E} = 0$ follows automatically from our construction of the space of momenta. Thus, by eliminating the gauge degrees of freedom in our configuration space by working with $\tilde{\vec{A}}$ rather than $\vec{A}$, we have succeeded in giving a constraint-free Hamiltonian formulation of Maxwell's equations.[2]

Similarly, in the case of Einstein's equation there is gauge arbitrariness in our choice of configuration field h_{ab}. If ψ is any diffeomorphism of Σ_t, then h_{ab} and $\psi^* h_{ab}$ represent the same physical configuration. This suggests that we should take the configuration space of general relativity to be the set equivalence classes, $\tilde{h}_{ab}$, of Riemannian metrics on Σ_t, where two metrics are considered equivalent if they can be carried into each other by a diffeomorphism. This configuration space is known as *superspace* (Wheeler 1968). Using superspace as the configuration space, we find that for any vector field w^a on Σ_t the conjugate momenta π^{ab} now must satisfy

$$\int \pi^{ab}(\delta h_{ab} + D_{(a}w_{b)}) = \int \pi^{ab}\,\delta h_{ab} \quad , \tag{E.2.42}$$

which implies that π^{ab} automatically satisfies

$$D_a(h^{-1/2}\pi^{ab}) = 0 \quad . \tag{E.2.43}$$

Thus, the constraint (E.2.34) is eliminated by the choice of superspace as the configuration space.

However, the constraint (E.2.33) remains. This constraint may be viewed as resulting from the gauge arbitrariness involved in the choice of how to "slice" spacetime into space and time. It is very closely analogous to the constraint which arises when one "parameterizes" an originally unconstrained theory in a fixed, background spacetime, i.e., when one introduces into the Lagrangian a time function—which defines the choice of hypersurfaces, Σ_t, with respect to a reference surface Σ—and treats this time function as a dynamical variable (Kuchař 1973, 1981). In the case of such parameterized theories, the constraint analogous to (E.2.33) is linear in the momentum conjugate to the time function. One then can

2. The relation between the constraint $\vec{\nabla}\cdot\vec{E} = 0$ and the gauge transformations $\vec{A}\to\vec{A} - \vec{\nabla}\chi$ can be obtained more systematically as follows. Given a function f on phase space, we may associate with it a vector field V on phase space by the requirement that for any function g on phase space, we have $V(g) = \{f, g\}$, where the *Poisson bracket* $\{f, g\}$ is defined by

$$\{f, g\} = \int_{\Sigma_t}\left(\frac{\delta f}{\delta q}\frac{\delta g}{\delta \pi} - \frac{\delta g}{\delta q}\frac{\delta f}{\delta \pi}\right).$$

(We have not defined infinite-dimensional manifolds here or vector fields on them, so these remarks are intended only as heuristic.) One may verify directly that the vector field V associated in this manner with the "constraint function" $f = \int_{\Sigma_t}\chi\vec{\nabla}\cdot\vec{E}$ (where χ is an arbitrary function on Σ_t) is just the infinitesimal generator of the one-parameter family of transformations on phase space associated with the gauge transformations $\vec{A}\to\vec{A} - \vec{\nabla}\chi$. Thus, in this sense, in electromagnetism, the constraint "generates" the gauge transformations. By restricting to the "constraint submanifold" $\vec{\nabla}\cdot\vec{E} = 0$ and to the space of orbits of V on this submanifold, we obtain a consistent, constraint-free Hamiltonian formulation on a "reduced phase space." Similarly, in the gravitational case, the vector field associated with the constraint function $2h^{1/2}\xi_b D_a(h^{-1/2}\pi^{ab})$, where ξ^a is an arbitrary vector field on Σ_t, generates the one-parameter family of diffeomorphisms on Σ_t associated with ξ^a. Thus, one is led to choose as the new configuration space the metrics on Σ_t modulo diffeomorphisms (or, more precisely, modulo diffeomorphisms which can be continuously deformed to the identity; see Friedman and Sorkin 1980).

"deparameterize" the theory by solving the constraint for this momentum. However, in the case of Einstein's equation, the constraint (E.2.33) is quadratic in the momentum, and a similar deparameterization does not appear to be possible. Thus, it does not appear possible to find a choice of configuration space for general relativity such that only the "true dynamical degrees of freedom" are present in its phase space. The presence of the constraint (E.2.33) appears to be an unavoidable feature of the Hamiltonian formulation of general relativity. This provides a serious obstacle to the formulation of a quantum theory of gravitation by the canonical quantization approach (see chapter 14).

Finally, we return to the issue of boundary terms in the Hamiltonian formulation. Consider, first, the case of a closed universe, i.e., $M = \mathbb{R} \times \Sigma$, where Σ is compact. Consider the region, U, of M bounded by two constant time hypersurfaces Σ_1 and Σ_2. Then the modified gravitational action S'_G, equation (E.1.42), will receive boundary contributions from Σ_1 and Σ_2. However, these boundary terms will be canceled by the contributions of the third term on the right-hand side of equation (E.2.28). Note that the fourth term in equation (E.2.28) yields no boundary contributions since $n^a \nabla_a n^c$ is orthogonal to the normal, n^c, to Σ_1 and Σ_2. In addition, the last term in equation (E.2.32) makes no contribution since there is no spatial boundary. Thus, in the case of a closed universe, our final answer for H_G is unchanged when all boundary terms are reinserted. Note that because of equations (E.2.33) and (E.2.34), the numerical value of H_G vanishes for every solution. This suggests that we should define the total energy of a closed universe to be zero. In other words, it suggests that there does not exist a nontrivial notion of total energy in a closed universe. However, this argument is not conclusive, since one always can "parameterize" a theory in the manner mentioned above so as to make its Hamiltonian vanish. If general relativity could be "deparameterized," a notion of total energy in a closed universe could well emerge.

Consider, now, the case of asymptotically flat spacetimes. Again, consider a region of M bounded by two hypersurfaces Σ_1 and Σ_2. As before, we wish to consider metric variations for which h_{ab} is held fixed on Σ_1 and Σ_2, but now the most natural spatial boundary condition is that the variations preserve asymptotic flatness rather than that the induced metric be held fixed on a distant spatial boundary. This new boundary condition requires the addition of further boundary terms into the gravitational action (E.1.42). Furthermore, the boundary terms[3] from equations (E.2.28) and (E.2.32) now will contribute to H_G. Instead of keeping careful account of all these terms, we shall proceed by calculating the boundary terms arising from variations of H_G and then modifying H_G to get rid of these terms. Introduce on Σ_t asymptotic Cartesian coordinates as described in problem 2 of chapter 11. We consider the case where t^a asymptotically becomes a time translation at spatial infinity, i.e., we take $N \to 1$ and $N_a \to 0$ as $r \to \infty$. Let S denote a coordinate sphere of radius r. Then, only the term $-h^{1/2}N\,{}^{(3)}R$ in equation (E.2.32) produces a non-

3. To avoid confusion, we remaind the reader that in equation (E.1.42) K is the trace of the extrinsic curvature of the boundary, whereas in equation (E.2.28) K_{ab} is the extrinsic curvature of Σ_t.

vanishing boundary term on S in the limit $r \to \infty$ when H_G is varied. But the calculation of this term is the three-dimensional analog of the calculation of the boundary term in the gravitational action given at the end of the previous section, except that we no longer require the metric to be held fixed at S. Thus, using equation (7.5.14) (or, even better, using the three-dimensional analog of the first line of eq. [E.1.38]) we find that for a one-parameter variation of h_{ab} and π^{ab} which preserves asymptotic flatness, we have (Regge and Teitelboim 1974)

$$\frac{dH_G}{d\lambda} = \int_{\Sigma_t} [A_{ab}\delta\pi^{ab} - B^{ab}\delta h_{ab}] - \delta C \quad . \tag{E.2.44}$$

Here A_{ab} and B^{ab} are given by the right-hand sides of equations (E.2.35) and (E.2.36), respectively, and the boundary term δC is given by

$$\delta C = \lim_{r\to\infty} \int_S r^a h^{bc}[D_c(\delta h_{ab}) - D_a(\delta h_{bc})] \quad , \tag{E.2.45}$$

where r^a is the unit normal to S and the natural volume element on S is understood. We can rewrite δC in coordinate component form as

$$\begin{aligned}\delta C &= \lim_{r\to\infty} \sum_{\nu=1}^{3} \int_S \left(\frac{\partial \delta h_{\mu\nu}}{\partial x^\nu} - \frac{\partial \delta h_{\nu\nu}}{\partial x^\mu}\right) r^\mu \\ &= \delta\left\{\lim_{r\to\infty} \sum_{\nu=1}^{3} \int_S \left(\frac{\partial h_{\mu\nu}}{\partial x^\nu} - \frac{\partial h_{\nu\nu}}{\partial x^\mu}\right) r^\mu\right\} \quad ,\end{aligned} \tag{E.2.46}$$

where we have discarded terms which do not contribute in the limit as $r \to \infty$. Thus, in order to get a Hamiltonian whose variation produces no boundary terms from spatial infinity, we define a new gravitational Hamiltonian H'_G by

$$H'_G = H_G + \alpha \quad , \tag{E.2.47}$$

where α is the term inside the braces in equation (E.2.46). The right-hand sides of equations (E.2.35) and (E.2.36) then truly are the functional derivatives of H'_G with respect to h_{ab} and π^{ab} for variations which preserve asymptotic flatness.

The numerical value of H'_G for a solution of Einstein's equation is just α. This suggests that α should be interpreted as proportional to the total energy of an asymptotically flat spacetime. This provides the motive for the definition given in chapter 11, equation (11.2.14). (The constant of proportionality between α and energy can be determined by evaluating α for the Schwarzschild solution.) Similarly, the definition of total momentum, equation (11.2.15), can be motivated by examining the boundary terms in H_G which occur when we take $N \to 0$ and require N_a to go to a translation as $r \to \infty$. Indeed, a notion of angular momentum (see problem 6 of chapter 11) arises from consideration of more general asymptotic behavior of the lapse and shift (Regge and Teitelboim 1974).

APPENDIX F

UNITS AND DIMENSIONS

Geometrized Units

In this book, we have used "geometrized units," where the gravitational constant G, and speed of light c, are set equal to one. All quantities which in ordinary units have dimension expressible in terms of length L, time T, and mass M, are given the dimension of a power of length in geometrized units. Since $G = c = 1$, all factors involving G and c may be omitted from formulas, and, indeed this is why it is convenient to use geometrized units in general relativity. However, if one wishes to evaluate quantities in ordinary, "nongeometrized" units, the factors of G and c must be reinserted. This is easily done as follows.

In "nongeometrized" units, the dimension of c is L/T and the dimension of G/c^2 is L/M. Hence, the "conversion factor" relative to geometrized units for a quantity with dimension of time is c, while the conversion factor for a quantity with dimension of mass is G/c^2. More generally, a quantity with dimension $L^n T^m M^p$ in ordinary units has dimension L^{n+m+p} in geometrized units and the conversion factor is $c^m(G/c^2)^p$. The dimensions and conversion factors for some frequently encountered quantities are given in Table F.1.

In order to convert a formula written in geometrized units to one which is valid in nongeometrized units, one first must identify the nongeometrized dimension of all quantities appearing in the equation. Then one simply obtains the conversion factor for each quantity from Table F.1 or computes it by the above formula. If one then *multiplies* each quantity appearing in the equation by its conversion factor, the resulting equation will be valid in nongeometrized units.

Planck Units

For calculations involving quantum effects in general relativity, it is natural to employ Planck units where $\hbar$ is set equal to 1 in addition to $G = c = 1$. In Planck units, all quantities which in ordinary units have dimension expressible in terms of L, T, and M now become dimensionless. In particular, all lengths are expressed as dimensionless multiples of the Planck length, $l_P = (G\hbar/c^3)^{1/2}$. To convert a formula valid in Planck units to one valid in ordinary units, we simply identify the nongeometrized dimension of all quantities appearing in the equation. Then we *multiply* each such quantity by its conversion factor, which equals its conversion factor for

Table F.1
CONVERSION FACTORS TO GEOMETRIZED UNITS

Quantity	Nongeometrized Dimension	Geometrized Dimension	Conversion Factor
Acceleration	LT^{-2}	L^{-1}	c^{-2}
Angular momentum	$L^2T^{-1}M$	L^2	G/c^3
Electric charge (cgs)	$L^{3/2}T^{-1}M^{1/2}$	L	$G^{1/2}/c^2$
Energy	$L^2T^{-2}M$	L	G/c^4
Energy density	$L^{-1}T^{-2}M$	L^{-2}	G/c^4
Force	$LT^{-2}M$	1	G/c^4
Length	L	L	1
Mass	M	L	G/c^2
Mass density	$L^{-3}M$	L^{-2}	G/c^2
Pressure	$L^{-1}T^{-2}M$	L^{-2}	G/c^4
Time	T	L	c
Velocity	LT^{-1}	1	c^{-1}

geometrized units divided by l_P^n, where L^n is the geometrized dimension of the quantity.

In Planck units, the fundamental scales of length, time, mass, and other quantities (with respect to which all physical quantities are expressed as dimensionless ratios) are just the inverses of the above conversion factors. For the convenience of the reader, we list some of these fundamental scales below, together with their values in cgs units, calculated using the values $c = 3.00 \times 10^{10}$ cm s^{-1}, $G = 6.67 \times 10^{-8}$ cm^3 g^{-1} s^{-2}, and $\hbar = 1.05 \times 10^{-27}$ erg-sec.

length: $l_P = (G\hbar/c^3)^{1/2} \approx 1.6 \times 10^{-33}$ cm,
time: $t_P = l_P/c \approx 5.4 \times 10^{-44}$ s,
mass: $m_P = l_P c^2/G \approx 2.2 \times 10^{-5}$ g,
energy: $E_P = l_P c^4/G \approx 2.0 \times 10^{16}$ ergs $\approx 1.3 \times 10^{19}$ GeV,
mass density: $\rho_P = l_P^{-2} c^2/G \approx 5.2 \times 10^{93}$ g cm^{-3},
temperature: $T_P = E_P/k = l_P c^4/Gk \approx 1.4 \times 10^{32}$ K.

REFERENCES

Abers, E.S., and Lee, B.W. 1973, "Gauge Theories," *Phys. Rept.*, **C9** 1–141.

Adams, R.A. 1975 *Sobolev Spaces* (New York: Academic Press).

Alley, C. O. 1979, "Relativity and Clocks," in *Proceedings of the 33rd Annual Symposium on Frequency Control* (Washington: Electronic Industries Association).

Ames, W. L., and Thorne, K. S. 1968, "The Optical Appearance of a Star That Is Collapsing through Its Gravitational Radius," *Astrophys. J.*, **151**, 659–670.

Arnett, W. D., and Bowers, R. L. 1977, "A Microscopic Interpretation of Neutron Star Structure," *Astrophys. J. Suppl.*, **33**, 415–436.

Arnowitt, R., Deser, S., and Misner, C. W. 1962, "The Dynamics of General Relativity" in *Gravitation: An Introduction to Current Research*, ed. L. Witten (New York: Wiley).

Ashtekar, A. 1980, "Asymptotic Structure of the Gravitational Field at Spatial Infinity," in *General Relativity and Gravitation,* Vol. **2**, ed. A. Held (New York: Plenum).

Ashtekar, A., and Geroch, R. P. 1974. "Quantum Theory of Gravitation," *Rept. Prog. Phys.*, **37**, 1211–1256.

Ashtekar, A., and Hansen, R. O. 1978, "A Unified Treatment of Null and Spatial Infinity in General Relativity. I. Universal Structure, Asymptotic Symmetries, and Conserved Quantities at Spatial Infinity," *J. Math. Phys.* **19**, 1542–1566.

Ashtekar, A., and Horowitz, G. T. 1982, "Energy-Momentum of Isolated Systems Cannot be Null," *Phys. Lett.*, **89A**, 181–184.

Ashtekar, A., and Magnon, A. 1975, "Quantum Fields in Curved Space-Times," *Proc. Roy. Soc. Lond.*, **A346**, 375–394.

Ashtekar, A., and Magnon-Ashtekar, A. 1979*a*, "On Conserved Quantities in General Relativity," *J. Math. Phys.*, **20**, 793–800.

Ashtekar, A., and Magnon-Ashtekar, A., 1979*b*,"Energy-Momentum in General Relativity," *Phys. Rev. Lett.*, **43**, 181–184.

Ashtekar, A., and Streubel, M. 1981, "Symplectic Geometry of Radiative Modes and Conserved Quantities at Null Infinity," *Proc. Roy. Soc. Lond.*, **A376**, 585–607.

Ashtekar, A., and Winicour, J. 1982, "Linkages and Hamiltonians at Null Infinity," *J. Math. Phys.*, **23**, 2410–2417.

Atiyah, M. F., and Ward, R. S. 1977, "Instantons and Algebraic Geometry," *Commun. Math. Phys.*, **55**, 117–124.

Avis, S. J., and Isham, C. J. 1980, "Generalized Spin Structures on Four Dimensional Spacetimes," *Commun. Math. Phys.*, **72**, 103–118.

Backer, D. C., Kulkarni, S. R., Heiles, C., Davis, M. M., and Goss, W. M. 1982 "A Millisecond Pulsar," *Nature,* **300**, 615–618.

Bardeen, J. M., Carter, B., and Hawking, S. W. 1973, "The Four Laws of Black Hole Mechanics," *Commun. Math. Phys.*, **31**, 161–170.

Bardeen, J. M., Press, W. H., and Teukolsky, S. A. 1972, "Rotating Black Holes: Locally Nonrotating Frames, Energy Extraction, and Scalar Synchronotron Radiation," *Astrophys. J.*, **178**, 347–369.

Bargmann, V. 1954, "On Unitary Ray Representations of Continuous Groups," *Ann. Math.*, **59**, 1–46.

Bargmann, V. 1947, "Irreducible Unitary Representations of the Lorentz Group," *Ann. Math.*, **48**, 568–640.

Bargmann, V., and Wigner, E. P. 1948, "Group Theoretical Discussion of Relativistic Wave Equations," *Proc. Nat. Acad. Sci. U.S.*, **34**, 211–223.

Beig, R., and Simon, W. 1980, "Proof of a Multipole Conjecture due to Geroch," *Commun. Math. Phys.*, **78**, 75–82.

Bekenstein, J. D. 1972, "Nonexistence of Baryon Number for Static Black Holes," *Phys. Rev.*, **D5**, 1239–1246.

Bekenstein, J. D. 1973*a*, "Extraction of Energy and Charge from a Black Hole," *Phys. Rev.*, **D7**, 949–953.

Bekenstein, J. D. 1973*b*, "Black Holes and Entropy," *Phys. Rev.*, **D7**, 2333–2346.

Bekenstein, J. D., 1974, "Generalized Second Law of Thermodynamics in Black-Hole Physics," *Phys. Rev.* **D9**, 3292–3300.

Belinskii, V. A., Khalatnikov, J. M., and Lifshitz, E. M. 1970, "Oscillatory Approach to a Singular Point in the Relativistic Cosmology," *Advan. in Phys.*, **19**, 525–573.

Bell, J. S., and Leinaas, J. M. 1983, "Electrons as Accelerated Thermometers," *Nucl. Phys.*, **B212**, 131–150.

Bers, L., John, F., and Schechter, M. 1964, *Partial Differential Equations* (Providence, RI: American Mathematical Society).

Bianchi, L. 1897, "Sugli Spazii a Tre Dimensioni che Ammettono un Gruppo Continuo di Movimenti," *Mem. di Mat. Soc. Ital. Sci.* **11**, 267-ff.

Birkhoff, G. D. 1923, *Relativity and Modern Physics*, (Cambridge, MA: Harvard University Press).

Birrell, N. D. 1981, "Interacting Quantum Field Theory in Curved Space-Time," in *Quantum Gravity 2*, ed. C. J. Isham, R. Penrose, and D. W. Sciama (Oxford: Clarendon Press).

Birrell, N. D., and Davies, P. C. W. 1982, *Quantum Fields in Curved Space* (Cambridge: Cambridge University Press).

Bishop, R. L., and Crittenden, R. J. 1964, *Geometry of Manifolds* (New York: Academic Press).

Bondi, H., van der Burg, M. G. J., and Metzner, A. W. K. 1962, "Grativational Waves in General Relativity. VII. Waves from Axi-symmetric Isolated Systems," *Proc. Roy. Soc. Lond.*, **A269**, 21–52.

Boulware, D. 1973, "Naked Singularities, Thin Shells, and the Reissner-Nordström Metric," *Phys. Rev.*, **D8**, 2363–2368.

Boyer, R. H., and Lindquist, R. W. 1967, "Maximal Analytic Extension of the Kerr Metric," *J. Math. Phys.*, **8**, 265–281.

Brill, D., and Cohen, J. M. 1966, "Rotating Masses and Their Effects on Inertial Frames," *Phys. Rev.*, **143**, 1011–1015.

Brown, L. S. 1977, "Stress-Tensor Trace Anomaly in a Gravitational Metric: Scalar Fields," *Phys. Rev.*, **D15**, 1469–1483.

Buchdahl, H. A. 1962, "On the Compatibility of Relativistic Wave Equations in Riemannian Spaces," *Nuovo Cim.*, **25**, 486–496.

Candelas, P. 1980, "Vacuum Polarization in Schwarzschild Spacetime," *Phys. Rev.*, **D21**, 2185–2202.

Cantor, M. 1973, "Global Analysis over Noncompact Spaces," Ph.D. thesis, University of California, Berkeley (unpublished).

Carter, B. 1968*a*, "Global Structure of the Kerr Family of Gravitational Fields," *Phys. Rev.*, **174**, 1559–1571.

Carter, B. 1968*b*, "Hamilton-Jacobi and Schrodinger Separable Solutions of Einstein's Equations," *Commun. Math. Phys.*, **10**, 280–310.

Carter, B. 1969, "Killing Horizons and Orthogonally Transitive Groups in Space-Time," *J. Math. Phys.*, **10**, 70–81.

Carter, B. 1971, "Axisymmetric Black Hole Has Only Two Degrees of Freedom," *Phys. Rev. Lett.*, **26**, 331–332.

Carter, B. 1973, "Black Hole Equilibrium States," in *Black Holes*, ed. C. DeWitt and B. S. DeWitt (New York: Gordon & Breach).

Chakrabarti, S. K., Geroch, R. P., and Liang, C. B. 1983, "Timelike Curves of Limited Acceleration in General Relativity," *J. Math. Phys.*, **24**, 597–598.

Chandrasekhar, S. 1939, *An Introduction to the Study of Stellar Structure* (Chicago: University of Chicago Press).

Chandrasekhar, S. 1964, "The Dynamical Instability of Gaseous Masses Approaching the Schwarzschild Limit in General Relativity," *Astrophys. J.*, **140**, 417–433.

Chandrasekhar, S. 1979, "On the Equations Governing the Perturbations of the Reissner-Nordström Black Hole," *Proc. Roy. Soc. Lond.*, **A365**, 453–465.

Chandrasekhar, S. 1983, *The Mathematical Theory of Black Holes* (Oxford: Clarendon Press).

Chandrasekhar, S., and Hartle, J. B. 1982, "On Crossing the Cauchy Horizon of a Reissner-Nordström Black Hole," *Proc. Roy. Soc. Lond.*, **A384**, 301–315.

Chandrasekhar, S., and Tooper, R. F. 1964, "The Dynamical Instability of the White-Dwarf Configurations Approaching the Limiting Mass," *Astrophys. J.*, **139**, 1396–1398.

Choquet-Bruhat, Y. 1962, "The Cauchy Problem," in *Gravitation: An Introduction to Current Research*, ed. L. Witten (New York: Wiley).

Choquet-Bruhat, Y., and Geroch, R. P. 1969, "Global Aspects of the Cauchy Problem in General Relativity," *Commun. Math. Phys.*, **14**, 329–335.

Choquet-Bruhat, Y., and York, J. W., Jr. 1980, "The Cauchy Problem," in *General Relativity and Gravitation,* Vol. **1**, ed. A. Held (New York: Plenum).

Christ, N. H. 1980, "Conservation-Law Violation at High Energy by Anomalies," *Phys. Rev.*, **D21**, 1591–1602.

Christodoulou, D. 1970, "Reversible and Irreversible Transformations in Black Hole Physics," *Phys. Rev. Lett.*, **25**, 1596–1597.

Christodoulou, D., and O'Murchadha, N. 1981, "The Boost Problem in General Relativity," *Commun. Math. Phys.*, **80**, 271–300.

Christodoulou, D., and Ruffini, R. 1971, "Reversible Transformations of a Charged Black Hole," *Phys. Rev.*, **D4**, 3552–3555.

Chrzanowski, P. 1975, "Vector Potential and Metric Perturbations of a Rotating Black Hole," *Phys. Rev.*, **D11**, 2042–2062.

Clarke, C. J. 1971, "Magnetic Charge, Holonomy and Characteristic Classes: Illustrations of the Methods of Topology in General Relativity," *Gen. Rel. and Grav.*, **2**, 43–51.

Coddington, E. A., and Levinson, N. 1955, *Theory of Ordinary Differential Equations* (New York: McGraw-Hill).

Cohen, J. M., and Kegeles, L. S. 1974, "Electromagnetic Fields in Curved Spaces: A Constructive Procedure," *Phys. Rev.*, **D10**, 1070–1084.

Coleman, S. 1973, "Renormalization and Symmetry: A Review for Non-specialists," in *Properties of the Fundamental Interactions*, ed. A. Zichichi (Bologna: Editorice Compositori).

Collins, C. B., and Ellis, G. F. R. 1979, "Singularities in Bianchi Cosmologies," *Phys. Rep.*, **56**, 65–105.

Courant, R., and Hilbert, D. 1962, *Methods of Mathematical Physics*, Vol. **2**: *Partial Differential Equations* (New York: Interscience).

Cowley, A. P., Crampton, D., Hutchings, J. B., Remillard, R., and Penfold, J. 1983, "Discovery of a Massive Unseen Star in LMC X-3," *Astrophys. J.*, **272**, 118–122.

Deser, S., van Nieuwenhuizen, P., and Boulware, D. 1975, "Uniqueness and Nonrenormalizability of Quantum Gravitation," in *General Relativity and Gravitation*, ed. G. Shaviv and J. Rosen (New York: Wiley).

DeWitt, B. S. 1967*a*, "Quantum Theory of Gravity. II. The Manifestly Covariant Theory," *Phys. Rev.*, **162**, 1195–1239.

DeWitt, B. S. 1967*b*, "Quantum Theory of Gravity. III. Applications of the Covariant Theory," *Phys. Rev.*, **162**, 1239–1256.

DeWitt, B. S. 1979, "Quantum Gravity: The New Synthesis," in *General Relativity, an Einstein Centenary Survey*, ed. S. W. Hawking and W. Israel (Cambridge: Cambridge University Press).

Dicke, R. H. 1962, "Mach's Principle and Invariance under Transformations of Units," *Phys. Rev.*, **125**, 2163–2167.

Dimock, J. 1979, "Scalar Quantum Field in an External Gravitational Field," *J. Math. Phys.*, **20**, 2549–2555.

Dixon, W. G. 1974, "Dynamics of Extended Bodies in General Relativity. III. Equations of Motion," *Phil. Trans. Roy. Soc. Lond.*, **A277**, 59–119.

Douglass, D. H., and Braginsky, V. B. 1979, "Gravitational Radiation Experiments," in *General Relativity, an Einstein Centenary Survey*, ed. S. W. Hawking and W. Israel (Cambridge: Cambridge University Press).

Dyson, F. W., Eddington, A. S. E., and Davidson, C. R. 1920, "A Determination of the Deflection of Light by the Sun's Gravitational Field from Observations Made at the Total Eclipse of May 29, 1919," *Phil. Trans. Roy. Soc. A*, **220**, 291–333.

Eddington, A. S. 1924, "A Comparison of Whitehead's and Einstein's Formulas," *Nature*, **113**, 192.

Ehlers, J. 1957, "Konstrucktionen und Charakterisierung der Einsteinschen Gravitationsfeldgleichungen," dissertation (Hamburg).

Einstein, A. 1915*a*, "Zur Allgemeinen Relativitätstheorie," *Preuss. Akad. Wiss. Berlin, Sitzber.*, 778–786.

Einstein, A. 1915*b*, "Der Feldgleichungen der Gravitation," *Preuss. Akad. Wiss. Berlin, Sitzber.*, 844–847.

Eisenhart, L. P. 1949, *Riemannian Geometry* (Princeton: Princeton Univerity Press).

Ellis, G. F. R., and MacCallum, M. A. H. 1969, "A Class of Homogeneous Cosmological Models," *Commun. Math. Phys.*, **12**, 108–141.

Ernst, F. J. 1968, "New Formulation of the Axially Symmetric Gravitational Field Problem," *Phys. Rev.*, **167**, 1175–1178.

Fadde'ev, L. D., and Popov, V. N. 1967, "Feynman Diagrams for the Yang-Mills Field," *Phys. Lett.*, **25B**, 29–30.

Fawcett, M. 1983, "The Energy-Momentum Tensor near a Black Hole," *Commun. Math. Phys.*, **89**, 103–115.

Feinberg, G. 1967, "Possibility of Faster-than-Light Particles," *Phys. Rev.*, **159**, 1089–1105.

Fierz, M., and Pauli, W. 1939, "Relativistic Wave Equations for Particles of Arbitrary Spin in an Electromagnetic Field," *Proc. Roy. Soc. Lond.*, **A173**, 211–232.

Finkelstein, D. 1958, "Past-Future Asymmetry of the Gravitational Field of a Point Particle," *Phys. Rev.*, **110**, 965–967.

Fischer, A., and Marsden, J. 1972, "The Einstein Evolution Equations as a First-Order Symmetric Hyperbolic Quasilinear System," *Commun. Math Phys.*, **28**, 1–38.

Fischer, A. E., and Marsden, J. E. 1979, "The Initial Value Problem and the Dynamical Formulation of General Relativity," in *General Relativity, an Einstein Centenary Survey*, ed. S. W. Hawking and W. Israel (Cambridge: Cambridge University Press).

Flanders, H. 1963, *Differential Forms with Applications to the Physical Sciences* (New York: Academic Press).

Fock, V. A. 1939, "Sur le Mouvement des Masses Finies d'Apres la Théorie de Gravitation Einsteinienne," *J. Phys. U.S.S.R.*, **1**, 81–116.

Fomalont, E. B., and Sramek, R. A. 1976, "Measurement of the Solar Gravitational Deflection of Radio Waves in Agreement with General Relativity," *Phys. Rev. Lett.*, **36**, 1475–1478.

Friedman, J. L., and Sorkin, R. D. 1980, "Spin 1/2 from Gravity," *Phys. Rev. Lett.*, **44**, 1100–1103. (Erratum, **45**, 148E.)

Friedmann, A. 1922, Über die Krümmung des Raumes," *Z. Phys.*, **10**, 377–386.

Fulling, S. A. 1973, "Nonuniqueness of Canonical Field Quantization in Riemannian Space-Time," *Phys. Rev.*, **D7**, 2850–2862.

Fulling, S. A. 1983, "Two-Point Functions and Renormalized Observables," in *Gauge Theory and Gravitation*, ed. K. Kikkawa, N. Nakaniski, and H. Narai (Berlin: Springer-Verlag).

Fulling, S. A., Narcowich, F., and Wald. R. M. 1982, "Singularity Structure of the Two-Point Function in Quantum Field Theory in Curved Spacetime. II," *Ann. Phys.*, **136**, 243–272.

Gannon, D. 1975, "Singularities in Nonsimply Connected Space-Times," *J. Math. Phys.*, **16**, 2364–2367.

Gannon, D. 1976, "On The Topology of Spacelike Hypersurfaces, Singularities, and Black Holes," *Gen. Rel. and Grav.*, **7**, 219–232.

Gårding, L. 1947, "Notes on Continuous Representations of Lie Groups," *Proc. Nat. Acad. Sci. U.S.*, **33**, 331–332.

Geroch, R. P. 1968*a*, "Spinor Structure of Space-Times in General Relativity. I," *J. Math. Phys.*, **9**, 1739–1744.
Geroch, R. P. 1968*b*, "Local Characterization of Singularities in General Relativity," *J. Math. Phys.*, **9**, 450–465.
Geroch, R. P. 1968*c*, "What Is a Singularity in General Relativity?" *Ann. Phys.*, **48**, 526–540.
Geroch, R. P. 1970*a*, "Spinor Structure of Space-Times in General Relativity. II," *J. Math. Phys.*, **11**, 343–348.
Geroch, R. P. 1970*b*, "Domain of Dependence," *J. Math. Phys.*, **11**, 437–449.
Geroch, R. P. 1971, "A Method for Generating Solutions of Einstein's Equation," *J. Math. Phys.*, **12**, 918–924.
Geroch, R. P. 1972*a*, "A Method for Generating New Solutions of Einstein's Equations. II," *J. Math. Phys.*, **13**, 394–404.
Geroch, R. P. 1972*b*, "Structure of the Gravitational Field at Spatial Infinity," *J. Math. Phys.*, **13**, 956–968.
Geroch, R. P. 1977, "Asymptotic Structure of Space-Time," in *Asymptotic Structure of Space-Time*, ed. F. P. Esposito and L. Witten (New York: Plenum).
Geroch, R. P. 1978*a*, *General Relativity from A to B* (Chicago: University of Chicago Press).
Geroch, R. P. 1978*b*, "Null Infinity Is Not a Good Initial Data Surface," *J. Math. Phys.*, **19**, 1300–1303.
Geroch, R. P., and Horowitz, G. T. 1978, "Asymptotically Simple Does Not Imply Asymptotically Minkowskian," *Phys. Rev. Lett.*, **40**, 203–206.
Geroch, R. P., and Horowitz, G. 1979, "Global Structure of Spacetimes," in *General Relativity, an Einstein Centenary Survey*, ed. S. W. Hawking and W. Israel (Cambridge: Cambridge University Press).
Geroch, R. P., and Jang, P. S. 1975, "Motion of a Body in General Relativity," *J. Math. Phys.*, **16**, 65–67.
Geroch, R. P., Kronheimer, E. H., and Penrose, R. 1972, "Ideal Points in Space-Time," *Proc. Roy. Soc. Lond.*, **A327**, 545–567.
Geroch, R. P., Liang, C. B., and Wald. R. M. 1982, "Singular Boundaries of Space-Times," *J. Math. Phys.*, **23**, 432–435.
Geroch, R. P., and Xanthopoulos, B. C. 1978, "Asymptotic Simplicity Is Stable," *J. Math. Phys.*, **19**, 714–719.
Geroch, R. P., and Winicour, J. 1981, "Linkages in General Relativity," *J. Math. Phys.*, **22**, 803–812.
Gibbons, G. W. 1979, "Cosmological Fermion-Number Non-Conservation," *Phys. Lett.*, **84B**, 431–434.
Gibbons, G. W., Hawking, S. W., Horowitz, G. T., and Perry, M. J. 1983, "Positive Mass Theorems for Black Holes," *Commun. Math. Phys.*, **88**, 295–308.
Gibbons, G. W., Hawking, S. W., and Siklos, S., 1983, ed. *The Very Early Universe* (Cambridge: Cambridge University Press).
Gibbons, G. W., and Perry, M. J. 1976, "Black Holes in Thermal Equilibrium," *Phys. Rev. Lett.*, **36**, 985–987.
Gibbons, G. W., and Perry, M. J. 1978, "Black Holes and Thermal Green's Functions," *Proc. Roy. Soc. Lond.*, **A358**, 467–494.
Goldstein, H. 1980, *Classical Mechanics* (Reading, MA: Addison-Wesley).
Guth, A. H. 1981, "Inflationary Universe: A Possible Solution to the Horizon and Flatness Problems," *Phys. Rev.*, **D23**, 347–356.
Güven, R. 1977, "Wave Mechanics of Electrons in Kerr Geometry," *Phys. Rev.*, **D16**, 1706–1711.
Haag, R., and Kastler, D. 1964, "An Algebraic Approach to Quantum Field Theory," *J. Math Phys.*, **5**, 848–861.
Hafele, J. C., and Keating, R. E. 1972, "Around the World Atomic Clocks: Observed Relativistic Time Gains," *Science*, **177**, 168–170.
Hahn, S. G., and Lindquist, R. W. 1964, "The Two Body Problem in Geometrodynamics," *Ann. Phys.*, **29**, 304–331.
Hajicek, P. 1973, "General Theory of Vacuum Ergospheres," *Phys. Rev.*, **D7**, 2311–2316.
Hansen, R. O. 1974, "Multipole Moments in Stationary Space-Times," *J. Math. Phys.*, **15**, 46–52.

Hartle, J. B. 1972, "Can a Schwarzschild Black Hole Exert Long-Range Neutrino Forces?" in *Magic without Magic*, ed. J. Klauder (San Francisco: Freeman).

Hartle, J. B. 1978, "Bounds on the Mass and Moment of Inertia of Non-Rotating Neutron Stars," *Phys. Rept.*, **46C**, 201–247.

Hartle, J. B., and Hawking, S. W. 1976, "Path-Integral Derivation of Black Hole Radiance," *Phys. Rev.*, **D13**, 2188–2203.

Hauser, J., and Ernst, F. J. 1981, "A Proof of the Geroch Conjecture," *J. Math. Phys.*, **22**, 1051–1063.

Hawking, S. W. 1967, "The Occurrence of Singularities in Cosmology. III. Causality and Singularities," *Proc. Roy. Soc. Lond.*, **A300**, 182–201.

Hawking, S. W. 1971, "Gravitational Radiation from Colliding Black Holes," *Phys. Rev. Lett.*, **26**, 1344–1346.

Hawking, S. W. 1975, "Particle Creation by Black Holes," *Commun. Math. Phys.*, **43**, 199–220.

Hawking, S. W. 1976, "Breakdown of Predictability in Gravitational Collapse," *Phys. Rev.*, **D14**, 2460–2473.

Hawking, S. W. 1977, "Zeta Function Regularization of Path Integrals in Curved Spacetime," *Commun. Math. Phys.*, **55**, 133–148.

Hawking, S. W. 1979, "The Path Integral Approach to Quantum Gravity," in *General Relativity, an Einstein Centenary Survey*, ed. S. W. Hawking and W. Israel (Cambridge: Cambridge University Press).

Hawking, S. W., and Ellis, G. F. R. 1973, *The Large Scale Structure of Space-Time* (Cambridge: Cambridge University Press).

Hawking, S. W., and Hartle, J. B. 1972, "Energy and Angular Momentum Flow into a Black Hole," *Commun. Math. Phys.*, **27**, 283–290.

Hawking, S. W., and Israel, W. 1979, ed., *General Relativity, an Einstein Centenary Survey* (Cambridge: Cambridge University Press).

Hawking, S. W., and Penrose, R. 1970, "The Singularities of Gravitational Collapse and Cosmology," *Proc. Roy. Soc. Lond.*, **A314**, 529–548.

Hawking, S. W., and Roček, M. 1981, ed., *Superspace and Supergravity* (Cambridge: Cambridge University Press).

Hicks, N. J. 1965, *Notes on Differential Geometry* (Princeton: Van Nostrand).

Hocking, J. G., and Young, G. S. 1961, *Topology* (Reading: Addison-Wesley).

Hoenselaers, C., Kinnersley, W., and Xanthopoulos, B. C. 1979, "Symmetries of the Stationary Einstein-Maxwell Equations. VI. Transformations Which Generate Asymptotically Flat Spacetimes with Arbitrary Multipole Moments," *J. Math. Phys.*, **20**, 2530–2536.

Horowitz, G. T. 1980, "Semiclassical Relativity: The Weak Field Limit," *Phys. Rev.*, **D21**, 1445–1461.

Horowitz, G. T., and Perry, M. J. 1982, "Gravitational Energy Cannot Become Negative," *Phys. Rev. Lett.*, **48**, 371–374.

Hulse, R. A., and Taylor, J. H. 1975, "Discovery of a Pulsar in a Binary System," *Astrophys. J.*, **195**, L51–L53.

Isham, C. J. 1978, "Spinor Fields in Four Dimensional Spacetime," *Proc. Roy. Soc. Lond.*, **A364**, 591–599.

Isham, C. J., Penrose, R., and Sciama, D. W. 1975, ed. *Quantum Gravity* (Oxford: Clarendon Press).

Isham, C. J., Penrose, R., and Sciama, D.W. 1981, ed., *Quantum Gravity 2* (Oxford: Clarendon Press).

Israel, W. 1967, "Event Horizons in Static Vacuum Space-Times," *Phys. Rev.*, **164**, 1776–1779.

Israel, W. 1968, "Event Horizons in Static Electrovac Space-Times," *Commun. Math. Phys.*, **8**, 245–260.

Israel, W. 1976, "Thermo-Field Dynamics of Black Holes," *Phys. Lett.*, **57A**, 107–110.

Jackson, J. D. 1962, *Classical Electrodynamics* (New York: Wiley).

Jacobson, N. 1962, *Lie Algebras* (New York: Wiley).

Jang, P. S., and Wald, R. M. 1977, "The Positive Energy Conjecture and the Cosmic Censor Hypothesis," *J. Math. Phys.*, **18**, 41–44.

Johnson, R. A. 1977, "The Bundle Boundary in Some Special Cases," *J. Math. Phys.*, **18**, 898–902.

Kasner, E. 1925, "Solutions of the Einstein Equations Involving Functions of Only One Variable," *Trans. Am. Math. Soc.*, **27**, 155–162.

Kay, B. S. 1978, "Linear Spin-Zero Quantum Fields in External Gravitational and Scalar Fields," *Commun. Math. Phys.*, **62**, 55–70.

Kay, B. S. 1982, "Quantum Fields in Curved Space-Time and Scattering Theory," in *Differential Geometric Methods in Mathematical Physics*, ed. H. D. Doebner, S. I. Andersson, and H. R. Petry (Berlin: Springer-Verlag).

Kelley, J. 1955, *General Topology* (Princeton: Van Nostrand-Reinhold).

Kerr, R. P. 1963, "Gravitational Field of a Spinning Mass as an Example of Algebraically Special Metrics," *Phys. Rev. Lett.*, **11**, 237–238.

Kerr, R. P., and Schild, A. 1965, "A New Class of Vacuum Solutions of the Einstein Field Equations," in *Proceedings of the Galileo Galilei Centenary Meeting on General Relativity, Problems of Energy and Gravitational Waves*, ed. G. Barbera (Florence: Comitato Nazionale per le Manifestazione Celebrative).

Kinnersley, W. 1969, "Type D Vacuum Metrics," *J. Math. Phys.*, **10,** 1195–1203.

Kirshner, R. P., Oemler, A., Schechter, P. L., and Shectman, S. A. 1981, "A Million Cubic Megaparsec Void in Boötes?" *Astrophys. J.*, **248,** L57–L60.

Klein, O. 1929, "Die Reflexion von Electronen an einem Potentialsprung nach der relativistischen Dynamic von Dirac," *Z. Physik,* **53,** 157–165.

Kobayashi, S., and Nomizu, K. 1963, *Foundations of Differential Geometry,* Vol. **1** (New York: Interscience).

Komar, A. 1959, "Covariant Conservation Laws in General Relativity," *Phys. Rev.,* **113,** 934–936.

Kramer, D., Stephani, H., MacCallum, M., and Herlt, E. 1980, *Exact Solutions of Einstein's Field Equations* (Cambridge: Cambridge University Press).

Kruskal, M. D. 1960, "Maximal Extension of Schwarzschild Metric," *Phys. Rev.*, **119**, 1743–1745.

Kuchař, K. 1973, "Canonical Quantization of Gravity," in *Relativity, Astrophysics, and Cosmology*, ed. W. Israel (Dordrecht: Reidel).

Kuchař, K. 1976, "Dynamics of Tensor Fields in Hyperspace. III," *J. Math. Phys.*, **17**, 801–820.

Kuchař, K. 1981, "Canonical Methods of Quantization," in *Quantum Gravity 2*, ed. C. J. Isham, R. Penrose, and D. W. Sciama (Oxford: Clarendon Press).

Kundu, P. 1981, "On the Analyticity of Stationary Gravitational Fields at Spatial Infinity," *J. Math. Phys.*, **22**, 2006–2011.

Landau, L. D., and Lifshitz, E. M. 1962, *The Classical Theory of Fields* (Oxford: Pergamon).

Leray, J. 1952, "Hyperbolic Differential Equations," duplicated notes from Princeton Institute for Advanced Study (unpublished).

Lichnerowicz, A. 1944, "L'integration des Equations de la Gravitation Relativiste et le Probleme des *n* Corps," *J. Math. Pures Appl.*, **23**, 37–63.

Lichnerowicz, A. 1955, *Théories Relativistes de la Gravitation et de Electromagnetisme* (Paris: Masson).

Lovelock, D. 1972, "The Four-Dimensionality of Space and the Einstein Tensor," *J. Math. Phys.*, **13**, 874–876.

Ludvigsen, M., and Vickers, J. A. G. 1982, "A Simple Proof of the Positivity of Bondi Mass," *J. Phys. A*, **15**, L67–L70.

MacCallum, M. A. H. 1979, "Anisotropic and Inhomogeneous Relativistic Cosmologies," in *General Relativity, an Einstein Centenary Survey*, ed. S. W. Hawking and W. Israel (Cambridge: Cambridge University Press).

Mazur, P. O. 1982, "Proof of Uniqueness of the Kerr-Newman Black Hole Solution," *J. Phys. A*, **15**, 3173–3180.

Milnor, J. W. 1963, "Spin Structures on Manifolds," *L'enseignement math.*, **9**, 198–203.

Misner, C. W. 1963, "The Flatter Regions of Newman, Unti, and Tamburino's Generalized Schwarzschild Space," *J. Math. Phys.*, **4**, 924–937.

Misner, C. W. 1967, "Taub-NUT Space as a Counterexample to Almost Anything," in *Relativity Theory and Astrophysics I: Relativity and Cosmology*, ed. J. Ehlers, Lectures in Applied Mathematics, Vol. **8** (Providence: American Mathematical Society).

Misner, C. W. 1969, "Mixmaster Universe," *Phys. Rev. Lett.*, **22**, 1071–1074.

Misner, C. W., Thorne, K. S., and Wheeler, J. A. 1973, *Gravitation* (San Francisco: Freeman)

Moncrief, V. 1975, "Gauge Invariant Perturbations of Reissner-Nordström Black Holes," *Phys. Rev.*, **D12**, 1526–1537.

Müller zum Hagen, H. 1970, "On the Analyticity of Stationary Vacuum Solutions of Einstein's Equation," *Proc. Camb. Phil. Soc.*, **68**, 199–201.

Müller zum Hagen, H., Robinson, D. C., and Seifert, H. J. 1973, "Black Holes in Static Vacuum Space-Times," *Gen. Rel. Grav.*, **4**, 53–78.

Müller zum Hagen, H., and Seifert, H. J. 1977, "On Characteristic Initial-Value and Mixed Problems," *Gen. Rel. and Grav.*, **8**, 259–301.

Newman, E. T., Couch, E., Chinnapared, K., Exton, A., Prakash, A., and Torrence, R. 1965, "Metric of a Rotating Charged Mass," *J. Math. Phys.*, **6**, 918–919.

Newman, E. T., and Penrose, R. 1962, "An Approach to Gravitational Radiation by a Method of Spin Coefficients," *J. Math. Phys.*, **3**, 566–578; erratum **4**, 998.

Noakes, D. R. 1983, "The Initial Value Formulation of Higher Derivative Gravity," *J. Math. Phys.*, **24**, 1846–1850.

Nordstrom, G. 1918, "On the Energy of the Gravitational Field in Einstein's Theory," *Proc. Kon. Ned. Akad. Wet.*, **20**, 1238–1245.

Paczyński, B. 1974, "Mass of Cygnus X-1," *Astron. and Astrophys.*, **34**, 161–162.

Paczyński, B. 1983, "Mass of LMC X-3," *Astrophys. J. Lett.*, **273**, L81–L84.

Page, D. N. 1976*a*, "Particle Emission Rates from a Black Hole: Massless Particles from an Uncharged, Nonrotating Hole," *Phys. Rev.*, **D13**, 198–206.

Page, D. N. 1976*b*, "Particle Emission Rates from a Black Hole. II. Massless Particles from a Rotating Hole," *Phys. Rev.*, **D14**, 3260–3273.

Page, D. N. 1980, "Is Black-Hole Evaporation Predictable?" *Phys. Rev. Lett.*, **44**, 301–304.

Page, D. N., and Hawking, S. W. 1976, "Gamma Rays from Primordial Black Holes," *Astrophys. J.*, **206**, 1–7.

Panangaden, P., and Wald, R. M. 1977, "Probability Distribution for Radiation from a Black Hole in the Presence of Incoming Radiation," *Phys. Rev.*, **D16**, 929–932.

Papapetrou, A. 1951, "Spinning Test Particles in General Relativity. I," *Proc. Roy. Soc. Lond.*, **A209**, 248–258.

Papapetrou, A. 1953, "Eine Rotationssymetrische Lösung in der Allgemeinen Relativitätstheorie," *Ann. Physik*, **12**, 309–315.

Papapetrou, A. 1966, "Champs Gravitationnels Stationnares a Symmetrie Axiale," *Ann. Inst. Henri Poincaré*, **A4**, 83–105.

Parker, L. 1969, "Quantized Fields and Particle Creation in Expanding Universes. I," *Phys. Rev.*, **183**, 1057–1068.

Parker, L. 1975, "Probability Distribution of Particles Created by a Black Hole," *Phys. Rev.*, **D12**, 1519–1525.

Peebles, P. J. E. 1971, *Physical Cosmology* (Princeton: Princeton University Press).

Peebles, P. J. E. 1980, *The Large-Scale Structure of the Universe* (Princeton: Princeton University Press).

Penrose, R. 1960, "A Spinor Approach to General Relativity," *Ann. Phys.*, **10**, 171–201.

Penrose, R. 1963, "Asymptotic Properties of Fields and Space-Times," *Phys. Rev. Lett.*, **10**, 66–68.

Penrose, R. 1965*a*, "Gravitational Collapse and Space-Time Singularities," *Phys. Rev. Lett.*, **14**, 57–59.

Penrose, R. 1965*b*, "Zero Rest-Mass Fields Including Gravitation: Asymptotic Behavior," *Proc. Roy. Soc. Lond.*, **A284**, 159–203.

Penrose, R. 1967, "Twistor Algebra," *J. Math. Phys.*, **8**, 345–366.

Penrose, R. 1968, "Structure of Space-Time," in *Battelle Rencontres, 1967*, ed. C. M. DeWitt and J. A. Wheeler (New York: Benjamin).

Penrose, R. 1969, "Gravitational Collapse: The Role of General Relativity," *Rev. del Nuovo Cimento*, **1**, 252–276.

Penrose, R. 1972, *Techniques of Differential Topology in Relativity* (Philadelphia: Siam).

Penrose, R. 1975, "Twistor Theory, the Aims and Achievements," in *Quantum Gravity*, ed. C. J. Isham, R. Penrose, and D. W. Sciama (Oxford: Clarendon Press).

Penrose, R. 1979, "Singularities and Time Asymmetry," in *General Relativity, an Einstein Centenary Survey*, ed. S. W. Hawking and W. Israel (Cambridge: Cambridge University Press).

Penrose, R. 1981, "Time-Asymmetry and Quantum Gravity," in *Quantum Gravity 2*, ed. C. J. Isham, R. Penrose, and D. W. Sciama (Oxford: Clarendon Press).

Penrose, R., and MacCallum, M. A. H. 1972, "Twistor Theory: An Approach to the Quantization of Fields and Space-Time," *Phys. Rept.*, **6C**, 241–316.

Penrose, R., and Rindler, W. 1984, *Spinors and Space-Time*, Vol. **1**: *Two-Spinor Calculus and Relativistic Fields* (Cambridge: Cambridge University Press).

Penrose, R., and Ward, R. S. 1980, "Twistors for Flat and Curved Space-Time," in *General Relativity and Gravitation*, Vol. **2**, ed. A. Held (New York: Plenum Press).

Penzias, A. A., and Wilson, R. W. 1965, "A Measurement of Excess Antenna Temperature at 4080 Mc/s," *Astrophys. J.*, **142**, 419–421.

Persides, S., and Papadopoulos, D. 1979, "A Covariant Formulation of the Landau-Lifshitz Complex," *Gen. Rel. and Grav.*, **11**, 233–243.

Petrov. A. Z. 1954, *Reports of the State University of Kazan*, **114**, book 8, 55.

Petrov, A. Z. 1969, *Einstein Spaces* (New York: Pergamon).

Pirani, F. A. E. 1965, "Introduction to Gravitational Radiation Theory," in *Lectures on General Relativity*, ed. S. Deser and K. W. Ford (Englewood Cliffs, NJ: Prentice-Hall).

Pound, R. V., and Rebka, G. A. 1960, "Apparent Weight of Photons," *Phys. Rev. Lett.*, **4**, 337–341.

Press, W. H., and Teukolsky, S. A. 1973, "Perturbations of a Rotating Black Hole. II. Dynamical Stability of the Kerr Metric," *Astrophys. J.*, **185**, 649–673.

Price, R. H. 1972*a*, "Nonspherical Perturbations of Relativistic Gravitational Collapse. I. Scalar and Gravitational Perturbations," *Phys. Rev.*, **D5**, 2419–2438.

Price, R. H. 1972*b*, "Nonspherical Perturbations of Relativistic Gravitational Collapse. II. Integer-Spin, Zero-Rest-Mass Fields," *Phys. Rev.*, **D5**, 2439–2454.

Reasenberg, R. D., Shapiro, I. I., MacNeil, P. E., Goldstein, R. B., Breidenthal, J. C., Brenkle, J. P., Cain, D. L., Kaufman, T. M., Kormarek, T. A., and Zygielbaum, A. I. 1979, "Viking Relativity Experiment: Verification of Signal Retardation by Solar Gravity," *Astrophys. J. Lett.*, **234**, 219–221.

Reed, M., and Simon, B. 1972, *Methods of Modern Mathematical Physics. I. Functional Analysis* (New York: Academic Press).

Reed, M., and Simon, B. 1979, *Methods of Modern Mathematical Physics. III. Scattering Theory* (New York: Academic Press).

Rees, M. J. 1978, "Emission from the Nuclei of Nearby Galaxies: Evidence for Massive Black Holes?" in *Structure and Properties of Nearby Galaxies*, ed. E. M. Berkhuijsen and R. Wielebinski (International Astronomical Union).

Regge, T., and Teitelboim, C. 1974, "Role of Surface Integrals in the Hamiltonian Formulation of General Relativity," *Ann. Phys.*, **88**, 286–318.

Reissner, H. 1916, "Über die Eigengravitation des elektrischen Felds nach der Einsteinshen Theorie," *Ann. Physik*, **50**, 106–120.

Reula, O. 1982, "Existence Theorem for Solutions of Witten's Equation and Nonnegativity of Total Mass," *J. Math. Phys.*, **23**, 810–814.

Reula, O., and Tod, P. 1984, "Positivity of the Bondi Energy," *J. Math. Phys.*, in press.

Riesz, F., and Sz-Nagy, B. 1955, *Functional Analysis* (New York: Ungar).

Robinson, D. C. 1975, "Uniqueness of the Kerr Black Hole," *Phys. Rev. Lett.*, **34**, 905–906.

Robinson, D. C. 1977, "A Simple Proof of the Generalization of Israel's Theorem," *Gen. Rel. Grav.*, **8**, 695–698.

Royden, H. L. 1963, *Real Analysis* (New York: Macmillan).

Rumpf, H. 1976, "Covariant Treatment of Particle Creation in Curved Space-Time," *Phys. Lett.*, **61B**, 272–274.

Ryan, M. P., Jr., and Shepley, L. C. 1975, *Homogeneous Relativistic Cosmologies* (Princeton: Princeton University Press).

Sachs, R. K. 1962*a*, "On the Characteristic Initial Value Problem in Gravitational Theory," *J. Math. Phys.*, **3**, 908–914.

Sachs, R. K. 1962*b*, "Gravitational Waves in General Relativity, VIII. Waves in Asymptotically Flat Space-Time," *Proc. Roy. Soc. Lond.*, **A270**, 103–126.

Sachs, R. K. 1962*c*, "Asymptotic Symmetries in Gravitational Theory," *Phys. Rev.*, **128**, 2851–2864.

Sargent, W. L. W., Young, P. J., Boksenberg, A., Shortridge, K., Lynds, C. R., and Hartwick, F. D. A. 1978, "Dynamical Evidence for a Central Mass Concentration in the Galaxy M87," *Astrophys. J.*, **221**, 731–744.

Schmidt, B. G. 1971, "A New Definition of Singular Points in General Relativity," *Gen. Rel. and Grav.*, **1**, 269–280.

Schoen, R., and Yau, S.-T. 1979, "Proof of the Positive-Action Conjecture in Quantum Relativity," *Phys. Rev. Lett.*, **42**, 547–548.

Schoen, R., and Yau, S.-T. 1981, "Proof of the Positive Mass Theorem. II," *Commun. Math. Phys.*, **79**, 231–260.

Schoen, R., and Yau, S.-T. 1982, "Proof That the Bondi Mass is Positive," *Phys. Rev. Lett.*, **48**, 369–371.

Schoen, R., and Yau, S.-T. 1983, "The Existence of a Black Hole due to Condensation of Matter," *Commun. Math. Phys.*, **90,** 575–579.

Schwarzschild, K. 1916*a*, "Über das Gravitationsfeld eines Massenpunktes nach der Einsteinschen Theorie," *Sitzber. Deut. Akad. Wiss. Berlin*, Kl. Math.-Phys. Tech., 189–196.

Schwarzschild, K. 1916*b*, Über das Gravitationsfeld einer Kugel aus inkompressibler Flussigkeit nach der Einsteinschen Theorie," *Sitzber. Deut. Akad. Wiss. Berlin*, Kl. Math.-Phys. Tech., 424–434.

Sen, A. 1982, "Quantum Theory of Spin 3/2 Field in Einstein Spaces," *Int. J. Theo. Phys.*, **21**, 1–35.

Sewell, G. 1982, "Quantum Fields on Manifolds: PCT and Gravitationally Induced Thermal States," *Ann. Phys.*, **141**, 201–224.

Smarr, L. 1979, "Gauge Conditions, Radiation Formulae and the Two Black Hole Collision," in *Sources of Gravitational Radiation*, ed. L. Smarr (Cambridge: Cambridge University Press).

Smoot, G. F., Gorenstein, M. V., and Muller, R. A. 1977, "Detection of Anisotropy in the Cosmic Blackbody Radiation," *Phys. Rev. Lett.*, **39**, 898–901.

Sorkin, R. D. 1981,"A Criterion for the Onset of Instability," *Astrophys. J.*, **249**, 254–257.

Starobinskii, A. A. 1973, "Amplification of Waves during Reflection from a Rotating Black Hole," *Zh. Eksp. Teor. Fiz.*, **64**, 48–57 (English transl.: *Soviet Phys. — JETP*, **37**, 28–32).

Steenrod, N. 1951, *The Topology of Fibre Bundles* (Princeton: Princeton University Press).

Steigman, G. 1976, "Observational Tests of Antimatter Cosmologies," in *Annual Review of Astronomy and Astrophysics*, **14**, 339–372.

Stelle, K. S. 1977, "Renormalization of Higher-Derivative Quantum Gravity," *Phys. Rev.*, **D16**, 953–969.

Streater, R. F., and Wightman, A. S. 1964, *PCT, Spin and Statistics, and All That* (New York: Benjamin).

Szekeres, G. 1960, "On the Singularities of a Riemannian Manifold," *Publ. Mat. Debrecen*, **7**, 285–301.

Tamburino, L., and Winicour, J. 1966, "Gravitational Fields in Finite and Conformal Bondi Frames," *Phys. Rev.*, **150**, 1039–1053.

Taub, A. H. 1951, "Empty Space-Times Admitting a Three-Parameter Group of Motions," *Ann. Math.*, **53**, 472–490.

Taubes, C. H., and Parker, T. 1982, "On Witten's Proof of the Positive Energy Theorem," *Commun. Math. Phys.*, **84**, 223–238.

Taylor, J. F., and McCulloch, P. M. 1980, "Evidence for the Existence of Gravitational Radiation from Measurements of the Binary Pulsar PSR 1913+16," in *Ninth Texas Symposium on Relativistic Astrophysics*, ed. J. Ehlers, J. J. Perry, and M. Walker (New York: New York Academy of Sciences).

Teitelboim, C. 1972," Nonmeasurability of the Quantum Numbers of a Black Hole," *Phys. Rev.*, **D5**, 2941–2954.

Teukolsky, S. A. 1972, "Rotating Black Holes: Separable Wave Equations for Gravitational and Electromagnetic Perturbations," *Phys. Rev. Lett.*, **29**, 1114–1118.

Teukolsky, S. A., and Press, W. H. 1974, "Perturbations of a Rotating Black Hole. III. Interaction of the Hole with Gravitational and Electromagnetic Radiation," *Astrophys. J.*, **193**, 443–461.

Thirring, H., and Lense, J. 1918, "Über den Einfluss der Eigenrotation der Zentralkörper auf die Bewegung der Planeten und Monde nach der Einsteinschen Gravitationstheorie," *Phys. Z.*, **19**, 156–163.

Thorne, K. S. 1978, "General-Relativistic Astrophysics," in *Theoretical Principles in Astrophysics and Relativity*, ed. N. R. Lebovitz, W. H. Reid, and P. O. Vandervoort (Chicago: University of Chicago Press).

Tomimatsu, A., and Sato, H. 1972, "New Exact Solution for the Gravitational Field of a Spinning Mass," *Phys. Rev. Lett.*, **29**, 1344–1345.

Tomimatsu, A., and Sato, H. 1973, "New Series of Exact Solutions for Gravitational Fields of Spinning Masses," *Prog. Theor. Phys. (Kyoto)*, **50**, 95–110.

Trautman, A. 1965, "Foundations and Current Problems of General Relativity," in *Lectures on General Relativity*, ed. S. Deser and K. W. Ford (Englewood Cliffs, NJ: Prentice-Hall).

Unruh, W. G. 1973, "Separability of the Neutrino Equations in a Kerr Background," *Phys. Rev. Lett.*, **31**, 1265–1267.

Unruh, W. G. 1974, "Second Quantization in the Kerr Metric," *Phys. Rev.*, **D10**, 3194–3205.

Unruh, W. G. 1976, "Notes on Black Hole Evaporation," *Phys. Rev.*, **D14**, 870–892.

Unruh, W. G. 1977, "Origin of the Particles in Black-Hole Evaporation," *Phys. Rev.*, **D15**, 365–369.

Unruh, W. G., and Wald, R. M. 1982, "Acceleration Radiation and the Generalized Second Law of Thermodynamics," *Phys. Rev.*, **D25**, 942–958.

Unruh, W. G., and Wald, R. M. 1984, "What Happens When an Accelerating Observer Detects a Rindler Particle," *Phys. Rev.*, **D29**, 1047–1056.

van Nieuwenhuizen, P. 1981, "Supergravity," *Phys. Reports*, **68C**, 189–398.

Vessot, R. F. C., and Levine, M. W. 1979, "A Test of the Equivalence Principle Using a Space-borne Clock," *Gel. Rel. Grav.*, **10**, 181–204.

Vessot, R. F. C., Levine, M. W., Mattison, E. M., Blomberg, E. L., Hoffman, T. E., Nystrom, G. U., Farrel, B. F., Decher, R., Eby, P. B., Baugher, C. R., Watts, J. W., Teuber, D. L., and Wills, F. O. 1980, "Test of Relativistic Gravitation with a Space-borne Hydrogen Maser," *Phys. Rev. Lett.*, **45**, 2081–2084.

Wald, R. M. 1972*a*, "Electromagnetic Fields and Massive Bodies," *Phys. Rev.*, **D6**, 1476–1479.

Wald, R. M. 1972*b*, "Gravitational Spin Interaction," *Phys. Rev.*, **D6**, 406–413.

Wald, R. M. 1973, "On Perturbations of a Kerr Black Hole," *J. Math. Phys.*, **14**, 1453–1461.

Wald, R. M. 1974*a*, "Gedanken Experiments to Destroy a Black Hole," *Ann. Phys.*, **82**, 548–556.

Wald, R. M. 1974*b*, "Black Hole in a Uniform Magnetic Field," *Phys. Rev.*, **D10**, 1680–1685.

Wald, R. M. 1974*c*, "Energy Limits on the Penrose Process," *Astrophys. J.*, **191**, 231–233.

Wald, R. M. 1975, "On Particle Creation by Black Holes," *Commun. Math. Phys.*, **45**, 9–34.

Wald, R. M. 1976, "Stimulated Emission Effects in Particle Creation near Black Holes," *Phys. Rev.*, **D13**, 3176–3182.

Wald, R. M. 1977*a*, *Space, Time and Gravity: The Theory of the Big Bang and Black Holes* (Chicago: University of Chicago Press).

Wald, R. M. 1977*b*, "The Back Reaction Effect in Particle Creation in Curved Spacetime," *Commun. Math. Phys.*, **54**,1–19.

Wald, R. M. 1978*a*, "Construction of Solutions of Gravitational, Electromagnetic, or Other Perturbation Equations from Solutions of Decoupled Equations," *Phys. Rev. Lett.*, **41**, 203–206.

Wald, R. M. 1978*b*, "Trace Anomaly of a Conformally Invariant Quantum Field in Curved Spacetime," *Phys. Rev.*, **D17**, 1477–1484.

Wald, R. M. 1979*a*, "Note on the Stability of the Schwarzschild Metric," *J. Math. Phys.*, **20**, 1056–1058; erratum, **21**, 218 (1980).

Wald, R. M. 1979*b*, "Existence of the S-Matrix in Quantum Field Theory in Curved Spacetime," *Ann. Phys.*, **118**, 490–510.

Wald, R. M. 1979*c*, "On the Euclidean Approach to Quantum Field Theory in Curved Spacetime," *Commun. Math. Phys.*, **70**, 221–242.

Wald, R. M. 1980, "Quantum Gravity and Time Reversibility," *Phys. Rev.*, **D21**, 2742–2755.

Walker, M., and Penrose, R. 1970, "On Quadratic First Integrals of the Geodesic Equations for Type [22] Spacetimes," *Commun. Math. Phys.*, **18**, 265–274.

Ward, R. S. 1981, "The Twistor Approach to Differential Equations," in *Quantum Gravity 2*, ed. C. J. Isham, R. Penrose, and D. W. Sciama (Oxford: Clarendon Press).

Warner, F. W. 1971, *Foundations of Differentiable Manifolds and Lie Groups* (Glenview, IL: Scott, Foresman and Co.).

Weinberg, S. 1972, *Gravitation and Cosmology* (New York: Wiley).

Weinberg, S. 1979, "Ultraviolet Divergences in Quantum Theories of Gravitation," in *General Relativity, an Einstein Centennary Survey*, ed. S. W. Hawking and W. Israel (Cambridge: Cambridge University Press).

Weyl, H. 1917, "Zur Gravitationstheorie," *Ann. Physik*, **54**, 117–145.

Wheeler, J. A. 1968, "Superspace and the Nature of Quantum Geometrodynamics," in *Battelle Recontres*, ed. C. M. DeWitt and J. A. Wheeler (New York: Benjamin).

Wightman, A. S. 1971, ed. *Troubles in the External Field Problem for Invariant Wave Equations* (New York: Gordon & Breach).

Wigner, E. P. 1939, "On Unitary Representations of the Inhomogeneous Lorentz Group," *Ann. Math.*, **40**, 149–204.

Wigner, E. P. 1959, *Group Theory and Its Application to the Quantum Mechanics of Atomic Spectra* (New York: Academic Press).

Will, C. M. 1981, *Theory and Experiment in Gravitational Physics* (Cambridge: Cambridge University Press).

Winicour, J. 1968, "Some Total Invariants of Asymptotically Flat Space-Times," *J. Math. Phys.* **9**, 861–867.

Witten, E. 1981, "A New Proof of the Positive Energy Theorem," *Commun. Math. Phys.*, **80**, 381–402.

Xanthopoulos, B. C. 1981, "Exterior Spacetimes for Rotating Stars," *J. Math Phys.*, **22**, 1254–1259.

Yip, P. 1983, "Spinors in Two Dimensions," *J. Math. Phys.*, **24**, 1206–1212.

Yodzis, P., Seifert, H.-J., and Müller zum Hagen, H. 1973, "On the Occurrence of Naked Singularities in General Relativity," *Commun. Math. Phys.*, **34**, 135–148.

Yodzis, P., Seifert, H.-J., and Müller zum Hagen, H. 1974, "On the Occurrence of Naked Singularities in General Relativity. II," *Commun. Math. Phys.*, **37**, 29–40.

York, J. W., Jr. 1971, "Gravitational Degrees of Freedom and the Initial-Value Problem," *Phys. Rev. Lett.*, **26**, 1656–1658.

Young, P. J., Westphal, J. A., Kristian, J., Wilson, C. P., and Landauer, F. P. 1978, "Evidence for a Supermassive Object in the Nucleus of the Galaxy M87 from SIT and CCD Area Photometry," *Astrophys. J.*, **221**, 721–730.

Zel'dovich, Ya. B. 1972, "Amplification of Cylindrical Electromagnetic Waves Reflected from a Rotating Body," *Zh. Eksp. Teor. Fiz.*, **62**, 2076–2081 (English transl. *Soviet Phys.—JETP*, **35**, 1085–1087 [1972]).

Zipoy, D. 1966, "Topology of Some Spheroidal Metrics," *J. Math. Phys.*, **7**, 1137–1143.

INDEX

(Italicized page numbers denote definitions)